Introductory Algebra

Richelle M. (Rikki) Blair
Lakeland Community College

PEARSON
Addison Wesley

Boston San Francisco New York
London Toronto Sydney Tokyo Singapore Madrid
Mexico City Munich Paris Cape Town Hong Kong Montreal

Publisher: Greg Tobin
Editor in Chief: Maureen O'Connor
Executive Project Manager: Kari Heen
Project Editor: Katie Nopper
Managing Editor: Ron Hampton
Senior Production Supervisor: Kathleen A. Manley
Senior Designer/Cover Designer: Dennis Schaefer
Photo Researcher: Beth Anderson
Software Development: Marty Wright and Mary Durnwald
Marketing Manager: Jay Jenkins
Marketing Coordinator: Tracy Rabinowitz
Senior Author Support/Technology Specialist: Joe Vetere
Senior Prepress Supervisor: Caroline Fell
Senior Manufacturing Buyer: Evelyn Beaton
Text Design: Leslie Haimes
Production Coordination, Composition, and Illustrations: Pre-Press Company, Inc.

Cover photo: Mark Tomalty/Masterfile

Photo credits: 44 PAL/PA WideWorld Photo, 55 PhotoDisc, 55 PhotoDisc, 55 PhotoDisc, 58 PhotoDisc, 58 Beth Anderson, 104 PhotoDisc Red, 104 Beth Anderson, 106 PAL/Hulton Archive, 108 Beth Anderson, 128 PAL/Courtesy of the Library of Congress, 206 PAL/NASA/Jet Propulsion Laboratory, 405 Brand X Pictures, 409 Blend Images (RF), 411 PhotoDisc, 428 PhotoDisc Red, 428 PhotoDisc Blue, 428 PhotoDisc, 430 PhotoDisc, 438 Brand X Pictures, 454 Brand X Pictures, 501 Corbis RF, 508 Brand X Pictures, 566 PhotoDisc Red, 566 Beth Anderson, 577 Brand X Pictures, 577 Beth Anderson

Library of Congress Cataloging-in-Publication Data

Blair, Richelle M.
Introductory algebra/Richelle M. Blair.
p. cm.
ISBN 0-201-65878-X (alk. paper)
1. Algebra—Textbooks. I. Title.

QA154.3.B5827 2006
512—dc22

2005048885

1 2 3 4 5 6 7 8 9 10—VHP—09 08 07 06 05

To my mother and father
for their love and support.

Contents

Preface

From the Author

Students enter Introductory Algebra with a variety of mathematics experiences, some positive and some negative. The goal of this textbook is to engage all students and develop their algebraic thinking and confidence in learning mathematics. Standards-based and traditional content have been blended by actively involving students in real-world problem–solving, reasoning, and skill development. The instructional strategies in this textbook are supported by research on the science of learning and teaching mathematics with the goal of addressing multiple learning styles of students and different teaching styles of faculty. Content and instructional strategies have been chosen based on what we know about how students best learn and how faculty best teach.

Focus on Student Learning

Research shows that developing an understanding of the concepts of equivalence and variable are key to students developing an understanding of algebraic systems. In this textbook, students will review arithmetic as needed and develop the following:

- An understanding of equivalence
- The ability to use variables in a variety of ways
- Problem-solving skills with real-world problems
- Confidence in solving mathematics problem situations
- Quantitative literacy and algebraic skills to be applied in other courses

Recent developments in the science of learning emphasize the importance of students being involved and taking control of their own learning. Active student learning activities have been shown to increase the degree to which students transfer learning to new concepts.[1] Learning new concepts involves connecting previously learned information to new information, reflecting, and transferring that knowledge to new problem situations. Each section of this textbook presents student activities to create an active learning environment and build cognitive connections. In addition, students are asked to do the following:

- Read, write, explore, discuss, reflect, and do mathematics as active learners
- Complete true/false, matching, and reasoning questions
- Analyze data and solve real-world problems
- Create 3×5-inch study cards of important definitions, formulas, and examples
- Complete a mathematics dictionary

Focus on Standards-Based Teaching

An instructor using this textbook is seen not only as an expert mathematician and problem solver, but also as a facilitator and manager of student learning in an active classroom, employing a variety of instructional activities. This textbook embraces the

[1]Bransford, Brown, and Cocking (Eds.). (2000). *How People Learn—Brain, Mind, Experience and School.* National Academy Press, Washington, D.C.

standards of the American Mathematical Association of Two-Year Colleges,[2] the National Council of Teachers of Mathematics,[3] and the Mathematical Association of America[4] by employing a variety of standards-based strategies:

- Activities for students to work (learn) together
- Activities for interactive lecturing
- Step-by-step examples
- Organizing, analyzing, and graphing real-world data
- Connecting mathematics to other disciplines
- Optional problems using a graphing calculator to explore, visualize, observe patterns, and generalize and confirm solutions, if desired

Organization and Features of the Textbook

Each section of each chapter is organized into three sub-sections: *Getting Started, Discussing Mathematics*, and *Doing Mathematics*. The innovative keystone feature of the textbook is the student activities in the Getting Started section. These activities and problems are designed to help the student review previously learned concepts and connect that knowledge to new concepts in the section. The Getting Started section can be used in a variety of ways, depending on the desired approach chosen by the instructor. Getting Started is followed by Discussing Mathematics, an explanation of concepts and examples, and Doing Mathematics, a selection of practice problems. The entire textbook is written in a comfortable and welcoming tone and style that students can read on their own.

Using the textbook is as easy as 1-2-3:

1 2 3 — *Getting Started* presents student activities to be used as classroom group work, lecture discussions, or assigned as homework for the next day.

1 **2** 3 — *Discussing Mathematics* presents definitions, examples of key concepts, and helpful hints.

1 2 **3** — *Doing Mathematics* presents practice problems.

Supplements for the Instructor

INSTRUCTOR'S SOLUTIONS MANUAL

(ISBN 0-321-27957-3) This supplement contains complete, worked-out solutions to all of the exercises in the text.

---

[2]AMATYC. (1995). *Crossroads in Mathematics: Standards for Introductory College Mathematics Before Calculus*. Memphis, TN: AMATYC. These standards are in the process of being revised and are to be released in November 2006.

[3]NCTM. (2000). *Principles and Standards for School Mathematics*. Reston, VA: NCTM.

[4]MAA. (2004). *Undergraduate Programs and Courses in the Mathematical Sciences: CUPM Curriculum Guide 2004*. Washington, DC: MAA.

PRINTED TEST BANK/INSTRUCTOR'S RESOURCE GUIDE

(ISBN 0-321-27958-1) This supplement contains several tests per chapter, teaching notes correlated to each chapter, and a variety of other helpful resources for this text.

TESTGEN WITH QUIZMASTER

(ISBN 0-321-27959-X) TestGen enables instructors to build, edit, print, and administer tests. Featuring a computerized bank of questions developed to cover all text objectives, TestGen is available on a dual-platform Windows/Macintosh CD-ROM.

Supplements for the Student

STUDENT'S SOLUTIONS MANUAL

(0-321-39954-4) This supplement contains complete, worked-out solutions to the odd-numbered exercises in the text.

ADDISON-WESLEY MATH TUTOR CENTER

The Addison-Wesley Math Tutor Center is staffed by qualified mathematics instructors who provide students with tutoring on examples and odd-numbered exercises from the textbook. Tutoring is available via toll-free telephone, toll-free fax, e-mail, or the Internet. White Board technology allows tutors and students to actually see problems worked while they "talk" in real time over the Internet during tutoring sessions. For more information, go to www.aw-bc.com/tutorcenter.

Acknowledgments

A textbook reflects the philosophy of its author, but is strengthened by the collective wisdom of the mathematics community. Many thanks to all those involved in the development of this textbook, especially the following reviewers:

Dan Abbey, Broward Community College
Randall Allbritton, Daytona Beach Community College
Peter Arvanites, Rockland Community College
James J. Ball, Indiana State University
Linda Beattie, Western New Mexico University
Diane Benner, Harrisburg Area Community College
Jose Calavia, Broward Community College, Central
Terry Cheng, Irvine Valley College
Walter Deckert, American River College
Dale C. Downs, Ivy Tech State College
Irene Durancyzk, University of Minnesota
Scott Eckert, Cuyamaca College
Lucy Edwards, Las Positas College
Jackie English, Northern Oklahoma College, Enid
Lori Fuller, Tunxis Community-Technical College
Edna Greenwood, Tarrant County College, Northwest Campus
Jean M. Horn, Northern Virginia Community College, Woodbridge
Evan Innerst, Canada College
Marilyn Jacobi, Gateway Community College

John C. Jacobs, formerly of Massachusetts Bay Community College
Patricia Jones, Methodist College
Maureen Kelley, Northern Essex Community College
Sandra Lanoue, Tulsa Community College
Lee LaRue, Paris Junior College
Valerie Lazzara, Palm Beach Community College
Wing Lee, Valdosta State University
Douglas Leigh, Broome Community College
Christine Heinecke Lehmann, Purdue University North Central
Jean-Marie Magnier, Springfield Technical Community College
Carol A. Marinas, Barry University
Kenneth J. Mead, Genessee Community College
Jean P. Millen, Georgia Perimeter College, Central Campus
Michael Montano, Riverside Community College, City Campus
Penny Morris, Polk Community College
Linda Murphy, Northern Essex Community College
Gail Neal, Danville Community College
Karen L. Pagel, Dona Ana Branch Community College
Marilyn Platt, Gaston College
Susan D. Poston, Chemeketa Community College
Janice L. Roy, Montcalm Community College
Roger A. Roy, University of Maine, Fort Kent
Joyce Saxon, Morehead University
Doris A. Schraeder, McLennan Community College
Mark Sigfrids, Kalamazoo Valley Community College
Gwen Terwilliger, University of Toledo
Ping-Jung Tintera, Texas A&M, Corpus Christi
Betty Vix Weinberger, Delgado Community College
Cora S. West, Florida Community College at Jacksonville
Mary L. Wolyniak, Broome Community College

The production of a textbook is a multifaceted endeavor, made possible through support and encouragement from the publisher. Special thanks to Greg Tobin, publisher at Addison-Wesley, for embracing the spirit and philosophy of the approach in this textbook. I am grateful for the ongoing advice of Deana Richmond during its development edit. Many thanks to Maureen O'Connor, editor in chief, for her support and direction; Kari Heen, executive project manager, for keeping the project on track; Lauren Morse and Katie Nopper, project editors, for their attention to the many tasks during development; Kathy Manley, senior production supervisor, and Dennis Schaefer, senior designer, for their creative ideas during production; Jay Jenkins, marketing manager, and Tracy Rabinowitz, marketing coordinator, for getting the word out about the textbook; and the entire Addison-Wesley developmental mathematics team for their continued support in making this text all that it can be.

Rikki Blair

richelle.blair@sbcglobal.net

Arithmetic Review Skills Test

This test is intended to assess skills at the beginning of the course.
Do not use a calculator on this test.

Round the given number to the indicated place-value. (R. 1)

1. Round to the nearest ones place: 549.89

(a) 549 (b) 550 (c) 549.9 (d) 500

2. Round to the nearest tenths place: 150.523

(a) 150 (b) 150.5 (c) 150.52 (d) 200

Find the numerator of the fraction to make the two fractions equivalent. (R. 1)

3. $\frac{2}{5} = \frac{\quad}{30}$

(a) 12 (b) 4 (c) 6 (d) 2

Rewrite the mixed number as an improper fraction. (R. 1)

4. $4\frac{1}{5}$

(a) $\frac{11}{5}$ (b) $\frac{21}{5}$ (c) $\frac{21}{4}$ (d) $\frac{11}{4}$

Rewrite the improper fraction as a mixed number. (R. 1)

5. $\frac{43}{7}$

(a) $6\frac{1}{7}$ (b) $7\frac{6}{7}$ (c) $5\frac{6}{7}$ (d) $7\frac{1}{7}$

Reduce the fraction to lowest terms. (R. 1)

6. $\frac{18}{48}$

(a) $\frac{6}{16}$ (b) $\frac{1}{4}$ (c) $\frac{3}{8}$ (d) $\frac{9}{24}$

Determine the number that is closest to the exact arithmetic answer by estimating. (R. 1)

7. Is 10.11×9 closer to 90 or closer to 900?

(a) 90 (b) 900

8. Is $\frac{35.7}{5}$ closer to 7 or closer to 0.7?

(a) 7 (b) 0.7

Multiply or divide the fractions and reduce to lowest terms. (R.2)

9. $\frac{28}{3} \cdot \frac{15}{7}$

(a) $\frac{420}{21}$ (b) $\frac{43}{10}$ (c) 20 (d) 22

10. $1\frac{4}{7} \cdot 2\frac{3}{4}$

(a) $\frac{32}{11}$ (b) 2 (c) $4\frac{1}{3}$ (d) $\frac{121}{28}$

11. $\frac{15}{7} \div \frac{27}{35}$

(a) $\frac{50}{34}$ (b) $\frac{525}{189}$ (c) $\frac{405}{245}$ (d) $\frac{25}{9}$

12. $1\frac{4}{7} \div 5\frac{2}{3}$

(a) $\frac{33}{68}$ (b) $\frac{187}{21}$ (c) $\frac{33}{119}$ (d) $\frac{33}{24}$

Add or subtract the fractions and reduce to lowest terms. Write answers that are improper fractions as mixed numbers. (R.2)

13. $\frac{7}{5} + \frac{3}{5}$

(a) $\frac{21}{25}$ (b) 2 (c) $2\frac{1}{3}$ (d) 1

14. $\frac{8}{5} + \frac{2}{7}$

(a) $\frac{10}{12}$ (b) $\frac{5}{6}$ (c) $1\frac{31}{35}$ (d) $\frac{10}{35}$

15. $\frac{11}{8} - \frac{2}{5}$

(a) $\frac{9}{3} = 3$ (b) $3\frac{1}{40}$ (c) $\frac{39}{40}$ (d) $1\frac{1}{40}$

Perform the operations and reduce to lowest terms. Follow Order of Operations. Write answers that are improper fractions as mixed numbers. (R.2)

16. $\frac{17}{2} - \frac{7}{2} \cdot \frac{4}{3}$

(a) $-2\frac{3}{4}$ (b) $3\frac{5}{6}$ (c) $6\frac{2}{3}$ (d) $13\frac{1}{3}$

17. $\left(\frac{11}{2} - \frac{7}{2}\right) \cdot \frac{4}{3}$

(a) $5\frac{1}{3}$ (b) $1\frac{1}{3}$ (c) $2\frac{2}{3}$ (d) $6\frac{2}{3}$

Perform the operations. (R.3)

18. 142.56 + 35.47

(a) 1780.3 (b) 17.803 (c) 1.7803 (d) 178.03

19. 142.56 − 35.47

(a) 107.09 (b) 10,709 (c) 1070.9 (d) 0.10709

20. 142.56 × 35.47

(a) 50.56603 (b) 5056.6032 (c) 50566.03 (d) 0.5056

21. 46.08 ÷ 4.5 × 15.45

(a) 1.58208 (b) 15.8208 (c) 158.208 (d) 1582.08

Perform the operations. (R.3)

22. What is 40% of 250?

(a) 1 (b) 10 (c) 100 (d) 1000

23. 20 is what percent of 400?

(a) 1% (b) 5% (c) 50% (d) 5.5%

Solve the following proportions. (R.4)

24. $\frac{3}{5} = \frac{x}{20}$

(a) $x = 60$ (b) $x = 6$ (c) $x = 4$ (d) $x = 12$

25. $\frac{x}{32} = \frac{3}{4}$

(a) $x = 8$ (b) $x = 12$ (c) $x = 24$ (d) $x = 4$

Convert the given unit of measurement to the unit of measurement specified using the equivalencies or approximate equivalencies below. Round to the nearest hundredth. (R.4)

1 yard (yd) = 3 feet (ft)
1.09 yard (yd) ≈ 1 meter (m)
1 mile ≈ 1.61 kilometers (km)

26. Convert 100 kilometers = _____ miles

(a) 6.211 miles (b) 62.11 miles (c) 1.61 miles (d) 16.1 miles

27. Convert 4.6 meters = _____ yards

(a) 5.01 yards (b) 50.1 yards (c) 4.6 yards (d) 1.09 yards

28. Write 2^3 in expanded and simplified (standard) form. (R.5)

(a) $2 \times 3 = 6$ (b) $2 \times 2 \times 2 = 8$ (c) $3 \times 3 = 9$ (d) $3 \times 2 \times 2 = 12$

29. Convert 1.02×10^{-2} from scientific notation to standard form. (R.5)

(a) 0.00102 (b) 0.00102 (c) 0.0102 (d) 0.102

30. Convert 2,530,000 from standard form to scientific notation. (R.5)

(a) 25.3×10^5 (b) 2.53×10^5 (c) 25.3×10^6 (d) 2.53×10^6

Reflection: Which four problems on this Skills Test did you find the most difficult? Review the sections in Chapter R related to those difficult problems.

CHAPTER

1

Symbols, Real Numbers, and Problem Solving

The equal symbol was first used by Robert Recorde (1510–1558). A translation of what he wrote: I will set a pair of parallel lines (as the equal sign) because no two things can be more equal.

1.1 Symbols and Notations

OBJECTIVES

- State multiple meanings of mathematics symbols.
- Express relationships between real numbers.
- Use objects as placeholders for real numbers.

1 2 3

Getting Started

In this section, we explore the meaning of various mathematics symbols, beginning with the = symbol. In transitioning from arithmetic to algebra, it should be recognized that the equal symbol has two interpretations.

- The equal symbol can represent an instruction to "get an answer": $6 + 2 = ?$.
- The equal symbol can communicate that each side of the equation represents the same quantity and that both sides are equivalent: $2 + 5 - 1 = 3 \cdot 2$.

A mathematical statement that includes the equal symbol, =, is called an **equation**. We will call everything to the left of the equal symbol the "left-hand side" of the equation and everything to the right of the equal symbol the "right-hand side" of the equation.

1. Perform the indicated operations. The equal symbol in these problems is an instruction to "get an answer."

(a) $3 + 4 =$ ______ (b) $\frac{11}{5} - \frac{4}{5} =$ ______

(c) $7 \times 3 =$ ______ (d) $14 \div 2 =$ ______

2. To explore the concept of equivalency, think about how many ways \$1.00 may be obtained with different combinations of coins. List two combinations of pennies, nickels, dimes, and quarters that total \$1.00 in the table. Multiply the number of coins by the value of each coin. Add the total value of each type of coin and write an equation showing your total as \$1.00. It is not necessary to use each type of coin in every combination.

	Pennies	Nickels	Dimes	Quarters	Write an equation representing the TOTAL of \$1.00
EXAMPLE	5 × \$0.01	3 × \$0.05	3 × \$0.10	2 × \$0.25	5 × \$0.01 + 3 × \$0.05 + 3 × \$0.10 + 2 × \$0.25 = \$1.00
(a)					
(b)					

3. The = symbol is used to indicate that two quantities are "equivalent." The ≠ symbol is used between two expressions that are "not equivalent." Complete each statement by writing the appropriate symbol, either = or ≠, on the line between the two quantities.

(a) $15 + 15 + 4$ ______ 9×4 (b) $3\frac{1}{2}$ ______ 3.5

(c) $50 \div 5$ ______ $7 + 5$ (d) $\frac{1}{2}$ ______ $\frac{4}{8}$

4. A balanced scale is a way to visualize equivalency between quantities. When the quantity on the left side is equivalent to the quantity on the right side, the scale is *balanced*. Fill in the box in each problem with the number that will keep the scale balanced.

(a) 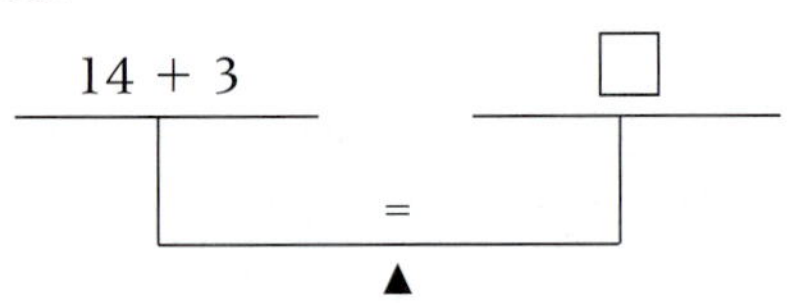

(b) 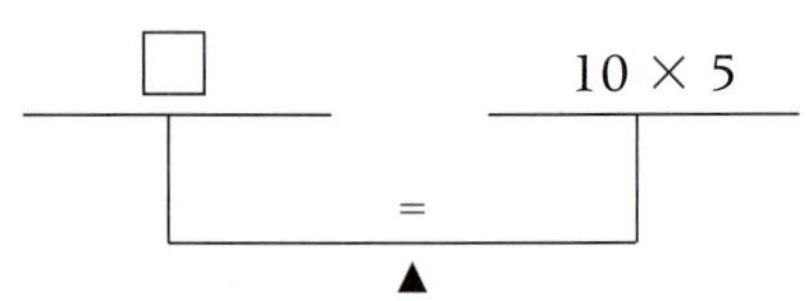

The set of real numbers can be visualized on the **real number line**. Positive numbers are written to the right of zero and negative numbers to the left of zero. The arrows on each end of the real number line indicate that the line continues forever.

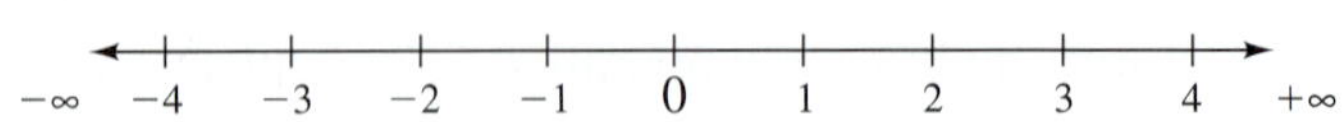

When a number is *negative*, the − symbol is always written. When a number is *positive*, the + symbol may or may not be written. The farther a number is to the right on the number line, the larger the number. The farther a number is to the left on the number line, the smaller the number.

5. Some mathematics symbols have multiple meanings. The + symbol can be an instruction to *add* or it may denote that a number is *greater than zero* (a positive number). Write (A) if the + symbol is an instruction to add. Write (P) if the + symbol denotes a positive number.

(a) $2 + 4$ ______ (b) $10 - (+3)$ ______

(c) $+12 - 4$ ______ (d) $1 + 3$ ______

6. The − symbol can be an instruction to *subtract* or it may denote that a number is *less than zero* (a negative number). Write (S) if the − symbol is an instruction to subtract. Write (N) if the − symbol denotes a negative number.

(a) $200 + (-6)$ ______ (b) $7 - 3$ ______

(c) $(-2) + 6$ ______ (d) $15 - 8$ ______

7. The > and < symbols are also used to express relationships between quantities. The > symbol means "greater than" and the < symbol means "less than." Complete each statement by writing the appropriate symbol, either < or >, on the line between the two quantities.

(a) 10 ______ $5 + 3$ (b) $\frac{1}{2}$ ______ 8

(c) -3 ______ -6 (d) 0 ______ -1

8. Multiple symbols exist to denote multiplication and division. Write whether the indicated operation in each problem is multiplication (M) or division (D).

(a) 10×5 ______ (b) $(10)(5)$ ______

(c) $5\overline{)10}$ ______ (d) $10 \div 5$ ______

(e) $10/5$ or $\frac{10}{5}$ ______ (f) $10 \cdot 5$ ______

9. Other mathematics symbols and notations have specific meanings and act as a kind of shorthand notation. Numbers or operations are understood but not always written explicitly.

(a) Does the number $5\frac{3}{4}$ mean $5 + \frac{3}{4}$ or $5 \times \frac{3}{4}$?

(b) An **exponent** is a symbol that denotes repeated multiplication. The expression 3^4 is read "3 to the fourth power." The exponent, 4, is an instruction to multiply the **base**, 3, four times:

$$3^4 = \underbrace{3 \cdot 3 \cdot 3 \cdot 3}_{\text{4 times}} = 81.$$

Rewrite the expression 5^3 without an exponent.

Discussing Mathematics

Mathematical symbols are used in many ways. They communicate instructions to perform an operation or process, denote whether a number is greater than or less than zero, denote relationships between quantities or expressions, and/or represent shorthand notation.

= Symbol	Examples
• The equal symbol, =, may communicate an instruction to perform an arithmetic operation and "get an answer."	$10 + 4 = ?$ Add 10 and 4. $10 + 4 = 14$ The = symbol instructs us to "get an answer."
• The equal symbol, =, may indicate that the expressions on both sides of the equation are "equivalent."	$10 + 6 - 3 = 15 - 2$ The = symbol is used to communicate equivalency. The left side of the equation is equivalent to the right side of the equation.

EXAMPLE 1: An equation can be visualized with a balanced scale. A scale is balanced only when the quantities on each side are equivalent.

$10 + 6 - 3 \quad = \quad 13$ Equation

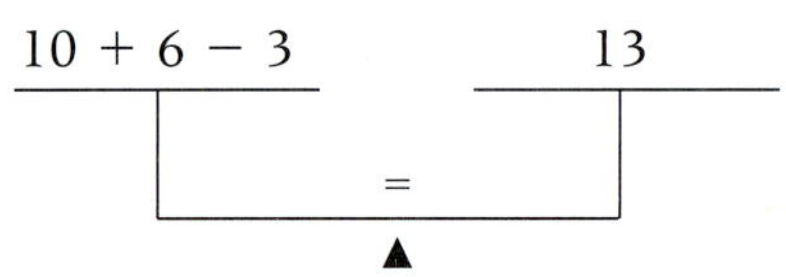

Each side is equivalent to 13. The scale is balanced.

EXAMPLE 2: Objects, like a ★, can be used to represent unknown quantities and used as placeholders to form an equation.

★ + ★ + ★ = 3★ Equation

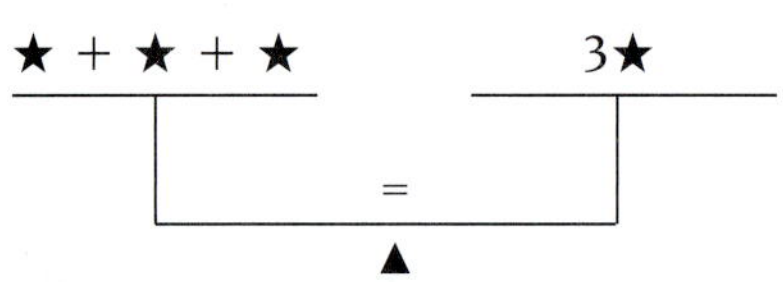

Each side is equivalent to 3 stars. The scale is balanced. The multiplication symbol is not always written when an object or variable appears in the expression: ★ + ★ + ★ = 3·★ = 3★.

THE REAL NUMBER LINE AND THE + AND − SYMBOLS

The real number line is a visualization of the set of real numbers. Positive numbers are written to the right of zero and negative numbers to the left of zero. Zero is considered to be neither positive nor negative. When a number is negative, the − symbol is always written in front of the number. When a number is positive, the + symbol may or may not be written. The farther a number is to the right on the number line, the larger the number. The farther a number is to the left on the number line, the smaller the number. **Infinity symbols**, $-\infty$ and $+\infty$, are symbols that represent a direction and communicate that a line continues forever.

The Real Number Line

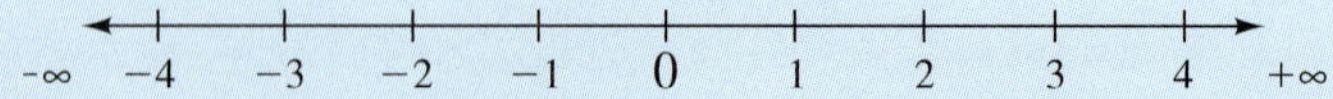

Numbers are called **opposites** if they are the same number of units from zero, but on the opposite side of zero.

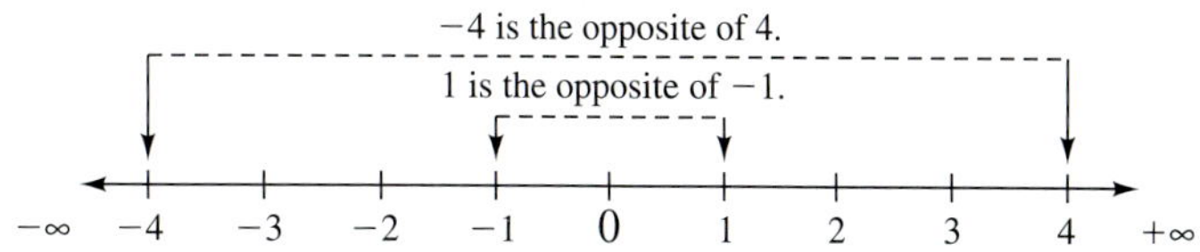

The symbols + and − have more than one meaning in mathematics.

+ Symbol	Examples
• The + symbol may be an instruction to add.	$5 + 6 = 11$ The + symbol is an instruction to add 5 and 6.
• The + symbol may communicate that a number is a positive number, a number to the right of zero on the number line.	$20 - (+3) = 17$ The + symbol indicates that 3 is a positive number, subtracted from 20.

− Symbol	Examples
• The − symbol may be an instruction to subtract.	$7 - 3 = 4$ The − symbol is an instruction to subtract 3 from 7.
• The − symbol may communicate that a number is a negative number, a number to the left of zero on the number line.	$15 + (-2) = 13$ The − symbol means that 2 is a negative number, added to 15.
• The − symbol may be an instruction to take the opposite of the number.	$-(4) = -4$ The − symbol is an instruction to take the opposite of 4. $-(-1) = 1$ The − symbol in front of the parentheses is an instruction to take the opposite of −1.

INEQUALITY SYMBOLS

Inequality symbols communicate a relationship between numbers, objects, or expressions.

INEQUALITY SYMBOLS		
Greater than	**Examples**	
• The > symbol means strictly "greater than."	$10 > 5$	10 is strictly greater than 5.
• The ≥ symbol means "greater than *or* equal to."	$5 \geq 2$ $3 \geq 3$	5 is greater than 2. 3 is equal to 3.
Less than	**Examples**	
• The < symbol means strictly "less than."	$-4 < 2$	−4 is strictly less than 2.
• The ≤ symbol means "less than *or* equal to."	$-5 \leq 1$ $9 \leq 9$	−5 is less than 1. 9 is equal to 9.

MULTIPLICATION AND DIVISION SYMBOLS

The operations of multiplication and division may be denoted by more than one symbol.

Multiplication symbols	Examples
• These symbols are instructions to multiply or find the product.	10×5 $10 \cdot 5$ $10*5$ $(10)(5)$ $10(5)$ $(10)5$

Division symbols	Examples
• These symbols are instructions to divide or find the quotient.	$10 \div 5$ $10/5$ $\frac{10}{5}$ $5\overline{)10}$ The fraction bar indicates the operation of division.

➢ *A multiplication symbol is not always written explicitly. When objects or letters are used as placeholders, there may be no symbol written. Multiplication is understood:*

$$3 \cdot a = 3a \quad or \quad 5x = 5 \cdot x.$$

➢ *The "×" symbol for multiplication is not used in algebra to avoid confusion when the letter "x" is used as a variable.*

OTHER SYMBOLS AND NOTATIONS

The symbols used in mixed numbers, exponents, square roots, and other operations are a shorthand notation for mathematics operations.

Square root symbol	Examples
• The square root symbol, $\sqrt{\ }$, is an instruction to find the square root. The expression under the symbol is called the **radicand**.	$\sqrt{36} = 6$ since $6^2 = 6 \cdot 6 = 36$ 36 is the radicand. The principal square root is the non-negative number that when multiplied by itself results in a product equal to the radicand.

Approximate symbol	Examples
• The ≈ symbol is used to denote a number that is approximate and/or has been rounded.	$\frac{2}{3} \approx 0.67$ $\pi \approx 3.14$

SUBSTITUTING VALUES FOR OBJECTS

Objects, such as ♦, ★, or □, or letters are used as placeholders for numbers in algebra. **Evaluating an expression** is the process of substituting numbers for the objects or letters and performing the operations.

EXAMPLE 3: Given ♦ = 3, □ = 5, and ★ = 1, substitute the values for the objects. Write the appropriate symbol, either =, <, or >, on the line between the two quantities.

$$2\cdot♦ + 6\cdot★ \quad \underline{\ ?\ } \quad 2\cdot□$$

$$2\cdot 3 + 6\cdot 1 \quad \underline{\ ?\ } \quad 2\cdot 5$$

$$6 + 6 \quad \underline{\ ?\ } \quad 10$$

$$12 \quad \underline{\ >\ } \quad 10$$

EXAMPLE 4: Given $x = 3$ and $y = 5$, substitute the values for the letters. Write the appropriate symbol, either =, <, or >, on the line between the two quantities.

$$2x + 3y \quad \underline{\ ?\ } \quad 4y + 1$$

$$2\cdot 3 + 3\cdot 5 \quad \underline{\ ?\ } \quad 4\cdot 5 + 1$$

$$6 + 15 \quad \underline{\ ?\ } \quad 20 + 1$$

$$21 \quad \underline{\ =\ } \quad 21$$

It is essential to understand and be able to apply the definitions of the words and the meanings associated with the mathematical symbols. Memorize the definitions and notations included in this section. They will be used throughout the book.

Doing Mathematics

- Write the definitions of the words and symbols listed for this section in the Dictionary in Appendix A.
- On 3 × 5-inch cards, write the following for future reference:
  (a) Three different mathematics symbols that are used to indicate multiplication
  (b) Three different mathematics symbols that are used to indicate division

Problems 1–2: Create two mathematics statements that are true using only the numbers 3, 5, 7, 8, and 10. You must use the operations of addition, subtraction, multiplication, and division at least once in each problem. You do not have to use all five numbers in each statement. You may use numbers more than once in a statement. Be sure that the quantities on each side of the equal symbol are equivalent.

1. ____________________ = ____________________

2. ____________________ = ____________________

Problems 3–12: Complete the statements below by writing the appropriate symbol, either =, <, or >, on the line between the two quantities.

3. $2 + 4$ ________ 2×3

4. $\frac{1}{2}$ ________ $\frac{1}{3}$

5. $(2)(4)$ ________ 4^2

6. $2\cdot 0$ ________ $0\cdot 2$

7. $1000 \div 5$ ________ 210

8. $2 + 3$ ________ 2×3

9. $20\cdot 5$ ________ $15 - 3$

10. $\sqrt{49}$ ________ 8

11. $100/20$ ________ $2 + 3$

12. $\frac{1}{4}$ ________ $\frac{6}{24}$

Problems 13–20: Given ♦ = 3, □ = 5, and ★ = 1, substitute the values for the objects. Perform the operations and write the appropriate symbol, either =, <, or >, on the line between the two quantities. Show your work.

13. 6·♦ + 2·□ ________ 28

14. 3·□ ________ 2·♦ + 4·★

15. 30 ________ 5·★ + 4·□

16. 4·★ ________ ★ + ★ + ★ + ★ + ★ + ★

17. 100 ÷ □ ________ 50

18. ★ + ♦ ________ □

19. 3·♦ − 2·♦ ________ ★

20. 5·★ − ♦ ________ 3·★

Problems 21–28: Given $x = 2$, $y = 4$, and $z = 0$, substitute the values for the letters. Perform the operations and write the appropriate symbol, either =, <, or >, on the line between the two quantities. Show your work.

21. $3x - 4z$ ________ y

22. $2y - 5z$ ________ x

23. x^2 ________ y^2

24. y^2 ________ z^2

25. $5y + 2x$ ________ $6y$

26. $10x - 3z$ ________ $5y$

27. $24 \div y$ ________ $3x$

28. $32 \div x$ ________ $4y$

Problems 29–33: Equivalence can be visualized with a scale. The goal is to keep the scale balanced, with the quantity on the left side equivalent to the quantity on the right side. Fill in the box in each problem below with the appropriate value to keep the scale balanced.

29.

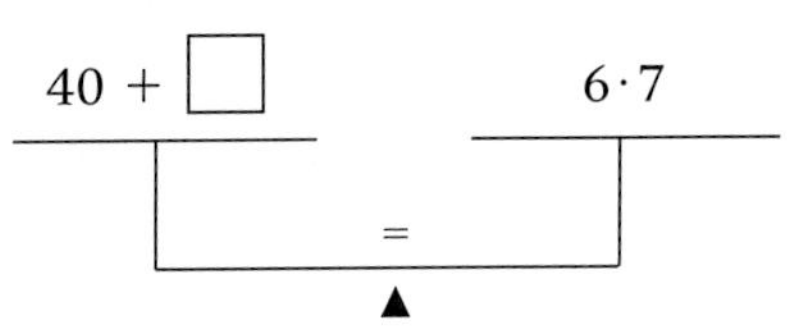

30.

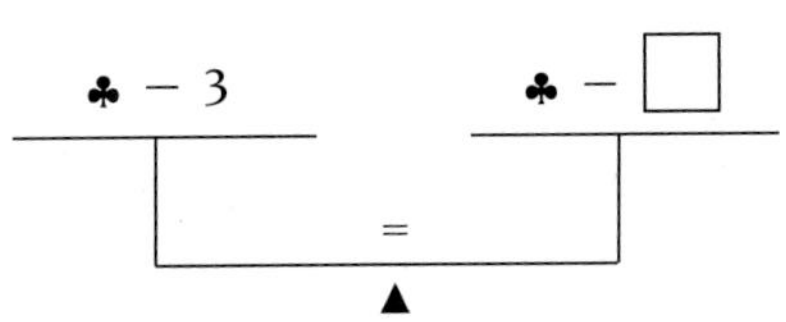

31.

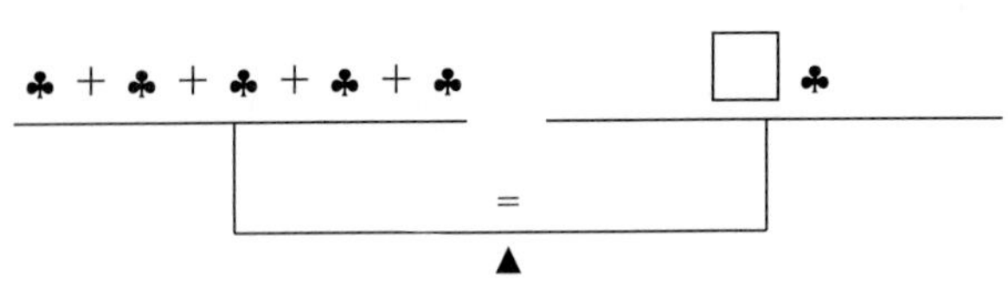

32.

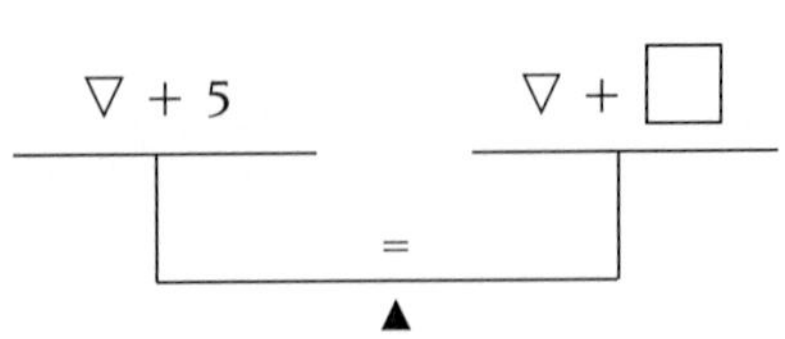

33.

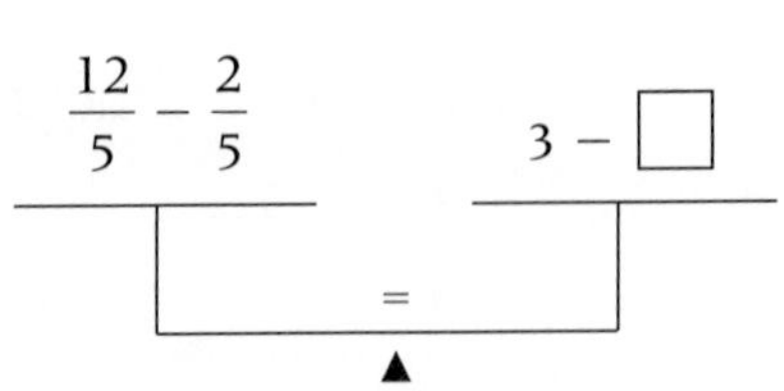

Problems 34–39: Fill in the blank with a number to create an equation in which the right side of the equation is equivalent to the left side.

34. $6 + \_\_\_\_\_\_ = 10$

35. $10 \cdot 5 = 25 \cdot \_\_\_\_\_\_$

36. $15 = 12 + \_\_\_\_\_\_$

37. $32 - 7 = 4 \cdot 5 + \_\_\_\_\_\_$

38. $\frac{90}{10} + \_\_\_\_\_\_ = 6 \cdot 2$

39. $\frac{6}{4} + \_\_\_\_\_\_ = 2$

40. Create an inequality statement of your own using the indicated inequality symbols.

(a) $\_\_\_\_\_\_\_\_\_\_ < \_\_\_\_\_\_\_\_\_\_$

(b) $\_\_\_\_\_\_\_\_\_\_ > \_\_\_\_\_\_\_\_\_\_$

(c) $\_\_\_\_\_\_\_\_\_\_ \leq \_\_\_\_\_\_\_\_\_\_$

(d) $\_\_\_\_\_\_\_\_\_\_ \geq \_\_\_\_\_\_\_\_\_\_$

Problems 41–43: Writing Exercises

41. Write (A) if the + symbol is an instruction to add (A). Write (P) if the + symbol denotes a positive number.

(a) $16 - (+4)$

(b) $\Delta + \blacklozenge$

(c) $(+15)(6)$

(d) $9 + 3$

(e) Write a paragraph using complete sentences that describes your understanding of the two meanings of the + symbol.

42. Write (S) if the − symbol is an instruction to subtract (S). Write (N) if the − symbol is a negative number.

(a) $10 - 5$

(b) $20 + (-4)$

(c) $\frac{-30}{10}$

(d) $x - y$

(e) Write a paragraph using complete sentences that describes your understanding of the two meanings of the − symbol.

43. What is the difference between the (−) and the ((−)) key on your calculator?

Problems 44–49: Write an equivalent expression without an exponent showing multiplication explicitly. Find the final product. Example: $2^4 = 2 \cdot 2 \cdot 2 \cdot 2 = 16$.

44. 5^2

45. 6^3

46. $(3.5)^3$

47. 2^5

48. 3^3

49. $(2.3)^2$

1.2 Properties of Real Numbers, Order of Operations

OBJECTIVES

- Recognize and write the subsets of the set of real numbers.
- Write examples of the Commutative, Associative, Identity, and Distributive Properties of real numbers.
- Use the Order of Operations to perform arithmetic operations.
- Use variables to generalize mathematical statements.

Getting Started

In this section, the sets of numbers contained on the real number line will be presented along with arithmetic rules called Properties of Real Numbers. Objects will be replaced with letters as variables to form generalizations that will guide our work throughout the textbook. The Order of Operations will be presented and used to perform operations on positive real numbers. The concepts will be extended to negative real numbers in Section 1.3.

1. The real number line helps us to visualize relationships between numbers. Positive real numbers appear to the right of zero and negative real numbers to the left of zero. A number farther to the right on the number line is larger than a number to its left and a number farther to the left is smaller. The letters x, b, t, and d denote the values of real numbers in their respective locations on this number line. The letter d is farther to the right on the number line than t.

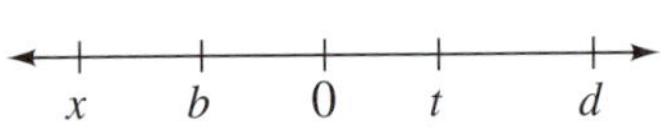

Therefore, $d > t$.

(a) Is the number represented by the letter x a positive or negative number?

(b) Is the number represented by the letter d a positive or negative number?

Write the appropriate symbol, either $<$ or $>$, on the line between the variables.

(c) x ______ b (d) t ______ d

(e) d ______ x (f) $-x$ ______ 0

Any group of numbers is called a **set** of numbers. The numbers in a set are called the **elements** or **members** of the set. When writing a set, place braces around all the elements, insert a comma between each element, and label each set with a name or a letter. For example, $A = \{1, 2, 3\}$ and $B = \{1, 2, 3, 4, 5, 6\}$. If a set has a limited number of elements, the set is called **finite**. If the number of the elements continues indefinitely, the set is called **infinite**. An infinite set can be written by listing at least three elements of the set to show the pattern and then writing three dots called an **ellipsis**. The ellipsis tells us that the pattern of the elements continues indefinitely.

The set of real numbers is composed of many sets of numbers.

- The infinite set of **counting numbers** (also called **natural numbers**) consists of the set $\{1, 2, 3, \ldots\}$ and will be labeled C.
- The infinite set of **whole numbers** consists of the set of counting numbers and zero, $\{0, 1, 2, 3, \ldots\}$, and will be labeled W.
- The set of **integers** consists of the set of whole numbers, their opposites, and zero, $\{\ldots -3, -2, -1, 0, 1, 2, 3, \ldots\}$, and will be labeled I.

  ➢ *Fractions and decimals,* between *the elements in the sets, are* not *elements of these three sets.*

2. Classify each real number as a member of the set of counting numbers (C), whole numbers (W), integers (I), or none (N). Use all the names that apply.

(a) 100 (b) -720

(c) -13 (d) $\dfrac{-2}{9}$

(e) 0 (f) 2.35

The three sets above, along with numbers that can be written as fractions or decimals comprise two other sets of real numbers.

- The set of **rational numbers** are fractions that can be written in the form $\dfrac{a}{b}$, where a and b $(b \neq 0)$ are integers, and will be labeled Q. Rational numbers have decimal equivalents that are terminating or repeating decimals.
- The set of **irrational numbers** are real numbers whose decimal equivalents are neither terminating nor repeating and will be labeled P.

3. Classify each real number as a member of the set of rational numbers (Q) or irrational numbers (P).

(a) $\dfrac{1}{2}$ (b) π

(c) $-2.341678\ldots$ (d) $0.272727\ldots$

4. The results of certain operations on real numbers create unique patterns that define arithmetic rules called *Properties*. Let's look at some problems, observe the patterns, and draw conclusions. An operation is said to be **commutative** if when the *order* of the numbers is changed, the result is the same answer.

Perform the indicated operations and answer the questions.

	Arithmetic problem	Does changing the *order* of the numbers result in the same answer?	
Ex.	$2 + 3 \stackrel{?}{=} 3 + 2$ $5 = 5$	Yes.	
(a)	$10 + 5 \stackrel{?}{=} 5 + 10$		Is addition commutative?
(b)	$5 \cdot 6 \stackrel{?}{=} 6 \cdot 5$		Is multiplication commutative?
(c)	$10 - 5 \stackrel{?}{=} 5 - 10$		Is subtraction commutative?
(d)	$24 \div 6 \stackrel{?}{=} 6 \div 24$		Is division commutative?

5. Grouping symbols, such as parentheses and brackets, indicate that the operation(s) inside the grouping symbols is (are) to be performed first. An operation is said to be **associative** if when the position of the *grouping* symbols is changed, the result is the same answer. Perform the indicated operations.

	Arithmetic problem	Does changing the *grouping* of the numbers result in the same answer?	
Ex.	$(7 + 1) + 4 \stackrel{?}{=} 7 + (1 + 4)$ $(8) + 4 \stackrel{?}{=} 7 + (5)$ $12 = 12$	Yes.	
(a)	$(2 + 9) + 3 \stackrel{?}{=} 2 + (9 + 3)$		Is addition associative?
(b)	$(4 \cdot 2) \cdot 5 \stackrel{?}{=} 4 \cdot (2 \cdot 5)$		Is multiplication associative?
(c)	$(12 - 9) - 3 \stackrel{?}{=} 12 - (9 - 3)$		Is subtraction associative?
(d)	$(24 \div 6) \div 2 \stackrel{?}{=} 24 \div (6 \div 2)$		Is division associative?

6. A number or variable may appear as a multiplier in front of parentheses enclosing the operation of addition or subtraction. Determine whether you obtain the same result when you perform the operation inside the parentheses and then multiply or when you multiply the factor by each term and then add or subtract.

	Arithmetic problem	Are the answers the same?
Ex.	$3(4 + 1) \stackrel{?}{=} 3 \cdot 4 + 3 \cdot 1$ $3(5) \stackrel{?}{=} 12 + 3$ $15 = 15$	Yes.
(a)	$5(4 + 7) \stackrel{?}{=} 5 \cdot 4 + 5 \cdot 7$	
(b)	$9(2 - 1) \stackrel{?}{=} 9 \cdot 2 - 9 \cdot 1$	

7. The *order* in which operations are performed in a problem is important when parentheses do not appear in the problem. Perform the operations: $2 + 3 \cdot 4 =$ ______.

The answer should be 14! Multiplication should be performed first and then addition: $2 + \underbrace{3 \cdot 4}$

$$2 + \quad 12 = 14.$$

To avoid confusion, a set of rules, called the Order of Operations, dictates which operation should be performed first and will be discussed later.

8. Perform the operations. You should not arrive at the same answer for both problems.

(a) $\dfrac{20 + 8}{4} =$ ______ (b) $20 + \dfrac{8}{4} =$ ______

9. Team up with one or more other students and write the equation $2 + 3 = 5$ on a piece of paper. Pass the paper to a second student. The second student extends the equation by adding, subtracting, multiplying, or dividing on each side of the equation the same number of his/her choice to create an equivalent equation. For example, the second student might write: $2 + 3 - 4 = 5 - 4$. Do not attempt to divide by zero. Pass the paper to another student until the equation has been rewritten at least four times. Each student checks to verify that the equation is equivalent.

1 2 3 Discussing Mathematics

THE SET OF REAL NUMBERS

All real numbers are represented on the real number line.

The Real Number Line

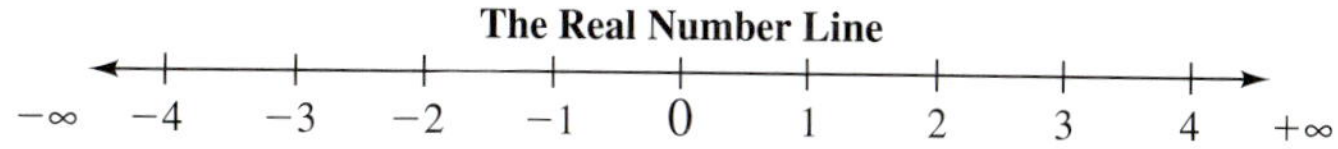

The letter ***R*** will be used to denote the **set of real numbers** composed of the set of rational numbers (containing counting numbers, whole numbers, and integers) and irrational numbers. A set is a **subset** of another set if every element in the first set is also an element of the second set. The subsets of real numbers can be visualized with nested boxes:

The Set of Real Numbers = *R*

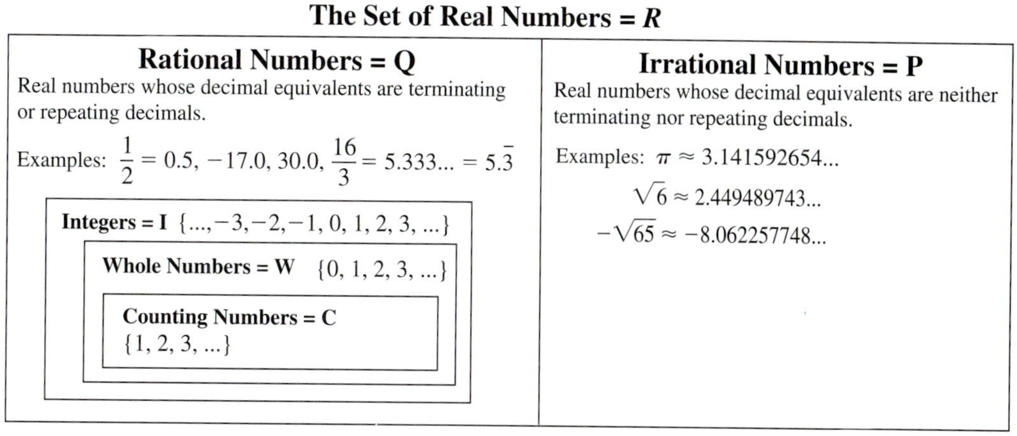

Any number that can be written in the form $\frac{a}{b}$, where a and b are integers and $b \neq 0$, is a **rational number**. When the fraction is changed to its decimal equivalent, the result is either a terminating or repeating decimal. The denominator of a rational number can never be zero because division by zero is undefined. A **terminating decimal** has a finite number of decimal places with a remainder of zero.

EXAMPLE 1: The decimal equivalent of $\frac{3}{4}$ is found by dividing the numerator (3) by the denominator (4).

$$\frac{3}{4} = 0.75$$

```
    0.75        0.75 is a terminating decimal.
 4)3.00
   28
    20
    20
     0   <--  Remainder of zero
```

A **repeating decimal** is a decimal in which a sequence of numbers in the quotient repeats and the same remainder repeats. The sequence of numbers repeating is called the **repetend**. A repeating decimal can be written with a line drawn above the repetend called a **vinculum**. Another notation for a repeating decimal uses an **ellipsis** after the repetend to show that the pattern continues.

EXAMPLE 2: The decimal equivalent of $\frac{8}{11}$ is found by dividing the numerator (8) by the denominator (11).

Repetend is 72 Vinculum

$$\frac{8}{11} = 0.7272\ldots = 0.\overline{72}$$

Ellipsis

$0.\overline{72}$ is a repeating decimal.

```
     0.7272 . . .
 11)8.0000 . . .
    77
     30
     22
      80
      77
       30
       22
        8
```

Irrational numbers are real numbers that cannot be written as a quotient of integers and can be identified by their decimal form, which is neither a repeating nor a terminating decimal. The number of decimal places is infinite and the decimal values do not repeat in a pattern. Irrational numbers are often rounded to a specified number of decimal places (consult Chapter R for a review of rounding).

EXAMPLE 3: Irrational numbers and their decimal approximations:

(a) $\sqrt{11} \approx 3.3166$ (rounded to four decimal places)

(b) $-\sqrt{5} \approx -2.24$ (rounded to two decimal places)

EXAMPLE 4: Conclusions can be formed about the values of the numbers represented by the letters m and c shown in the respective locations on the number line:

(a) The letter m represents a negative number (left of zero) on the line above. A negative symbol in front of the letter m is an instruction to "take the opposite of the value represented by m." The expression $-m$ is a positive number and $-m > 0$.

(b) The letter c represents a positive number (right of zero) on the line above. A negative symbol in front of the letter c is an instruction to "take the opposite of the value represented by c." The expression $-c$ is a negative number and $-c < 0$.

PROPERTIES OF REAL NUMBERS

Mathematical rules that are true for all real number values are called Properties. Properties communicate patterns that are true for specific numbers and generalizations follow using objects or letters for all real numbers. The objects and letters are called **variables**.

Identity Property of Addition and Multiplication

Adding the number 0 to any real number results in a sum *identical* to the original real number. The number 0 is called the **additive identity**. Similarly, the number 1 is called the **multiplicative identity** because multiplying any real number by 1 results in a product *identical* to the original real number.

IDENTITY PROPERTIES (KEY WORD: IDENTICAL)

For any real number x:

Identity Property of Addition: $x + 0 = x$	The number zero is called the Additive Identity.
Identity Property of Multiplication: $x \cdot 1 = x$	The number one is called the Multiplicative Identity.

EXAMPLE 5: To generalize the Identity Property of Addition:

$$15 + 0 = 15$$

Replace the number 15 with the letter x.

$$x + 0 = x$$

Generalization: The **Identity Property of Addition** states that the sum of any real number x and zero is *identical* to the original real number, x.

EXAMPLE 6: To generalize the Identity Property of Multiplication:

$$50 \cdot 1 = 50$$

Replace the number 50 with the letter x.

$$x \cdot 1 = x$$

Generalization: The **Identity Property of Multiplication** states that the product of any real number x and 1 is *identical* to the original real number, x.

Commutative Property of Addition and Multiplication

Changing the *order* of the addends in an addition problem or changing the order of the factors in a multiplication problem does not change the result. The operations of addition and multiplication are said to be commutative and can be generalized to form the Commutative Properties of Addition and Multiplication.

COMMUTATIVE PROPERTIES (KEY WORD: ORDER)

For all real numbers x and y:

Commutative Property of Addition: $x + y = y + x$

Commutative Property of Multiplication: $x \cdot y = y \cdot x$

➢ **Watch out!** *Changing the order of the numbers or variables in expressions involving subtraction or division does not produce the same results:* $10 - 5 \neq 5 - 10$; $20 \div 4 \neq 4 \div 20$.

EXAMPLE 7: To generalize the Commutative Property of Addition:

$20 + 4 = 4 + 20$ — Replace 20 with the variable x and 4 with the variable y (dashed line).

$x + y = y + x$ — Generalization: The **Commutative Property of Addition** states that the sum of two real numbers is the same, regardless of the *order* in which they are added.

EXAMPLE 8: To generalize the Commutative Property of Multiplication:

$3 \cdot 7 = 7 \cdot 3$ — Replace 3 with the variable x and 7 with the variable y (dashed line).

$x \cdot y = y \cdot x$ — Generalization: The **Commutative Property of Multiplication** states that the product of two real numbers is the same, regardless of the *order* in which they are multiplied.

Associative Property of Addition and Multiplication

Changing the *grouping* of the addends in an addition problem or of the factors in a multiplication problem does not change the result and can be generalized into the Associative Properties of Addition and Multiplication.

ASSOCIATIVE PROPERTIES (KEY WORD: GROUPING)

For all real numbers x, y, and z:

Associative Property of Addition: $(x + y) + z = x + (y + z)$

Associative Property of Multiplication: $(x \cdot y) \cdot z = x \cdot (y \cdot z)$

➢ **Watch out!** *Changing the grouping in expressions involving subtraction or division does not produce the same results:* $(7 - 3) + 2 \neq 7 - (3 + 2)$; $(10 \div 5) \cdot 3 \neq 10 \div (5 \cdot 3)$.

EXAMPLE 9: To generalize the Associative Property of Addition (and similarly for multiplication):

$$(5 + 7) + 1 = 5 + (7 + 1)$$

Replace 5 with the variable x, 7 with the variable y, and 1 with the variable z.

$$(x + y) + z = x + (y + z)$$

Generalization: The **Associative Property of Addition** states that the sum of real numbers is the same, regardless of the *grouping* of the addends.

Distributive Properties

When the operation of multiplication is indicated in front of a grouping symbol and the operations of addition or subtraction are enclosed in the grouping symbol, the Distributive Property is employed. The number in front of the parentheses is multiplied and distributed across the numbers inside the parentheses. The process of distributing is like "spreading out" the multiplier outside of the parentheses. Note that the multiplication symbol in front of parentheses is usually not written explicitly.

DISTRIBUTIVE PROPERTY OF MULTIPLICATION OVER ADDITION (OR SUBTRACTION) (KEY WORDS: SPREADING OUT)

For all real numbers x, y, and z:

$$x(y + z) = xy + xz \qquad x(y - z) = xy - xz$$

EXAMPLE 10: To generalize the Distributive Property of Multiplication over Addition (and similarly for subtraction):

$$3(5 + 2) = 3 \cdot 5 + 3 \cdot 2$$

Replace 3 with the variable x, 5 with the variable y, 2 with the variable z.

$$x(y + z) = x \cdot y + x \cdot z$$

Generalization: Distributive Property of Multiplication over Addition.

EXAMPLE 11: The Commutative Property can be applied to expressions involving parentheses and combined with the Distributive Property. The factor may appear in front or after the parentheses with the same result.

$$\begin{aligned} 3(5 + 2) &= (5 + 2)3 \\ 3 \cdot 5 + 3 \cdot 2 &= 5 \cdot 3 + 2 \cdot 3 \\ 15 + 6 &= 15 + 6 \\ 21 &= 21 \end{aligned}$$

A generalization of the Distributive Property and the Commutative Property to expressions involving letters and parentheses:

$$x(y + z) = (y + z)x$$

EXAMPLE 12: Apply the Distributive Property to perform the operations: $2(x + 3)$. Until we know the value of x, it is not possible to compute the sum inside the parentheses. The expression can be rewritten applying the Distributive Property, "spreading out" the 2 by multiplying 2 by x and 3:

$$2(x + 3) = 2 \cdot x + 2 \cdot 3 = 2x + 6$$

ORDER OF OPERATIONS

The **Order of Operations** dictates which operation to perform first, second, and so forth, when more than one operation appears in a problem. The Order of Operations can be remembered by memorizing an acronym, **PEMDAS**, or the sentence, **Please Excuse My Dear Aunt Sally.**

ORDER OF OPERATIONS				
	Operation	**Process**	**Acronym**	**Sentence**
First	**P**arentheses	Perform any operations enclosed in parentheses, brackets, other grouping symbols, or absolute value. ➢ *A fraction bar may act as a grouping symbol. Perform operation(s) in the numerator and/or denominator.*	**P**	**P**lease
Second	**E**xponents	Simplify any expression involving an exponent.	**E**	**E**xcuse
Third	**M**ultiplication *or* **D**ivision (left to right)	Start at the left side of the expression and identify the first multiplication *or* division operation in the problem. ➢ *If division appears first, perform the division. If multiplication appears first, perform the multiplication.*	**M** **D**	**M**y **D**ear
Fourth	**A**ddition *or* **S**ubtraction (left to right)	Start at the left side of the expression and identify the first addition *or* subtraction operation in the problem. ➢ *If subtraction appears first, perform subtraction. If addition appears first, perform the addition.*	**A** **S**	**A**unt **S**ally

➢ ***Note!*** *The operation of multiplication is not* always *performed before division. Multiplication* or *division is performed in the order in which the operation occurs in the problem, left to right. The operation of addition is not* always *performed before subtraction. Addition* or *subtraction is performed in the order in which the operation occurs in the problem, left to right.*

EXAMPLE 13: Simplify $20 - 3 \cdot 5 + 4$ using Order of Operations.

$= 20 - 15 + 4$ Multiply $3 \cdot 5$.

$= 5 + 4$ Subtract 15 from 20.

$= 9$ Add $5 + 4$.

EXAMPLE 14: Simplify $12 \div 3 \cdot 2 - 2 + 6$ using Order of Operations.

$= 4 \cdot 2 - 2 + 6$ Divide $12 \div 3$ (multiplication *or* division left to right).

$= 8 - 2 + 6$ Multiply $4 \cdot 2$.

$= 6 + 6$ Subtract 2 from 8 (addition *or* subtraction left to right).

$= 12$ Add $6 + 6$.

EXAMPLE 15: Simplify $2^3 + \frac{1}{3} \cdot 45 + 6 - 28$ using Order of Operations.

$= 8 + 15 + 6 - 28$	Simplify exponent in 2^3.
	Multiply $\frac{1}{3} \cdot 45$.
$= 23 + 6 - 28$	Add $8 + 15$.
$= 29 - 28$	Add $23 + 6$.
$= 1$	Subtract 28 from 29.

EXAMPLE 16: Simplify $\frac{20 + 8}{4} + 3^2$ using Order of Operations.

$= \frac{28}{4} + 3^2$	Add $20 + 8$ (fraction bar acts as a grouping symbol).
$= \frac{28}{4} + 9$	Simplify exponent in 3^2.
$= 7 + 9$	Divide $28 \div 4$.
$= 16$	Add $7 + 9$.

EXAMPLE 17: Simplify $3(10 - 2)^2 + 4^2 - 1$ using Order of Operations.

$= 3(8)^2 + 4^2 - 1$	Simplify operation in parentheses.
$= 3(64) + 16 - 1$	Simplify exponents.
$= 192 + 16 - 1$	Multiply $3(64)$.
$= 208 - 1$	Add $192 + 16 + 9$.
$= 207$	Subtract.

CALCULATORS

A scientific or graphing calculator is an instructional tool to explore numeric patterns and to help us perform complicated mathematical processes, but it should not be used for simple operations with one- or two-digit numbers. There are many different manufacturers and models of calculators and most have these common characteristics.

- All calculators perform arithmetic operations. Most scientific calculators and graphing calculators have been programmed with the Order of Operations.
- Most keys on a scientific or graphing calculator perform more than one operation or function.
- All graphing calculators have a screen on which operations and results are displayed.
- Graphing calculators may be used to produce graphs of data and equations.

Your instructor will provide guidelines about the use of the calculator in your course.

Doing Mathematics

- Write the definitions of the words and symbols listed for this section in the Dictionary in Appendix A.
- On a 3 × 5-inch card, write the following for future reference:
  (a) The five subsets of the real numbers
  (b) The Identity Properties of Addition and Multiplication
  (c) The Commutative Properties of Addition and Multiplication
  (d) The Associative Properties of Addition and Multiplication
  (e) The Distributive Property
  (f) The Order of Operations

Problems 1–8: The real number line helps us to visualize relationships between numbers. Let the letters w, k, g, and h represent the values of real numbers in their respective locations on the number line:

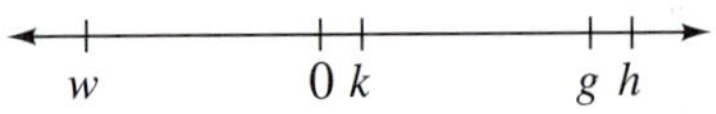

Write the appropriate inequality symbol, < or >, on the line between the numbers and variables.

1. k ______ 0

2. g ______ 0

3. w ______ g

4. k ______ g

5. $-w$ ______ 0

(*Hint:* See Example 4)

6. $-h$ ______ 0

7. $-k$ ______ h

8. h ______ w

Problems 9–22: Classify each real number as a member of the set(s) of counting numbers (C), whole numbers (W), integers (I), rational numbers (Q), or irrational numbers (P). Use all the names that apply.

9. 0

10. $-\sqrt{15}$

11. 1502

12. $-\frac{31}{5}$

13. 0.35

14. π

15. $\sqrt{36}$

16. -10

17. $\sqrt{2}$

18. $0.\overline{6}$

19. $\frac{17}{100}$

20. $2\frac{3}{4}$

21. $5.\overline{09}$

22. 24.5

23. Use a calculator to find a decimal approximation for each irrational number. Write all the digits displayed on your calculator (different calculators may display a different number of digits), followed by an ellipsis.

(a) $\sqrt{21}$ (b) $-\sqrt{107}$

(c) $\dfrac{\sqrt{10}}{3}$ (d) $\sqrt{1/2}$

24. Use a calculator to find a decimal approximation for each irrational number. Write all the digits displayed on your calculator (different calculators may display a different number of digits), followed by an ellipsis.

(a) $\sqrt{55}$ (b) $-\sqrt{17}$

(c) $\dfrac{\sqrt{30}}{7}$ (d) $\sqrt{1/6}$

Problems 25–32: Indicate whether the statement is true or false. If false, rewrite the statement as a true statement.

25. Every integer is a real number.

26. All real numbers are rational numbers.

27. Repeating and terminating decimals are irrational numbers.

28. Every whole number is an integer.

29. Every rational number can be written as a repeating decimal.

30. The set of rational numbers is a subset of the real numbers.

31. The number 1 is the additive identity.

32. The set of whole numbers is a subset of the set of irrational numbers.

Problems 33–38: Use the Commutative Property of Addition (or Multiplication) to write an equivalent expression on the right side of the equal symbol. Do not perform the operations. Example: $2 + 3 = 3 + 2$.

33. $51 + 6 =$ **34.** $a \cdot b =$

35. $\dfrac{17}{4} \cdot \dfrac{1}{3} =$ **36.** $\dfrac{4}{3} + \dfrac{5}{2} =$

37. $x + y =$ **38.** $125 \cdot 42 =$

Problems 39–44: Use the Associative Property of Addition (or Multiplication) to write an equivalent expression on the right side of the equal symbol. Do not perform the operations. Example: $2 + (1 + 9) = (2 + 1) + 9$.

39. $2 + (3 + 5) =$ **40.** $9 \cdot (5 \cdot 7) =$

41. $a \cdot (b \cdot c) =$ **42.** $(x + y) + z =$

43. $(7 \cdot 6) \cdot 3 =$ **44.** $5.32 + (7.67 + 9.11) =$

Problems 45–50: Use the Distributive Property over Addition (or Subtraction) to write an equivalent expression on the right side of the equal symbol. Do not perform the operations. Example: $4(2 + 9) = 4 \cdot 2 + 4 \cdot 9$.

45. $10(2 + 6) =$

46. $7(5 - 2) =$

47. $(15 + 78)2 =$

48. $\frac{2}{3}\left(9 - \frac{6}{5}\right) =$

49. $a(b + c) =$

50. $(y - z)x =$

Problems 51–56: Use the Distributive Property over Addition (or Subtraction) to write equivalent statements on the right-hand side of the equation in the problems below. Do not perform the operations. Example: $4 \cdot 2 + 4 \cdot 9 = 4(2 + 9)$.

51. $2 \cdot 5 + 2 \cdot 3 =$

52. $b \cdot a - c \cdot a =$

53. $537 \cdot 6 - 537 \cdot 2 =$

54. $(1.5)(2.35) + (1.5)(6.43) =$

55. $\frac{1}{2} \cdot x - \frac{1}{2} \cdot y =$

56. $x \cdot y + x \cdot z =$

Problems 57–72: Determine which Property is illustrated in the statements below: Identity Property of Addition (or Multiplication), Commutative Property of Addition (or Multiplication), Associative Property of Addition (or Multiplication), or Distributive Property over Addition (or Subtraction).

57. $5 + \frac{1}{2} = \frac{1}{2} + 5$

58. $a + b = b + a$

59. $(a + b)c = ac + bc$

60. $3 + (2.56 + 4) = (3 + 2.56) + 4$

61. $3 \cdot (2 \cdot 4) = (3 \cdot 2) \cdot 4$

62. $x \cdot 1 = x$

63. $x \cdot y = y \cdot x$

64. $\frac{5}{9} \cdot 2 = 2 \cdot \frac{5}{9}$

65. $(3 + 4)2 = 3 \cdot 2 + 4 \cdot 2$

66. $5 \cdot 6 - 5 \cdot 2 = 5(6 - 2)$

67. $4(7 - 3) = 28 - 12$

68. $152.37 = 152.37 \cdot 1$

69. $17 + 5 = 5 + 17$

70. $(2 + 9) + 10 = 2 + (9 + 10)$

71. $-\frac{3}{4} + 0 = -\frac{3}{4}$

72. $3 \cdot 5 + 3 \cdot 7 = 3(5 + 7)$

Problems 73–92: Use the Order of Operations and perform the indicated operations without using a calculator. Write the result of each step.

73. $6.59 + 5.2(4)$

74. $\frac{4}{49} + \frac{1}{7} \cdot \frac{45}{7}$

75. $(5 + 3)2 + 7$

76. $6 + 2 \cdot 10 \div 5$

77. $6 + 10 \div 5 \cdot 2$

78. $(5.1)^2 + 10 - 3 + 2 \cdot 6$

79. $3(5 + 2) - 2^3 + 5(6 - 2)$

80. $\dfrac{50 + 5}{5}$

81. $(6 + 10) \div 4 \cdot 2$

82. $6 + 12 \div 3 \cdot 4 - 2 + 4^2$

83. $(4^3 - 4) \div 10$

84. $(6 - 4)^2 \div 2 \cdot 5$

85. $\dfrac{3^3 - 2}{3 + 2}$

86. $\dfrac{6 + 3 \cdot 2}{6^2 - 30}$

87. $100 - 2 + 15 \cdot \dfrac{1}{5} - 50 \div 25$

88. $15 - \dfrac{1}{2} \cdot 4 + 12 \div 6 - 3$

89. $2(5 - 2)^2 + 3^2 - 2^2$

90. $4(7 - 5)^2 + 5^2 - 3^2$

91. $10^2 - (6 - 3)^2 + 5$

92. $6^2 - (4 - 2)^2 + 7$

Problems 93–94: Perform the operations using a calculator.

93. $\dfrac{245.367 + 543.789}{37.35} + (45.782)^2$

94. $(10)^3 - 259 + 10.35 \cdot \dfrac{1}{5} - 146.49 \div 25$

95. Circle the calculator screen that represents the correct entry and answer to the problem: $\dfrac{20 + 8}{4}$. Explain why you made your choice.

(a) 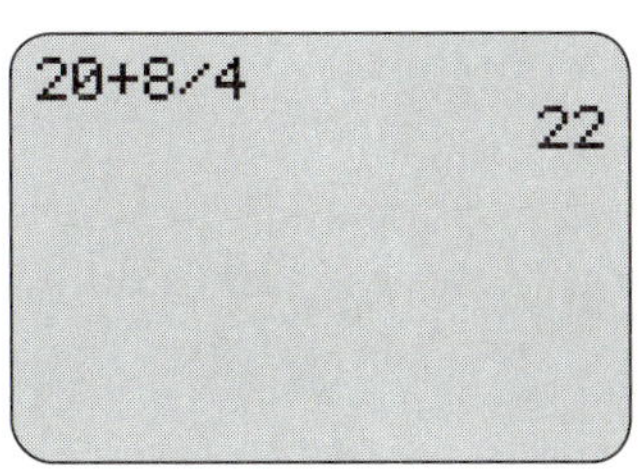

(b) 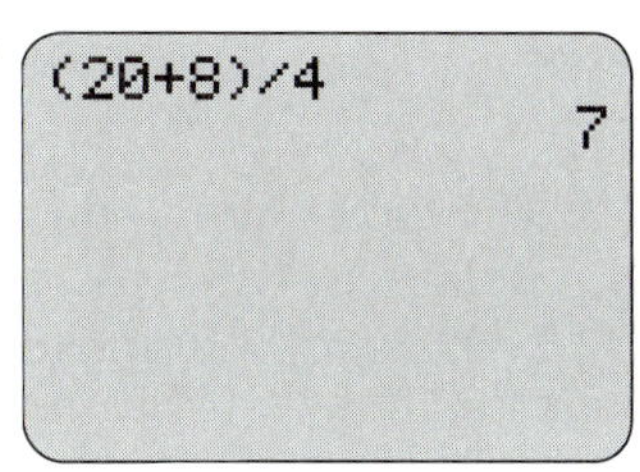

1.3 Addition and Subtraction of Real Numbers

OBJECTIVE

- Add and subtract real numbers.

1 2 3 Getting Started

In the previous section, we defined the set of real numbers and used the real number line to visualize the set. In this section, we focus on operations with positive and negative numbers beginning with two real-world problems.

1. You are going to keep track of the charges you make on your credit card by using a table. *Purchases will be shown as negative numbers and payments will be shown as positive numbers.* Compute the account balance after each purchase or payment and complete the table.

Date	Activity on the account (purchases, returns, and payments)	Balance (in $)
	Starting balance	0.00
Jan. 15	Purchased a shirt costing $45.00	−45.00
Jan. 26	Purchased a leather jacket costing $150.75	−195.75
Feb. 3	Returned the shirt that cost $45.00	
Feb. 15	Made a payment of $100.00	
Feb. 19	Purchased tennis shoes costing $80.25	

2. (a) Summarize the mathematical process used to compute the balance on your account when you made a purchase.

(b) Summarize the mathematical process used to compute the balance on your account when you made a payment.

(c) Summarize the mathematical process used to compute the balance on your account when you returned a purchase.

3. The next month you made additional purchases and payments on the same account. Represent the activity on the account by writing addition and subtraction *symbolically*. Complete the table.

Date	Activity on the account (purchases, returns, and payments)	Summarize the activity symbolically	Balance (in $)
Feb. 19	Balance brought forward		−131.00
Feb. 20	Purchased gas costing $15.00	−131.00 + (−15.00)	−146.00
Feb. 27	Purchased a pair of jeans costing $55.50		
Mar. 3	Purchased dinner costing $45.75		
Mar. 12	Returned the jeans that cost $55.50		
Mar. 15	Made a payment of $100.00		
Mar. 16	Balance in account		

4. When playing golf, players count the number of times they hit the ball on each hole. The object is to score a PAR or less on each hole. PAR is a predetermined number of strokes for a given hole. If the golfer's score on a particular hole is above PAR, the number of strokes above PAR is recorded as positive and written with a + sign. If the golfer's score is below PAR, the number of strokes under PAR is recorded as negative and written with a − sign. The sum of the total number of strokes above or below PAR is used in golf tournaments to summarize who is ahead in the tournament.

(a) Compute the signed number score from PAR on each hole: Golfer's score − PAR and record the result on the bottom row of the scorecard.

Hole	1	2	3	4	5	6	7	8	9	Sum of scores
PAR	4	3	5	4	4	3	5	4	5	37
Golfer's Score	7	4	4	3	4	4	4	7	4	
Signed number score: Golfer's score − PAR	7 − 4 = +3	4 − 3 = +1	4 − 5 = −1							

(b) Find the sum of the golfer's scores on all nine holes.

(c) How many strokes from PAR (either above or below) was the golfer's score?

5. Describe two real-world situations involving addition or subtraction of negative numbers.

(a)

(b)

Discussing Mathematics

Positive and negative numbers are called **signed numbers**. Often, we perform the operations with signed numbers in our heads, without thinking consciously of the mathematical steps involved. The mathematical statements in problems 1 and 2 can be grouped into three categories. Guidelines for performing operations in each category will be discussed.

CATEGORY 1: ADDITION—ADDENDS WITH *LIKE* SIGNS	
Jan. 26	$-45.00 + (-150.00) = -195.00$
CATEGORY 2: ADDITION—ADDENDS WITH *UNLIKE* SIGNS	
Feb. 15	$-150.75 + (100.00) = -50.75$
CATEGORY 3: SUBTRACTION	
Feb. 3	$-195.75 - (-45) = -150.75$

ADDITION OF REAL NUMBERS

One way to visualize the sum of signed numbers is to draw the sum on the number line. Making a mental picture of the process might help to arrive and/or confirm your answers. After practice, you may not choose to actually draw a number line for each problem. Recall that the numbers being added together are called **addends**.

VISUALIZING ADDITION ON THE NUMBER LINE	
Step 1:	Begin at zero on a number line and count the appropriate number of units of the first addend, moving in the direction of its sign. If the number is positive, move to the right. If the number is negative, move to the left. Rest your pencil on that number.
Step 2:	Count the appropriate number of units of the second addend, moving in the direction of its sign. Your pencil is now on the number representing the sum and final answer.

EXAMPLE 1: The credit card transaction on Jan. 26 (p. 24) is an example of two addends with LIKE signs. Add $-45 + (-150)$ on the number line.

Step 1: Begin at zero and move 45 units in the NEGATIVE direction. Rest your pencil on -45.

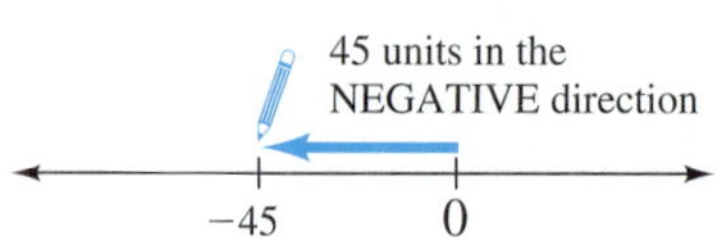

Step 2: Move 150 units from -45 in the NEGATIVE direction. Your pencil is on -195. $-45 + (-150) = \mathbf{-195}$

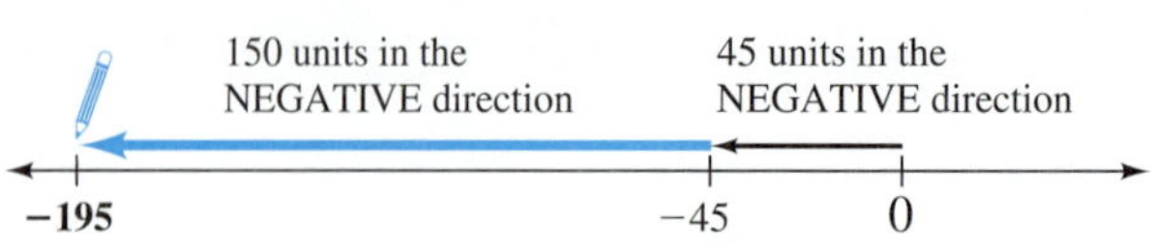

EXAMPLE 2: The credit card transaction on Feb. 15 (p. 24) is an example of two addends with UNLIKE signs. Add $-150.75 + (+100.00)$ on the number line.

Step 1: Begin at zero and move 150.75 units in the NEGATIVE direction. Rest your pencil on -150.75.

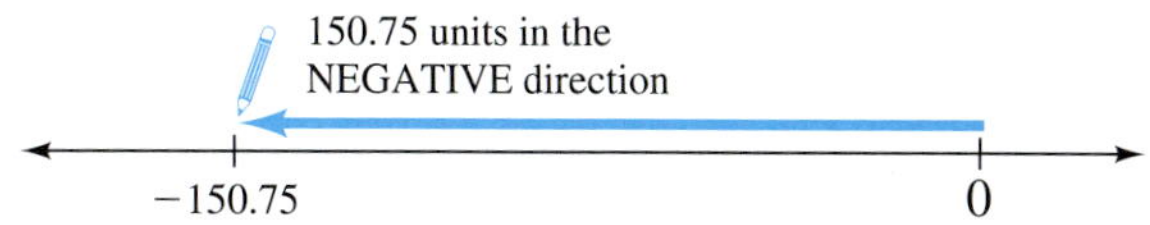

Step 2: Move 100 units from -150.75 moving in the POSITIVE direction. Your pencil is on -50.75. $-150.75 + (+100) = \mathbf{-50.75}$

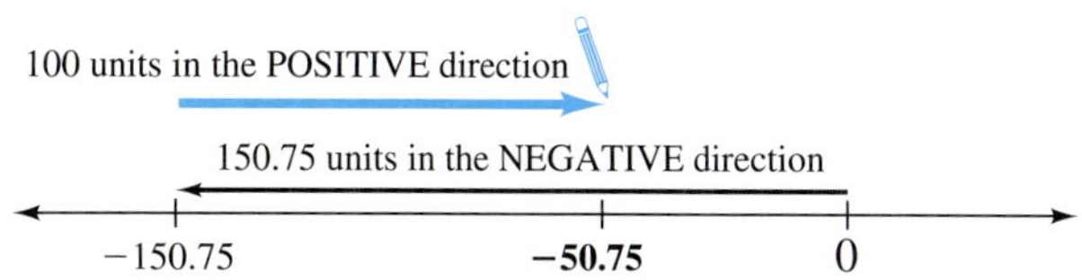

➢ *Keep in mind that the + symbol for a positive number is not usually written explicitly. However, when adding or subtracting signed numbers, it may be helpful to clearly write the + symbol on each addend.*

EXAMPLE 3: Add $5.6 + (-3.1)$ on the number line.

Rewrite the problem with the sign of each addend clearly: $(+5.6) + (-3.1)$.

Step 1: Begin at zero and move 5.6 units in the POSITIVE direction. Rest your pencil on 5.6.

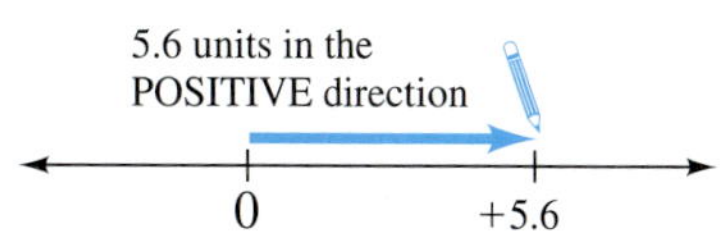

Step 2: Move 3.1 units from $+5.6$ in the NEGATIVE direction. Your pencil is on 2.5. $(+5.6) + (-3.1) = \mathbf{2.5}$

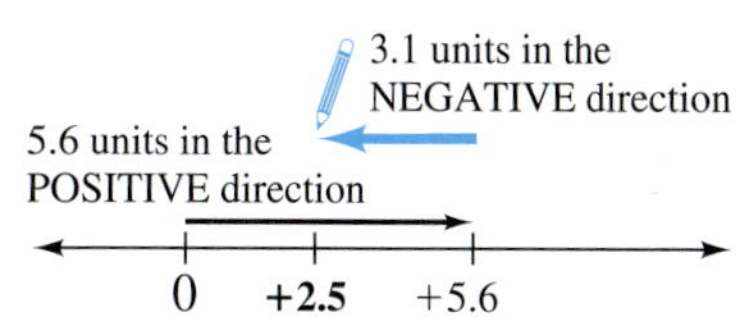

To generalize the process for adding (and subtracting) signed numbers, the concept of absolute value must be introduced. The distance of a number from zero is called the **absolute value** and denoted by vertical bars on each side of the number.

EXAMPLE 4: $|-195| = 195$ because -195 is 195 units from zero.

$|+45| = 45$ because $+45$ is 45 units from zero.

ADDING SIGNED NUMBERS WITH LIKE SIGNS USING ABSOLUTE VALUE (CATEGORY 1)
• Add the absolute value of the addends. • The sign of the sum is the same as the sign of the addends.

ADDING SIGNED NUMBERS WITH UNLIKE SIGNS USING ABSOLUTE VALUE (CATEGORY 2)
• Subtract the absolute value of the addends. • The sign of the sum is the same as the sign of the number farthest from zero (largest absolute value).

EXAMPLE 5: Adding signed numbers on the number line and symbolically.

		Addition on the number line	Symbolically
(a)	$(+5) + (+3) = 8$ LIKE signs	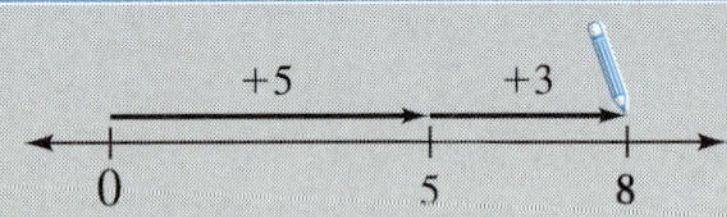 	$(+5) + (+3)$ $= +(\lvert +5 \rvert + \lvert +3 \rvert)$ $= +(5 + 3) = +8$
(b)	$(-5) + (-3) = -8$ LIKE signs	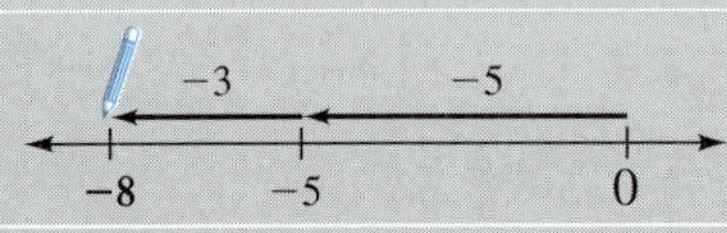 	$(-5) + (-3)$ $= -(\lvert -5 \rvert + \lvert -3 \rvert)$ $= -(5 + 3) = -8$
(c)	$(+5) + (-3) = 2$ UNLIKE signs	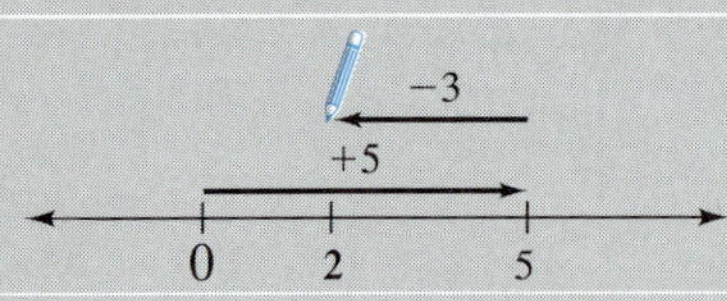 	$(+5) + (-3)$ $= +(\lvert +5 \rvert - \lvert -3 \rvert)$ $= +(5 - 3) = +2$
(d)	$(-5) + (+3) = -2$ UNLIKE signs	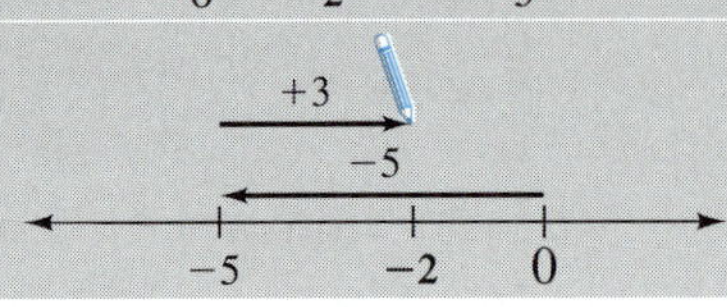 	$(-5) + (+3)$ $= -(\lvert -5 \rvert - \lvert +3 \rvert)$ $= -(5 - 3) = -2$

When the problem involves more than two addends, rearranging the addends may make the arithmetic a bit easier. The Commutative and Associative Properties of addition indicate that the order and grouping of the addends can be changed without affecting the sum.

EXAMPLE 6: The sum of the 9 golf scores of the difference from PAR in Getting Started (p. 25) can be reordered with negative numbers together and positive numbers together:

Original numbers:

$(+3) + (+1) + (-1) + (-1) + 0 + (+1) + (-1) + (+3) + (-1) =$

Reordered:

$(-1) + (-1) + (-1) + (-1) + (+3) + (+1) + 0 + (+1) + (+3) =$

Regrouped:

$[(-1) + (-1) + (-1) + (-1)] + [(+3) + (+1) + 0 + (+1) + (+3)] =$

$$\underbrace{\underset{\text{Sum of the negative numbers}}{[(-4)]} + \underset{\text{Sum of the positive numbers}}{[(+8)]} =}_{+4}$$

SUBTRACTION OF SIGNED NUMBERS

In a subtraction problem, the **subtrahend** is subtracted from the **minuend** to form the result, called the **difference**.

5	Minuend
$-\underline{3}$	Subtrahend
2	Difference

Subtraction is the inverse operation of addition and means to "add the opposite of the number." To subtract on the number line, move in *the opposite direction* of the sign of the subtrahend and add.

EXAMPLE 7: Subtract $5 - (-3)$ on the number line.

Rewrite the problem with the + symbol written $(+5) - (-3)$ explicitly.

Step 1: Begin at zero and move 5 units in the POSITIVE direction. Rest your pencil on $+5$.

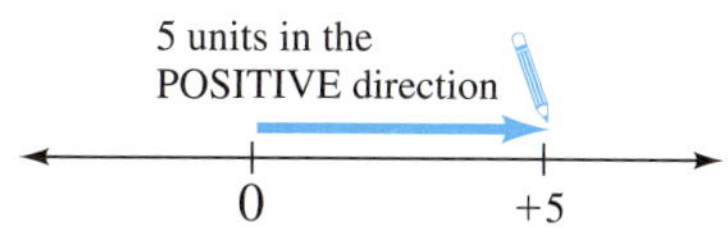

Step 2: The subtrahend is -3. Since -3 is being subtracted from $+5$, move 3 units from $+5$ in the *opposite direction* of -3 (move to the right). Your pencil is on $+8$. $5 - (-3) = \mathbf{+8}$.

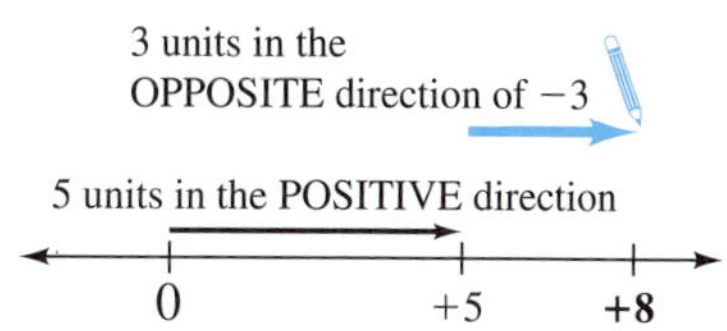

EXAMPLE 8: When considering the return of the shirt on Feb. 3 (p. 24), think of returning an item as "taking away" the fact that we purchased it.

$$-195.75 - (-45)$$

$$= -195.75 + (+45)$$

$$= -150.75$$

To subtract, change the operation to addition *and* change the sign of the subtrahend.
Add.

SUBTRACTING SIGNED NUMBERS (CATEGORY 3)
• Change the sign of the subtrahend. • Change the operation of subtraction to addition. • Follow the guidelines for addition of signed numbers.

Subtraction is sometimes described as "change the sign of the subtrahend and add."

EXAMPLE 9: Subtract the signed numbers.

		First: Change the sign of the subtrahend. Second: Change the operation of subtraction to addition.	**Third: Add the numbers according to the guidelines for addition.**
(a)	$10 - 5$ The problem can be rewritten as $10 - (+5)$.	$= 10 \quad - \quad (+5)$ $= 10 \quad + \quad (-5)$	$= 5$
(b)	$-10 - (5)$ The problem can be rewritten as $-10 - (+5)$.	$= -10 \quad - \quad (+5)$ $= -10 \quad + \quad (-5)$	$= -15$
(c)	$10 - (-5)$	$= 10 \quad - \quad (-5)$ $= 10 \quad + \quad (+5)$	$= 15$
(d)	$-10 - (-5)$	$= -10 \quad - \quad (-5)$ $= -10 \quad + \quad (+5)$	$= -5$

EVALUATING EXPRESSIONS

Positive and/or negative values can be assigned to variables. To evaluate an expression, substitute the value of the variable into the expression, enclosing the number value in parentheses. Then perform the operations following the Order of Operations and the Guidelines for Addition and Subtraction.

EXAMPLE 10: Evaluate the expression $a - b$, given $a = -10$ and $b = -6$.

$$a - b$$

$$(-10) - (-6)$$

Substitute the given values of the variables into the respective positions in the expression, enclosing the values of the variables in parentheses.

Subtract: change the sign of the subtrahend and add.

$$= (-10) + (+6)$$

$$= -4$$

EXAMPLE 11: Evaluate the expression $a^2 - (b - c)$, given $a = -4$, $b = 5$, and $c = -12$.

$$a^2 - (b - c)$$

$$(-4)^2 - ((5) - (-12))$$

Substitute the given values of the variables into the respective positions in the expression, *enclosing the values of the variables in parentheses.*

$$= (-4)^2 - ((5) + (+12))$$

To subtract in parentheses, change the sign of the subtrahend and add.

$$= (-4)^2 - (+17)$$

$$= 16 - (+17)$$

Simplify exponents.

$$= 16 + (-17)$$

Subtract: change the sign of the subtrahend and add.

$$= -1$$

Doing Mathematics

- Write the definitions of the words and symbols listed for this section in the Dictionary in Appendix A.
- On a 3 × 5-inch card, write the Guidelines for Addition and Subtraction of signed numbers for future reference.

1. Record charges made on a credit card in the table. Purchases should be shown as negative numbers, the return of an item shown as subtracting a negative number $(-(-\text{amount}))$, and payments as positive numbers.

Date	Activity on account (purchases, returns, and payments)	Summarize the activity symbolically	Balance (in $)
10/1	Balance brought forward		−35.00
10/3	Purchased a book costing $55.30		
10/4	Purchased a calculator costing $90.65		
10/15	Returned the book that cost $55.30		
10/31	Made a payment of $100		
11/9	Purchased tennis racquet costing $120.85		
11/20	Purchased tennis balls costing $10.05		
11/30	Made a payment of $100		

2. The temperature in Cleveland, Ohio, on January 2 was 25° F at 4 A.M. The temperature dropped 5° at 6 A.M. Then the temperature rose 3° each hour for two hours.

(a) Write a mathematical statement summarizing the changes symbolically.

(b) What was the temperature at 8 A.M.?

3. The high temperature in Anchorage, Alaska on February 17 was −6° C. The low temperature for the day was −16° C.

(a) Write a mathematical statement summarizing the difference between the high and the low symbolically.

(b) What was the overall change in temperature on that day?

Problems 4–10: Use the number line to add or subtract.

4. $4 + 2 =$

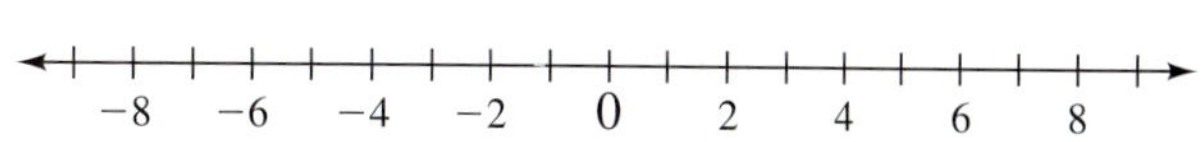

5. $-4 + (-2) =$

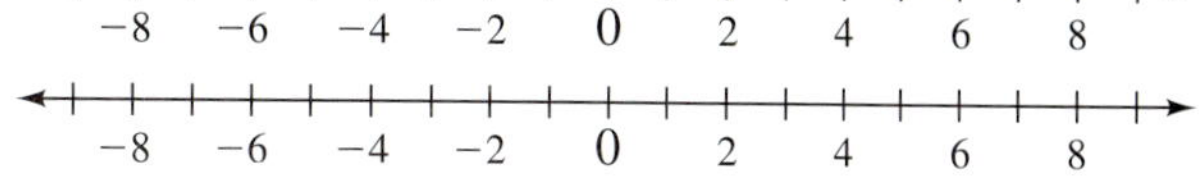

6. $-4 + 2 =$

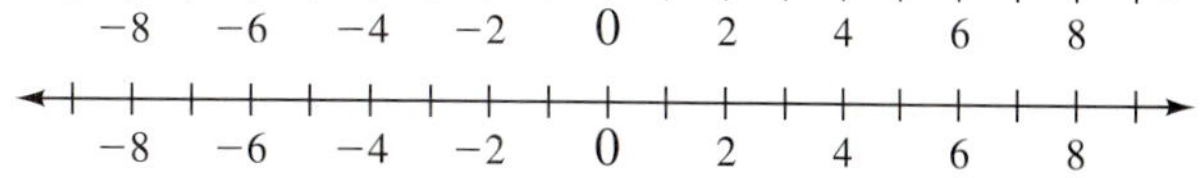

7. $4 - 2 =$

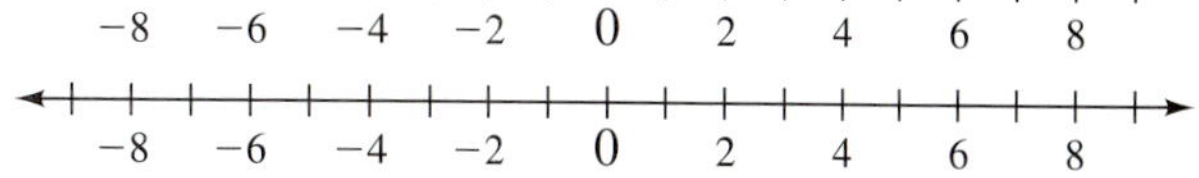

8. $4 - (-2) =$

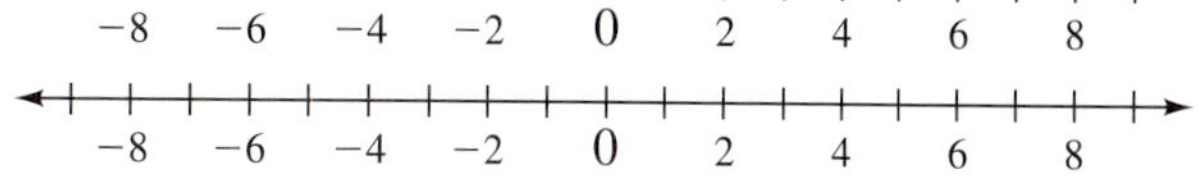

9. $-4 - (-2) =$

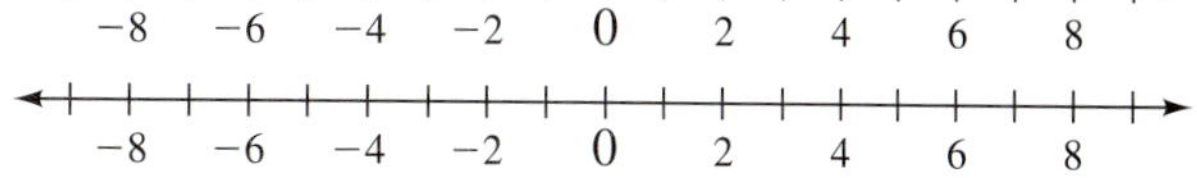

10. $-4 - 2 =$

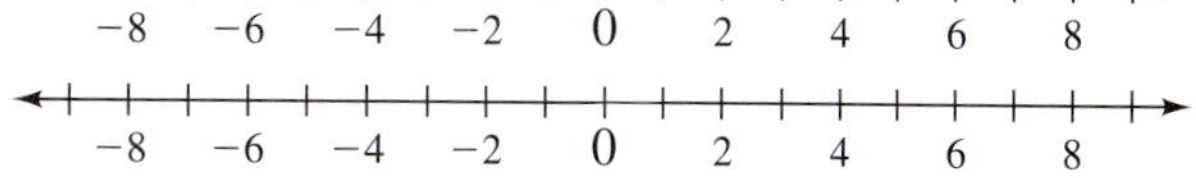

Problems 11–39: Perform the operations.

11. $|-5|$

12. $|-7|$

13. $-|-10|$

14. $-|-2|$

15. $-9 + 10$

16. $-5 + 7$

17. $-5 + (-10)$

18. $-16 + (-6)$

19. $-3.5 + 15.6$

20. $-2.6 + 10.8$

21. $+15 + (-15)$

22. $+3 + (-3)$

23. $-15 + (16)$

24. $-9 + (5)$

25. $2.1 - 5.3$

26. $7.5 - 1.3$

27. $-16 + 3$

28. $-55 + 5$

29. $6 - (-9)$

30. $10 - (-7)$

31. $-7 - 5$

32. $-11 - 2$

33. $-7 - (-5)$

34. $-11 - (-2)$

35. $50 + (-60) - (-10)$

36. $80 + (-5) + (-10)$

37. $-6.2 + (-3.5) - 2.6$

38. $(-2) + (-3) + (5) + (-10) + (-6) + 20$

39. $(-1) - (-3) + (-2) - 6 + (-10) + (-3) - (-5)$

Problems 40–49: The variables x, a, s, and w represent real numbers in the respective locations on the number line. Complete the statements by writing the appropriate symbol, $<$ or $>$, on the line.

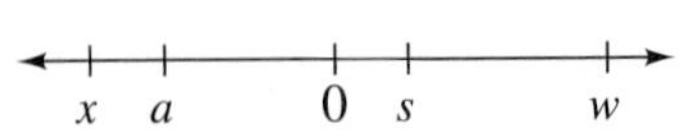

40. $s + w$ ______ 0

41. $|x|$ ______ 0

42. $|s|$ ______ 0

43. $-a$ ______ 0

44. a ______ s

45. $s + w$ ______ a

46. $w - s$ ______ 0

47. $-x$ ______ $-a$

48. $-a$ ______ w

49. $x - a$ ______ 0

Problems 50–54: Evaluate each expression, given that $a = -2$, $b = -3$, and $c = 11$.

50. $a + b + c$

51. $a - b - c$

52. $c - a + b$

53. $a^2 - c$

54. $a^2 + b$

55. List three examples of the use of negative numbers in daily life.

(a) ______________________

(b) ______________________

(c) ______________________

1.4 Multiplication and Division of Real Numbers

OBJECTIVE

- Multiply and divide real numbers.

1 2 3 Getting Started

In this section, we discuss the operations of multiplication and division of signed numbers and identify patterns for the products (result of multiplication) and quotients (result of division).

1. Look at the algebraic signs of the real numbers in the arithmetic problems and identify a pattern in the algebraic signs of the products and quotients.

(a)	$(-5)(6) =$	-30	(b)	$(5)(6) =$	30
(c)	$(5)(-6) =$	-30	(d)	$(-5)(-6) =$	30
(e)	$\frac{-10}{5} =$	-2	(f)	$\frac{10}{5} =$	2
(g)	$\frac{10}{-5} =$	-2	(h)	$\frac{-10}{-5} =$	2

(i) Look at (a), (c), (e), and (g), above. When two real numbers have UNLIKE signs, what is the sign of the product or quotient?

(j) Look at (b), (d), (f), and (h), above. When two real numbers have LIKE signs, what is the sign of the product or quotient?

➢ *Multiplication is the result of "repeated addition." The product of a negative real number and a positive real number is always a negative real number.*

2. Why is the product of two negative numbers positive? Look for a pattern in Column 2.

(a) Use the pattern to complete the last three rows in Column 2.

	Column 1	Column 2: Product	
In Column 1 the number -6 is multiplied by integers, starting with 5 and decreasing by 1.	$5 \cdot (-6) =$	-30	(b) What is the pattern of the numbers in Column 2?
	$4 \cdot (-6) =$	-24	
	$3 \cdot (-6) =$	-18	
	$2 \cdot (-6) =$	-12	
	$1 \cdot (-6) =$	-6	
	$0 \cdot (-6) =$	0	
	$-1 \cdot (-6) =$	$+6$	
	$-2 \cdot (-6) =$	$+12$	
	$-3 \cdot (-6) =$		
	$-4 \cdot (-6) =$		
	$-5 \cdot (-6) =$		

(c) What is the sign of the product of two negative real numbers?

3. Recall the vocabulary of division $\frac{\text{Dividend}}{\text{Divisor}} = \text{Quotient}$ and that the product of the quotient and the divisor is always equal to the dividend: Divisor · Quotient = Dividend. Use multiplication to check the quotients. Complete the blanks.

(a) Given $\frac{10}{-5} = -2$: Divisor · Quotient = Dividend

_______ · _______ = _______

(b) Given $\frac{-10}{5} = -2$. Divisor · Quotient = Dividend

_______ · _______ = _______

(c) Given $\frac{-10}{-5} = 2$. Divisor · Quotient = Dividend

_______ · _______ = _______

(d) What is the algebraic sign of the quotient of two real numbers with UNLIKE signs?

(e) What is the algebraic sign of the quotient of two real numbers with LIKE signs?

4. Use the observations from the problems above to perform these operations.

(a) $(3)(-7)$

(b) $(-3)(-7)$

(c) $\frac{-20}{4}$

(d) $\frac{-20}{-4}$

1 2 3

Discussing Mathematics

MULTIPLICATION AND DIVISION OF REAL NUMBERS

Multiplication is the result of repeated addition.

- The product $(5)(6)$ is equivalent to 6 added together 5 times.

$$(5)(6) = (6) + (6) + (6) + (6) + (6) = 30$$

- The product $(5)(-6)$ is equivalent to -6 added together 5 times.

$$(5)(-6) = (-6) + (-6) + (-6) + (-6) + (-6) = -30$$

- The product $(-5)(-6)$ is equivalent to $-(5)(-6)$, the opposite of -6 added together 5 times.

$$\begin{aligned}(-5)(-6) &= -(5)(-6) \\ &= -[(-6) + (-6) + (-6) + (-6) + (-6)] = -[-30] = 30\end{aligned}$$

Similar examples can be shown to illustrate that division is the result of repeated subtraction. These patterns can be generalized into the following Guidelines for Multiplication and Division of real numbers and are grouped into two categories, depending on whether the signs of the numbers being multiplied or divided are the same (LIKE signs) or are different (UNLIKE signs).

GUIDELINES FOR MULTIPLICATION AND DIVISION

	MULTIPLICATION	DIVISION
The product or quotient of real numbers with **LIKE** signs is *positive*.	$(+)(+) = (+)$	$\frac{(+)}{(+)} = (+)$
	$(-)(-) = (+)$	$\frac{(-)}{(-)} = (+)$
The product or quotient of real numbers with **UNLIKE** signs is *negative*.	$(+)(-) = (-)$	$\frac{(+)}{(-)} = (-)$
	$(-)(+) = (-)$	$\frac{(-)}{(+)} = (-)$

EXAMPLE 1: Multiply.

(a) $(5)(3) = 15$
(b) $(-5)(-3) = 15$
} The product of real numbers with LIKE signs is positive.

(c) $(5)(-3) = -15$
(d) $(-5)(3) = -15$
} The product of real numbers with UNLIKE signs is negative.

EXAMPLE 2: Divide.

(a) $\frac{15}{5} = 3$

(b) $\frac{-15}{-5} = 3$

The quotient of real numbers with LIKE signs is positive.

(c) $\frac{15}{-5} = -3$

(d) $\frac{-15}{5} = -3$

The quotient of real numbers with UNLIKE signs is negative.

ORDER OF OPERATIONS

The Order of Operations and properties discussed in Section 1.2 applies to all real numbers.

ORDER OF OPERATIONS	
First	**P**arentheses or Grouping Symbols (if nested, begin with the innermost operation)
Second	**E**xponents
Third	**M**ultiplication or **D**ivision (left to right)
Fourth	**A**ddition or **S**ubtraction (left to right)

➢ *Be careful to apply the Order of Operations appropriately when simplifying signed numbers expressions with parentheses and exponents.*

EXAMPLE 3: Simplify each problem using the Order of Operations.

(a) $(-5)^2$ — Simplify exponent.

$= (-5)\cdot(-5)$ — The quantity inside the parentheses is multiplied two times.

$= +25$

(b) -5^2 — Simplify exponent.

$= -(5)(5)$ — Square the base 5. The negative sign is an instruction to take the opposite of 25.

$= -25$

(c) $(-4)^2 \div (-8)(-3)$

$= 16 \div (-8)(-3)$ — Simplify exponent.

$= (-2)(-3)$ — Divide.

$= 6$ — Multiply.

(d) $\frac{-4^2 - 2}{3}$

$= \frac{-16 - 2}{3}$ — Simplify exponent.

$= \frac{-18}{3}$ — Subtract (the fraction bar acts as a grouping symbol).

$= -6$ — Divide.

EXAMPLE 4: Simplify each problem using the Order of Operations.

(a) $\left(-\frac{3}{4}\right)\cdot\left(-\frac{5}{2}\right)-\left(-\frac{1}{2}\right)^3$

$=\left(-\frac{3}{4}\right)\cdot\left(-\frac{5}{2}\right)-\left(-\frac{1}{8}\right)$ Simplify exponent.

$=\left(\frac{15}{8}\right)-\left(\frac{-1}{8}\right)$ Multiply.

$=\left(\frac{15}{8}\right)+\left(\frac{+1}{8}\right)$ Subtract (change sign and add).

$=\frac{16}{8}=2$ Reduce the fraction.

(b) $-3^2[2^3(-6-(-4))]$

$=-3^2[2^3(-6+(+4))]$ Subtract inside the innermost parentheses.

$=-3^2[2^3(-2)]$ Change the sign and add.

$=-3^2[8(-2)]$ Simplify exponent inside brackets.

$=-3^2[-16]$ Multiply inside brackets.

$=-9[-16]$ Simplify exponent outside of brackets.

$=144$ Multiply.

EVALUATING EXPRESSIONS

Use great care when applying the Order of Operations involving variables. *Enclose the value of a variable in parentheses and substitute.*

EXAMPLE 5: Evaluate each expression for the given values.

(a) Evaluate $-a^2-(b-c)$, given $a=-5$, $b=-3$, and $c=2$:

$=-(-5)^2-((-3)-(2))$ Substitute the values for the variables.

$=-(25)-(-5)$ Simplify.

$=-25+(+5)=-20$

(b) Evaluate a^2-b^2, given $a=-5$ and $b=-3$:

$=(-5)^2-(-3)^2$ Substitute the values for the variables.

$=(+25)-(+9)=16$ Simplify.

1 2 3 Doing Mathematics

- Write the definitions of the words and symbols listed in this section in the Dictionary in Appendix A.
- On a 3 × 5-inch card, write the Guidelines for Multiplication and Division of signed numbers for future reference.

Problems 1–24: Perform the operations.

1. $(4)(2)$

2. $(-4)(2)$

3. $(4)(-2)$

4. $(-4)(-2)$

5. $\dfrac{4}{2}$

6. $\dfrac{-4}{2}$

7. $\dfrac{4}{-2}$

8. $\dfrac{-4}{-2}$

9. $(-6)(-2)$

10. $(+5.32)(-2.1)$

11. $(-6)^2$

12. -6^2

13. $-(-6)^2$

14. $(-2)^3$

15. $\left(\dfrac{-3}{5}\right)\left(\dfrac{+4}{3}\right)$

16. $-10 \div (+5)$

17. $\dfrac{-2}{5} \div \left(\dfrac{-8}{15}\right)$

18. $0 \div (-2.4)$

19. $\left(\dfrac{-1}{2}\right)^3$

20. $\dfrac{-16.75}{-0.25}$

21. $\dfrac{-61}{2} - \left(\dfrac{5}{2}\right)$

22. $-\dfrac{2^2}{3}$

23. $-63(10)$

24. $5(-80)$

Problems 25–40: Perform the operations. Compute the answers, step-by-step.

25. $10 + 1 - 5 \cdot 2 \div 5$

26. $2^2 + 8 \div (-2)$

27. $(-2)^2 \cdot 12 \div 6 + (-5)$

28. $-2^2 + 3(2 - 6)$

29. $6(4 - 6)^2 \cdot 3 + 7$

30. $(3 - 5)^2 + (-9) \cdot (-7) - 100$

31. $(-5)(-6)(-1)(-2)$

32. $\dfrac{12 \div (-4) \cdot (-2)^3}{5 + 7}$

33. $(+7) + (-2) + (-3) + 3 + 5 + (-2) + (-10)$

34. $(-1)(-1)(-1) - (-2)(-2)(-2)$

35. $\left[\dfrac{4}{25} - \left(\dfrac{-1}{5}\right)^2\right]^2$

36. $-0.35(16.32 - 18.52)$

37. $\dfrac{[-3^2 - (-4)]}{(-2)^2 + 1} - (-5)^2$

38. $3(5 - 11) - (-2)^3 \cdot (3)$

39. $-2^4[-2^3(-2^2)]$

40. $-5[-(1 + 3)^2 - (-2^3 - (11 - 4))]$

Problems 41–48: The variables x, a, s, and w represent real numbers in the respective locations on the number line. Complete each statement by writing the appropriate symbol, $<$ or $>$, on the line.

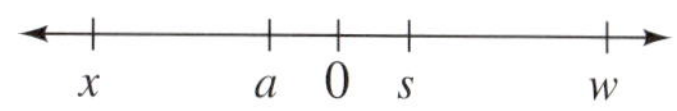

41. $s \cdot w$ ______ 0

42. $s \cdot a$ ______ 0

43. $\dfrac{-a}{s}$ ______ 0

44. $a \cdot \dfrac{-w}{x}$ ______ s

45. $s + w$ ______ a

46. $w - a$ ______ 0

47. $-x \cdot s$ ______ 0

48. $-a \cdot (-s)$ ______ w

Problems 49–55: Evaluate each expression, given that $a = 2$, $b = -3$, and $c = -1$:

49. $a \cdot b \cdot c$

50. $\dfrac{a \cdot b}{c}$

51. $\dfrac{a + b}{c}$

52. $b^2 + a$

53. $a^2 + b^2$

54. $-a^2 \cdot b^2$

55. $a^2 \cdot b^2 \cdot c^2$

1.5 Problem-Solving Strategies and Mathematics Study Skills

OBJECTIVES

- Identify, describe, and use different problem-solving strategies to solve problems.
- Discuss important study skills needed to achieve success in studying mathematics.

Getting Started

The goal in this section is to solve real-world problems, identify multiple methods or strategies used to solve the problems, and discuss mathematics study skills that contribute to success in this course. You already have all the mathematics skills you need to solve the problems given. As you work on each problem, think about which mathematical processes you are using and be able to describe the method you employed.

1. You have been asked to purchase favors for the children who are going to attend a birthday party at your home. You have decided to purchase yo-yos and boxes of crayons as the favors. Each yo-yo costs \$3 and a box of crayons costs \$2. You will

spend a total of \$49 on the favors. Every child attending will receive one favor and 20 children will be at the party.

(a) How many yo-yos should you buy?

(b) How many boxes of crayons should you buy?

(c) Describe the method or strategy you employed to solve the problem.

2. You own a company that manufactures videos. The fixed operating expense to produce any number of the videos is \$1000. Each video will sell for \$8. The word *revenue* refers to the income produced from the sale of the videos and is calculated by multiplying the number of videos sold by \$8. How much profit will be made on the sale of 125 videos? 500 videos? (Hint: Profit = Revenue − Operating Expenses)

(a) Calculate the profit if 125 videos are sold.

(b) Calculate the profit if 500 videos are sold.

(c) *Break even* means that the revenue and the operating expenses are equal and the profit is zero. How many videos must be sold to break even?

(d) Describe the method or strategy you employed to answer the questions above.

3. Compare your methods of solution with another student. List two other strategies that could have been used to solve problems 1 and 2.

(a) ______________________________

(b) ______________________________

Good study habits and skills are important when studying any course. Some skills needed to be successful in a mathematics course are different from the skills needed to succeed in other courses. Answer questions 4–10 to identify some of your personal strengths or weaknesses regarding your mathematics study skills.[1] Choose the statement that best describes your habits of studying and learning mathematics.

4. The time I spend studying math is

(a) 1 to 2 hours a week.

(b) 3 to 5 hours a week.

(c) More than 5 hours a week.

5. My attendance record in math class is as follows:

(a) I attend class only on test days.

(b) I often miss class.

(c) I seldom miss class.

6. When I am in math class,

(a) I seldom write all the steps to a problem.

(b) I often write all the steps to a problem.

(c) I almost always write all the steps to a problem.

7. When I become confused in math class,

(a) I seldom ask questions of the instructor.

(b) I often ask questions of the instructor.

(c) I almost always ask questions of the instructor.

[1] Questions 4–10 were adapted from Paul D. Nolting, *Winning at Math: Your Guide to Learning Mathematics Through Successful Study Skills* (Bradenton: Academic Success Press, 1997), 23–27.

8. With math homework assignments,
   (a) I almost always fall behind.
   (b) I often fall behind.
   (c) I seldom fall behind.

9. When I try to read my math textbook and get stuck,
   (a) I almost always stop reading.
   (b) I often stop reading.
   (c) I seldom stop reading.

10. When I think about my abilities in mathematics,
    (a) I never think about my math abilities.
    (b) I believe that I cannot do well in math courses.
    (c) I usually believe that I can do well in math courses.

11. Score your answers above by awarding 1 point for each question answered (a), 2 points for each question answered (b), and 3 points for each question answered (c).
    - A score of 10 or below means that you have weak study skills.
    - A score between 11 and 17 means that you have good study skills, but there is room to improve.
    - A score above 17 means that you have excellent study skills.

12. List two actions that you can employ to improve your study skills and maximize your success in this course:
    (a) ______________________________
    (b) ______________________________

Discussing Mathematics

PROBLEM-SOLVING STRATEGIES

Each person may approach a mathematics problem differently and employ a different problem-solving strategy. While one strategy may be more efficient than another, it is important to find the solution to the problem by any method. Some problem-solving strategies are listed below for solving problems 1 and 2.

Draw a Picture or Graph Pictures and graphs can be useful to visualize and figure out the solution to a problem.

Draw 20 squares representing the 20 favors to be purchased.

Write \$2 in each square and find the total in all squares. The total spent is \$40, which is too low. We need to spend 9 more dollars in order to spend the entire \$49.

\$2	\$2	\$2	\$2	\$2	\$2	\$2	\$2	\$2	\$2
\$2	\$2	\$2	\$2	\$2	\$2	\$2	\$2	\$2	\$2

Add \$1 in each of 9 squares to bring the total to \$49.

\$2 **+\$1**	\$2 **+\$1**	\$2 **+\$1**	\$2 **+\$1**	\$2 **+\$1**	\$2 **+\$1**	\$2 **+\$1**	\$2 **+\$1**	\$2 **+\$1**	\$2
\$2	\$2	\$2	\$2	\$2	\$2	\$2	\$2	\$2	\$2

Look at the total number of squares containing \$3 and the total number of squares containing \$2: 9 squares contain \$3 and 11 squares contain \$2. Therefore, 9 yo-yos at \$3 and 11 boxes of crayons at \$2 need to be purchased, for a total of \$49.

Guess and Check An educated guess can be the start of the process of finding the answer to a problem. If the first guess doesn't lead to solution, a second guess is made based on the outcome of the first guess. Don't give up. Revise your first guess and try again. The guess and check strategy may not always be the most effective strategy, but it can lead to solution. One student suggested the following method. There are other processes you might have chosen.

Since each yo-yo costs \$3, divide 3 into \$49.

$$\frac{49}{3} = 16.\overline{3} \approx 16$$

Use 16 as your first guess for the number of yo-yos.

20 favors are needed, therefore, 4 boxes of crayons must be purchased if 16 yo-yos are purchased.

Compute the total cost of 16 yo-yos (at \$3) and 4 boxes of crayons at \$2. We have only \$49 to spend.

$$16 \cdot \$3 + 4 \cdot \$2 = \$48 + \$8 = \$56$$

The total \$ is too high.

Adjust your guess to 10 yo-yos and 10 boxes of crayons. Compute the total cost.

$$10 \cdot \$3 + 10 \cdot \$2 = \$30 + \$20 = \$50$$

Almost! The total amount is still too high. Make another guess of 9 yo-yos and 11 boxes of crayons.

Since the total was too high, adjust your guess again to 9 yo-yos and 11 boxes of crayons. Compute the total cost.

$$9 \cdot \$3 + 11 \cdot \$2 = \$27 + \$22 = \$49$$

The right answer!

Make a Table or List and Look for a Pattern The information given in a problem or obtained from arithmetic operations can be organized in a table to identify a pattern that leads to a solution.

Number of \$3 favors	Number of \$2 favors	Computations	Total \$ spent
16	4	$16 \cdot \$3 + 4 \cdot \$2$	\$56
12	8	$12 \cdot \$3 + 8 \cdot \$2$	\$52
10	10	$10 \cdot \$3 + 10 \cdot \$2$	\$50
9	11	$9 \cdot \$3 + 11 \cdot \$2$	\$49 Correct answer

Create an Equation One or more equations can produce the solution to a problem. You are not expected to know how to create or solve the equations above at this time. We will come back to these equations later in the text.

Problem 1—the yo-yo problem: To create one equation representing the problem situation, let the variable x represent the number of yo-yos to be purchased at \$3 each. The number of boxes of crayons is $(20 - x)$.

Given: yo-yos cost \$3 each; boxes of crayons cost \$2 each.

$\$3 \cdot \text{number of yo-yos} + \$2 \cdot \text{number of boxes of crayons} = \$49$

$$3x + 2(20 - x) = 49.$$

Solution: $x = 9$ yo-yos and $y = 11$ boxes of crayons to be purchased.

Problem 2—the video problem: A pattern in the computations and profit in problem 2 can be observed and generalized into an equation to compute other values. Variables can be used as placeholders for unknowns in the problem: the variable x represents the unknown number of videos to be sold and the variable P represents the profit in this table. The generalized equation can be used to calculate other profits by substituting a value for x into the equation.

x = number of videos sold	Computations	P = Profit (in dollars)
50	$8(50) - 1000$	-600
100	$8(100) - 1000$	-200
125	$8(125) - 1000$	0 (break even)
150	$8(150) - 1000$	200
1000	$8(1000) - 1000$	7000
Generalize x	$8(x) - 1000$	$P = 8x - 1000$

➢ ***Which problem-solving strategy is best?*** *The information given in a problem may determine an effective strategy to use. Multiple strategies are appropriate and acceptable. At this point in the course, you are not expected to create and solve equations. We will do this later in this text. The goal is to choose a strategy and work toward solving the problem. As we develop additional algebraic skills, we will be better equipped to identify and use more efficient strategies.*

GEORGE POLYA'S PROBLEM-SOLVING METHOD

Many great minds have thought about the most effective strategies for solving mathematics problems. Professor George Polya[2] outlined a four-step problem-solving method[3] with the express purpose of helping students solve problems:

I. **Understanding the problem:** You have to *understand* the problem. What is the unknown? What are the data?

II. **Devising a plan:** Find the connection between the data and the unknowns. Have you seen it before? Or have you seen the same problem in a slightly different form? Do you know a related problem?

III. **Carrying out the plan:** Can you see clearly that the step(s) are correct? Check each step.

IV. **Looking back:** Examine the solution. Is it reasonable? Can you check the result? Can you derive the result differently? Revise the plan and try it again.

Professor Polya was a proponent for trying any and all methods to solve a problem. One of his famous quotes is the following: "A great discovery can solve a great problem but there is a grain of discovery in the solution of any problem. Your problem may be modest, but it challenges your curiosity and brings into play your inventive facilities, and if you solve it by your own means, you may experience the tension and enjoy the triumph of discovery."

EXAMPLE 1: Use Professor Polya's four-step process to solve this problem:

Janice is having a fence installed around the boundary of her property. The dimensions of the property are shown in the figure below. If the fence cost $5.00 per linear foot plus $100 charge for installation, how much will the fence cost?

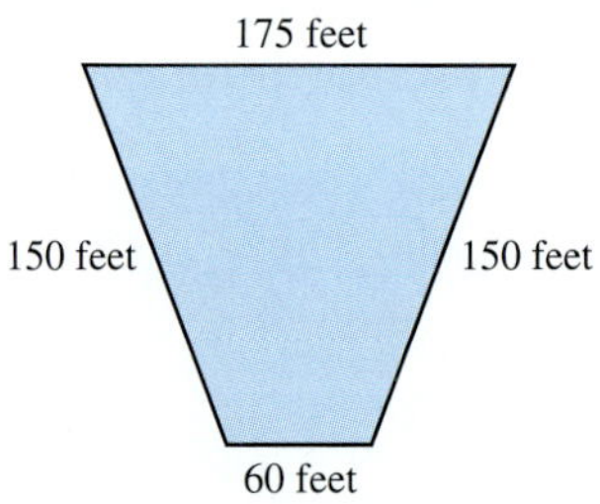

[2]George Polya (1887–1985) received his doctorate in mathematics from the University of Budapest in 1912. His work in mathematics included papers on series, number theory, combinatorics, voting systems, astronomy, probability, and integral functions. After moving to the United States, he held faculty positions at Boston University and Stanford University.

[3]George Polya, *How to Solve It: A New Aspect of Mathematical Method* (Princeton: Princeton University Press, 1973) xvi–xvii.

I. Understanding the problem	• The dimensions of each of the four sides of the property are given. • The total cost of the project includes the cost of the fencing, plus installation. • The unkown is the cost of the fence to be installed on all four sides.
II. Devising a plan	• Choose a problem-solving strategy using the picture and understanding of the perimeter of a geometric figure. • The sum of the lengths of the four sides of the property is called the perimeter. • Find the perimeter of the property by adding the lengths of the four sides. Multiply the sum by \$5. Add the cost of installation.
III. Carrying out the plan	• Calculate the cost of the fencing: Perimeter = 175 + 150 + 60 + 150 = 535 feet Total cost = \$5·535 + \$100 = \$2775
IV. Looking back	• Check the answer: $5(175 + 150 + 60 + 150) + 100 = 5 \cdot 535 + 100 = 2775$ • The total cost of the fence is \$2775.

STUDY SKILLS IN MATHEMATICS

Recognizing important mathematics study habits and skills is the first step to improving your skills and maximizing your performance in this course. Read the following carefully and think about what you can do to improve your study habits.

- **Studying mathematics requires concentration and effort.** "In a math course, you must be able to do three things: (1) *understand* the material, (2) *process* the material, and (3) *apply* what you have learned to correctly solve a problem."[4] In other subjects you need to understand the material, but you may not have to actually solve a problem. For example, in political science you learn about politics, but you don't become a political candidate yourself. In mathematics, you must be able to solve all the problems *yourself* to be successful. You *must* practice.
- **Mathematics *is* a foreign language.** Words and symbols have special meanings in mathematics. Definitions and formulas must be memorized. Speaking, writing, translating, and understanding the language of mathematics is a learned skill.
- **Mathematics should be approached aggressively.** A college mathematics course usually moves about twice as fast as a similar high school class. In college courses, persistence is not enough. Just attending class is not enough. You need to prove that you know the material to earn a passing grade. Lectures will rarely cover every detail of a particular section or chapter. You will be expected to supplement the lectures by reading the text, working exercises, discussing the con-

[4]Nolting, *Winning at Math,* 32.

cepts with other students, asking questions of the instructor in and out of class, talking to tutors, and looking for information in other texts.

- **It is important to do all the homework to pass a mathematics course.** In mathematics, don't wait until the night before the test to cram. Practice solving problems every day and keep up with the assignments to develop confidence. *Don't let yourself get behind.* It's almost impossible to catch up in a mathematics class once you fall behind. Plan on two hours of homework time for each hour spent in the classroom.

WHAT CAN YOU DO TO BE PREPARED FOR CLASS AND IMPROVE YOUR STUDY SKILLS?

- Arrive to class on time each day.
- Come to class prepared. Bring pencils, paper, erasers, a ruler, graph paper, colored pens, a calculator, and a positive attitude.
- Introduce yourself to other students in your class. Arrange time outside of class to study together.
- With your instructor's permission, make a tape recording of lectures and replay them to clarify concepts that may be unclear.
- If your instructor permits the use of a calculator in class, make sure you are knowledgeable about how to use it.
- Rewrite or review your notes as soon as possible after class is over, filling in gaps if you can. Make a list of concepts that are unclear. Ask your instructor about concepts that need clarification.
- Read ahead in the book.
- Study at least two hours outside of class for each hour in class.
- Don't put off doing homework until the last minute. Complete all assignments each day and be sure you can do each problem without looking at your notes. It may be necessary to complete the problems more than one time.
- Try to study for exams in an environment similar to the environment in which you'll take the exam. You can't change your test environment, but you can change your study environment.

HOW CAN YOU IMPROVE YOUR MATHEMATICS TEST-TAKING SKILLS?

Before the Test

- Start studying at least three days before the test.
- Get a good night's sleep.
- Review your notes the morning of the test. Do not try to learn new materials the day of the test.

During the Test

- When you receive your test, turn it over and write down any important formulas, definitions, or facts that you might need during the test. This "memory data dump" provides a mental and physical list to refer to during the test.

- Read the directions carefully.
- Look over the entire test before you do any problems. Put a checkmark next to the problems that can be completed quickly. Go back and complete the problems with the checkmarks, skipping more difficult problems.
- Go back and complete more difficult problems.
- Do not spend too much time on any one problem.
- After completing as many problems as possible, review the entire test. Write something down for each problem, even if it is only the first step. Perhaps you will receive partial credit for writing one or two steps that are correct.

After the Test

- Test-taking should be a learning experience. Use returned tests to identify concepts where you need practice and to improve study skills for the next test.
- Rework each problem that you missed. Check with your instructor to make sure your reworked answers are correct.
- Keep your returned tests and corrected answers to use to study for the final exam.

STUDENT RIGHTS AND RESPONSIBILITIES

The first step in improving mathematics skills is to *take charge* of your own learning and develop confidence in your abilities. To do this, begin by recognizing your rights and responsibilities as a student:

What Are My RIGHTS as a Student?

- You have the right to be treated fairly and with respect.
- You have the right to expect your teacher to come to class prepared and on time.
- You have the right to expect that all the material in the course will be presented in a clear, concise manner, which facilitates your learning.
- You have the right to a good learning environment.
- You have the right to ask any questions you may have.
- You have the right to know what is expected of you in the classroom and course.

You also Have RESPONSIBILITIES that Accompany These Rights.

- You have an obligation to treat other students and your instructor politely and respectfully.
- You have an obligation to come to class prepared for the topics to be discussed.
- You have an obligation to be on time, to participate in the class discussions, and to promote a good learning environment for yourself and others.
- You have a responsibility to listen to your instructor and other students.
- You have a responsibility to ask questions when you have them.
- You have an obligation to make every effort to learn the material.

Many books have been written about problem solving, improving study skills, reducing mathematics anxiety, and becoming a better mathematics student. A list of some resources is found in Appendix C.

Doing Mathematics

- On 3 × 5-inch cards, write the following for future reference:
  - (a) Polya's Problem-Solving Method
  - (b) Four problem-solving strategies outlined in this section and a short description of each
  - (c) At least three actions you can employ to improve your study skills and maximize your success in this course

Problems 1–13: Solve each problem using any method. Show your work. Describe the problem-solving strategies you used. You are not expected to create or solve equations. A guess and check strategy is appropriate.

1. You are going on vacation and plan to rent a car. There is a fixed rental fee of $45 to rent the car, plus a charge of $0.15 for each mile driven.

(a) What is the cost of driving the car 500 miles?

(b) How many miles can you drive if you have $210 to spend on car rental?

(c) Describe the problem-solving strategy you used to solve the problem.

2. You are going to rent a party room for a reception. There is a fixed rental fee of $500 for the room, plus a charge of $12.00 per person for snacks.

(a) What is the cost of a party with 100 persons attending?

(b) How many persons can you invite if you have $2900 to spend on the party?

(c) Describe the problem-solving strategy you used to solve the problem.

3. A farmer would like to use 400 square feet of his acreage to make a square pig pen. (Hint: Area of a square = Length of the Side · Length of the Side)

Area =
400 square feet

(a) Determine the length of each side of the pen.

(b) Describe the problem-solving strategy you used to solve the problem.

4. Denise is planning a rectangular garden and needs 120 square feet of area in which to plant the flowers she has chosen. She wants one side of the garden to be 15 feet long. (Hint: Area of a rectangle = Length · Width)

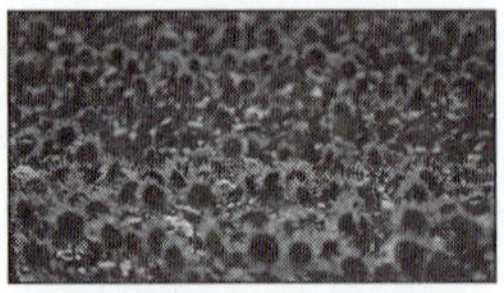

Area =
120 square feet

(a) Determine the dimension of the other side of the garden.

(b) Describe the problem-solving strategy you used to solve the problem.

5. A stripe is to be painted around the edge of a new drug store. The cost of painting the stripe is \$9.25 per foot. The dimensions (in feet) of the building are given in the figure. (Hint: Perimeter is the measure of the length around the outer boundary of the figure.)

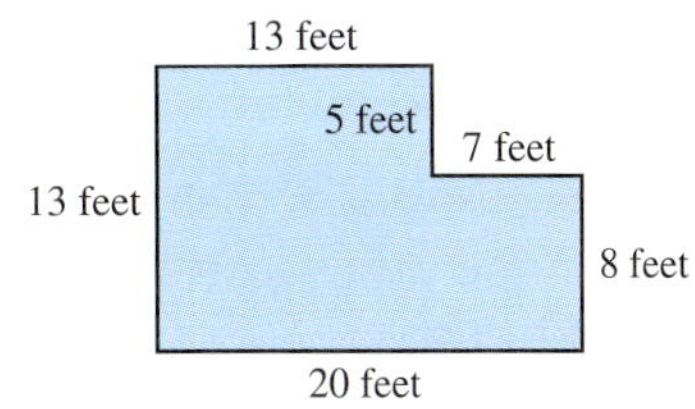

(a) What is the perimeter of the building (around the drug store) in feet?

(b) What is the cost of painting the stripe?

(c) Describe the problem-solving strategy you used to solve the problem.

6. Aunt Jane has baked 64 very large cookies to give to 10 relatives who are coming to Thanksgiving dinner. She plans to put them in large boxes that will hold 8 cookies or smaller boxes that will hold 4 cookies. She plans to give all the cookies away and needs to have 10 filled boxes.

(a) How many large boxes and how many small boxes are needed to have 10 total boxes and distribute exactly 64 cookies?

(b) Describe the problem-solving strategy you used to solve the problem.

7. Kevin signed a cell phone contract that awards 1750 minutes per month with a discount rate. The discount rates are \$0.01 per minute for each local call and \$0.02 per minute for each long-distance call. Kevin is able to spend \$25.00 per month for his cell phone.

(a) How many minutes of local calls and how many minutes of long-distance calls can he make using the entire 1750 total minutes and spending the entire \$25?

(b) Describe the problem-solving strategy you used to solve the problem.

8. Mary has accepted a job selling calculators. She receives a monthly salary of \$300 plus a commission of 6% of her monthly sales.

(a) What is her total salary for the month of May if her sales total \$500?

(b) What would her monthly sales in June need to be to double her commission in May?

(c) Describe the problem-solving strategy you used to solve the problem.

9. John has accepted a telemarketing job. He receives a monthly salary of \$250 plus a commission of \$0.25 for each call he completes.

(a) What is his total salary for the month of September if he makes 800 calls?

(b) What is his total salary for the month of October if he makes 1,000 calls?

(c) Describe the problem-solving strategy you used to solve the problem.

10. Jason received scores of 93, 87, and 76 on his first three mathematics tests. He would like to receive at least a grade of B in the course and needs at least 320 points on the four tests to be given during the semester to earn a B.

(a) What is the lowest grade he can receive on the fourth test and still receive a grade of B for the course?

(b) Describe the problem-solving strategy you used to solve the problem.

11. Judy received scores of 98, 89, and 81 on her first three physics tests. She would like to receive a grade of A in the course and needs at least 360 points on the four tests to be given during the semester to earn a grade of A.

(a) What is the lowest grade she can receive on the fourth test and still receive a grade of A for the course?

(b) Describe the problem-solving strategy you used to solve the problem.

12. Stephanie owns a business that builds computers according to customers' specifications. The fixed operating costs of her business are $4,000 per week. Each computer sells for $2,000.

(a) How many computers must she sell to break even?

(b) If her profit for one week was $10,000, how many computers did she sell?

(c) Describe the problem-solving strategy you used to solve the problem.

13. Connect the 9 dots below with 4 straight lines without lifting your pencil. Hint: You do not need to stay within the boundaries of the 9 dots. Hint: Think outside the box.

(a) Describe the problem-solving strategy you used to solve the problem.

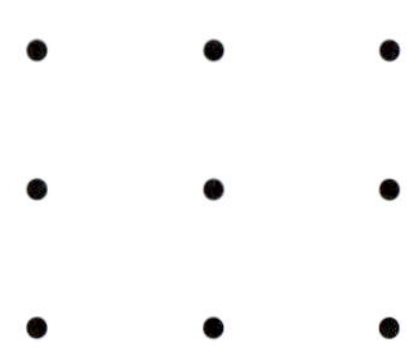

Problems 14–15: Writing Exercises

14. Reflect upon the problem-solving methods you used in problems 1–13.

(a) Which method did you use most often?

(b) What did you like about that method?

(c) What would make one method preferable over another method?

15. Obtain the book *Overcoming Math Anxiety*[5] by Sheila Tobias from your college library. If you are unable to find this particular book, locate any book about "math anxiety." Choose *one chapter* from the book. Read it and write a reaction paper of the chapter with three full paragraphs written in complete sentences:

(a) The first paragraph should be a summary of the chapter you read.

(b) The second paragraph should be your answer to these questions: What ideas in the chapter did you agree with? What ideas did you disagree with?

(c) The third paragraph should answer the question: What strategies can you employ to be more successful in your algebra course?

[5]Sheila Tobias, *Overcoming Math Anxiety* (New York: W. W. Norton, 1995).

Chapter 1 Summary

SECTION 1.1:

The Real Number Line

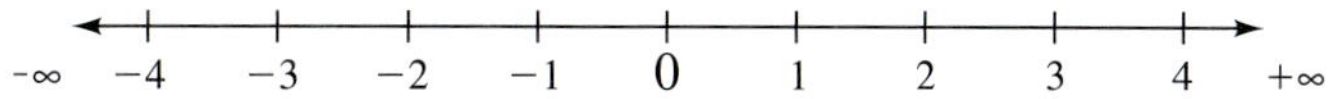

SECTION 1.2:

The Set of Real Numbers = *R*

Rational Numbers = Q

Real numbers whose decimal equivalents are terminating or repeating decimals.

Examples: $\frac{1}{2} = 0.5$, -17.0, 30.0, $\frac{16}{3} = 5.333... = 5.\bar{3}$

Integers = I $\{..., -3, -2, -1, 0, 1, 2, 3, ...\}$

Whole Numbers = W $\{0, 1, 2, 3, ...\}$

Counting Numbers = C $\{1, 2, 3, ...\}$

Irrational Numbers = P

Real numbers whose decimal equivalents are neither terminating nor repeating decimals.

Examples: $\pi \approx 3.141592654\ ...$

$\sqrt{6} \approx 2.449489743\ ...$

$-\sqrt{65} \approx -8.062257748\ ...$

Properties of Real Numbers: For all real numbers *x*, *y*, and *z*

	Key Word:
Identity Properties of Addition and Multiplication: $x + 0 = x$ and $x \cdot 1 = x$	Identical
Commutative Properties of Addition and Multiplication: $x + y = y + x$ and $x \cdot y = y \cdot x$	Order
Associative Properties of Addition and Multiplication: $(x + y) + z = x + (y + z)$ and $(x \cdot y) \cdot z = x \cdot (y \cdot z)$	Grouping
Distributive Property of Multiplication over Addition and Subtraction: $x(y + z) = xy + xz$ and $x(y - z) = xy - xz$	Spreading out

Order of Operations

First	**P**arentheses or Grouping Symbols
Second	**E**xponents
Third	**M**ultiplication or **D**ivision (left to right)
Fourth	**A**ddition or **S**ubtraction (left to right)

SECTION 1.3:

Guidelines for Addition and Subtraction

Addends with *LIKE* signs: Add the absolute value of the addends.

- The sign of the sum is the same as the sign of the addends.

Addends with *UNLIKE* signs: Subtract the absolute value of the numbers.

- The sign of the sum is the same as the sign of the number farthest from zero (largest absolute value).

All subtraction problems:*

1. Change the sign of the subtrahend.
2. Change the operation of subtraction to addition.
3. Follow the Guidelines for Addition of Signed Numbers.

*The three steps can be summarized as follows: "Change the sign and add."

SECTION 1.4:

Guidelines for Products and Quotients of Signed Numbers

The product or quotient of real numbers with *UNLIKE* signs is negative.

The product or quotient of real numbers with *LIKE* signs is positive.

SECTION 1.5: Many problem-solving strategies are appropriate, including Polya's Four-Step Problem-Solving Method.

Chapter 1 Review

- Review the definitions you have written for this chapter in the Dictionary.
- Review any properties and/or processes you have written on 3 × 5-inch cards.

SECTION 1.1

Problems 1–2: Create two mathematics statements that are true using only the numbers 2, 6, 8, 9, and 11. You must use the operations of addition, subtraction, multiplication, and division at least once in each problem. You do not have to use all five numbers in each statement. You may use numbers more than once in a statement. Be sure that the quantities on each side of the = symbol are equivalent.

1. ________ = ________ **2.** ________ = ________

Problems 3–8: Fill in the blank with a number to create an equation where the left side of the equation is equivalent to the right side.

3. $5.35 + \_\_\_\_ = 6.00$

4. $\frac{36}{4} = 3 \cdot \_\_\_\_$

5. $23 + 7 = 50 - 25 + \_\_\_\_$

6. $1 = \frac{5}{7} + \_\_\_\_$

7. $100 \div \_\_\_\_ = 25$

8. $15 \cdot \_\_\_\_ = 45$

Problems 9–14: Write an equivalent expression without an exponent. Find the final product.

9. 4^3

10. 25^2

11. $(1.3)^3$

12. 3^5

13. 2^4

14. $(1.1)^4$

SECTION 1.2

Problems 15–22: Let the letters x, y, and z represent real numbers in their respective locations on the number line:

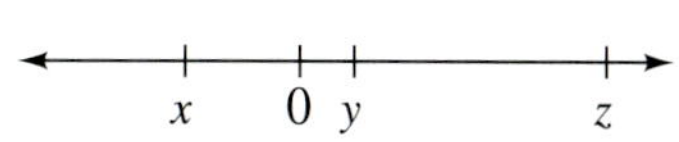

Write the appropriate inequality symbol, < or >, on the line between the numbers and variables.

15. x ____ 0

16. y ____ 0

17. x ____ z

18. y ____ z

19. $-x$ ______ 0

20. $-y$ ______ 0

21. $-x$ ______ $-z$

22. $-x$ ______ $-y$

Problems 23–32: Classify each real number as a member of the set(s) of counting numbers (C), whole numbers (W), integers (I), rational numbers (Q), or irrational numbers (P). Use all the names that apply.

23. 135

24. 0.4444. . .

25. $\sqrt{11}$

26. 20

27. $\frac{-150}{7}$

28. $-\frac{11}{4}$

29. $-(-7)$

30. 2.645751. . .

31. 15.68

32. -5237

Problems 33–40: Indicate whether the statement is true or false. If false, rewrite the statement to make it true.

33. Every whole number is a real number.

34. All real numbers are integers.

35. Repeating and terminating decimals are rational numbers.

36. Every integer is a whole number.

37. Every repeating decimal is an irrational number.

38. The set of irrational numbers is a subset of the set of real numbers.

39. The set of whole numbers is a subset of the set of rational numbers.

40. Every irrational number is a nonterminating, nonrepeating decimal.

Problems 41–50: Determine which Property is illustrated in the statements below: Identity Property of Addition (or Multiplication), Commutative Property of Addition (or Multiplication), Associative Property of Addition (or Multiplication), or Distributive Property over Addition (or Subtraction).

41. $(5 + 3) + 2 = 5 + (3 + 2)$

42. $20 \times 2 = 2 \times 20$

43. $(3 - 7)5 = 3 \cdot 5 - 7 \cdot 5$

44. $5.66 \cdot 1 = 5.66$

45. $\frac{3}{4} + \frac{7}{10} = \frac{7}{10} + \frac{3}{4}$

46. $3 \cdot (8 \cdot 4) = (3 \cdot 8) \cdot 4$

47. $6 + 0 = 6$

48. $(a + b)c = a \cdot c + b \cdot c$

49. $x \cdot y = y \cdot x$

50. $\left(\frac{3}{4} + \frac{7}{10}\right) + \frac{2}{5} = \frac{3}{4} + \left(\frac{7}{10} + \frac{2}{5}\right)$

Problems 51–64: Use the Order of Operations and perform the indicated operations *without* using a calculator. Write the result of each step.

51. $8 + 12 \div 6 \times 3$

52. $8 + 2 \times 6 \div 2$

53. $(8 + 2) \times 6 \div 2$

54. $2^3 + 9 - 3 + 4 \cdot 5$

55. $4(5 + 2) - 3^3 + 5(6 - 2)$

56. $\frac{30 - 6}{6}$

57. $36 \div 4 \times \frac{2}{3} - 3 + 7$

58. $8 + 12 \div 3 \times 4 - 3 + 4^2$

59. $(3^3 - 9) \div \frac{1}{3}$

60. $3^3 - 9 \div 9$

61. $\frac{(3.1)^2 - 0.61}{1 + 2}$

62. $\frac{4 + 4 \times 5}{6}$

63. $\frac{11}{2} - \frac{9}{2} \div \frac{1}{2}$

64. $\frac{11}{2} + \frac{9}{2} \cdot \frac{1}{2}$

SECTION 1.3

Problems 65–69: Use the number line to add or subtract.

65. $-6 + 2 =$

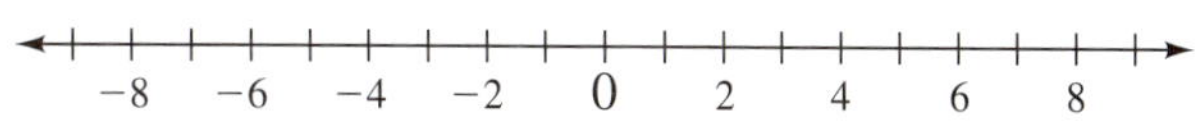

66. $-6 + (-2) =$

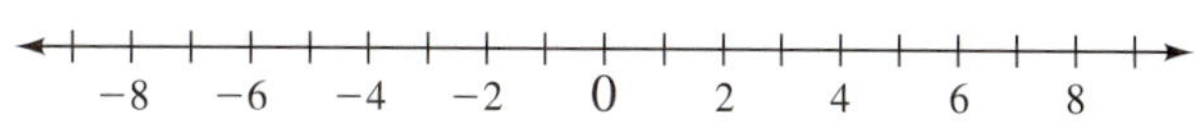

67. $-6 - 2 =$

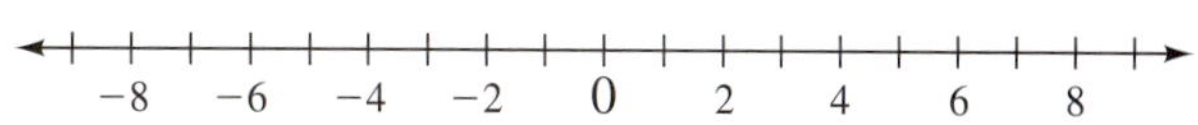

68. $-6 - (-2) =$

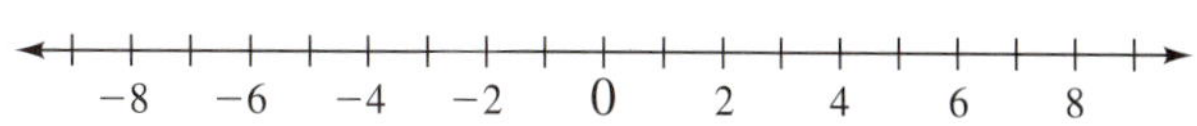

69. $-(-6) - 2 =$

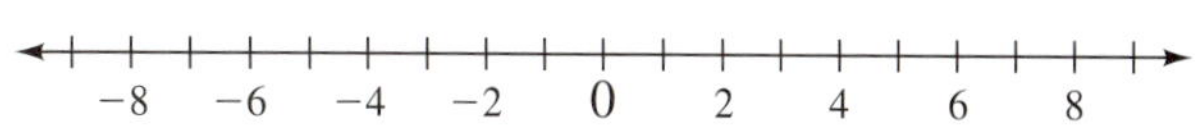

Problems 70–85: Perform the operations.

70. $|25|$

71. $|-10|$

72. $-20 + (-2)$

73. $10 - (-8)$

74. $-13.5 + 5.6$

75. $-5 + 5$

76. $-10 - (-6)$

77. $-1.5 + (-2.5)$

78. $-19 + 10$

79. $20 - (-7)$

80. $56 - (-6)$

81. $-82 + (-3)$

82. $5 + (-6) - (-1)$

83. $8 + (-2) + (-9)$

84. $2 - (-3) + (-6) - (-10) - 1 + 22$

85. $(-3) + (-2) + (+5) - 6 + (-10) + (-3) + (-1)$

Problems 86–89: The variables x, y, and z represent real numbers in the respective locations on the number line. Complete the statements below by writing the appropriate symbol, $<$ or $>$, on the line.

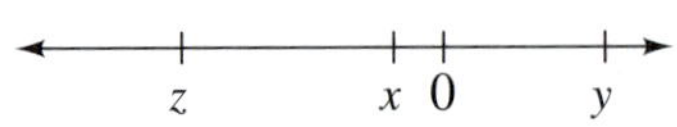

86. $-z$ ______ 0

87. $x + z$ ______ 0

88. x ______ z

89. $z - x$ ______ 0

Problems 90–93: Evaluate each expression, given that $x = -1$, $y = 5$, and $z = -2$.

90. $x + y + z$

91. $x - y - z$

92. $z - x - y$

93. $-(-x - y)$

SECTION 1.4

Problems 94–111: Perform the operations. Compute the answers, step-by-step.

94. -2^2

95. $(-2)^3$

96. $(-2)^2$

97. $-(-2)^2$

98. $(-2)^3$

99. -2^3

100. $8 - 4^2$

101. $8^2 - 4^2$

102. $(8 - 4)^2$

103. $-(8 - 4)^2$

104. $10 \div (-5) + 2^3 \cdot (-3) - (-5)$

105. $-11(9 - 5) + (-4)(5)(-7) \div 2 \cdot (-5)$

106. $-18 \div (2 \times 3) + (-6) + (-15)$

107. $6(3 - 9)^2 \cdot 3 + (-10)$

108. $\left[\frac{5}{9} - \left(-\frac{4}{9}\right)\right] \times (-21) \div (-7)$

109. $\dfrac{-12 \div (-3) \cdot (-2)^3}{6 + 2}$

110. $\dfrac{3 + 3}{3} + 6 - (-13) + \dfrac{5 - (-6)}{3} \div \dfrac{1}{3}$

111. $\left[\frac{5}{8} - \left(\frac{-9}{8}\right)\right]^2 \cdot \frac{4}{7} \div \frac{21}{2}$

Problems 112–115: Perform the operations.

112. $\dfrac{[(-5)^2 - (-5)]}{-2(5)} - 4^2$

113. $-5(6 - 2)^2 - 7^2(-1)$

114. $2[-3^2(-3^3\{-2^2\})]$

115. $3[-(5 - 2)^2 - (3^3 - \{6 - 9\})]$

Problems 116–123: The variables x, y, and z represent real numbers in the respective locations on the number line. Complete each statement by writing the appropriate symbol, $<$ or $>$, on the line.

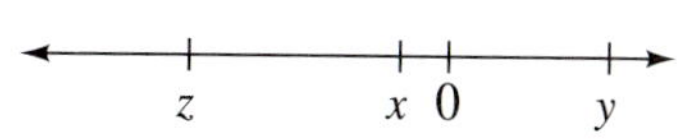

116. $x \cdot z$ ______ 0

117. $x \cdot y$ ______ 0

118. $(x)^2$ ______ 0

119. $y \cdot x$ ______ $z \cdot 0$

120. $x + z$ ______ 0

121. $z - x$ ______ 0

122. $y - x$ ______ 0

123. $x - z$ ______ 0

Problems 124–127: Evaluate each expression, given that $x = -1$, $y = 5$, and $z = -2$.

124. $x \cdot y \cdot (-z)$

125. $x^2 + y^2 + z^2$

126. $\dfrac{x^3 - z^2}{x}$

127. $\dfrac{y + z}{x} + \dfrac{z - y}{x}$

Problems 128–131: Evaluate each expression, given that $x = 1$, $y = -3$, and $z = -2$.

128. $x \cdot y \cdot (-z)$

129. $x^2 + y^2 + z^2$

130. $\dfrac{x^3 - z^2}{x}$

131. $\dfrac{y + z}{x} + \dfrac{z - y}{x}$

SECTION 1.5

Problems 132–136: Solve each problem by any method. Show your work.

132. The meter on a taxi calculates the cost of the taxi ride as a base fare of \$3.50 plus \$0.10 for each mile driven.

(a) What is the cost of a 20-mile taxi ride?

(b) How many miles can you travel if you only have \$10 to spend?

(c) Describe the problem-solving strategy you used to solve the problem.

133. A farmer would like to use 500 square feet of his acreage to make a rectangular pen for his animals. The length of the pen must be 25 feet.

(a) Determine the width of the pen.

(b) Describe the problem-solving strategy you used to solve the problem.

134. You have decided to replace the fencing around your property. Fencing costs \$9 per foot and a gate costs \$45. The dimensions (in feet) of the property are given in the figure. A 5-foot gate is to be installed on the shortest side:

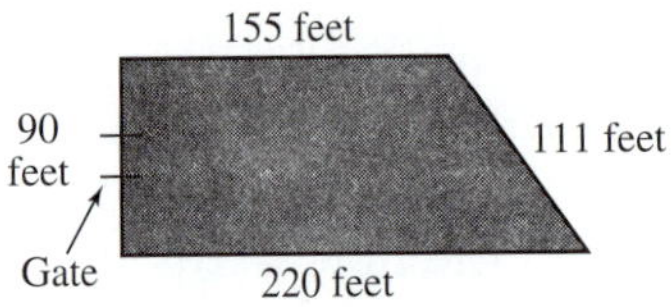

(a) What is the total perimeter (in feet) of your property?

(b) What is the cost of the fencing and gate?

(c) Describe the problem-solving strategy you used to solve the problem.

135. Jeremy received scores of 80, 86, and 92 on three biology tests. He would like to receive a grade of A in the course and needs at least 360 points on the four tests to be given during the semester.

(a) What is the lowest grade he can receive on the fourth test and still receive a grade of A for the course?

(b) Describe the problem-solving strategy you used to solve the problem.

136. John has been offered a job selling cars and his base salary is to be \$1000 plus 3% commission on the total monthly sales.

(a) What is John's total salary for the month of September if his total sales are \$100,000?

(b) What must his total monthly sales be to make a total monthly salary of \$4500?

(c) Describe the problem-solving strategy you used to solve the problem.

Practice Test—Chapter 1

SECTION 1.1

1. Given □ = 3, ★ = 2, and ♣ = 5, substitute the values for the objects. Perform the operations and write the appropriate symbol, either =, $<$, or $>$, on the line between the two quantities.

(a) 4·♣ + 7·□ ______ 41

(b) $\frac{\square}{\clubsuit}$ ______ $\clubsuit^2$

(c) 7·★ ______ 2·□ + 4·♣

(d) ♣·□ ______ ★

SECTION 1.2

2. Classify each real number as a member of the set(s) of counting numbers (C), whole numbers (W), integers (I), rational numbers (Q), or irrational numbers (P). Use all the names that apply.

(a) 1.5

(b) $\sqrt{5}$

(c) 0

(d) $\frac{3}{4}$

(e) -6

(f) $1\frac{2}{3}$

(g) $0.\overline{3}$

(h) 100

3. Use the Commutative Property of Addition (or Multiplication) to write an equivalent expression on the right side of the equal symbol. Do not perform the operations.

(a) $25 + 30 =$

(b) $a + b =$

(c) $9 \times 10 =$

(d) $x \cdot y =$

4. Use the Associative Property of Addition (or Multiplication) to write an equivalent expression on the right side of the equal symbol. Do not perform the operations.

(a) $1 + (2 + 7) =$

(b) $(a + b) + c =$

(c) $(3 \times 5) \times 10 =$

(d) $(x \cdot y) \cdot z =$

5. Use the Distributive Property to write an equivalent expression on the right side of the equal symbol. Do not perform the operations.

(a) $20(8 + 6) =$

(b) $a(b - c) =$

(c) $3 \cdot 2 + 3 \cdot 5 =$

(d) $x \cdot y + x \cdot z =$

SECTION 1.3

6. Use the number line to add or subtract.

(a) $4 - (-2) =$

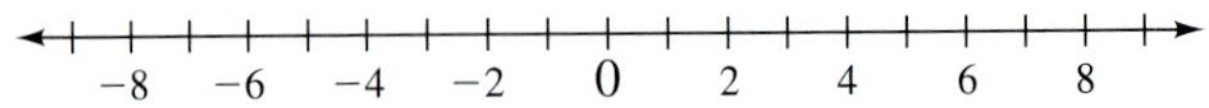

(b) $-4 + (-2) =$

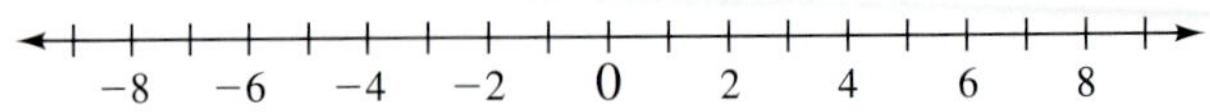

7. Use the Order of Operations to perform the operations indicated without using a calculator.

(a) $12 + 8 \div 4 - 2$

(b) $4(7 + 3)^2 - 5^2$

SECTION 1.4

8. The variables a, b, c, and d represent real numbers in the respective locations on the number line.

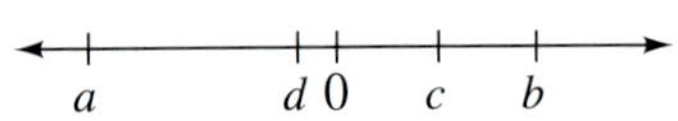

(a) Is $-b$ positive or negative?

(b) Is the sum of a and d a negative or positive number?

(c) Is $d \cdot c$ a positive or negative number?

(d) Is a^2 a positive or negative number?

9. Evaluate each expression, given that $a = -3$, $b = -2$, and $c = -1$.

(a) $a \cdot b \cdot c$

(b) $a^2 + b^2$

(c) $-a^2 \cdot b^2$

(d) $-(a + b)$

Problems 10–14: Use the Order of Operations to perform the indicated operations.

10. $-2^2 + 3(2 - 6)$

11. $-18 \times 6 \div 3 + (-6)$

12. $3(5 + 2) - |-6|$

13. $\dfrac{[-3^2 - (-4)]}{(-2)^2 + 1} - (-4)^2$

14. $-2^4[-2^3(-2)^2]$

SECTION 1.5

15. Sarah owns a small business. There was a loss of \$5 on Monday and a loss of \$12 on Tuesday. On Wednesday there was a profit of \$19 and on Thursday there was a loss of \$18. Find the total profit or loss for the four days.

16. During exercise the recommended maximum heart rate in beats per minute is given by the formula $M = 0.8(220 - A)$, where M is the maximum heart rate and A is the person's age. What is the recommended maximum heart rate in beats per minute for a person who is 40 years old?

Answers to Getting Started—Chapter 1

SECTION 1.1

1. (a) $3 + 4 = 7$

(b) $\frac{11}{5} - \frac{4}{5} = \frac{7}{5}$

(c) $7 \times 3 = 21$

(d) $14 \div 2 = 7$

2. Answers will vary.

3. (a) $\neq$

(b) $=$

(c) $\neq$

(d) $=$

4. (a)

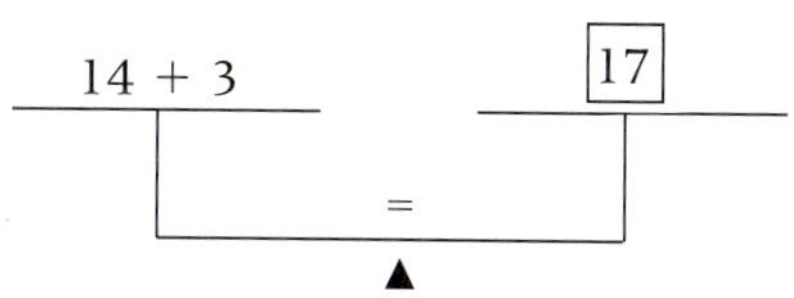

(b)

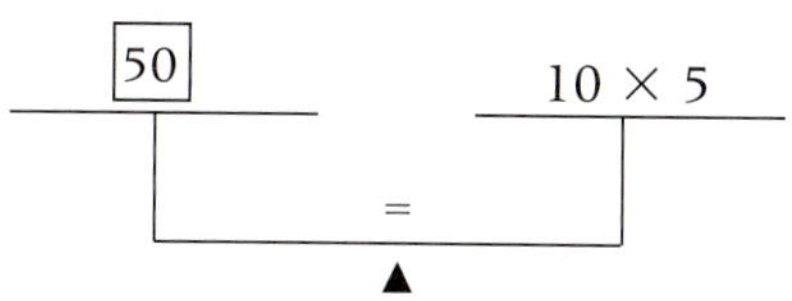

5. (a) $2 + 4$; A

(b) $10 - (+3)$; P

(c) $+12 - 4$; P

(d) $1 + 3$; A

6. (a) $200 + (-6)$; N

(b) $7 - 3$; S

(c) $(-2) + 6$; N

(d) $15 - 8$; S

7. (a) $10 > 5 + 3$

(b) $\frac{1}{2} < 8$

(c) $-3 > -6$

(d) $0 > -1$

8. (a) M

(b) M

(c) D

(d) D

(e) D

(f) M

9. (a) $5\frac{3}{4}$ means $5 + \frac{3}{4}$

(b) $5^3 = 5 \cdot 5 \cdot 5 = 125$

SECTION 1.2

1. (a) Negative

(b) Positive

(c) $x < b$

(d) $t < d$

(e) $d > x$

(f) $-x > 0$

2. (a) C, W, I

(b) I

(c) I

(d) N

(e) W, I

(f) N

3. (a) Q

(b) P

(c) P

(d) Q

4. (a) Yes. $15 = 15$. Yes

(b) Yes. $30 = 30$. Yes

(c) No. $5 \neq -5$. No

(d) No. $4 \neq \frac{1}{4}$. No

5. (a) Yes. $14 = 14$. Yes

(b) Yes. $40 = 40$. Yes

(c) No. $0 \neq 6$. No

(d) No. $2 \neq 8$. No

6. (a) Yes. $55 = 55$

(b) Yes. $9 = 9$

7. $2 + 3 \cdot 4 = 2 + 12 = 14$

8. (a) 7

(b) 22

9. Answers will vary.

SECTION 1.3

1.

Jan. 15	−45.00
Jan. 26	−195.75
Feb. 3	−150.75
Feb. 15	−50.75
Feb. 19	−131.00

2. Answers will vary:

(a) a *purchase*: add a negative value

(b) a *payment*: add a positive value

(c) *returned a purchase*: take away the fact that the purchase was made (add the value back into the balance).

3.

Feb. 19		−131.00
Feb. 20	−131.00 + (−15.00)	−146.00
Feb. 27	−146.00 + (−55.50)	−201.50
Mar. 3	−201.50 + (−45.75)	−247.25
Mar. 12	−247.25 − (−55.50)	−191.75
Mar. 15	−191.75 + 100.00	−91.75
Mar. 16		−91.75

4. (a) See table.

Golfer's Score	7	4	4	3	4	4	4	7	4	41
Signed number score	+3	+1	−1	−1	0	+1	−1	+3	−1	+4

(b) 41 (c) 4 above PAR.

5. Answers will vary.

(a) Checking account balance

(b) Change in temperature

SECTION 1.4

1. (i) The sign of a product or quotient of two real numbers with UNLIKE signs is negative.

(j) The sign of the product or quotient of two real numbers with LIKE signs is positive.

2. (a)

$-3\cdot(-6)$	+18
$-4\cdot(-6)$	+24
$-5\cdot(-6)$	+30

(b) Each number is increasing by 6 units.

(c) The sign of the product of two negative real numbers is a positive number.

3. (a) $(-5)\cdot(-2) = +10$

(b) $5\cdot(-2) = -10$

(c) $(-5)\cdot 2 = -10$

(d) The sign of the quotient of real numbers with UNLIKE signs is negative.

(e) The sign of the quotient of real numbers with LIKE signs is positive.

4. (a) −21

(b) 21

(c) −5

(d) 5

SECTION 1.5

1. (a) 9 yo-yos

(b) 11 boxes of crayons

(c) Answers will vary.

2. (a) Profit if 125 videos are sold: $0.00

(b) Profit if 500 videos are sold: $3000.00

(c) Break even = 125 videos

(d) Answers will vary.

3. Many problem-solving strategies are available. Some include: guess and check, draw a picture, make a table, or write an equation.

4–12. Answers will vary.

CHAPTER

2

Variables, Polynomials, Solving Linear Equations and Inequalities

The earliest known treatise on algebra is credited to Diophantus of Alexandria (third century A.D.). The term algebra *is from an Arabic word,* al-jabr. *The ancient mathematician Abu Ja'far Muhammand, later known as Al Kharazmi, introduced the writing down of calculations in place of using an abacus.*

2.1 What *Is* Algebra Anyway?

OBJECTIVES

- Recognize different uses of variables and symbols.
- Define algebra.
- Identify monomials, binomials, trinomials, and polynomials.
- State the degree of a polynomial.
- Write a polynomial in ascending or descending order.

1 2 3 Getting Started

Algebra has been called "arithmetic with letters," but it is much more than that. Algebra is a powerful language with its own symbols and vocabulary and is sometimes called the "language of science." Algebra is an organized set of procedures and principles that are used to solve problems. Algebra is an organized mathematical system.

Variables are used in algebra for different reasons. A variable might be used in an equation, as an unknown, where the equation describes a problem situation. Let's explore examples of different uses of variables.

1. To illustrate the use of a variable or symbol as placeholders for *assigned values,* substitute the given numerical values for the objects and variables into the expressions: $\blacklozenge = 7, x = 10, y = -3$. Write the appropriate symbol, either =, <, or >, on the line between the quantities. Show your work.

(a) $\blacklozenge + \blacklozenge + \blacklozenge$ ______ $3\blacklozenge$ (b) $2\blacklozenge - 3x$ ______ $5y$

2. The value of a variable can change. *Assign your own values* to these objects using any real number (other than zero or one): ♦ = ______, x = ______, y = ______. Substitute the values into the given expressions and write the appropriate symbol, either =, <, or >, on the line between the quantities. Show your work.

(a) ♦ + ♦ + ♦ ______ 3♦ (b) 2♦ − 3x ______ 5y

3. Variables are used as *objects in a mathematical structure or set of rules* as in the Properties of Real Numbers. Assign each variable in the Distributive Property a value and perform the operations on each side of the equation.

$$3(x + y) = 3x + 3y$$

4. Variables are used in an *equation or formula* to represent values that vary. The formula for the area of a rectangle is $A = l \cdot w$, where A is a variable that represents the area, l is a variable that represents the length, and w is a variable that represents the width. Use the formula to find the area of each rectangle.

(a)

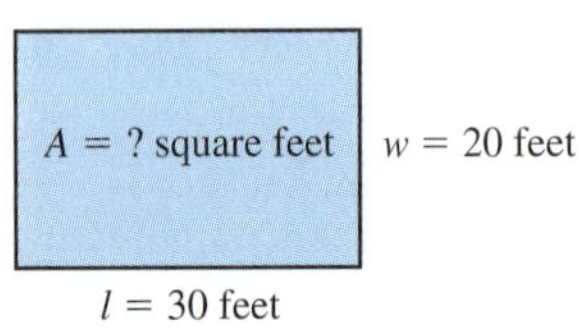

(b)

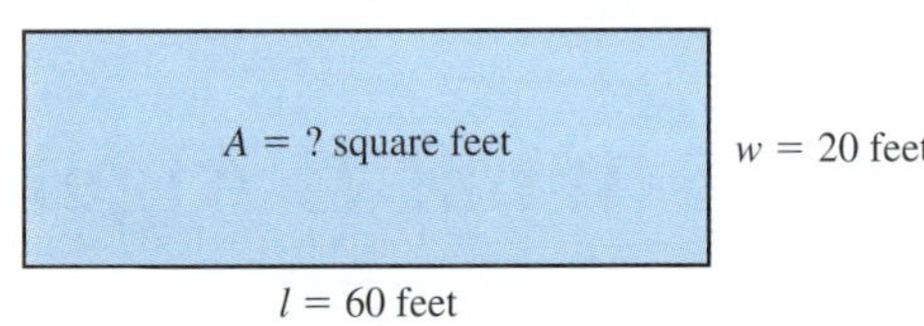

5. A variable can represent an *unknown value*. Find the value of the unknown variable that satisfies each equation.

(a) $x + 2 = 10$ (b) $12 - y = 3$

6. Variables may be used as placeholders *denoting a relationship* between quantities. Given the variables in the respective locations on the number line, write the appropriate symbol, either =, <, or >, on the line between the quantities.

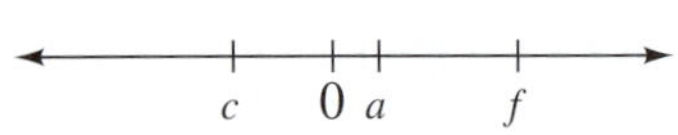

(a) $a \cdot f$ ______ c (b) c ______ 0

1 2 3 Discussing Mathematics

VARIABLES

Symbols, objects, and letters are used frequently as variables in the discipline of algebra in the following ways.

	Variables Are Used	Examples
I.	As placeholders for *assigned values.*	Evaluate $a^2 - b^2$ when $a = -5$ and $b = -3$.
II.	As objects in *a mathematical system or set of rules.* Symbols and variables can represent an infinite number of values. Operations are performed upon variables.	Identity Property of Addition: ♦ + 0 = ♦ Commutative Property of Multiplication: $x \cdot y = y \cdot x$
III.	In *equations or formulas* for values that vary.	$A = l \cdot w$ or $P = 2l + 2w$
IV.	As placeholders for *unknown values.*	Solve for x: $x + 4 = 10$
V.	As placeholders *denoting relationships* between quantities.	$d > c$
VI.	To *generalize or describe a pattern.*	$P = 8x - 1000$ Number of Sides $= 3n + 1$

WHAT IS ALGEBRA?

Algebra has been defined as a mathematical system used to generalize arithmetical operations by permitting letters or other symbols to represent numbers. It is a complete mathematical system that helps us to organize our thought processes, communicate relationships, and solve problems.

WHAT IS ALGEBRA?

I. A powerful language with its own symbols and vocabulary
II. A mathematical system used to generalize arithmetic
III. An organized set of procedures and principles
IV. A problem-solving tool

ALGEBRA IS USED TO

I. Generalize calculations
II. Create a record of our thinking
III. Outline a process that may be difficult to manage mentally
IV. Represent our thinking to others

CREATING A SYSTEM—THE VOCABULARY OF POLYNOMIALS

As we begin to extend our understanding of the algebraic structure, we also need to extend our mathematical vocabulary. Numbers or quantities that are multiplied are called **factors**. Mathematical **terms** are numbers, variables, or variables multiplied by a number or variable that are separated by addition or subtraction. A term that does not contain a variable is called a **constant**. A term containing variables might include one or several factors. The number preceding the variable is called the **numerical coefficient** of the term. When a number does not precede the variable, the numerical coefficient is understood to be 1.

A polynomial in one variable is an expression composed of one term or the sum of two or more terms containing the same variable. The **degree of a polynomial** is the value of the largest exponent on the variable of all the terms in the polynomial. Polynomials with one, two, or three terms have special names, as shown.

POLYNOMIAL

A **polynomial** is an algebraic expression composed of one term or the sum of two or more terms. Polynomials with one, two, or three terms have special names.

A polynomial with one term is called a **monomial**.

A polynomial with two terms is called a **binomial**.

A polynomial with three terms is called a **trinomial**.

EXAMPLE 1: Polynomials and their corresponding terms, names, numerical coefficients, and degrees:

Polynomial	Number of terms	Name	Terms	Numerical coefficient or constant	Degree
$2x$	one	monomial	$2x$	2	1
$2y + 5y^2$	two	binomial	$2y$ $5y^2$	2 5	2
$x^2 - 5x + 9$	three	trinomial	x^2 $-5x$ 9	1 -5 constant	2
$-3x^5 + 2x^2 - x + 7$	four	polynomial	$-3x^5$ $2x^2$ $-x$ 7	-3 2 -1 constant	5

➢ *Polynomials in one variable can be written in either ascending or descending order according to the exponents on the variables. In practice, polynomials are usually written in descending order.*

EXAMPLE 2: The terms in the polynomial $-3x^5 + 2x^2 - x + 7$ are written in **descending order** with the exponents on the variables decreasing as you look at the terms from left to right.

The terms in the polynomial $7 - x + 2x^2 - 3x^5$ are written in **ascending order** with the exponents on the variables increasing as you look at the terms from left to right.

1 2 3 Doing Mathematics

- Write the definitions of the words and symbols listed for this section in the Dictionary in Appendix A.
- On a 3 × 5-inch card, write the vocabulary of polynomials for future reference.

1. Review the table about uses of variables on page 67 of this section. Write I, II, III, IV, V, or VI to describe the use of variables in each statement. Use as many Roman numerals and categories that are appropriate.

	Problems using variables	Write I, II, III, IV, V, or VI
(a)	$15 > x$	
(b)	Commutative Property of Addition: $A + B = B + A$	
(c)	What is the value of x in this equation? $20 = 10 \cdot x$	
(d)	The number of sides in the geometric figure $= 2n + 1$	
(e)	The area of a square: $A = s^2$	
(f)	Evaluate $a^2 + b^2 = c^2$ when $a = 5$ and $b = 3$.	

2. Let x and y represent real numbers.

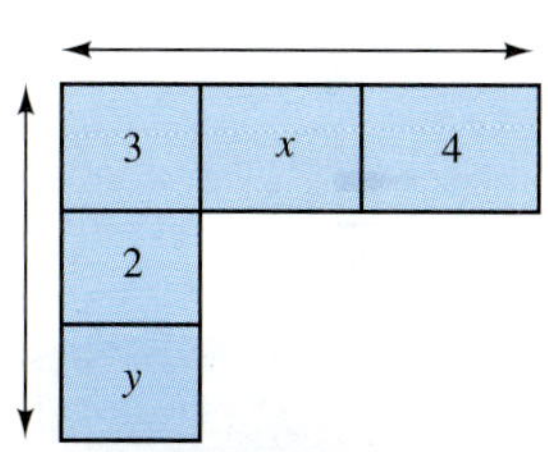

(a) Find a value for x and y that will create the same sum for the numbers in the horizontal and vertical boxes. $x =$ ______ $y =$ ______ (There are an infinite number of solutions.)

(b) Write an equation representing the relationship between the horizontal sum and the vertical sum.

Problems 3–12: Substitute the given numerical values for the objects and variables into each expression: $\bigstar = 2$, $x = -5$, $y = -7$. Write the appropriate symbol, either $=$, $<$, or $>$, on the line between the quantities. Show your work.

3. $x + x$ ______ $2x$

4. $\bigstar \cdot x - \bigstar \cdot y$ ______ $\bigstar(x - y)$

5. $2x + 3y$ ______ $5xy$

6. $8\bigstar - 5x$ ______ $3x\bigstar$

7. $xy \cdot 1$ ______ xy

8. $3y$ ______ $y + y$

9. $6\bigstar$ ______ $6x$

10. $2x + 3x$ ______ $5x$

11. $x(\bigstar + y)$ ______ $x\bigstar + xy$

12. $2y$ ______ $2\bigstar$

Problems 13–20: Solve each equation by inspection, determining the value of the variable that satisfies the equation. Verify the solution.

13. $x + 12 = 20$

14. $y - 2 = 15$

15. $\dfrac{50}{x} = 25$

16. $2y = -80$

17. $12 = y - 3$

18. $\dfrac{75}{x} = -3$

19. $-4y = -16$

20. $x + 5 = 28$

Problems 21–30: Substitute the given numerical values for the objects and variables into each expression: $\blacklozenge = 7, x = -4, y = -3$. Perform the operations and determine the value of each expression. Show your work.

21. $x^2 + 5x - 9$

22. $x(\blacklozenge + y)$

23. $y^2 + 2xy - 5y + 20$

24. $(x - y) \cdot \blacklozenge$

25. $x^2 + 5\blacklozenge$

26. $-(3x)(4y)$

27. $2x(x + y)$

28. $x + y + \blacklozenge$

29. $-x(\blacklozenge - y)$

30. $3x + 4y^2$

Problems 31–37: Complete the table.

Polynomial	Monomial, binomial, trinomial, or polynomial	List the terms	Numerical coefficient of each term or constant	Degree of the polynomial
31. $-5x - 7$				
32. $1.5x^2$				
33. $2x^3 - 4x^2 + 5$				
34. $7y^3 - 5y^2 + 9y - 10$				
35. $\frac{1}{2}x - 5 + \frac{2}{3}x^2$				
36. 100				
37. $-3x$				

Problems 38–41: Rewrite each expression with terms in descending order.

38. $10x - 16 + 3x^5 - 4x^2 + x^3 - 9x^4$

39. $-15 + 3y^2 - 5y + 2y^3 + 18y^6$

40. $1.5x - 5x^2 - 2.75 + 4.25x^3$

41. $\frac{3}{2}x^2 - \frac{1}{2}x + 2x^3 - 5$

Problems 42–45: Rewrite each expression with terms in ascending order.

42. $10x + 16 + 3x^5 - 4x^2 + x^3 - 9x^4$

43. $-15 + 3y^2 - 5y + 2y^3 + 18y^6$

44. $1.5x - 5x^2 + 2.75 + 4.25x^3$

45. $-x^4 + \frac{16}{3} - \frac{2}{3}x^2 + x$

Problems 46–49: The perimeter of a geometric figure is the distance around the boundary of the figure. The perimeter of a rectangle is found by adding the measures of the four sides. The formula for the perimeter of a rectangle is $P = 2l + 2w$, where P represents the perimeter of the rectangle, l represents the length, and w represents the width.

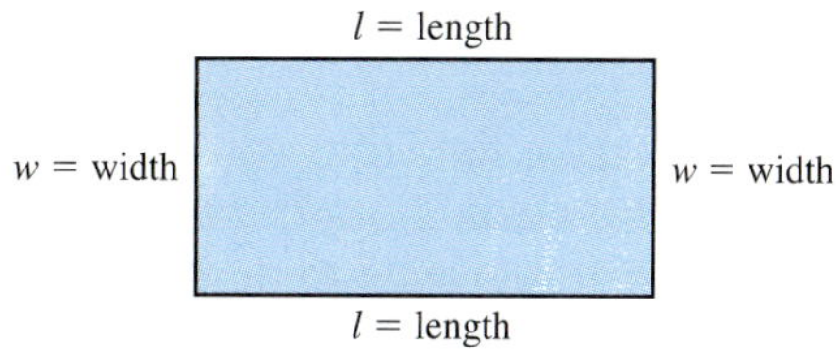

Draw a picture of each rectangle described below. Use the picture or formula to find the missing dimension.

46. $P =$ ______ feet, $l = 7$ feet, $w = 5$ feet

47. $P = 160$ feet, $l = 20$ feet, $w =$ ______ feet

48. $P =$ ______ feet, $l = 5.62$ feet, $w = 3.78$ feet

49. $P = 200$ feet, $l =$ ______ feet, $w = 60$ feet

Problems 50–51: Writing Exercises

50. List three formulas you can remember. Write the meaning of each variable used in the formula.

(a) ______________________

(b) ______________________

(c) ______________________

51. Describe a real-world situation in which you could have used a variable to represent the situation to solve the problem.

2.2 Collecting LIKE Terms

OBJECTIVES

- Determine equivalence between expressions.
- Identify and collect LIKE terms.
- Add and subtract polynomials, horizontally and vertically.
- Use the Distributive Property.

1 2 3 Getting Started

In this section, we focus on the processes of adding and subtracting LIKE terms and rewriting algebraic expressions in equivalent forms.

1. Imagine that you have these coins in your pocket. How would you describe these coins to a friend?

One way to describe the coins is to count the coins that are alike: 2 quarters, 3 dimes, and 2 nickels. Another way to describe the coins would be to state the total amount of money of $0.90. Variables can represent each type of coin: quarter = q, dime = d, nickel = n and equivalent forms.

$$\underbrace{\text{1 quarter} + \text{1 quarter}} + \underbrace{\text{1 dime} + \text{1 dime} + \text{1 dime}} + \underbrace{\text{1 nickel} + \text{1 nickel}} = \text{total money}$$

$$\begin{array}{ccccccc} \text{2 quarters} & + & \text{3 dimes} & + & \text{2 nickels} & = & \$0.90 \\ 2q & + & 3d & + & 2n & = & \$0.90 \\ 2(0.25) & + & 3(0.10) & + & 2(0.05) & = & \$0.90 \end{array}$$

All four equations represent the same coins, which are equivalent to $0.90.

2. Adding and subtracting algebraic expressions is a way of rewriting equivalent forms of expressions. The process involves counting LIKE objects. Determine whether the expressions are equivalent and insert the appropriate symbol, either = or ≠, on the line between the two expressions.

(a) 4 apples + 5 apples ________ 9 apples

(b) 100 bears − 10 bears ________ 90 bears

(c) 3 dogs + 5 cats ________ 8 monkeys

(d) 13 cats − 5 dogs ________ 8 monkeys

(e) Summarize the pattern(s) you observed in the statements that are equivalent.

3. In algebra, we use a process called **collecting LIKE terms** to add and subtract algebraic expressions whose variable factors (letters and exponents) are exactly the same. Look for a relationship between the left side of each equation and the equivalent form on the right side of the equation in (a)–(e).

(a) $4y + 3y = 7y$

(b) $5y - 3y = 2y$

(c) $x + x + y + y + y = 2x + 3y$

(d) $2x^2 + 3x^2 = 5x^2$

(e) $7xy - 4xy = 3xy$

(f) Describe the pattern(s) you observed in the equivalent statements.

4. Values can be substituted for variables to help confirm equivalence of expressions containing variables. The process does not prove the statement for all values, but it can be used to observe patterns. Substitute the numerical values, $y = 3$ and $x = 5$, in each expression. Determine whether the expressions are equivalent for these values and insert the appropriate symbol, either $=$ or $\neq$, on the line between the two expressions. Show your work.

(a) $3y - 2y$ ______ y

(b) $x + x + y + y + y$ ______ $2x + 3y$

(c) $5y - 3x$ ______ $2y$

(d) $6x^2 - 4x^2$ ______ $2x$

(e) Summarize the pattern(s) you observed in the equivalent statements.

(f) Summarize the pattern(s) you observed in the statements that are not equivalent.

➢ *The process of collecting LIKE terms employs the Distributive Property to rewrite the expression in an equivalent form.*

EXAMPLE 1:

$$\underbrace{6x + 4x}$$
$$(6 + 4)x$$
$$10x$$

$6x$ and $4x$ are LIKE terms. Rewrite LIKE terms using the Distributive Property. Add the numbers in the parentheses.

EXAMPLE 2:

$$\underbrace{9y^2 - 6y^2} + 5y$$
$$(9 - 6)y^2 + 5y$$
$$3y^2 + 5y$$

$9y^2$ and $6y^2$ are LIKE terms. Rewrite LIKE terms using the Distributive Property. Subtract the numbers in the parentheses. Write the term $5y$. $3y^2$ and $5y$ are not LIKE terms.

5. Combine LIKE terms using the Distributive Property.

(a) $2x + 3x$

(b) $15y + 5y - 3x$

(c) $6x^2 - 4x^2$

(d) $11ab - 2ab + 3c + 7c$

1 **2** 3

Discussing Mathematics

Our understanding of the equal symbol as an equivalence statement, combined with our understanding of variables as placeholders and as objects in an algebraic system, allows us to create equivalent mathematical statements. Collecting LIKE terms is a way to write an equivalent statement, simplify an expression, and perform the operations.

COLLECTING LIKE TERMS

Terms whose variable factors are exactly the same (letters and exponents) are called **LIKE terms**. The process of adding or subtracting algebraic expressions is called **collecting LIKE terms**. Collecting LIKE terms employs the Commutative Property of Addition to change the order of terms, the Associative Property of Addition to change the grouping of terms, and the Distributive Property to add or subtract the numerical coefficients of LIKE terms.

COLLECTING LIKE TERMS

1. Identify the LIKE terms. Apply the Commutative and Associative Properties of Addition to rearrange and group the LIKE terms. Terms are most often written in descending order of exponents.
2. Apply the Distributive Property, enclosing the numerical coefficients of LIKE terms in parentheses.
3. Add or subtract the numbers in the parentheses.

EXAMPLE 3: Add by collecting LIKE terms.

$2x + 3y + 5x + 16x + 15y$

$= \underbrace{(2x + 5x + 16x)} + \underbrace{(3y + 15y)}$ Apply the Commutative and Associative Properties to rearrange the terms and group terms containing x and terms containing y.

$= (2 + 5 + 16)x + (3 + 15)y$ Rewrite LIKE terms using the Distributive Property.

$= 23x + 18y$ Add the numbers in the parentheses.

➢ *Beware! Remember that addition is commutative and associative but subtraction is not. In Chapter 1, we summarized the subtraction process in the statement "change the sign and add." The Commutative and Associative Properties can be applied only when the operations of subtraction are rewritten as addition.*

EXAMPLE 4: Add and subtract, writing each subtraction as addition.

$10x + 3y - 5x - y - 2x$

$= 10x + 3y + (-5x) + (-y) + (-2x)$ Rewrite each subtraction as addition.

$= \underbrace{(10x + (-5x) + (-2x))} + \underbrace{(3y + (-y))}$ Apply the Commutative and Associative Properties to rearrange the terms to group LIKE terms.

$= (10 + (-5) + (-2))x + (3 + (-1))y$ Apply the Distributive Property.

$= 3x + 2y$ Add the numbers in the parentheses.

➢ *Can subtraction be performed without rewriting the operation(s) as addition?* Yes, attach the − symbol to each subtracted term as the terms are rearranged and grouped. The addition symbols are not written explicitly, but understood. The streamlined process looks like this: $10x + 3y - 5x - y - 2x$

$= \underbrace{10x - 5x - 2x} + \underbrace{3y - y}$

$= (10 - 5 - 2)x + (3 - 1)y = 3x + 2y$

➢ *Beware of terms that are UNLIKE and should not be added:* $5t^2m + 3tm^2 - 2tm$.

ADDING AND SUBTRACTING POLYNOMIALS INVOLVING PARENTHESES

Parentheses may enclose a monomial, binomial, trinomial, or any polynomial. To simplify an expression involving parentheses, apply the Order of Operations. Perform operations inside parentheses first. If the parentheses enclose a monomial or the operations are unable to be performed inside the parentheses, it is necessary to "remove the parentheses" in order to proceed with the problem.

ADDING AND SUBTRACTING POLYNOMIALS

1. If parentheses appear in the problem, remove them.
   - When an addition sign appears in front of the parentheses (or no operation is explicitly written), keep the sign of each term the same, and remove the parentheses. (Equivalent to multiplying by $+1$.)
   - When a subtraction sign appears in front of the parentheses, change the sign of each term inside the parentheses, and remove the parentheses. (Equivalent to multiplying by -1.)
   - When a monomial appears in front of the parentheses, multiply the monomial by each term inside the parentheses, and remove the parentheses.
2. Identify and collect LIKE terms.
   (a) Use the Commutative and Associative Properties of Addition to rearrange and group the LIKE terms.
   (b) Use the Distributive Property, enclosing the numerical coefficients of LIKE terms in parentheses.
   (c) Add or subtract the numbers in the parentheses.

➢ *Terms are most often written in descending order of exponents. It is customary to write variables within a term in alphabetical order and terms in a polynomial in descending order.*

EXAMPLE 5: Simplify: $(8x^2 + 3) + (2x^2 - 7)$.

$= 8x^2 + 3 + 2x^2 - 7$ Remove parentheses.

$= (8x^2 + 2x^2) + (3 - 7)$ Rearrange and group LIKE terms.

$= (8 + 2)x^2 + (3 - 7)$ Apply the Distributive Property.

$= 10x^2 - 4$ Perform operations in parentheses.

EXAMPLE 6: Simplify: $(-4x^3 + 5x^2 - 5) - (-2x^3 - x^2 + 5x + 8)$.

$= -4x^3 + 5x^2 - 5 + 2x^3 + x^2 - 5x - 8$ Remove parentheses.

$= (-4x^3 + 2x^3) + (5x^2 + x^2) - 5x + (-5 - 8)$ Rearrange and group LIKE terms.

$= (-4 + 2)x^3 + (5 + 1)x^2 - 5x + (-5 - 8)$ Perform the operations in parentheses.

$= -2x^3 + 6x^2 - 5x - 13$

➢ *When a monomial appears in front of parentheses and multiplication is indicated, the monomial is multiplied (or distributed) by each term inside the parentheses. The result is written without parentheses.*

EXAMPLE 7: Simplify: $4(x + 3) - 2x$.

$4(x + 3) - 2x$	Distribute 4 across the terms in parentheses.
$= 4x + 12 - 2x$	Multiply 4 by x and multiply 4 by $+3$.
$= 4x - 2x + 12$	Rearrange and group LIKE terms.
$= (4 - 2)x + 12$	Perform operations in parentheses.
$= 2x + 12$	

EXAMPLE 8: Simplify: $-4(x^2 - 3x + 6) - (2x^2 - 6x + 10)$.

$-4(x^2 - 3x + 6) - (2x^2 - 6x + 10)$	Multiply each term in the first parentheses by -4.
$= -4x^2 + 12x - 24 - 2x^2 + 6x - 10$	Multiply by -1, changing the sign of each term in the second parentheses.
$= (-4x^2 - 2x^2) + (12x + 6x) + (-24 - 10)$	Rearrange and group LIKE terms.
$= (-4 - 2)x^2 + (12 + 6)x + (-24 - 10)$	Perform operations in parentheses.
$= -6x^2 + 18x - 34$	

ADDING AND SUBTRACTING POLYNOMIALS VERTICALLY

Polynomials can be added or subtracted vertically with a process similar to adding or subtracting multidigit numbers. Columns are formed with LIKE terms in each column. Numerical coefficients in each column are added or subtracted.

EXAMPLE 9: Add the polynomials vertically $(9x^4 + 2x^3 - 5) + (2x^4 - x^3 + 4x - 9)$.

Arrange LIKE terms in vertical columns. If a term is missing, leave a space.

$$\begin{array}{rrrrr} & 9x^4 & + 2x^3 & & - 5 \\ + & 2x^4 & - x^3 & + 4x & - 9 \\ \hline & 11x^4 & + x^3 & + 4x & - 14 \end{array}$$

Add the numerical coefficients in each column, collecting LIKE terms.

➤ *In subtraction problems, the sign of each term in the subtrahend is changed, and LIKE terms are combined.*

EXAMPLE 10: Subtract $-2y^4 + 11y^3 + 5y^2 - y - 18$ from $14y^4 - 23y^3 + 10y^2 - 6y + 20$ vertically. Be sure to subtract the subtrahend from the minuend.

Step 1: Arrange the polynomials with LIKE terms in vertical columns.

$$\begin{array}{rrrrrr} & 14y^4 & - 23y^3 & + 10y^2 & - 6y & + 20 \\ - & -2y^4 & + 11y^3 & + 5y^2 & - y & - 18 \\ \hline \end{array}$$

Step 2: Copy the minuend. Rewrite the subtrahend, *changing the sign of each term* and then add.

$$\begin{array}{rrrrrr} & 14y^4 & - 23y^3 & + 10y^2 & - 6y & + 20 \\ + & +2y^4 & - 11y^3 & - 5y^2 & + y & + 18 \\ \hline & 16y^4 & - 34y^3 & + 5y^2 & - 5y & + 38 \end{array}$$

Step 3: Add the numerical coefficients in each column, combining LIKE terms.

1 2 3 Doing Mathematics

- Write the definitions of the words and symbols listed for this section in the Dictionary in Appendix A.
- On a 3 × 5-inch card, write the steps involved in the process of adding and subtracting polynomials for future reference.

Problems 1–12: Add and subtract by collecting LIKE terms.

1. $3x + 2x - 5x + 12x - 7x$

2. $-5xy + 14x - 9xy + 22x - (-7x)$

3. $1.25x^2 + 2.35x - (-5.25x^2) - 10.00x^2 + (-7.33x)$

4. $x - 2x - x - 12x - 7x$

5. $\frac{3}{2}y + \frac{1}{2}y - \frac{8}{2}y - \left(\frac{-3}{2}y\right)$

6. $3x^3 + 2x^2 - 5x^3 + 12x^2 - 7x^2$

7. $-2.3ab + 5.2a - 1.2b - (-12.6ab) + (-5.5b)$

8. $6x - \frac{5}{9}x + \frac{11}{9}x - 6x - \left(-\frac{3}{9}x\right)$

9. $33xyz + 21x - 5xy + 42xyz - (-7xz)$

10. $4.7cd + 5.1cd - 2c + 4d - (-3c) - 2.3cd$

11. $\frac{3}{4}x + 2x - \frac{5}{4}x + 12x - \frac{2}{4}x$

12. $\frac{1}{3}y + 5x + \frac{2}{3}y - 4y$

Problems 13–16: Substitute the numerical values $x = 2$, $y = -3$, and $z = -1$ for the variables. Determine whether the expressions are equivalent for these values and insert the appropriate symbol, either $=$ or $\neq$, on the line between the two expressions.

13. $5x - 3x$ ______ $2x$

14. $8xy - 6xy$ ______ 2

15. $y(x - z)$ ______ $yx - y$

16. $-2x^2 - (-3x^2)$ ______ x^2

Problems 17–22: Apply the Distributive Property to remove parentheses.

17. $5(x - 6)$

18. $7(x + 5)$

19. $-(x^2 + 2x + 1)$

20. $3(2y^2 - 5y + 4)$

21. $-2(5x^2 - 6)$

22. $-2(2y - 3)$

Problems 23–38: Perform the operations, removing parentheses and collecting LIKE terms.

23. $(10x - 3) + (15x + 7)$

24. $(3x + 2) + 2(4x - 5)$

25. $(9x^2 + 1) - 2(6x^2 - 3)$

26. $(5xy - 3) - (2xy + x)$

27. $\left(\frac{4}{7}x - \frac{1}{5}\right) + \left(\frac{10}{7}x - \frac{4}{5}\right)$

28. $-9(x + 2) - (x + 5)$

29. $(5x^2 - 2x + 9) + (x^2 - 6x - 25)$

30. $(x^2 + 4x + 7) - (x - 15)$

31. $(4x^2 - x - 6) - 3(x^2 - 2x + 5)$

32. $(3x^4 - 4x^2 - 1) + (2x^4 - 2x^2 - 3x + 5)$

33. $(2x - 5) + (1.5x + 6) + (2.5x - 1) - (x - 2)$

34. $(y^3 - y^2 - y - 1) - (y^3 - y^2 - y - 1)$

35. $(2ab - b + 1) + \left(\frac{3}{2}ab + a + 1\right) - \left(\frac{1}{2}ab + a + 1\right)$

36. $-(x^2 + x - 1) - (3x^2 - x + 1) + (x^2 - 5x + 4)$

37. $-2(x^2 + x - 1) - 5(x^2 - x + 2) + 3(x^2 - 7x + 4)$

38. $(10x^5 - 2) - (2x^3 + x - 3)$

Problems 39–42: Add the polynomials, vertically.

39.
$$\begin{array}{rrrrr} & 2x^3 & -\ 5x^2 & +\ x & -\ 9 \\ + & 8x^3 & +\ 2x^2 & -\ x & +\ 8 \\ \hline \end{array}$$

40.
$$\begin{array}{rrrrr} & -x^3 & & +\ 2x & +\ 10 \\ + & 7x^3 & +\ 2x^2 & -\ x & -\ 11 \\ \hline \end{array}$$

41.
$$\begin{array}{rrrrr} & -2x^4 & -\ 5x^2 & +\ x & -\ 9 \\ + & & -2x^2 & -\ 3x & -\ 8 \\ \hline \end{array}$$

42.
$$\begin{array}{rrrrrr} & 2x^5 & & -\ 5x^2 & +\ x & \\ + & & -8x^3 & -\ x^2 & -\ x & +\ 8 \\ \hline \end{array}$$

Problems 43–46: Subtract the polynomials, vertically. Be sure to change the sign of each term in the subtrahend.

43.
$$\begin{array}{rrrrr} & 2x^3 & -\ 5x^2 & +\ x & -\ 9 \\ - & 8x^3 & +\ 2x^2 & -\ x & +\ 8 \\ \hline \end{array}$$

44.
$$\begin{array}{rrrrr} & -x^3 & & +\ 2x & +\ 10 \\ - & 7x^3 & +\ 2x^2 & -\ x & -\ 11 \\ \hline \end{array}$$

45.
$$\begin{array}{rrrrr} & -2x^4 & -\ 5x^2 & +\ x & -\ 9 \\ - & & -\ 2x^2 & -\ 3x & -\ 8 \\ \hline \end{array}$$

46.
$$\begin{array}{rrrrrr} & 2x^5 & & -\ 5x^2 & +\ x & \\ - & & -8x^3 & -\ x^2 & -\ x & +\ 8 \\ \hline \end{array}$$

Problem 47: Writing Exercise

47. Summarize in your own words the meaning of the expression *LIKE terms* in algebra.

2.3 Solving Linear Equations

OBJECTIVES

- Solve linear equations using a balanced scale.
- Solve linear equations using the Equality Properties of Equations.
- Solve linear equations involving parentheses and involving variables on both sides of the equation.

1 2 3 Getting Started

In this section, we outline a mathematical procedure for solving linear (first-degree) equations involving one variable. **Solving an equation** is the process of finding the value of the variable that satisfies the equation. It may be possible to solve some equations by inspection. However, other equations will be more difficult.

The **scale method** of solving an equation helps us visualize the process of isolating the variable using a balanced scale. The **symbolic method** uses operations with variables to solve equations.

EXAMPLE 1: Solve the equation $x - 7 = 10$ using the scale and symbolic method.

The scale method: The equation is represented by a scale with the fulcrum as the = symbol. To solve the equation and isolate the variable, we need to "undo" the subtraction of 7 on the left side of the equation.

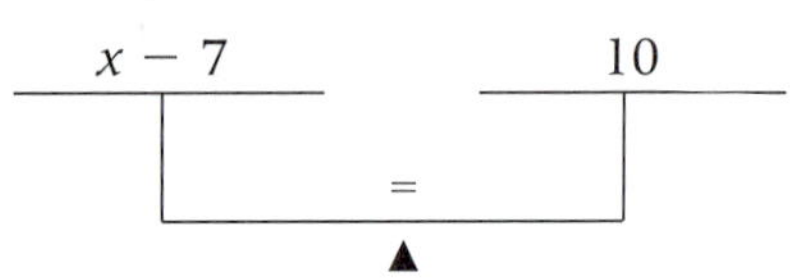

Addition is the *inverse* operation of subtraction. The variable can be isolated and the scale balanced by *adding* 7 on each side of the scale.

Add 7 to each side of the scale:

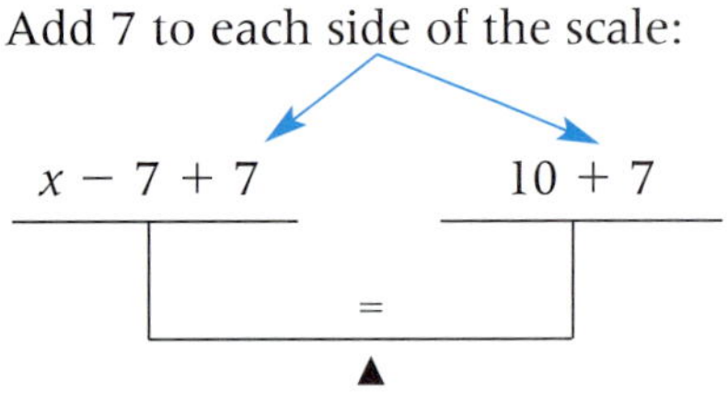

Simplify each side of the scale. The solution is $x = 17$.

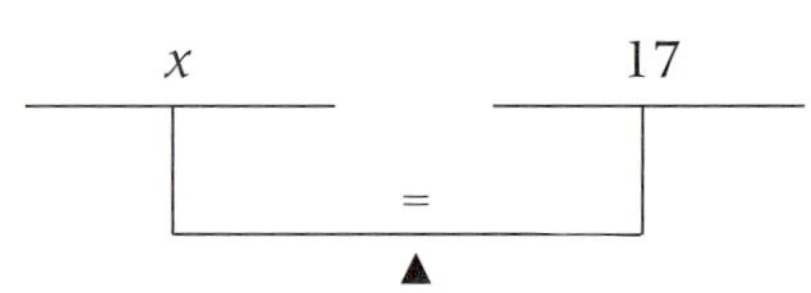

$$x - 7 = 10$$
$$(17) - 7 \stackrel{?}{=} 10$$
$$10 = 10$$

Verify the solution by substituting $x = 17$ into the original equation.

✓ It checks.

The symbolic method: The equation-solving process represented symbolically.

$$x - 7 = 10$$
$$x - 7 + 7 = 10 + 7 \quad \text{Add 7 to each side of the equation.}$$
$$x = 17$$

1. Solve $x - 19 = 5$ using both the scale and symbolic method. Add 19 to each side of the scale and equation.

The scale method:

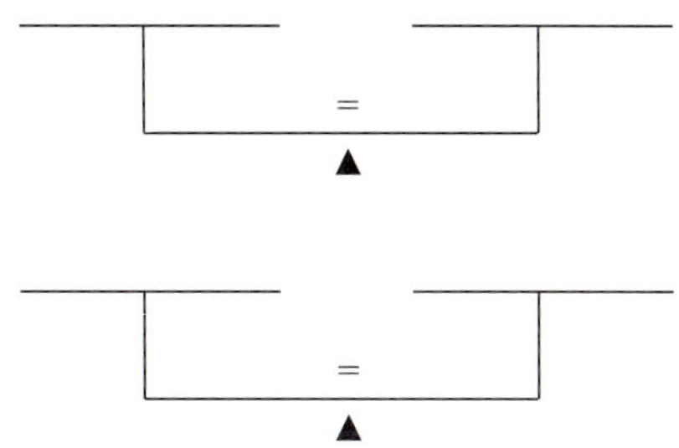

The symbolic method:

$x - 19 \qquad = 5$

$x - 19 + \_\_\_\_\_ = 5 + \_\_\_\_\_$

$x = \_\_\_\_\_$

Verify the solution:

2. Subtraction is the *inverse* operation of addition. Solve $x + 6 = 5$ with the scale and symbolic method by *subtracting* 6 on each side of the scale and equation.

The scale method:

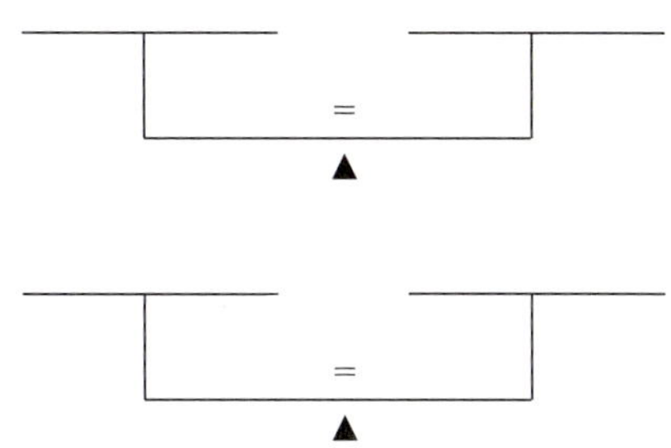

The symbolic method:

$x + 6 \qquad = 5$

$x + 6 - \_\_\_\_\_ = 5 - \_\_\_\_\_$

$x = \_\_\_\_\_$

Verify the solution:

EXAMPLE 2: Solve the equation $5x = 15$.

The scale method:

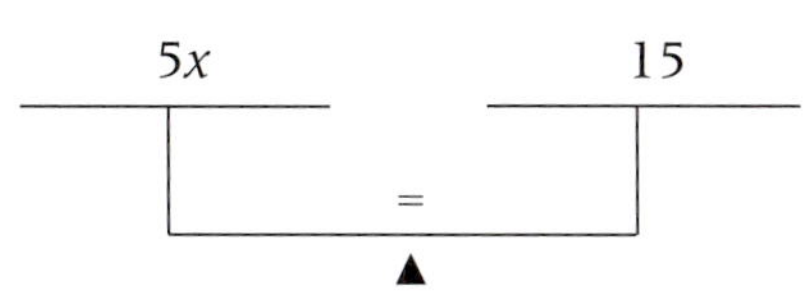

Division is the *inverse* operation of multiplication. The variable is isolated by *dividing* by 5 on each side of the scale.

Divide by 5 on each side of the scale:

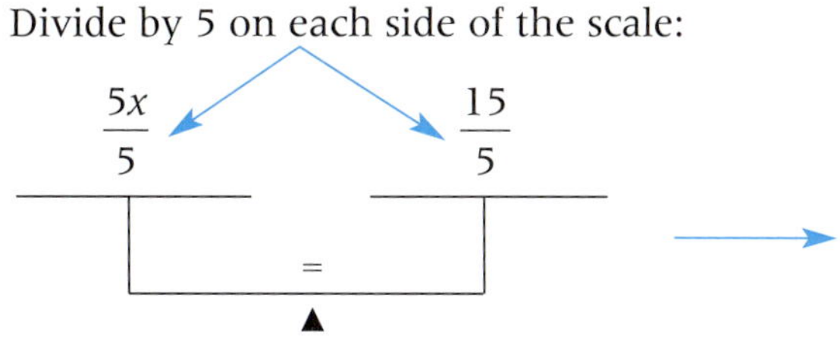

Simplify each side of the scale. The solution is $x = 3$.

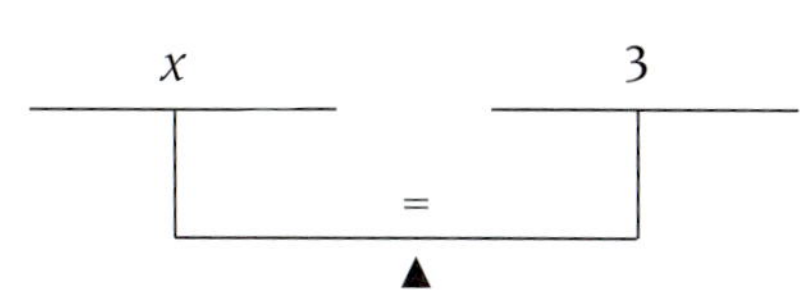

$$5x = 15$$
$$5(3) \stackrel{?}{=} 15$$
$$15 = 15$$

Verify the solution by substituting $x = 3$ into the original equation.

✓ It checks.

The symbolic method:

$$5x = 15$$
$$\frac{5x}{5} = \frac{15}{5}$$
$$x = 3$$

Divide by 5 on each side of the equation.

3. Solve $6x = 24$ by dividing by 6 on each side of the scale or equation.

The scale method:

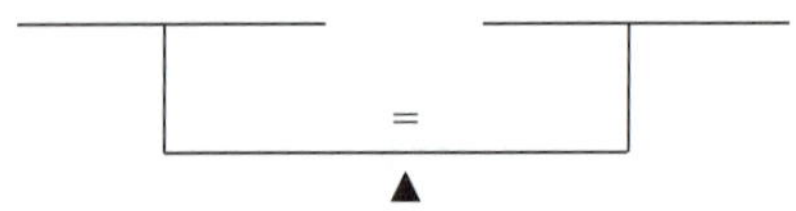

The symbolic method:

$$6x = 24$$

$$\frac{6x}{\text{---------}} = \frac{24}{\text{---------}}$$

$x =$ ______ Verify the solution:

4. Multiplication is the *inverse* operation of division. Solve $\frac{x}{7} = 6$ by *multiplying* by 7 on each side of the scale or equation.

The scale method:

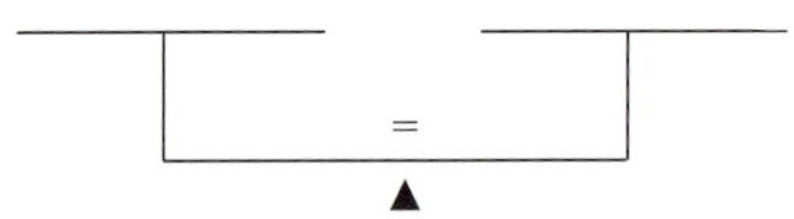

The symbolic method:

$$\frac{x}{7} = 6$$

$$\_\_\_\_\_\_ \cdot \frac{x}{7} = \_\_\_\_\_\_ \cdot 6$$

$x =$ ______ Verify the solution:

5. Summarize in your own words the steps for solving a linear equation.

Discussing Mathematics

Solving a linear equation is the process of identifying the solution that satisfies the equation. The balanced scale helped us to visualize the process of isolating the variable to solve an equation. Now we will formalize the symbolic process of solving an equation.

SOLVING LINEAR EQUATIONS

The **Equality Principles of Equations** are generalizations of the processes of adding, subtracting, multiplying, or dividing constants and variables on each side of an equation. In each case, the *inverse* operation is performed to isolate the variable. In generalizing, the object ♦, denotes a real number or algebraic expression.

EQUALITY PRINCIPLES OF EQUATIONS

Let ♦ = any real number or algebraic expression.

Addition Principle:	If $x = y$, then $x + ♦ = y + ♦$.
Subtraction Principle:	If $x = y$, then $x - ♦ = y - ♦$.
Multiplication Principle:	If $x = y$, then $x \cdot ♦ = y \cdot ♦$.
Division Principle:	If $x = y$, then $\frac{x}{♦} = \frac{y}{♦}$, $(♦ \neq 0)$.

➢ *The Addition, Subtraction, and Multiplication Principles are trivial when ♦ is equal to zero. Addition and Subtraction of zero is unnecessary. Multiplication by zero eliminates the equation. Division by zero is undefined.*

EXAMPLE 3: Solve $y - 4 = 17$.

$y - 4 + 4 = 17 + 4$ — Addition is the *inverse* of subtraction. Add 4 on each side of the equation.

$y = 21$ — Solution.

$y - 4 = 17$ — Verify the solution by substituting $y = 21$ into the original equation.

$(21) - 4 \stackrel{?}{=} 17$ — ✓ It checks.

$17 = 17$

EXAMPLE 4: Solve $25 + y = 60$.

$25 + y - 25 = 60 - 25$ — Subtraction is the *inverse* of addition. Subtract 25 on each side of the equation.

$y = 35$ — Solution.

$25 + y = 60$ — Verify the solution by substituting $y = 35$ into the original equation.

$25 + (35) \stackrel{?}{=} 60$ — ✓ It checks.

$60 = 60$

EXAMPLE 5: Solve $-7y = 35$.

$\frac{-7y}{-7} = \frac{35}{-7}$ — Division is the *inverse* of multiplication. Divide by -7 on each side of the equation.

$y = -5$ — Solution. Verify the solution by substituting $y = -5$ into the original equation. ✓ It checks.

EXAMPLE 6: Solve $\frac{x}{4} = 9$.

$\overset{1}{\cancel{4}} \cdot \frac{x}{\underset{1}{\cancel{4}}} = 4 \cdot 9$ — Multiplication is the *inverse* operation of division. Multiply by 4 on each side of the equation. Divide out common factors.

$x = 36$ — Solution. Verify the solution by substituting $x = 36$ into the original equation. ✓ It checks.

SOLVING LINEAR EQUATIONS WHEN VARIABLES APPEAR ON BOTH SIDES OF AN EQUATION

In Examples 1 through 6, the variable appeared on only one side of the equation. Now we consider equations in which terms containing variables will appear on each side of the equation. Variables, like constants, can be added, subtracted, multiplied, or divided on each side of an equation. It may be necessary to perform more than one inverse operation to isolate the variable and determine the solution of the equation.

STEPS FOR SOLVING A LINEAR EQUATION

1. Simplify each side of the equation. Remove parentheses, if necessary. Collect LIKE terms.
2. Add (or subtract) terms on each side of the equation so that all terms containing the variable are on one side and all constant terms are on the other side.
3. Simplify each side of the equation by combining LIKE terms.
4. Divide each side of the equation by the coefficient of the variable or multiply each side by the reciprocal of the coefficient of the variable.
5. Verify by substituting the solution into the original equation.

➢ *When solving equations, always choose the operation that "undoes" the operation of the term or factor that you want to change.*

EXAMPLE 7: Solve $4x = 2x - 10$.

$4x - 2x = 2x - 2x - 10$	Subtract $2x$ from each side of the equation.
$2x = -10$	Collect LIKE terms.
$\dfrac{2x}{2} = \dfrac{-10}{2}$	Divide by 2 on each side of the equation.
$x = -5$	Solution.
$4(-5) \stackrel{?}{=} 2(-5) - 10$	Verify by substituting $x = -5$ into the original equation, $4x = 2x - 10$.
$-20 \stackrel{?}{=} -10 - 10$	
$-20 = -20$	✓ It checks.

EXAMPLE 8: Solve $16 - 2y = 6y$.

$16 - 2y + 2y = 6y + 2y$	Add $2y$ to each side of the equation.
$16 = 8y$	Collect LIKE terms.
$\dfrac{16}{8} = \dfrac{8y}{8}$	Divide by 8 on each side of the equation.
$2 = y$	Solution. Verify by substituting $y = 2$ into the original equation. ✓ It checks.

➢ *When using the Equality Principles to solve the equations, simplify each side of the equation before and after adding, subtracting, multiplying, or dividing on each side of the equation. Order of Operations must always be employed.*

EXAMPLE 9: Solve $18a - 11a = 10 + 2a$.

$7a = 10 + 2a$	Simplify the left side of the equation.
$7a - 2a = 10 + 2a - 2a$	Subtract $2a$ on each side of the equation.
$5a = 10$	Collect LIKE terms.
$\dfrac{5a}{5} = \dfrac{10}{5}$	Divide by 5 on each side of the equation.
$a = 2$	Verify the solution by substituting $a = 2$ into the original equation. ✓ It checks.

EXAMPLE 10: Solve $5x - 2(x + 3) = -18$.

$5x - 2x - 6 = -18$	Remove parentheses by distributing -2.
$3x - 6 = -18$	Combine LIKE terms.
$3x - 6 + 6 = -18 + 6$	Add 6 to each side of the equation to isolate the variable.
$3x = -12$	
$\frac{3x}{3} = \frac{-12}{3}$	Divide by 3 on each side of the equation.
$x = -4$	Verify the solution by substituting $x = -4$ into the original equation. ✓ It checks.

EXAMPLE 11: Solve $-4.1x + 10.2 = 1 - (11.2x + 5)$.

$-4.1x + 10.2 = 1 - 11.2x - 5$	Remove parentheses by multiplying by -1.
$-4.1x + 10.2 = -11.2x - 4$	Simplify the right side of the equation.
$-4.1x + 11.2x + 10.2 = -11.2x + 11.2x - 4$	Add $11.2x$ to each side of the equation.
$7.1x + 10.2 = -4$	Collect LIKE terms.
$7.1x + 10.2 - 10.2 = -4 - 10.2$	Subtract 10.2 from each side of the equation.
$7.1x = -14.2$	Simplify each side of the equation.
$\frac{7.1x}{7.1} = \frac{-14.2}{7.1}$	Divide by 7.1 on each side of the equation.
$x = -2$	Verify the solution by substituting $x = -2$ into the original equation. ✓ It checks.

EXAMPLE 12: Solve $x - \frac{3}{4} = \frac{1}{2}$.

$x - \frac{3}{4} + \frac{3}{4} = \frac{1}{2} + \frac{3}{4}$	Add $\frac{3}{4}$ to each side of the equation.
$x = \frac{1}{2} + \frac{3}{4}$	Find the common denominator and add.
$x = \frac{2}{4} + \frac{3}{4} = \frac{5}{4}$	Verify the solution by substituting $x = \frac{5}{4}$ into the original equation. ✓ It checks.

MODELING REAL-WORLD SITUATIONS

Equations help us represent and solve real-world situations. A variable is used to represent an unknown and an equation is formed. The Equality Principles of Equations are then used to solve the equations and answer specific questions.

EXAMPLE 13: Marge owns a secondhand clothing store. The fixed operating costs of her business are \$1500 per week. Every item in the store sells for \$6.50. Let x represent the number of items sold and let P represent the profit.

(a) If her profit for one week was \$1750, how many items did she sell? Write an equation describing Marge's total monthly profit.

$$\text{Profit} = \text{Revenue} - \text{Operating Costs}$$
$$\text{Profit} = \$6.50 \cdot (\text{number of items sold}) - \$1500$$
$$P = 6.50x - 1500$$

$1750 = 6.50x - 1500$ Substitute 1750 for P. Solve for x.

$1750 + 1500 = 6.50x - 1500 + 1500$ Add 1500 on each side of the equation.

$3250 = 6.50x$

$\frac{3250}{6.50} = \frac{6.50}{6.50}x$ Divide each side of the equation by 6.50.

$500 = x$

Marge sold 500 items.

(b) How many items must be sold for Marge to break even? Break even occurs when $P = 0$.

$0 = 6.50x - 1500$ Substitute 0 for P. Solve for x.

$0 + 1500 = 6.50x - 1500 + 1500$ Add 1500 on each side of the equation.

$1500 = 6.50x$

$\frac{1500}{6.50} = \frac{6.50}{6.50}x$ Divide each side of the equation by 6.50.

$230.77 = x$

Marge must sell 231 items to break even. It is impossible to sell a fraction of a clothing item.

1 2 3 Doing Mathematics

- Write the definitions of the words and symbols listed for this section in the Dictionary in Appendix A.
- On a 3 × 5-inch card, write the process of solving a linear equation for future reference.

1. Determine whether the given value is a solution of the equation. Show your work.

(a) Is $x = 5$ a solution of $x + 4 = 19$?

(b) Is $x = 2$ a solution of $3x + 1 = 7$?

2. Determine whether the given value is a solution of the equation. Show your work.

(a) Is $x = -3$ a solution of $-4x = -7(x + 3)$?

(b) Is $x = \frac{1}{4}$ a solution of $4000 = 6400x$?

Problems 3–53: Solve each equation. Verify the solution by substituting.

3. $x + 5 = 10$

4. $y + 3 = 7$

5. $y - 7 = 12$

6. $x - 2 = 20$

7. $5x = -100$

8. $2y = -14$

9. $-3y = 15$

10. $-10x = 60$

11. $\frac{x}{10} = 4$

12. $\frac{y}{5} = 2$

13. $\frac{y}{3} = -4$

14. $\frac{x}{6} = -10$

15. $\frac{x}{-2} = 3$

16. $\frac{x}{-4} = 2$

17. $2x - 5 = 15$

18. $5y - 10 = 15$

19. $3y + 1 = 19$

20. $4y + 5 = 25$

21. $4y + 2 = 1$

22. $3x + 1 = 5$

23. $\frac{y}{3} - 6 = 20$

24. $\frac{x}{2} - 3 = 5$

25. $\frac{x}{4} + 1 = 9$

26. $\frac{x}{5} + 6 = 10$

27. $\frac{2x}{9} + 2 = 8$

28. $\frac{3x}{5} + 1 = 7$

29. $12x + 28x = 8$

30. $6y + 2y = 80$

31. $100y - 50y = 400$

32. $48x - 20x = 56$

33. $5x - 50 = 3x$

34. $20 - 4x = 6x$

35. $14 + 2x = 9x$

36. $9y - 27 = 6y$

37. $3x - 2(x + 5) = 19$

38. $\frac{10}{3}x + 5 = \frac{4}{3}x + 2$

39. $\frac{7}{2}x + 10 = \frac{1}{2}x + 22$

40. $5y - 6 = 2(2y + 5)$

41. $2(m - 5) = -6(m - 1)$

42. $10.2x - 5.0 = 15.8 - (1.8x - 3.2)$

43. $7x + 13 = -3(x + 9)$

44. $4\left(\frac{x}{2} - 5\right) = -3x + 5$

45. $9\left(\frac{x}{3} - 1\right) = -4x - 15$

46. $x - 3x + 6 = 2(x - 3)$

47. $-3y = 5y - (6 + 2y)$

48. $-3x - 10 = -5(x + 6)$

49. $2x^2 + 2x - 6 = -2(-x^2 + x - 15)$

50. $\frac{1}{2}(2x + 6) = -2(x + 6)$

51. $2(4x + 13) - 6x = 4x - 2(1 - x)$

52. $2(1.12a - 4.64) = 5.74a + 5.42$

53. Challenge problem: $-2y(y - 3) - (3y^2 - 1) = -8 - (5y^2 - 1)$

54. Mary has accepted a part-time job selling calculators. She will receive a fixed monthly salary of \$300 plus a commission of 6% of her monthly sales. Let the variable x represent Mary's monthly sales. Let T represent her total monthly pay.

(a) Use the variables x and T to write an equation that describes the total monthly pay Mary will receive.

(b) What is Mary's total pay for the month of May if her sales total \$500?

(c) Mary would like to double her salary in July, compared to May. What would Mary's monthly sales in July need to be to double her May pay?

55. Tom owns a business that builds computers according to a customer's specifications. The fixed operating costs of his business are \$2700 per week. In addition to fixed operating costs, each computer costs \$600 to produce. Each computer sells for \$1500. Let c represent the number of computers sold. Let P represent the profit Tom makes from selling computers.

(a) Use the variables c and P and write an equation that describes Tom's weekly profit.

(b) How many computers must Tom sell in one week to break even?

(c) How many computers did Tom sell if his profit was at least \$10,000?

56. Joan has accepted a telephone survey job. She receives a fixed monthly salary of \$200 plus a commission of \$0.20 for each call she completes. Let c represent the number of calls Joan completes. Let T represent her total monthly pay.

(a) Write an equation describing Joan's total monthly pay.

(b) What is Joan's total salary for the month of May if she completes 800 calls?

(c) How many calls does Joan need to complete in one month to make \$2000?

57. Jesse owns a video store that rents video games. The fixed operating costs of his business are \$875 per week. Each video rents for \$3.50.

(a) Let g represent the number of video games rented and let P represent the profit. Write an equation describing Jesse's total weekly profit.

(b) How many games must be rented per week for Jesse to break even?

(c) If Jesse's profit for one week was \$525, how many games were rented?

58. Jessica has signed a cellphone contract offering a discount rate of \$0.03 per minute for each call. Jessica estimates that she will make twice as many local calls as long distance calls. She is able to spend \$27.00 per month for her cell phone.

(a) Let the variable c represent the number of minutes of long distance calls. Let $2c$ represent the number of minutes of local calls. Write an equation describing the problem situation.

(b) How many minutes of long distance calls and local calls can she make, if her bill is exactly \$27?

2.4 Solving and Graphing Linear Inequalities, Set-Builder and Interval Notation

OBJECTIVES

- Graph linear inequality statements on the number line.
- Write linear inequality statements using set-builder notation and interval notation.
- Solve linear inequalities.

1 2 3 Getting Started

Inequality symbols, $<$ (less than), $>$ (greater than), $\leq$ (less than or equal to), and $\geq$ (greater than or equal to) denote relationships between quantities. In this section, we consider inequalities in problem situations and develop techniques for solving statements of inequality involving variables.

1. Suppose that your grade in this course is to be determined from the average of the scores of four 100-point tests. You received scores of 91, 70, and 79 on the first three tests. In order to receive a grade of B or higher, a minimum of 80% of the total 400 points possible, or 320 points, must be earned.

(a) What is the minimum (lowest) score on the fourth test that will enable you to earn a grade of B or higher in the course?

(b) What is the maximum (highest) score that you can earn on the fourth test?

(c) List five other scores for the fourth test that will earn you a grade of B or higher.

(d) Graph the scores you listed in parts (a)–(c) on the number line below. Label the points.

(e) Is it possible to receive a grade of A in the course (minimum 360 points)? Why or why not?

2. Determine whether the direction of the inequality symbol is the same or reversed when a number, chosen arbitrarily, is *added or subtracted* on each side of the inequality. Write the result of the operation, along with the appropriate symbol, either $<$ or $>$, on the lines.

EXAMPLE :	9	$<$	15	Original inequality statement.
	$9 + 2$		$15 + 2$	2 is added to each side.
	$\underline{11}$	$\underline{<}$	$\underline{17}$	Write the result and symbol.

(a) $9 \quad < \quad 15$

$9 + 8 \qquad 15 + 8$ 8 is added.

_____ __ _____

(b) $9 \quad < \quad 15$

$9 - 8 \qquad 15 - 8$ 8 is subtracted.

_____ __ _____

(c) $10 \quad > \quad 7$

$10 + (-2) \qquad 7 + (-2)$ -2 is added.

______ __ ______

(d) $10 \quad > \quad 7$

$10 - (-2) \qquad 7 - (-2)$ -2 is subtracted.

______ __ ______

(e) After adding or subtracting a positive or negative number on both sides of the inequalities, is the direction of the inequality symbol the same or reversed?

3. Determine whether the direction of the inequality symbol is the same or reversed when a number, chosen arbitrarily, is *multiplied or divided* on each side of the inequality. Write the result of the operation, along with the appropriate symbol, either $<$ or $>$, on the lines.

(a) $5 \quad < \quad 8$

$5(3) \qquad 8(3)$ Multiply by 3.

____ __ ____

(b) $45 \quad > \quad 25$

$\frac{45}{5} \qquad \frac{25}{5}$ Divide by 5.

__ __ __

(c) After multiplying or dividing by a positive number on both sides of the inequalities, is the direction of the inequality symbol the same or reversed?

(d) $7 \quad > \quad 1$

$7(-2) \qquad 1(-2)$ Multiply by -2.

____ __ ____

(e) $45 \quad > \quad 25$

$\frac{45}{-5} \qquad \frac{25}{-5}$ Divide by -5.

__ __ __

(f) After multiplying or dividing by a negative number on both sides of the inequalities, is the direction of the inequality symbol the same or reversed?

Solving inequalities is similar to solving equations. The Equality Principles can be utilized in much the same way to *isolate the variable*. The solution of an inequality statement involving variables is the set of all values of the variable that satisfy the statement and may be written in set-builder notation, in interval notation, or on a graph.

➢ *Adding or subtracting any real number on each side of an inequality statement results in a true statement. The direction of the inequality remains the same.*

➢ *Multiplying or dividing by the same positive real number on each side of an inequality statement results in a true statement. The direction of the inequality remains the same.*

EXAMPLE 1: Solve the inequality statement: $y - 6 > 13$.

$y - 6 > 13$

$y - 6 + 6 > 13 + 6$ Isolate the variable by adding 6 to each side of the inequality statement.

$y > 19$ Solution set written in set-builder notation: $\{y \mid y > 19\}$.

4. Solve each inequality statement. Write the solution in set-builder notation.

(a) $y - 10 > 5$

(b) $y + 5 \le 8$

(c) $4x \le 12$

(d) $\frac{x}{5} > 3$

➢ *Multiplying or dividing by the same negative real number on each side of an inequality statement results in a false statement. To write a true statement, the direction of the inequality must be reversed.*

EXAMPLE 2: Solve the inequality statement: $-10x \leq 50$.

$-10x \leq 50$ — Divide by -10 on each side of the inequality.

$\dfrac{-10x}{-10} \boxed{\geq} \dfrac{50}{-10}$ — Reverse the inequality symbol to write a true statement.

$x \geq -5$ — ➢ *Draw a* **reminder square** *around the inequality symbol as a reminder to reverse the inequality symbol. Solution set written in set-builder notation:* $\{x \mid x \geq -5\}$.

5. Solve each inequality statement. Write the solution set in set-builder notation.

(a) $-3x \leq 12$ (b) $\dfrac{x}{-5} \leq 2$

Discussing Mathematics

The solution set of an inequality statement is often visualized with a graph or written in set-builder or interval notation.

SET-BUILDER NOTATION

Statements of inequality are often written symbolically using set-builder and/or interval notation. **Set-builder notation** uses set braces, the variable, the boundary number(s), and the appropriate inequality symbol:

The set $x < 5$ written in set-builder notation: $\{x \mid x < 5\}$

and is read "the set of all x, such that, x is strictly less than 5."

GRAPHS OF INEQUALITY STATEMENTS AND INTERVAL NOTATION

Graphs help us visualize solution sets of inequality statements. Darkened lines are drawn from the boundary point on top of the numbers in the solution set. An arrow indicates an infinite set.

➢ **Open dots and parentheses** *are used when graphing strict inequalities ($<$ or $>$) and indicate that the boundary number of the set is NOT included in the solution set.*

➢ **Closed dots or square brackets** *are used to graph nonstrict inequalities ($\leq$ or $\geq$) and indicate that the boundary number IS included in the solution set.*

Interval Notation uses parentheses or brackets to enclose the boundaries of a set, without displaying the variable. Parentheses denote that the boundary number is *not* included. Brackets denote that the boundary number *is* included. Infinite sets use infinity symbols, $+\infty$ or $-\infty$, enclosed with a parenthesis.

The graph of the strict inequality, $x > 5$:

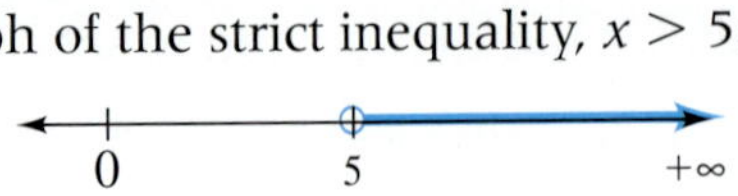

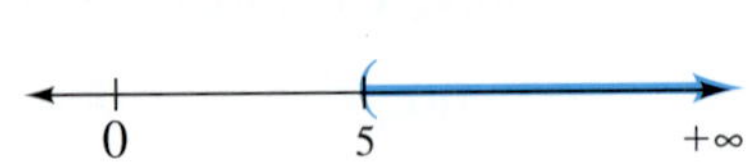

Interval notation for the strict inequality, $x > 5$:

$(5, \infty)$

Parenthesis indicates 5 is not included. | 5 is left boundary point. | A right boundary number does not exist.

The graph of the nonstrict inequality, $x \geq 5$:

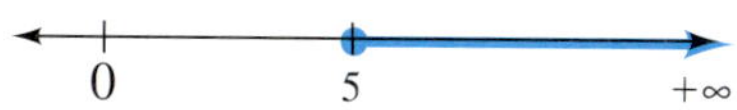

OR

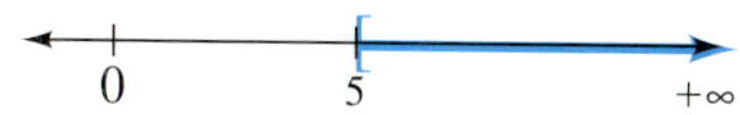

Interval notation for the nonstrict inequality, $x \geq 5$:

$[5, \infty)$

Bracket indicates 5 is included. | 5 is left boundary point. | A right boundary number does not exist.

➢ *Parentheses and open dots are two symbols that mean the same thing. Closed dots and brackets also mean the same thing. In this text, we will always use parentheses (instead of open dots) and square brackets (instead of closed dots).*

Summary of inequality statements in equivalent set-builder notation, graph format, and interval notation:

Set-builder notation	Graph	Interval notation
$\{x \mid x > 5\}$		$(5, \infty)$
$\{x \mid x \geq 5\}$		$[5, \infty)$
$\{x \mid x < 5\}$	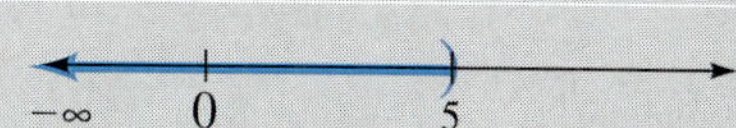	$(-\infty, 5)$
$\{x \mid x \leq 5\}$		$(-\infty, 5]$

ADDITION PRINCIPLE OF INEQUALITIES

The Addition Principle of Inequalities generalizes our observations from Getting Started problems. The ♦ symbol denotes a quantity added (or subtracted) on each side of an inequality statement.

➢ *Adding or subtracting the same number on each side of an inequality statement results in a true statement. The direction of the inequality stays the same.*

ADDITION (AND SUBTRACTION)* PRINCIPLE OF INEQUALITIES

Let a, b, and ♦ represent any real number, variable, or algebraic expression containing one or more terms. The statement is true for $<$, $\leq$, $>$, and $\geq$.

$$\text{If } a < b, \text{ then } a + \blacklozenge < b + \blacklozenge.$$

*Subtraction of any real number can be rewritten as addition.

EXAMPLE 3: Solve the inequality statement: $y - 10 > 5$.

$$y - 10 > 5$$

$$y - 10 + 10 > 5 + 10 \quad \text{Add 10 to each side of the inequality.}$$

$$y > 15$$

The solution written in three forms:

Graph:

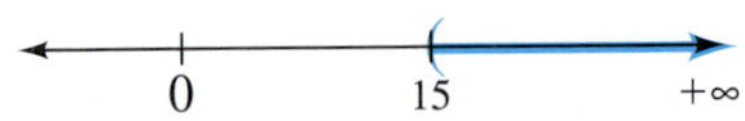

Set-builder notation: $\{y \mid y > 15\}$

Interval notation: $(15, \infty)$

To verify the solution, choose a value in the interval $y > 15$ and substitute the value into the original inequality. A value in the solution interval results in a *true* statement.

Choose $y = 20$: $20 - 10 \overset{?}{>} 5$

$10 > 5$ True

A value that is not in the solution interval should result in a *false* statement when substituted into the original inequality.

Choose $y = 10$: $10 - 10 \overset{?}{>} 5$

$0 > 5$ False

This method serves as a quick check, but does not constitute a formal proof. It is not possible to check all values in the solution interval.

EXAMPLE 4: Solve the inequality statement: $5x + 3 \leq 4x - 10$.

$$5x + 3 \leq 4x - 10$$

$$5x - 4x + 3 \leq 4x - 4x - 10 \quad \text{Subtract } 4x \text{ from each side.}$$

$$x + 3 \leq -10 \quad \text{Simplify.}$$

$$x + 3 - 3 \leq -10 - 3 \quad \text{Subtract 3 from each side.}$$

$$x \leq -13$$

The solution set written in three forms:

Graph:

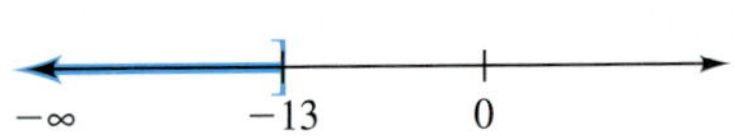

Set-builder notation: $\{x \mid x \leq -13\}$

Interval notation: $(-\infty, -13]$

To verify the solution, choose a value in the interval $x \leq -13$ and substitute the value into the original inequality. A true statement should result. Similarly, a value that is not in the solution interval should result in a false statement when substituted into the original inequality.

MULTIPLICATION AND DIVISION PRINCIPLES OF INEQUALITIES

Our observations in Getting Started also lead us to generalizations of the Multiplication and Division Principles of Inequalities.

- *Multiplying or dividing by the same positive quantity results in a true statement where the direction of the inequality is the same.*
- *Multiplying or dividing by the same negative quantity results in a false statement. The direction of the inequality must be reversed to write a true statement. A* **reminder square** *is used to remind us to reverse the direction of the inequality when multiplying or dividing by a negative quantity.*

Reminder Square

MULTIPLICATION AND DIVISION PRINCIPLES OF INEQUALITIES

Let a and b represent any real number, variable, or algebraic expression. Each statement is true for $<$, $\leq$, $>$, and $\geq$.

Let ♦ represent any positive real number, variable, or algebraic expression (positive real numbers are numbers greater than zero; zero is not a positive number).

If $a < b$, then $a \cdot \blacklozenge < b \cdot \blacklozenge$. If $a < b$, then $\frac{a}{\blacklozenge} < \frac{b}{\blacklozenge}$.

Let ♦ represent any negative real number, variable, or algebraic expression (negative real numbers are numbers less than zero; zero is not a negative number).

If $a < b$, then $a \cdot \blacklozenge \boxed{>} b \cdot \blacklozenge$. If $a < b$, then $\frac{a}{\blacklozenge} \boxed{>} \frac{b}{\blacklozenge}$.

EXAMPLE 5: Solve the inequality statement: $-\frac{y}{3} > 6$.

$$\left(-\frac{3}{1}\right) \cdot -\frac{y}{3} \boxed{<} 6 \cdot \left(-\frac{3}{1}\right)$$ Multiply both sides of the inequality by -3 $\left(\text{or } -\frac{3}{1}\right)$.

$$y < -18$$ Reverse the direction of the inequality.

The solution set in three forms:

Graph:

Set-builder notation: $\{y \mid y < -18\}$

Interval notation: $(-\infty, -18)$

To verify the solution, choose a value in the interval $y < -18$ and substitute the value into the original inequality.

Choose a value that is not in the solution interval.

Choose $y = -21$: $-\frac{-21}{3} \stackrel{?}{>} 6$

$-(-7) \stackrel{?}{>} 6$

$7 > 6$ True

Choose $y = -12$: $-\frac{-12}{3} \stackrel{?}{>} 6$

$-(-4) \stackrel{?}{>} 6$

$4 > 6$ False

SOLVING LINEAR INEQUALITIES

SOLVING LINEAR INEQUALITIES

1. Remove parentheses first.
2. Use the Addition Principles of Inequalities to collect LIKE terms on one side of the inequality and collect constants on the other side.
3. Use the Multiplication or Division Principle of Inequalities to isolate the variable. Draw a reminder square and reverse the direction of the inequality symbol when multiplying or dividing by a negative quantity.
4. Write the solution as an inequality statement, in graph format, in set-builder notation, or in interval notation.

EXAMPLE 6: Solve the inequality statement: $7x - 2 \geq 10x + 19$.

$7x - 10x - 2 \geq 10x - 10x + 19$ — Subtract $10x$ from each side of the inequality.

$-3x - 2 \geq 19$ — Collect LIKE terms.

$-3x - 2 + 2 \geq 19 + 2$ — Add 2 to each side of the inequality.

$-3x \geq 21$

$\frac{-3x}{-3} \boxed{\leq} \frac{21}{-3}$ — Divide each side by -3. Reverse the direction of the inequality symbol.

$x \leq -7$

The solution set in three forms:

Graph:

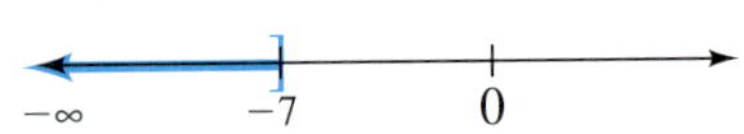

Set-builder notation: $\{x|x \leq -7\}$

Interval notation: $(-\infty, -7]$

To verify the solution, choose a value in the interval $x \leq -7$ and substitute the value into the original inequality. A true statement should result. Similarly, a value that is not in the solution interval should result in a false statement when substituted into the original inequality.

➢ *Example 6 can be solved differently to avoid dividing by a negative value. Subtracting 7x from each side, instead of 10x, isolates the variable on the right side of the inequality with a positive sign. The inequality symbol does not need to be reversed in this process:*

$$7x - 2 \geq 10x + 19$$

$$7x - 7x - 2 \geq 10x - 7x + 19$$

$$-2 \geq 3x + 19$$

$$-2 - 19 \geq 3x + 19 - 19$$

$$-21 \geq 3x$$

$$-\frac{21}{3} \geq \frac{3x}{3}$$

$-7 \geq x$ or $x \leq -7$ — The answer may be written in either form.

EXAMPLE 7: Solve the inequality statement: $8(x-2)-2(x-3) \le 8x$.

$8x - 16 - 2x + 6 \le 8x$ Remove parentheses.

$6x - 10 \le 8x$ Collect LIKE terms.

$6x - 6x - 10 \le 8x - 6x$ Subtract $6x$ from each side of the inequality.

$-10 \le 2x$

$\frac{-10}{2} \le \frac{2x}{2}$ Divide by 2 on each side of the inequality.

$-5 \le x$

➢ *It is NOT necessary to reverse the direction of the inequality symbol since we divided by +2.*

The solution set in three forms:

Graph:

Set-builder notation: $\{x|x \ge -5\}$

Interval notation: $[-5, \infty)$

To verify the solution, choose a value in the interval $x \ge -5$ and substitute the value into the original inequality. A true statement should result.

SOLVING REAL-WORLD PROBLEMS INVOLVING INEQUALITIES

Many real-world problems can be represented with statements of inequality and solved using the principles presented in this section. Let's consider the test score problem from Getting Started.

EXAMPLE 8: Suppose that your grade in this course is to be determined from the average of the scores of four 100-point tests. You received scores of 92, 69, and 83 on the first three tests. In order to receive a grade of B or higher, a minimum of 80% of the total 400 points possible, or 320 points, must be earned. Let x represent the unknown score on the fourth test.

$92 + 69 + 83 + x \ge 320$ The sum of the four scores must be greater than or equal to 320 in order to obtain a grade of B or higher.

$244 + x \ge 320$

$244 + x \ge 320$ Solve the inequality.

$244 - 244 + x \ge 320 - 244$

$x \ge 76$ The solution set of the inequality is: $\{x|x \ge 76\}$.

The highest score possible on the fourth test is 100. Therefore, the solution set to this problem situation is restricted between and including 76 and 100: $\{x|76 \le x \le 100\}$. Any score less than or equal to 100 AND greater than or equal to 76 on the fourth test will earn a grade of B or higher in the course.

➢ *When a mathematical statement or solution to a problem includes two inequality symbols, the statement is called a compound inequality. Compound inequalities are written with the numbers in the order they appear on the number line:* $76 \le x \le 100$.

Doing Mathematics

- Write the definitions of the words and symbols listed for this section in the Dictionary in Appendix A.
- On a 3 × 5-inch card, write the process of solving a linear inequality for future reference.

Problems 1–4: Graph each inequality statement. Use parentheses or brackets (not open or closed dots) on boundary points. Draw a darkened line to indicate the numbers in the set. Write each set in interval notation.

1. $x > -1$

(a) Graph: 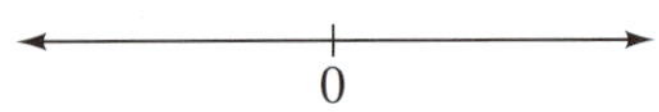

(b) Interval notation:

2. $x \geq -1$

(a) Graph: 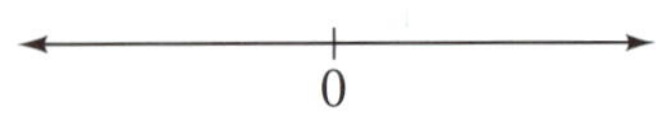

(b) Interval notation:

3. $x < -1$

(a) Graph: 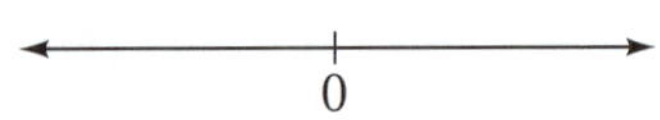

(b) Interval notation:

4. $x \leq -1$

(a) Graph: 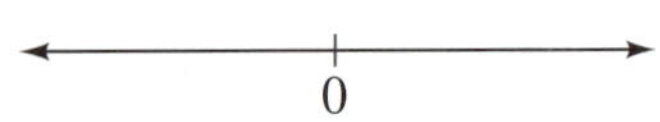

(b) Interval notation:

Problems 5–8: Graph each interval. Write the interval in set-builder notation.

5. $[-2, \infty)$

(a) Graph: 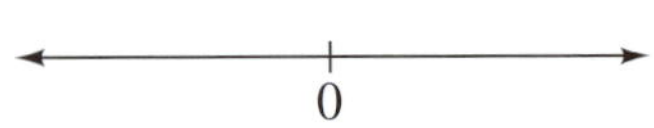

(b) Set-builder notation:

6. $(5, \infty)$

(a) Graph: 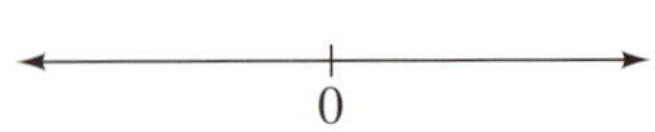

(b) Set-builder notation:

7. $(-\infty, -3]$

(a) Graph: 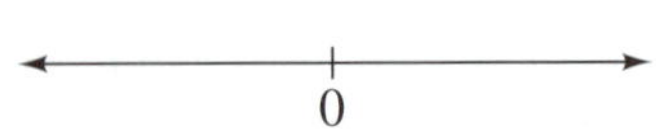

(b) Set-builder notation:

8. $(-\infty, 0)$

(a) Graph: 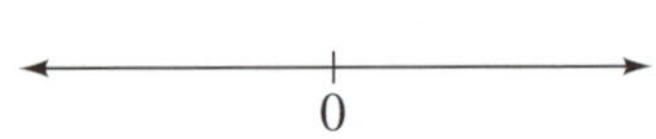

(b) Set-builder notation:

Problems 9–28: Solve each inequality. Draw a graph of the solution set and write the solution in set-builder and interval notation.

9. $x + 5 > 12$

(a) Graph: 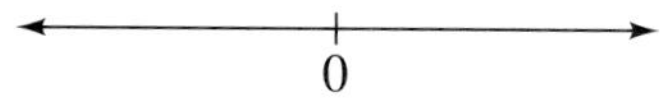

(b) Set-builder notation:

(c) Interval notation:

10. $y - 4 < 20$

(a) Graph: 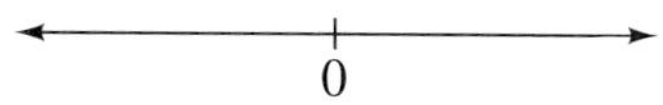

(b) Set-builder notation:

(c) Interval notation:

11. $12 \leq 2x + 6$

(a) Graph: 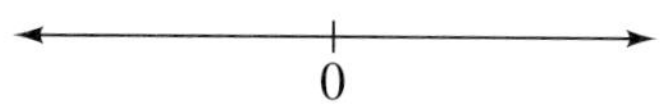

(b) Set-builder notation:

(c) Interval notation:

12. $36 \geq -2z + 30$

(a) Graph: 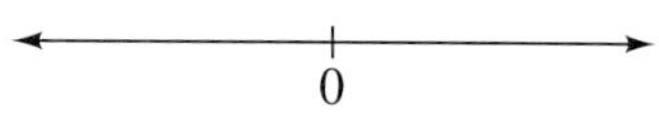

(b) Set-builder notation:

(c) Interval notation:

13. $x - 5 + 9 \geq 0$

(a) Graph: 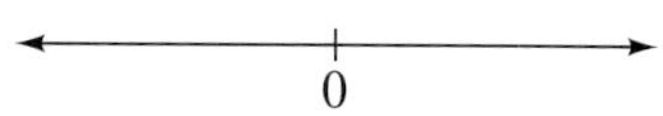

(b) Set-builder notation:

(c) Interval notation:

14. $20 - 13y < -6$

(a) Graph: 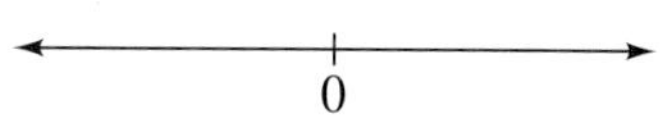

(b) Set-builder notation:

(c) Interval notation:

15. $7x + 6 - 2x \geq 2 - 3x$

(a) Graph: 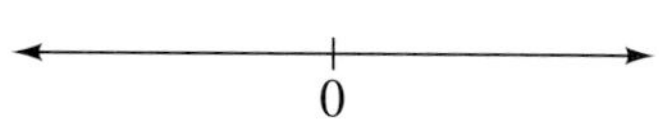

(b) Set-builder notation:

(c) Interval notation:

16. $10 + 9x - 3 \geq 4 - 3x$

(a) Graph: 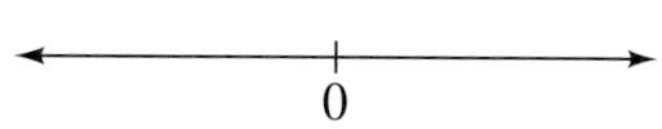

(b) Set-builder notation:

(c) Interval notation:

17. $13x + 5 - 18x < 4 - x$

(a) Graph: 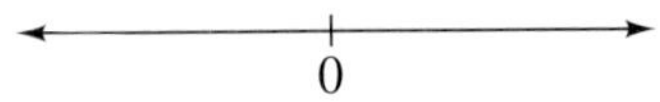

(b) Set-builder notation:

(c) Interval notation:

18. $6x + 7 - 4x < 7x - 3$

(a) Graph: 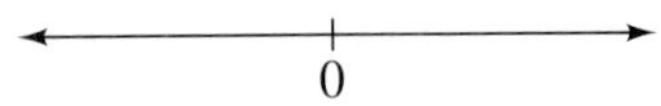

(b) Set-builder notation:

(c) Interval notation:

19. $4(6 - 2a) > 14(3 - a)$

(a) Graph: 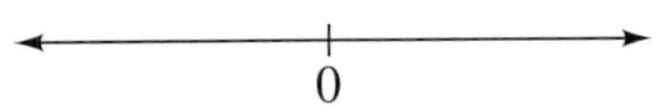

(b) Set-builder notation:

(c) Interval notation:

20. $3(5x + 6) \le 6(3x + 2)$

(a) Graph: 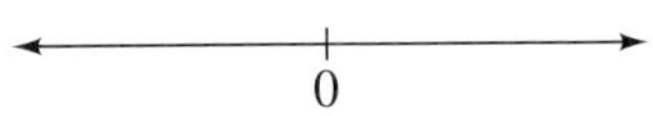

(b) Set-builder notation:

(c) Interval notation:

21. $26 \ge 2(7y - 8) + 7(2 - y)$

(a) Graph: 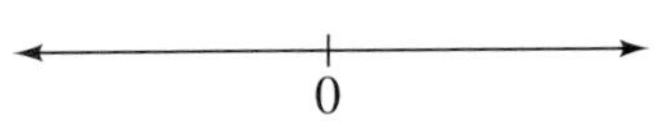

(b) Set-builder notation:

(c) Interval notation:

22. $0 \ge 4(3 - y) - 3(4 - y)$

(a) Graph: 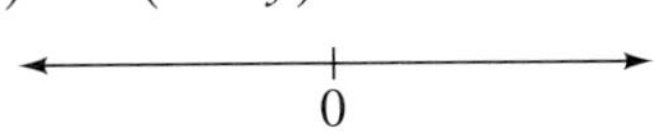

(b) Set-builder notation:

(c) Interval notation:

23. $5 - 3(4x + 3) \le 4(x - 1)$

(a) Graph: 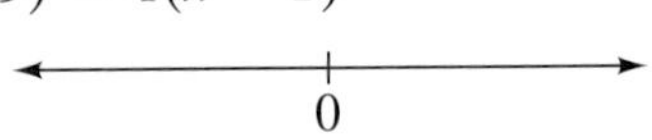

(b) Set-builder notation:

(c) Interval notation:

24. $2(2 - x) - (5 - 4x) < 13$

(a) Graph: 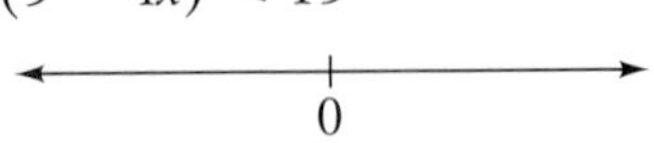

(b) Set-builder notation:

(c) Interval notation:

25. $-2.5(4x - 1.4) \geq 2x + 21.5$

(a) Graph: 0

(b) Set-builder notation:

(c) Interval notation:

26. $-0.4y + 2 < -1.2y - 0.4$

(a) Graph: 0

(b) Set-builder notation:

(c) Interval notation:

27. $\frac{2x}{7} \leq \frac{4}{7}$

(a) Graph: 0

(b) Set-builder notation:

(c) Interval notation:

28. $\frac{x}{5} + \frac{4x}{5} \geq 1$

(a) Graph: 0

(b) Set-builder notation:

(c) Interval notation:

Problems 29–30: Challenge problems. Solve each inequality, draw a graph of the solution sets, and write the solution in set-builder and interval notation.

29. $\frac{3}{4}x - \frac{1}{2} > \frac{1}{4}x + \frac{1}{2}$

(a) Graph: 0

(b) Set-builder notation:

(c) Interval notation:

30. $\frac{3}{5}x - \frac{1}{15} > \frac{5}{9}x + \frac{1}{5}$

(a) Graph: 0

(b) Set-builder notation:

(c) Interval notation:

31. Suppose that your grade in this course is to be determined from the average of the scores of four 100-point tests. You received the scores of 90, 96, and 87 on the first three tests. In order to receive a grade of A or higher, a minimum of 90% of the total 400 points possible, or 360 points, must be earned.

(a) What is the minimum score you need to earn on the fourth test in order to receive a grade of A or higher in the course?

(b) Write an inequality statement describing the problem situation.

(c) Solve the inequality in part (b), showing your work.

(d) Graph the solution set: 

(e) Write the solution in set-builder notation.

(f) Write the solution in interval notation.

(g) Does the problem situation place any restriction(s) on the answer?

32. A Priority package weighing 1 pound or less can be sent through the U.S. Mail locally or to zones 1, 2, or 3 for $3.85. A package weighing more than 1 pound but less than or equal to 2 pounds can be sent for $3.95. A package weighing more than 2 pounds but less than or equal to 3 pounds can be sent for $4.75. A package weighing more than 3 pounds but less than or equal to 4 pounds can be sent for $5.30. A package weighing more than 4 pounds but less than or equal to 5 pounds can be sent for $5.85. The rates for packages over 5 pounds are based on the weight of the package and the distance from the origin ZIP code to the destination ZIP code. (http://www.usps.gov)

(a) Let x represent the weight in pounds of a package. Write inequality statements that represent the weight of a package. Compute the cost of mailing a Priority package.

Weight	Symbolic representation of the weight of the package	Mailing cost of the package
1 pound or less	$0 < x \leq 1$	$3.85
More than 1 pound and up to and including 2 pounds	$1 < x \leq 2$	$3.95
More than 2 pounds and up to and including 3 pounds	$2 < x \leq 3$	$4.75
More than 3 pounds and up to and including 4 pounds		
More than 4 pounds and up to and including 5 pounds		
Over 5 pounds		Need ZIP code to compute

(b) If the cost of mailing a package was $5.30, what is the maximum number of pounds that the package could weigh?

(c) If the package weighed 4.2 pounds, how much would it cost to mail it?

33. Beth is a salesperson for a cellular phone company, Phones-On-The-Go. As an incentive to its salespersons, the company initiated a contest with a prize of a two-week trip to Hawaii to any salesperson selling more than 600 phones in a six-month period. In the months of January, February, March, April and May, Beth sold 60, 79, 105, 90, and 87 phones, respectively.

(a) Let x = the number of phones to be sold in June. Write the problem situation as an inequality statement.

(b) Solve the inequality and write the solution in set-builder notation.

(c) How many phones does Beth need to sell to win a trip to Hawaii?

(d) Do you think she will win the trip? Why or why not?

34. The Coffee Cage at Smarts College has fixed operating costs of \$300 per month. How much revenue must be generated to earn a profit of at least \$500? (Profit = Revenue − Operating Costs)

(a) Let R represent the revenue. Write an inequality statement describing the problem situation.

(b) Solve the inequality and write the solution set in interval notation:

35. Optional Internet Exercise: Visit the Internet Web site for the US Postal Service (http://www.usps.gov) and obtain the Domestic Rates and Fees for Customers for Express Mail, next-day and second-day PO to Addressee.

(a) Let x represent the weight in pounds of a package. Write inequality statements that represent the weight of Express Mail—Custom Designed packages. Compute the cost of mailing a package.

Weight	Symbolic representation of the weight	Mailing cost
$\frac{1}{2}$ pound or less	$0 < x \leq \frac{1}{2}$	
More than $\frac{1}{2}$ pound and up to and including 1 pound		
More than 1 pound and up to and including 2 pounds		
More than 2 pounds and up to and including 3 pounds		
More than 3 pounds and up to and including 4 pounds		
More than 4 pounds and up to and including 5 pounds		

(b) How much does it cost to send an Express Mail package weighing 3.25 pounds?

36. Create and solve a real-world problem situation involving an inequality statement.

2.5 Formulas and Literal Equations

OBJECTIVES

- Generalize problem situations, create, and solve equations.
- Solve problems involving formulas.
- Solve formulas and literal equations for a specified variable.

1 2 3 Getting Started

In this section, we pull together the concepts discussed in this chapter using variables to generalize a problem situation and solve equations. We begin with the formula of the area of a rectangle: $A = l \cdot w$ with length (l) and width (w). This formula can be solved for any of the three variables. To solve for w:

$$A = l \cdot w \qquad \text{Assume } l \neq 0. \text{ (Formula 1)}$$

$$\frac{A}{l} = \frac{l \cdot w}{l} \qquad \text{Divide by } l \text{ on each side of the equation.}$$

$$\frac{A}{l} = w \qquad \text{(Formula 2)}$$

1. A rectangular playground has an area of 400 square feet. Use Formula 2 and compute the corresponding width for the given lengths in the figures below.

(a) 

width = ___?___

Given: length = 25 feet

(b) 

width = ___?___

Given: length = 50 feet

(c) Solve the formula $A = l \cdot w$ for l.

2. Complete the table and compute the missing dimensions of a rectangular playground with an area of 400 square feet. Use the appropriate formula: $l = \frac{A}{w}$, $w = \frac{A}{l}$, $A = l \cdot w$.

	Length (l)	Width (w)	Area (A)
Ex.	200 feet	$w = \frac{400}{200} = 2$	400 square feet
(a)	100 feet		400 square feet
(b)		8 feet	400 square feet
(c)	25 feet		400 square feet
(d)		20 feet	400 square feet
(e)	40 feet		400 square feet

3. The formula for the area of a triangle is: $A = \frac{1}{2}bh$, where b represents the length of the base and h represents the height of the triangle. Complete the steps outlined as follows to solve the formula $A = \frac{1}{2}bh$ for b.

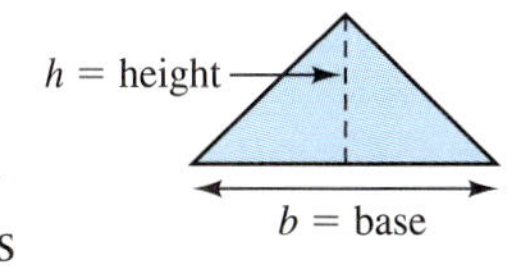

(a) ________________________ Multiply each side of the equation by 2 and simplify.

____

(b) ________________________ Divide each side of the equation by h.

(c) ________________________ Equivalent formula written in an alternate form.

4. There are many values of the base and height of a triangle that will produce an area of 400 square feet. Use the formula $\frac{2A}{h} = b$ to determine the base of each triangle.

(a)

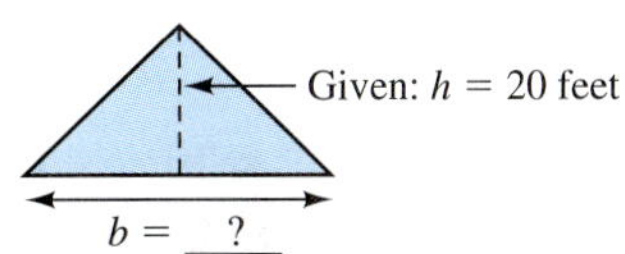

(b)

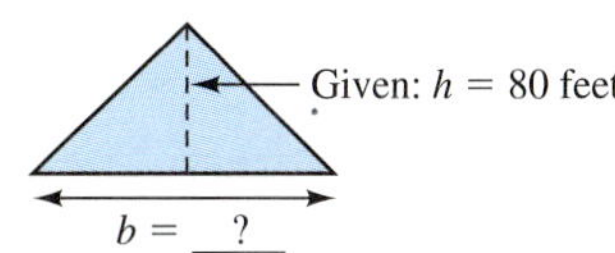

Discussing Mathematics

A **literal equation** is an equation containing two or more variables. Formulas are literal equations that are generalizations of frequently used problem situations with variables representing unknown quantities. A formula may be solved for any variable that appears in the formula and rewritten in an equivalent form.

EXAMPLE 1: The formula $P = 2l + 2w$ represents the perimeter of a rectangle with perimeter (P), width (w), and length (l). Solve the perimeter formula for w.

$$P = 2l + 2\boxed{w}$$ To help focus on the variable w, draw a box around the term containing w.

$$P - 2l = 2l - 2l + 2\boxed{w}$$ Subtract $2l$ from each side of the equation.

$$P - 2l = 2w$$

$$\frac{P - 2l}{2} = \frac{2w}{2}$$ Divide by 2 on each side of the equation

$$\frac{P - 2l}{2} = w$$ Equivalent form of the perimeter formula, solved for w.

EXAMPLE 2: Solve the equation $4 - x = y + 5$ for y.

$4 - x = \boxed{y} + 5$	Draw a box around the term containing the variable y.
$4 - 5 - x = \boxed{y} + 5 - 5$	Subtract 5 on each side of the equation to isolate the variable y.
$-1 - x = y$	The result may also be written as $y = -x - 1$.

PYTHAGOREAN THEOREM

Pythagoras (560–480 BC) dedicated his life to learning about science, philosophy, mathematics, and music and developed the Pythagorean Theorem. He established a school where his students and teachers believed the world could be explained through mathematics and science.

The Pythagorean Theorem is a useful formula describing the relationship between the sides of a right triangle. A **right triangle** is a triangle in which two sides are perpendicular and form a right angle (90°). The third side, opposite the right angle, is the longest side called the **hypotenuse**.

PYTHAGOREAN THEOREM

Given that a and b represent the lengths of the legs of a right triangle with c as the hypotenuse:

$$a^2 + b^2 = c^2 \quad \text{or} \quad c^2 = a^2 + b^2.$$

The Pythagorean Theorem solved for each variable:

$$c = \sqrt{a^2 + b^2}$$
$$a = \sqrt{c^2 - b^2}$$
$$b = \sqrt{c^2 - a^2}$$

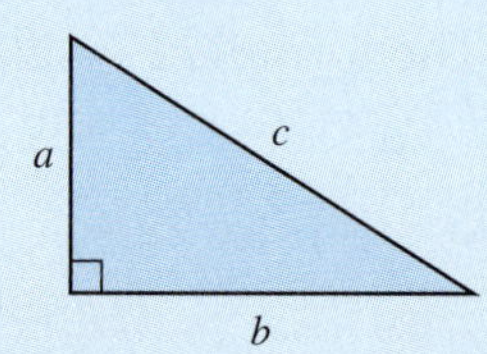

EXAMPLE 3: Joe is building a shed with the height of the roof 8 feet and the distance from the center of the roof to the outside wall 6 feet (see figure). Joe wants to find the length of the roof rafter.

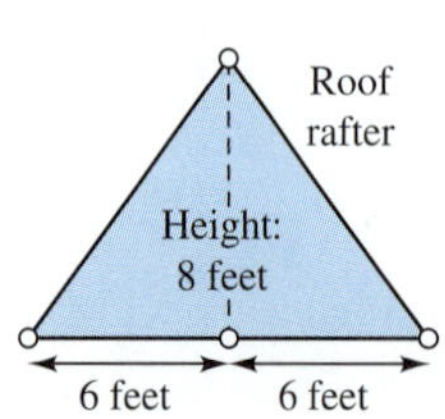

Joe's drawing of one triangle:

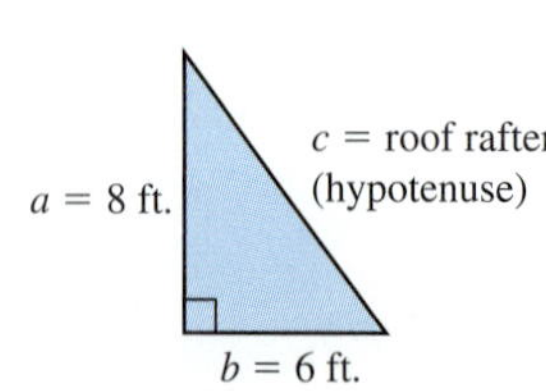

To find the length of the roof rafter (the hypotenuse), let $b = 6$ and $a = 8$ in the Pythagorean Theorem formula $c = \sqrt{a^2 + b^2}$ and solve for c.

$$c = \sqrt{8^2 + 6^2} = \sqrt{64 + 36} = \sqrt{100} = 10 \text{ feet}$$

Memorize these **frequently used formulas**.

Rectangles where l = length, w = width: Area: $A = lw$ Perimeter: $P = 2l + 2w$ 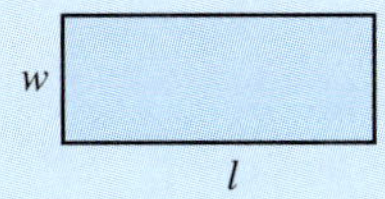 	Circles where r = radius: Area: $A = \pi r^2$ Circumference: $C = 2\pi r$ 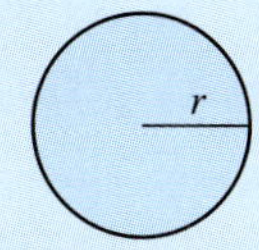
Triangles where a, b, and c are sides of any triangle, b = base, and h = height: Area: $A = \frac{1}{2}bh$ Perimeter: $P = a + b + c$ 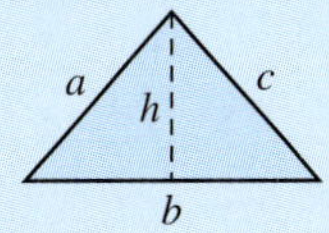 	The Pythagorean Theorem: $a^2 + b^2 = c^2$

1 2 3 Doing Mathematics

- Write the definitions of the words, symbols, and formulas for this section in the Dictionary in Appendix A.
- On a 3 × 5-inch card, write the Pythagorean Theorem and the frequently used formulas for future reference.

1. Consider the set of all rectangles with an area of 500 square feet. Use the equivalent forms of the area formula to complete the table: $l = \frac{A}{w}$ and $w = \frac{A}{l}$. Round your answers to the nearest tenth.

	Length (l)	Width (w)	Area (A) = 500 square feet
(a)	100 feet		500 square feet
(b)		50 feet	500 square feet
(c)	75 feet		500 square feet
(d)		10 feet	500 square feet
(e)	25 feet		500 square feet
(f)		60 feet	500 square feet
(g)	How many rectangles can be created with area 500 square feet?		

2. Solve the formula for the perimeter of a rectangle, $P = 2l + 2w$, for the variable l.

3. Solve the formula for the perimeter of a triangle, $P = a + b + c$, for the variable a.

4. Consider the set of all rectangles with a perimeter of 200 square feet. Use the equivalent forms of the perimeter formula to complete the table:
$l = \dfrac{P - 2w}{2}$ and $w = \dfrac{P - 2l}{2}$.

	Length (l)	Width (w)	Perimeter (P) = 200 feet
(a)	25 feet		200 feet
(b)		50 feet	200 feet
(c)	80 feet		200 feet
(d)		10 feet	200 feet
(e)	30 feet		200 feet
(f)		75 feet	200 feet
(g)	How many rectangles can be created with perimeter 200 feet?		

5. You are going to construct a flower garden in the shape of a right triangle with the two perpendicular sides bordered by a fence. You plan to edge the third side (the hypotenuse of the triangle) with brick. The dimensions are given in the figure.

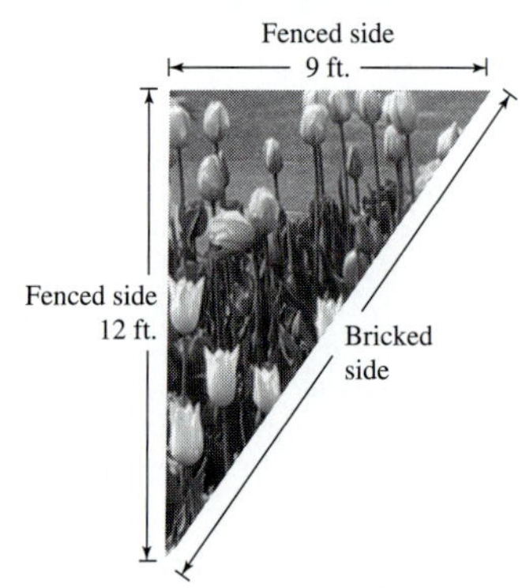

(a) Use the Pythagorean Theorem to find how many feet of the garden will be edged in brick. Show your work.

(b) Find the perimeter of the garden ($P = a + b + c$).

Problems 6–9: Find the length of side a of a right triangle with the given dimensions.

6. $c = 25, b = 15$

7. $c = 5, b = 3$

8. $c = 5, b = 4$

9. $c = 10, b = 8$

10. How many feet of fence should be purchased to enclose a circular flower garden with a radius of 5 feet? What is the area of the garden? Use $\pi \approx 3.14$.

11. How many feet of fence should be purchased to enclose a circular flower garden with a radius of 7 feet? What is the area of the garden? Use $\pi \approx 3.14$.

12. Business profit can be calculated using the formula $P = R - C$, where P = Profit, R = Revenue, and C = Operating Costs.

(a) Solve $P = R - C$ for C.

(b) Louise owns a bakery. Last week her revenue was \$400. Find her operating costs last week if she broke even (when profit is zero).

13. The formula expressing the relationship between temperature in Fahrenheit (F) and Celsius (C) is given by the formula $F = \frac{9}{5}C + 32$, also written $F = 1.8C + 32$.

(a) Solve this formula for C. Show your work.

(b) Given the Fahrenheit temperatures in the table, find the corresponding Celsius temperatures. Round to the nearest tenth.

Fahrenheit (F)	Celsius (C)
−20°	
40°	
45°	
60°	

Problems 14–33: Solve each formula or equation for the indicated variable.

14. $d = rt$ Solve for r.

15. $I = Prt$ Solve for r.

16. $\frac{d}{r} = t$ Solve for d.

17. $V = lwh$ Solve for w.

18. $C = 2\pi r$ Solve for r.

19. $y = mx + b$ Solve for m.

20. $y = mx + b$ Solve for x.

21. $d = 2r$ Solve for r.

22. $V = \pi r^2 h$ Solve for h.

23. $A + B + C = 180°$ Solve for A.

24. $A = P + Prt$ Solve for t.

25. $x + y = 5$ Solve for x.

26. $x - y = 14$ Solve for y.

27. $3x + 5y = 10$ Solve for x.

28. $4x + y - 12 = 0$ Solve for y.

29. $y - 4 = 2(x + 5)$ Solve for y.

30. $y + 1 = -3(x - 4)$ Solve for x.

31. $5x - 4y = 12$ Solve for x.

32. $3x + 4y = 20$ Solve for y.

33. $3x - 11 = 4y + 9$ Solve for x.

Chapter 2 Summary

SECTION 2.1:

Algebra is

I. A powerful language with its own symbols and vocabulary
II. A mathematical system used to generalize arithmetic
III. An organized set of procedures and principles
IV. A problem-solving tool

Polynomial

A **polynomial** is an algebraic expression composed of one term or the sum of two or more terms.

A polynomial with one term is called a **monomial**.

A polynomial with two terms is called a **binomial**.

A polynomial with three terms is called a **trinomial**.

SECTION 2.2:

LIKE terms are terms whose variable factors are exactly the same (letters and exponents).

SECTION 2.3:

Equality Principles of Equations

Let ♦ = any real number or algebraic expression containing one or more terms.

Addition Principle:	If $x = y$, then $x + ♦ = y + ♦$.
Subtraction Principle:	If $x = y$, then $x - ♦ = y - ♦$.
Multiplication Principle:	If $x = y$, then $x \cdot ♦ = y \cdot ♦$.
Division Principle:	If $x = y$, then $\frac{x}{♦} = \frac{y}{♦}$, $(♦ \neq 0)$.

SECTION 2.4:

Set-Builder Notation, Graphs, and Interval Notation

The variable x represents any real number and the variable a represents a single, specific number on the number line:

Set-Builder Notation	Graph	Interval Notation
$\{x \mid x \geq a\}$	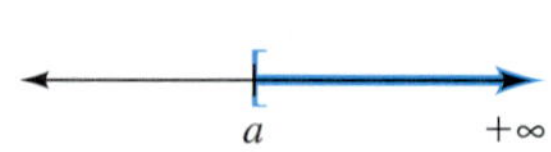	$[a, \infty)$
$\{x \mid x > a\}$	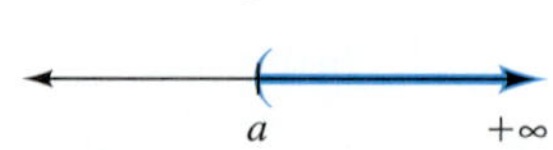	(a, ∞)
$\{x \mid x \leq a\}$		$(-\infty, a]$
$\{x \mid x < a\}$	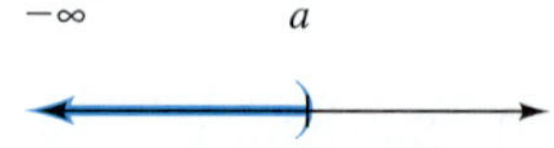	$(-\infty, a)$

Addition (and Subtraction)[2] Principle of Inequalities

Let a, b, and ♦ represent any real number, variable, or algebraic expression containing one or more terms. The statement is true for $<$, $\le$, $>$, and $\ge$.

$$\text{If } a < b, \text{ then } a + ♦ < b + ♦.$$

[2]Subtraction of any real number can be rewritten as addition.

Multiplication and Division Principle of Inequalities

Let a and b represent any real number, variable, or algebraic expression. Each statement is true for $<$, $\le$, $>$, and $\ge$.

Let ♦ represent any positive real number, variable, or algebraic expression (positive real numbers are numbers greater than zero; zero is not a positive number):

$$\text{If } a < b, \text{ then } a \cdot ♦ < b \cdot ♦. \qquad \text{If } a < b, \text{ then } \frac{a}{♦} < \frac{b}{♦}.$$

Let ♦ represent any negative real number, variable, or algebraic expression (negative real numbers are numbers less than zero; zero is not a negative number):

$$\text{If } a < b, \text{ then } a \cdot ♦ \boxed{>} b \cdot ♦. \qquad \text{If } a < b, \text{ then } \frac{a}{♦} \boxed{>} \frac{b}{♦}.$$

SECTION 2.5:

Frequently Used Formulas

Rectangles where l = length, w = width	Area: $A = lw$ Perimeter: $P = 2l + 2w$
Circles where r = radius	Area: $A = \pi r^2$ Circumference: $C = 2\pi r$
Triangles where a, b, and c are sides of the triangle, b = base, and h = height	Area: $A = \frac{1}{2}bh$ Perimeter: $P = a + b + c$
Pythagorean Theorem where a and b represent the lengths of the perpendicular sides of a right triangle, and c is the hypotenuse	$c^2 = a^2 + b^2$ $a = \sqrt{c^2 - b^2}$, $b = \sqrt{c^2 - a^2}$ $c = \sqrt{a^2 + b^2}$

Chapter 2 Review

- Review the definitions you have written for this chapter in the Dictionary.
- Review any properties or processes you have written on 3 × 5-inch cards.

SECTION 2.1

1. Let x and y represent real numbers.

(a) Find a value for x and y that will create the same sum for the numbers in the horizontal and vertical boxes.
$x =$ ______, $y =$ ______

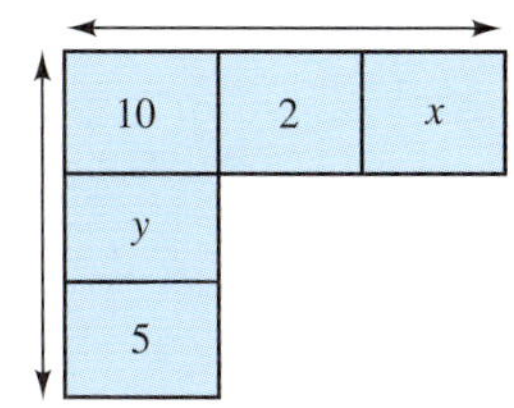

(b) How many values for x and y can you find that will create the same sum?

(c) Write an equation representing the relationship between the horizontal sum and the vertical sum.

Problems 2–7: Substitute the given numerical values for the objects and variables into each expression: ♦ = 6, $x = -4$, ★ = 1, $y = -2$. Perform the operations and determine the value of each expression. Show your work.

2. (7♦)(2x★)

3. 7★ + 10♦ − $3x$

4. $-3x$(♦ − y)

5. $-2y$(★ + ♦)

6. $y^2 + 2xy - 5y + 20$

7. $4(3x^2 + 5x - 9)$

Problems 8–15: Complete the table.

Polynomial	Monomial, binomial, trinomial, or polynomial	List the terms	Numerical coefficient of each term or constant	Degree of the polynomial
EXAMPLE: $2x + 1$	Binomial	$2x$ 1	2 Constant	1
8. $5x^2 - 7z$				
9. $11.5a$				
10. $2x^2 - 4x + 5$				
11. $7x^3 + x - 5$				
12. $8y + 10$				
13. $x + y + z + w$				
14. $-26x$				
15. $x^3 - 6x^2 + 9x - 1$				

Problems 16–17: Rewrite each expression with terms in descending order.

16. $-9x^4 - 16x^2 + 7x^6 - 4x + x^3$

17. $1.5y - 5y^3 + 3.75y^2 + 4.25y^5$

Problems 18–19: Rewrite each expression with terms in ascending order.

18. $-10x^2 - 11x + 2x^5 - 8x^3 + 7x^4$

19. $-15 + 3y^2 - 5y + 2y^3 + 18y^6$

Problems 20–21: The formula for the area of a rectangle is $A = l \cdot w$, where A represents the area of the rectangle, l represents the length, and w represents the width. Compute the missing dimension.

20. $A = 100$ square feet,
$l = 20$ feet, $w =$ ____ feet

21. $A =$ ____ square feet,
$l = 8$ feet, $w = 6$ feet

Problems 22–25: Solve each equation. Verify the solution.

22. $z + 9 = 29$

23. $y - 2 = 15$

24. $4x = -16$

25. $\frac{y}{3} = 10$

SECTION 2.2

Problems 26–32: Add or subtract by collecting LIKE terms.

26. $-5x + 2x - 10x + 12x - 7x$

27. $xy + 12x - 19xy + 2x - (-6x)$

28. $1.2x^2 + 5.3x - (-5.2x^2) - 10.2x^2 + (-7.3x)$

29. $\frac{7}{2}z - \frac{1}{2}z - \frac{11}{2}z - \left(\frac{-3}{2}z\right)$

30. $-2.3ab + 5.2a - 1.2b - (-12.6ab) + (-5.5b)$

31. $x - x - x - x + x$

32. $-3xyz + 2xz - 5xy + 4xyz - (-7xz)$

Problems 33–36: Substitute the numerical values $x = -2$, $y = 3$, $z = -1$ for the variables. Determine whether the expressions are equivalent for these values and insert the appropriate symbol, either = or ≠, on the line between the two expressions. Show your work.

33. $5x + 3x$ ____ $8x$

34. $7x^2 - 5x^2$ ____ $2x$

35. $9xy - 2xz$ ____ $7yz$

36. $x(y - z)$ ____ $xy - xz$

Problems 37–45: Perform the operations by removing parentheses and collecting LIKE terms.

37. $(x^2 + 2x - 3) + (15x + 7)$

38. $(6y - 3) - (2y - 3)$

39. $\left(\frac{24}{7}x - \frac{6}{5}\right) + \left(\frac{4}{7}x - \frac{4}{5}\right)$

40. $-9(2x - 1) - (3x + 5)$

41. $(5x^2 - 2x + 9) + (x^2 - 6x - 25)$

42. $(-4x^4 + 8x^2 - 1) + (x^2 - 2x^2 - x + 5)$

43. $(2x - 5) + (1.5x + 6) + (2.5x - 1) - (x - 2)$

44. $-(3x^2 + x - 1) - 2(3x^2 - 2x + 1) + (3x^2 - 5x + 4)$

45. $-4(x^2 + x - 1) + 5(2x^2 - x + 9) - 3(x^2 - 7x + 4)$

Problems 46–49: Add or subtract the polynomials vertically.

46.
$$\begin{array}{rr} & 10x^3 - 7x^2 + 2x - 9 \\ + & 18x^3 + 2x^2 - 5x + 8 \\ \hline \end{array}$$

47.
$$\begin{array}{rr} & -4x^3 \qquad\qquad + 2x - 12 \\ + & 7x^3 - 3x^2 - x - 11 \\ \hline \end{array}$$

48.
$$\begin{array}{rr} & 2x^3 - 5x^2 + 2x - 9 \\ - & x^3 + 2x^2 - 3x - 8 \\ \hline \end{array}$$

49.
$$\begin{array}{rr} & -x^3 \qquad\qquad + 2x + 10 \\ - & 7x^3 + 2x^2 - x - 11 \\ \hline \end{array}$$

50. Write an equation using variables describing the total value of the coins in David's pocket if he has q quarters, d dimes, n nickels, and p pennies totaling \$1.48.

51. Ann is taking inventory of the furniture in her stores. This table is a summary of the inventory:

Store 1	Number of items in stock
Couch	3
Dining room chair	15
Family room chair	5
Lamp	21
Store 2	
Couch	2
Dining room chair	11
Family room chair	4
Lamps	11

Write an equation using variables that summarizes the total cost of all pieces of furniture. Use c = cost of one couch, d = cost of one dining room chair, f = cost of one family room chair, p = cost of one lamp, and T = total cost of all furniture.

SECTION 2.3

Problems 52–55: Determine whether the given variable is a solution of the equation. Show your work.

52. Is $x = -3$ a solution of $x - 9 = -12$?

53. Is $y = 5$ a solution of $2y + 6 = 11$?

54. Is $z = 0$ a solution of $2z = 2$?

55. Is $x = -4$ a solution of $\frac{3x}{-2} = 6$?

Problems 56–63: Solve each equation for the given variable. Verify each solution.

56. $12x - 28x = 32$

57. $100y + 50y = 300$

58. $\frac{5}{2}x - 25 = \frac{3}{2}x + 2$

59. $-3(x - 6) = 4(x - 1) + 1$

60. $20.2x - 5.0 = 25.8 - (1.8x - 3.2)$

61. $3\left(\frac{x}{3} - 5\right) = -3x + 5$

62. $-4x - 50 = -5(x + 6)$

63. $2x^2 - 3x - 6 = -2(-x^2 + x - 15)$

64. Sandy has accepted a job selling cars. She will receive a monthly salary of \$500 plus a commission of 2% of monthly sales. Let the variable x represent Sandy's monthly sales. Let the variable S represent her monthly salary.

(a) Use the variables x and S and write an equation that describes the total monthly salary Sandy will receive.

(b) What is Sandy's total salary for the month of September if her sales total \$60,000?

(c) Sandy would like to triple her salary in November, compared to September. What would Sandy's monthly sales in November need to be to triple her September salary?

SECTION 2.4

Problems 65–68: Graph each inequality statement. Use parentheses or brackets on boundary points. Draw a darkened line to indicate the numbers in the set. Write each set in interval notation.

65. $x > -3$

(a) Graph: 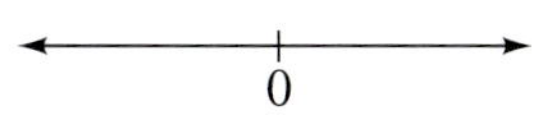

(b) Interval notation:

66. $x \geq -3$

(a) Graph: 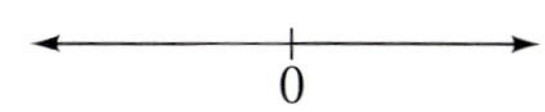

(b) Interval notation:

67. $x < -3$

(a) Graph: 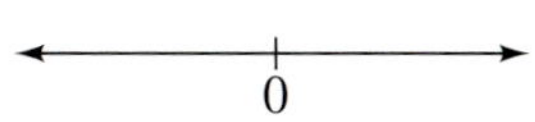

(b) Interval notation:

68. $x \leq -3$

(a) Graph: 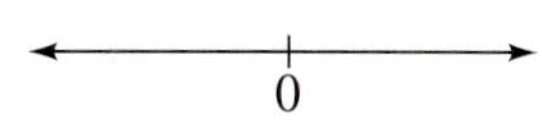

(b) Interval notation:

Problems 69–72: Graph each interval. Write the interval in set-builder notation.

69. $(-1, \infty)$

(a) Graph: 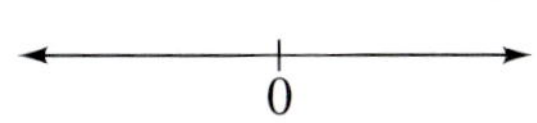

(b) Set-builder notation:

70. $[0, \infty)$

(a) Graph: 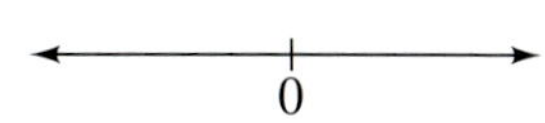

(b) Set-builder notation:

71. $(-\infty, 1)$

(a) Graph: 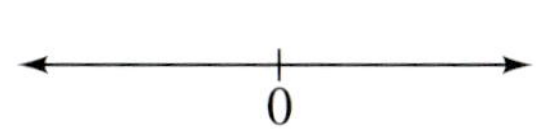

(b) Set-builder notation:

72. $(-\infty, 3)$

(a) Graph: 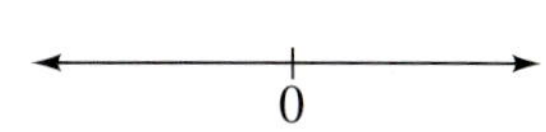

(b) Set-builder notation:

73. Optional Internet Exercise: United Parcel Service (UPS) rates for packages are available on the Internet or from your local UPS office. Suppose you were going to send a package weighing 4.5 pounds from the UPS office in your city to a friend in Anchorage, Alaska. Visit the UPS Web site or call UPS to complete the table. Let x represent the weight in pounds of a package.

Weight	Symbolic representation of the weight	Cost to mail a package from the UPS office in your city to Anchorage, Alaska
1 pound or less	$0 < x \leq 1$	
More than 1 pound and up to and including 2 pounds	$1 < x \leq 2$	
More than 2 pounds and up to and including 3 pounds	$2 < x \leq 3$	
More than 3 pounds and up to and including 4 pounds	$3 < x \leq 4$	
More than 4 pounds and up to and including 5 pounds	$4 < x \leq 5$	

74. Bill is a salesperson for a local car dealership. As an incentive to its salespersons, the dealership initiated a contest with a prize of a weekend trip to Las Vegas to any salesperson selling more than 25 cars in a six-month period. In the months of March, April, May, June, and July, Bill sold 3, 6, 5, 6, and 4 cars, respectively.

(a) Let x = the number of cars to be sold in August. Write the problem situation as an inequality statement.

(b) How many cars does Bill need to sell to win the trip to Las Vegas?

(c) Do you think he will win the trip? Why or why not?

Problems 75–82: Solve each inequality. Draw a graph of the solution set and write the solution in set-builder and interval notation.

75. $x + 6 < 12$

(a) Graph: 0

(b) Set-builder notation:

(c) Interval notation:

76. $-2z + 15 \leq 27$

(a) Graph: 0

(b) Set-builder notation:

(c) Interval notation:

77. $20 - 2x > -6$

(a) Graph: 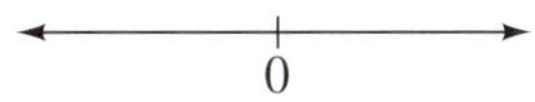

(b) Set-builder notation:

(c) Interval notation:

78. $2(5y + 6) > 6(2y + 4)$

(a) Graph: 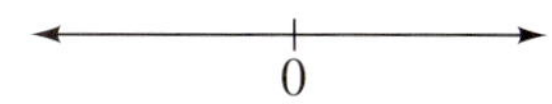

(b) Set-builder notation:

(c) Interval notation:

79. $-3(5 - x) < 6(2 - x)$

(a) Graph: 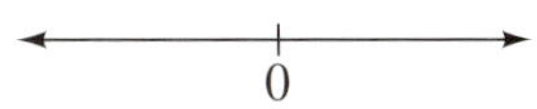

(b) Set-builder notation:

(c) Interval notation:

80. $2.6x - 0.2 < 1.4x + 2.2$

(a) Graph: 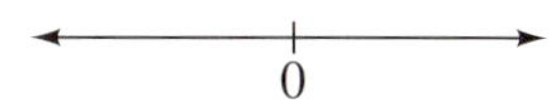

(b) Set-builder notation:

(c) Interval notation:

81. $\dfrac{4 - 3x}{2} \geq 4$

(a) Graph: 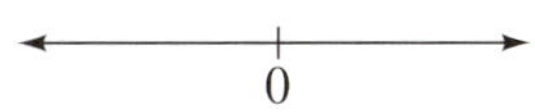

(b) Set-builder notation:

(c) Interval notation:

82. $\dfrac{5 - x}{3} \leq 2$

(a) Graph: 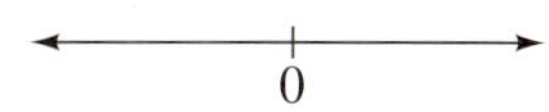

(b) Set-builder notation:

(c) Interval notation:

SECTION 2.5

83. Consider the set of all rectangles with an area of 300 square feet. Use the equivalent forms of the area formula to complete the table: $l = \dfrac{A}{w}$ and $w = \dfrac{A}{l}$.

	Length (l)	**Width (w)**	**Area (A)**
(a)	100 feet		300 square feet
(b)		50 feet	300 square feet
(c)	25 feet		300 square feet
(d)		10 feet	300 square feet
(e)	50 feet		300 square feet
(f)		30 feet	300 square feet
(g)	How many rectangles can be created with area 300 square feet?		

84. Consider the set of all rectangles with a perimeter of 300 feet. Use the equivalent forms of the perimeter formula to complete the table: $l = \dfrac{P - 2w}{2}$ and $w = \dfrac{P - 2l}{2}$.

	Length (l)	Width (w)	Perimeter (P)
(a)	100 feet		300 feet
(b)		50 feet	300 feet
(c)	25 feet		300 feet
(d)		10 feet	300 feet
(e)	50 feet		300 feet
(f)		30 feet	300 feet
(g)	How many rectangles can be created with perimeter 300 feet?		

Problems 85–95: Solve each equation or formula for the indicated variable.

85. $I = Prt$ Solve for t.

86. $V = lwh$ Solve for h.

87. $Ax + By = C$ Solve for y.

88. $A + B + C = 180°$ Solve for C.

89. $P = a + b + c$ Solve for b.

90. $x + 2y = 7$ Solve for x.

91. $4x + 5y = 15$ Solve for x.

92. $8x + 5y - 12 = 0$ Solve for y.

93. $y - 3 = -2(x - 1)$ Solve for x.

94. $7x - 5y = 12$ Solve for y.

95. $x - 11 = 4y + 9$ Solve for x.

Practice Test—Chapter 2

SECTION 2.1

1. Substitute the given numerical values for the objects and variables into the expressions: ♦ = 5, $x = 3$, ★ = −1, $y = -2$. Perform the operations.

(a) (8 ♦ − x)(−4★) (b) $2(x^2 - x + 1)$

(c) 6★ − 7 ♦ − $3x$ (d) $x^2 + xy - y + 9$

2. Complete the table.

Polynomial	Monomial, binomial, trinomial, or polynomial	List the terms	Numerical coefficient of each term or constant	Degree of the polynomial
(a) $125x + y^2 + z$				
(b) $25a^3$				
(c) $4x^5 - y$				
(d) $x^4 + 2x^3 - 3x + 9$				

3. Rewrite each expression with terms in descending order.

(a) $x^4 + 1 - 9x^3 + 2x^5 - 3x + x^2$ (b) $5y^4 - y + 7y^2 - 2y^3$

4. Rewrite each expression with terms in ascending order.

(a) $8x^2 + 3x - 2x^4 - 4x^3 + 7$ (b) $10 + 3y^5 - 5y + 2y^3$

5. Solve each equation. Verify each solution.

(a) $x - 3 = 11$ (b) $y + 2 = 10$

(c) $-5x = -30$ (d) $\frac{y}{6} = 7$

SECTION 2.2

6. Add and subtract by collecting LIKE terms.

(a) $4x - 2x + 6x - 5y + 10y$

(b) $10x^2y^2 - 5xy + x^2y^2 + 9xy$

(c) $\frac{1}{2}xy - \frac{1}{2}xy + 4xy + 2x - 5y + 7xy - x - y$

(d) $xy + xz - 3xy - 4zy - 7xy$

7. Perform the operations. Remove parentheses and collect LIKE terms.

(a) $(3x^2 + x - 1) + (5x + 4)$

(b) $-5(2x - 1) - (4x + 7)$

(c) $(x - 1) + (5x + 3) - (2x - 1) - (x - 9)$

(d) $(5x^2 - 2x + 9) - (x^2 - 6x - 25)$

8. Add (or subtract) each polynomial, vertically.

(a)
$$\begin{array}{rrrrr} & 12x^3 & + 5x^2 & - 4x & + 3 \\ + & 8x^3 & - 2x^2 & - 5x & + 8 \\ \hline \end{array}$$

(b)
$$\begin{array}{rrrrr} & 2x^3 & - 6x^2 & + 2x & - 3 \\ - & x^3 & & - 6x & - 7 \\ \hline \end{array}$$

SECTION 2.3

Problems 9–13: Solve each equation for the given variable. Verify each solution.

9. $2y + 5y = 12 + 3y$

10. $\frac{5}{2}x = \frac{3}{2}x - 4$

11. $3(y + 1) = 2(y - 5)$

12. $4x - 2(x - 3) = x + 1$

13. $2.3x + 2(1.1 - x) = 1.8 - 0.7x$

14. Marat has accepted a job selling stereos. He will receive a monthly salary of \$400 plus a commission of 4% of monthly sales. Let the variable x represent Marat's monthly sales. Let the variable S represent his monthly salary.

(a) Use the variables x and S and write an equation that describes the total monthly salary Marat will receive.

(b) What is Marat's total salary for the month of November if his monthly sales total \$80,000?

(c) Marat would like to double his salary in December, compared to November. What would Marat's monthly sales in December need to be to double his November salary?

SECTION 2.4

15. Graph each inequality statement. Write each set in interval notation.

(a) $x > -6$

Graph:

Interval notation:

(b) $x \le 5$

Graph:

Interval notation:

16. Graph each interval. Write the interval in set-builder notation.

(a) $(-\infty, -1]$

Graph:

Set-builder notation:

(b) $[0, \infty)$

Graph:

Set-builder notation:

Problems 17–20: Solve each inequality. Draw a graph of the solution set and write the solution in set-builder and interval notation.

17. $-2z - 9 > 5$

(a) Graph:

(b) Set-builder notation:

(c) Interval notation:

18. $-2(x - 1) \ge 6(3 - x)$

(a) Graph:

(b) Set-builder notation:

(c) Interval notation:

19. $24 - 5x < -6.5$

(a) Graph:

(b) Set-builder notation:

(c) Interval notation:

20. $\dfrac{3x - 2}{7} \le 1$

(a) Graph:

(b) Set-builder notation:

(c) Interval notation:

SECTION 2.5

Problems 21–25: Solve each equation or formula for the indicated variable.

21. $I = Prt$ Solve for P.

22. $C = \pi d$ Solve for d.

23. $Ax + By = C$ Solve for y.

24. $3(x - 4) = y + 5$ Solve for x.

25. $5x - 2y = 10$ Solve for y.

Answers to Getting Started—Chapter 2

SECTION 2.1

1. (a) ♦ + ♦ + ♦ = 3♦

21 = 21

(b) 2♦ − $3x$ < $5y$

−16 < −15

2. Numerical answers will vary.

3. Answers will vary.

Let $x = 2$ and $y = 5$:

$$3(2 + 5) = 3(2) + 3(5)$$
$$3(7) = 6 + 15$$
$$21 = 21$$

4. (a) $A = 30 \cdot 20 = 600$ square feet

(b) $A = 60 \cdot 20 = 1200$ square feet

5. (a) $x = 8$

(b) $y = 9$

6. (a) >

(b) <

SECTION 2.2

1. One way to describe the coins is to count the coins that are alike: 2 quarters, 2 nickels, and 3 dimes. Another way to describe the coins would be to state the total amount of money of $0.90.

2. (a) 4 apples + 5 apples = 9 apples

(b) 100 bears − 10 bears = 90 bears

(c) 3 dogs + 5 cats ≠ 8 monkeys

(d) 13 cats − 5 dogs ≠ 8 monkeys

(e) Equivalent statements: LIKE objects are added or subtracted.

3. (f) Numerical coefficients are added or subtracted for LIKE terms.

4. (a)

$$3y - 2y = y$$
$$3(3) - 2(3) = 3$$
$$9 - 6 = 3$$
$$3 = 3$$

(b)

$$x + x + y + y + y = 2x + 3y$$
$$5 + 5 + 3 + 3 + 3 = 2(5) + 3(3)$$
$$19 = 10 + 9$$
$$19 = 19$$

(c)

$$5y - 3x \neq 2y$$
$$5(3) - 3(5) \neq 2(3)$$
$$15 - 15 \neq 6$$
$$0 \neq 6$$

(d)

$$6x^2 - 4x^2 \neq 2x$$
$$6(5)^2 - 4(5)^2 \neq 2(5)$$
$$6(25) - 4(25) \neq 10$$
$$150 - 100 \neq 10$$
$$50 \neq 10$$

(e) Equivalent statements: LIKE terms are collected.

(f) Statements that are not equivalent: LIKE terms are NOT collected.

5. (a) $2x + 3x = 5x$

(b) $15y + 5y - 3x = 20y - 3x$

(c) $6x^2 - 4x^2 = 2x^2$

(d) $11ab - 2ab + 3c + 7c = 9ab + 10c$

SECTION 2.3

1.

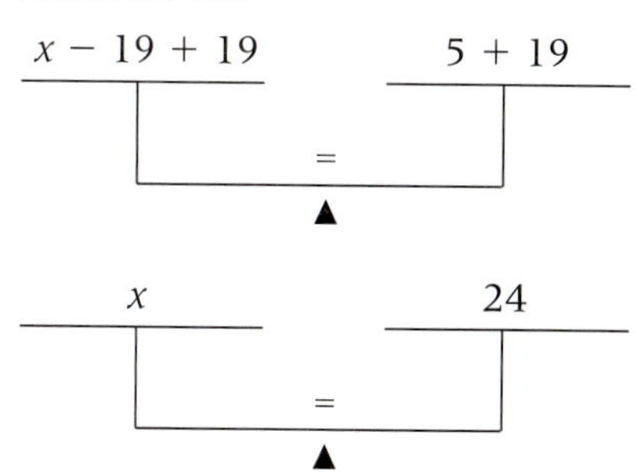

$$x - 19 = 5$$
$$x - 19 + 19 = 5 + 19$$
$$x = 24$$

Verify the solution:

$$(24) - 19 \stackrel{?}{=} 5$$
$$5 = 5$$

2.

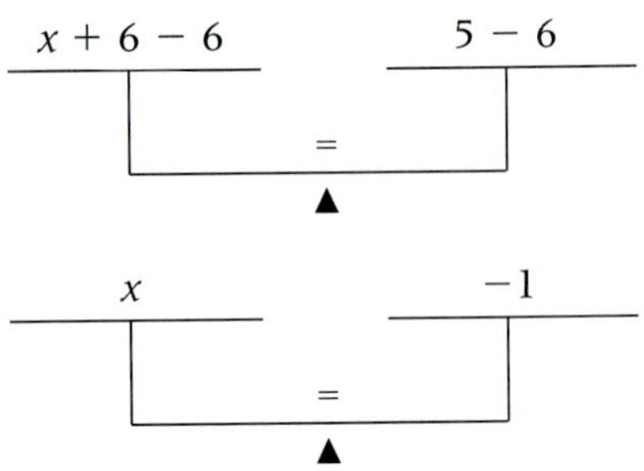

$$x + 6 = 5$$
$$x + 6 - 6 = 5 - 6$$
$$x = -1$$

Verify the solution:

$$(-1) + 6 \stackrel{?}{=} 5$$
$$5 = 5$$

3.

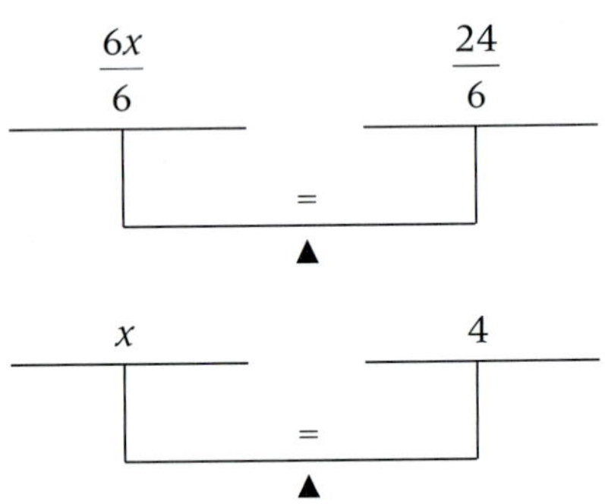

$$6x = 24$$
$$\frac{6x}{6} = \frac{24}{6}$$
$$x = 4$$

Verify the solution:

$$6(4) \stackrel{?}{=} 24$$
$$24 = 24$$

4.

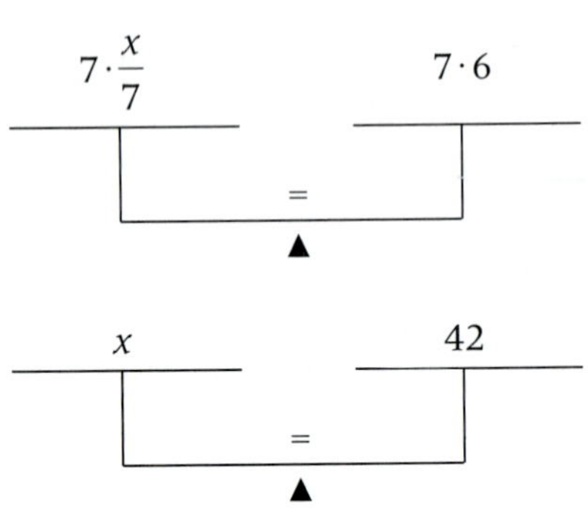

$$\frac{x}{7} = 6$$
$$\overset{1}{7}\cdot\frac{x}{\underset{1}{7}} = 7\cdot 6$$
$$x = 42$$

Verify the solution:

$$\frac{42}{7} \stackrel{?}{=} 6$$
$$6 = 6$$

5. Answers will vary. Isolate the variable by adding, subtracting, multiplying, or dividing on each side of the equation. Verify the solution.

SECTION 2.4

1. (a) 80

(b) 100

(c) Any number less than or equal to 100 and greater than or equal to 80.

(d) Answers will vary.

(e) No. You would need 120 points on a test with a maximum of 100 points.

2. (a) $17 < 23$

(b) $1 < 7$

(c) $8 > 5$

(d) $12 > 9$

(e) Same

3. (a) $15 < 24$

(b) $9 > 5$

(c) Same

(d) $-14 < -2$

(e) $-9 < -5$

(f) Reversed

4. (a)
$$y - 10 > 5$$
$$y - 10 + 10 > 5 + 10$$
$$y > 15 \quad \{y|y > 15\}$$

(b)
$$y + 5 \le 8$$
$$y + 5 - 5 \le 8 - 5$$
$$y \le 3 \quad \{y|y \le 3\}$$

(c) $4x \le 12$

$$\frac{4x}{4} \le \frac{12}{4}$$

$$x \le 3 \quad \{x|x \le 3\}$$

(d) $$\frac{x}{5} > 3$$

$$5 \cdot \frac{x}{5} > 3 \cdot 5$$

$$x > 15 \quad \{x|x > 15\}$$

5. (a) $-3x \le 12$

$$\frac{-3x}{-3} \boxed{\ge} \frac{12}{-3}$$

$$x \ge -4 \quad \{x|x \ge -4\}$$

(b) $$\frac{x}{-5} \le 2$$

$$(-5) \cdot \frac{x}{-5} \boxed{\ge} 2(-5)$$

$$x \ge -10 \quad \{x|x \ge -10\}$$

SECTION 2.5

1. (a) width = 16 feet

(b) width = 8 feet

(c) $l = \frac{A}{w}$

2. (a) 4 ft

(b) 50 ft

(c) 16 ft

(d) 20 ft

(e) 10 ft

3. (a) $2 \cdot A = 2 \cdot \frac{1}{2}bh$

$$2A = bh$$

(b) $$\frac{2A}{h} = \frac{bh}{h}$$

(c) $$\frac{2A}{h} = b$$

4. (a) $b = 40$ feet

(b) $b = 10$ feet

CHAPTER

3

Relations, Functions, and the Cartesian Plane

René Descartes (1596–1650), a famous mathematician and philosopher, is credited with creating the Cartesian coordinate system and is called the "father of analytical geometry."

3.1 The Cartesian Plane, Quadrants, and Ordered Pairs

OBJECTIVES

- Construct a Cartesian plane and plot ordered pairs.
- Determine coordinates of ordered pairs on a given graph.
- Choose an appropriate scale and plot a set of ordered pairs.
- Graph a linear equation on a Cartesian plane.
- Graph points and linear equations on a graphing calculator (optional).

1 2 3

Getting Started

When a relationship exists between numbers, a picture or graph can often make that relationship easier to see. The graphs of Hourly Earnings[1] and Unemployment Rate,[2] which appeared on the front page of the *Wall Street Journal*, are visual representations of societal trends and patterns.

[1] Hourly Earnings, *Wall Street Journal*, March 10, 1999, page 1.
[2] Unemployment Rate, *Wall Street Journal*, March 8, 1999, page 1.

HOURLY EARNINGS

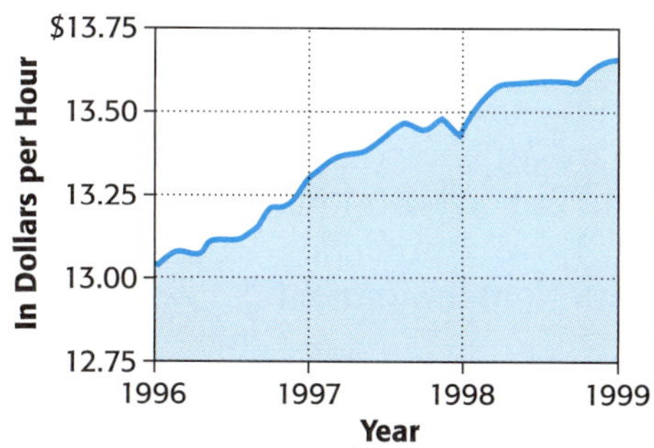

AVERAGE HOURLY PAY of production workers in the private business sector rose in February to $13.66 from $13.64 the preceding month, the Labor Department reports.

UNEMPLOYMENT RATE

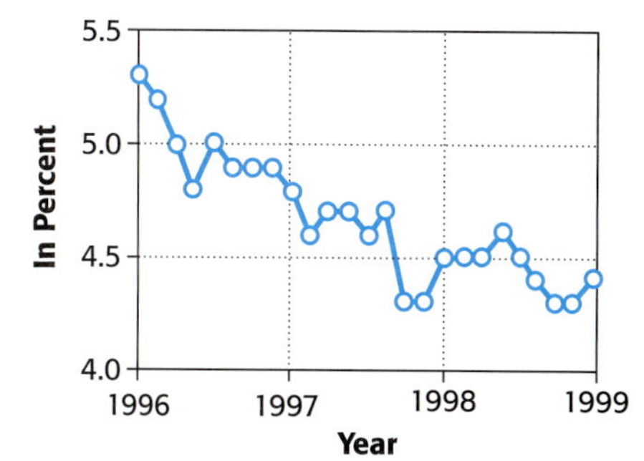

UNEMPLOYMENT in February rose to a seasonally adjusted 4.4% of the civilian labor force from 4.3% the preceding month, according to the Labor Department.

In this section, we introduce a flat, two-dimensional grid called the **Cartesian plane**.

CONSTRUCTING A CARTESIAN PLANE

1. To construct a Cartesian plane, follow these steps on the grid provided.

(a) Draw a vertical line on top of an existing line in the middle of the grid below. This line represents a number line. Draw arrows on each end of the line.

(b) Draw a horizontal line on top of an existing line in the middle of the grid. This line represents a number line. Draw arrows on each end of the line.

(c) Darken the two lines so that they appear darker and thicker than the other grid lines.

(d) Label the horizontal line the "x-axis" on the right end of the line.

(e) Label the vertical line the "y-axis" on the top of the line.

(f) The center point where the x- and y-axes intersect is called the **origin**. This point represents the zero value on each number line. Write the number 0 on the intersection of the x- and y-axes.

(g) Label the x-axis as you would label a number line. Start at 0 and label each grid line on the x-axis with positive integers as you move to the right. Label each grid line with negative integers as you move left.

(h) Next, start at 0 and label each grid line on the y-axis with positive integers as you move up and negative integers as you move down. In other words, label your y-axis as you would label a number line turned sideways.

(i) The x- and y-axes divide the plane into four regions, which are called **quadrants**. Quadrants are numbered using Roman numerals (I, II, III, and IV). Label the upper right region with Roman numeral I in the upper right-hand corner of the grid. Move counterclockwise (around to the left) and label the upper left region with Roman numeral II. Continue the pattern and complete the numbering of Quadrants III and IV with III in the bottom left quadrant and IV in the bottom right quadrant.

You have successfully constructed a Cartesian plane, if your grid looks like the Cartesian plane below. The integer labels on the x- and y-axes may not be easily seen if the numbers are written exactly on top of the grid lines. The numbers are easier to read if they are written to the right and/or below the grid lines.

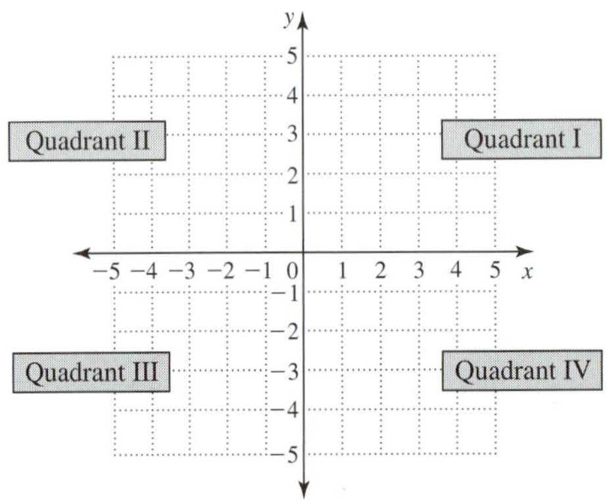

ORDERED PAIRS

Points are plotted on the Cartesian plane using two numbers called coordinates. The first number is called the **first coordinate** or ***x*-coordinate**.[3] The second number is called the **second coordinate** or ***y*-coordinate**. These coordinates state the address of each point on the plane and are called ordered pairs. An x- and y-coordinate pair is written in parentheses with a comma between the coordinates (x, y). An **ordered pair** consists of the following:

- A first number representing the number of units and the direction from 0 along the x-axis
- A second number representing the number of units and the direction from 0 along the y-axis

The order in which the numbers are written is important. The x-coordinate is always written first. The y-coordinate is always written second.

Ordered pair: (first coordinate, second coordinate) $= (x, y)$

[3] The x-coordinate is also called the abscissa. The y-coordinate is also called the ordinate.

PLOTTING ORDERED PAIRS

Plotting an ordered pair is the process of drawing a dot on the address of the coordinates of an ordered pair. To plot the ordered pair $(-2, 3)$ do the following.

First: Begin by placing your pencil at the origin.

Second: The x-coordinate is -2, a negative number. Move your pencil along the x-axis left (in the negative direction) 2 units from the origin. Rest your pencil at -2 on the x-axis.

Third: The y-coordinate is 3, a positive number. Move your pencil 3 units up (in the positive direction) parallel to the y-axis.

Fourth: Draw a dot on the point and label the point with the address $(-2, 3)$. This process is illustrated on the grid below:

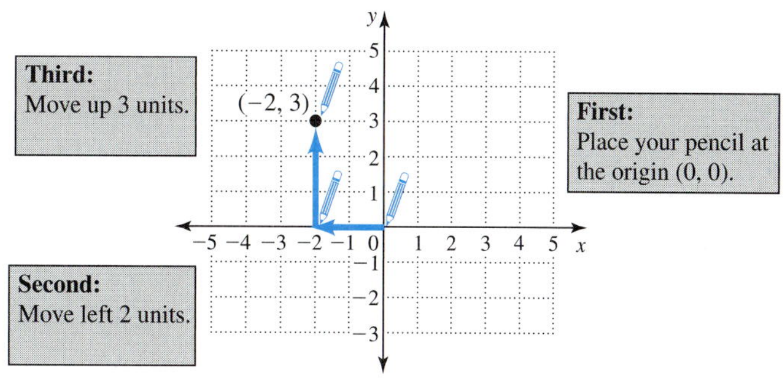

2. Locate the ordered pairs on the grid below and write the quadrant in which each point appears.

(a) $(1, 2)$ Quadrant ______ (b) $(-3, 4)$ Quadrant ______

(c) $(3, -4)$ Quadrant ______

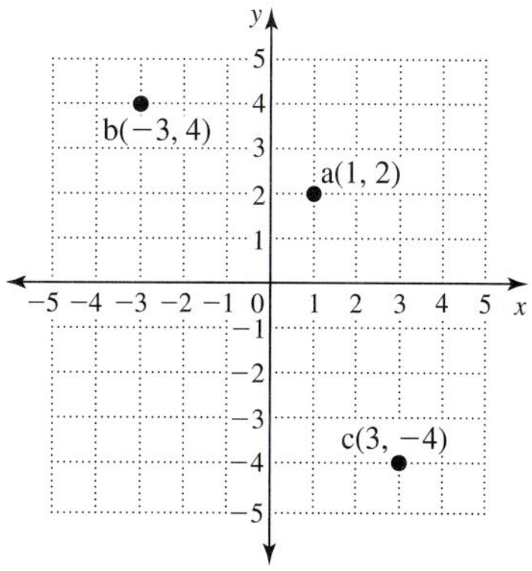

Plot these ordered pairs on the Cartesian plane above. Label each point with the appropriate letter, and write the quadrant in which each point appears.

(d) $(2, 1)$ Quadrant ______ (e) $(-4, -2)$ Quadrant ______

(f) $(3, 5)$ Quadrant ______ (g) $(-2, -4)$ Quadrant ______

(h) $(0, 2)$ Quadrant ______ (i) $(-4, 0)$ Quadrant ______

TIC-TAC-MATH

3. To practice plotting points, let's play a game called Tic-Tac-Math. In this game, you and a partner will take turns plotting points on a grid. One person will use an X to mark a choice and the other person will use an O to mark a choice. This game is just like Tic-Tac-Toe, except that each player must say the address of the chosen point out loud before marking it with an X or an O.

Round One

Play the game as follows on the grid below.

Step 1: The first person starts the game by choosing the first point in Quadrant I. He/She states the coordinates of the point and places an X on the point.

Step 2: The second person chooses a second point in any quadrant, states the coordinates of the point, and places an O on the point.

Step 3: Repeat the process by taking turns choosing a point anywhere on the plane. If a player forgets to state the coordinates of the chosen point out loud, he/she loses a turn.

Object of the game: The object of the first round of this game is to get three consecutive Xs, or three consecutive Os, in a row, either horizontally, vertically, or diagonally. As you play, try to block your opponent.

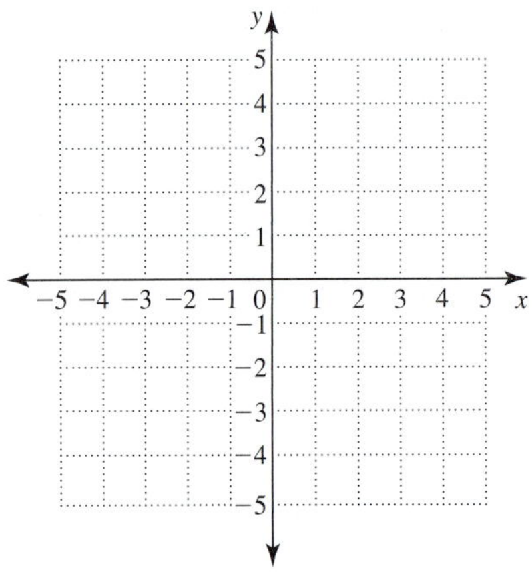

4. Let's play a second round of Tic-Tac-Math and change the rules as outlined below.

Round Two

Play the game as follows on the grid below.

Step 1: The first person starts the game by choosing the first point in Quadrant III. He/She states the coordinates of the point and places an X on the point.

Step 2: The second person chooses a second point in any quadrant, states the coordinates of the point, and places an O on the point.

Step 3: Repeat the process by taking turns choosing a point anywhere on the plane. If a player forgets to state the coordinates of the chosen point out loud, he/she loses a turn.

Object of Round Two: The object of Round Two of this game is to get four consecutive Xs, or four consecutive Os, in a row, either horizontally, vertically, or diagonally. As you play, try to block your opponent.

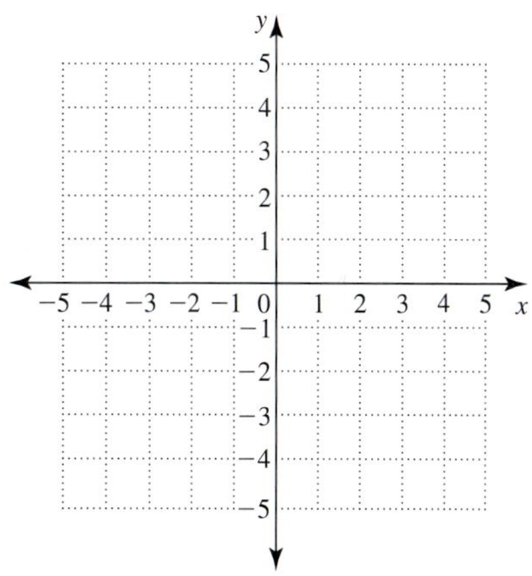

5. Summarize what has been discussed.

(a) What is a Cartesian plane?

(b) What is a quadrant?

(c) What is an ordered pair?

1 2 3 Discussing Mathematics

CHOOSING A SCALE

Graphs on the Cartesian plane are visualizations of data, individual points, positions, and relationships between numbers. In order to display some graphs completely and effectively, a scale other than one unit for each grid line may be used. The scale for the x- and y-axes depends on the ordered pairs to be plotted. The scale does not have to be the same value for both the x- and y-axes. The x and y values may suggest a scale, if the coordinate values are equally spaced. Otherwise, a scale is chosen that allows all or most of the points to be displayed.

CHOOSING A SCALE AND PLOTTING A LIST OF ORDERED PAIRS

1. Make a list of the ordered pairs.
2. Determine the smallest and largest first coordinate (x) value. Look at the pattern of the first coordinate values and determine a scale for the grid lines on the horizontal axis.
3. Repeat step 2 using the second coordinate (y) values. Determine a scale for the grid lines on the vertical axis.
4. Use your scales to label the gridlines on the horizontal and vertical axes.
5. Label each axis with a title describing the first and second coordinate values.
6. Plot the individual ordered pairs.

EXAMPLE 1: Plot the points in the table of values listing the number of hours worked and the salary earned for an employee at a pizza shop, where x represents the number of hours worked and y represents the salary earned.

x = number of hours worked	y = salary earned @ $6 per hour	The entries in the table can be written as ordered pairs
0	$0	(0, 0)
10	$60	(10, 60)
20	$120	(20, 120)
30	$180	(30, 180)
40	$240	(40, 240)

All the ordered pairs can be plotted in Quadrant I. Each x-coordinate value increases by 10 units. A scale of 10 units will accommodate all the ordered pairs along the x-axis. Label each grid line on the x-axis with a multiple of 10. Each y-coordinate value increases by 60 units. A scale of 60 units will accommodate all the ordered pairs along the y-axis. Label each grid line on the y-axis with a multiple of 60. Label the axes with word titles and plot the ordered pairs.

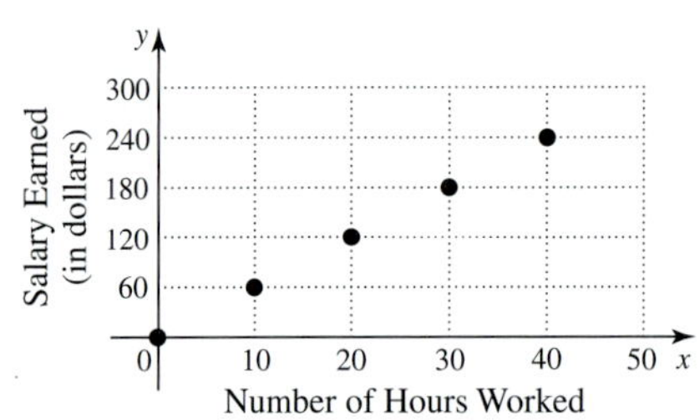

GRAPHING WITH A GRAPHING CALCULATOR

The graphing calculator is a helpful tool in graphing ordered pairs and equations. The viewing window and scale of each axis for each problem must be chosen carefully. A special notation is used to communicate the viewing window on a graphing calculator:

The smallest first coordinate value displayed on the screen is called **Xmin**.	The largest first coordinate value displayed on the screen is called **Xmax**.	The scale of the horizontal axis is called **XScl**.	The smallest second coordinate value displayed on the screen is called **Ymin**.	The largest second coordinate value displayed on the screen is called **Ymax**.	The scale of the vertical axis is called **Yscl**.
The values of Xmin and Xmax are written in interval notation, followed by the scale: **[Xmin, Xmax] XScl**			The values of Ymin and Ymax are written in interval notation, followed by the scale: **[Ymin, Ymax] Yscl**		

To display the ordered pairs in Example 1 on a graphing calculator, this viewing window could be chosen: [Xmin, Xmax] Xscl [Ymin, Ymax] Yscl

[−5, 50] Xscl 10 [−5, 300] Yscl 60

The graph is in Quadrant I only, however, the smallest x value needs to be slightly less than 0 and the largest x value slightly larger than 40. Similarly, the smallest y value needs to be less than 0 and the largest y value larger than 240.

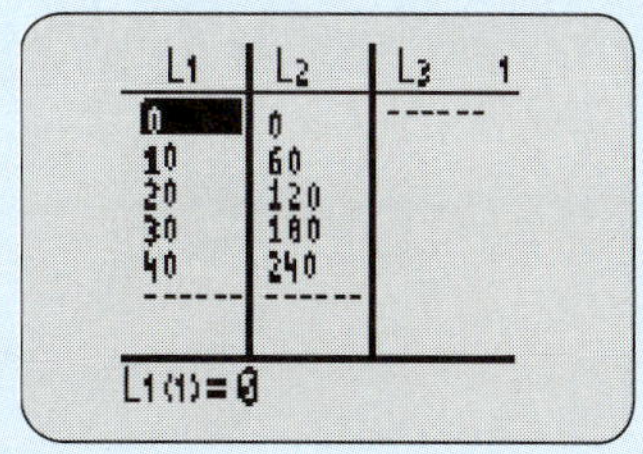

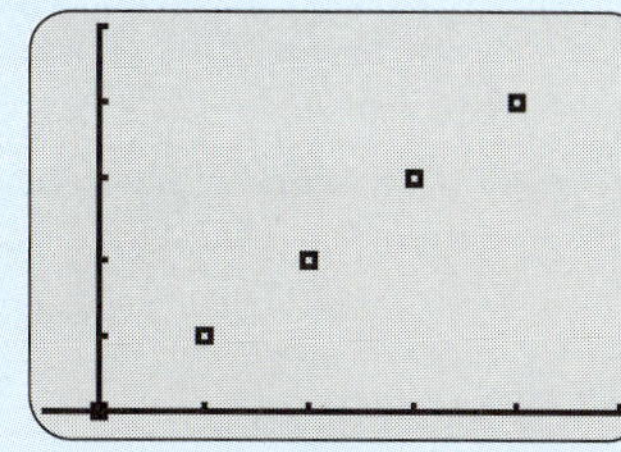

Window: [−5, 50] Xscl 10 [−5, 300] Yscl 60

Your instructor can show you how to create the graph and table on your calculator of the salary problem outlined in Example 1.

GRAPHING AN EQUATION

A graph is a visualization of an equation. The graph of a first-degree equation in two variables is a line. Ordered pairs that satisfy the equation are computed, points plotted, and the points connected to form the graph.

EXAMPLE 2: Do the ordered pairs (2, 11) and (−1, 0) satisfy the equation $y = 5x + 1$?

$y = 5x + 1$ — Substitute $x = 2$ and $y = 11$ into the equation.

$11 \stackrel{?}{=} 5(2) + 1$

$11 = 11$ — Yes. The ordered pair (2, 11) satisfies the equation.

$0 \stackrel{?}{=} 5\cdot(-1) + 1$ — Substitute $x = -1$ and $y = 0$ into the equation.

$0 \neq -4$ — No. The ordered pair (−1, 0) does not satisfy the equation.

GRAPHING AN EQUATION

1. Choose at least two negative values, zero, and two positive values for x and record them in a table of values.
2. Substitute the x values into the equation and compute the corresponding y values. Record the y values in the table.
3. Write each x and y pair in the table as an ordered pair.
4. Plot the ordered pairs.
5. Connect the dots, extending the line beyond the dots to show that the pattern continues. Draw arrows on each end of the line.

EXAMPLE 3: Graph the equation $y = 2x - 3$.

x	Computations	y	Ordered pairs
-2	$y = 2(-2) - 3 = -4 - 3 =$	-7	$(-2, -7)$
-1	$y = 2(-1) - 3 = -4 - 3 =$	-5	$(-1, -5)$
0	$y = 2(0) - 3 = 0 - 3 =$	-3	$(0, -3)$
1	$y = 2(1) - 3 = 0 - 3 =$	-1	$(1, -1)$
2	$y = 2(2) - 3 = 0 - 3 =$	1	$(2, 1)$

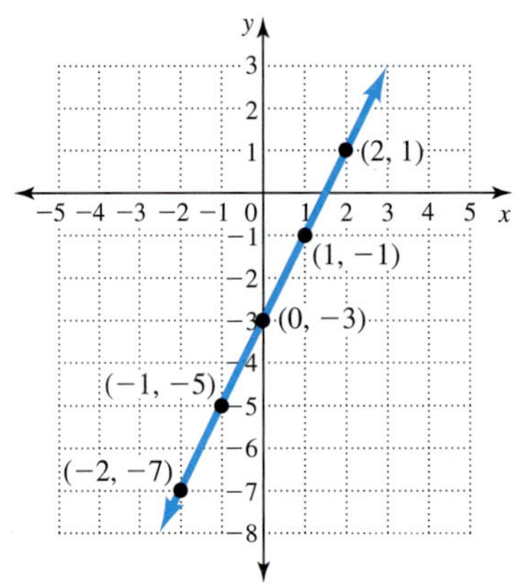

Doing Mathematics

- Write the definitions of the words and symbols listed for this section in the Dictionary in Appendix A.
- On a 3 × 5-inch card, write the following for future reference:
  (a) Constructing a Cartesian plane
  (b) Choosing a scale for the vertical and/or horizontal axes (axis)
  (c) Graphing an equation

Problems 1–2: Plot the ordered pairs on the Cartesian plane. Label each point with the corresponding letter. Indicate the Quadrant in which each point is located.

1. (a) $(1, 3)$ (b) $(-1, -2)$ (c) $(1, -4)$
(d) $(-4, -1)$ (e) $(-1, 2)$ (f) $(-1, -3)$
(g) $(4, -1)$ (h) $(4, 4)$ (i) $(-4, 1)$
(j) $(3, 1)$ (k) $(1, -2)$ (l) $(0, 0)$

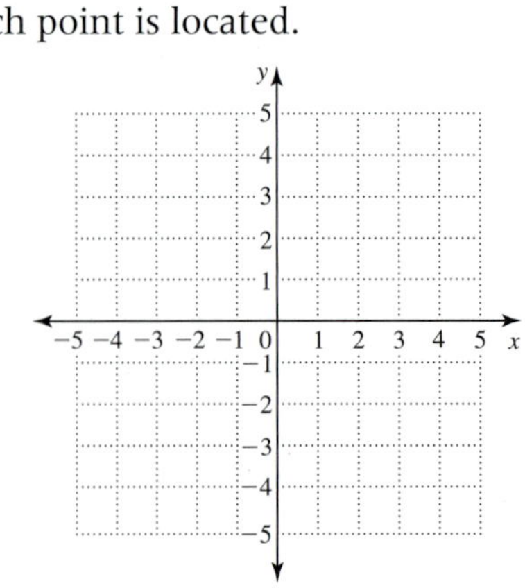

2. (a) $(2, -4)$ (b) $(-3, 2)$ (c) $(2, 4)$
   (d) $(0, 2)$ (e) $(-3, -1)$ (f) $(2, -1)$
   (g) $(-4, -3)$ (h) $(-4, 4)$ (i) $(-2, 0)$
   (j) $(3, -3)$ (k) $(4, 2)$ (l) $(-3, -4)$

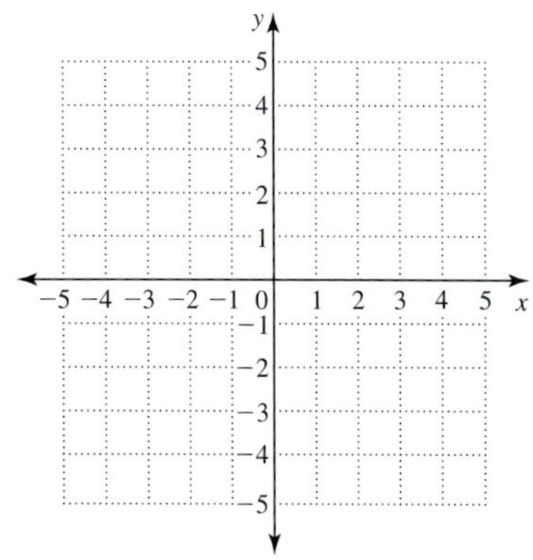

Problems 3–6: Given the grid and points:

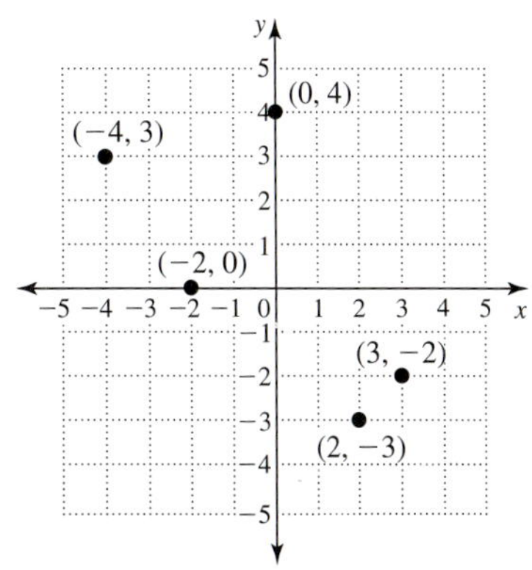

3. What is the first coordinate value of the point that has a second coordinate value of 3?

4. What is the first coordinate value of the point that has a second coordinate value of 0?

5. What is the second coordinate value of the point that has a first coordinate value of 2?

6. What is the second coordinate value of the point that has a first coordinate value of 0?

7. Determine the Dow Jones Industrial averages for the dates listed. Estimate answers.

   (a) December 1, 2000: ______
   (b) January 1, 2001: ______
   (c) February 1, 2001: ______
   (d) March 1, 2001: ______
   (e) March 21, 2001: ______

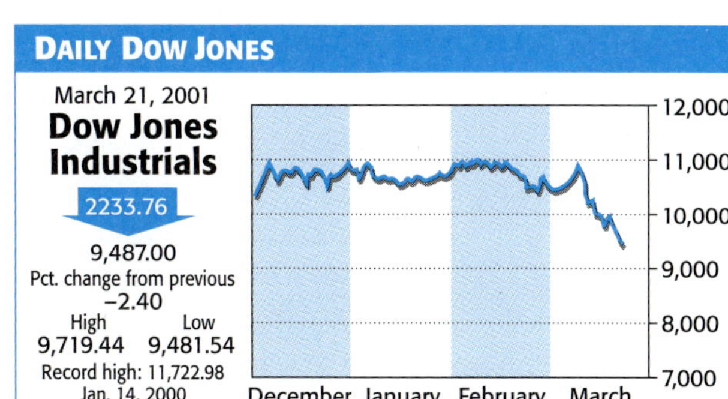

8. The following figure shows the fall enrollment at U.S. institutions of higher education in recent years, along with projected enrollments through fall 2007.[4] Enrollment statistics are taken (or projected) at the beginning of the fall of each year.

(a) Estimate the projected enrollment of women in U.S. institutions of higher education in the fall term of 2007.

(b) Estimate the projected enrollment of men in U.S. institutions of higher education in the fall term of 2007.

(c) What trends do you notice in the graph?

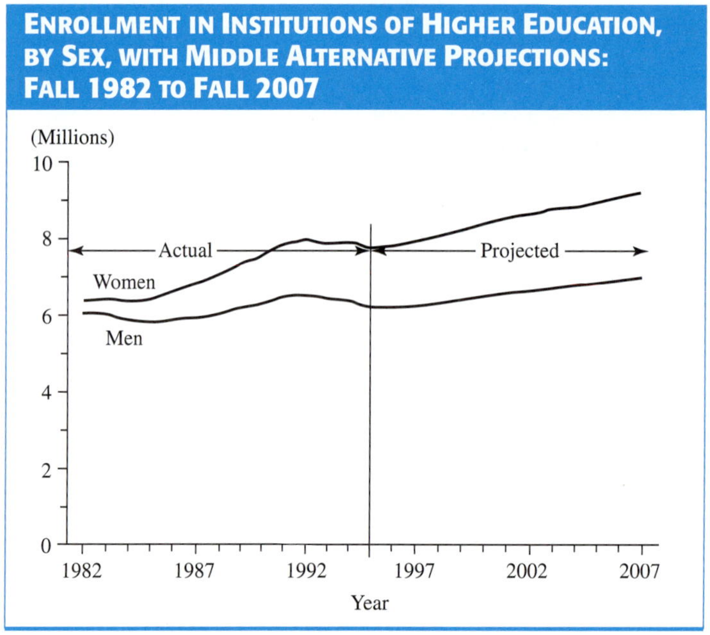

Problems 9–10: Identify an appropriate scale and plot the ordered pairs.

9. (a) List the sets of ordered pairs in the third column.

x = first coordinate	y = second coordinate	Ordered pairs
0	10	(0, 10)
2	15	
4	20	
6	25	

(b) Smallest first coordinate value: ______; Largest first coordinate value: ______ Scale for horizontal axis: ______

(c) Smallest second coordinate value: ______; Largest second coordinate value: ______; Scale for vertical axis: ______

(d) Label the horizontal and vertical axes with numbers according to the scale you chose above. Plot the ordered pairs.

[4] National Center for Education Statistics.

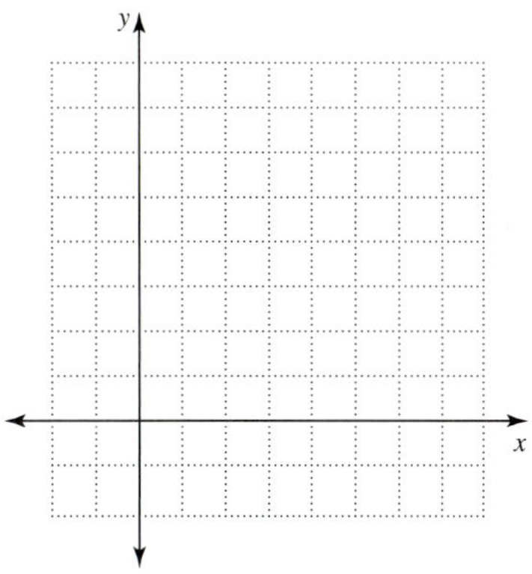

10. (a)List the sets of ordered pairs in the third column.

x = first coordinate	y = second coordinate	Ordered pairs
−100	−30	(−100, −30)
−50	−15	
0	0	
50	15	
100	30	

(b) Smallest first coordinate value: ______; Largest first coordinate value: ______ Scale for horizontal axis: ______

(c) Smallest second coordinate value: ______; Largest second coordinate value: ______; Scale for vertical axis: ______

(d) Label the horizontal and vertical axes with numbers according to the scale you chose above. Plot the ordered pairs.

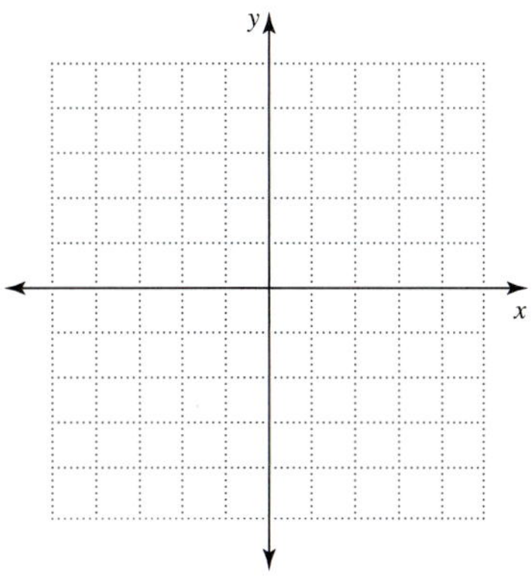

11. Tuition at a nearby community college is $45 per credit hour.

(a) Complete the table.

n = Number of credit hours taken	C = Tuition cost ($)
3	135
6	270
9	
15	

(b) Smallest first coordinate value: ______; Largest first coordinate value: ______
Scale for horizontal axis: ______

(c) Smallest second coordinate value: ______; Largest second coordinate value: ______; Scale for vertical axis: ______

(d) Label the horizontal and vertical axes with numbers according to the scale you chose above. Label both the horizontal and vertical axes with a title to explain what each axis represents. Plot the ordered pairs.

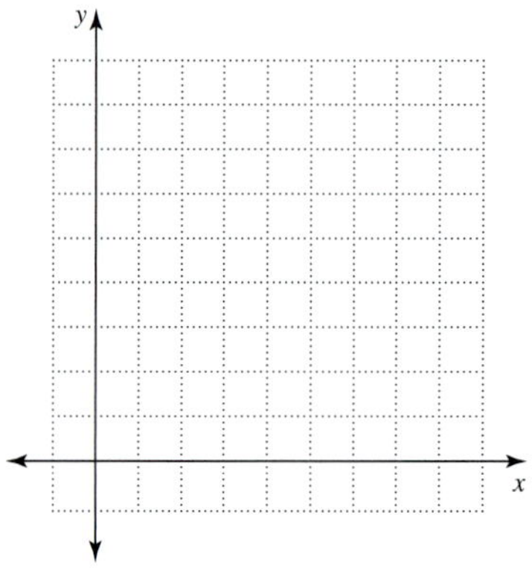

12. The table lists the number of grams of potassium nitrate (KNO_3) that dissolve per 100 grams of water at a given Celsius temperature.

t = Temperature (°C)	P = Number of grams of potassium nitrate
0	10
10	20
20	32
40	60
60	110
80	160
100	242

(a) Smallest first coordinate value: _______; Largest first coordinate value: _______
Scale for horizontal axis: _______

(b) Smallest second coordinate value: _______; Largest second coordinate value: _______; Scale for vertical axis: _______
(*Hint:* The scale for the vertical axis is not obvious from the y values.
You can choose any value that will display all coordinates on the grid you are using. One way to choose a scale is to divide the largest coordinate value by the number of gridlines to be labeled and then round that number: $\frac{242}{8} = 30.25 \approx 30$. Try a scale of 30 on the vertical axis. Feel free to experiment with different scales. Be sure all data points are displayed.)

(c) Label the horizontal and vertical axes with numbers according to the scale you chose above. Label both the horizontal and vertical axes with a title to explain what each axis represents. Plot the ordered pairs.

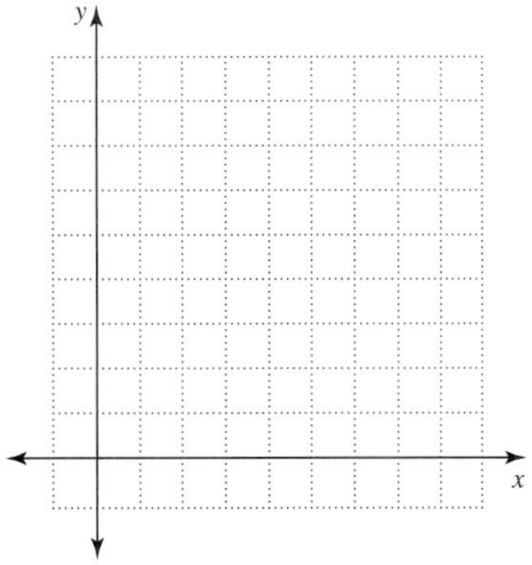

13. The number of weeks of pregnancy (gestation period) is related to the length of the adult mammal as shown in the table below.

Animal	l = Length of adult (in meters)	N = Number of weeks of pregnancy
A. Rabbit	0.5	4
B. Skunk	0.7	7
C. Coyote	1.3	9
D. Bear	1.75	28
E. Gorilla	2	36
F. Giraffe	5.7	60

(a) Smallest first coordinate value: _______; Largest first coordinate value: _______
Scale for horizontal axis: _______

(b) Smallest second coordinate value: _______; Largest second coordinate value: _______; Scale for vertical axis: _______

(c) Label the horizontal and vertical axes with numbers according to the scale you chose in (a) and (b). Label both the horizontal and vertical axes with a title to explain what each axis represents. Plot the ordered pairs and label each point with the letter above identifying the mammal name.

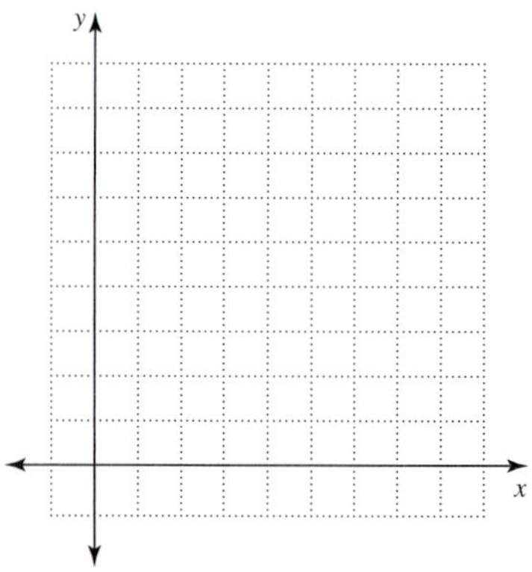

(d) Describe the relationship of the length of the pregnancy (gestation period) to the length of the adult.

14. This table gives the percentage of cigarette-smoking males and females (18 years and over) in the United States for the years 1965–1994.[5] Let x represent the number of years since 1965, y_1 represent the % of cigarette-smoking males, and y_2 represent the % of cigarette-smoking females.

Year	x = Number of years since 1965	y_1 = % of cigarette-smoking males (18 years and over)	y_2 = % of cigarette-smoking females (18 years and over)
1965	0	51.6	34.0
1974	9	42.9	32.5
1979	14	37.2	30.3
1983	18	34.7	29.9
1985	20	32.1	28.2
1987	22	31.0	26.7
1990	25	28.0	23.1
1991	26	27.5	23.6
1992	27	28.2	24.8
1993	28	27.5	22.7
1994	29	27.8	23.3

[5] Centers for Disease Control and Prevention.

(a) Determine a scale and plot both sets of ordered pairs on the same grid. Mark the points for the males with a black pen and mark the points for the females with a blue pen. Connect the black dots with a black pen and connect the blue dots with a blue pen.

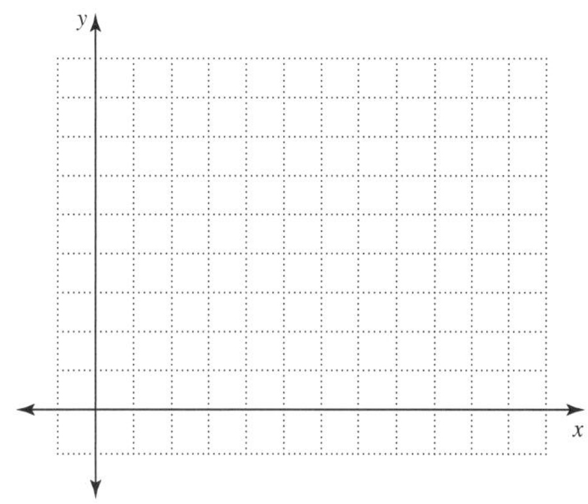

(b) Describe any pattern that you noticed with the points on the graph.

15. Go to the library and find a copy of the *Wall Street Journal, USA Today,* or a local newspaper. Find three examples of lists of ordered pairs and/or graphs of real-life problem situations. Copy the lists and/or graphs on an $8\frac{1}{2} \times 11$-inch piece of paper and keep the paper. List at least three ordered pairs displayed on the graph:
(,) (,) (,)

Problems 16–19: Do the ordered pairs satisfy the equation $y = 5 - 2x$?

16. $(0, 0)$

17. $(-1, -7)$

18. $(-1, 7)$

19. $(1, 3)$

Problems 20–27: Graph each equation.

(a) Substitute each x-value into the equation and find the corresponding y-value. Record the values and computations in the table. Write each ordered pair in the fourth column.

(b) Plot the points.

(c) Connect the dots. Extend the line and draw arrows on each end to show that the pattern continues.

20. $y = x - 4$

x	Computations	y	Ordered pairs
−2			
−1			
0			
1			
2			

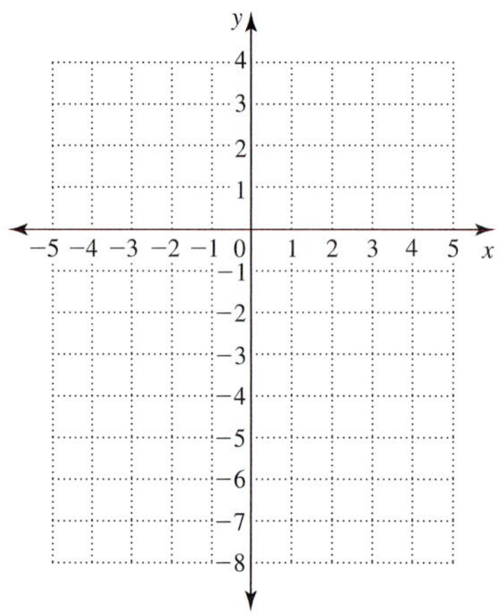

21. $y = x - 1$

x	Computations	y	Ordered pairs
−2			
−1			
0			
1			
2			

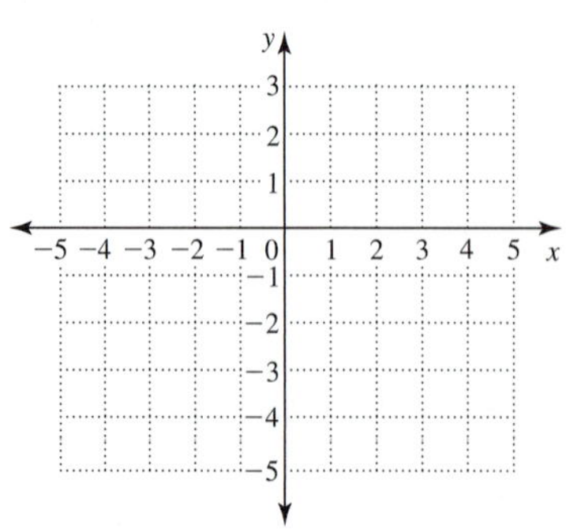

22. $y = -2x + 3$

x	Computations	y	Ordered pairs
−2			
−1			
0			
1			
2			

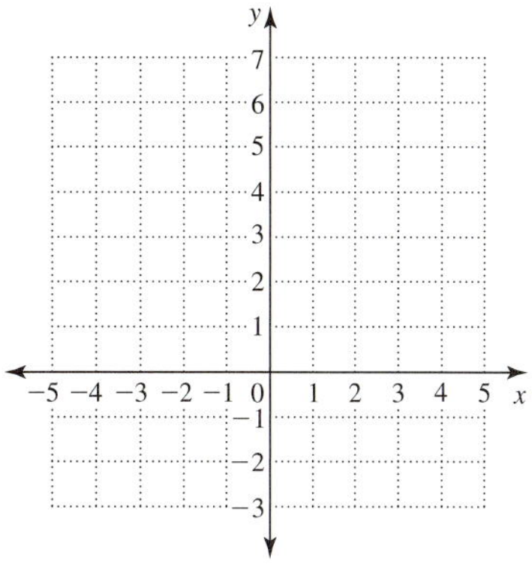

23. $y = -3x + 1$

x	Computations	y	Ordered pairs
−2			
−1			
0			
1			
2			

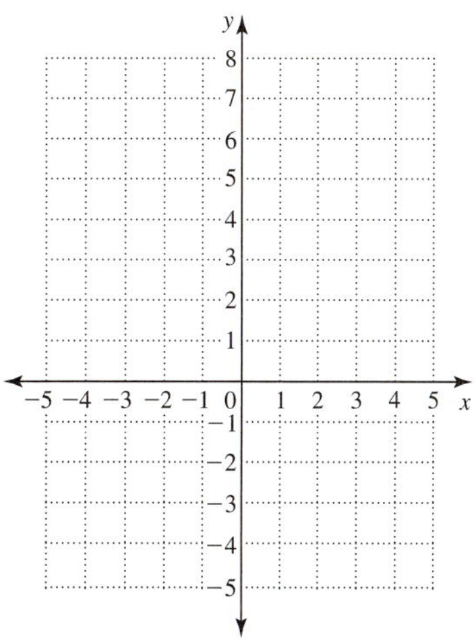

24. $y = -x + 4$

x	Computations	y	Ordered pairs
−2			
−1			
0			
1			
2			

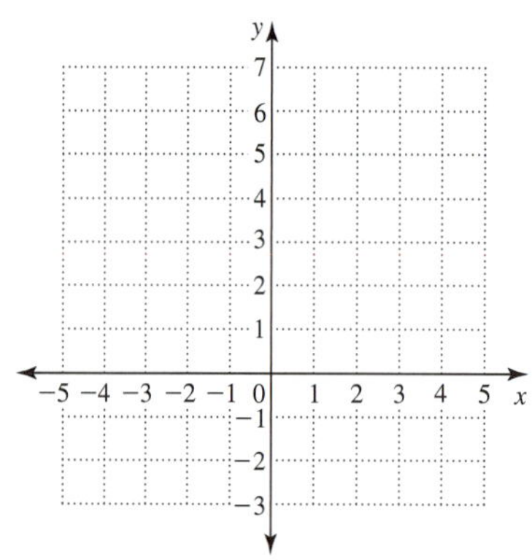

25. $y = -x + 6$

x	Computations	y	Ordered pairs
−2			
−1			
0			
1			
2			

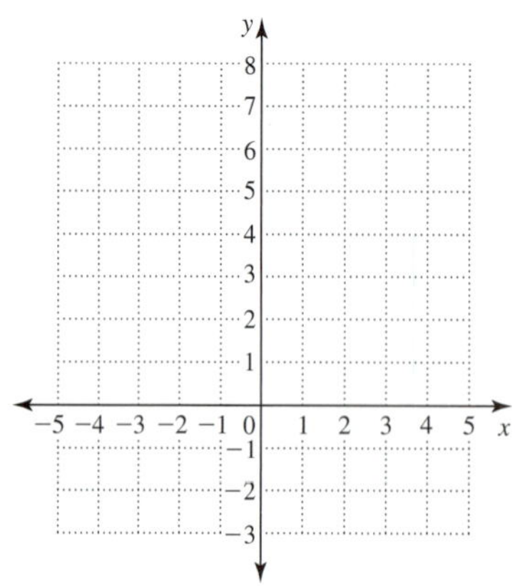

26. $y = 4 + 2x$

x	Computations	y	Ordered pairs
−2			
−1			
0			
1			
2			

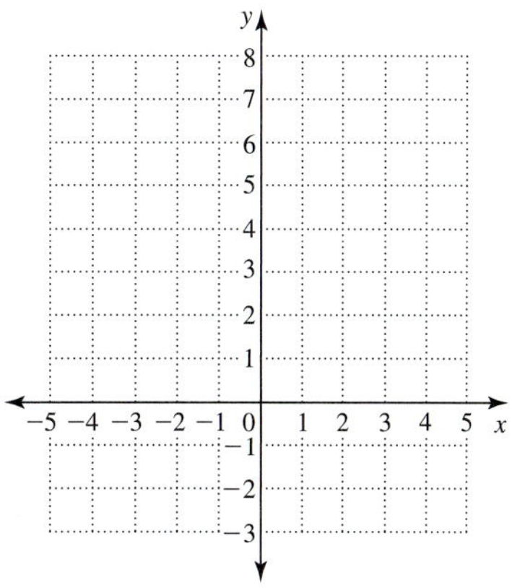

27. $y = 6 + 3x$

x	Computations	y	Ordered pairs
−2			
−1			
0			
1			
2			

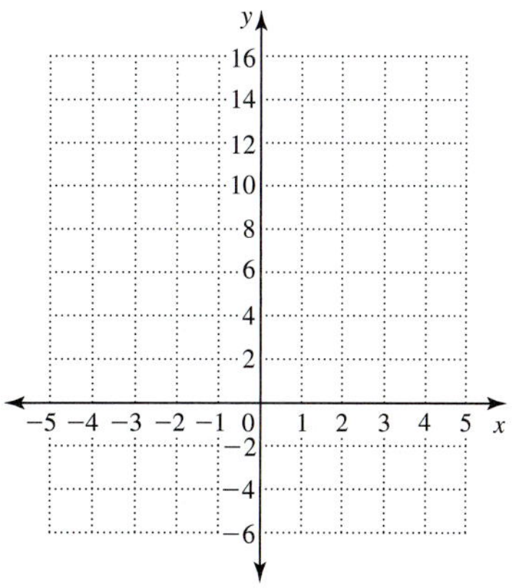

3.2 Relations, Functions, and Function Notation

OBJECTIVES

- Identify and graph a relation and a function.
- Use an Evaluating Machine to evaluate and solve a relation or function.
- Write and evaluate a function in function notation.
- Use the vertical line test to identify a function.

1 2 3 Getting Started

In this section, the concepts of relation and function are introduced and corresponding graphs discussed. A **relation** is any set of ordered pairs. Consider this list of Test Scores from an Algebra class. Each score and corresponding name defines an ordered pair. The first coordinate of each ordered pair is the test score. The second coordinate is the name(s) of the student(s) receiving the given score.

Test scores	Name of student receiving the score	Ordered pairs
70	Jenny	(70, Jenny)
89	Tom	(89, Tom)
59	Kristen	(59, Kristen)
70	Jane	(70, Jane)
95	Bill	(95, Bill)
89	Mike	(89, Mike)
95	Kathy	(95, Kathy)
70	Debbie	(70, Debbie)
91	Nicole	(91, Nicole)
89	Ashley	(89, Ashley)

Evaluating Machines are used to visualize relations. The first coordinate value is called the **input value**. The input proceeds through a process to determine the second coordinate value, called the **output value**.

Evaluating Machine

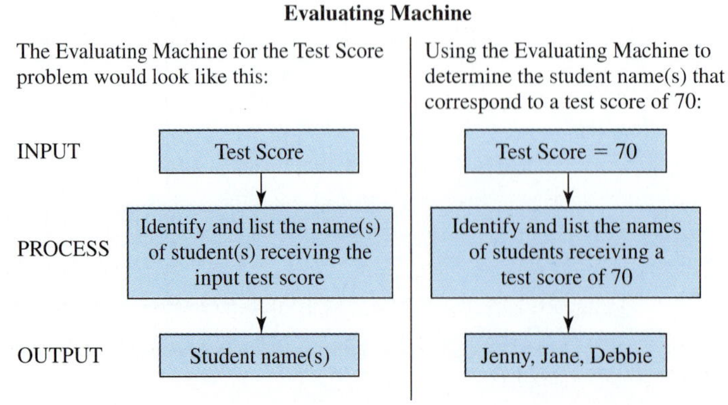

1. Use the Evaluating Machine for the Test Score problem to write the output values for the given input values.

(a) List the output (student name(s)) for the input test score of 59.

(b) List the output (student name(s)) for the input test score of 89.

(c) List the output (student name(s)) for the input test score of 91.

(d) List the output (student name(s)) for the input test score of 95.

2. We can construct an Evaluating Machine for the video profit equation in Chapter 1. The problem situation was described by the equation $P = 8x - 1000$, where the variable x represents the number of videos sold and P represents the profit. The profit depends on the number of videos sold. The input variable x, number of videos sold, is called the **independent variable**. The output variable P, profit, is called the **dependent variable**. The equation, $P = 8x - 1000$, states the relationship between the variables x and P.

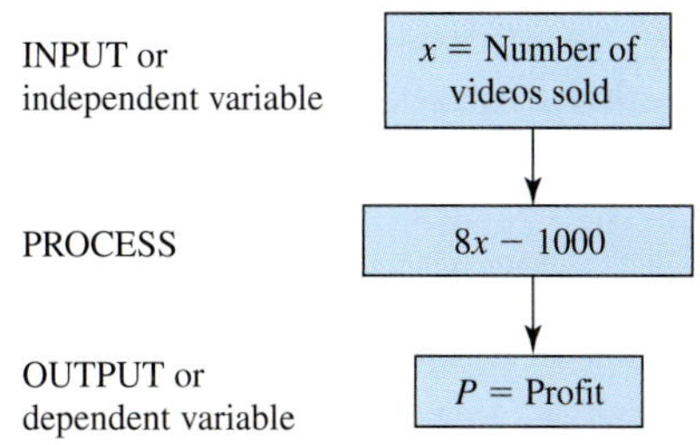

(a) Use the Evaluating Machine to compute the output when the input is 175.

(b) Use the Evaluating Machine to compute the output when the input is 400.

(c) How many output values are possible for each input value in the Evaluating Machine in the profit equation $P = 8x - 1000$?

3. A **relation** is any set of ordered pairs. A relation may have more than one output value for any input value. The test score problem is an example of a relation, where some input values have more than one output. The video problem is an example of a function. Every input value has *one and only one* output value. The equation $P = 8x - 1000$ is a function. A **function** is a relation in which every input value (first coordinate value or x-value) is paired with exactly one output value (second coordinate value or y-value).

(a) Is every relation a function? Explain your answer.

(b) Is every function a relation? Explain your answer.

Discussing Mathematics

RELATIONS AND FUNCTIONS

Each input value in a relation has one or more output value(s). Some relations are also called functions. A **function** is a relation where every input is paired with exactly ONE output. In an ordered pair of a function, (x, y), there is one and only one y value for each given x value.

DEFINITION OF A RELATION AND A FUNCTION

A relation is any set of ordered pairs.

A function is a relation where each input value has *one and only one* output value.

VERTICAL LINE TEST

Relations and functions can be graphed on a Cartesian plane. The input values are normally represented on the horizontal axis. The output values are normally represented on the vertical axis. The **vertical line test** is used to determine whether or not a relation is a function. Vertical lines drawn on the graph of a function will intersect in one and only one point.

VERTICAL LINE TEST

- If all vertical lines intersect the graph of a relation at only one point, the relation *is* also a function. One and only one output value exists for each input value.
- If any vertical line intersects the graph of a relation at more than one point, the relation "fails" the test and is NOT a function. More than one output value exists for some (or all) input value(s).

EXAMPLE 1: Use the vertical line test to determine whether or not each graph is a relation or a function.

Graph of the relation defined by the set of ordered pairs:

$\{(-1, 3), (3, 2), (-1, -3), (3, -2), (4, 4), (-4, -5), (3, 1), (0, 2)\}$

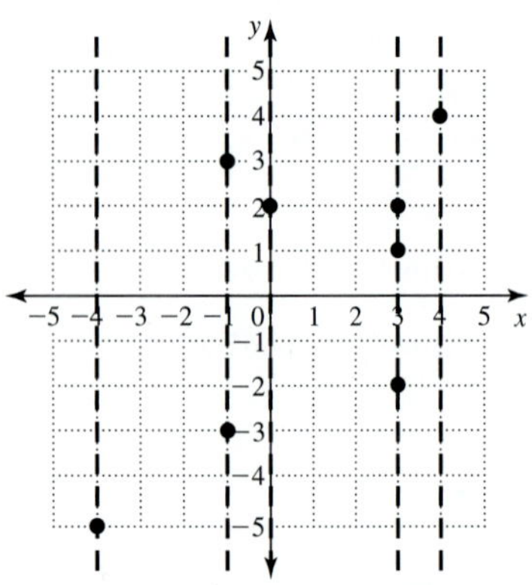

Two vertical lines intersect more than one point.
The relation is *not* a function.

Graph of $P = 8x - 1000$:

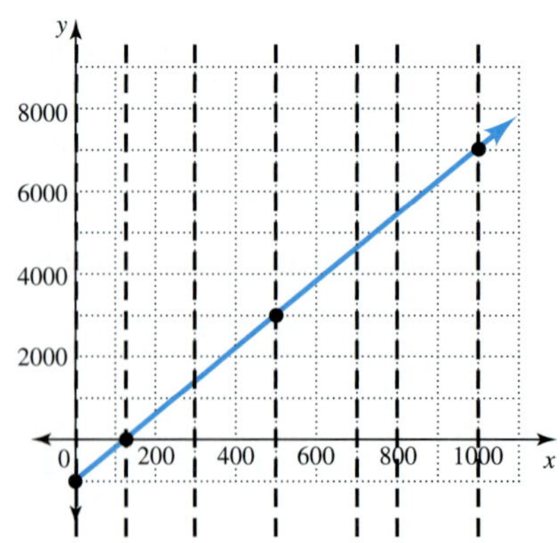

Each vertical line intersects the graph in one and only one point.
The relation *is* a function.

EVALUATING AND SOLVING

Evaluating machines can be used to evaluate or solve a function. Assigning a value to the input and using the process of the function to compute the corresponding output is called **evaluating**. Assigning a value to the output and using the process of the function backwards to compute the input is called **solving**.

EXAMPLE 2: Use an Evaluating Machine to *evaluate* the expression $3x + 2$ when $x = 5$.

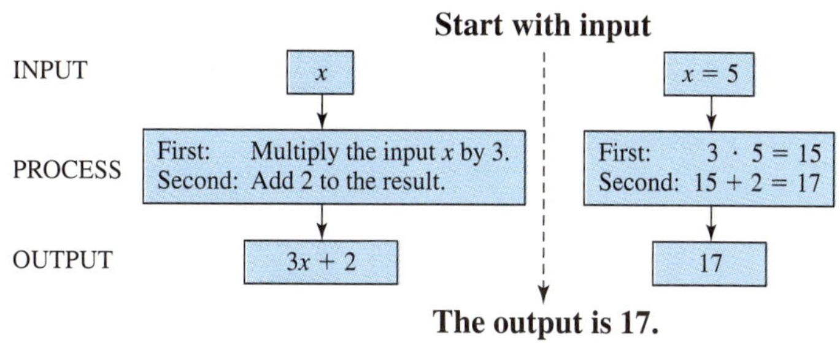

EXAMPLE 3: Use an Evaluating Machine backwards to *solve* the equation $y = 3x + 2$ when $y = 17$.

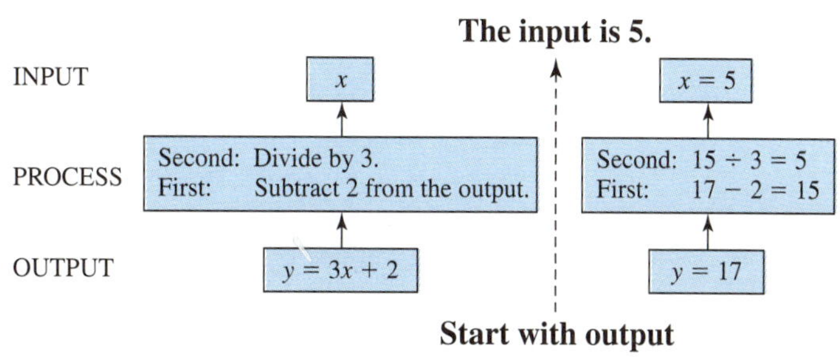

FUNCTION NOTATION

Functions are usually written using a special notation. **Function notation** is a shorthand notation communicating that the dependent variable is a function of the independent variable. Function notation replaces the dependent variable with an expression using both the dependent and independent variable. When the profit function (Getting Started, problem 2) is written with function notation, $P(x)$ replaces P:

$$P = 8x - 1000$$
$$\downarrow$$
$$P(x) = 8x - 1000$$

Function Notation
Dependent variable
(independent variable)

PROCESS involving
the independent variable

➢ ***$P(x)$ does not imply multiplication. The variable P is not being multiplied by x.***

➢ *The variable f is the most common letter used to denote the dependent variable of a function and x is the most common letter used to denote the dependent variable.*

EVALUATING, SOLVING, AND GRAPHING A FUNCTION USING FUNCTION NOTATION

Function notation is a useful shorthand notation for evaluating and solving functions, and writing ordered pairs of the form $(x, P(x))$.

EXAMPLE 4: Evaluate the profit function $P(x) = 8x - 1000$ when $x = 50$.

$$P(x) = 8x - 1000$$

$$P(50) = 8(50) - 1000$$ Substitute 50 for x.

$$P(50) = 400 - 1000$$ Perform the operations on the right side of the equation.

$$P(50) = -600$$

Some ordered pairs associated with the profit function $P(x) = 8x - 1000$:

Number of videos sold: x	Profit: $P(x) = 8x - 1000$	Ordered pairs: $(x, P(x))$
50	$P(50) = 8(50) - 1000$ $P(50) = -600$	$(50, -600)$
563	$P(563) = 8(563) - 1000$ $P(563) = 3504$	$(563, 3504)$
700	$P(700) = 8(700) - 1000$ $P(700) = 4600$	$(700, 4600)$

EXAMPLE 5: Given the function $f(x) = -3x^2 + 5$, evaluate $f(-2)$.

$$f(x) = -3x^2 + 5$$

$$f(-2) = -3(-2)^2 + 5$$ Substitute -2 for x.

$$f(-2) = -3(4) + 5$$ Perform the operations on the right side of the equation.

$$f(-2) = -12 + 5$$

$$f(-2) = -7$$ Corresponding ordered pair: $(-2, -7)$.

EXAMPLE 6: Find the number of videos sold if the profit is \$3504 in the function $P(x) = 8x - 1000$.

$$P(x) = 8x - 1000$$

$$3504 = 8x - 1000$$ Substitute \$3504 for $P(x)$. Solve for x.

$$3504 + 1000 = 8x - 1000 + 1000$$ Add 1000 on each side of the equation.

$$4504 = 8x$$ Divide by 8 on each side of the equation.

$$\frac{4504}{8} = \frac{8x}{8}$$

$$563 = x$$

The number of videos sold is 563 when the profit is \$3504.

EXAMPLE 7: Graph the function $f(x) = 4x + 3$.

Evaluate the function for different values of x and record the corresponding $f(x)$. Write the coordinates as ordered pairs.

x	Computations	$f(x)$	Ordered pairs $(x, f(x))$
-1	$f(-1) = 4(-1) + 3 = -4 + 3 =$	-1	$(-1, -1)$
0	$f(0) = 4(0) + 3 = 0 + 3 =$	3	$(0, 3)$
1	$f(1) = 4(1) + 3 = 4 + 3 =$	7	$(1, 7)$

Label the vertical axis the $f(x)$-axis. Plot the points. Connect the dots.

Extend the line and draw arrows on each end of the line.

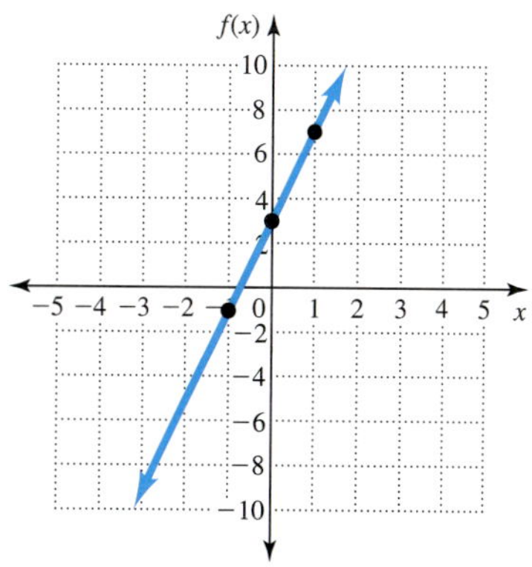

EXAMPLE 8: A graph can be used to estimate input and/or output values. Look at the graph of $P(x) = 8x - 1000$ and find the output value when the input value is 300.

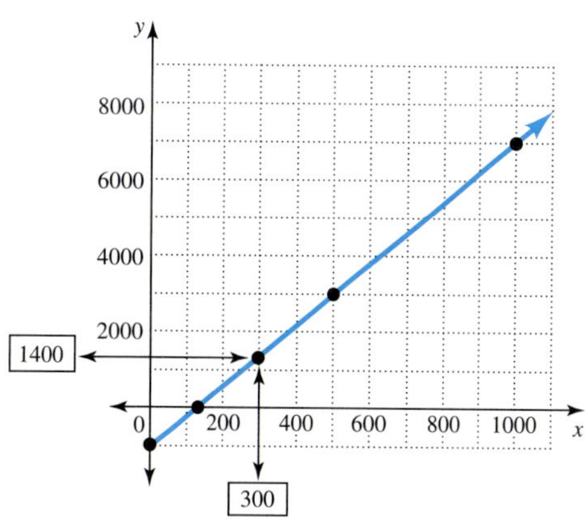

The output value appears to be approximately 1500.

To find the exact value when the input is 300, substitute $x = 300$ and evaluate the function: $P(300) = 8(300) - 1000 = \$1400$

The exact output value is $1400 when the input value is 300.

➢ *Function notation will be used throughout this course and in future mathematics courses. Any letter may be used for either the independent or dependent variable. In this text, a function will often be labeled $f(x)$.*

1 2 3 Doing Mathematics

- Write the definitions of the words and symbols listed for this section in the Dictionary in Appendix A.
- On 3 × 5-inch cards, write the following processes and/or definitions for future reference:
  (a) An example of an Evaluating Machine
  (b) The process of evaluating a function
  (c) The process of solving a function
  (d) The process of graphing a function

Problems 1–4: Draw the horizontal and vertical axes. Label the axes appropriately for the points to be plotted. Plot the ordered pairs. Use the definition of relation and function and the vertical line test to determine whether each set represents a function.

1. $\{(0, 0), (0, 1), (1, -2), (1, 3), (3, -2), (3, 4)\}$
Function? Yes or No

2. $\{(-1, -4), (-2, 1), (0, 3), (1, 2)\}$
Function? Yes or No

3.

Input	Output
3	4
3	5
3	6
3	7
3	8

Function? Yes or No

4.

Input	Output
4	3
5	3
6	3
7	3

Function? Yes or No

Problems 5–8: Use the vertical line test to determine whether the relation is also a function.

5.

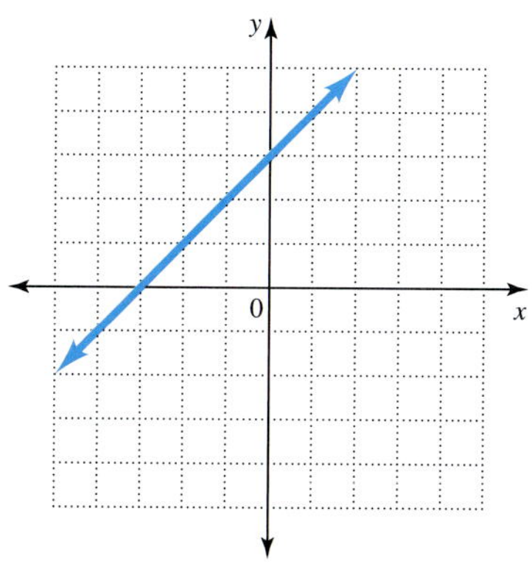

6.

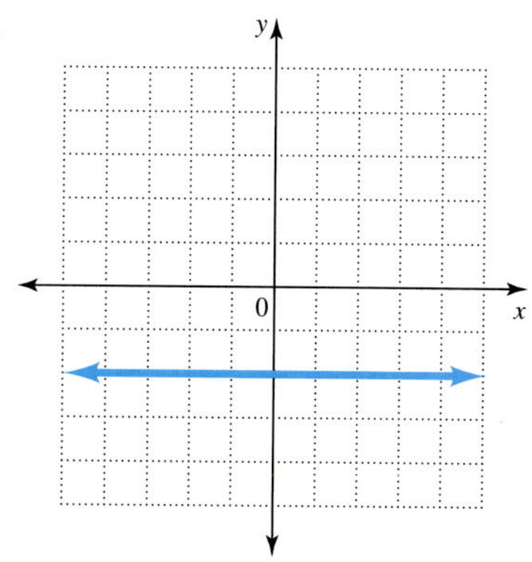

7.

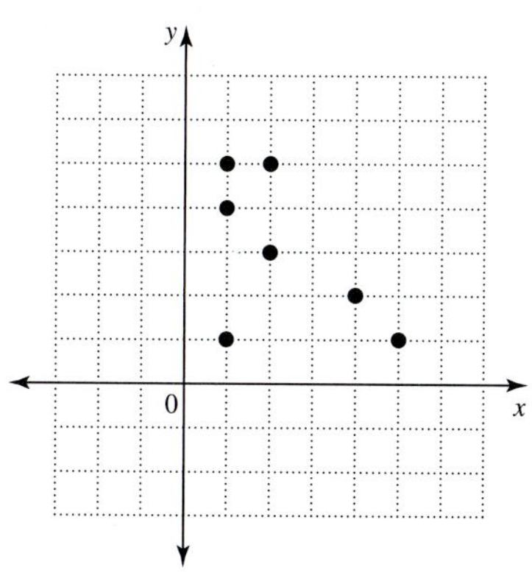

8.

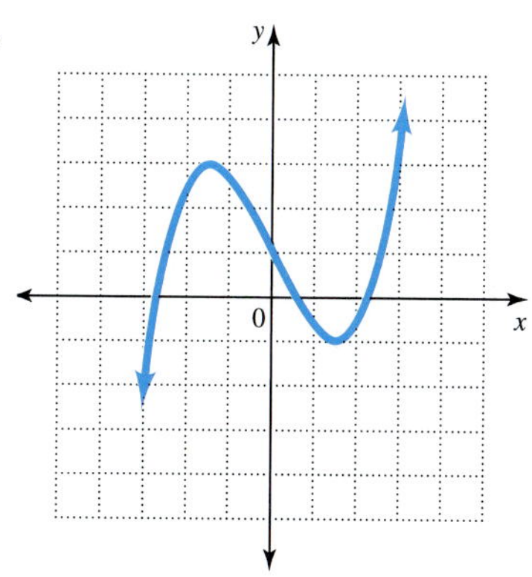

Problems 9–16: Graph each function. Substitute each x value into the function and compute the corresponding $f(x)$ value. Record the ordered pairs in the table. Plot the points. Connect the dots. Extend the line and draw arrows on each end of the line.

9. Graph the function $f(x) = 2x - 1$.

x	$f(x)$	Ordered pairs
−2		
−1		
0		
1		
2		

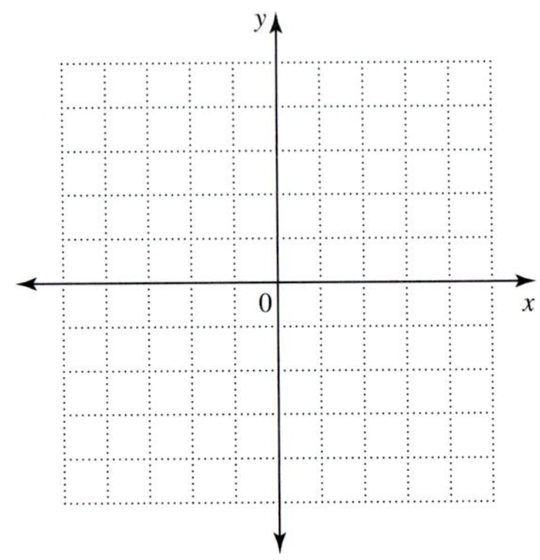

10. Graph the function $f(x) = x + 2$.

x	$f(x)$	Ordered pairs
−2		
−1		
0		
1		
2		

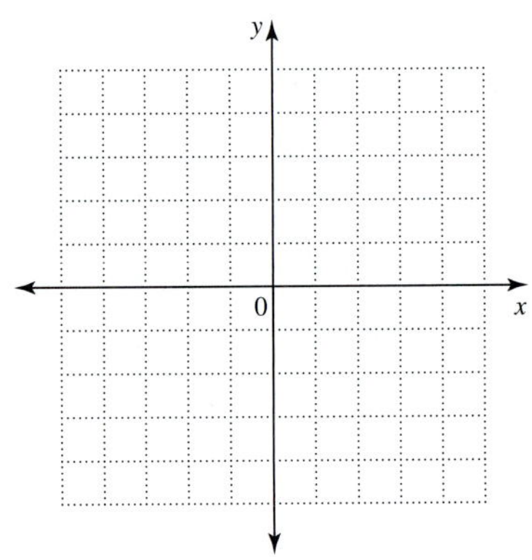

11. Graph the function $f(x) = 2x$.

x	$f(x)$	Ordered pairs
−2		
−1		
0		
1		
2		

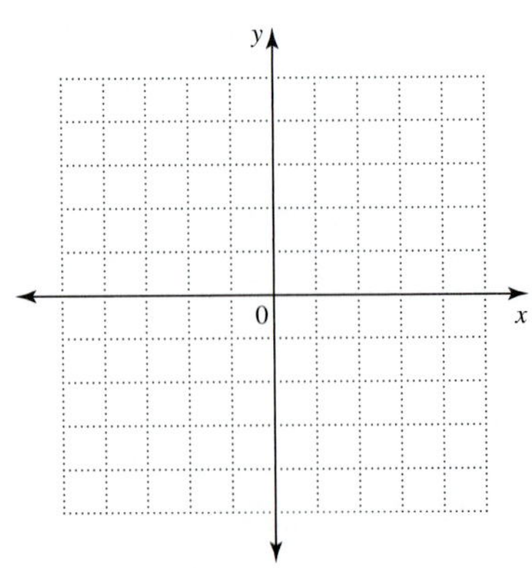

12. Graph the function $f(x) = 3x$.

x	$f(x)$	Ordered pairs
-2		
-1		
0		
1		
2		

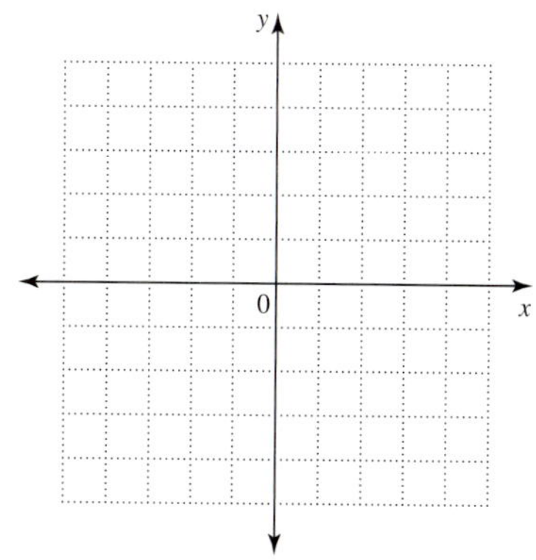

13. Graph the function $f(x) = -x$.

x	$f(x)$	Ordered pairs
-2		
-1		
0		
1		
2		

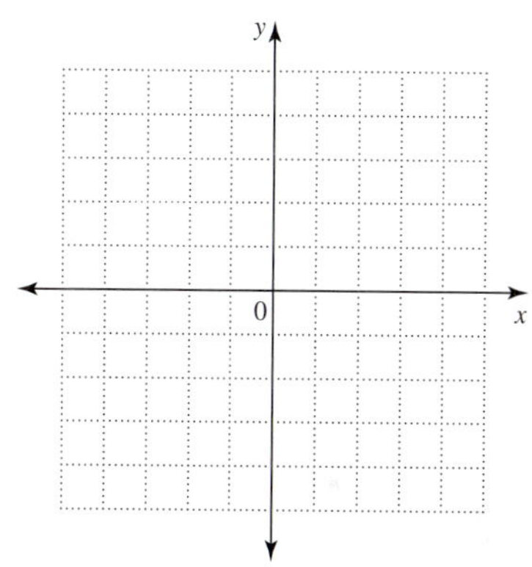

14. Graph the function $f(x) = 1 - x$.

x	$f(x)$	Ordered pairs
-2		
-1		
0		
1		
2		

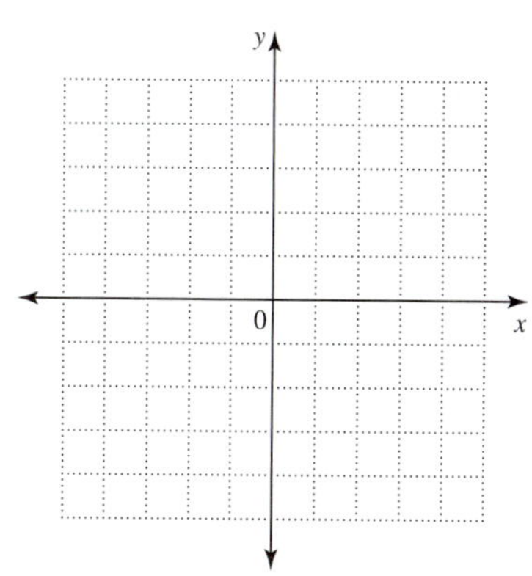

15. Graph the function $f(x) = 6$.

x	$f(x)$	Ordered pairs
-2		
-1		
0		
1		
2		

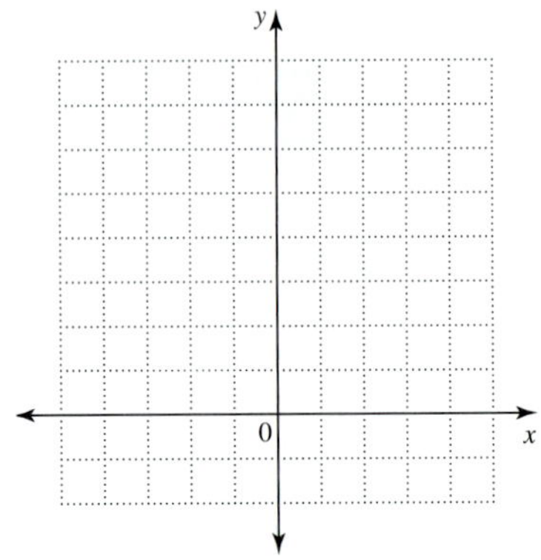

16. Graph the function $f(x) = -2$.

x	$f(x)$	Ordered pairs
-2		
-1		
0		
1		
2		

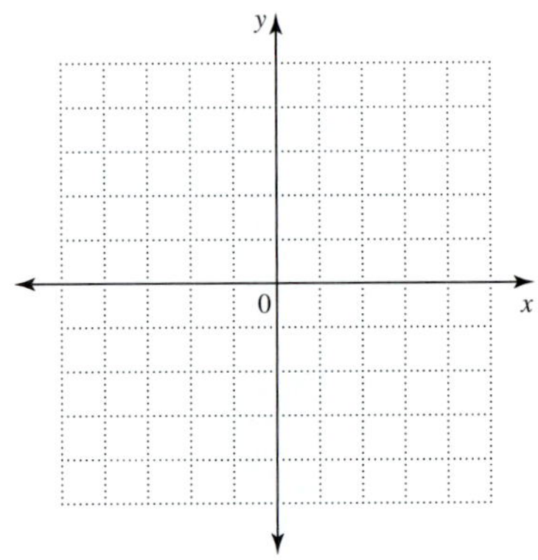

17. Draw an Evaluating Machine for evaluating the expression $-x^2$. The input is x and the output is $-x^2$.

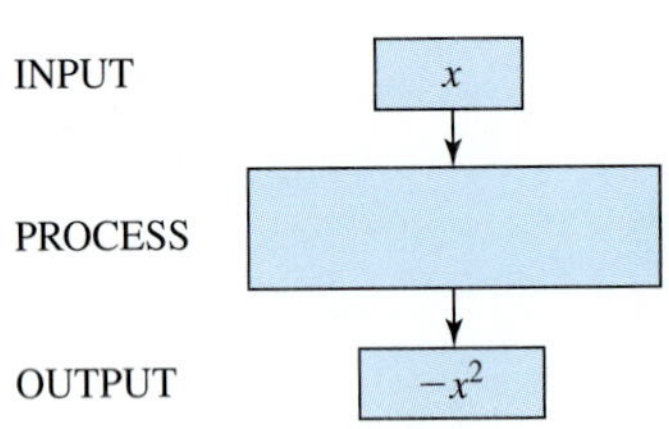

Use the Evaluating Machine to find the output value when

(a) the input value is $x = 3$.
(b) the input value is $x = -3$.
(c) the input value is $x = 20$.
(d) the input value is $x = -20$.

18. Draw an Evaluating Machine for evaluating the expression $(-x)^2$. The input is x and the output is $(-x)^2$.

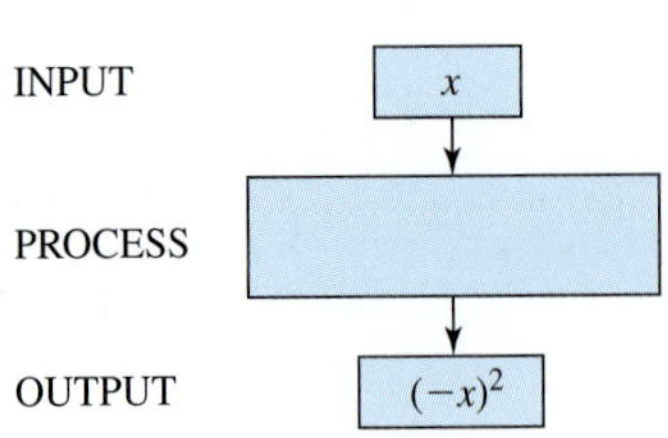

Use the Evaluating Machine to find the output value when

(a) the input value is $x = 3$.
(b) the input value is $x = -3$.
(c) the input value is $x = 20$.
(d) the input value is $x = -20$.

19. Draw an Evaluating Machine for solving the equation $y = 5(x - 6)$.

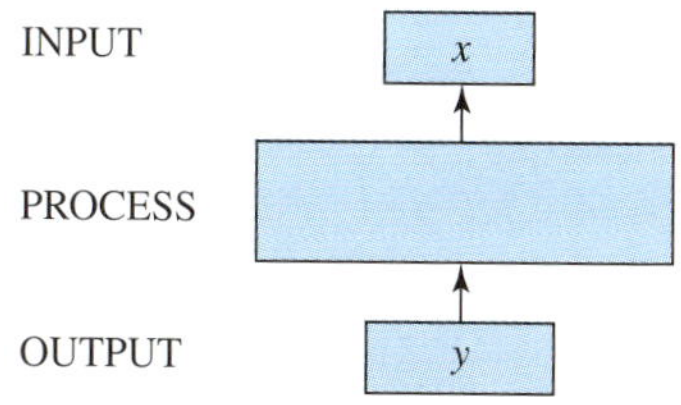

Use the Evaluating Machine to find the input value when

(a) the output value is $y = 10$:

(b) the output value is $y = 20$:

(c) the output value is $y = -15$:

(d) the output value is $y = -30$:

20. Draw an Evaluating Machine for solving the equation $y = 2x + 3$.

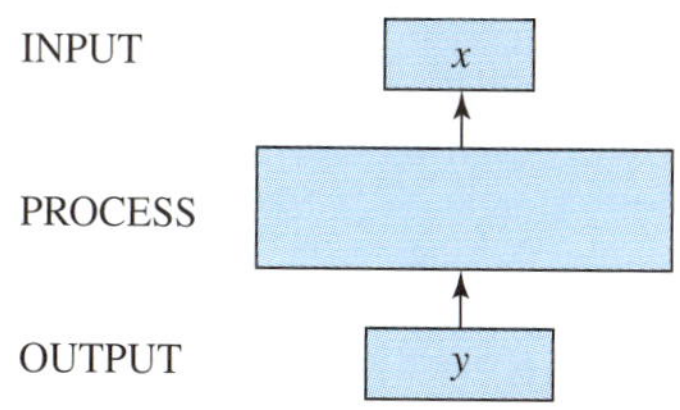

Use the Evaluating Machine to find the input value when

(a) the output value is $y = 13$:

(b) the output value is $y = 27$:

(c) the output value is $y = -5$:

(d) the output value is $y = -17$:

Problems 21–24: Evaluate each function for the given input values.

21. Given $f(x) = 2x + 1$,

(a) evaluate $f(5)$.

(b) evaluate $f(-2)$.

22. Given $g(x) = x^2 - 4x + 5$,

(a) evaluate $g(-3)$.

(b) evaluate $g(2)$.

23. Given $f(y) = y^2 + 3$,

(a) evaluate $f(10)$.

(b) evaluate $f(-2)$.

24. Given $h(a) = a - 17$,

(a) evaluate $h(-4)$.

(b) evaluate $h(0)$.

Problems 25–28: Solve the function $f(x) = 5x + 1$ for the given output values.

25. Solve for x, given $f(x) = 11$.

26. Solve for x, given $f(x) = 1$.

27. Solve for x, given $f(x) = -4$.

28. Solve for x, given $f(x) = -14$.

29. Mike has been offered a job selling calculators. His annual salary is \$15,000 plus 35% commission on the dollar amount of his sales of calculators.

(a) Write the word or words that describe the input or independent variable.

(b) Write the word or words that describe the output or dependent variable.

(c) Complete the table.

S = Dollars amount of sales	Computations: Total salary = Base salary + Commission	T = Total salary ($)	Ordered pairs
1000	15,000 + 0.35(1000)	15,350	(1000, 15350)
5000			
6000			
10,000			
s	Generalize	$T(s)$ =	

(d) On a separate piece of graph paper, draw a Cartesian plane. Label the horizontal axis with the independent variable and the vertical axis with the dependent variable. Use Xscl 500 and Yscl 2500. Plot the points. Connect the dots.

(e) Draw a vertical line on the graph at $s = 7500$. Estimate the second coordinate value of the point where the vertical line intersects the graph. Write the coordinates of the point of intersection.

(f) What is an estimate of Mike's total salary if his dollar amount of sales is 7500?

(g) Find the exact value of Mike's total salary if his dollar amount of sales is 7500. Evaluate $T(7500)$.

(h) Find the exact value of Mike's total salary if his dollar amount of sales is 9500. Evaluate $T(9500)$.

(i) Calculate the dollar amount of sales Mike would need to make a total salary of $17,870. Substitute $17,870 for $T(s)$ and solve for s.

(j) Calculate the dollar amount of sales Mike would need to make a total salary of $20,250. Substitute $20,250 for $T(s)$ and solve for s.

30. You are going on vacation and plan to rent a car. There is a fixed rental fee of $25 to rent the car, plus a charge of $0.10 for each mile driven. What is the cost to rent the car?

(a) Write the word or words that describe the input or independent variable.

(b) Write the word or words that describe the output or dependent variable.

(c) Complete the table.

m = Number of miles driven	Computations	to rent the car
100	$25 + 0.10(100)	
200		
500		
700		
m	Generalize	$C(m)$ =

(d) On a separate piece of graph paper, draw a Cartesian plane. Label the horizontal axis with the independent variable and the vertical axis with the dependent variable. Use Xscl 100 and Yscl 15. Plot the points. Connect the dots.

(e) Draw a vertical line on the graph at $m = 300$. Estimate the second coordinate value of the point where the vertical line intersects the graph. Write the coordinates of the point of intersection.

(f) What is an estimate of the cost to rent the car when the number of miles driven is 300?

(g) Find the exact cost to rent the car if the number of miles driven is 300. Evaluate $C(300)$.

(h) Find the exact cost to rent the car if the number of miles driven is 1000. Evaluate $C(1000)$.

(i) Calculate the exact number of miles driven if the cost to rent the car is \$135. Substitute \$135 for $C(m)$ and solve for m.

(j) Calculate the exact number of miles driven if the cost to rent the car is \$200. Substitute \$200 for $C(m)$ and solve for m.

31. Bud is the Quality Control Manager at a local machine shop. One of his jobs is to determine how close the weight of a manufactured machine part is to 5 pounds. He uses a standard 5-pound weight to calibrate (or measure) the weights of the parts. The manufactured machine part is weighed and compared to the desired weight. Bud and his staff have weighed six parts and recorded the weights in the table.

Standard 5-pound weight	Weights of manufactured machine parts
5 lb	5.0 lb
5 lb	5.1 lb
5 lb	4.9 lb
5 lb	4.8 lb
5 lb	5.2 lb
5 lb	4.9 lb

(a) Graph the sets of ordered pairs on the grid.

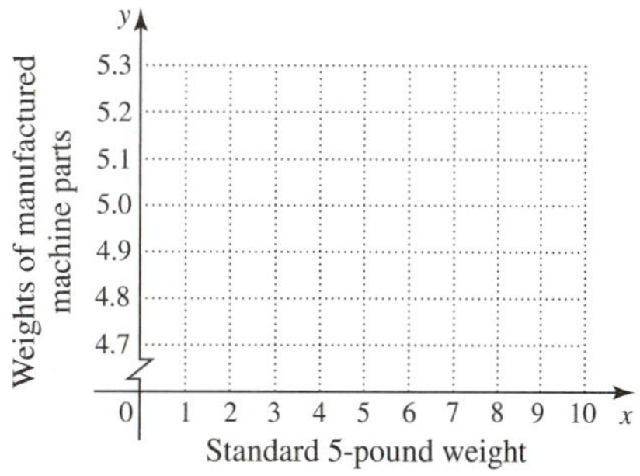

(b) Do the sets of ordered pairs create a relation that is also a function? Explain your answer.

(c) The specifications for the manufacturing process of the machine part state that the weight of any acceptable part must be greater than or equal to 4.9 and less than or equal to 5.1 pounds. Which machine parts are acceptable?

32. Eddie is the Quality Control Manager at a local machine shop. One of his jobs is to determine how close the length of a manufactured machine part is to 4 inches. He uses a 4-inch measuring rod to calibrate (or measure) the lengths of the parts. The manufactured machine part is measured and compared to the desired length. Eddie and his staff have measured six parts and recorded the lengths in the table.

Standard 4-inch measuring rod	Lengths of manufactured machine parts
4 inches	3.7 inches
4 inches	4.1 inches
4 inches	4.0 inches
4 inches	3.9 inches
4 inches	4.2 inches
4 inches	3.8 inches

(a) Graph the sets of ordered pairs on the grid.

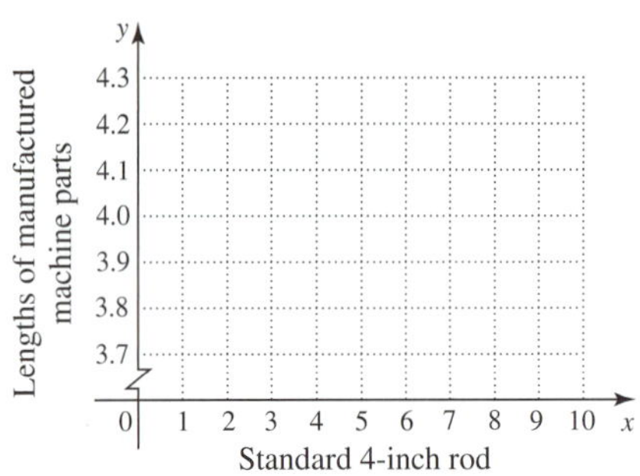

(b) Do the sets of ordered pairs create a relation that is also a function? Explain your answer.

(c) The specifications for the manufacturing process of the machine part state that the length of any acceptable part must be greater than or equal to 3.9 and less than or equal to 4.1 inches. Which machine parts are acceptable?

33. Create your own real-world function. Describe the independent variable, the dependent variable, and the process. Draw an Evaluating Machine for your function and show five ordered pairs.

34. Use the Internet or any other source to find a set of data points for a topic you are interested in (temperatures in a city, levels of snowfall, records of sporting events).

(a) Determine whether the data set represents a function.

(b) Graph the ordered pairs. Label the axes and the graph.

(c) Can you describe any pattern that occurs in the graph?

3.3 Linear Relations and Functions, Graphing and Intercepts

OBJECTIVES

- Define a linear equation.
- Identify and compute the x- and y-intercepts of the graph of a linear function.
- Graph linear relations and functions using the intercept method.
- Identify the zero of a linear relation or function.
- Write the function form of a horizontal line and the general form of a vertical line.

1 2 3 Getting Started

In this section, linear relations and functions will be considered. Notice that the word "linear" includes the word "line." The graph of a linear equation is a line. Two important points on a graph are the x- and y-intercepts. We begin by revisiting the video profit function and its graph to look for the intercepts.

1. Video profit problem: You own a company that manufactures videos with fixed costs for operating expenses of \$1000. Each video will sell for \$8. The problem is represented by the profit function, $P(x) = 8x - 1000$, and the corresponding graph.

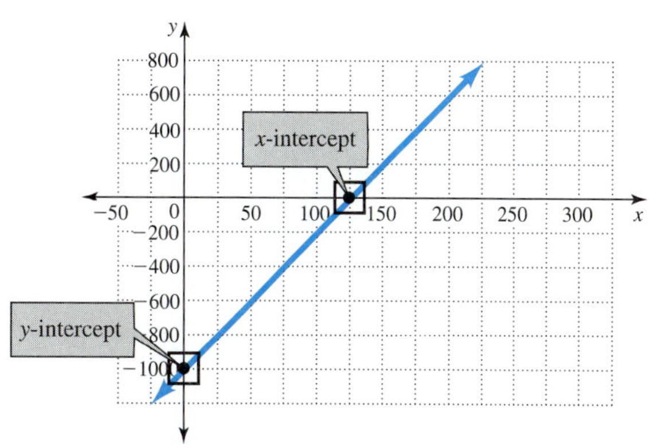

(a) The point where a graph crosses the x-axis or the horizontal axis is called the x-**intercept**. Write the coordinates of the x-intercept of the graph.

(b) The second coordinate value of every x-intercept is zero: $(x, 0)$. Why?

(c) The point where a graph crosses the y-axis or the vertical axis is called the y-**intercept**. Write the coordinates of the y-intercept of the graph.

(d) The first coordinate value of every y-intercept is zero: $(0, y)$. Why?

2. Let's graph the function $y = x - 1$ and identify the x- and y-intercepts.

(a) Substitute each x value into the equation and find the corresponding y value. Record the values in the table.

(b) Plot the points. Connect the dots. Extend the line and draw arrows on each end of the line.

x	y
−2	
−1	
0	
1	
2	
3	

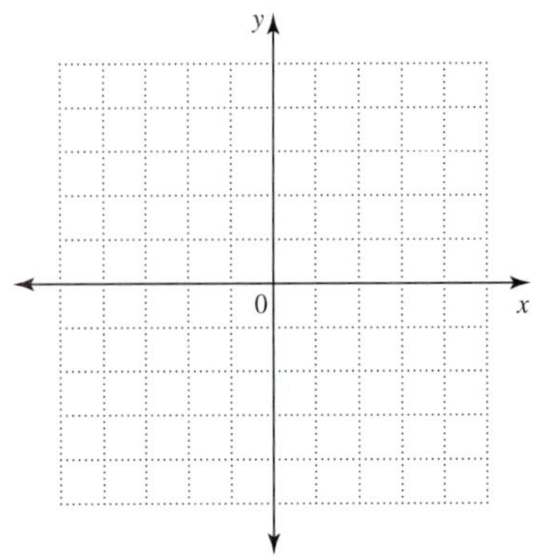

(c) Write the coordinates of the x-intercept.

(d) Write the coordinates of the y-intercept.

3. Linear relations may form horizontal or vertical lines. Beginning on February 1 of last year, the average temperature in Nice Spot, USA was 80 degrees for 10 days in a row.

(a) Complete the table of values. The day of the month represents the x-coordinate and the average temperature represents the y-coordinate.

(b) Graph the ordered pairs.

Date	x = Day of the month	y = Average high temperature
Feb 1	1	
Feb 2	2	
Feb 3	3	
Feb 4	4	
Feb 5	5	
Feb 6	6	
Feb 7	7	
Feb 8	8	
Feb 9	9	
Feb 10	10	

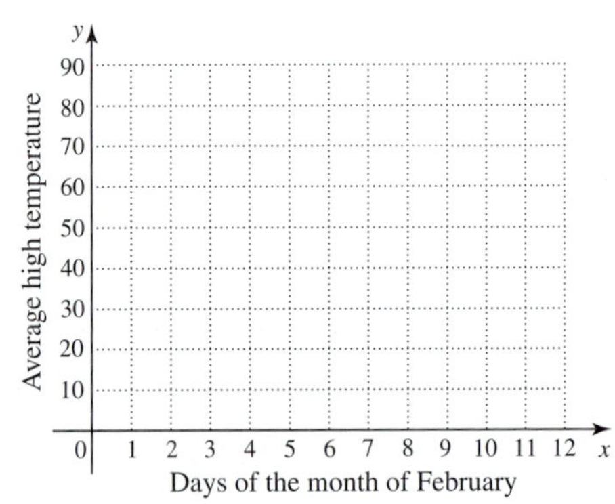

➢ *The points lie on a horizontal line composed of distinct points that are not connected.*

(c) Does the graph of the relation represent a function? Explain your answer.

4. Let's graph another problem. You are going to carpet your family room. You will choose from eight different carpets. Each of the eight carpets has a different total cost, as listed in the table. The area of the family room represents the x-coordinates and the cost of the carpets represents the y-coordinates.

(a) Graph the ordered pairs.

x = Area of the family room	y = Cost of carpet
300	300
300	375
300	450
300	550
300	600
300	675
300	750
300	800

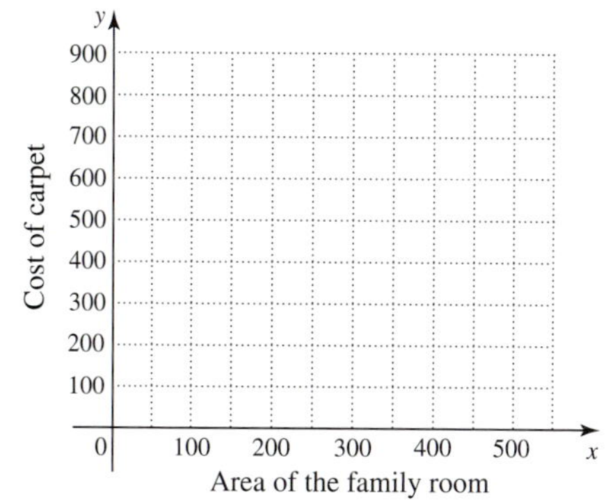

➢ *The points lie on a vertical line composed of distinct points that are not connected.*

(b) Does the graph represent a function? Explain your answer.

5. Describe a line that does not have an x-intercept.

6. Describe a line that does not have a y-intercept.

1 2 3

Discussing Mathematics

LINEAR FUNCTIONS

The graph of a first-degree equation in two variables is a horizontal line, a vertical line, or a line with a slant or slope. A horizontal line or a line that slants up or down is a function. Vertical lines are relations, not functions. When a linear relation is a function, the dependent variable, y, may be replaced with $f(x)$.

LINEAR RELATIONS AND FUNCTIONS

- Linear relations are of the form $y = ax + b$ where x is the independent variable, y is the dependent variable, and a and b are constants.
- A linear function is written in the form $f(x) = ax + b$.

Linear functions may use variables other than x and/or y. In the video profit equation, $P(x)$ replaces $f(x)$ in the linear function where $a = 8$ and $b = -1000$:

$$f(x) = ax + b$$

$$P(x) = 8x - 1000$$

In the formula for converting Celsius temperature to Fahrenheit temperature is a linear function, $F(C)$ replaces $f(x)$, C replaces x, $a = 1.8$, and $b = 32$:

$$f(x) = ax + b$$

$$F(C) = 1.8C + 32$$

GRAPHING USING THE INTERCEPT METHOD

Graphs of many linear relations intersect one or both of the axes.

INTERCEPTS OF A LINEAR FUNCTION $f(x) = ax + b$

- The point where a graph crosses the vertical axis (the y-axis) is called the y-intercept. The first coordinate (or x value) of every y-intercept is zero.
- The point where a graph crosses the horizontal axis (the x-axis) is called the x-intercept. The second coordinate (or y value) of every x-intercept is zero.

When $x = 0$ in the function form of a linear function $f(x) = ax + b$, a generalized expression for the y-intercept results:

$$f(x) = ax + b$$

Let $x = 0$: $$f(0) = a(0) + b$$
$$f(0) = b$$
$$y = b$$

The coordinates of the y-intercept of a linear function $f(x) = ax + b$ are $(0, b)$, where b denotes a constant.

The x-intercept of a relation or function is especially important and is given another name, **the zero of the relation** or **the zero of the function**. The x-intercept is called the "zero" because substituting the x-coordinate value of the x-intercept in a relation or function results in zero.

The graph of a linear function is uniquely defined by two points and can be constructed by plotting only the x- and y-intercepts. The process of graphing a line by plotting the intercepts is called the **Intercept Method** of graphing.

GRAPHING A FUNCTION $f(x) = ax + b$ USING THE INTERCEPT METHOD

1. Calculate the y-intercept (let $x = 0$ and solve for y).
2. Calculate the x-intercept (let $y = 0$ and solve for x).
3. Plot the x- and y-intercepts on a Cartesian plane.
4. Connect the points with a line.
5. Compute and plot the coordinates of two more points to verify the graph.

EXAMPLE 1: Graph the linear function $f(x) = 2x - 6$ using the Intercept Method.

$f(0) = 2(0) - 6$ — Compute the y-intercept: let $x = 0$ and solve for $f(0) = y$. The y-intercept has coordinates $(0, -6)$.

$f(0) = -6$

$0 = 2x - 6$ — Compute the x-intercept: let $y = 0$ and solve for x. The x-intercept has coordinates $(3, 0)$.

$3 = x$ — Plot the intercepts $(3, 0)$ and $(0, -6)$. Connect the dots with a line. Choose any two additional values for x, compute the corresponding y value. Record the ordered pairs in a table, along with the intercepts.

x	y
0	−6
3	0
Two additional points to verify the graph:	
1	−4
2	−2

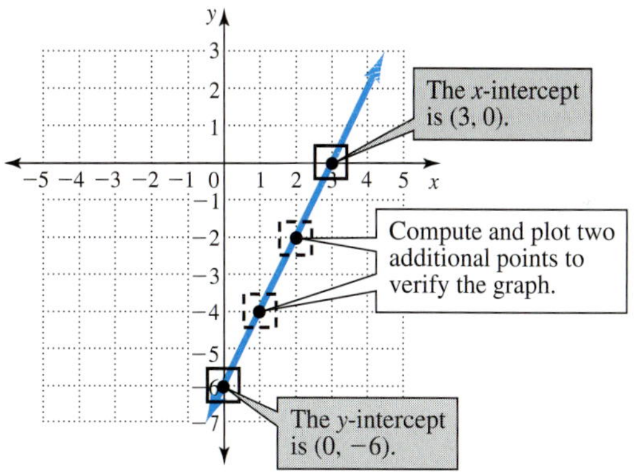

EXAMPLE 2: Graph the linear equation $2x + 3y = 6$ using the Intercept Method.

$2(0) + 3y = 6$ — Compute the y-intercept: let $x = 0$ and solve for y. The y-intercept has coordinates $(0, 2)$.

$y = 2$

$2x + 3(0) = 6$ — Compute the x-intercept: let $y = 0$ and solve for x. The x-intercept has coordinates $(3, 0)$.

$x = 3$ — Plot the intercepts $(3, 0)$ and $(0, 2)$. Connect the dots with a line. Choose any two additional values for x, compute the corresponding y value. Record the ordered pairs in a table, along with the intercepts.

x	y
0	2
3	0
Two additional points to verify the graph:	
-1	$\frac{8}{3}$
1	$\frac{4}{3}$

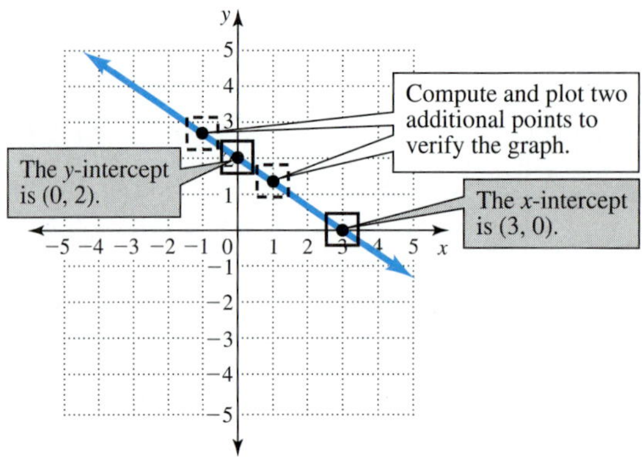

HORIZONTAL LINES

Constant functions are relations where the dependent variable is constant and the ordered pairs lie on a horizontal line. The graph of a constant function is a continuous horizontal line. The numerical coefficient of the independent variable is zero. In general, the equation of a constant function is of the form:

$$f(x) = ax + b$$

When $a = 0$: $f(x) = 0 \cdot x + b.$

$$f(x) = b, \text{ where } b \text{ is a constant.}$$

FUNCTION FORM OF A HORIZONTAL LINE

$f(x) = b$, where b is a real number constant.

$f(x) = b$ is equivalent to $y = b$.

➢ *To plot a horizontal line, plot point $(0, b)$ and draw a horizontal line through that point.*

EXAMPLE 3: Graph $f(x) = 4$ (Also written $y = 4$.)

Plot the point $(0, b) = (0, 4)$ and draw a horizontal line through that point.

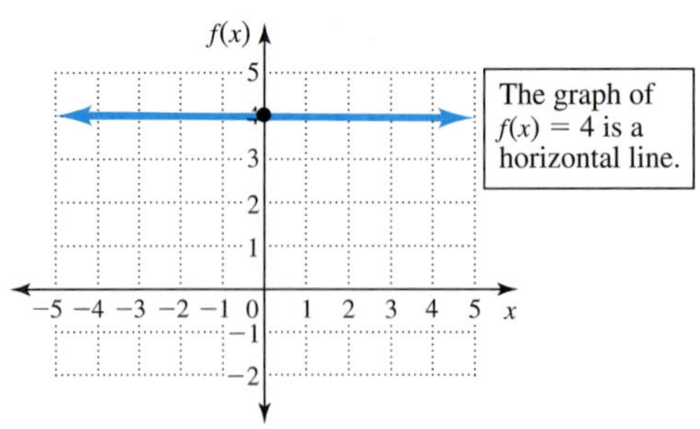

For every x value, the corresponding y is 4. There is no x-intercept for a horizontal line (except when $b = 0$).

VERTICAL LINES

Vertical lines are relations, not functions, and never written with function notation. The value of the independent variable is a constant. The numerical coefficient of the dependent variable is zero. In general, the equation of a horizontal line is of the form:

GENERAL FORM OF A VERTICAL LINE

$x = k$, where k is a constant.

➢ *To plot a vertical line, plot point $(k, 0)$ and draw a vertical line through that point.*

EXAMPLE 4: Graph $x = 3$.

Plot the point $(k, 0) = (3, 0)$ and draw a horizontal line through that point.

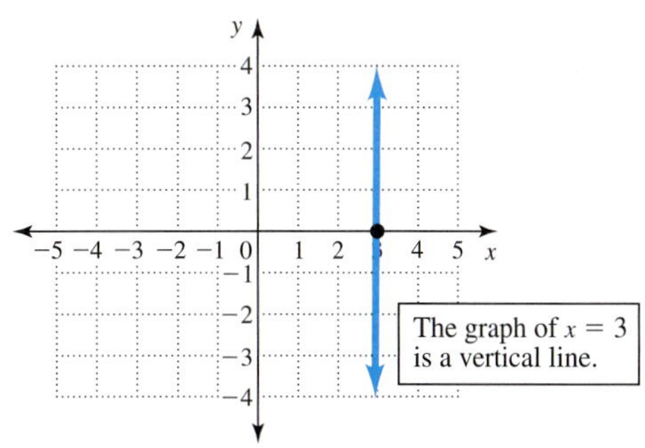

For every y value, the corresponding x-value is 3. There is no y-intercept for a vertical line (except when $k = 0$).

1 2 3 Doing Mathematics

- Write the definitions of the words and symbols listed for this section in the Dictionary in Appendix A.
- On a 3 × 5-inch card, write the following facts or processes for future reference:
  - (a) The general form of a linear relation and function
  - (b) Graphing a line using the intercept method
  - (c) Graphing a horizontal line
  - (d) Graphing a vertical line

Problems 1–2: Circle the statements that are linear functions.

1. (a) $f(x) = 3x + 1$ (b) $f(x) = x^2 + 2x + 6$

(c) $y = 7 - 3x$ (d) $x = 4$

2. (a) $y = -10$ (b) $f(x) = x^3 - 4x^2 + x - 1$

(c) $x = -3$ (d) $y = -4x + 6$

3. Given the graph of $y(x) = 2x + 4$,

(a) write the coordinates of the x-intercept.

(b) write the coordinates of the y-intercept.

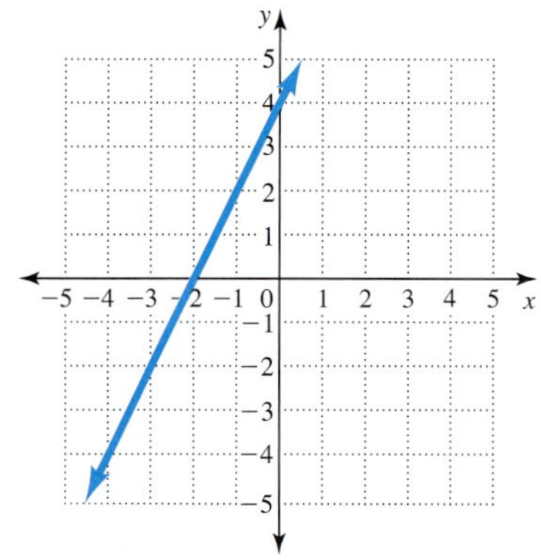

4. Given the graph of $f(x) = -3x - 9$,

(a) write the coordinates of the x-intercept.

(b) write the coordinates of the y-intercept.

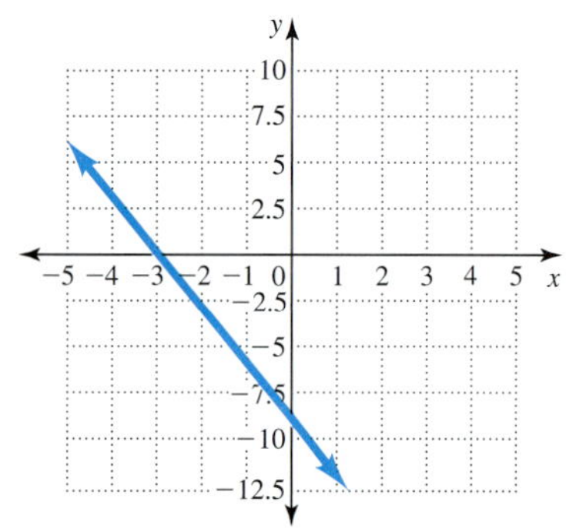

5. Given the graph of $f(x) = 5$,

(a) write the coordinates of the x-intercept.

(b) write the coordinates of the y-intercept.

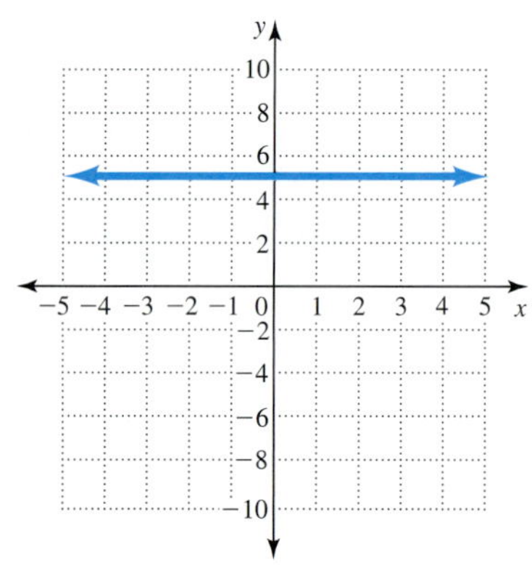

Problems 6–15: Graph each function using the Intercept Method.

6. Graph $f(x) = 3x + 2$.

x	$f(x)$
0	
	0
Two additional points to verify the graph:	

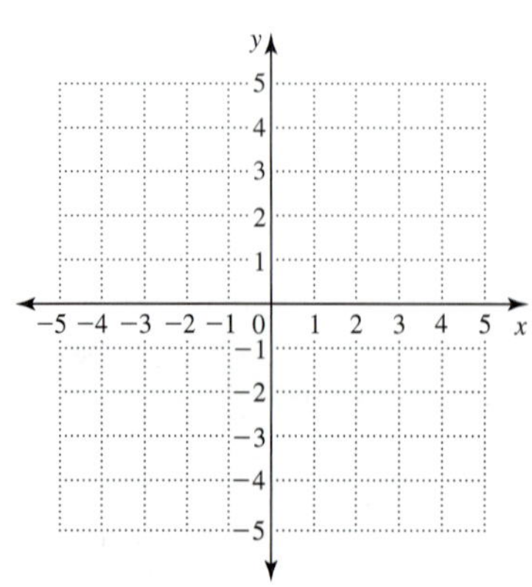

7. Graph $f(x) = 2x - 1$.

x	$f(x)$
0	
	0
Two additional points to verify the graph:	

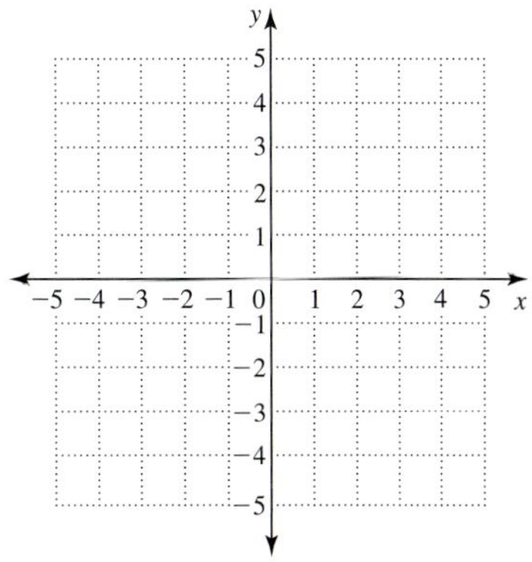

8. Graph $f(x) = -x - 3$.

x	$f(x)$
0	
	0
Two additional points to verify the graph:	

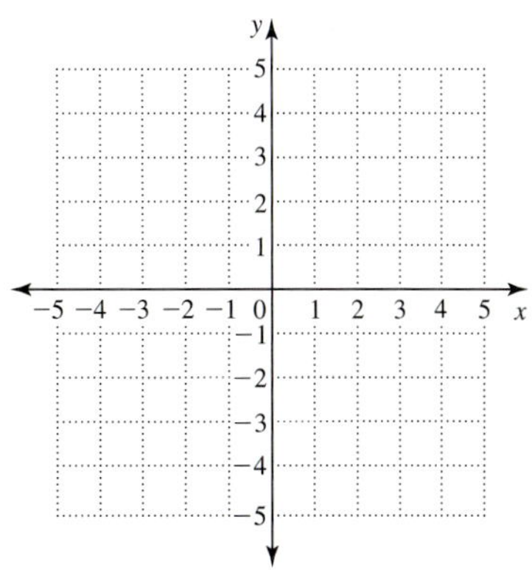

9. Graph $f(x) = -x + 4$.

x	$f(x)$
0	
	0
Two additional points to verify the graph:	

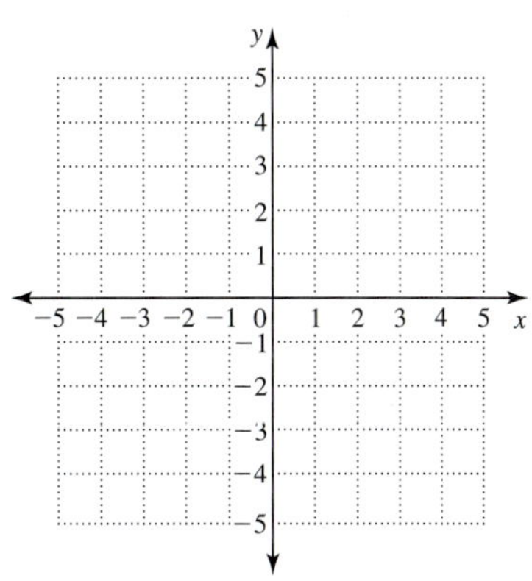

10. Graph $4x - 3y = 12.$

x	y
0	
	0
Two additional points to verify the graph:	

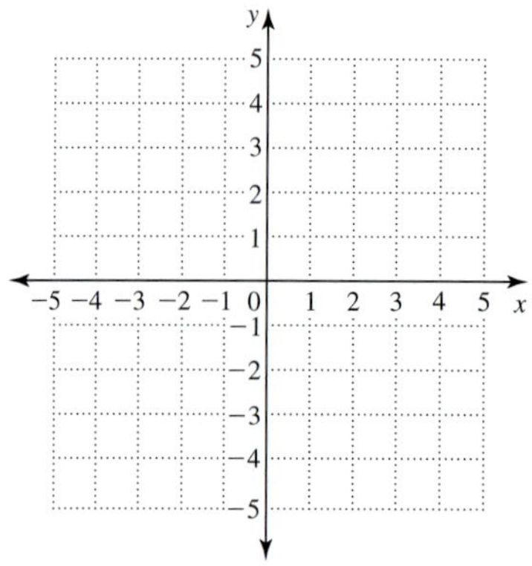

11. Graph $2x - 3y = 6.$

x	y
0	
	0
Two additional points to verify the graph:	

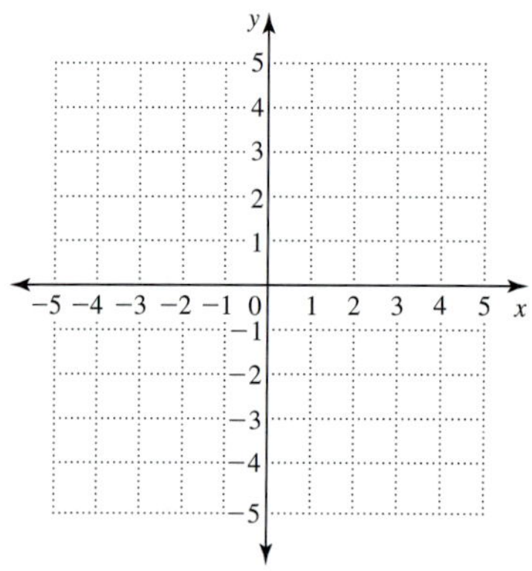

12. Graph $y = 2x - 8.$

x	y
0	
	0
Two additional points to verify the graph:	

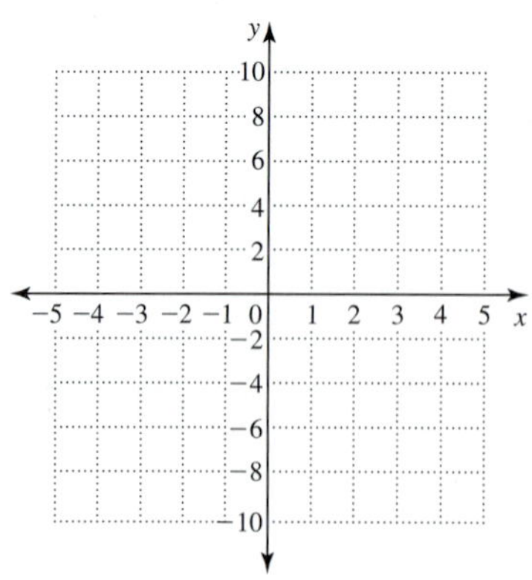

13. Graph $y = -2x + 2$.

x	y
0	
	0
Two additional points to verify the graph:	

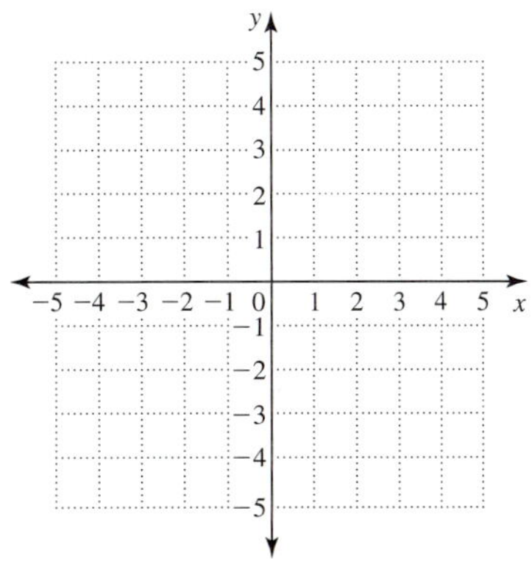

14. Graph $y = \frac{1}{2}x - 1$.

x	y
0	
	0
Two additional points to verify the graph:	

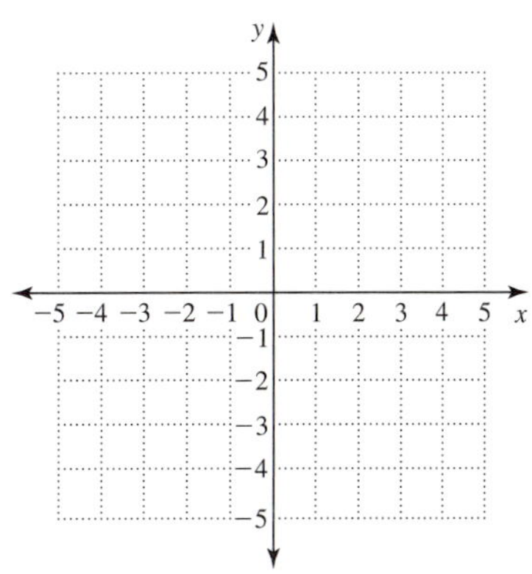

15. Graph $y = -\frac{3}{4}x$.

x	y
0	
	0
Two additional points to verify the graph:	

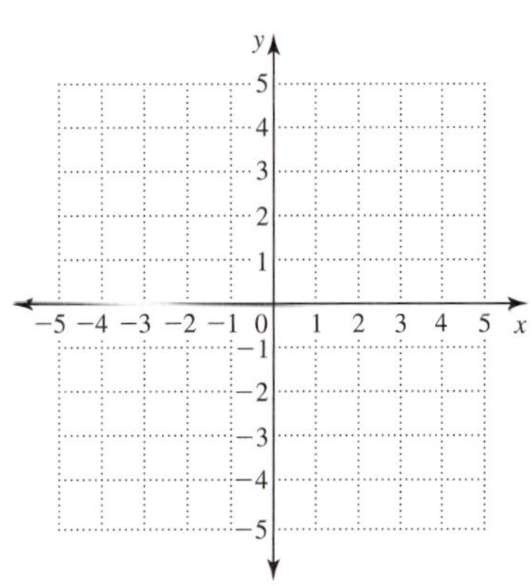

Problems 16–19: Graph each horizontal or vertical line.

(a) Does the graph represent a function? Explain your answer.

(b) If the relation is a function, rewrite the equation in function notation.

16. Graph $y = 2$.

x	y
−3	
−2	
−1	
0	
1	
2	
3	

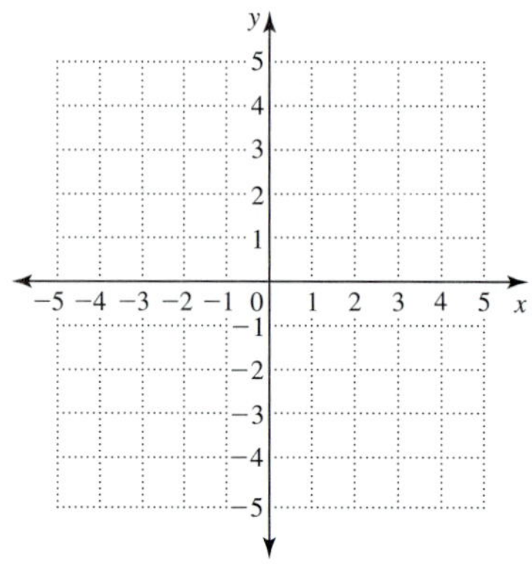

17. Graph $y = -1$.

x	y
−3	
−2	
−1	
0	
1	
2	
3	

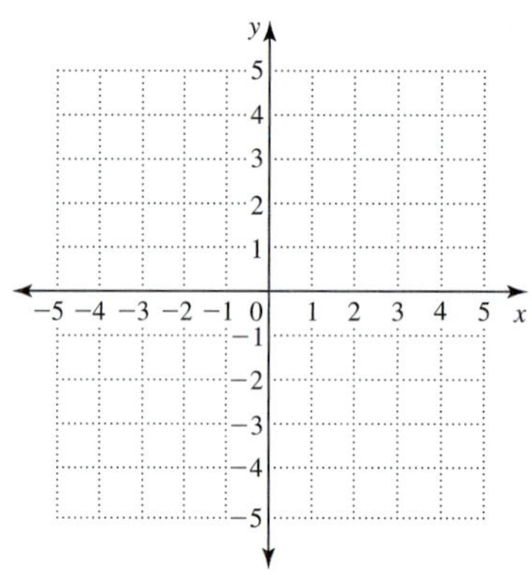

18. Graph $x = -3$.

x	y
	−3
	−2
	−1
	0
	1
	2
	3

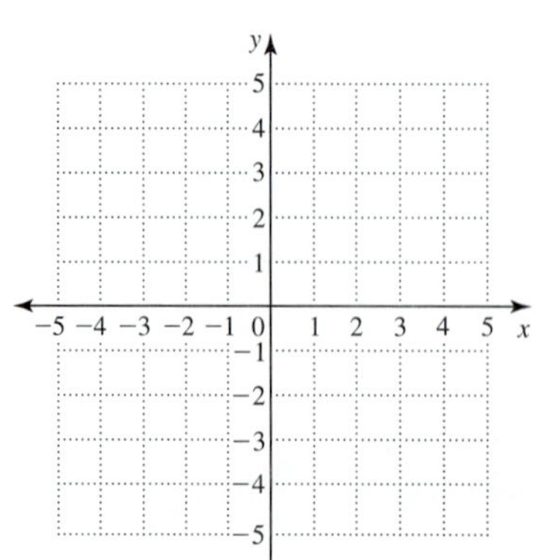

19. Graph $x = 0$.

x	y
	-3
	-2
	-1
	0
	1
	2
	3

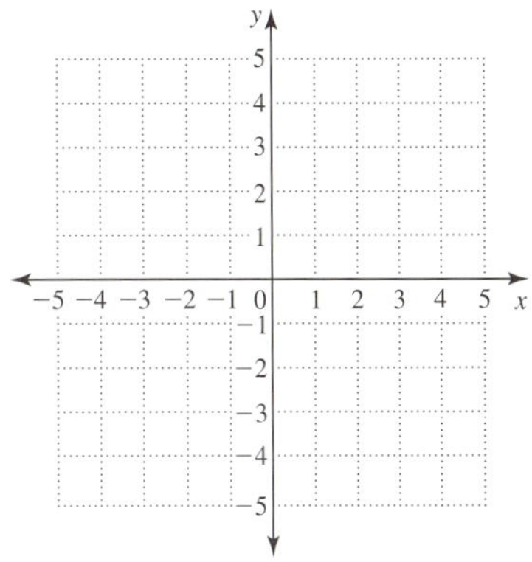

20. Ann owns a furniture store. Her average revenue per month is \$200 for each customer. Her fixed operating expenses per month are \$1500.

(a) Complete the table and generalize the profit function. Let n represent the number of purchasing customers in a given month. Let P represent monthly profit. Hint: Profit = Revenue − Operating expenses.

n = Number of purchasing customers in a given month	Computations	P = Monthly profit (in dollars)
0		
10		
15		
20		
25		
n		$P(n) =$

(b) Determine a scale, label the axes, and graph the ordered pairs on a separate piece of graph paper.

(c) How many customers need to make a purchase for Ann to break even?

(d) Use the graph to estimate the profit when 12 customers make a purchase in a given month.

(e) Calculate the exact profit when 12 customers make a purchase in a given month. Evaluate $P(12)$.

21. Joe is a tennis pro and charges clients \$40 for an hour-long tennis lesson. He must pay the tennis club \$15 per hour to use their tennis courts for lessons.

(a) Complete the table and generalize the salary function. Let n represent the number of lessons per week. Let S represent Joe's salary for one week.

n = Number of lessons per week	Computations	S = Salary per week (in dollars)
0		
5		
8		
10		
12		
n		$S(n) =$

(b) Determine a scale, label the axes, and graph the ordered pairs on a separate piece of graph paper.

(c) How many tennis lessons must Joe teach in one week to break even?

(d) What is Joe's salary for the week if he teaches 20 lessons?

(e) Use the graph to estimate the number of lessons he needs to teach in one week to have a salary of $275 after expenses.

(f) Calculate the exact number of lessons he needs to teach in one week to have a salary of $750 after expenses.

22. The U.S. Census Office has collected data on the number of persons living alone for all age groups. The table presents the number of persons ages 30–34 living alone.[1] The equation of the line approximating this data set is

$$N(t) = -0.033t + 2.07.$$

Year	t = Number of years since 1995	N = Number of persons living alone (ages 30–34, in millions)
1995	0	2.07
1997	2	2.00
1998	3	1.97
	t	$N(t) = -0.033t + 2.07$

(a) Determine a scale, label the axes, and graph the ordered pairs on a separate piece of graph paper.

(b) Use the graph to estimate the number of persons living alone who are 30–34 years of age in 1999.

(c) Use the function to calculate the exact number of persons living alone who are 30–34 years of age in 2005.

Problems 23–24: Writing Exercises

23. Explain why the graph of a horizontal line is a function.

24. Explain why the graph of a vertical line is not a function.

[1] U.S. Census Bureau.

3.4 Rate of Change, Slope-Intercept Form

OBJECTIVES

- Find the slope of a line.
- Write the slope of an increasing, decreasing, horizontal, or vertical line.
- Graph a line using the y-intercept and slope.
- Find the equation of a line given the y-intercept and slope.

1 2 3

Getting Started

In this section, we explore the slope of a line and its relationship to the corresponding equation of the line. The slope describes the "slant," steepness of a line, or angle of inclination of a line. The **slope** is the ratio of the vertical displacement to the horizontal displacement. We begin our discussion of slope by comparing the "vertical drop" of ski slopes in Colorado. Vertical drop is measured from the top of the ski slope with a vertical line, perpendicular to the horizon.

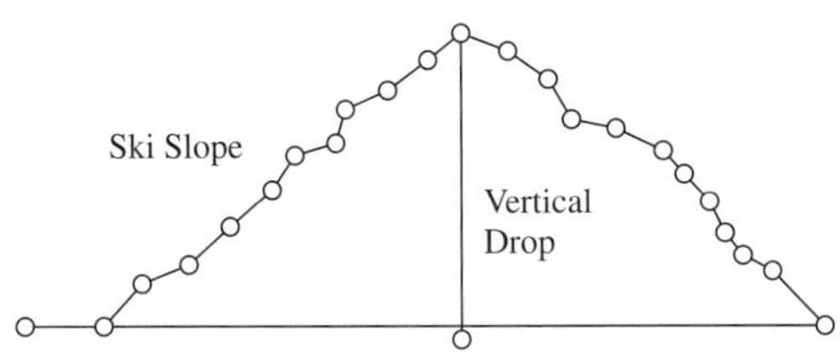

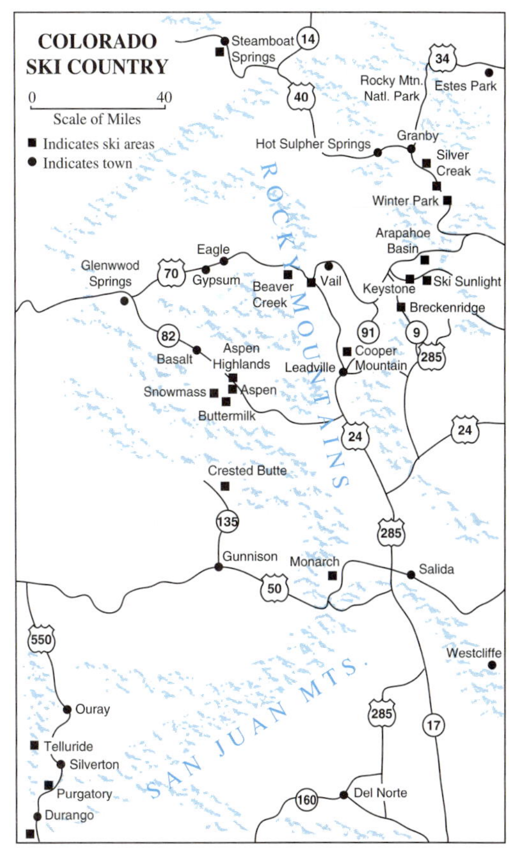

1. The table lists the vertical drop of the longest ski runs in Colorado.[1]

Name of ski resort	Vertical drop
Aspen	3267 ft
Beaver Creek	3340 ft
Breckenridge	2583 ft
Copper Mountain	2760 ft
Snowmass	3596 ft
Steamboat Springs	3600 ft
Telluride	3155 ft
Vail	3150 ft
Winter Park	2200 ft

(a) Which ski resort would you choose if you were an expert skier? Why?

(b) Which ski resort would you choose if you were a beginner skier? Why?

[1] *Fodor's 89: Ski Resorts of North America*, Fodor's Travel Publications, Inc., 1989.

Vertical drop is only one way to compare ski resorts. Assume a ski trail is a straight line from the top of the mountain to the base of the mountain. A sketch of the slopes at Steamboat Springs and Winter Park Ski Resorts is shown below. Measuring as a straight line, the longest ski run at Steamboat is 3 miles or 15,840 feet. The longest run at Winter Park is 3.9 miles or 20,592 feet. The Pythagorean Theorem is used to calculate the horizontal distance of each ski slope as the base (b) of each right triangle.

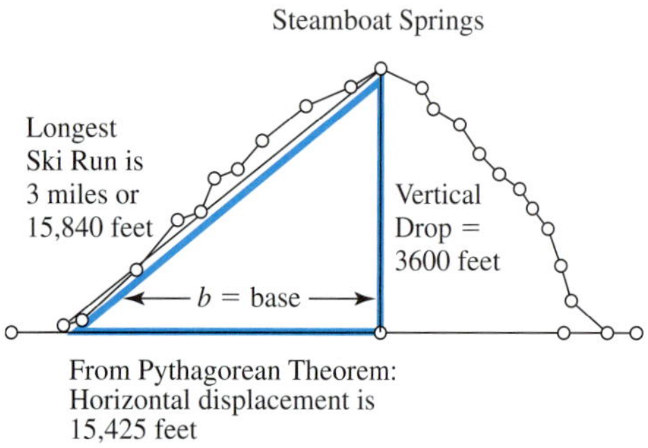

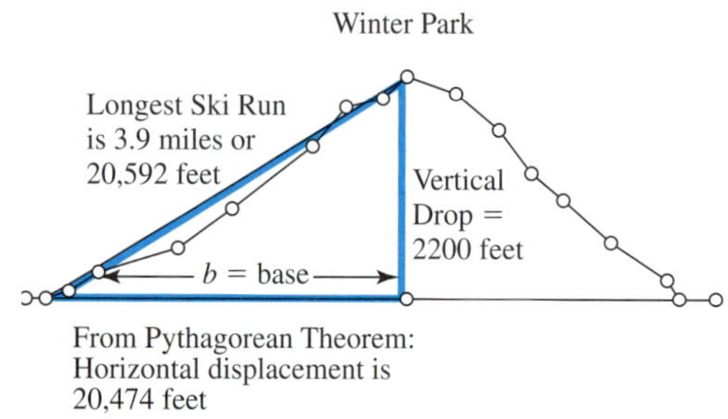

2. A **ratio** communicates a relationship between the two quantities as a fraction. The slope or steepness of a ski run can be described with a ratio of the vertical drop to the horizontal distance: Slope $= \dfrac{\text{Vertical Drop in feet}}{\text{Horizontal Distance in feet}}$.

(a) Write the ratio of the vertical drop to the horizontal distance in the longest ski run at Steamboat Springs as a fraction and as a decimal, rounding your answer to the nearest hundredth.

Ratio as a fraction — Ratio as a decimal

$$\frac{\text{Vertical Drop in feet}}{\text{Horizontal Distance in feet}} = \underline{\qquad\qquad} =$$

(b) Write the ratio of the vertical drop to the horizontal distance in the longest ski run at Winter Park as a fraction and as a decimal, rounding your answer to the nearest hundredth.

Ratio as a fraction — Ratio as a decimal

$$\frac{\text{Vertical Drop in feet}}{\text{Horizontal Distance in feet}} = \underline{\qquad\qquad} =$$

3. Compare the ratio of the longest ski run at Steamboat Springs to the ratio of the longest ski run at Winter Park in problem 2.

(a) Which ski slope is steeper? How do you know?

(b) Which ski slope should you choose if you are a beginner and would like the flattest slope?

1 2 3 Discussing Mathematics

SLOPE OF A LINE

The slope or steepness of a line or the rate of change of a linear relation can be represented by the ratio of the change in output values (y-coordinates) to the change in input values (x-coordinates). The slope of a line is always constant.

When computing the slope of a line between two points, the coordinates of the first point are labeled (x_1, y_1) and the coordinates of the second point are labeled (x_2, y_2). The difference of the y-coordinates is called the **vertical displacement** or RISE and is labeled RISE $= \Delta y = y_2 - y_1$ (or $y_1 - y_2$). The difference of the x-coordinates of two points is called the **horizontal displacement** or RUN and is labeled RUN $= \Delta x = x_2 - x_1$ (or $x_1 - x_2$).

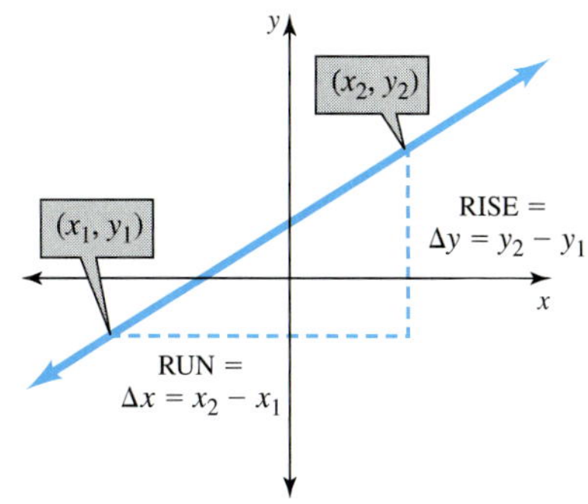

The **slope** of a line, denoted with the variable m, is the ratio of the vertical displacement to the horizontal displacement.

SLOPE FORMULA

The slope of a line between two points (x_1, y_1) and (x_2, y_2) on the line is

$$m = \frac{\text{Vertical Displacement}}{\text{Horizontal Displacement}} = \frac{\text{RISE}}{\text{RUN}} = \frac{\Delta y}{\Delta x} = \frac{y_2 - y_1}{x_2 - x_1}, \ (x_1 \neq x_2).$$

➢ *The slope of a line can be positive, negative, zero, or undefined.*

➢ *The ratios* $\frac{y_2 - y_1}{x_2 - x_1}$ *and* $\frac{y_1 - y_2}{x_1 - x_2}$ *are equivalent. You may subtract the coordinates in either order. However, you must subtract the x-coordinates in the denominator in the same order as the y-coordinates in the numerator.*

EXAMPLE 1: Find the slope of the line between the two points $(-2, -1)$ and $(3, 1)$.

Label the first point $(-2, -1)$ as (x_1, y_1).

Label the second point $(3, 1)$ as (x_2, y_2).

$$m = \frac{\text{RISE}}{\text{RUN}} = \frac{y_2 - y_1}{x_2 - x_1} = \frac{1 - (-1)}{3 - (-2)} = \frac{2}{5}$$

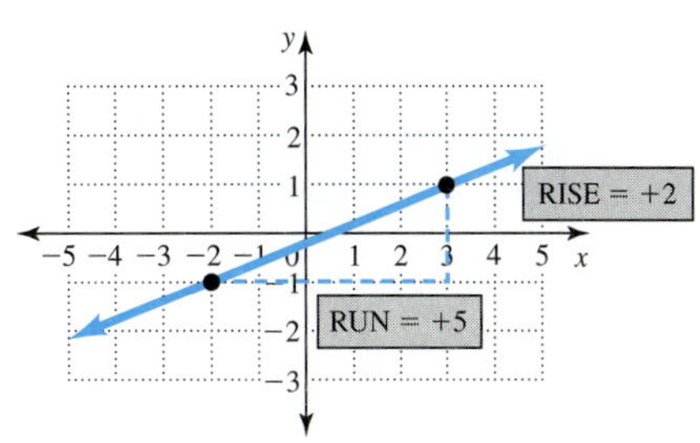

➢ *When the points are labeled in the reverse order, with* $(x_2, y_2) = (-2, -1)$ *and* $(x_1, y_1) = (3, 1)$, *the slope is the same:* $m = \frac{y_2 - y_1}{x_2 - x_1} = \frac{-1 - (1)}{-2 - (3)} = \frac{-2}{-5} = \frac{2}{5}$

The value of the slope for all linear equations of the form $y = ax + b$ is the coefficient of the independent variable x. The constant, b, is the y-intercept of the graph of the line. The equation is called the Slope-Intercept Form of a line when the variable a is replaced with m.

SLOPE-INTERCEPT FORM OF A LINE

$$y = mx + b \quad \text{or} \quad f(x) = mx + b$$

where m is the slope and b is the y-coordinate of the y-intercept.

EXAMPLE 2: Compare the slopes of $y = 2x + 1$ and $y = -2x + 1$.

Equation 1: $y = 2x + 1$

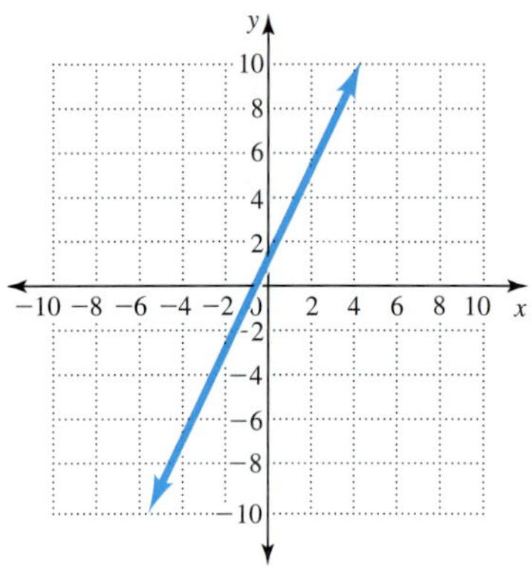

x	y_1
−4	−7
−3	−5
−2	−3
−1	−1
0	1
1	3
2	5

$m = 2$ and $b = 1$
The slope and the coefficient of x is $+2$.
The y-intercept is $(0, 1)$.

The graph of Equation 1 is a line that slants up. The line is increasing as it is viewed from left to right. **The slope of the increasing line is a positive number.**

Equation 2: $y = -2x + 1$

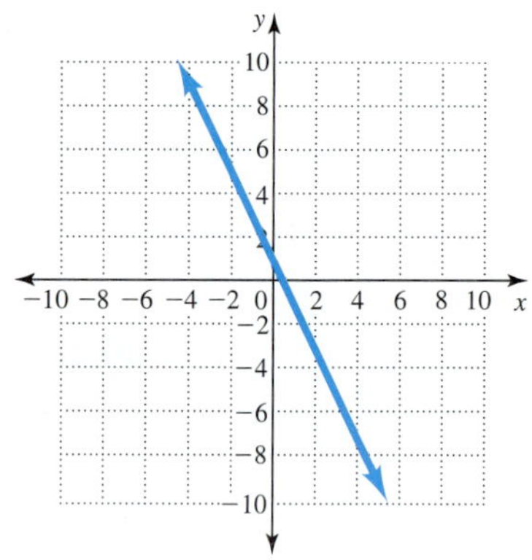

x	y_1
−4	9
−3	7
−2	5
−1	3
0	1
1	−1
2	−3

$m = -2$ and $b = 1$
The slope and the coefficient of x is -2.
The y-intercept is $(0, 1)$.

The graph of Equation 1 is a line that slants down. The line is decreasing as it is viewed from left to right. **The slope of the decreasing line is a negative number.**

EXAMPLE 3: Compare the slopes of the vertical line (Equation 1) and the horizontal line (Equation 2).

Equation 1: $x = 4$

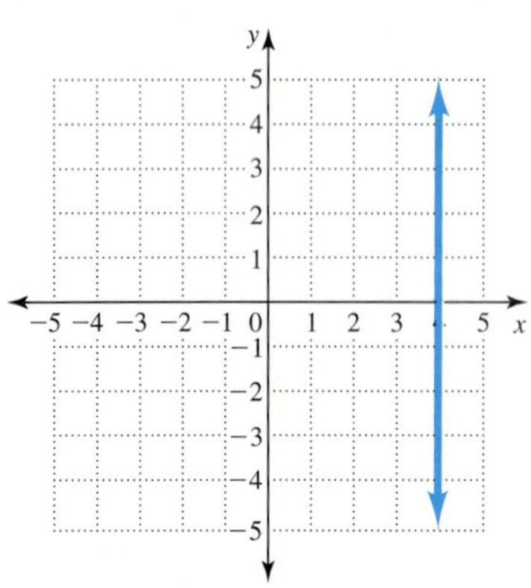

x	y
4	−2
4	−1
4	0
4	1
4	2

The graph of $x = 4$ is a vertical line.
This equation is not a function and the graph cannot be constructed on most graphing calculators.

Calculate the slope, using the points

$$(x_1, y_1) = (4, 1) \text{ and}$$
$$(x_2, y_2) = (4, -2).$$

$$m = \frac{y_2 - y_1}{x_2 - x_1} = \frac{-2 - 1}{4 - 4} = \frac{-3}{0}$$

$$m = \text{undefined}$$

The y-intercept does not exist.

The x-intercept is $(4, 0)$.

The denominator of the slope of a vertical line is always zero because the x-coordinates of all points are the same.

The slope of any vertical line is undefined.

Equation 2: $y = 3$

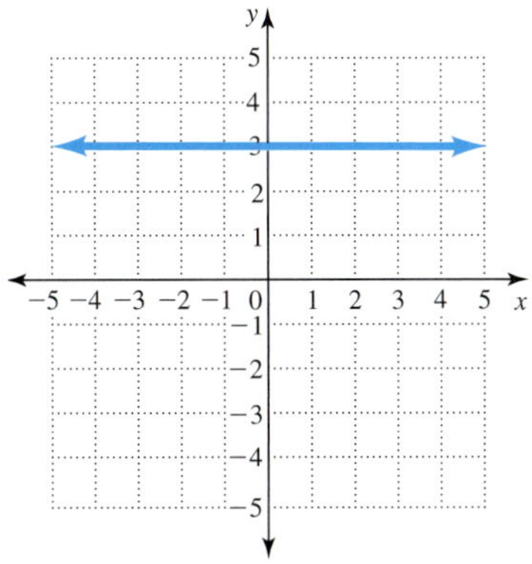

x	y_1
−2	3
−1	3
0	3
1	3
2	3
3	3
4	3

The graph of $y = 3$ is a horizontal line.

Calculate the slope, using the points

$$(x_1, y_1) = (-2, 3) \text{ and}$$
$$(x_2, y_2) = (0, 3).$$

$$m = \frac{y_2 - y_1}{x_2 - x_1} = \frac{3 - 3}{0 - (-2)} = \frac{0}{2} = 0$$

$$m = 0$$

An x-intercept does not exist.

The y-intercept is (0, 3).

The numerator of the slope of a horizontal line is always equal zero because the y-coordinates of all points are the same.

The slope of a horizontal line is 0.

Here is a summary of the behaviors and corresponding slope of increasing and decreasing lines:

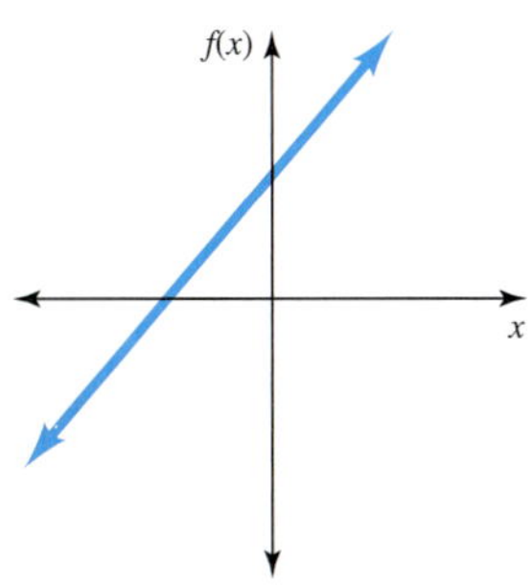

The slope of an **increasing line** is a positive number: $m > 0$. The larger the absolute value of the slope, the steeper the line.

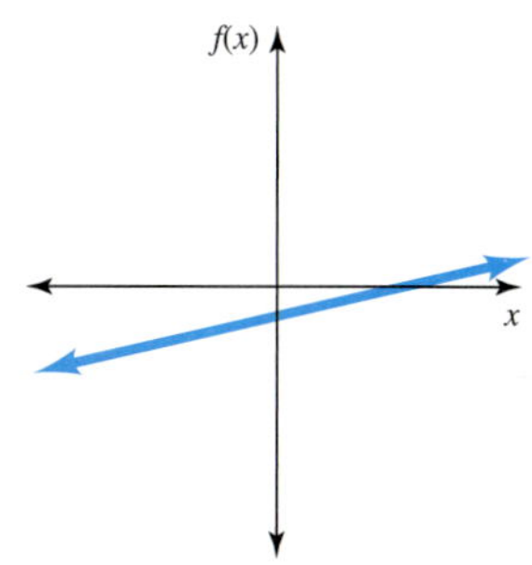

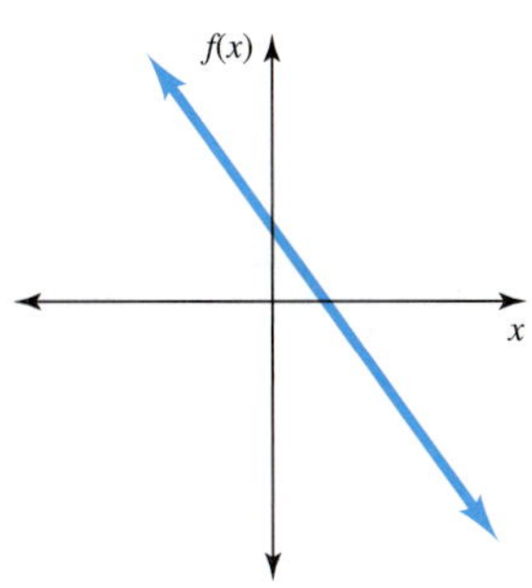

The slope of a **decreasing line** is a negative number: $m < 0$. The larger the absolute value of the slope, the steeper the line.

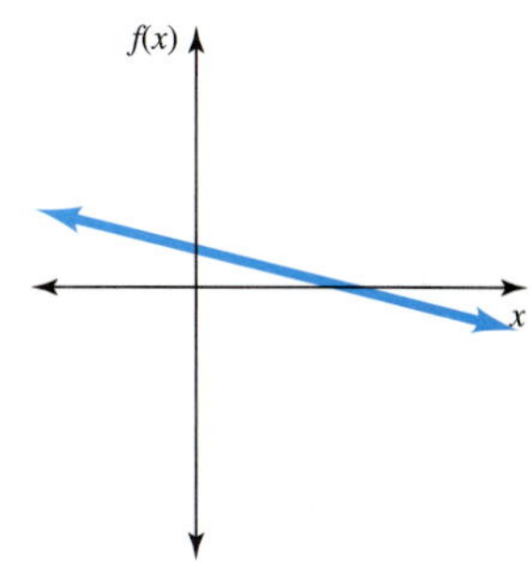

The slope of a **vertical line** $x = k$ is undefined. k is the x-coordinate of the x-intercept.

The slope of a **horizontal line** $y = b$ is zero. b is the y-coordinate of the y-intercept.

GRAPHING A LINE USING THE SLOPE-INTERCEPT METHOD

The Slope-Intercept Method for graphing a line, begins by plotting the y-intercept on a grid. Then move vertically and horizontally from the y-intercept according to the RISE and RUN specified in the slope.

GRAPHING A LINE USING THE SLOPE-INTERCEPT METHOD

1. Identify the y-intercept and the RISE and the RUN from the slope. Place your pencil at the y-intercept and count the appropriate number of units for the RISE and the RUN. Draw a second point at that location.
2. Connect the y-intercept and the second plotted point with a line.
3. Compute and plot the coordinates of two other points to verify the graph.

EXAMPLE 4: Graph the function $y = \frac{-3}{2}x + 2$ using the Slope-Intercept Method.

1. Identify the slope and the y-intercept: $y = mx + b$

$$y = \frac{-3}{2}x + 2$$

Slope $= m = \frac{-3}{2}$ $\quad$ y-intercept $= b = 2$

2. Place your pencil at the y-intercept $(0, 2)$. Use the RISE and count 3 units down (the negative direction). Use the RUN and count 2 units right (the positive direction). Place a point at that location, $(2, -1)$.
3. Connect the y-intercept and the second point $(2, -1)$ with a line.

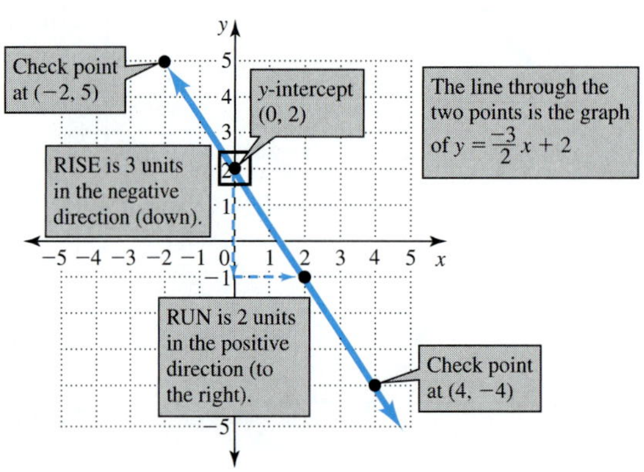

4. Compute and plot two other points to verify the graph.

x	y
−2	5
4	−4

FINDING THE EQUATION OF A LINE, GIVEN THE y-INTERCEPT AND THE SLOPE

The Slope-Intercept Form of a line is used to create the equation of a line when the slope and the y-intercept are given.

FINDING THE EQUATION OF A LINE, GIVEN THE y-INTERCEPT AND THE SLOPE

Given the y-intercept, $(0, b)$, and the slope, m, substitute the given values for m and b into the Slope-Intercept Form of a line: $y = mx + b$ or $f(x) = mx + b$

EXAMPLE 5: Find the equation of the line given the slope $= -\frac{1}{2}$ and y-intercept $= (0, 9)$.

$$y = mx + b$$
$$y = -\frac{1}{2}x + 9$$

Substitute $m = -\frac{1}{2}$ and $b = 9$ into the Slope-Intercept Form of a line.

REAL-WORLD PROBLEMS

EXAMPLE 6: Tom owns a business that manufactures and sells notebooks. The table shows the number of notebooks sold and the profit earned. Tom would like to create an equation to predict the profit when 5000 notebooks are sold.

x = Number of notebooks sold	$P(x)$ = Profit (in dollars)
0	−800
100	−300
500	1700
1000	4200
5000	?

Tom graphed the points.

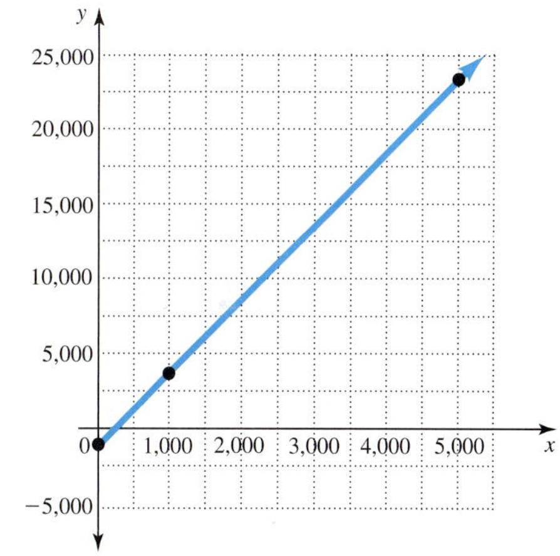

The points appeared to be linear. He used the Slope Formula and the Slope-Intercept Form of a line to find the profit equation, and then the profit for 5000 notebooks. Label $(0, -800) = (x_1, y_1)$ and $(500, 1700) = (x_2, y_2)$ and compute slope.

$$m = \frac{y_2 - y_1}{x_2 - x_1} = \frac{1700 - (-800)}{500 - 0} = \frac{2500}{500} = \frac{5}{1}$$

$$y = mx + b$$
$$P(x) = 5x - 800$$

The y-intercept (when $x = 0$) from the data points is $(0, -800)$.

Label the independent variable (x) and the dependent variable $(P(x))$. Substitute $m = 5$ and $b = -800$ in the Slope-Intercept Form of the line.

$$P(5000) = 5(5000) - 800$$
$$= 25{,}000 - 800 = 24{,}200$$

To compute the profit when 5000 notebooks are sold, evaluate the profit function when $x = 5000$, find $P(5000)$.

When 5000 notebooks are sold, the profit is $24,200.

Doing Mathematics

- Write the definitions of the words and symbols listed for this section in the Dictionary in Appendix A.
- On a 3 × 5-inch card, write the following formulas or concepts for future reference:
  (a) The slope formula
  (b) The Slope-Intercept form of a line
  (c) How to graph a line using the Slope-Intercept Method
  (d) The behavior of a line relative to its slope (increasing, decreasing, horizontal, vertical)

1. The vertical drops of some ski slopes in Vermont[2] are listed in the table. Circle the three ski resorts most likely to have the steepest ski runs, based only on vertical drop.

Name of ski resort	Vertical drop in feet	Name of ski resort	Vertical drop in feet
Ascutney	1530	Mount Snow	1200
Bolton Valley	1600	Okemo	1967
Bromley	1334	Pico	1967
Burke Mountain	2000	Stowe	2350
Haystack	1400	Stratton	2003
Jay Peak	2153	Suicide Six	650
Killington	3160	Sugarbush	2600

2. One of the longest ski runs at Killington is 3578 feet with a vertical drop of 3160 feet.

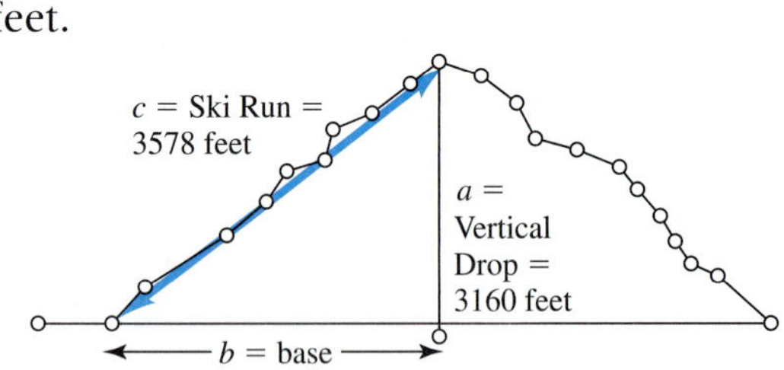

(a) Solve for the horizontal distance b, where $a = 3160$ and $c = 3578$. Round your answer to the nearest hundredth.

$$b = \sqrt{c^2 - a^2}$$

$$b^2 = \sqrt{3578^2 - 3160^2} =$$

(b) Find the slope of the ski run. Slope $= \dfrac{\text{Vertical Drop}}{\text{Horizontal Distance}} =$

---

[2] *Fodor's 89: Ski Resorts of North America*, Fodor's Travel Publications, Inc., 1989, p. 100.

3. The longest ski run at Stowe, called the Juggernaut, is 52,800 feet with a vertical drop of 2350 feet.

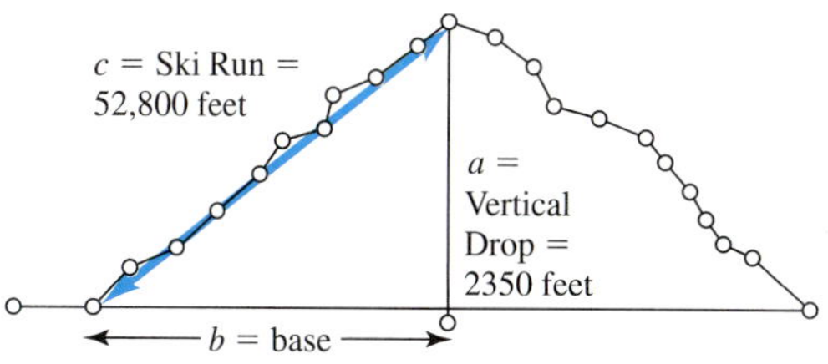

(a) Solve for the horizontal distance b, where $a = 2350$ and $c = 52{,}800$. Round your answer to the nearest hundredth.

$$b = \sqrt{c^2 - a^2}$$
$$b^2 = \sqrt{52{,}800^2 - 2350^2} =$$

(b) Find the slope of the ski run. Slope $= \dfrac{\text{Vertical Drop}}{\text{Horizontal Distance}} =$

Problems 4–7: The vertical drops and horizontal distances of four types of ski trails in the Rocky Mountains are listed in the table. Calculate the slope of each ski trail. Round your answer to the nearest hundredth.

Type of ski trail	Vertical drop (in feet)	Horizontal distance (in feet)	Slope
4. Beginner	1098	4097	
5. Intermediate	1098	2718	
6. Expert	1388	2137	
7. Super Expert	800	1061	

8. Why is the ski trail in Problem 4 called the Beginner trail?

9. Why is the ski trail in Problem 7 called the Super Expert trail?

Problems 10–11: Determine whether each graph is increasing, decreasing, horizontal (constant), or vertical.

10. (a)

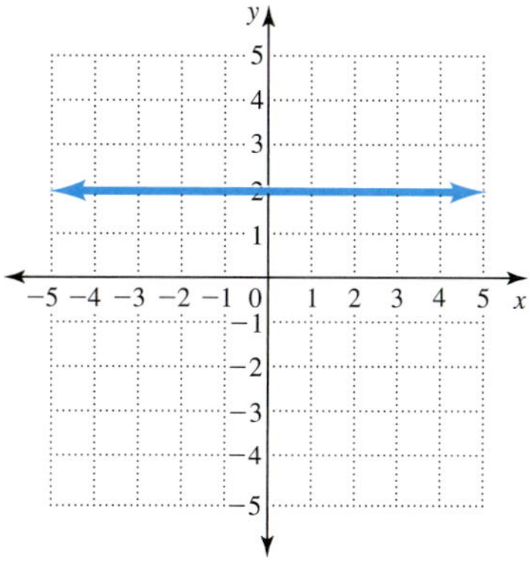

(b)

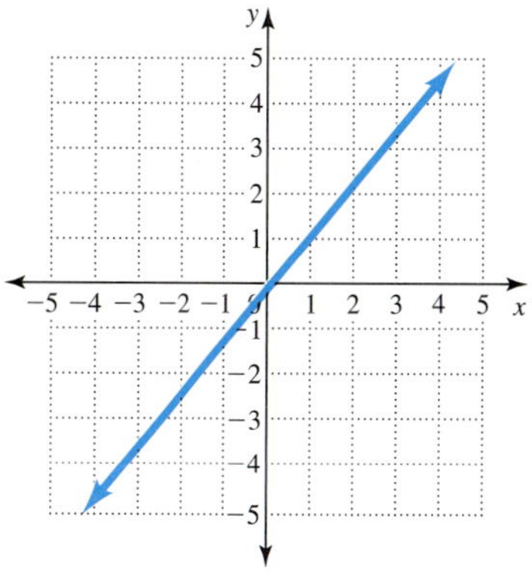

(c)

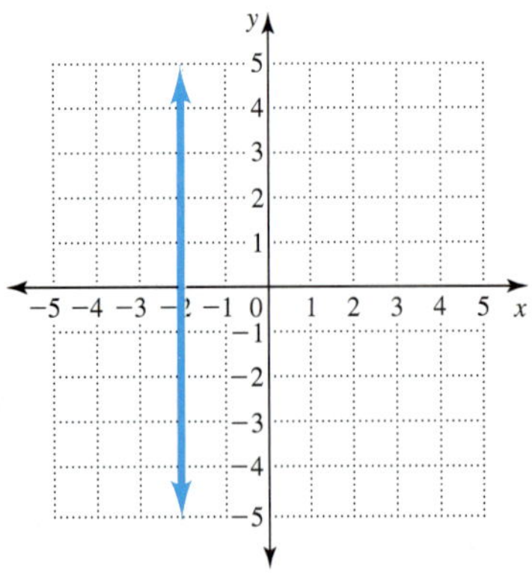

(d)

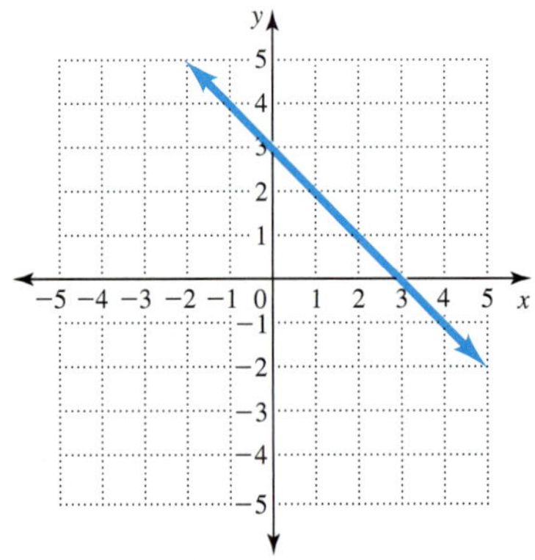

11. (a)

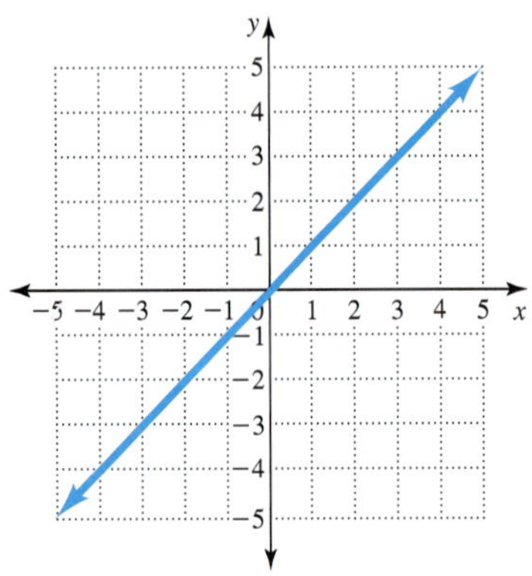

(b)

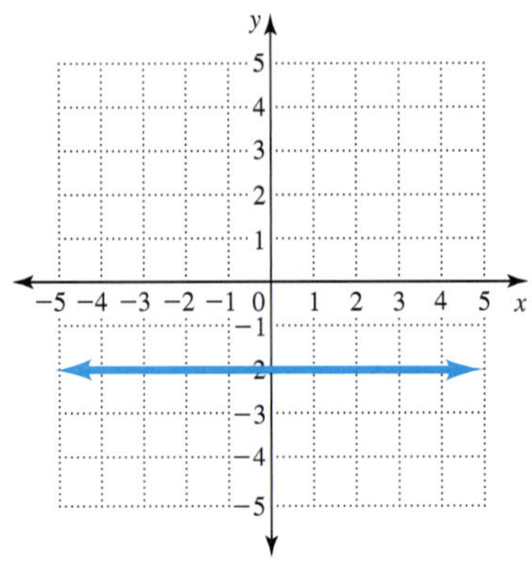

(c)

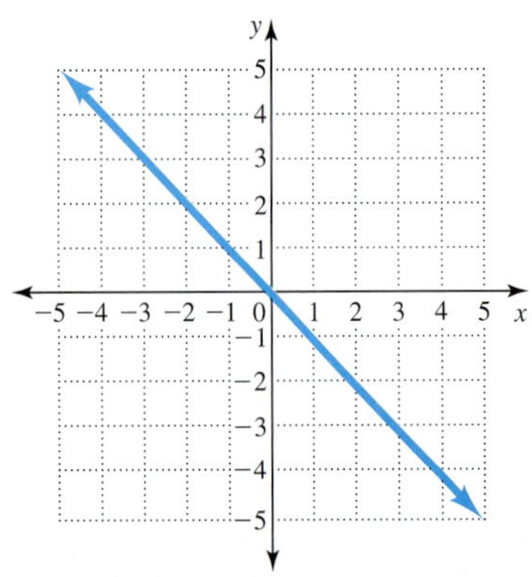

(d)

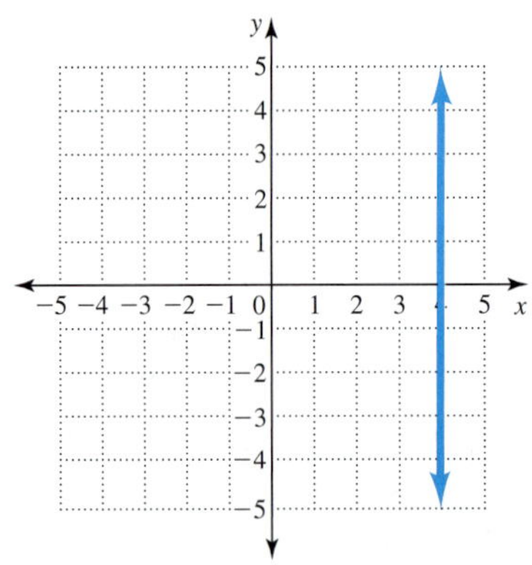

Problems 12–13: Determine whether the slope is positive, negative, zero, or undefined.

12. (a)

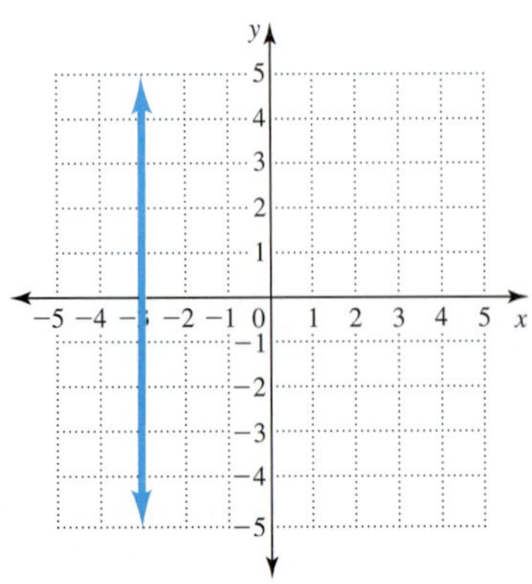

(b)

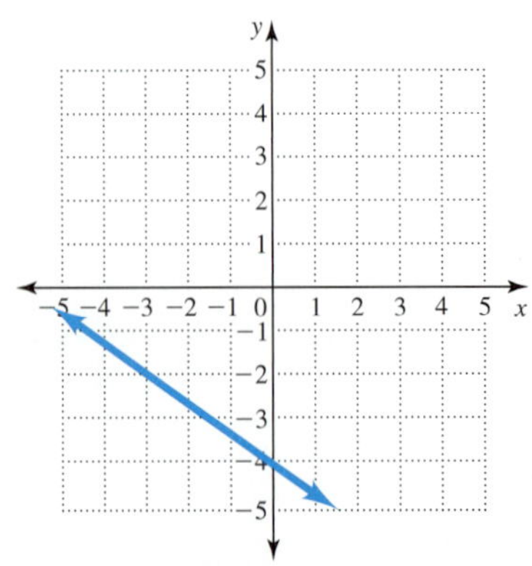

(c)

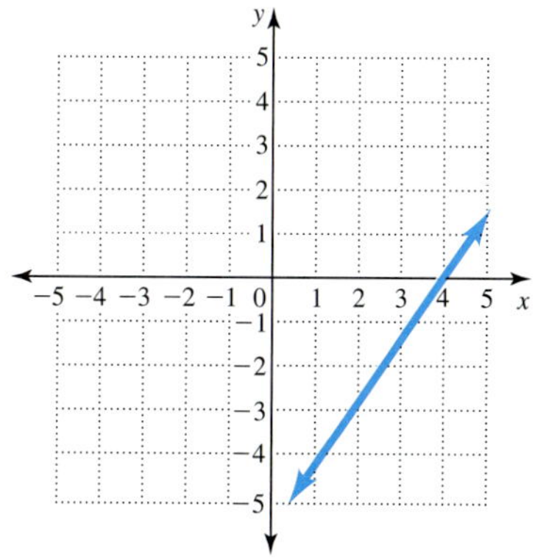

(d)

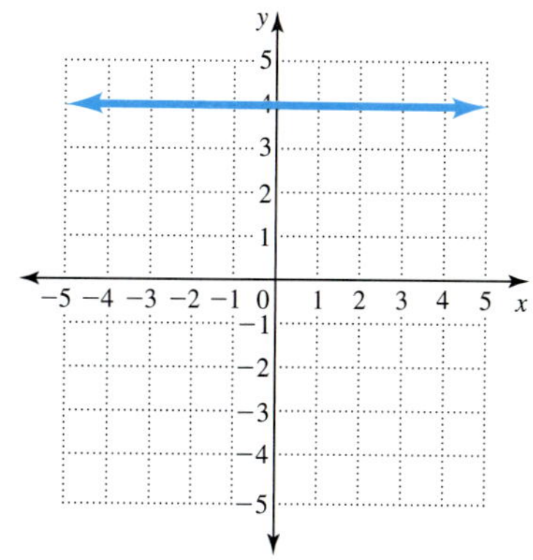

13. (a)

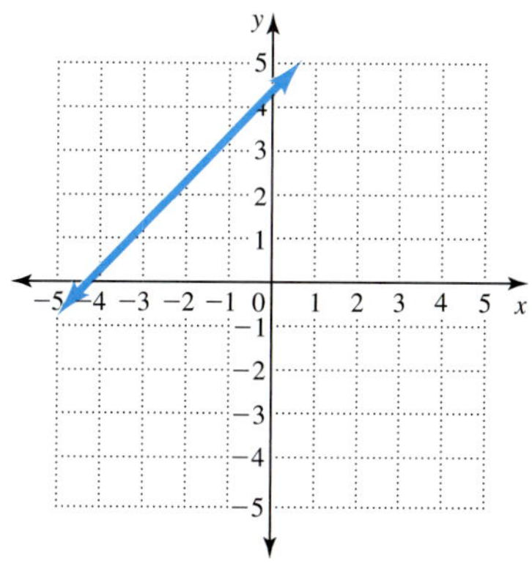

(b)

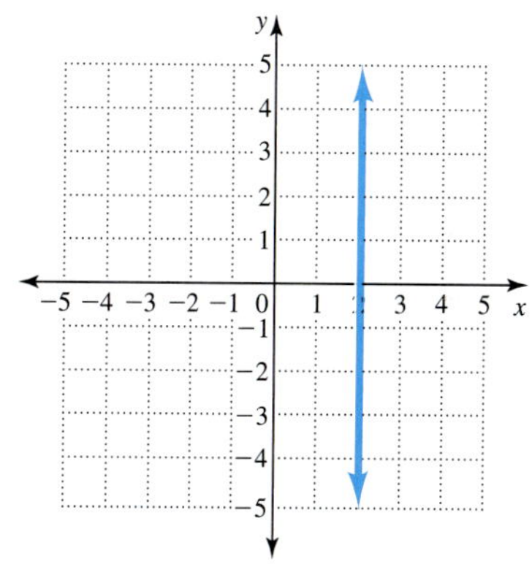

(c)

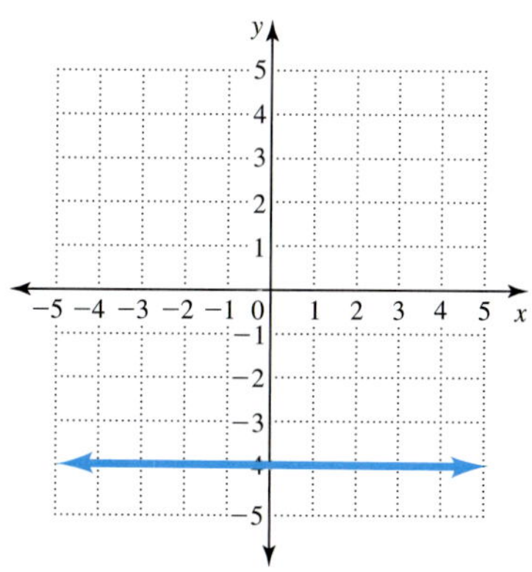

(d)

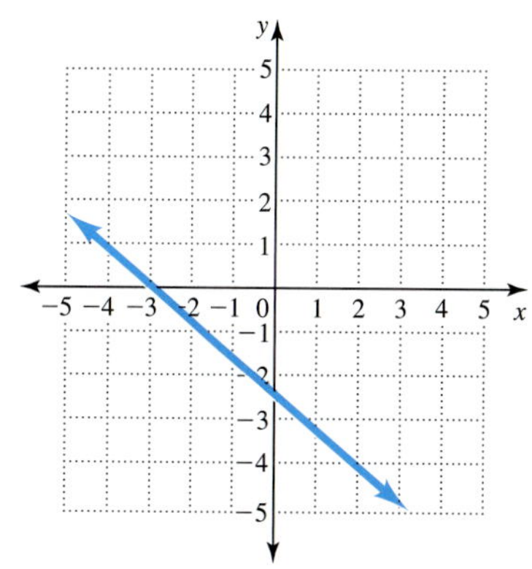

Problems 14–29: Identify the coordinates of the y-intercept of the line and the slope from the equation. Graph each equation or function using the Slope-Intercept Method. Verify the graph by computing and plotting two other points on the line.

14. $f(x) = x + 3$

(a) y-intercept of the line: ______ (b) Slope of the line: ______

x	$f(x)$

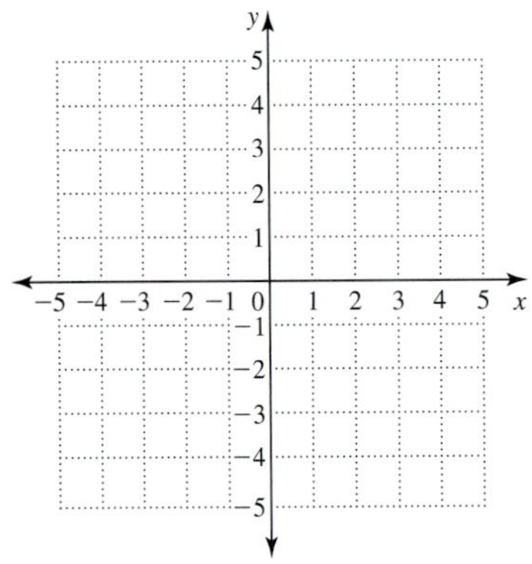

15. $f(x) = x - 1$

(a) y-intercept of the line: ______ (b) Slope of the line: ______

x	$f(x)$

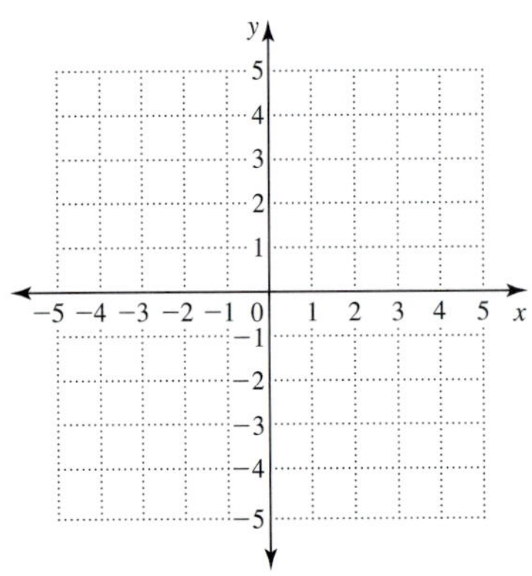

16. $y = 2x + 1$

(a) y-intercept of the line: ______ (b) Slope of the line: ______

x	y

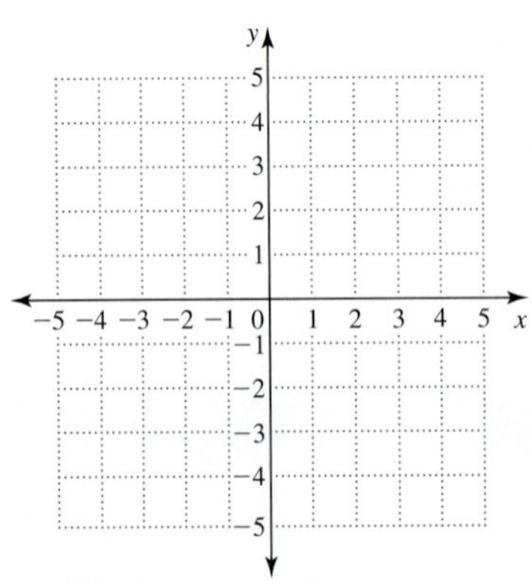

17. $y = 3x + 2$

(a) y-intercept of the line: ______ (b) Slope of the line: ______

x	y

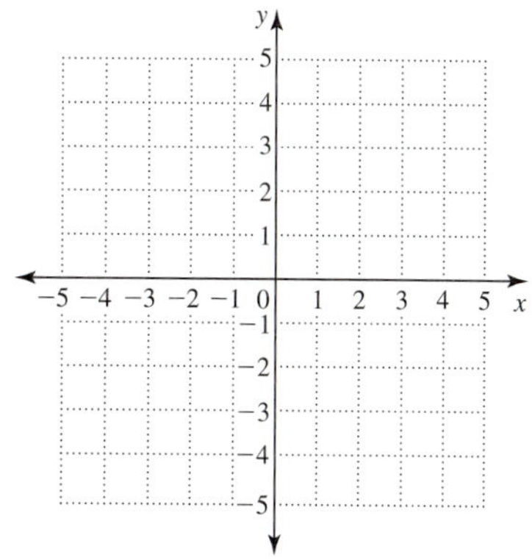

18. $y = -3x + 1$

(a) y-intercept of the line: ______ (b) Slope of the line: ______

x	y

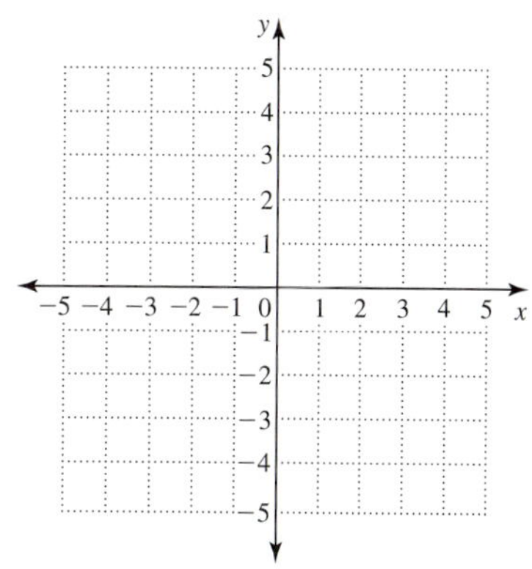

19. $y = -2x - 3$

(a) y-intercept of the line: ______ (b) Slope of the line: ______

x	y

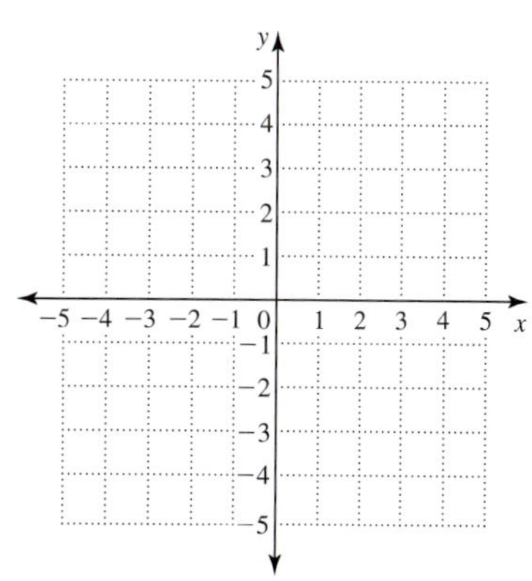

20. $y = 4$

(a) y-intercept of the line: ______ (b) Slope of the line: ______

x	y

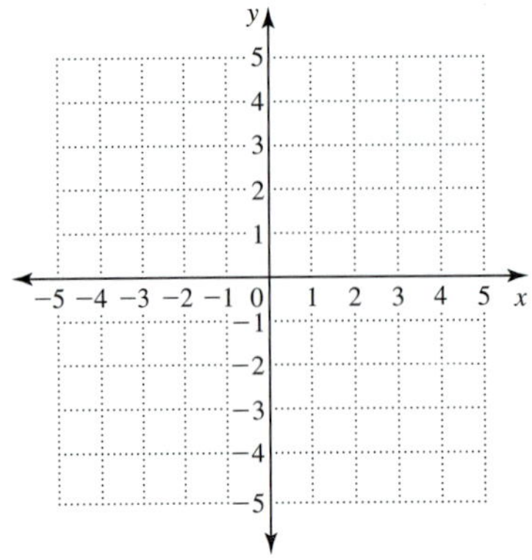

21. $y = -2$

(a) y-intercept of the line: ______ (b) Slope of the line: ______

x	y

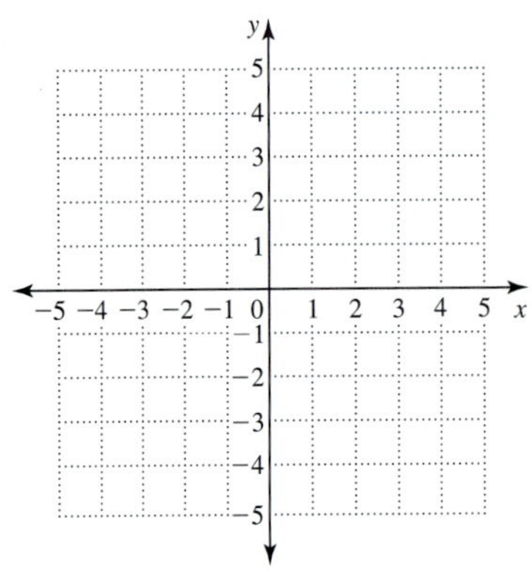

22. $x = -2$

(a) y-intercept of the line: ______ (b) Slope of the line: ______

x	y

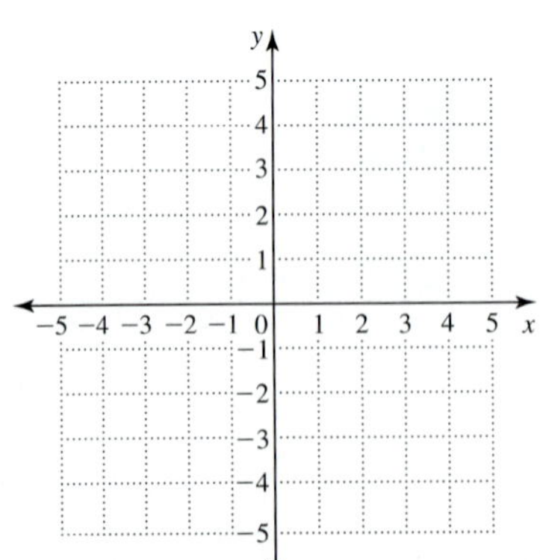

23. $x = 4$

(a) y-intercept of the line: ______ (b) Slope of the line: ______

x	y

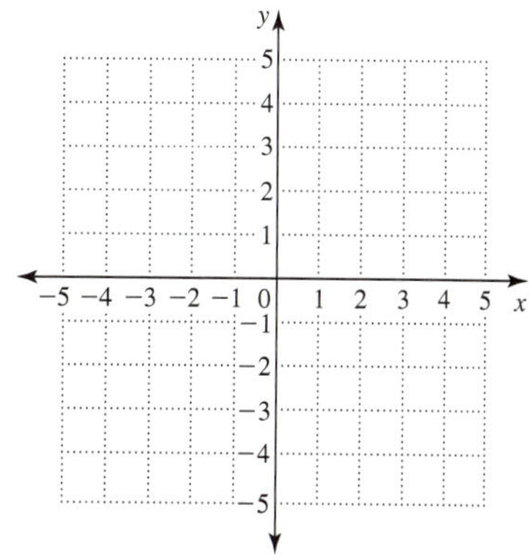

24. $y = \frac{2}{3}x - 1$

(a) y-intercept of the line: ______ (b) Slope of the line: ______

x	y

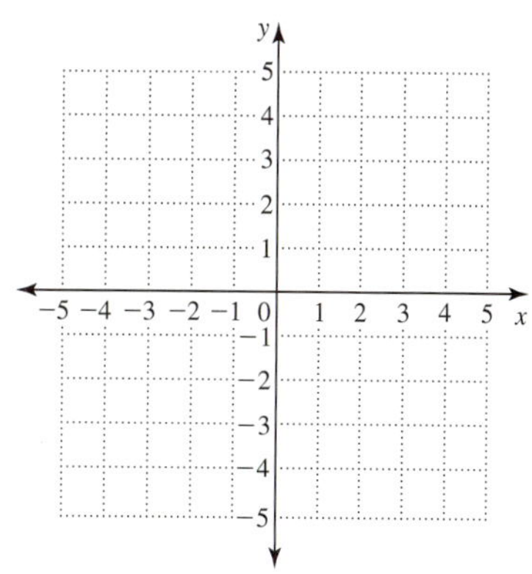

25. $y = \frac{3}{4}x - 2$

(a) y-intercept of the line: ______ (b) Slope of the line: ______

x	y

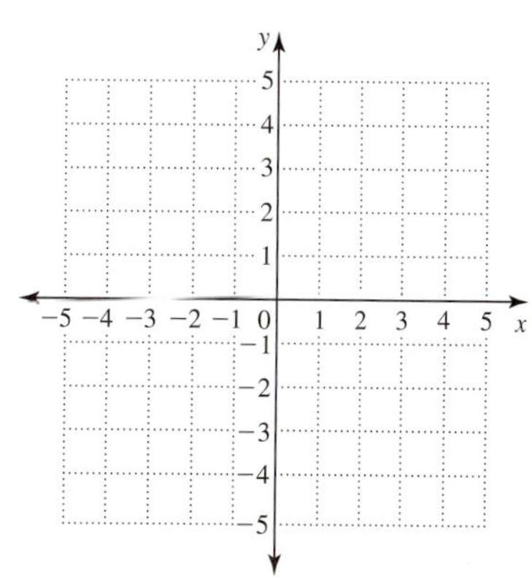

26. $y = -\frac{2}{5}x + 3$

(a) y-intercept of the line: ______ (b) Slope of the line: ______

x	y

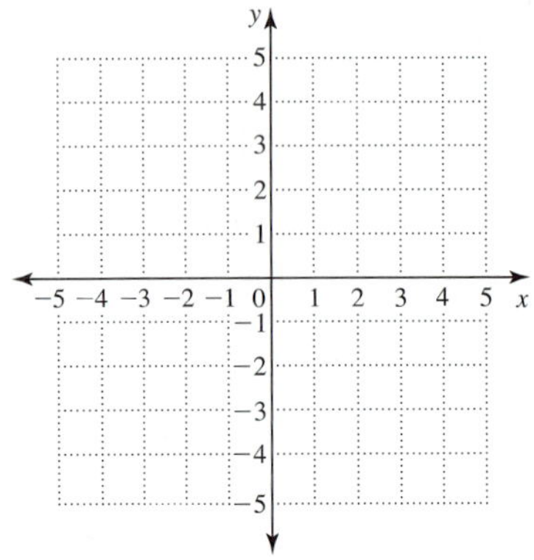

27. $y = -\frac{3}{2}x + 1$

(a) y-intercept of the line: ______ (b) Slope of the line: ______

x	y

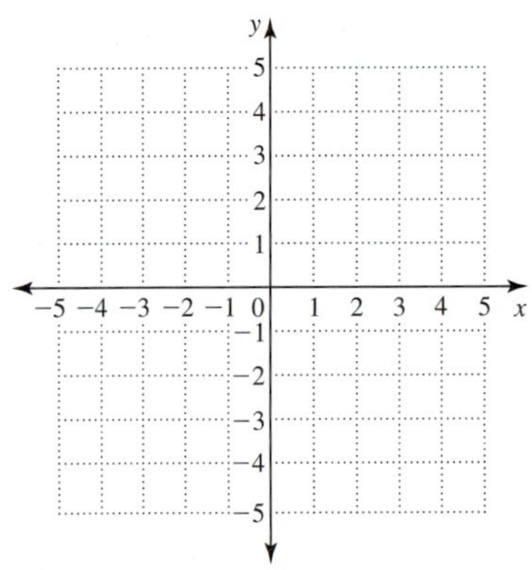

28. $f(x) = -3x + 2$

(a) y-intercept of the line: ______ (b) Slope of the line: ______

x	$f(x)$

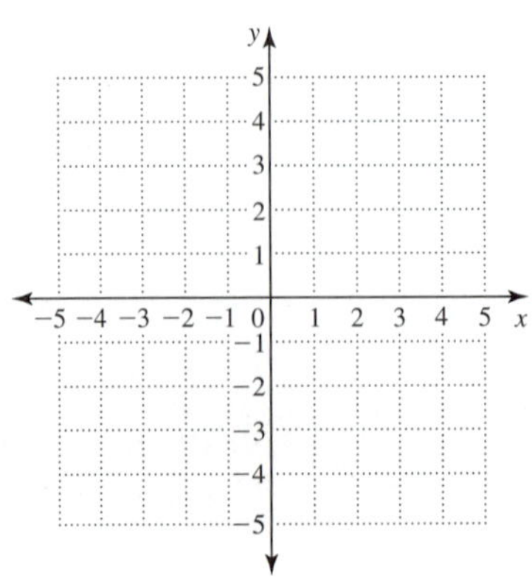

29. $f(x) = -2x + 1$

(a) y-intercept of the line: ______ (b) Slope of the line: ______

x	$f(x)$

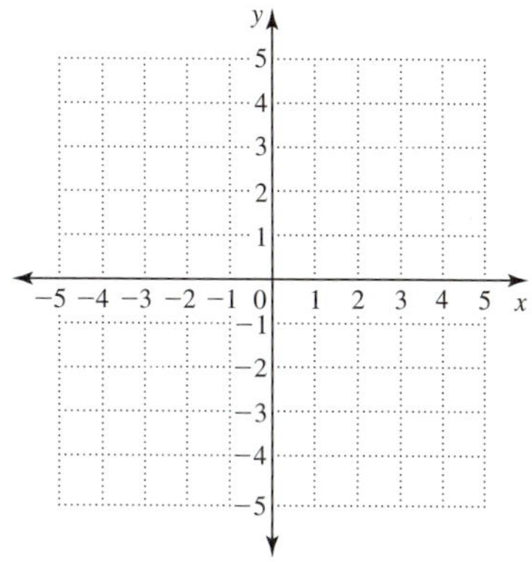

Problems 30–41: Write the equation of each line, given the slope and y-intercept.

30. Slope $= 5$; y-intercept $= (0, -2)$

31. Slope $= 4$; y-intercept $= (0, -1)$

32. Slope $= \frac{2}{3}$; y-intercept $= (0, 1)$

33. Slope $= \frac{2}{7}$; y-intercept $= (0, 3)$

34. Slope $= 2.3$; y-intercept $= (0, -4.2)$

35. Slope $= 1.75$; y-intercept $= (0, -3.5)$

36. Slope $= -6$; y-intercept $= (0, 5)$

37. Slope $= -4$; y-intercept $= (0, 3)$

38. Slope $= -1$; y-intercept $= \left(0, \frac{1}{2}\right)$

39. Slope $= -1$; y-intercept $= \left(0, -\frac{5}{7}\right)$

40. Slope $= -1$; y-intercept $= (0, 0)$

41. Slope $= 3$; y-intercept $= (0, 0)$

42. Celsius degrees and the equivalent Fahrenheit degrees are an example of a linear function.

$x =$ Temperature in Celsius degrees	$y =$ Temperature in Fahrenheit degrees
−50	−58
−4	24.8
0	32
2	35.6
20	68

(a) Choose any two points from the table and calculate the slope.

(b) Identify the y-intercept from the table.

(c) Use the slope and the y-intercept to write the equation of the line in Slope-Intercept Form.

(d) Use the equation to find the Fahrenheit temperature if the Celsius temperature is 13 degrees.

43. Patrick took a trip to visit a friend at college in another state. His speedometer was broken, but the odometer still worked. Patrick kept track of the number of miles he traveled each hour.

x = time traveled (in hours)	y = distance traveled (in miles)
0	0
1	65
2	130
3	195

(a) The graph is linear. Choose any two points from the table and calculate the slope of the line.

(b) Identify the y-intercept from the table.

(c) Use the slope and y-intercept to write the equation of the line in Slope-Intercept Form.

(d) Use the equation to predict how far Patrick had traveled after 6 hours if his speed remained the same throughout the trip.

Problems 44–47: The total number of Associate degrees conferred in the indicated fields of study in the academic years 1997, 1998, and 1999 are given in each table.[3] Each set of data can be approximated with a linear function.

(a) Calculate the slope of the line using the second and third points in the table.

(b) Identify the y-intercept from the table.

(c) Use the slope and y-intercept to write the equation of the line in Slope-Intercept Form.

(d) Use the equation to predict the number of degrees to be conferred in 2007–2008 ($t = 10$).

44. Associate degrees conferred in Agriculture and natural resources:

Academic year	x = Number of years since 1998	y = Degrees conferred (in thousands)
1997–1998	0	4.2
1998–1999	1	4.3
1999–2000	2	4.4

[3] National Center for Education Statistics.

45. Associate degrees conferred in Physical Sciences:

Academic year	x = Number of years since 1998	y = Degrees conferred (in thousands)
1997–1998	0	2.3
1998–1999	1	2.4
1999–2000	2	2.5

46. Associate degrees conferred in Accounting:

Academic year	x = Number of years since 1998	y = Degrees conferred (in thousands)
1997–1998	0	14.8
1998–1999	1	14.3
1999–2000	2	13.6

47. Associate degrees conferred in Law and Legal Studies:

Academic year	x = Number of years since 1998	y = Degrees conferred (in thousands)
1997–1998	0	7.8
1998–1999	1	7.4
1999–2000	2	7.3

Problem 48: Writing Exercise

48. Write a definition of slope in your own words. Use the words vertical displacement and horizontal, RISE, RUN, and ratio in the definition.

3.5 Point-Slope Formula

OBJECTIVES

- Use the Point-Slope Formula to find the equation of a line.
- Write the equation of a line in General Form.

1 2 3

Getting Started

In this section, the Point-Slope Formula is derived from the Slope Formula and used to write the equation of a line when either two points, or the slope and one point, are given.

1. The total number of Associate degrees awarded in the fifty states and Washington, DC, is projected to be 684 thousand in the year 2009.[1] This projection is based on the fact that the number of degrees awarded is increasing at a constant rate of approximately 4 thousand each year. We will find the linear equation describing the problem situation and use it to determine how many degrees will be awarded in the year 2013.

(a) The number of degrees awarded is changing at a constant rate. Write the rate of change (slope) as a ratio.

$$\text{Rate of change} = \frac{\left(\begin{array}{c}\text{change in the number of Associate degrees}\\ \text{awarded (in thousands)}\end{array}\right)}{\text{(each year)}} = \frac{\_\_\_}{1}$$

(b) Write an ordered pair from the given information by filling in the blank with the number of Associate degrees awarded.

$$\left(\text{year}, \begin{array}{c}\text{Number of Associate degrees}\\ \text{awarded (in thousands)}\end{array}\right)$$

$$(\underline{2009}, \underline{\qquad\qquad\qquad})$$

To visualize the problem, let's construct a rough graph (not to scale) of the line described in this problem situation. The rate of change of this problem situation is positive and the graph of the line is increasing. Let the x-axis represent the year and the y-axis represent the number of degrees awarded in thousands. Plot ordered pair (2009, 684) on the grid. The slope (rate of change) is a constant rate of 4 thousand per year. Sketch a somewhat steep, increasing line through the point (2009, 684). Let (x, y) represent a second unknown point on the line.

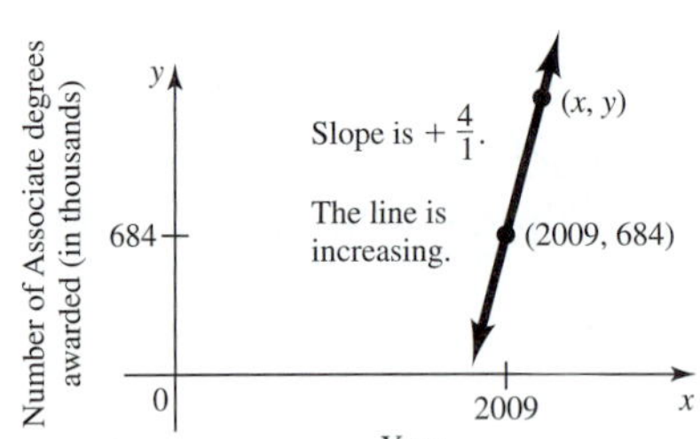

(c) To find the equation of the line, label $(x_1, y_1) = (2009, 684)$ and $(x_2, y_2) = (x, y)$. Substitute $x_1 = 2009$, $y_1 = 684$, $x_2 = x$, and $y_2 = y$ and $m = \frac{4}{1}$ into the slope formula: $m = \frac{y_2 - y_1}{x_2 - x_1}$

$$\_\_\_ = \_\_\_ \qquad \text{Equation 1}$$

[1] National Center for Education Statistics.

Problems 5–8:

(a) Given the graph and table of values for the function, determine whether the line is increasing or decreasing.

(b) From the equation, determine the slope.

(c) Write the y-intercept.

5. $y = -3x - 4$

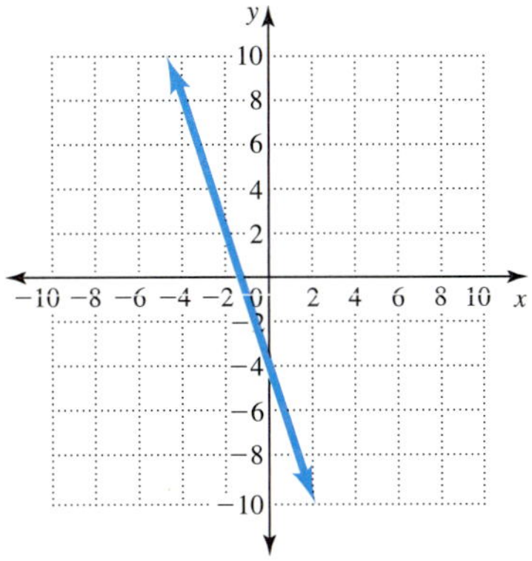

x	y_1
−4	8
−3	5
−2	2
−1	−1
0	−4
1	−7
2	−10

6. $y = 2x + 5$

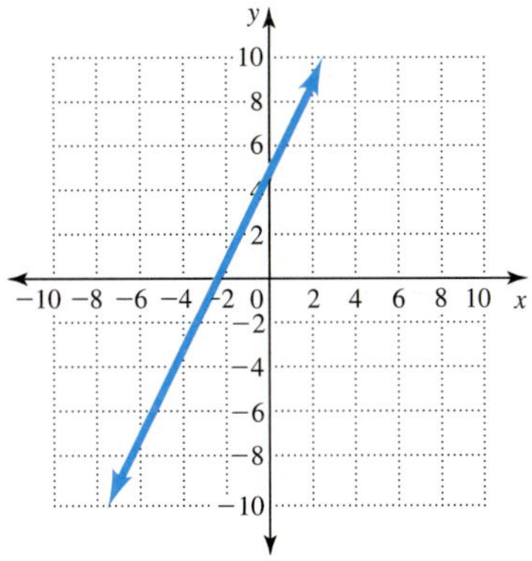

x	y_1
−4	−3
−3	−1
−2	1
−1	3
0	5
1	7
2	9

7. $y = 3x - 4$

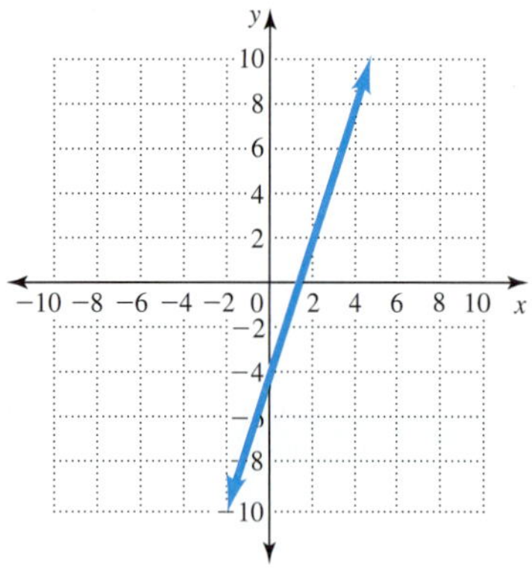

x	y_2
−4	−16
−3	−13
−2	−10
−1	−7
0	−4
1	−1
2	2

8. $y = -2x + 5$

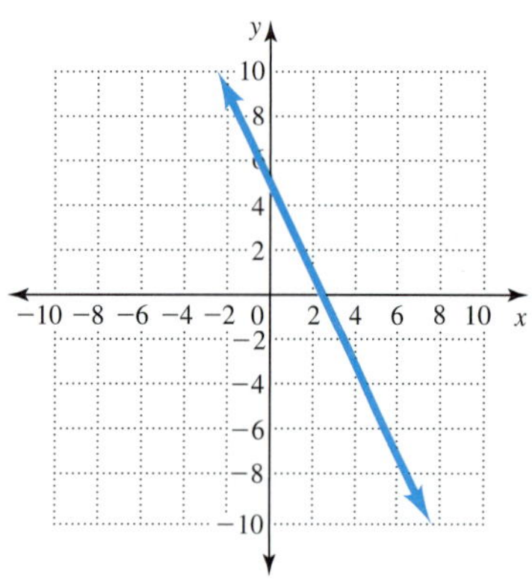

x	y_1
−4	13
−3	11
−2	9
−1	7
0	5
1	3
2	1

Problems 9–36: Find the equation of each line with the given characteristics. Write each equation in Slope-Intercept Form ($y = mx + b$).

9. Slope 5 and y-intercept $(0, 1)$.

10. Slope 4 and y-intercept $(0, 3)$.

11. Slope -5 and y-intercept $(0, -1)$.

12. Slope -3 and y-intercept $(0, -2)$.

13. Slope 5, passing through the point $(2, 5)$.

14. Slope 4, passing through the point $(3, 2)$.

15. Slope -2, passing through the point $(2, 3)$.

16. Slope -3, passing through the point $(1, 7)$.

17. Slope 6, passing through the point $(-1, -1)$.

18. Slope 2, passing through the point $(-2, -5)$.

19. Slope $\frac{1}{2}$, passing through the point $(1, 2)$.

20. Slope $\frac{1}{3}$, passing through the point $(-1, 2)$.

21. Slope $\frac{-5}{3}$, passing through the point $(1, -3)$.

22. Slope $\frac{-2}{5}$, passing through the point $(-5, -1)$.

23. Slope $= -6$, passing through the point $(0, 4)$.

24. Slope $= -1$, passing through the point $(0, 9)$.

25. Slope $= 0$, passing through the point $(4, 1)$.

26. Slope $= 0$, passing through the point $(-4, 2)$.

27. Through the points $(5, 3)$ and $(-1, -3)$.

28. Through the points $(-2, 7)$ and $(4, 1)$.

29. Through the points $(2, -1)$ and $(0, -1)$.

30. Through the points $(-1, 5)$ and $(-4, 5)$.

31. Through the points $(-2, 5)$ and $(-6, -1)$.

32. Through the points $(2, 4)$ and $(4, -1)$.

33. Through the points $(-5, -1)$ and $(-5, 4)$.

34. Through the points $(-2, 5)$ and $(-2, -1)$.

35. Passing through the point $(-2, 7)$ with an x-intercept of -3

36. Through the origin and the point $(5, 3)$.

Problems 37–42: Rewrite the equations of the lines in problems 9–14 in General Form $(Ax + By = C)$.

37. Problem 9

38. Problem 10

39. Problem 11

40. Problem 12

41. Problem 13

42. Problem 14

43. Given the two equations and their corresponding graphs:

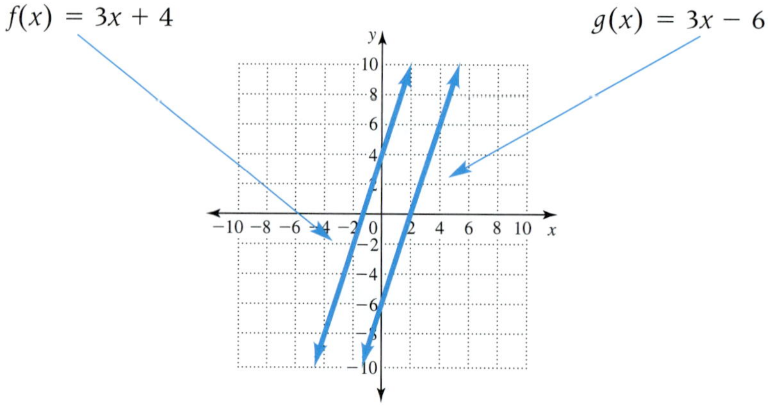

(a) What is the slope of $f(x)$?

(b) What is the y-intercept of $f(x)$?

(c) What is the slope of $g(x)$?

(d) What is the y-intercept of $g(x)$?

(e) The two lines are parallel. What can you conclude about the slopes of two parallel lines?

44. Given the two equations and their corresponding graphs:

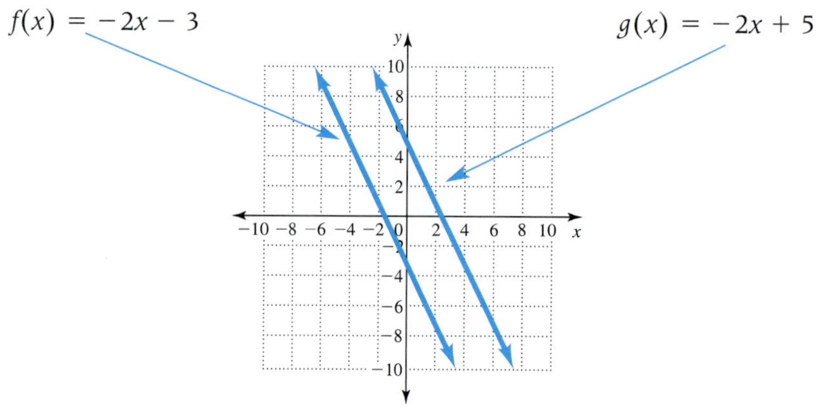

(a) What is the slope of $f(x)$?

(b) What is the y-intercept of $f(x)$?

(c) What is the slope of $g(x)$?

(d) What is the y-intercept of $g(x)$?

(e) The two lines are parallel. What can you conclude about the slopes of two parallel lines?

45. The total number of Associate degrees awarded to men in the fifty states and Washington, DC, is projected to be 244 thousand in the year 2008.[2] This projection is based on the fact that the number of Associate degrees awarded is increasing at a constant rate of approximately 1 thousand each year.

(a) Write the ordered pair from the given information.

(year , number of Associate degrees awarded in thousands)

(_____, _____)

[2] National Center for Education Statistics.

(b) Fill in the numerical information for the rate of change (slope) and write the ratio.

$$\text{Slope} = \frac{\begin{pmatrix}\text{change in the number of Associate degrees}\\ \text{awarded in thousands}\end{pmatrix}}{(\text{each year})} = \underline{\quad}$$

(c) Use the slope and the point (2008, 244) to create a linear function.

(d) Compute the number of Associate degrees (in thousands) projected to be awarded to men in the year 2012. Show your work.

46. The total number of Associate degrees awarded to women in the fifty states and Washington, DC, is projected to be 433 thousand in the year 2008. This projection is based on the fact that the number of Associate degrees awarded is increasing at a constant rate of approximately 3 thousand each year.

(a) Write the ordered pair from the given information.

(year, number of Associate degrees awarded in thousands)

(____, ____)

(b) Fill in the numerical information for the rate of change and write the ratio.

$$\text{Slope} = \frac{\begin{pmatrix}\text{change in the number of Associate degrees}\\ \text{awarded in thousands}\end{pmatrix}}{(\text{each year})} = \underline{\quad}$$

(c) Use the slope and the point (2008, 433) to create a linear function.

(d) Compute the number of Associate degrees (in thousands) projected to be awarded to women in the year 2012. Show your work.

47. The following table shows two data points of the concentration of carbon dioxide (CO_2) in parts per million (ppm) in the atmosphere at given years.[3]

Year	x = the number of years since 1970	(CO_2) concentration (in ppm)
1970	0	325.3
1990	20	354.0

(a) The complete data set for the twenty-year period approximates a linear function. Use the points in the table to find the slope of the line. Round your answer to the nearest tenth.

(b) Use the Point-Slope Formula to find the equation of the line.

(c) Use the equation to predict the concentration of atmospheric carbon dioxide in parts per million in the year 2005 ($x = 35$).

[3] World Resources Institute, 1992–1993.

48. Christopher Basich throws a 12-pound shot put for his high school track team. The table lists his longest throws during the months of March, April, and June 1999.

Month	Input: Month (by number)	Output: Distance (in feet)
March	3	38
April	4	48
June	6	59

(a) Graph the ordered pairs. Connect the first and third point with a straight line. The data approximates a linear function.

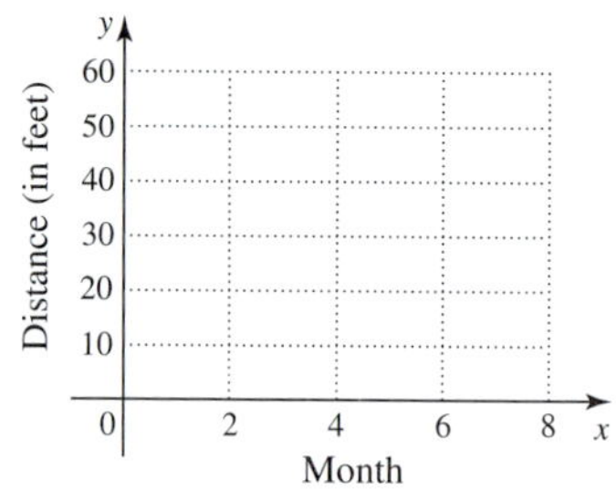

(b) Find the slope of the line using the first point in the table, $(x_1, y_1) = (3, 38)$, and the third point in the table, $(x_2, y_2) = (6, 59)$.

(c) Use the Point-Slope Formula to find the equation of the line.

(d) Use the equation to predict the distance Christopher will throw in July ($x = 7$) if he continues to improve at this constant rate.

(e) Is it realistic to assume that Christopher's shot put distances will increase at a constant rate indefinitely?

49. The NASA Lewis Materials Adherence Experiment (MAE) on the Sojourner Rover Mars Pathfinder measured the dust deposited on the rover's glass cover.[4] The data collected from the experiment is given in the graph. NASA used "Sol" as the independent variable representing the day of measurement.

Even though more than one measurement was recorded on some days and not all points lie on a straight line, NASA considered the data as approximating a linear function and used the equation of the line to predict dust deposits after the fourteenth day. A line was drawn between the ordered pairs at days 2 and 13.

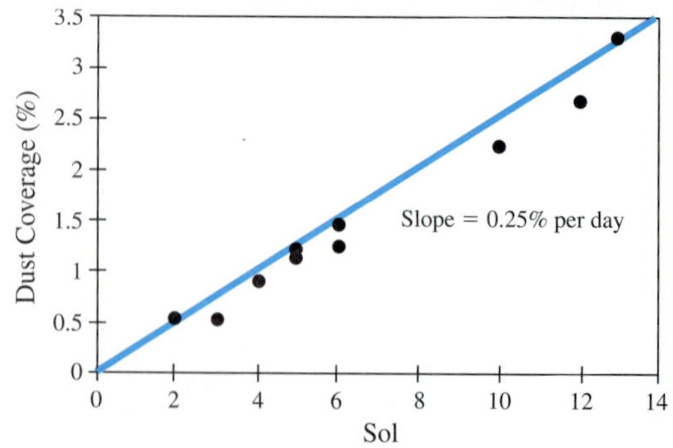

[4] NASA Glenn Research Center.

Day of measurement = Sol	Dust Coverage (%)
2	0.50
13	3.25

(a) Use the points in the table to find the slope of the line between the two points.

(b) Use the Point-Slope Formula to create the equation of the line.

(c) Use the equation to predict the % of Dust Coverage on the twentieth day ($x = 20$).

50. The U.S. Census Bureau has collected data on the number of persons living alone for all age groups. The table presents the data of the number of persons living alone for ages 30–34. Find an equation of the line approximating this data and use it to predict the number of persons living alone who are 30–34 years of age in 2005.

Year	t = Number of years since 1995	Number of persons living alone for ages 30–34 (in millions)
1995	0	2.07
1996	1	2.04
1997	2	2.00
1998	3	1.97
1999	4	1.94

(a) Let t represent the number of years since 1995. Label the first point in the table, (0, 2.07) as (x_1, y_1), and the last point in the table, (4, 1.94) as (x_2, y_2). Use these two points to find the slope of the line.

(b) Use the third point, the slope, and the Point-Slope Formula to find the equation of the line.

(c) Use the equation to predict the number of persons living alone, 30–34 years of age, in the year 2005.

51. The U.S. Census Bureau has collected data on the number of persons living alone for all age groups. The table presents the data of the number of persons living alone for ages 35–44. Find an equation of the line approximating this data and use it to predict the number of persons living alone who are 35–44 years of age in 2005.

Year	t = Number of years since 1995	Number of persons living alone for ages 30–34 (in millions)
1995	0	3.85
1996	1	3.97
1997	2	4.07
1998	3	4.16
1999	4	4.24

(a) Let t represent the number of years since 1995. Label the first point in the table, (0, 3.85) as (x_1, y_1), and the last point in the table, (4, 4.24) as (x_2, y_2). Use these two points to find the slope of the line.

(b) Use the third point, the slope, and the Point-Slope Formula to find the equation of the line.

(c) Use the equation to predict the number of persons living alone, 35–44 years of age, in the year 2005.

Chapter 3 Summary

SECTION 3.1: The **Cartesian plane** is used to graph relations and functions.

SECTION 3.2:

Definition of a Relation and a Function

A relation is any set of ordered pairs.

A function is a relation where each input value has *one* and *only* one output value.

Vertical Line Test

- If all vertical lines intersect the graph of a relation at only one point, the relation IS also a function. One and only one output value exists for each input value.
- If any vertical line intersects the graph of a relation at more than one point, the relation "fails" the test and is NOT a function. More than one output value exists for some (or all) input value(s).

SECTION 3.3:

Linear Relations and Functions

- Linear relations are of the form $y = ax + b$ where x is the independent variable, y is the dependent variable, and a and b are constants.
- A linear function is written in the form $f(x) = ax + b$.

Intercepts of a Linear Function $f(x) = ax + b$

- The point where a graph crosses the vertical axis (the y-axis) is called the y-intercept. The first coordinate (or x value) of every y-intercept is zero.
- The point where a graph crosses the horizontal axis (the x-axis) is called the x-intercept. The second coordinate (or y value) of every x-intercept is zero.

Graphing a function $f(x) = ax + b$ using the Intercept Method

1. Calculate the x-intercept (let $y = 0$ and solve for x).
2. Calculate the y-intercept (let $x = 0$ and solve for y).
3. Plot the x- and y-intercepts on a Cartesian plane.
4. Connect the points with a line.
5. Compute and plot the coordinates of two more points to verify the graph.

Function form of a Horizontal Line

$f(x) = b$, where b is a real number constant.

$f(x) = b$ is equivalent to $y = b$.

General form of a Vertical Line

$x = k$, where k is a constant.

SECTION 3.4:

Slope Formula

The slope of a line between two points (x_1, y_1) and (x_2, y_2) on the line is:

$$m = \frac{\text{Vertical Displacement}}{\text{Horizontal Displacement}} = \frac{\text{RISE}}{\text{RUN}} = \frac{\Delta y}{\Delta x} = \frac{y_2 - y_1}{x_2 - x_1}, \ (x_1 \neq x_2).$$

- The slope of an *increasing* line is a positive value $m > 0$.
- The slope of a *decreasing* line is a negative value $m < 0$.
- The slope of a *vertical* line is undefined: m is undefined in $x = k$.
- The slope of a *horizontal* line is zero: $m = 0$ in $y = b$.

Slope-Intercept Form of a Line

$$y = mx + b \quad \text{or} \quad f(x) = mx + b$$

where m is the slope and b is the y-coordinate of the y-intercept.

Graphing a Line, using the Slope-Intercept Method

1. Identify the y-intercept and the RISE and the RUN from the slope. Place your pencil at the y-intercept and count the appropriate number of units for the RISE and the RUN. Draw a second point at that location.
2. Connect the y-intercept and the second plotted point with a line.
3. Compute and plot the coordinates of two other points to verify the graph.

SECTION 3.5:

Point-Slope Formula of a Line

$$y - y_1 = m(x - x_1),$$

where m is the slope and (x_1, y_1) is a given point on the line.

General Form of a Linear Equation

$$Ax + By = C,$$

where A and B are coefficients of respectively, and C is a constant. In most cases, A represents a positive integer.

To Find the Equation of a Line

Determine which information is given in the problem and apply the appropriate formula:

- **Given the slope (m) and the y-intercept** $(0, b)$: Substitute into the Slope-Intercept Form $f(x) = mx + b$. (Section 3.4)
- **Given the slope (m) and one point** (x_1, y_1): Substitute into the Point-Slope Formula $y - y_1 = m(x - x_1)$.
- **Given two points**: Find the slope $m = \frac{y_2 - y_1}{x_2 - x_1}$ and substitute into the Point-Slope Formula $y - y_1 = m(x - x_1)$.

Chapter 3 Review

- Review the definitions you have written for this chapter in the Dictionary.
- Review any properties or processes you have written on 3 × 5-inch cards.

SECTION 3.1

1. Label each axis with integer values. Plot the ordered pairs on the Cartesian plane. Label each point with the corresponding letter. For letters (a)–(i), indicate the Quadrant in which each point is located.

(a) $(2, 3)$ (b) $(4, 3)$ (c) $(-3, 3)$ (d) $(-1, -3)$

(e) $(-4, -1)$ (f) $(-1, 2)$ (g) $(1, -2)$ (h) $(1, -3)$

(i) $(-1, -2)$ (j) $(1, 0)$ (k) $(-1, 0)$ (l) $(0, 5)$

(m) $(0, 0)$

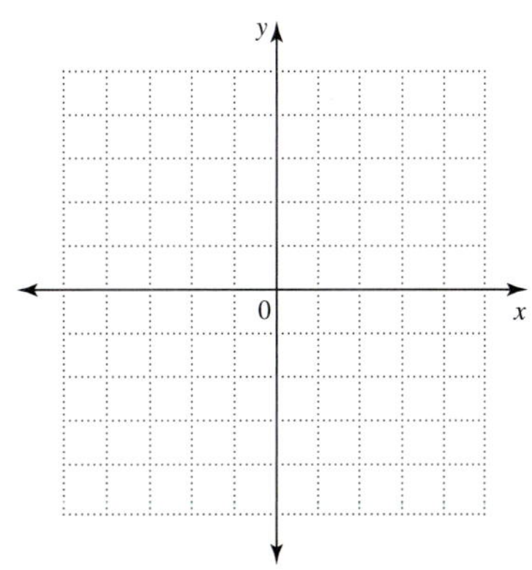

Problems 2–5: Given the grid and points:

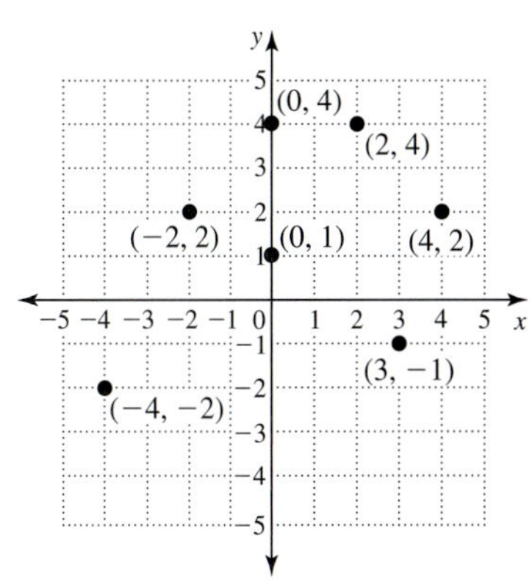

2. What are the first coordinate values of the points with a second coordinate value of 2?

3. What is the second coordinate value of the point with first coordinate value of -4?

4. What is the first coordinate value of the point with second coordinate value of -1?

5. What is the second coordinate value of the point(s) with first coordinate value of 0?

6. Tuition at Downtown University is $185 per credit hour.

(a) Complete the table and list the ordered pairs in the third column.

x = Number of credit hours	y = Tuition cost	Ordered pairs
1	185	
5	925	
10		
15		

(b) Smallest first coordinate value:
Largest first coordinate value:
Scale for horizontal axis:

(c) Smallest second coordinate value:
Largest second coordinate value:
Scale for vertical axis:

(d) Label each axis with word titles and your scale. Plot the ordered pairs.

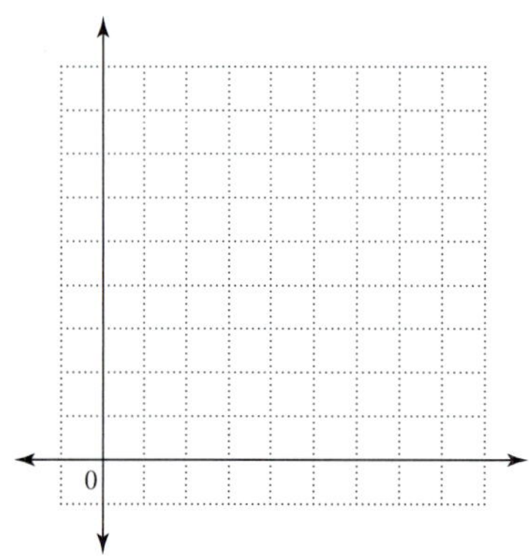

7. The life span of an insect can be modified by the temperature of the environment. Fruit flies, both male and female, live longer in cooler environments, as shown in the following table.[1]

(a) Write the ordered pairs.

Degrees (in Celsius)	Life span (in days)	Ordered pairs
10	120.5	
15	92.4	
20	40.2	
25	28.5	
30	13.6	

[1] R. W. Keck, and R. R. Patterson, *Biomath, Problem Solving for Biology Students*, San Francisco: Pearson Education, Inc., publishing as Pearson Benjamin/Cummings, 2000, 131.

(b) Smallest first coordinate value:
Largest first coordinate value:
Scale for horizontal axis:

(c) Smallest second coordinate value:
Largest second coordinate value:
Scale for vertical axis:

(d) Label each axis with word titles and your scale. Plot the ordered pairs.

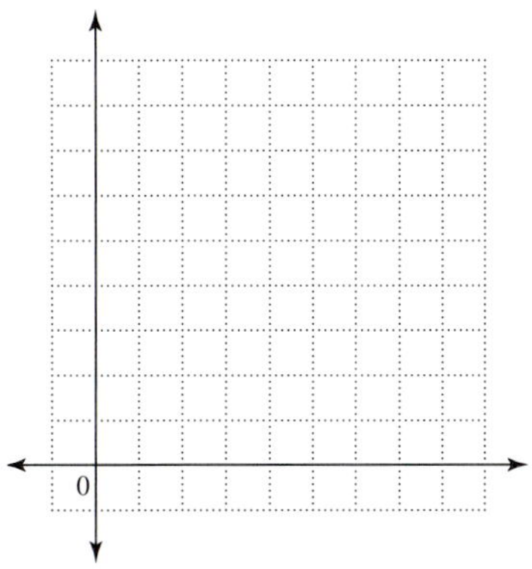

SECTION 3.2

8. Pam has been offered a job as a salesperson for a medical equipment distributor. Her base salary is \$20,000 plus 30% commission on total dollar sales of equipment.

(a) Write a word or words that describe the input or independent variable.

(b) Write a word or words that describe the output or dependent variable.

(c) Compute Pam's total salary for the given dollar amount of sales.

s = Sales of equipment (in \$)	Computations: Base salary + Commission	T = Total salary (in \$)
1000		
5000		
10,000		
15,000		
20,000		
s	Generalize	$T(s) =$

(d) Identify a scale, label the axes, and graph the ordered pairs from the table on a separate piece of graph paper.

(e) Use the graph to estimate the total salary if the sales are \$16,000.

(f) Evaluate the exact value of the total salary if the sales are \$16,000.

(g) Calculate the exact sales needed to earn a total salary of \$22,100.

9. You are going on vacation and plan to rent a car. There is a fixed rental fee of \$50 to rent the car, plus a charge of \$0.10 for each mile driven.

(a) Write the word or words that describe the input or independent variable.

(b) Write the word or words that describe the output or dependent variable.

(c) Complete the table of values.

m = Numbers of miles driven	Computations	C = Cost to rent the car
50		
100		
150		
200		
250		
m	Generalize	$C(m) =$

(d) Identify a scale, label the axes, and graph the ordered pairs from the table on a separate piece of graph paper.

(e) Use the graph to estimate the cost to rent the car if 300 miles are driven.

(f) Evaluate the exact cost to rent the car if 300 miles are driven.

(g) Calculate the exact number of miles driven if the cost to rent the car is \$68.

Problems 10–11: Plot each set of ordered pairs. Draw and label the axes. Use the definition of a relation and function and the vertical line test to determine whether each set represents a function.

10. $\{(0, 0), (-1, 2), (1, -2), (2, 3), (1, 4), (-3, 1)\}$
Function? Yes or No

11.

Input	Output
3	1
4	2
5	3
6	4
7	5

Function? Yes or No

12. Graph the function: $f(x) = 2x - 1$. Substitute each x value into the function and compute the corresponding $f(x)$ value. Record the ordered pairs in the table. Plot the points. Connect the dots. Extend the line and draw arrows on each end of the line.

x	$f(x)$	Ordered pairs
-2		
-1		
0		
1		
2		

Problems 13–16: Evaluate the functions for the given input values.

13. Given $f(x) = -2x + 5$,

(a) evaluate $f(-10)$.

(b) evaluate $f\left(\frac{1}{2}\right)$.

14. Given $g(x) = x^2 - 4x + 2$,

(a) evaluate $g(2)$.

(b) evaluate $g(-1)$.

15. Given $f(x) = 2x^2 - 7x$,

(a) evaluate $f(4)$.

(b) evaluate $f(-3)$.

16. Given $F(C) = 1.8C + 32$,

(a) evaluate $F(10)$.

(b) evaluate $F(-5)$.

SECTION 3.3

17. Circle the equations that are linear functions.

(a) $y = -4x^2 + 6$

(b) $f(x) = 7x + 1$

(c) $x = 4$

(d) $f(x) = x^3 + 2x + 6$

18. Circle the equations that are linear functions.

(a) $f(x) = 2x - 4$ (b) $y = 4x^2 + x - 1$

(c) $y = \frac{2}{3}x - 15$ (d) $x = -2$

19. Given the graph of $f(x) = -2x + 3$,

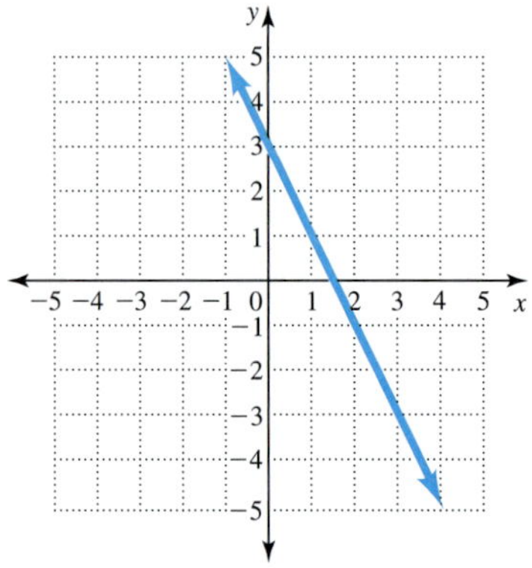

(a) Identify the coordinates of the x-intercept:

(b) Identify the coordinates of the y-intercept:

20. Given the graph of $f(x) = x - 4$,

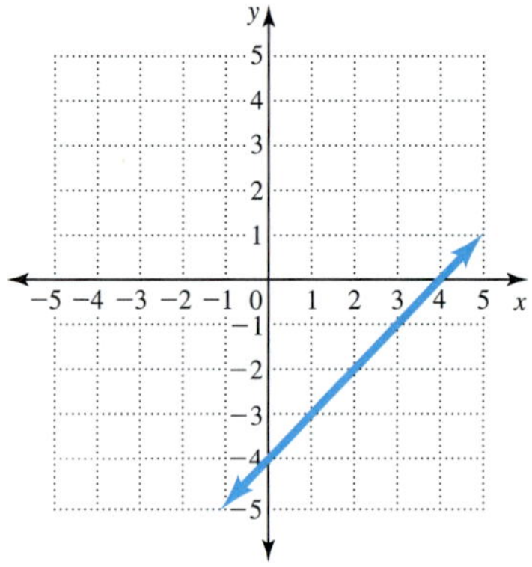

(a) Identify the coordinates of the x-intercept:

(b) Identify the coordinates of the y-intercept:

21. Richard is a quality control manager for a company that packages machine oil into plastic 1.0-liter containers. The volume of the oil in the containers is checked randomly and recorded. The table lists the volume of 5 containers selected at random.

Desired volume of 1.0-liter	Volume of selected container (in liters)
1.0	0.9
1.0	1.0
1.0	1.1
1.0	1.2
1.0	1.0

(a) Graph the ordered pairs on the grid.

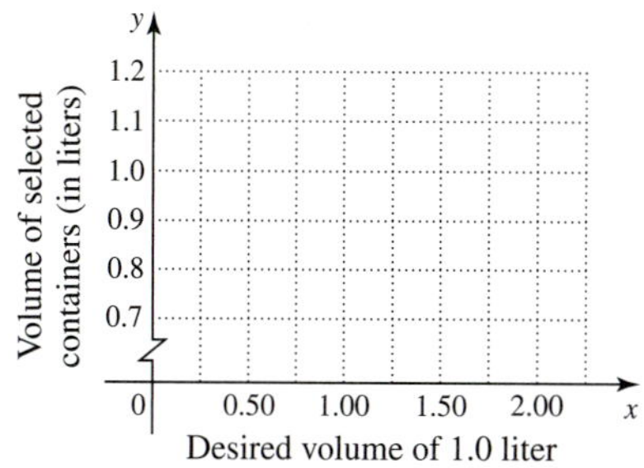

(b) Do the ordered pairs define a function? Why or why not?

Problems 22–23: Graph each equation using the Intercept Method.

22. $3x - y = 3$

x	*y*
0	
	0
Two additional points to verify the graph:	

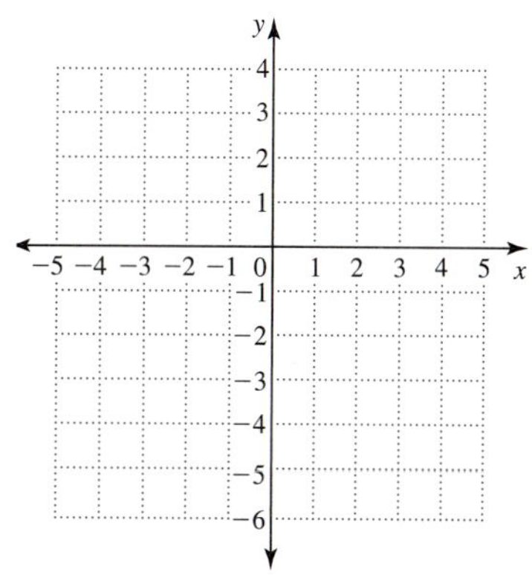

23. $3x + y = -6$

x	*y*
0	
	0
Two additional points to verify the graph:	

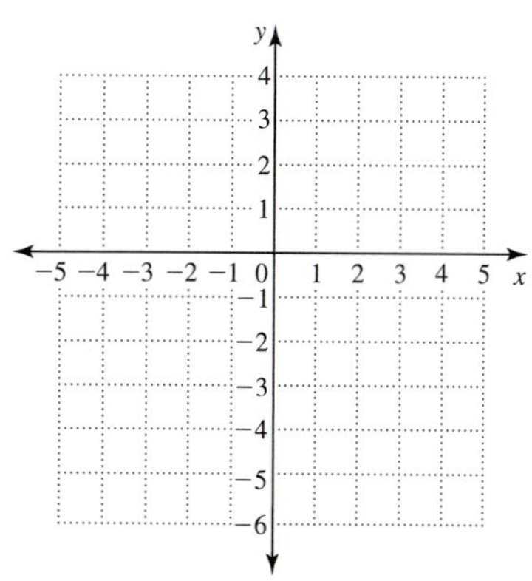

Problems 24–25: Graph each equation.

(a) Does the graph of the equation represent a function? How do you know?

(b) If the relation is a function, rewrite the equation in function notation.

24. $y = -2$

x	y

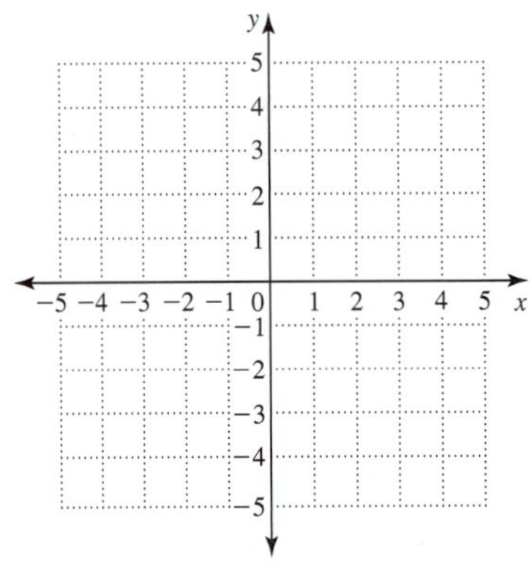

25. $x = 4$

x	y

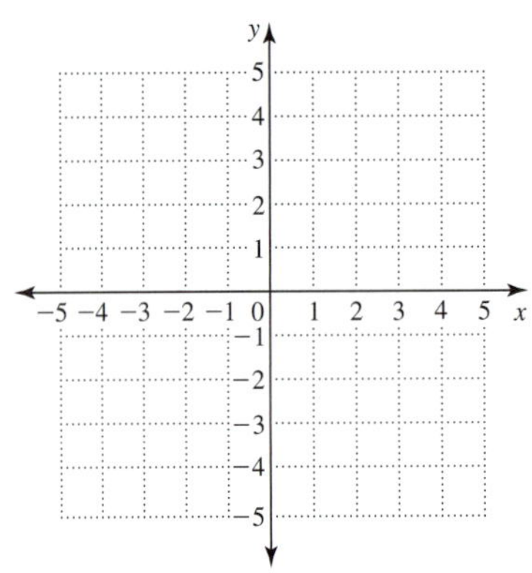

SECTION 3.4

Problems 26–29: The slope of a ski trail is determined by the ratio of the vertical drop and the horizontal distance. Find the slope of the four trails. Write the slope as a fraction and as a decimal, rounded to the nearest hundredth.

	Type of ski trail	Vertical drop (in feet)	Horizontal distance (in feet)	Slope	
				Fraction	Decimal
26.	Beginner	1080	4100		
27.	Intermediate	1090	2500		
28.	Expert	1310	2115		
29.	Super Expert	820	999		

30. Why is the ski trail in #29 called the "Super Expert trail?"

31. Determine whether each graph is increasing, decreasing, horizontal (constant), or vertical or whether the slope of the line is positive, negative, zero or undefined.

(a) Graph: Increasing Decreasing Horizontal Vertical

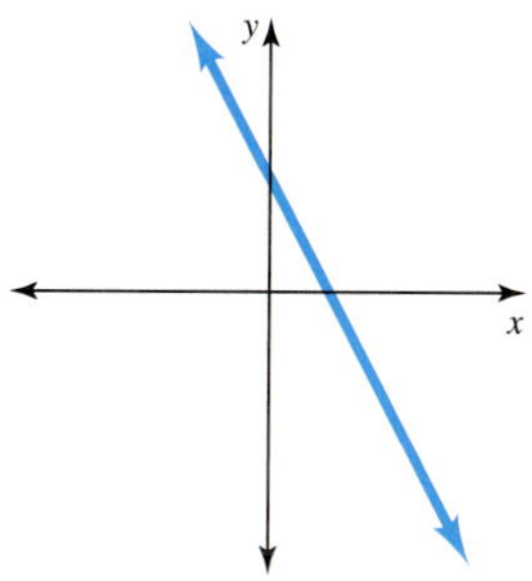

(b) Graph: Increasing Decreasing Horizontal Vertical

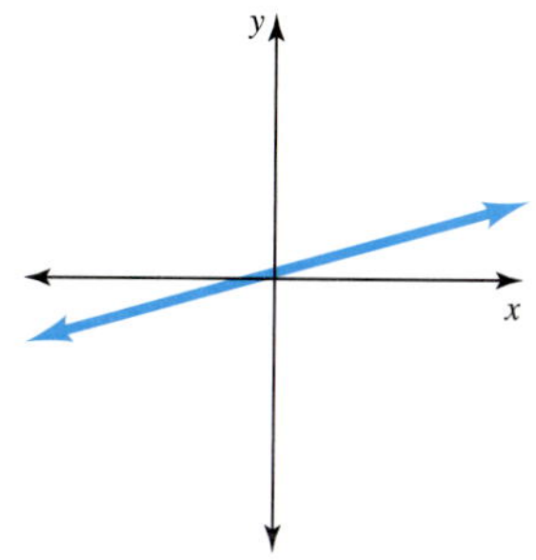

(c) Slope: Positive Negative Zero Undefined

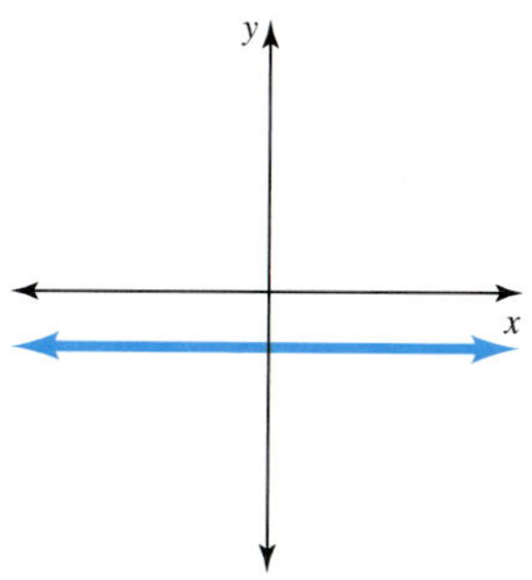

(d) Slope: Positive Negative Zero Undefined

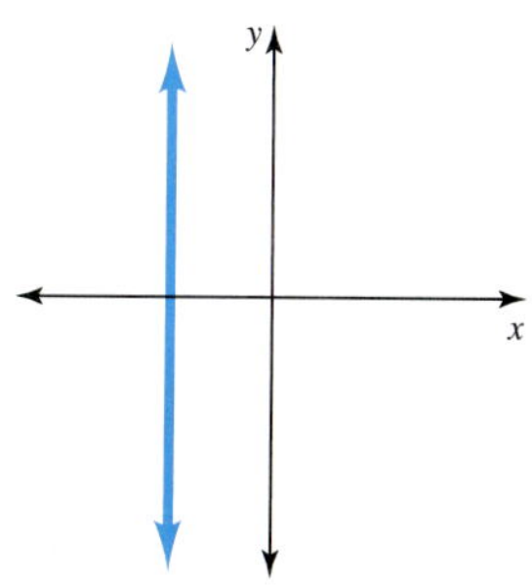

32. Fahrenheit temperatures and equivalent Celsius temperatures are an example of a linear function.

x = Fahrenheit temperature in degrees	y = Celsius temperature in degrees
−4	−20
0	−17.78
14	−10
32	0
50	10

(a) Choose any two points on the line and calculate the slope. Write the slope as a fraction.

(b) Identify the y-intercept from the table.

(c) Use the slope and y-intercept to write the equation in Slope-Intercept Form.

(d) Use the equation of the line to find the Celsius temperature if the Fahrenheit temperature is 63 degrees. Show your work.

33. Laurie drove to visit a friend at college in another state. Her speedometer was broken, but the odometer still worked and she kept track of how many miles she traveled each hour. The distances listed in the table describe a linear function.

x = Time traveled (in hours)	y = Distance traveled (in miles)
0	0
0.5	34
2	136
2.5	170
4	272

(a) Choose a scale and graph the points on a separate piece of graph paper. Connect the points.

(b) If the graph is linear, choose any two points on the line and calculate the slope.

(c) Identify the y-intercept from the table.

(d) Use the slope and the y-intercept to write the equation of the line in Slope-Intercept Form.

(e) Use the equation to predict how far Laurie had traveled after 6 hours if her speed remained the same throughout the trip.

Problems 34–35: The total number of Bachelor degrees conferred in the indicated fields of study in the academic years 1997, 1998, and 1999 are given in the table. Each set of data can be approximated with a linear function.[2]

(a) Calculate the slope of the line using the second and third points in the table.

(b) Identify the y-intercept from the table.

(c) Use the slope and y-intercept to write the equation of the line in Slope-Intercept Form.

(d) Use the equation to predict the number of degrees to be conferred in 2007–2008 ($t = 10$).

34. Bachelor degrees conferred in Economics to men and women:

Academic Year	x = Number of years since 1998	y = Degrees conferred (in thousands)
1997–1998	0	17.0
1998–1999	1	17.6
1999–2000	2	18.4

[2] National Center for Education Statistics.

35. Bachelor degrees conferred in Mathematics to men:

Academic Year	x = Number of years since 1998	y = Degrees conferred (in thousands)
1997–1998	0	6.6
1998–1999	1	6.5
1999–2000	2	6.4

Problems 36–39: Identify the coordinates of the y-intercept of the line and the slope from the equation. Graph each equation using the Slope-Intercept-Method. Verify the graph by computing and plotting two other points on the line.

36. $y = 2x - 3$

(a) y-intercept of the line: ______ (b) Slope of the line: ______

x	y

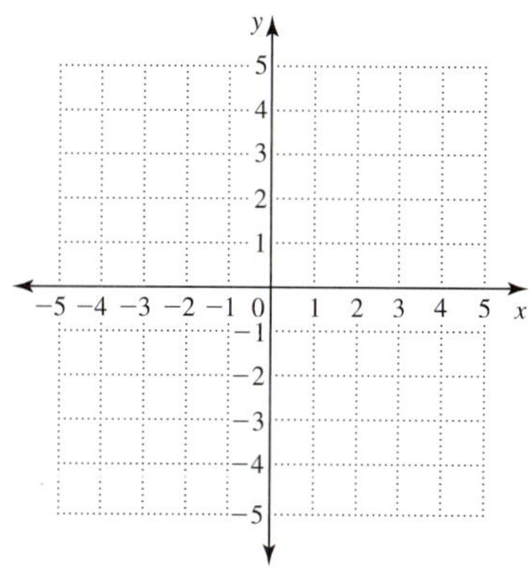

37. $y = -x + 1$

(a) y-intercept of the line: ______ (b) Slope of the line: ______

x	y

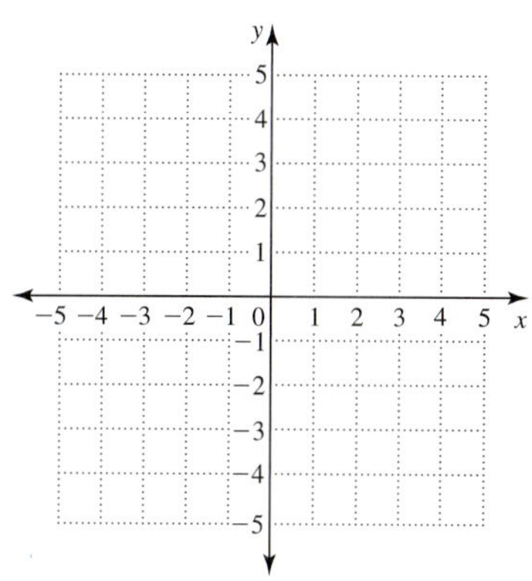

38. $y = 5$

(a) y-intercept of the line: ______ (b) Slope of the line: ______

x	*y*

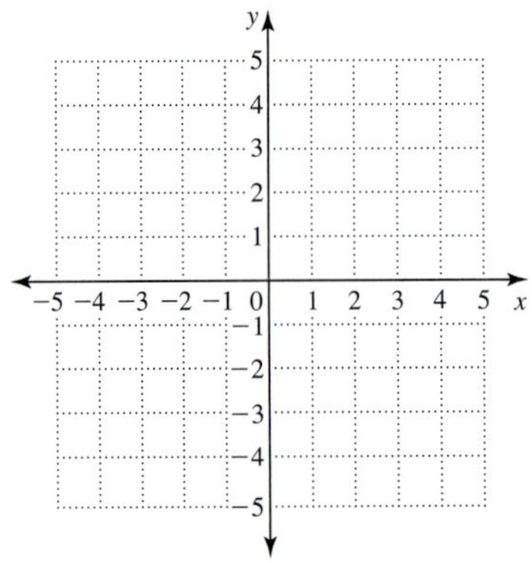

39. $x = 2$

(a) y-intercept of the line: ______ (b) Slope of the line: ______

x	*y*

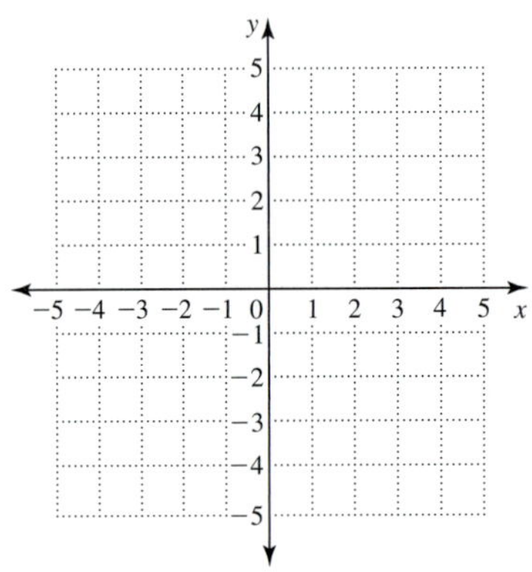

SECTION 3.5

40. The table shows the total miles (in billions) driven by trucks in the United States by year.

Year	t = Number of years since 1970	Total miles (in billions)
1980	10	399
1985	15	500

(a) Assume the data to be linear. Use the first point in the table, $(10, 399) = (x_1, y_1)$ and the second point $(15, 500) = (x_2, y_2)$ to find the slope.

(b) Use the point-slope form to find the equation of the line.

(c) Use the equation to predict the total truck miles driven in 1993 (Year 1993, $x = 23$).

41. The total number of Bachelor degrees awarded to both men and women in the fifty states and Washington, DC, is projected to be 1456 thousand in the year 2010.[3] This projection is based on the fact that the number of Bachelor degrees awarded is increasing at a constant rate of approximately 15 thousand each year. Use this information to create a linear function and compute the number of Bachelor degrees projected to be awarded in the year 2013.

(a) Write the ordered pair from the given information.

(year, number of Bachelor degrees awarded in thousands)

(______, ____________)

(b) Write the rate of change (slope) as a ratio.

(c) Use the slope and point to create a linear function.

(d) Compute the number of Bachelor degrees projected to be awarded to men and women in the year 2013.

Problems 42–49: Find the equation of each line with the given characteristics. Write the equations in Slope-Intercept Form.

42. Through the points $(4, 4)$ and $(-1, -1)$

43. Through the points $(-2, 7)$ and $(1, -3)$

44. Through the point $(2, -3)$ with x-intercept of 5

45. Through the point $(1, 6)$ with y-intercept of $(0, -1)$

46. Slope $\frac{1}{2}$ and y-intercept of $(0, 4)$

47. Slope $-\frac{2}{3}$ and y-intercept of $(0, -3)$

48. Slope $\frac{-5}{7}$ and x-intercept of $(0, 0)$

49. Slope -6 and x-intercept of $(9, 0)$

Problems 50–53: Rewrite the equation of each line in problems 42–45 in General Form ($Ax + By = C$).

50. Problem 42

51. Problem 43

52. Problem 44

53. Problem 45

[3] National Center for Education Statistics.

Practice Test—Chapter 3

SECTION 3.1

1. Given the grid and points:

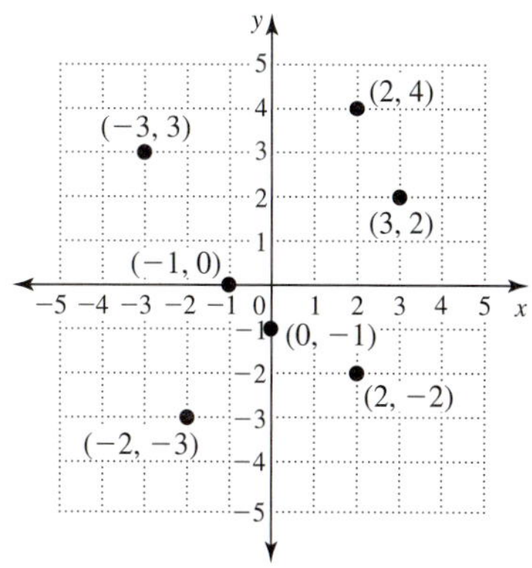

(a) What is the first coordinate value of the point with second coordinate value of 2?

(b) What is the second coordinate value of the point with first coordinate value of −2?

(c) What is the first coordinate value of the point with second coordinate value 0?

(d) What is the second coordinate value of the point(s) with first coordinate value of 2?

2. The mortality profile for squirrels is listed in the following table.[1] Ecologists use a life table to develop survivorship curves, beginning with a 1000-cohort sample in the first year.

Age period (in years)	Number of squirrels alive at the end of the age period
1	1000
2	796
4	151
6	11
8	0

(a) Smallest first coordinate value:
Largest first coordinate value:
Scale for horizontal axis:

(b) Smallest second coordinate value:
Largest second coordinate value:
Scale for vertical axis:

(c) Label each axis with your scale and word titles on a separate piece of graph paper. Plot the ordered pairs. Connect the points with a curve.

(d) Estimate the number of squirrels alive at the end of 3 years.

[1] Keck and Patterson, *Biomath*, 174.

SECTION 3.2

3. The College Student Government is selling raffle tickets for \$5 each. The cost of printing the raffle tickets was \$25. Let n represent the number of raffle tickets sold. Let P represent the total profit from the sale of tickets.
Profit = funds received from sale of raffle tickets − printing costs.

(a) Write the word or words that describe the input or independent variable.

(b) Write the word or words that describe the output or dependent variable.

(c) Let n represent the number of raffle tickets sold. Let P represent the total profit. Complete the following table of values.

n = Number of raffle tickets sold	P = Total profit
20	
40	
60	
80	
100	
n	$P(n) =$

(d) Identify a scale, label the axes, and graph the ordered pairs from the table on a separate piece of graph paper. Connect the points with a line.

(e) Use the graph to estimate the profit if the number of tickets sold is 90.

(f) Evaluate the exact profit if the number of tickets sold is 90.

(g) Calculate the exact number of tickets sold if the profit is \$400.

Problems 4–5: Draw and label the axes. Plot each set of ordered pairs. Use the definition of a relation and function and the vertical line test to determine whether each set represents a function.

4. $\{(-2, -3), (-2, 2), (-1, -1), (0, 1), (1, 1), (3, 4)\}$
Function? Yes or No

5.

Input	Output
0	1
2	2
4	3
6	4
8	5

Function? Yes or No

6. Evaluate each function for the given input values.

(a) Given $f(x) = 5x + 2$, evaluate $f(3)$.

(b) Given $f(x) = 5x + 2$, evaluate $f(-1)$.

(c) Given $g(x) = 3x^2 + 2x - 5$, evaluate $g(-3)$.

(d) Given $g(x) = 3x^2 + 2x - 5$, evaluate $g(0)$.

SECTION 3.3

7. Circle the statements that are linear functions.

(a) $y = 2x + 5$

(b) $x = 7$

(c) $f(x) = -x$

(d) $f(x) = x^3 - 1$

(e) $y = 2x^2 - 7$

(f) $y = \frac{5}{9}x + \frac{2}{3}$

8. Graph $f(x) = x + 3$ using the Intercept Method.

x	$f(x)$
0	
	0
Two additional points to verify the graph:	

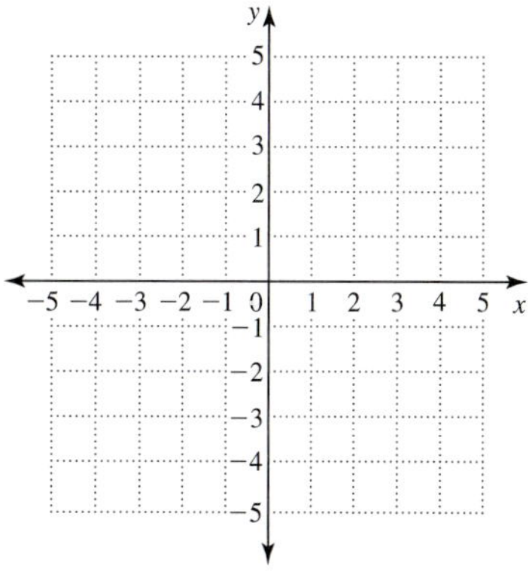

9. Graph $y = 3$.

x	y

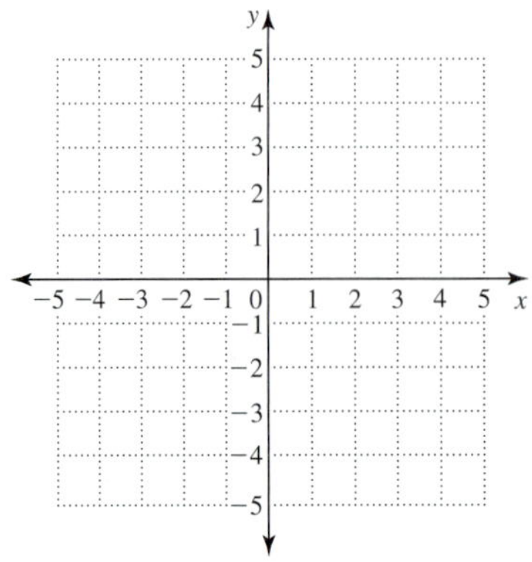

SECTION 3.4

Problems 10–12: Identify the coordinates of the y-intercept of the line and the slope from the equation. Graph each equation or function using the Slope-Intercept Method. Verify the graph by computing and plotting two other points on the line.

10. $f(x) = 3x - 1$

(a) y-intercept of the line: ______ (b) Slope of the line: ______

x	$f(x)$

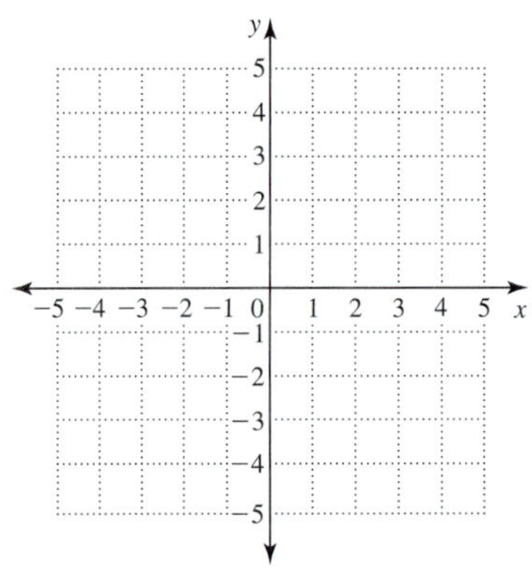

11. $f(x) = -\frac{3}{2}x - 1$

(a) y-intercept of the line: ______ (b) Slope of the line: ______

x	$f(x)$

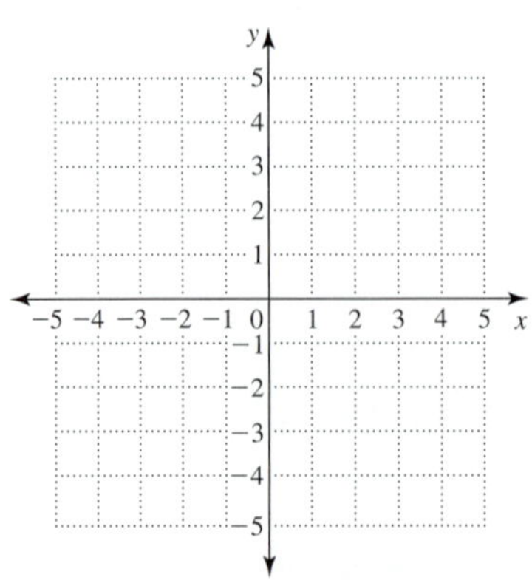

12. $y = -2x$

(a) y-intercept of the line: ______ (b) Slope of the line: ______

x	y

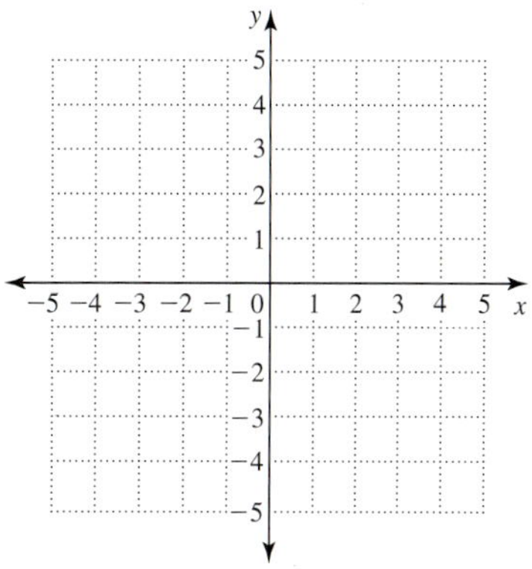

13. Write the equation of the line with slope = 5 and y-intercept of (0, 2).

14. Write the equation of the line with slope $= \frac{3}{2}$ and y-intercept of $(0, -1)$.

15. Write the equation of the line with slope undefined and x-intercept of $(-1, 0)$.

16. Write the equation of the line with slope = 0 and y-intercept of $(0, -6)$.

SECTION 3.5

17. The NASA Lewis Materials Adherence Experiment (MAE) on the Sojourner Rover[2] on Mars measured the dust deposited on the rover's solar array for two and a half weeks of operation. The graph of the data collected approximated a straight line and contained the two points listed in the table.

Number of days	Percent of dust coverage
2	0.5
6	1.5

(a) Calculate the slope of the line through the two points.

(b) Find the equation of the line through the two data points. Write the equation of the line in Slope-Intercept Form.

(c) Use the equation to predict the percent of dust coverage at the end of the twelfth day.

[2] NASA Glenn Research Center.

Problems 18–21: Find the equation of each line with the given characteristics. Write each equation in Slope-Intercept Form and General Form.

18. Passing through the points $(1, 2)$ and $(-1, 3)$.

19. Passing through the points $(-1, -3)$ and $(4, 1)$.

20. Through the point $(1, 4)$ with an x-intercept of -2.

21. Slope $\frac{1}{2}$ passing through the point $(0, 1)$.

Answers to Getting Started—Chapter 3

SECTION 3.1

1. Grid on page 130.
2. (a) (1, 2) Quadrant: I

   (b) (−3, 4) Quadrant: II

   (c) (3, −4) Quadrant: IV

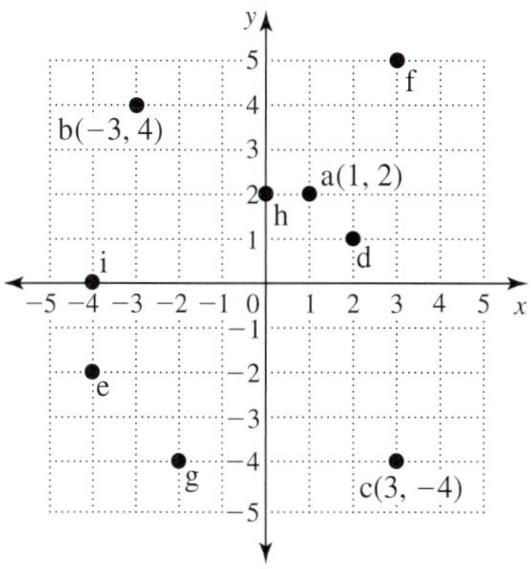

   (d) (2, 1) Quadrant: I

   (e) (−4, −2) Quadrant: III

   (f) (3, 5) Quadrant: I

   (g) (−2, −4) Quadrant: III

   (h) (0, 2) None

   (i) (−4, 0) None

5. Answers will vary

SECTION 3.2

1. (a) Kristen

   (b) Tom, Mike, Ashley

   (c) Nicole

   (d) Bill, Kathy

2. (a) Input is 175, output is 400:

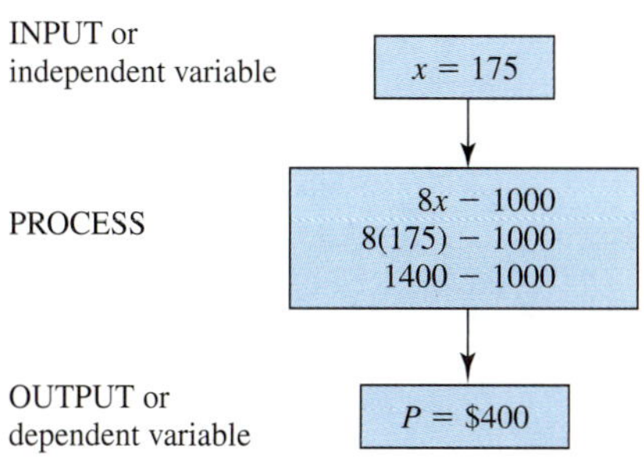

   (b) Input is 400, output is 2200:

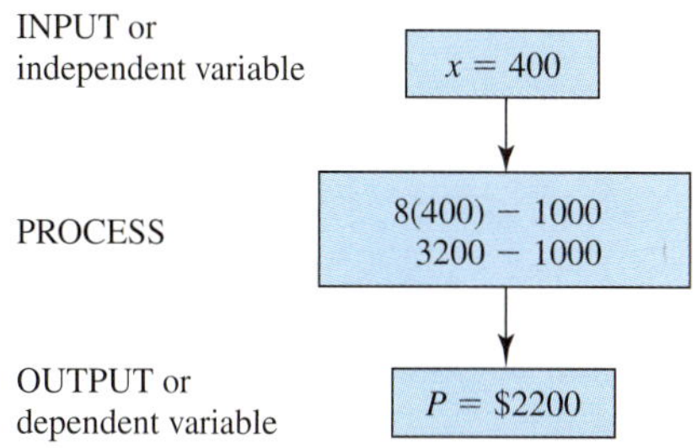

   (c) One output value for each input value.

3. (a) No. A function is a relation in which every first coordinate value is paired with exactly one second coordinate value. Some relations are functions. Others are not.

   (b) Yes, every function is a relation.

SECTION 3.3

1. (a) The coordinates of the x-intercept: (125, 0).

   (b) Every point on the x-axis has a second coordinate with a value of 0.

   (c) The coordinates of the y-intercept: (0, −1000).

   (d) Every point on the y-axis has a first coordinate with a value of 0.

2. (a)

x	**y**
−2	−3
−1	−2
0	−1
1	0
2	1
3	2

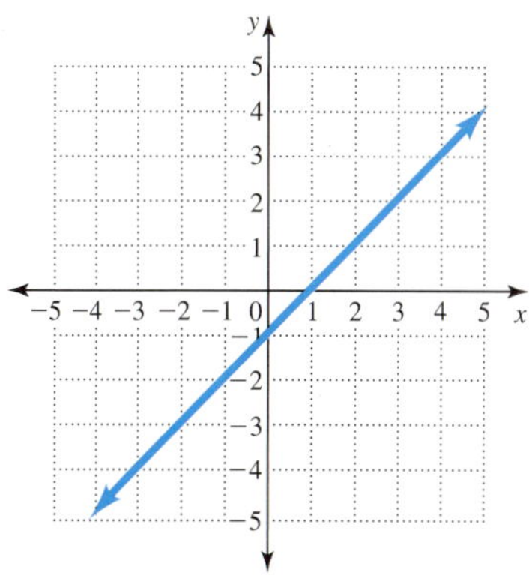

   (c) x-intercept: (1, 0)

   (d) y-intercept: (0, −1)

3. (a–b)

Date:	x	y
Feb 1	1	80
Feb 2	2	80
Feb 3	3	80
Feb 4	4	80
Feb 5	5	80
Feb 6	6	80
Feb 7	7	80
Feb 8	8	80
Feb 9	9	80
Feb 10	10	80

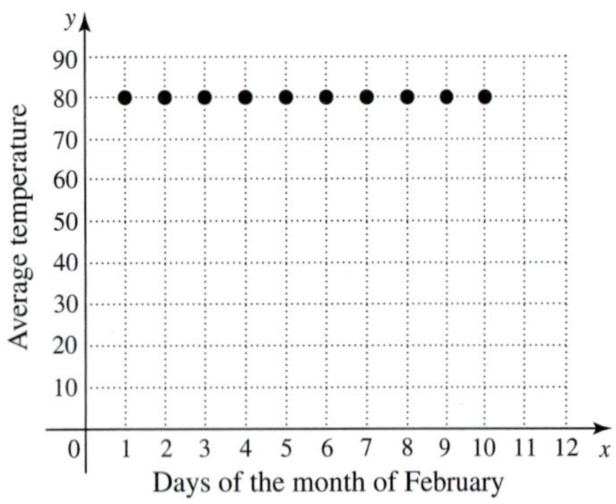

(c) Yes. For each value of the independent variable there is one and only one value for the dependent variable.

4. (a)

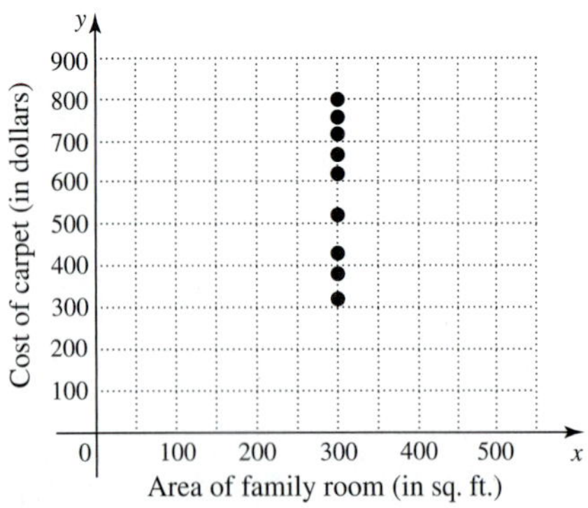

(b) No. There are eight values of the dependent variable for the one value of 300 of the independent variable.

5. A horizontal line that is not the x-axis has no x-intercept.

6. A vertical line that is not the y-axis has no y-intercept.

SECTION 3.4

1. (a) Steamboat Springs. Steamboat has the largest vertical drop. Snowmass is a very close second.

(b) Winter Park. Winter Park has the smallest vertical drop.

2. (a) $\dfrac{\text{Vertical Drop in feet}}{\text{Horizontal Distance in feet}} = \dfrac{3600}{15425} \approx 0.23$

(b) $\dfrac{\text{Vertical Drop in feet}}{\text{Horizontal Distance in feet}} = \dfrac{2200}{20474} \approx 0.11$

3. (a) Steamboat Springs is steeper because the ratio (or slope) is a larger number.

(b) Choose Winter Park if you are a beginner because the slope of the ski run is flatter or less steep.

SECTION 3.5

1. (a) Slope $= \dfrac{4}{1}$

(b) (2009, 684)

(c) $\dfrac{4}{1} = \dfrac{y - 684}{x - 2009}$

(d) $1(y - 684) = 4(x - 2009)$

(e) $y - 684 = 4x - 8036$

$y = 4x - 7352$

(f) Let $x = 2013$: $y = 4x - 7352$

$y = 4(2013) - 7352$

$y = 8052 - 7352$

$y = 700$ thousand

Cumulative Test Chapters 1–3

1. Use the number line to add or subtract.

(a) $-5 - (-3) =$

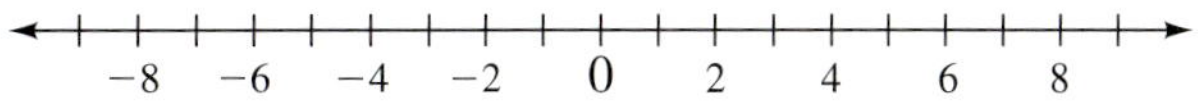

(b) $-6 + 4 =$

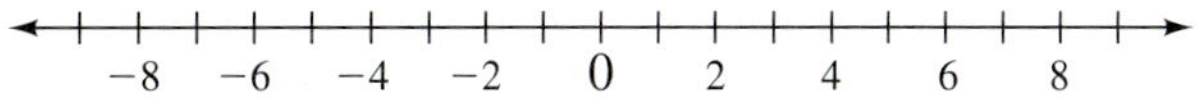

2. The variables a, b, c, and d represent real numbers in the respective locations on the number line.

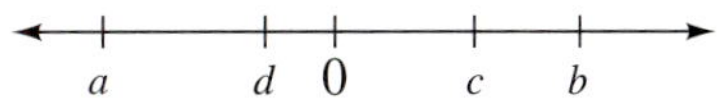

(a) Is $-d$ positive or negative?

(b) Is $a + c$ positive or negative?

(c) Is $a \cdot d$ positive or negative?

(d) Is $-c^2$ positive or negative?

3. Evaluate each expression, given that $a = -2$ and $b = 4$.

(a) $a^2 - b^2$

(b) $-a^2 \cdot b^2$

Problems 4–7: Use the Order of Operations to perform the indicated operations.

4. $-3^2 - 2(4 - 6)$

5. $-12 \times 6 \div 3 - (-2)$

6. $\dfrac{[-10 - (-4)]^2}{(-2)^2} - (-5)^2$

7. $-3^2[-3^3(-3)^2]$

8. Justin's landscaping business had a loss of \$500 on Monday and a loss of \$120 on Tuesday. On Wednesday there was a profit of \$819 and on Thursday there was a loss of \$88. Find the total profit or loss for the four days.

9. Substitute the given numerical values for the objects and variables into the expressions: $\blacklozenge = -2$, $x = -5$, $\bigstar = -1$, $y = -3$. Perform the operations.

(a) $(8\blacklozenge - x)(-4\bigstar)$

(b) $2(x^2 - x + 1)$

Problems 10–11: Perform the operations.

10. $(x^2 + x - 1) - 3(5x + 4)$

11. $(5x^2 - 1) - (5x + 3) - 3(2x^2 - 1) - (x - 8)$

12. Add the polynomials vertically.

$$\begin{array}{r} 12x^3 + 5x^2 - 4x + 3 \\ +\quad 8x^3 - 2x^2 - 5x + 8 \\ \hline \end{array}$$

Problems 13–14: Solve each equation.

13. $3(y + 1) = 2(y - 5)$

14. $5.5x + 2(1.5 - x) = 30 + 0.5x$

Problems 15–16: Solve each inequality. Draw a graph of the solution set and write the solution in set-builder and interval notation.

15. $-9 > 5 - 7z$

(a) Graph:

(b) Set-builder notation:

(c) Interval notation:

16. $-2(x - 1) \geq 4(3 - x)$

(a) Graph:

(b) Set-builder notation:

(c) Interval notation:

Problems 17–18: Solve each equation or formula for the indicated variable.

17. $Ax + By = C$ Solve for y.

18. $3(x - 4) = y + 5$ Solve for y.

19. The City Youth Group is selling t-shirts for \$5 each. The Youth Group purchased the t-shirts for \$4 each and shipping was \$20. Let n represent the number of t-shirts sold. Let P represent the total profit.

(a) Complete the following table of values.

n = Number of t-shirts sold	**P = Total profit**
20	
40	
60	
80	
100	
n	$P(n) =$

(b) Identify a scale, label the axes, and graph the ordered pairs from the table on a separate piece of graph paper.

(c) Use the graph to estimate the profit if the number of t-shirts sold is 55.

(d) Evaluate the exact profit if the number of t-shirts sold is 55.

(e) Calculate the exact number of t-shirts sold to break even.

20. Evaluate each function for the given input values.

(a) Given $f(x) = -x - 2$, evaluate $f(-3)$.

(b) Given $f(x) = -x + 2$, evaluate $f(1)$.

(c) Given $g(x) = 3x^2 + x - 5$, evaluate $g(-2)$.

(d) Given $g(x) = 3x^2 - 5$, evaluate $g(0)$.

21. Graph $f(x) = -x + 3$ using the Intercept Method.

x	$f(x)$
0	
	0
Two additional points to verify the graph:	

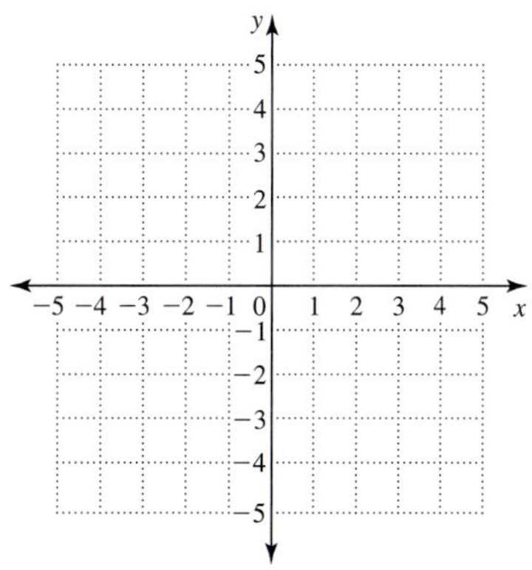

Problems 22–25: Find the equation of each line with the given characteristics. Write the equation in Slope-Intercept Form and General Form.

22. Passing through the points $(1, -2)$ and $(-1, 3)$

23. Slope -3 with y-intercept $(0, -2)$

24. Through the point $(1, 4)$ with y-intercept $(0, -2)$

25. Write the equation of the line with slope $-\frac{1}{2}$ and x-intercept $(-1, 0)$

26. The total number of Associate degrees conferred in Social Science and History in the academic years 1997, 1998, and 1999 are given in the table.[1] The data can be approximated with a linear function.

(a) Calculate the slope of the line using the second and third points in the table.

(b) Identify the y-intercept from the table.

(c) Use the slope and y-intercept to write the equation of the line in Slope-Intercept Form.

(d) Use the equation to predict the number of degrees to be conferred in 2007–2008 ($t = 10$).

Academic Year	x = Number of years since 1998	y = Degrees conferred (in thousands)
1997–1998	0	4.2
1998–1999	1	4.6
1999–2000	2	5.1

[1] National Center for Education Statistics.

CHAPTER

4

Linear Systems, Graphical Solutions of Linear Equations and Inequalities

Chinese mathematicians explored and solved systems of linear equations as early as 300–0 B.C. They began solving systems by a series of adjusted guesses. Later, they used a method very similar to the elimination method presented in Section 4.3.

4.1 Solving Linear Equations Graphically and Analytically

OBJECTIVES

- Solve linear equations using the X-Intercept and Intersection-of-Graphs Graphical Methods.
- Use graphical methods to verify solutions of linear equations found analytically.

1 2 3 Getting Started

In this chapter, we explore problems involving two equations called systems of equations. Before discussing systems, we need to consider graphical solutions of one linear equation. When an equation is solved on paper using algebraic techniques, the equation is solved analytically. When an equation is rewritten as an equation with two variables and graphed, the graph also can be used to determine the solution of the original equation graphically. The X-Intercept and Intersection-of-Graphs Methods are presented in this section.

There is a direct relationship between the solution of an equation in one variable and the associated function in two variables. A graphical method of solution may be used to confirm an analytical solution or used to solve an equation when analytical methods are difficult.

1. Follow these steps to solve the linear equation $0 = 2x + 2$ using the X-Intercept Graphical Method.

$$0 = 2x + 2$$

Substitute the variable y for zero in the equation to form the associated function.

$$y = 2x + 2$$

(a) Complete a table of values. Plot the points and connect them with a line.

x	$y = 2x + 2$
−3	
−2	
−1	
0	
1	
2	
3	

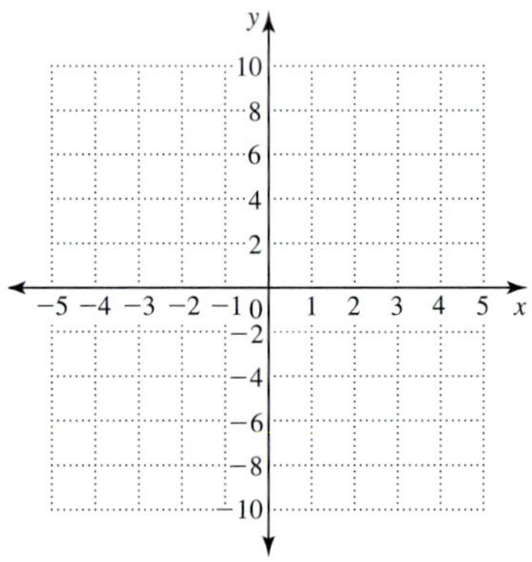

(b) Write the x-intercept of the graph.

(c) The graphical solution of the original equation is the x-coordinate of the x-intercept. Write the solution.

(d) Solve $0 = 2x + 2$ analytically.

(e) Is the x-coordinate of the x-intercept the same as the analytical solution of the equation? If not, recheck your work.

2. A second graphical solution method is called the Intersection-of-Graphs Method. Follow these steps to solve the equation $4x - 1 = 2x + 3$ using the Intersection-of-Graphs Method.

$$\underbrace{4x - 1}_{y_1} = \underbrace{2x + 3}_{y_2}$$

Label the left side of the equation y_1 and the right side of the equation y_2.

(a) Complete the tables of values and graph y_1 and y_2 on the same grid.

x	y_1	y_2
−2		
−1		
0		
1		
2		
3		

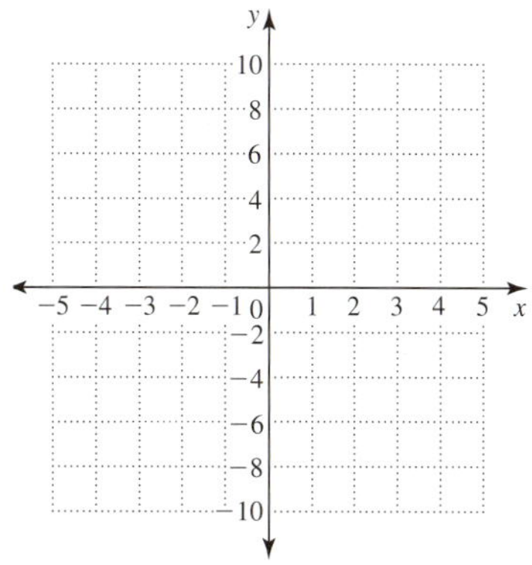

(b) Identify the point of intersection of the two lines.

(c) The graphical solution of the original equation is the x-coordinate of the point of intersection. Write the solution.

(d) Solve the equation analytically, showing all work.

$$4x - 1 = 2x + 3$$

(e) Is the x-coordinate of the point of intersection the same as the analytical solution of the equation? If not, recheck your work.

Discussing Mathematics

Most equations can be solved by analytical and graphical methods. Graphical and analytical methods complement each other, providing alternate methods for solving equations and checking our work.

THE *X*-INTERCEPT GRAPHICAL METHOD

The *X*-Intercept Method identifies the *x*-coordinate of the *x*-intercept of the graph as the **zero of a relation or function**, the value of the independent variable when the dependent variable is zero. The zero is where the graph crosses the *x*-axis and is the solution of an equation.

THE *X*-INTERCEPT GRAPHICAL METHOD OF SOLVING AN EQUATION

1. Rewrite the equation with all nonzero terms on one side of the equation equal to zero.
2. Substitute y for zero in the transformed equation to form the associated function.
3. Graph the function by making a table of values, plotting and connecting the points with a line.
4. Identify the x-intercept of the graph. The solution of the equation is the x-coordinate of the x-intercept.
5. Verify the answer by substituting the solution into the original equation.

EXAMPLE 1: Solve the equation $2(x + 3) = 4$ using the *X*-Intercept Graphical Method.

$2(x + 3) - 4 = 0$ Rewrite the equation with all nonzero terms on one side of the equation equal to zero.

$2(x + 3) - 4 = y$ Substitute y for zero to form the associated function.

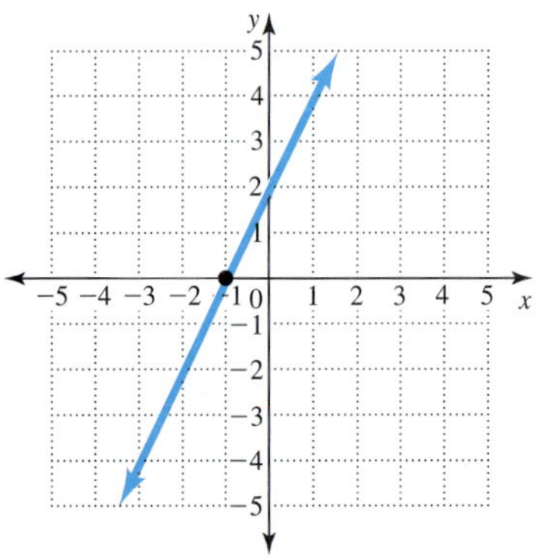

x	y_1
−3	−4
−2	−2
−1	0
0	2
1	4
2	6
3	8

Graph $y_1 = 2(x + 3) - 4$.

Note: Parentheses may be removed and the equation simplified to graph the equivalent equation $y_1 = 2x + 2$.

The x-intercept of the graph occurs at $x = -1$ and can be observed on the graph and as the ordered pair $(-1, 0)$ in the table.

➢ *The solution of the equation is the x-coordinate of the x-intercept, $x = -1$.*

Let $x = -1$: $2(x + 3) = 4$

$2((-1) + 3) \stackrel{?}{=} 4$

$4 = 4$

Verify the answer by substituting the solution into the original equation. ✓ It checks.

EXAMPLE 2: Solve the equation $3(x + 1) - 5 = 2x + 1$ using the X-Intercept Graphical Method.

$$3(x + 1) - 5 - 2x - 1 = 0$$
$$3x + 3 - 5 - 2x - 1 = 0$$
$$x - 3 = 0$$

Rewrite the equation with all nonzero terms on one side of the equation equal to zero. Remove parentheses to simplify.

$$x - 3 = y$$

Substitute y for zero to form the associated function.

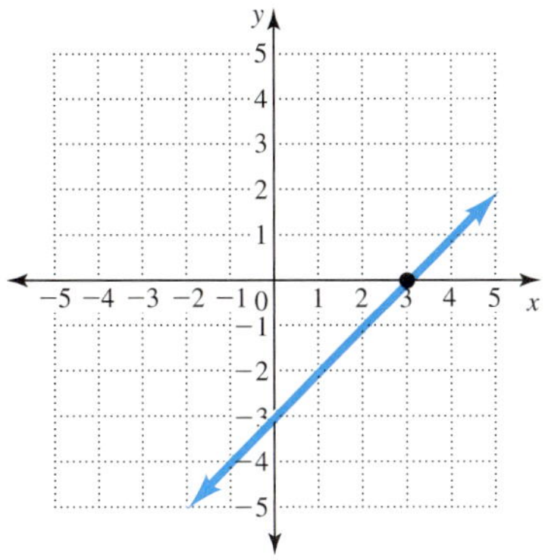

x	y_1
−3	−6
−2	−5
−1	−4
0	−3
1	−2
2	−1
3	0

Graph $y_1 = x - 3$.

The x-intercept of the graph occurs at $x = 3$ and can be observed on the graph and as the ordered pair (3, 0) in the table.

➢ *The solution of the equation is the x-coordinate of the x-intercept, $x = 3$.*

Let $x = 3$:
$$3(x + 1) - 5 = 2x + 1$$
$$3(3 + 1) - 5 \stackrel{?}{=} 2(3) + 1$$
$$7 = 7$$

Verify the answer by substituting the solution into the original equation. ✓ It checks.

INTERSECTION-OF-GRAPHS METHOD

A second graphical method is called the **Intersection-of-Graphs Method**. The associated function of each side of the original equation is graphed. The x-coordinate of the point of intersection is the solution of the original equation.

THE INTERSECTION-OF-GRAPHS METHOD FOR SOLVING AN EQUATION

1. Label the left side of the equation y_1. Label the right side of the equation y_2.
2. Make tables of values for y_1 and y_2.
3. Graph equations y_1 and y_2 on the same grid.
4. Identify the point of intersection of the two graphs where $y_1 = y_2$. The x-coordinate of the point of intersection is the solution of the original equation.
5. Verify the answer by substituting the solution into the original equation.

EXAMPLE 3: Solve the equation $6(x + 1) + 3 = 3(x + 2)$ using the Intersection-of-Graphs Method.

$$\underbrace{6(x+1)+3}_{y_1 = 6(x+1)+3} = \underbrace{3(x+2)}_{y_2 = 3(x+2)}$$

Label the left side of the equation y_1. Label the right side of the equation y_2.

Make tables of values for y_1 and y_2.

x	y_1	y_2
−4	−15	−6
−3	−9	−3
−2	−3	0
−1	3	3
0	9	6
1	15	9
2	21	12

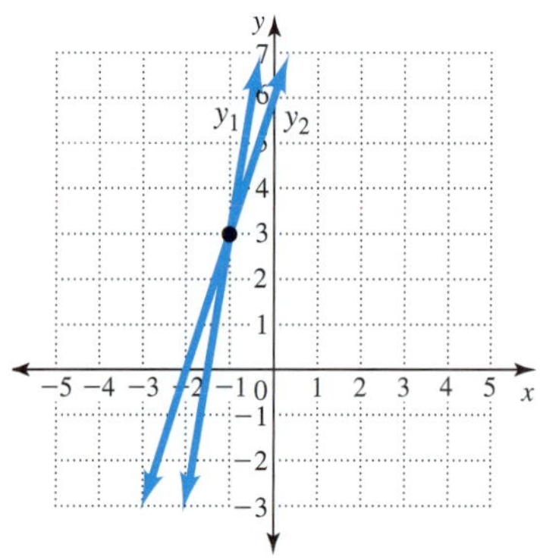

Graph equations y_1 and y_2 on the same grid.

The point of intersection of the two graphs is where $y_1 = y_2$ at $(-1, 3)$.

➢ *The x-coordinate of the point of intersection is the solution of the original equation* $x = -1$.

Let $x = -1$: $6(x + 1) + 3 = 3(x + 2)$

$$6((-1) + 1) + 3 \stackrel{?}{=} 3((-1) + 2)$$

$$6(0) + 3 \stackrel{?}{=} 3(1)$$

$$3 = 3$$

Verify the answer by substituting the solution into the original equation.

✓ It checks.

➢ *The choice of which graphical method to use to solve an equation depends on the equation and personal preference. Either the X-Intercept or Intersection-of-Graphs Graphical Method can be used to solve most equations.*

1 2 3 Doing Mathematics

- Write the definitions of the words and symbols listed for this section in the Dictionary in Appendix A.
- On 3 × 5-inch cards, write the following processes for future reference:
  (a) The X-Intercept Method for solving an equation graphically
  (b) The Intersection-of-Graphs Method for solving an equation graphically

Problems 1–8: Solve each equation analytically. Confirm each solution using the X-Intercept Graphical Method. Verify each answer by substituting the solution into the original equation.

1. Given $2(x + 1) = 0$.

(a) Solve the equation analytically.

(b) Graph the associated function.

x	y
−3	
−2	
−1	
0	
1	
2	
3	

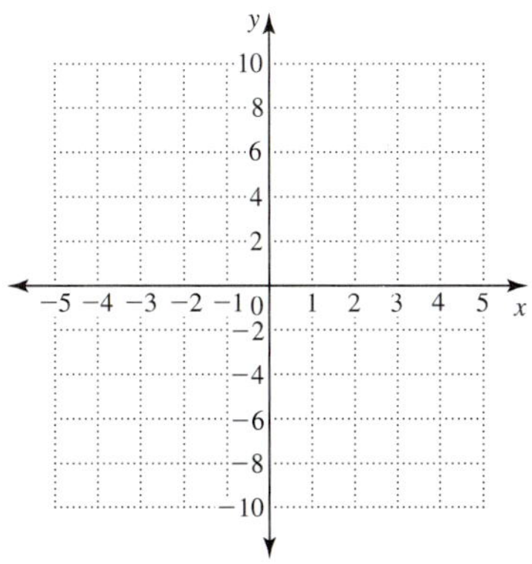

(c) What is the x-coordinate of the x-intercept of the graph?

(d) Does the graphical solution confirm the analytical solution? Why or why not?

2. Given $3(x - 2) = 0$.

(a) Solve the equation analytically.

(b) Graph the associated function.

x	y
−3	
−2	
−1	
0	
1	
2	
3	

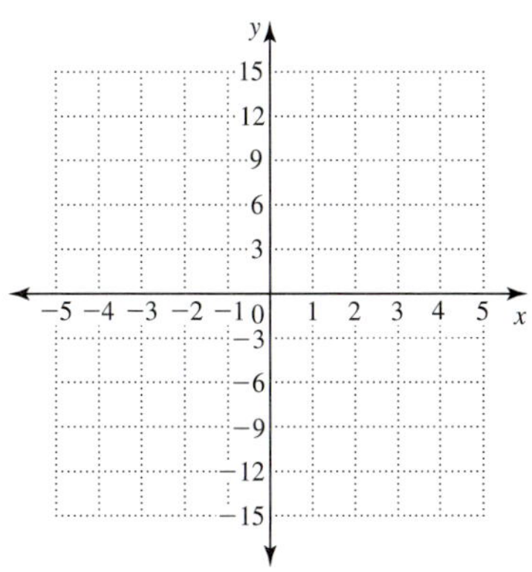

(c) What is the x-coordinate of the x-intercept of the graph?

(d) Does the graphical solution confirm the analytical solution? Why or why not?

3. Given $2x - 5 = -3$.

(a) Solve the equation analytically.

(b) Graph the associated function.

x	y
-3	
-2	
-1	
0	
1	
2	
3	

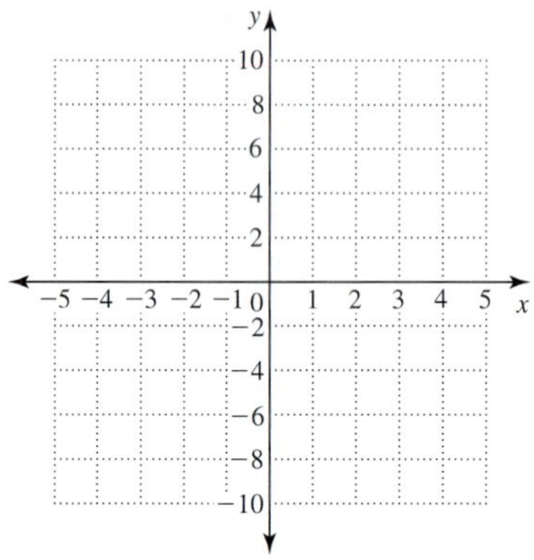

(c) What is the x-coordinate of the x-intercept of the graph?

(d) Does the graphical solution confirm the analytical solution? Why or why not?

4. Given $3x + 4 = -2$.

(a) Solve the equation analytically.

(b) Graph the associated function.

x	y
-3	
-2	
-1	
0	
1	
2	
3	

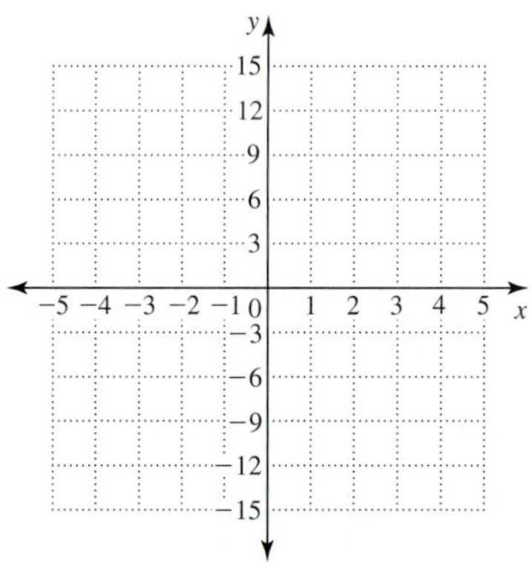

(c) What is the x-coordinate of the x-intercept of the graph?

(d) Does the graphical solution confirm the analytical solution? Why or why not?

5. Given $-(x + 1) = -1$.

(a) Solve the equation analytically.

(b) Graph the associated function.

x	y
-3	
-2	
-1	
0	
1	
2	
3	

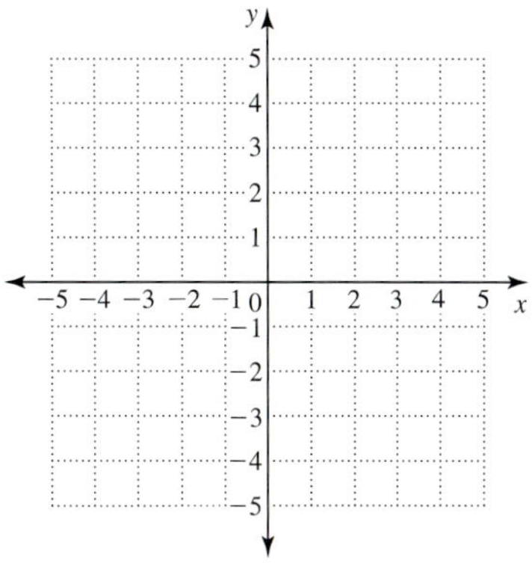

(c) What is the x-coordinate of the x-intercept of the graph?

(d) Does the graphical solution confirm the analytical solution? Why or why not?

6. Given $-2(x - 3) = -4$.

(a) Solve the equation analytically.

(b) Graph the associated function.

x	y
-1	
1	
3	
5	
7	
9	

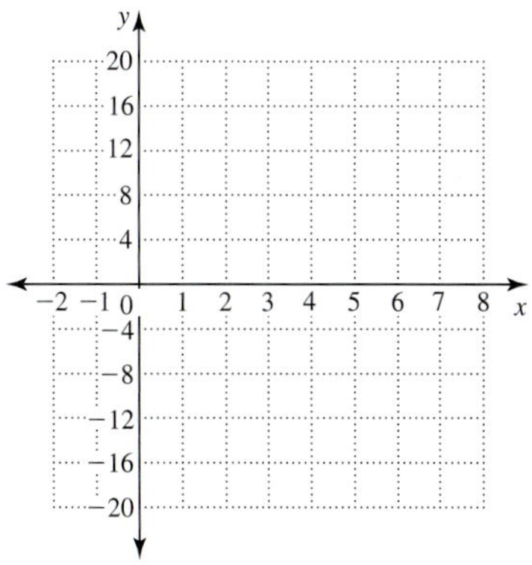

(c) What is the x-coordinate of the x-intercept of the graph?

(d) Does the graphical solution confirm the analytical solution? Why or why not?

7. Given $1 = -17 - 6x$.

(a) Solve the equation analytically.

(b) Graph the associated function.

x	y
−5	
−3	
−1	
1	
3	

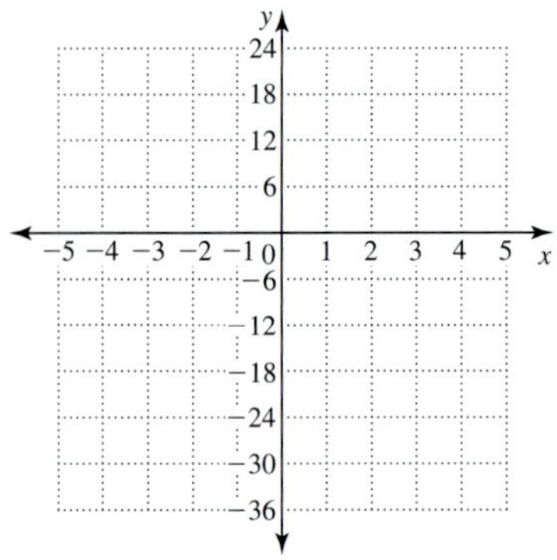

(c) What is the x-coordinate of the x-intercept of the graph?

(d) Does the graphical solution confirm the analytical solution? Why or why not?

8. Given $13 = -7 + 4x$.

(a) Solve the equation analytically.

(b) Graph the associated function.

x	y
2	
3	
4	
5	
6	
7	

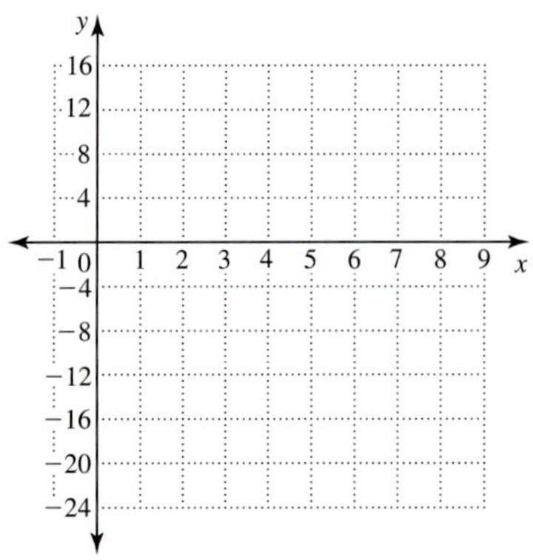

(c) What is the x-coordinate of the x-intercept of the graph?

(d) Does the graphical solution confirm the analytical solution? Why or why not?

Problems 9–16: Solve each equation analytically. Confirm the solution graphically, using the Intersection-of-Graphs Method. Label the left side of the equation y_1 and the right side of the equation y_2. Complete the tables of values for y_1 and y_2. Verify each answer by substituting the solution into the original equation.

9. Given $8x - 3 = 3(2x + 3) - x$.

(a) Solve the equation analytically.

(b) Graph both equations on the same grid.

x	y_1	y_2
−1		
0		
1		
2		
3		
4		
5		

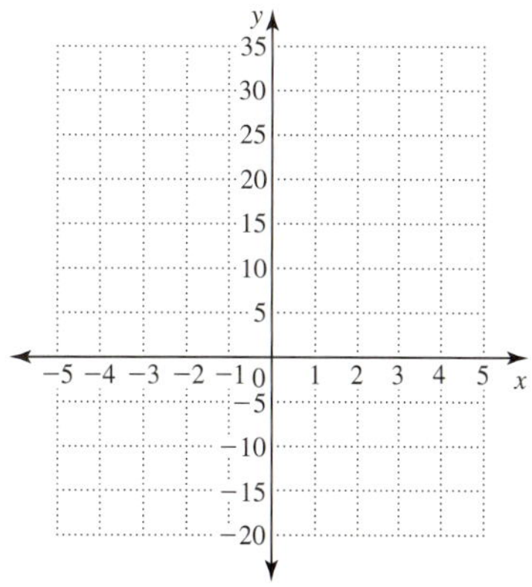

(c) What is the x-coordinate of the point of intersection of the graphs?

(d) Does the graphical solution confirm the analytical solution? Why or why not?

10. Given $2(6x - 1) = 2(3 + 4x) + 2x$.

(a) Solve the equation analytically.

(b) Graph both equations on the same grid.

x	y_1	y_2
2		
3		
4		
5		
6		

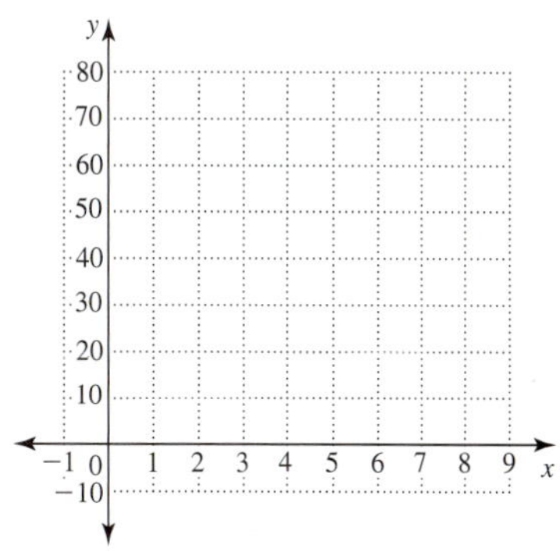

(c) What is the x-coordinate of the point of intersection of the graphs?

(d) Does the graphical solution confirm the analytical solution? Why or why not?

11. Given $2(3x + 5) = 2(x - 1)$.

(a) Solve the equation analytically.

(b) Graph both equations on the same grid.

x	y_1	y_2
−4		
−3		
−2		
−1		
0		

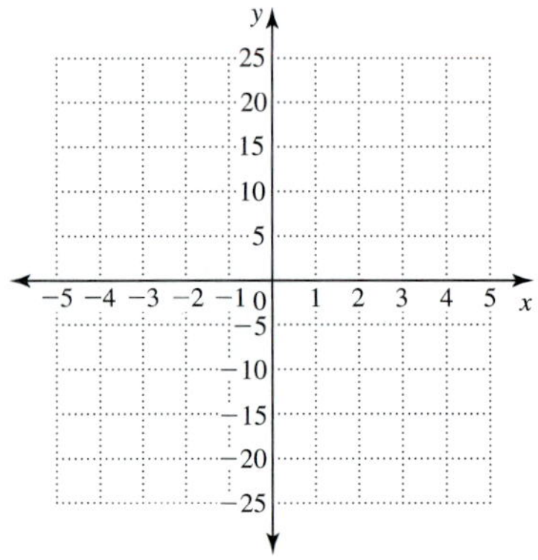

(c) What is the x-coordinate of the point of intersection of the graphs?

(d) Does the graphical solution confirm the analytical solution? Why or why not?

12. Given $6.3x - (1.3x + 2) = 4x + 1$.

(a) Solve the equation analytically.

(b) Graph both equations on the same grid.

x	y_1	y_2
−1		
0		
1		
2		
3		
4		
5		

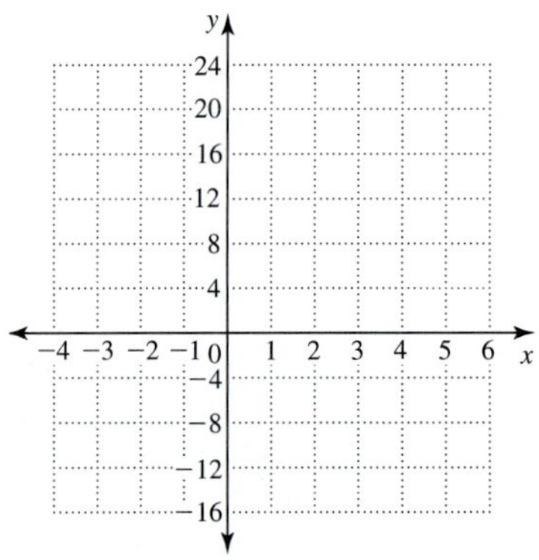

(c) What is the x-coordinate of the point of intersection of the graphs?

(d) Does the graphical solution confirm the analytical solution? Why or why not?

13. Given $-2.1x - (4.2x + 3.5) = -3x + 13$.

(a) Solve the equation analytically.

(b) Graph both equations on the same grid.

x	y_1	y_2
−7		
−6		
−5		
−4		
−3		

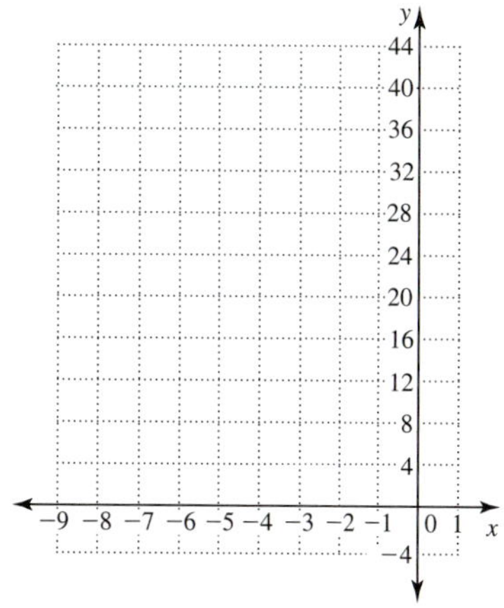

(c) What is the x-coordinate of the point of intersection of the graphs?

(d) Does the graphical solution confirm the analytical solution? Why or why not?

14. Given $5x + 2 = 2(x - 3.5)$.

(a) Solve the equation analytically.

(b) Graph both equations on the same grid.

x	y_1	y_2
−5		
−4		
−3		
−2		
−1		

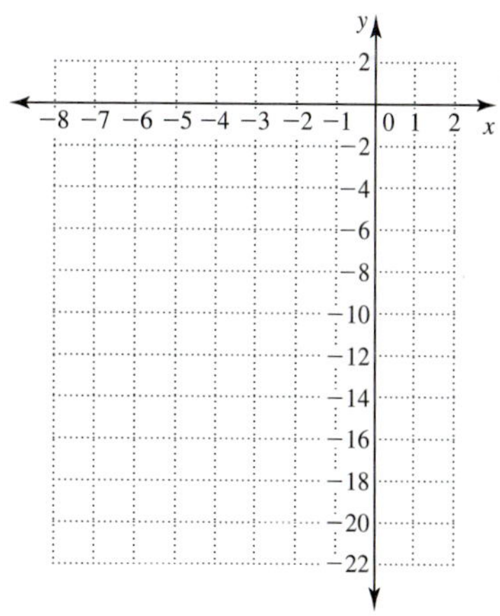

(c) What is the x-coordinate of the point of intersection of the graphs?

(d) Does the graphical solution confirm the analytical solution? Why or why not?

15. Given $(x + 3) - (x - 4) = 2x - 3$.

(a) Solve the equation analytically.

(b) Graph both equations on the same grid.

x	y_1	y_2
1		
2		
3		
4		
5		
6		
7		

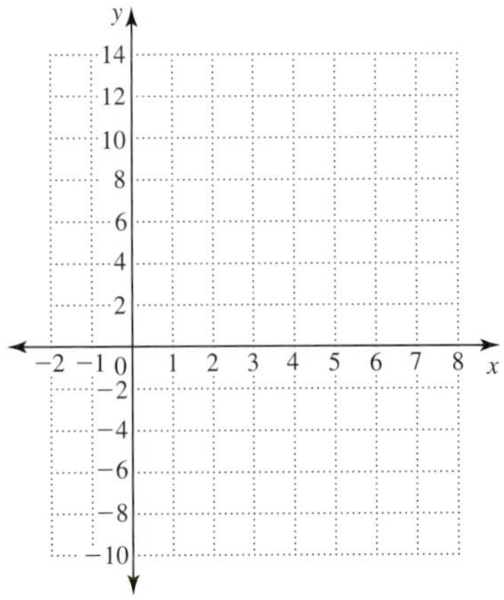

(c) What is the x-coordinate of the point of intersection of the graphs?

(d) Does the graphical solution confirm the analytical solution? Why or why not?

16. Given $(x - 2) - (x + 3) = 2x - 11$.

(a) Solve the equation analytically.

(b) Graph both equations on the same grid.

x	y_1	y_2
1		
2		
3		
4		
5		

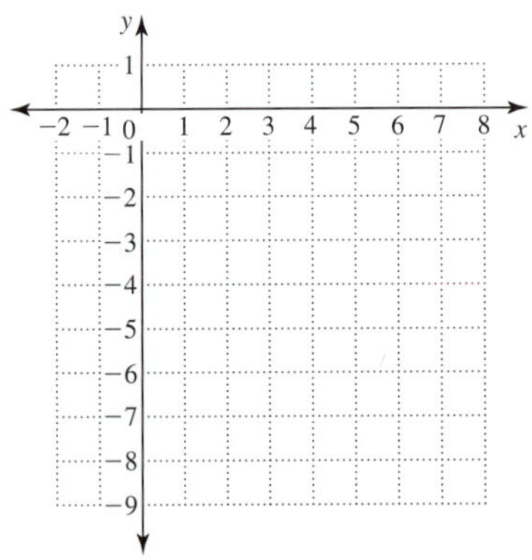

(c) What is the x-coordinate of the point of intersection of the graphs?

(d) Does the graphical solution confirm the analytical solution? Why or why not?

Problems 17–22: Solve each equation analytically and graphically using *either* the X-Intercept or the Intersection-of-Graphs Method. Verify each answer by substituting the solution into the original equation.

17. Given $3(0.5x + 0.2) = 0.5x - 0.4$.

(a) Solve the equation analytically.

(b) Solve the equation graphically.

x	y_1	y_2
−2		
−1		
0		
1		
2		

(c) What is the graphical solution?

(d) Does the graphical solution confirm the analytical solution? Why or why not?

18. Given $3(1.5x - 2) = 2.5x - 2$.

(a) Solve the equation analytically.

(b) Solve the equation graphically.

x	y_1	y_2
−2		
−1		
0		
1		
2		
3		

(c) What is the graphical solution?

(d) Does the graphical solution confirm the analytical solution? Why or why not?

19. Given $21 - 2(x - 4) = 5(1 - 2x)$.

(a) Solve the equation analytically.

(b) Solve the equation graphically.

➤ *Hint: If you choose the X-Intercept Method, use Xscl 1 and Yscl 1. If you choose the Intersection-of-Graphs Method, use Xscl 1 and Yscl 5.*

x	y_1	y_2
−4		
−3		
−2		
−1		
0		
1		

(c) What is the graphical solution?

(d) Does the graphical solution confirm the analytical solution? Why or why not?

20. Given $10 - 3(x + 2) = 2(6 - x) - 7$.

(a) Solve the equation analytically.

(b) Solve the equation graphically.

x	y_1	y_2
−3		
−2		
−1		
0		
1		
2		

(c) What is the graphical solution?

(d) Does the graphical solution confirm the analytical solution? Why or why not?

21. Given $-0.4x + 2.6 = 1.6x - 1.4$.

(a) Solve the equation analytically.

(b) Solve the equation graphically.

x	y_1	y_2
−1		
0		
1		
2		
3		

(c) What is the graphical solution?

(d) Does the graphical solution confirm the analytical solution? Why or why not?

22. Given $\frac{2}{3}x + \frac{1}{5} = x - \frac{4}{5}$.

(a) Solve the equation analytically.

(b) Solve the equation graphically.

x	y_1	y_2
−1		
0		
1		
2		
3		
4		

(c) What is the graphical solution?

(d) Does the graphical solution confirm the analytical solution? Why or why not?

23. The Armstrong Publishing Company publishes a magazine for college students. Each magazine sells for $2.00. The fixed costs for operating expenses to produce the magazine are $5000.

(a) Complete the table of values. Let x represent the number of magazines sold and generalize the problem situation into a profit function, $P(x)$.

x = Number of magazines sold	Computations	$P(x)$ = Profit function
0		
2000		
2500		
3000		
4000		
5000		
x		$P(x) =$

(b) Substitute the number zero for $P(x)$ in the function and solve the equation for x analytically.

(c) How many magazines must be sold to break even?

(d) Confirm your answer graphically using the X-Intercept Method. Determine a scale for the horizontal and vertical axes: Xscl = ______, Yscl = ______.

(e) Label the horizontal and vertical axes. Plot the points from the table on a separate piece of graph paper. Connect the points.

(f) Identify the x-intercept of the graph.

(g) How does the x-intercept of the graph relate to the breakeven point of the profit function?

24. MONY Records produces and markets compact discs for aspiring musicians. Each CD sells for $15.00. The fixed costs, including compensation for the artist, to produce the CD are $45,000.

(a) Complete the table of values. Let x represent the number of CDs sold and generalize the problem situation into a profit function, $P(x)$.

x = Number of CDs sold	Computations	$P(x)$ = Profit function
1000		
2000		
3000		
4000		
5000		
x		$P(x) =$

(b) Substitute the number zero for $P(x)$ in the function and solve the equation for x analytically.

(c) How many CDs must be sold to break even?

(d) Confirm the answer graphically using the X-Intercept Method. Determine a scale for the horizontal and vertical axes: Xscl = ______, Yscl = ______.

(e) Label the horizontal and vertical axes. Plot the points from the table on a separate piece of graph paper. Connect the points.

(f) Identify the x-intercept of the graph.

(g) How does the x-intercept of the graph relate to the breakeven point of the profit function?

25. The student government at Smart College is selling candy to raise funds. Five hundred (500) candy bars will be purchased at \$0.75 per bar. The selling price of each bar will be \$1.50. The shipping cost for all the candy bars is \$6.

(a) Complete the table of values. Let x represent the number of candy bars sold and generalize the problem situation into a profit function, $P(x)$.

x = Number of candy bars sold	Computations	$P(x)$ = Profit function
0		
200		
300		
400		
500		
x		$P(x) =$

(b) Substitute the number zero for $P(x)$ in the function and solve the equation for x analytically.

(c) How many candy bars must be sold to break even?

(d) Confirm the answer graphically using the X-Intercept Method. Determine a scale for the horizontal and vertical axes: Xscl = ______, Yscl = ______.

(e) Label the horizontal and vertical axes. Plot the points from the table on a separate piece of graph paper. Connect the points.

(f) Approximate the x-intercept of the graph.

(g) How does the x-intercept of the graph relate to the breakeven point of the profit function?

26. The *X & the Unknowns* band is selling t-shirts at their concerts. One thousand (1000) t-shirts with the band's logo will be purchased at $12.00 each. The selling price of each t-shirt will be $24. The shipping cost for all the t-shirts is $360.

(a) Complete the table of values. Let x represent the number of t-shirts sold and generalize the problem situation into a profit function, $P(x)$.

x = Number of t-shirts sold	Computations	$P(x)$ = Profit function
0		
315		
515		
815		
1000		
x		$P(x) =$

(b) Substitute the number zero for $P(x)$ in the function and solve the equation for x analytically.

(c) How many t-shirts must be sold to break even?

(d) Confirm the answer graphically using the X-Intercept Method. Determine a scale for the horizontal and vertical axes: Xscl = ______, Yscl = ______.

(e) Label the horizontal and vertical axes. Plot the points from the table on a separate piece of graph paper. Connect the points.

(f) Approximate the x-intercept of the graph.

(g) How does the x-intercept of the graph relate to the breakeven point of the profit function?

27. Beth and Kathy work part-time at CDs4U and make $6 and $8 per hour respectively. One week Kathy worked 5 hours. The total of Beth and Kathy's earnings that week was $130. Let x represent the number of hours Beth worked.

(a) Solve the equation describing the problem situation analytically: $6x + 8(5) = 130$.

(b) Confirm the answer graphically using the Intersection-of-Graphs Method. Label the left and right sides of the equation y_1 and y_2, respectively:

$$\underbrace{6x + 40}_{y_1} = \underbrace{130}_{y_2}$$

(c) Complete the tables of values and graph y_1 and y_2 on the same grid.

x	y_1	y_2
0		
5		
10		
15		
20		
25		

(d) Identify the point of intersection of the two lines.

(e) How many hours did Beth work?

(f) How does the point of intersection of the two graphs relate to the number of hours Beth worked?

28. Jesse and Carl work part-time as student reading tutors and make \$5.25 and \$6.25 per hour respectively. Last week Jesse worked 4 hours. The total of Jesse and Carl's earnings that week was \$71. Let x represent the number of hours Carl worked.

(a) Solve the equation describing the problem situation analytically: $5.25(4) + 6.25x = 71$

(b) Confirm your answer graphically using the Intersection-of-Graphs Method. Label the left and right sides of the equation y_1 and y_2, respectively:

$$\underbrace{21 + 6.25x}_{y_1} = \underbrace{71}_{y_2}$$

(c) Complete the tables of values and graph y_1 and y_2 on the same grid.

x	y_1	y_2
0		
2		
4		
6		
8		
10		

(d) Identify the point of intersection of the two lines.

(e) How many hours did Carl work?

(f) How does the point of intersection of the two graphs relate to the number of hours Carl worked?

29. Bill and Don went shopping. Bill spent \$10 more than Don. Together they spent \$210.

(a) Let x represent the amount spent by Don. Write an expression in terms of x that represents the amount spent by Bill.

(b) Write an equation in terms of x that represents the amount spent by Bill and Don.

(c) Solve the equation for x analytically.

(d) How much did Don spend?

(e) How much did Bill spend?

(f) Confirm your answer using either the X-Intercept or Intersection-of-Graphs Graphical Method. Draw graphs on a separate piece of graph paper.

30. Crystal and Leah shared expenses on a trip. Crystal spent \$150 more than Leah. Together they spent \$950.

(a) Let x represent the amount spent by Leah. Write an expression in terms of x that represents the amount spent by Crystal.

(b) Write an equation in terms of x that represents the amount spent by Crystal and Leah.

(c) Solve the equation for x analytically.

(d) How much did Leah spend?

(e) How much did Crystal spend?

(f) Confirm your answer using either the X-Intercept or Intersection-of-Graphs Graphical Method. Draw graphs on a separate piece of graph paper.

Problems 31–33: Writing Exercises

31. Describe why the analytical solution of a linear equation is equivalent to the zero of the equation.

32. Describe the relationship between the solution of an equation and the x-intercept of the graph of the corresponding function.

33. Describe the relationship between the solution of an equation and the point of intersection of the graphs of the corresponding functions.

4.2 Graphical Solutions of Linear Systems in Two Variables

OBJECTIVES

- Determine whether an ordered pair is a solution of a system of equations.
- Write a system of two linear equations in two variables from a problem situation.
- Solve a linear system in two variables using the Intersection-of-Graphs Method and identify the system as consistent, dependent, or inconsistent.

Getting Started

Many problem situations are defined with more than one equation. Two or more equations describing a problem situation are called a **system of equations**. A system may have one solution, an infinite number of solutions or no solution. In this section, we use the Intersection-of-Graphs Method for solving linear systems in two variables.

1. The solution of a system of two equations is an ordered pair, (x, y), that satisfies both equations. Given the system $\begin{cases} x + y = 6 \\ x - y = -2 \end{cases}$, substitute each ordered pair into both equations to determine whether the ordered pair is a solution of the system.

(a) $(2, 4)$ (b) $(1, 5)$

2. Let's create a system of two linear equations for the following real-world problem: You have been asked to purchase tennis balls and t-shirts as gifts for participants in a tennis tournament. Each can of tennis balls costs \$2 and each t-shirt costs \$5. You must spend exactly \$125. Each of the 40 participants will receive either a can of tennis balls or a t-shirt. How many cans of tennis balls should you purchase? How many t-shirts should you purchase?

(a) Let the variable x represent the number of cans of tennis balls to be purchased. Let the variable y represent the number of t-shirts to be purchased. Use the variables x and y to write an equation expressing the total number of gifts in the boxes below. Label the new equation Equation 1.

Number of cans of tennis balls		Number of t-shirts		Total number of gifts
	+		=	

(b) Each can of tennis balls costs \$2 and x represents the number of cans of tennis balls purchased. Write an expression using the variable x to represent the total amount spent on tennis balls.

(c) Each t-shirt costs \$5 and y represents the number of t-shirts purchased. Write an expression using the variable y to represent the total amount spent on t-shirts.

(d) The total amount to be spent is \$125. Write an equation using the variables x and y and your answers to (b) and (c) to represent the sum of the total amount of money spent on tennis balls and t-shirts in the boxes. Label the new equation Equation 2.

Total amount spent on tennis balls		Total amount spent on t-shirts		Total amount spent on gifts
	+		=	

(e) Solve Equation 1 for y. Substitute y_1 for y.

(f) Solve Equation 2 for y. Substitute y_2 for y.

➢ *The variables x and y represent the same quantities in both Equations 1 and 2 above. The solution to the problem situation is the ordered pair (x, y) that satisfies BOTH linear equations.*

(g) Use the Intersection-of-Graphs Method to find the point of intersection of the graphs of Equations 1 and 2. Complete the tables of values for equations y_1 and y_2. Graph both equations on the same grid.

x	$y_1 = 40 - x$	$y_2 = \dfrac{125 - 2x}{5}$
10		
15		
20		
25		
30		

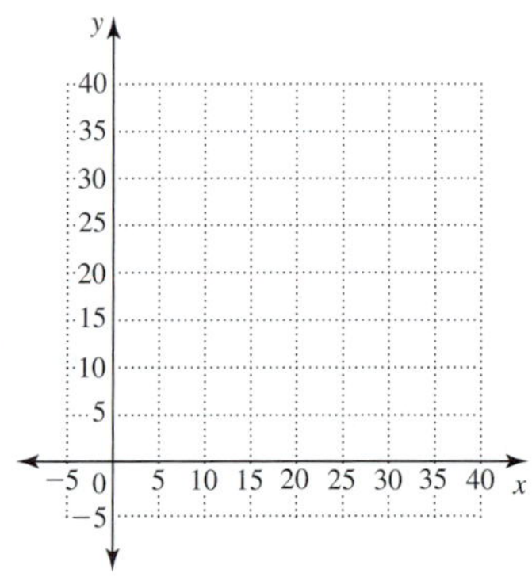

(h) Identify the point of intersection of the two lines from the graph and/or the tables of values. Write the coordinates of the point of intersection.

(i) How many cans of tennis balls should be purchased?

(j) How many t-shirts should be purchased?

(k) Verify the solution by substituting the coordinates of the point of intersection into the original Equations 1 and 2. Show your work.

3. Solve the linear system using the Intersection-of-Graphs Method.

$$x + y = 5 \qquad \text{Equation 1}$$
$$2x + 2y = 10 \qquad \text{Equation 2}$$

(a) Solve Equation 1 for y. Substitute y_1 for y.

(b) What is the slope of y_1? ______ What is the y-intercept of y_1? ______

(c) Solve Equation 2 for y. Substitute y_2 for y.

(d) What is the slope of y_2? ______ What is the y-intercept of y_2? ______

(e) Complete the tables of values for equations y_1 and y_2. Graph both equations on the same grid.

x	y_1	y_2
−2		
−1		
0		
1		
2		

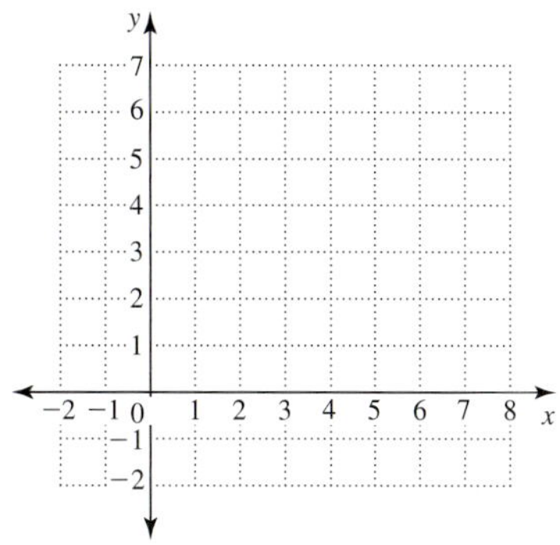

(f) Compare the slopes of the lines and the y-intercepts of the graphs. What can you conclude about the two lines?

(g) How many point(s) of intersection exist?

(h) How many points are included in the solution of the system?

4. Solve the linear system using the Intersection-of-Graphs Method.

$$2x + y = 5 \qquad \text{Equation 1}$$
$$2x + y = 1 \qquad \text{Equation 2}$$

(a) Solve Equation 1 for y. Substitute y_1 for y.

(b) What is the slope of y_1? ______ What is the y-intercept of y_1? ______

(c) Solve Equation 2 for y. Substitute y_2 for y.

(d) What is the slope of y_2? ______ What is the y-intercept of y_2? ______

(e) Complete the tables of values for equations y_1 and y_2. Graph both equations on the same grid.

x	y_1	y_2
−1		
0		
1		
2		

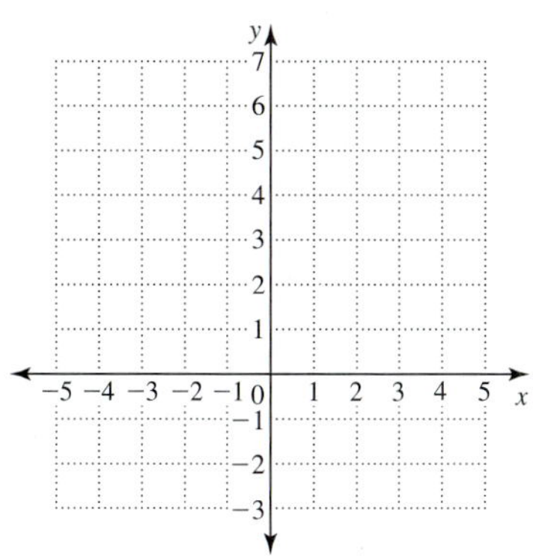

(f) Compare the slopes of the lines and the y-intercepts of the graphs. What can you conclude about the two lines?

(g) How many point(s) of intersection exist?

(h) How many points are included in the solution of the system?

1 2 3

Discussing Mathematics

SOLUTIONS TO SYSTEMS OF EQUATIONS

Many problem situations are best represented with two or more equations called a system of equations. The **solution of a system of equations** is the set(s) of ordered pairs that satisfy all the equations in the system.

EXAMPLE 1: Are the ordered pairs $(2, -6)$ and $(3, -3)$ solutions of the system?

$$3x - y = 12$$
$$2x + y = 3$$

(a) Substitute $x = 2$ and $y = -6$ into each equation.

$3x - y = 12$	$2x + y = 3$
$3(2) - (-6) \stackrel{?}{=} 12$	$2(2) + (-6) \stackrel{?}{=} 3$
$6 + 6 \stackrel{?}{=} 12$	$4 + (-6) \stackrel{?}{=} 3$
$12 = 12$	$-2 \neq 3$

The ordered pair $(2, -6)$ does not satisfy both equations. The ordered pair $(2, -6)$ is not a solution of the system.

(b) Substitute $x = 3$ and $y = -3$ into each equation.

$3x - y = 12$	$2x + y = 3$
$3(3) - (-3) \stackrel{?}{=} 12$	$2(3) + (-3) \stackrel{?}{=} 3$
$9 + 3 \stackrel{?}{=} 12$	$6 + (-3) \stackrel{?}{=} 3$
$12 = 12$	$3 = 3$

The ordered pair $(3, -3)$ does satisfy both equations. The ordered pair $(3, -3)$ is a solution of the system.

SOLVING A LINEAR SYSTEM OF EQUATIONS USING THE INTERSECTION-OF-GRAPHS METHOD

The solution of a linear system in two variables may be one ordered pair, an infinite number of ordered pairs, or there may be no solution. Graphing a system of equations provides us with a visual representation of the problem situation and the solution to the system.

VISUALIZING THE GRAPHS OF A LINEAR SYSTEM

1. A system of equations having at least one solution and one point of intersection is called **consistent**.

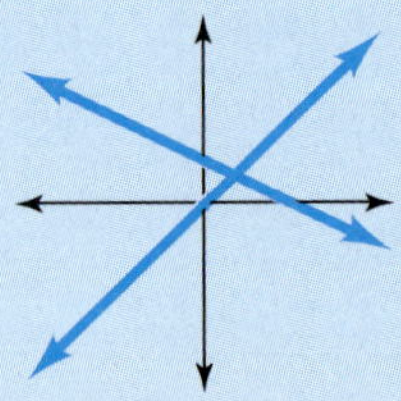

One point of intersection.

2. A consistent system of equations with an infinite number of solutions, where the graphs have an infinite number of points of intersection, is called **dependent**.

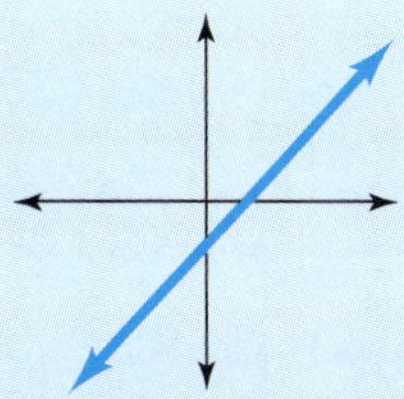

The two lines *coincide*. They intersect at all points along the line.

3. A system of equations with no solution, where the graphs do not intersect, is called **inconsistent**.

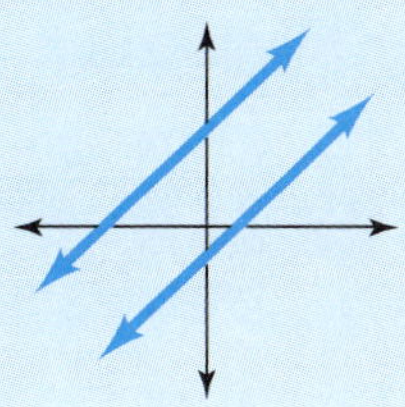

The two lines are parallel and *never* intersect.

Solving a system of two linear equations using the Intersection-of-Graphs Method involves the following steps.

THE INTERSECTION-OF-GRAPHS METHOD FOR SOLVING A LINEAR SYSTEM

1. If necessary, assign variables as the unknowns in the problem to create equations describing the problem situation. Use a picture or table, as needed. Label the equations of the system Equation 1 and Equation 2.
2. Rewrite each equation in function form, solving for y. Label the first equation y_1 and the second equation y_2.
3. Complete tables of values for y_1 and y_2. Graph both equations on the same grid.
4. Identify the point(s) of intersection of the two graphs, where $y_1 = y_2$. The ordered pairs denoting the point(s) of intersection of the graphs is the solution of the system.
5. Verify the solution by substituting the coordinates of the point(s) of intersection into the original equations of the system. If the system describes a problem situation, label the answer(s).

A CONSISTENT LINEAR SYSTEM

Let's consider a *consistent* linear system in two variables with one ordered pair as the solution.

EXAMPLE 2: Tickets to the Harbor Festival cost \$6 for adults and \$2 for children. 155 tickets were sold and \$810 in proceeds was received from ticket sales. How many adults attended and how many children attended? Solve this problem using the Intersection-of-Graphs Method.

Assign variables as unknowns, draw a picture, and create equations describing the problem situation.

Let x represent the number of adult tickets sold.

Let y represent the number of children's tickets sold.

Equation 1:

Number of adult tickets sold		Number of children's tickets sold		Total tickets sold
x	+	y	=	155

Equation 2:

Proceeds from sale of adult tickets		Proceeds from sale of children's tickets		Total proceeds
$6x$	+	$2y$	=	810

Solve the first equation for y and label the equation y_1.
Solve the second equation for y and label the equation y_2.

Equation 1:

$$x + y = 155$$
$$y = 155 - x$$
$$y_1 = 155 - x$$

Equation 2:

$$6x + 2y = 810$$
$$2y = 810 - 6x$$
$$y = \frac{810 - 6x}{2}$$
$$y = \frac{2(405 - 3x)}{2}$$
$$y_2 = 405 - 3x$$

Complete tables of values for y_1 and y_2.

x	y_1	y_2
100	55	105
105	50	90
110	45	75
115	40	60
120	35	45
125	30	30
130	25	15

Graph both equations on the same grid.

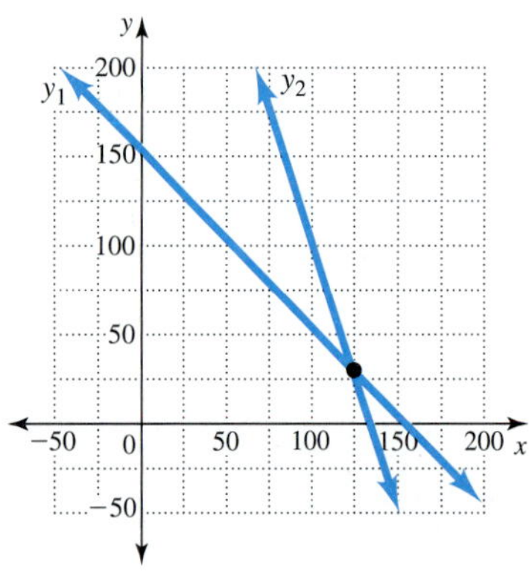

The two graphs intersect at the point (125, 30).

Identify the point(s) of intersection and the solution of the system.

The number of adult tickets purchased: 125

The number of children's tickets purchased: 30

Verify the solution by substituting the coordinates of the point of intersection into the original equations of the system. Check $x = 125$ and $y = 30$:

Equation 1:	Equation 2:
$x + y = 155$	$6x + 2y = 810$
$125 + 30 \stackrel{?}{=} 155$	$6(125) + 2(30) \stackrel{?}{=} 810$
$155 = 155$	$750 + 60 \stackrel{?}{=} 810$
	$810 = 810$

The ordered pair (125, 30) satisfies both equations.

A DEPENDENT LINEAR SYSTEM

The graphs of the equations of a *dependent* linear system coincide and the lines intersect in an infinite number of points.

EXAMPLE 3: Solve the system using the Intersection-of-Graphs Method.

$$3x + 6y = 15$$
$$x + 2y = 5$$

Equation 1:	Equation 2:
$3x + 6y = 15$	$x + 2y = 5$
$6y = 15 - 3x$	$2y = 5 - x$
$y_1 = \dfrac{15 - 3x}{6} = \dfrac{5 - x}{2}$	$y_2 = \dfrac{5 - x}{2}$

Solve the first equation for y and label the equation y_1.
Solve the second equation for y and label the equation y_2.

x	y_1	y_2
−3	4	4
−2	3.5	3.5
−1	3	3
0	2.5	2.5
1	2	2
2	1.5	1.5
3	1	1

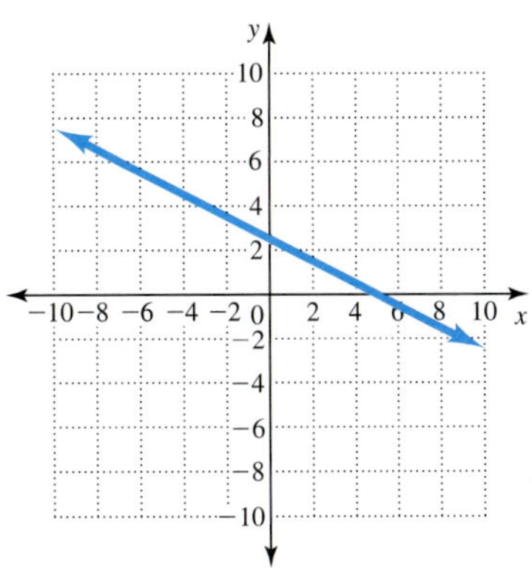

Graph both equations on the same grid.

y_1 and y_2 are the same line.

The lines, y_1 and y_2, coincide and intersect at all points along the line.

$$y_1 = y_2$$ The equations are equivalent.

$$\frac{5 - x}{2} = \frac{5 - x}{2}$$ The system is dependent.

AN INCONSISTENT LINEAR SYSTEM

The graphs of an inconsistent linear system do not intersect. The two lines represent parallel lines with equal slopes.

EXAMPLE 4: Solve the system using the Intersection-of-Graphs Method.

$$3x + y = 5$$
$$3x + y = 1$$

Equation 1:	Equation 2:	
$3x + y = 5$	$3x + y = 1$	Solve the first equation for y and label the equation y_1.
$y = 5 - 3x$	$y = 1 - 3x$	Solve the second equation for y and label the equation y_2.
$y_1 = -3x + 5$	$y_2 = -3x + 1$	

x	y_1	y_2
−3	14	10
−2	11	7
−1	8	4
0	5	1
1	2	−2
2	−1	−5
3	−4	−8

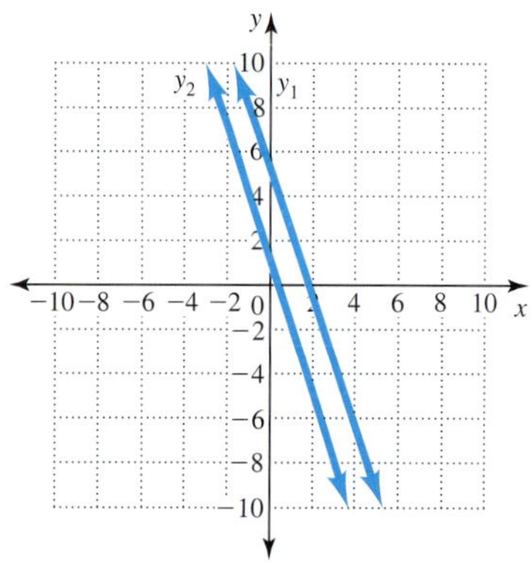

Complete tables of values for y_1 and y_2. Graph both equations on the same grid.

The slope of $y_1 = -3x + 5$ is -3. The slope of $y_2 = -3x + 1$ is -3. Both lines have the same slope, are parallel, and do not intersect. There are no ordered pairs in the table of values for which $y_1 = y_2$. There is no solution. The system is inconsistent.

Doing Mathematics

- Write the definitions of the words and symbols listed for this section in the Dictionary in Appendix A.
- On a 3 × 5-inch card, write the steps for solving a linear system in two variables using the intersection-of-graphs method for future reference.

Problems 1–8: Determine whether the given ordered pair is a solution to the system.

1. $x + y = 12$
$x - y = -2$ $(5, 7)$

2. $2x + y = 1$
$x - 3y = 4$ $(-1, 3)$

3. $x - y = 2$
$5x - y = -2$ $(0, 4)$

4. $y = -2x + 3$
$x - y = 3$ $(2, -1)$

5. $x + y = -5$
$x - y = 11$ $(3, -8)$

6. $-3x + 2y = 12$
$2x + 5y = 11$ $(2, -3)$

7. $2x - y = 12$
$x - 3y = 6$ $(0, 6)$

8. $y = -4x + 1$
$x + y = -2$ $(1, -3)$

Problems 9–36: Solve each system using the Intersection-of-Graphs Method. If the system is consistent, write the coordinates of the solution. Otherwise, write dependent or inconsistent.

9. $x + y = 6$
$x - y = -2$

10. $x - y = 0$
$3x - y = 4$

11. $2x - 4y = -2$
$-x + 2y = 5$

12. $x + 4y = 12$
$\frac{1}{2}x + 2y = 6$

13. $y = 1 - 5x$
$10x + 2y = 9$

14. $-3x + y = -12$
$2x + y = 3$

15. $2x + y = 3$
$6x + 3y = 9$

16. $x - y = 3$
$2x + y = 3$

17. $y = 2x + 14$
$x + y = 8$

18. $3x + 4y = 1$
$6x + 8y = 2$

19. $3x + 4y = 12$
$6x + 8y = 24$

20. $x + y = 2$
$y = 5 - x$

21. $x + y = 3$
$-x - y = 1$

22. $2x + y = 4$
$x - y = 2$

23. $5x - y = 11$
$6x - y = 12$

24. $4.2x - 2.8y = 10.6$
$-2.1x + 1.4y = -5.3$

25. $2x + 3y = 6$
$y = -\frac{2}{3}x + 2$

26. $y = 1 + 2x$
$3x - y = 2$

27. $3x + y = -8$
$5x - 2y = -6$

28. $-2x + 3y = 10$
$2x + 3y = 14$

29. $x - 2y = -11$
$5x + 2y = 5$

30. $y = 3x - 5$
$y = 5x - 11$

31. $x - 3y = -16$
$y = 12 - x$

32. $3x - y = 4$
$-3x + y = 1$

33. $2x + 5y = 7$
$-2x - 5y = -1$

34. $x = y + 5$
$x = 2y + 1$

35. $3(x + y) = 2(x - 4y) + 13$
$5(2x + y) = 3x + 19$

36. $3x - 2(y + 7) = 2$
$4(x + 6) + 7y = 26$

37. You have been asked to purchase favors for teenagers who are going to attend a birthday party at your home. You have decided to purchase t-shirts and baseball caps. Each t-shirt costs \$8 and each baseball cap costs \$5. You must spend a total of \$99. Each of the 15 guests will receive one favor. How many t-shirts and baseball caps should you buy?

(a) Let the variable x represent the number of t-shirts to be purchased. Let the variable y represent the number of baseball caps to be purchased.

Equation 1:

Number of t-shirts	+	Number of baseball caps	=	Total number of favors

Equation 2:

Total amount spent on t-shirts	+	Total amount spent on baseball caps	=	Total funds available

(b) Solve Equation 1 for y. Substitute y_1 for y.

(c) Solve Equation 2 for y. Substitute y_2 for y.

(d) Graph y_1 and y_2 on the same grid on a separate piece of graph paper.

(e) Write the coordinates of the point(s) of intersection of the two graphs.

(f) How many t-shirts should be purchased?

(g) How many baseball caps should be purchased?

(h) Verify the solution by substituting the coordinates of the point of intersection into the original equations of the system.

38. Sandi and Sharon work at CDs4U and earn \$7 per hour and \$10 per hour, respectively. One week the total of their hours worked was 28 hours. Their combined earnings were \$244. How many hours did each work?

(a) Let x represent the number of hours that Sandi worked and let y represent the number of hours that Sharon worked. Draw a picture of two equations representing the problem situation.

(b) Write an equation representing the total hours worked by Sandi and Sharon.

(c) Rewrite the equation in part (b) in function form, solving for y. Substitute y_1 for y.

(d) Write a second equation that represents the sum of Sandi and Sharon's earnings.

(e) Rewrite the equation in part (d) in function form, solving for y. Substitute y_2 for y.

(f) Graph y_1 and y_2 on the same grid on a separate piece of graph paper.

(g) Write the coordinates of the point(s) of intersection of the two graphs.

(h) How many hours did Sandi work?

(i) How many hours did Sharon work?

(j) Verify the solution by substituting the coordinates of the point of intersection into the original equations of the system.

39. Tickets to the circus cost \$10 for adults and \$5 for children. 500 people attended the circus and \$4000 was collected in ticket sales. How many adults and how many children attended the circus?

(a) Let x represent the number of adult tickets sold. Let y represent the number of children's tickets sold. Draw a picture of two equations representing the problem situation.

(b) Write an equation representing the total number of tickets sold.

(c) Rewrite the equation in part (b) in function form, solving for y. Substitute y_1 for y.

(d) Write a second equation that represents the total amount collected from the sale of all tickets.

(e) Rewrite the equation in part (d) in function form, solving for y. Substitute y_2 for y.

(f) Graph y_1 and y_2 on the same grid on a separate piece of graph paper.

(g) Write the coordinates of the point(s) of intersection of the two graphs.

(h) How many adult tickets were sold?

(i) How many children's tickets were sold?

(j) Verify the solution by substituting the coordinates of the point of intersection into the original equations of the system.

40. You have been asked to help purchase items to re-stock an auto parts store. You must purchase a total of 50 oil filters and air filters. Each oil filter costs \$4.00 and each air filter costs \$6.00. You must spend a total of \$240. How many oil filters and air filters should you buy?

(a) Let the variable x represent the number of oil filters to be purchased. Let the variable y represent the number of air filters to be purchased.

Equation 1:

Number of oil filters		Number of air filters		Total number of parts
	+		=	

Equation 2:

Total amount spent on oil filters		Total amount spent on air filters		Total funds available
	+		=	

(b) Rewrite Equation 1 in function form, solving for y. Substitute y_1 for y.

(c) Rewrite Equation 2 in function form, solving for y. Substitute y_2 for y.

(d) Graph y_1 and y_2 on the same grid on a separate piece of graph paper.

(e) Write the coordinates of the point(s) of intersection of the two graphs.

(f) How many oil filters should be purchased?

(g) How many air filters should be purchased?

(h) Verify the solution by substituting the coordinates of the point of intersection into the original equations of the system.

41. Damien and Jondarr teach science classes to children on Saturday mornings and earn \$30 and \$20 per hour respectively. Last month the total of their hours worked was 40 hours. Their combined earnings were \$960. How many hours did each work?

(a) Let x represent the number of hours that Damien worked and let y represent the number of hours that Jondarr worked. Draw a picture of two equations representing the problem situation.

(b) Write an equation representing the total hours worked by Damien and Jondarr.

(c) Rewrite the equation in part (b) in function form, solving for y. Substitute y_1 for y.

(d) Write a second equation that represents the sum of Damien and Jondarr's earnings.

(e) Rewrite the equation in part (d) in function form, solving for y. Substitute y_2 for y.

(f) Graph y_1 and y_2 on the same grid on a separate piece of graph paper.

(g) Write the coordinates of the point(s) of intersection of the two graphs.

(h) How many hours did Damien work?

(i) How many hours did Jondarr work?

(j) Verify the solution by substituting the coordinates of the point(s) of intersection into the original equations of the system.

42. Hot dogs and french fries were sold by a street vendor. Hot dogs cost \$2.00 each and each order of fries cost \$1.00. On a given day a total of 300 hot dogs and orders of fries were sold. The revenue collected was \$500. How many hot dogs and how many orders of french fries were sold on that day?

(a) Let x represent the number of hot dogs sold and let y represent the number of orders of french fries sold. Draw a picture of two equations representing the problem situation.

(b) Write an equation representing the total number of hot dogs and orders of fries sold.

(c) Rewrite the equation in part (b) in function form, solving for y. Substitute y_1 for y.

(d) Write a second equation to represent the total revenue collected from the sale of all hot dogs and fries.

(e) Rewrite the equation in part (d) in function form, solving for y. Substitute y_2 for y.

(f) Graph y_1 and y_2 on the same grid on a separate piece of graph paper.

(g) Write the coordinates of the point(s) of intersection of the two graphs.

(h) How many hot dogs were sold?

(i) How many orders of french fries were sold?

(j) Verify the solution by substituting the coordinates of the point(s) of intersection into the original equations of the system.

43. Explain how to determine whether a system of linear equations in two variables is consistent, inconsistent, or dependent using the Intersection-of-Graphs Method.

4.3 Solving Linear Systems Using Substitution and Elimination

OBJECTIVE

- Solve linear systems in two variables using substitution and elimination.

1 2 3 Getting Started

In this section, the methods of substitution and elimination are presented as analytical methods for solving linear systems.

1. Consider an investment problem: You have \$2000 to invest for one year. You plan to invest part of the money in an account that earns 6% annual interest and the rest of the money in an account that earns 10% annual interest. By the end of the year, you would like to earn \$152 in interest from the two accounts. How much money do you need to invest in each account?

(a) Let the variable x represent the amount invested at 6% and the variable y represent the amount invested at 10%. Use the variables x and y to write an equation representing the total amount invested.

Equation 1:

Amount invested at 6%	+	Amount invested at 10%	=	Total amount invested

(b) Simple interest is computed using the formula: $I = P \cdot r \cdot t$, where I represents the interest earned, P represents the amount invested (the principal), r represents the rate of interest, and t represents the number of years the principal is invested. In this problem, the money is to be invested for 1 year and $t = 1$.

To represent the amount of interest earned at 6%, substitute the variable x for the principal (P), 6% = 0.06 for the rate of interest (r), and 1 for t.

$$I = P \cdot r \cdot t$$

$$\downarrow \quad \downarrow \quad \downarrow$$

$$I = x \cdot (0.06) \cdot 1$$

$$I = 0.06x$$ The interest earned at 6% is $0.06x$.

To represent the interest earned at 10%, substitute the variable y for the principal (P), 0.10 for the rate of interest (r), and 1 for t.

$$I = P \cdot r \cdot t$$

$$\downarrow \quad \downarrow \quad \downarrow$$

$$I =$$

(c) Using the expressions from part (b), write an equation representing the total interest earned on the sum of the two investments.

Equation 2:

Interest earned at 6%	+	Interest earned at 10%	=	Total interest earned
				\$152

(d) Equations 1 and 2 represent a system of equations in two variables. Write the two equations of the system.

Equation 1:

Equation 2:

(e) Now let's solve the system using the Method of Substitution. Solve Equation 1 for y. Label the resulting equation, Equation 3.

(f) The Method of Substitution involves substituting an expression from one equation into the other equation. Because $y = (2000 - x)$, substitute $(2000 - x)$ for y into Equation 2.

$$0.06x + 0.10y = 152$$

$$\downarrow$$

$$0.06x + 0.10(\qquad\qquad) = 152$$

(g) Solve the resulting equation for x.

(h) To solve for the corresponding value of y of the solution, substitute the value for x into any equation of the system.

➢ *(Hint: Equation 3 is the easiest equation to use for this substitution.)*

(i) Write the solution of the system as an ordered pair. $(x, y) =$

(j) How much money should be invested at 6%?

(k) How much money should be invested at 10%?

(l) Verify the answer by substituting the solution into the original equations of the system.

Equation 1: $x + y = 2000$

Equation 2: $0.06x + 0.10y = 152$

1 2 3

Discussing Mathematics

METHOD OF SUBSTITUTION

The graphical solution presented in Section 4.2 provides a visual representation of the equations and the solution to the system. The **Method of Substitution** is one way to solve a system of equations analytically.

METHOD OF SUBSTITUTION FOR SOLVING A LINEAR SYSTEM IN TWO VARIABLES

1. Assign variables as the unknowns in the problem to create the equations describing the problem situation. Use a picture or table, as needed. Label the equations of the system Equation 1 and Equation 2.
2. Solve one of the equations of the system for one of the variables. If one variable has a numerical coefficient of $+1$ or -1, solve for that variable. Otherwise, solve for any one of the variables in either equation. Label the result Equation 3.
3. Substitute the expression for the variable obtained in Step 2 into the other equation. Solve the resulting equation for the remaining variable.
4. Substitute the result of Step 3 into Equation 3 and solve for the second variable. This process is called back-substitution. Write the solution as an ordered pair.
5. Verify the answer by substituting the solution into each equation of the original system. If the system describes a problem situation, label the answer(s) with the appropriate units of measure.

EXAMPLE 1: Solve the system using the Method of Substitution.

$x + y = 20$ Equation 1

$3x + 2y = 49$ Equation 2

$x + y = 20$ — Solve Equation 1 for y and label the result Equation 3.

$x - x + y = 20 - x$

Equation 3: $y = 20 - x$

Equation 2: $3x + 2y = 49$ — Since $y = 20 - x$, substitute $(20 - x)$ for y into Equation 2.

$3x + 2(20 - x) = 49$

$3x + 40 - 2x = 49$ — Solve the resulting equation for x.

$x + 40 = 49$

$x = 9$

$y = 20 - x$ — Back-substitute $x = 9$ into Equation 3.

$y = 20 - 9$ — Solve for y.

$y = 11$ — Write the answer as an ordered pair: (9, 11).

Verify the answer by substituting $x = 9$ and $y = 11$ into each equation of the original system.

Equation 1:

$(9) + (11) \stackrel{?}{=} 20$

$20 = 20$ ✓

Equation 2:

$3(9) + 2(11) \stackrel{?}{=} 49$

$49 = 49$ ✓

✓ The ordered pair (9, 11) satisfies both equations.

METHOD OF ELIMINATION

The Method of Substitution can produce complicated expressions, which are cumbersome to simplify. Another method, the **Method of Elimination**, may be more efficient than substitution in some problems. In the Method of Elimination, the goal is to *eliminate* one of the variables in the system by adding the two equations together and solving for the resulting variable.

METHOD OF ELIMINATION FOR SOLVING A LINEAR SYSTEM IN TWO VARIABLES

1. Create the equations describing the problem situation. Label the first equation Equation 1. Label the second equation Equation 2. Write each equation in standard form: $Ax + By = C$.
2. Look at the equations in the system and determine which variable can most easily be eliminated when the two equations are added together. It may be necessary to multiply one or both equations by a nonzero constant. Multiply EVERY term in the equation(s) by the constant(s).
3. Rewrite the system and add the equations together. The sum of the equations is a third equation (Equation 3) with one variable. Solve Equation 3 for the remaining variable.
4. Substitute the result of Step 4 into either of the original equations of the system and solve for the second variable. This process is called back-substitution. Write the solution as an ordered pair.
5. Verify the answer by substituting the solution into each equation of the original system. If the system describes a problem situation, label the answer(s) with the appropriate units of measure.

EXAMPLE 2: Solve the system using the Method of Elimination.

$x + y = 5$ Equation 1

$x - y = 3$ Equation 2

$$\begin{aligned} x + y &= 5 \\ x - y &= 3 \end{aligned}$$

Determine which variable can most easily be eliminated when the two equations are added together. Note that the y terms are eliminated when the equations are added vertically.

$$\begin{aligned} x + y &= 5 \\ x - y &= 3 \end{aligned}$$

Add the equations together, collecting LIKE terms vertically.

$$\begin{aligned} 2x &= 8 \\ x &= 4 \end{aligned}$$

Solve the resulting equation for x.

$$x + y = 5$$

Back-substitute $x = 4$ in Equation 1.

$$\begin{aligned} 4 + y &= 5 \\ y &= 1 \end{aligned}$$

Write the answer as an ordered pair (4, 1).

Verify the answer by substituting $x = 4$ and $y = 1$ into each equation of the original system. ✓ The ordered pair (4, 1) satisfies both equations.

➢ *It may be necessary to use the Multiplication Principle when solving a system of equations with the method of elimination. Multiplying each side of an equation by the same number produces an equivalent equation.*

If $x = y$, then $x \cdot \blacklozenge = y \cdot \blacklozenge$, where $\blacklozenge$ represents any real number.

For example, multiplying each term in the equation $2x - y = 4$ by the constant (2) produces a multiple of the original equation, equivalent to the original equation.

Original Equation:	$2x - y = 4$	Equivalent Equations
Multiply each term by 2	$(2)2x - (2)y = (2)4$	
Equivalent Equation:	$4x - 2y = 8$	

The new equivalent equation can then be added to the other equation in the system to eliminate one of the variables as illustrated in the next example.

EXAMPLE 3: Solve the system of equations using the Method of Elimination.

$x + 2y = 7$ Equation 1
$2x = y + 4$ Equation 2

$2x - y = 4$ Write Equation 2 in standard form: $Ax + By = C$.

$x + 2y = 7$
$2x - y = 4$
Rewrite the equations of the system and identify a variable to eliminate. If Equation 2 is multiplied by 2, the y terms are eliminated when the equations are added together.

$(2)2x - (2)y = (2)4$ Multiply each term of Equation 2 by 2.
$4x - 2y = 8$

$x + 2y = 7$ Write Equation 1 and the new Equation 2 as a system.
$4x - 2y = 8$
Add the equations together, collecting LIKE terms vertically.

$5x = 15$ Solve the resulting equation for x.
$x = 3$

$x + 2y = 7$ Back-substitute $x = 3$ in Equation 1.
$3 + 2y = 7$
$2y = 4$
$y = 2$ Write the answer as an ordered pair (3, 2).

Verify the answer by substituting $x = 3$ and $y = 2$ into each equation of the original system. ✓ The ordered pair (3, 2) satisfies both equations.

RECOGNIZING A DEPENDENT OR INCONSISTENT SYSTEM WHEN USING THE METHOD OF SUBSTITUTION OR ELIMINATION

- When the substitution or elimination process results in an identity (a true statement), the system is dependent.
- When the substitution or elimination process results in a false statement, the system is inconsistent.

EXAMPLE 4: When using the Method of Substitution to solve this system, an identity results.

$$x + y = 5 \qquad \text{Equation 1}$$
$$2x + 2y = 10 \qquad \text{Equation 2}$$

$$x + y = 5$$ Solve Equation 1 for y.

$$y = 5 - x$$

Equation 2: $$2x + 2y = 10$$ Since $y = 5 - x$, substitute $(5 - x)$ into Equation 2 for y.

$$2x + 2(5 - x) = 10$$ Solve the resulting equation for x.

$$2x + 10 - 2x = 10$$

$$10 = 10$$ Identity

The result is an identity, a statement true for all real numbers. The solution is an infinite number of ordered pairs. The system is dependent.

EXAMPLE 5: Use the Method of Substitution to solve the following system.

$$x + y = 8 \qquad \text{Equation 1}$$
$$x + y = 1 \qquad \text{Equation 2}$$

$$x + y = 8$$ Solve Equation 1 for y.

$$y = 8 - x$$

Equation 2: $$x + y = 1$$ Since $y = 8 - x$, substitute $(8 - x)$ into Equation 2 for y.

$$x + (8 - x) = 1$$ Solve the resulting equation for x.

$$x + 8 - x = 1$$

$$8 = 1$$ False

The result is a false statement. The system is inconsistent. There is no solution.

➢ *We have discussed three methods used to solve systems of two equations: graphing, substitution, and elimination. The solution method chosen depends on the system under consideration. Analytical solution(s) can be confirmed with a graphical method, if desired.*

SOLVING MIXTURE PROBLEMS USING LINEAR SYSTEMS

Mixture problems form a system of equations and can be solved using any of the methods discussed in this chapter.

EXAMPLE 6: Dr. Kling is a chemist at HomeTown College. He needs 45 liters of a 30% acid solution for an experiment in class. Unfortunately, the acid solutions in the storeroom are a 10% acid solution and a 40% acid solution. How many liters of each solution should he mix to produce 45 liters of 30% solution?

Let x represent the number of liters of the 10% solution needed to make 45 liters of 30% acid solution. Let y represent the number of liters of the 40% solution.

Equation 1:

Number of liters of 10% solution		Number of liters of 40% solution		Total number of liters in the final solution
x	+	y	=	45 liters

Calculate the number of liters of acid in a given solution:

Number of liters of acid in a solution = (percentage of acid) × (total number of liters).

Equation 2:

Liters of acid in 10% acid soultion		Liters of acid in 40% acid solution		Total liters of acid in the final Solution (30%)
Multiply the number of liters (x) by 10% 10% of x $0.10(x)$	+	Multiply the number of liters (y) by 40%	=	Multiply the number of liters, 45, by 30%
$0.1x$		$0.4y$		13.5

Solve the system using the Method of Substitution:

$$x + y = 45 \qquad \text{Equation 1}$$

$$0.1x + 0.4y = 13.5 \qquad \text{Equation 2}$$

Equation 3: $y = 45 - x$ — Solve Equation 1 for y. Label the result Equation 3.

$0.1x + 0.4y = 13.5$ — Since $y = 45 - x$, substitute $(45 - x)$ into Equation 2 for y.

$$0.1x + 0.4(45 - x) = 13.5$$

$0.1x + 18 - 0.4x = 13.5$ — Solve the resulting equation for x.

$$-0.30x = -4.5$$

$$x = \frac{-4.5}{-0.30}$$

$$x = 15 \text{ liters}$$

$y = 45 - x$ — Back-substitute $x = 15$ into Equation 3.

$$y = 45 - 15$$

$y = 30$ liters — Write the answer as an ordered pair (15, 30).

Verify the answer by substituting $x = 15$ and $y = 30$ into each equation of the original system to check that the values satisfy each of the original equations.

✓ The ordered pair (15, 30) satisfies both equations.

Solution: Dr. Kling should mix 15 liters of 10% acid solution together with 30 liters of 40% acid solution to obtain 45 liters of 30% acid solution.

Doing Mathematics

- Write the definitions of the words and symbols listed for this section in the Dictionary in Appendix A.
- On a 3 × 5-inch card, write the following processes for future reference:
  - (a) The steps for solving a linear system in two variables using the Method of Substitution.
  - (b) The steps for solving a linear system in two variables using the Method of Elimination.

Problems 1–16: Solve each system using the Method of Substitution. If the system is consistent, write the coordinates of the solution. Otherwise, write dependent or inconsistent. Verify the answer by substituting the solution into each equation of the original system. Confirm graphically, if desired.

1. $y = x - 1$
$4x - 3y = 12$

2. $y = 2x + 1$
$5x - 4y = -16$

3. $y - 3x = 2$
$x = \frac{1}{2}y$

4. $x = 4y$
$7y - 2x = -2$

5. $y = 3 - 2x$
$6x + 3y = 9$

6. $y = -2x + 3$
$8x + 4y = 5$

7. $\frac{1}{4}x - \frac{1}{5}y = 1$
$5x - 4y = 10$

8. $\frac{1}{2}x + \frac{1}{3}y = 1$
$3x + 2y = 6$

9. $x + y = 2$
$x + y = 5$

10. $3x - 2y = 6$
$15x - 10y = 30$

11. $x = y - 2$
$2x - 2y = 1$

12. $x = y + 3$
$2x - 2y = 5$

13. $y = 3x - 2$
$2x - 2y = 1$

14. $y = 5x + 2$
$x + 2y = 4$

15. $-3x + y = -12$
$2x + y = 3$

16. $2x + 3y = 4$
$4x + y = 3$

Problems 17–32: Solve each system using the Method of Elimination. If the system is consistent, write the coordinates of the solution. Otherwise, write dependent or inconsistent. Verify the answer by substituting the solution into each equation of the original system. Confirm graphically, if desired.

17. $x + y = 6$
$x - y = -2$

18. $x + 3y = 7$
$-x + 4y = 7$

19. $2x + y = 1$
$y = x - 4$

20. $2x + y = 0$
$4x - 2y = 2$

21. $x = y + 1$
$5x - 5y = 5$

22. $3x = 3y + 15$
$x - y = 5$

23. $x + 4y = 12$
$\frac{1}{2}x + 2y = 6$

24. $2x + 5y = 7$
$-2x = 5y + 7$

25. $x + y = 3$
$y = 1 - x$

26. $2 - x = y$
$x + y = 2$

27. $2x - 4y = -2$
$-x + 2y = 5$

28. $3y = 4x + 1$
$4x = 3y + 1$

29. $x - y = 0$
$3x - y = 4$

30. $3x + y = 8$
$x - \frac{1}{5}y = 0$

31. $5x - 4y = 9$
$x - 2y = -3$

32. $\frac{2}{3}x + \frac{1}{4}y = 3$
$\frac{1}{3}x - \frac{1}{4}y = 0$

33. You have \$4500 to invest for one year. You will use the interest earned to buy college textbooks next year, which will cost \$375. You plan to invest in two different accounts, because it is too risky to invest all of the money in one high interest account. One account earns 5% annual interest and the other account earns 11% annual interest. Determine how much money to invest in each account to earn \$375.

(a) Write two equations to form a system describing the problem. Let *x* represent the amount invested at 5%. Let *y* represent the amount invested at 11%.

Equation 1:

Amount invested at 5%	+	Amount invested at 11%	=	Total amount invested

Equation 2:

Interest earned at 5%	+	Interest earned at 11%	=	Total interest earned

(b) Rewrite Equations 1 and 2 as a system.

Equation 1:

Equation 2:

(c) Solve the system using either substitution or elimination.

(d) How much money should be invested at 5%?

(e) How much money should be invested at 11%?

(f) Verify the answer by substituting the solution into the original equations of the system. Confirm your answer graphically.

34. Congratulations! You won the lottery for $100,000 (after taxes). You plan to invest the total amount in two different accounts. One account earns 3% annual interest and the other account earns 7% annual interest. You need to use the $4000 total interest to help pay for your college expenses. Determine how much money to invest in each account.

(a) Write two equations to form a system describing the problem. Let x represent the amount invested at 3%. Let y represent the amount invested at 7%.

Equation 1:

Amount invested at 3%		Amount invested at 7%		Total amount invested
	+		=	

Equation 2:

Interest earned at 3%		Interest earned at 7%		Total interest earned
	+		=	

(b) Rewrite Equations 1 and 2 as a system.

Equation 1:

Equation 2:

(c) Solve the system using either substitution or elimination.

(d) How much money should be invested at 3%?

(e) How much money should be invested at 7%?

(f) Verify the answer by substituting the solution into the original equations of the system. Confirm your answer graphically.

35. Dave has 7 bills in his wallet in \$5 and \$10 bills for a total of \$55. How many bills of each kind does he have?

(a) Write two equations to form a system describing the problem. Let x represent the number of \$5 bills. Let y represent the number of \$10 bills.

Equation 1:

Number of \$5 bills	+	Number of \$10 bills	=	Total number of bills

Equation 2:

Value of \$5 bills (Hint: multiply number of bills by \$5)	+	Value of \$10 bills (Hint: multiply number of bills by \$10)	=	Total amount of money in Dave's wallet

(b) Rewrite Equations 1 and 2 as a system.

Equation 1:

Equation 2:

(c) Solve the system using either substitution or elimination.

(d) How many \$5 bills are in Dave's wallet?

(e) How many \$10 bills are in Dave's wallet?

(f) Verify the answer by substituting the solution into the original equations of the system. Confirm your answer graphically.

36. Elbert is a coin collector. He collects only "state" quarters and "wheat" pennies. His collection of 1100 coins is worth \$131 at face value. How many coins of each type does Elbert have?

(a) Write two equations to form a system describing the problem. Let x represent the number of quarters. Let y represent the number of pennies.

Equation 1:

Number of quarters	+	Number of pennies	=	Total number of coins

Equation 2:

Value of quarters (Hint: multiply number of quarters by \$0.25)	+	Value of pennies (Hint: multiply number of pennies by \$0.01)	=	Total amount of money

(b) Rewrite Equations 1 and 2 as a system.

Equation 1:

Equation 2:

(c) Solve the system using either substitution or elimination.

(d) How many quarters does Elbert have?

(e) How many pennies does Elbert have?

(f) Verify the answer by substituting the solution into the original equations of the system. Confirm your answer graphically.

37. The college bookstore will bundle 9 highlighter pens into a package to sell for \$5.25. Blue pens cost \$0.50 each and green pens cost \$0.75 each. How many blue pens and how many green pens should be put in each bundle?

(a) Draw a picture and write two equations to form a system describing the problem. Let x represent the number of blue pens. Let y represent the number of green pens.

Equation 1:

Equation 2:

(b) Solve the system using either substitution or elimination.

(c) How many blue pens should be in each bundle?

(d) How many green pens should be in each bundle?

(e) Verify the answer by substituting the solution into the original equations of the system. Confirm your answer graphically.

38. The Alwaysready Battery Company will bundle 8 AA batteries into a package to sell for \$9.00. Superduper batteries cost \$1.25 each and Reallygood batteries cost \$0.75 each. How many batteries of each type should be put in each bundle?

(a) Draw a picture and write two equations to form a system describing the problem. Let x represent the number of Superduper batteries. Let y represent the number of Reallygood.

Equation 1:

Equation 2:

(b) Solve the system using either substitution or elimination.

(c) How many Superdupers should be in each bundle?

(d) How many Reallygoods should be in each bundle?

(e) Verify the answer by substituting the solution into the original equations of the system. Confirm your answer graphically.

39. Lisa is a nurse who has two sedative solutions. One solution contains 25% sedative and the second solution contains 10% sedative. Lisa needs to give a patient a sedative solution that contains 20% sedative. How much 25% sedative solution should she mix with the 10% sedative solution to obtain 75 cubic centimeters of solution containing 20% sedative?

(a) Draw a picture and write two equations to form a system describing the problem. Let x represent the number of cubic centimeters of the 25% solution. Let y represent the number of cubic centimeters of the 10% solution.

Equation 1:

Equation 2:

(b) Solve the system using either substitution or elimination.

(c) How many cubic centimeters of 25% sedative solution should be used?

(d) How many cubic centimeters of 10% sedative solution should be used?

(e) Verify the answer by substituting the solution into the original equations of the system. Confirm your answer graphically.

40. Megan is a pharmacist who needs to mix two antibacterial solutions. One solution contains 20% active ingredient and the second solution contains 5% active ingredient. Megan needs to give a patient a solution that contains 15% of the active ingredient. How much 20% solution should she mix with the 5% solution to obtain 250 cubic centimeters of solution containing 15% of the active ingredient?

(a) Draw a picture and write two equations to form a system describing the problem. Let x represent the number of cubic centimeters of the 20% solution. Let y represent the number of cubic centimeters of the 5% solution.

Equation 1:

Equation 2:

(b) Solve the system using either substitution or elimination.

(c) How many cubic centimeters of 20% solution should be used?

(d) How many cubic centimeters of 5% solution should be used?

(e) Verify the answer by substituting the solution into the original equations of the system. Confirm your answer graphically.

41. Jessica is considering changing her cellphone contract. The new phone contract will cost \$0.10 per minute during peak hours (8 A.M.–5 P.M.) and \$0.05 per minute during off-peak hours (5 P.M.–8 A.M.). The plan will allow for a total of 500 minutes per month and the 500 minutes will cost \$30.00 per month. How many peak minutes (\$0.10 per minute) and how many nonpeak minutes (\$0.05) are allowed by the plan?

(a) Draw a picture and write two equations to form a system describing the problem situation. Let x represent the number of peak minutes at \$0.10 per minute. Let y represent the number of off-peak minutes at \$0.05 per minute.

Equation 1:

Equation 2:

(b) Solve the system using either substitution or elimination.

(c) How many peak minutes are possible with this plan?

(d) How many off-peak minutes are possible with this plan?

(e) Verify the answer by substituting the solution into the original equations of the system. Confirm your answer graphically.

42. Jessica has been offered a new cellphone contract. This phone contract will cost \$0.09 per minute during peak hours (8 A.M.–5 P.M.) and \$0.05 per minute during off-peak hours (5 P.M.–8 A.M.). The plan will allow for a total of 500 minutes per month and the 500 minutes will cost \$30.00 per month. How many peak minutes and how many nonpeak minutes are allowed by the plan?

(a) Draw a picture and write two equations to form a system describing the problem situation. Let x represent the number of peak minutes at \$0.09 per minute. Let y represent the number of off-peak minutes at \$0.05 per minute.

Equation 1:

Equation 2:

(b) Solve the system using either substitution or elimination.

(c) How many peak minutes are possible with this plan?

(d) How many off-peak minutes are possible with this plan?

(e) Verify the answer by substituting the solution into the original equations of the system. Confirm your answer graphically.

(f) Compare your answers for this problem with the answer to #41. Which plan is best for Jessica? What factors would affect her decision?

Problems 43–60: Solve each system by any method. If the system is consistent, write the coordinates of the solution. Otherwise, write dependent or inconsistent. Verify the answer by substituting the solution into each equation of the original system. Confirm graphically, if desired.

43. $x - y = 6$
$3x + y = 2$

44. $-2x - 3y = 5$
$4x + 2y = 10$

45. $3x + 4y = 12$
$6x + 8y = 5$

46. $7x + 8y = 2$
$14x + 16y = 2$

47. $y = 2x + 5$
$4x + y = 2$

48. $3x - 1 = y$
$5x + 2y = 9$

49. $\frac{1}{2}x + 2y = 3$
$x + 2y = 5$

50. $\frac{1}{2}x + \frac{1}{3}y = 5$
$x - y = 0$

51. $x = -\frac{3}{2}$
$4x + y = 5$

52. $y = \frac{3}{4}$
$5x + 4y = 13$

53. $2x + 4 = y$
$4x - 2y = 6$

54. $x = 4 - 3y$
$3y = 4 - x$

55. $0.2x - 0.1y = 0.8$
$3x + y = 6$

56. $0.3x + 5y = 1.2$
$x + 4y = 4$

57. $2x - y = 2$
$y = 2x - 4$

58. $3x - y = 1$
$y = 4x + 1$

59. $x = y$
$3x - 3y = 0$

60. $2x - 2y = 5$
$5x - 5y = 8$

4.4 Graphing and Solving Linear Inequalities in Two Variables

OBJECTIVE

- Solve and graph linear inequalities in two variables.

1 2 3 Getting Started

In this section, we discuss linear inequality statements in two variables. A **linear inequality in two variables** is of the form $Ax + By > C$ (or $<$, $\geq$, or $\leq$), where A and B are not both zero. Recall that the inequality symbols $<$ and $>$ are called strict inequalities and $\leq$ and $\geq$ are called nonstrict inequalities. The solution of a linear inequality in two variables is an infinite number of ordered pairs that satisfy the inequality, and is usually displayed with shading on the Cartesian plane.

1. Consider the inequality statement: $2x + 3y \leq 6$. Substitute the given coordinates into the inequality statement to determine whether the ordered pairs are solutions of the inequality.

(a) Is $(1, 4)$ a solution of the inequality?

(b) Is $(-3, 2)$ a solution of the inequality?

(c) Write another ordered pair that satisfies $2x + 3y \leq 6$.

2. The infinite number of ordered pairs in the solution of the nonstrict inequality $2x + 3y \leq 6$ can be visualized on the Cartesian plane. The points on the line $2x + 3y = 6$ form a boundary line for the solution of the inequality statement. The line divides the plane into one half-plane above the line and one half-plane below the line. The solution of the inequality is all points that satisfy the inequality, and will fall in one of the half-planes.

(a) Graph the *equality*, $2x + 3y = 6$, using the Intercept Method. Compute the x- and y-intercepts and graph the equation.

x	y
0	
	0

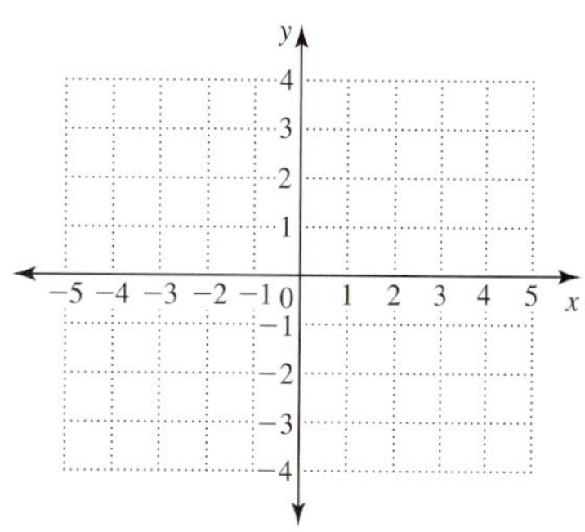

(b) Does the point $(0, 2)$ satisfy the nonstrict inequality ($\leq$)? Are all points on the line in the solution set of $2x + 3y \leq 6$? Why or why not?

(c) The point $(-3, 2)$ satisfies the inequality (problem 1b). Shade the half-plane on the grid containing the point $(-3, 2)$.

(d) If a test-point satisfies the inequality, then all points in that half-plane containing the test-point satisfy the inequality. Choose another point in the shaded area. Do the coordinates of your chosen point satisfy the inequality?

(e) Describe the solution set of all points that satisfy the inequality $2x + 3y \leq 6$.

3. Let's find the solution of the strict inequality $3x - 4y > 12$ by following these steps:

(a) Graph the *equation* $3x - 4y = 12$ using the Intercept Method.

x	y
0	
	0

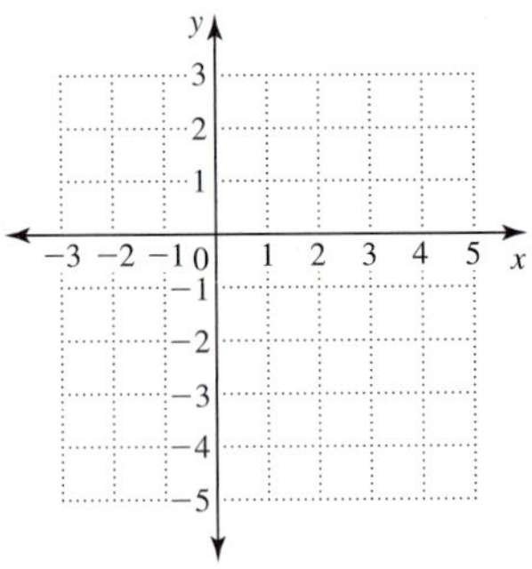

(b) Are the points on the line included in the solution of the strict inequality $3x - 4y > 12$? The points on the line $3x - 4y = 12$ form a boundary line for the solution of the inequality statement. The points on the line are not included and the line should be drawn as a dotted line.

➢ *The dotted line divides the plane into two half-planes: one half-plane above the line and one half-plane below the line. The solution of the inequality is all points in one of the half-planes. To determine which half-plane is the solution of the inequality, test a point in each half-plane. The half-plane containing the point(s) that satisfies the inequality is the solution of the inequality.*

(c) Write the coordinates of one point above the line as a test point. Substitute the coordinates into the inequality and determine whether or not this test point satisfies the original inequality $3x - 4y > 12$.

(d) Write the coordinates of a second point below the line as a second test point. Substitute the coordinates into the inequality and determine whether or not this test point satisfies the original inequality $3x - 4y > 12$.

(e) Which half-plane determines the solution set of the inequality $3x - 4y > 12$?
Circle one: Half-plane above the line Half-plane below the line

(f) Shade the appropriate half-plane on the grid to show the solution set of the inequality.

1 2 3

Discussing Mathematics

GRAPHING THE SOLUTION OF AN INEQUALITY

The **solution of an inequality in two variables** is the set of all points that satisfy the inequality. The collection of points comprising the solution set is best illustrated with a graph of the related equation and the appropriate shading above or below the line.

GRAPHING THE SOLUTION OF A LINEAR INEQUALITY IN TWO VARIABLES

1. Rewrite the inequality as an equality statement. Graph the equation.
2. Determine whether the line should be solid or dotted.
   - If the original inequality statement is a strict inequality (either $<$ or $>$), draw the line as a dotted line.
   - If the original inequality statement is a nonstrict inequality (either $\leq$ or $\geq$), draw the line as a solid line.
3. Choose a test-point above the line. Determine whether the coordinates of the test-point satisfy the original inequality
4. Choose a second test-point below the line. Determine whether the coordinates of the test-point satisfy the original inequality.
5. Shade the half-plane containing the test-point that satisfied the inequality from step 3 or 4.

➢ *The point (0, 0) is a good point to choose as a test-point with easy computations.*

EXAMPLE 1: Graph the inequality $x - 2y < 4$.

$x - 2y = 4$ Write the inequality as an *equality* statement and graph the equation by any method. The intercept-method is often very efficient.

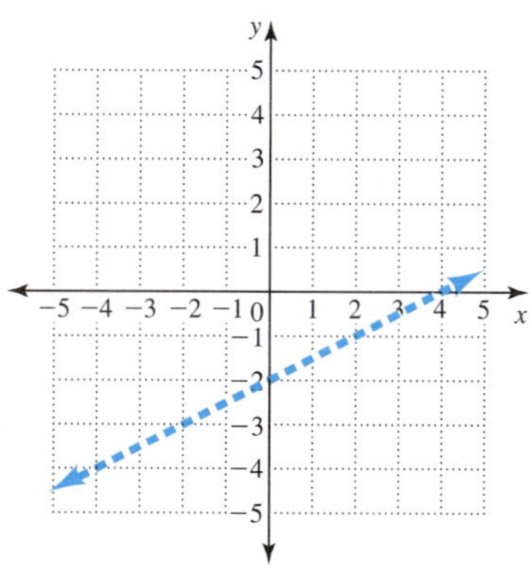

x	y_1
0	−2
4	0

Because the original inequality statement is a strict inequality ($<$), draw a dotted line.

Choose the test-point (0, 0) above the line:

$$x - 2y < 4$$

$$0 - 2(0) \stackrel{?}{<} 4$$

$$0 < 4$$

The result is a true statement. The point (0, 0) satisfies the original inequality.

Choose a second test-point (2, −4) below the line:

$$x - 2y < 4$$

$$2 - 2(-4) \stackrel{?}{<} 4$$

$$10 \not< 4$$

The result is a false statement. The point (2, −4) does not satisfy the original inequality.

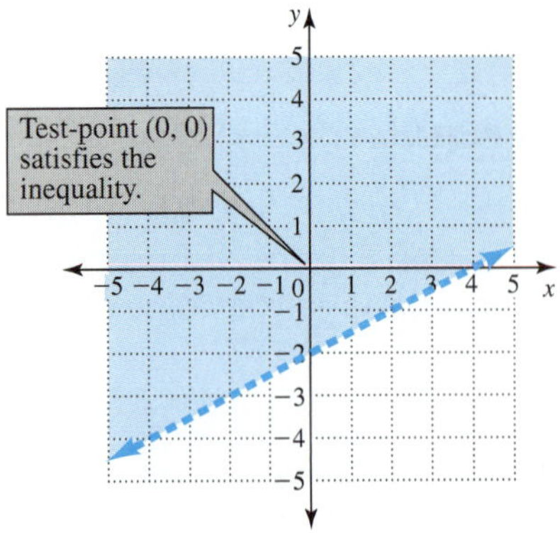

Shade the half-plane above the line containing the point (0, 0) that satisfied the inequality.

The line is not included in the solution as indicated by the dotted line.

Inequalities may form horizontal or vertical lines where the appropriate half-plane is either above or below, or to the left or right, of the lines formed from the related equality statements.

EXAMPLE 2: Graph the inequality $y \geq -3$.

The graph of the equality $y = -3$ is a horizontal line.
The original nonstrict inequality $(\geq)$ is drawn as a solid line.

Choose the test-point (0, 0) above the line:

$$y \geq -3$$
$$0 \overset{?}{\geq} -3$$
$$0 \geq -3$$

The point (0, 0) satisfies the original inequality.

Choose the test-point $(2, -5)$ below the line:

$$y \geq -3$$
$$-5 \overset{?}{\geq} -3$$
$$-5 \ngeq -3$$

The point $(2, -5)$ does not satisfy the original inequality.

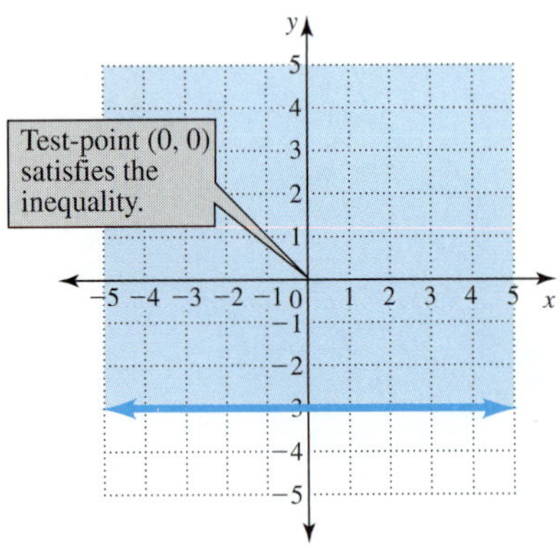

Shade the half-plane above the line containing the point (0, 0) that satisfied the inequality.

The solution set of the original inequality is the entire half-plane shaded above the line and all points on the solid line.

EXAMPLE 3: Graph the inequality $x < -1$.

The graph of the equality is a vertical line through $x = -1$.
The original strict inequality ($<$) is drawn as a dotted line.

Choose the test-point $(-3, 1)$ to the left of the line:

$$x < -1$$
$$-3 \overset{?}{<} -1$$
$$-3 < -1$$

The point $(-3, 1)$ satisfies the original inequality.

Choose the test-point $(0, 0)$ to the right of the line:

$$x < -1$$
$$0 \overset{?}{<} -1$$
$$0 \not< -1$$

The point $(0, 0)$ does not satisfy the original inequality.

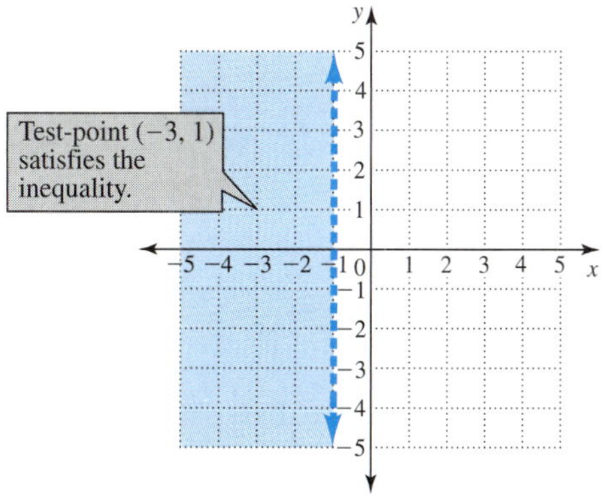

Shade the half-plane to the left the line containing the point from step 3 that satisfied the inequality.

The solution set of the original inequality is the entire half-plane shaded to the left of the line.

USING SLOPE-INTERCEPT FORM OF A LINE TO SOLVE LINEAR INEQUALITIES

Generalizations may be made about the solution of a linear inequality written in slope-intercept form, $y > mx + b$, $y \geq mx + b$, $y < mx + b$, or $y \leq mx + b$. The graphs below are sketches of any equation of the form $y = mx + b$, where m and b are positive numbers. Similar conclusions can be drawn when $m < 0$.

All points ABOVE the line where $y > mx + b$ $(m > 0, b > 0)$.	All points ON the line where $y = mx + b$ $(m > 0, b > 0)$.	All points BLEOW the line where $y < mx + b$ $(m > 0, b > 0)$.
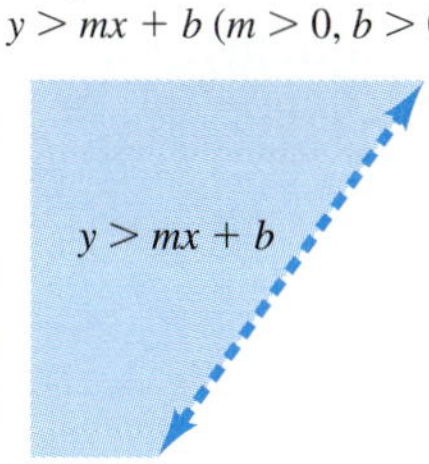	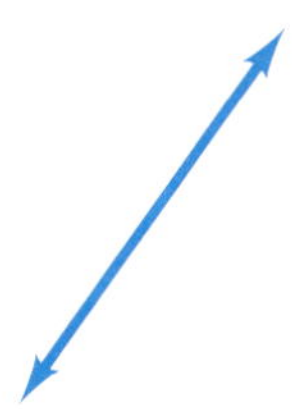	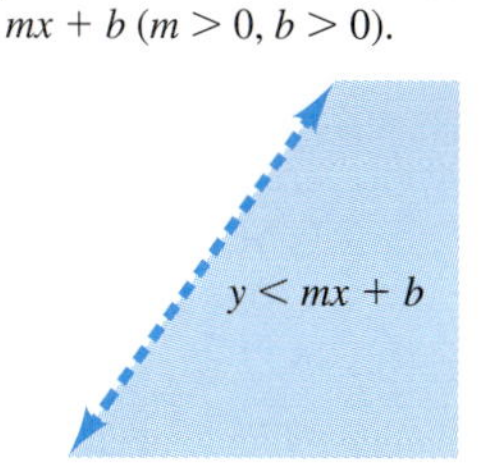
The half-plane ABOVE the line is the region where $y > mx + b$.	$y = mx + b$	The half-plane BELOW the line is the region where $y < mx + b$.

SHADING THE SOLUTION OF INEQUALITY STATEMENTS IN SLOPE-INTERCEPT FORM

$y > mx + b$ $y \geq mx + b$	Shade ABOVE the line.	The line is ***dotted*** for $y > mx + b$. The line is ***solid*** for $y \geq mx + b$.
$y < mx + b$ $y \leq mx + b$	Shade BELOW the line.	The line is ***dotted*** for $y < mx + b$. The line is ***solid*** for $y \leq mx + b$.

3 Doing Mathematics

- Write the definitions of the words and symbols listed for this section in the Dictionary in Appendix A.
- On a 3 × 5-inch card, write the process of solving and graphing a linear inequality in two variables for future reference.

Problems 1–6: Determine whether the given ordered pairs are solutions to the given inequality.

1. $x + y \le 12$
(a) $(0, 0)$
(b) $(5, 7)$
(c) $\left(\frac{1}{3}, 4\right)$

2. $x - y \le 8$
(a) $(0, 0)$
(b) $(2, 3)$
(c) $\left(9, \frac{1}{3}\right)$

3. $2x + y < 1$
(a) $(0, 0)$
(b) $\left(\frac{1}{2}, 6\right)$
(c) $(-1, -1)$

4. $x + 3y < -5$
(a) $(0, 0)$
(b) $\left(\frac{1}{3}, -2\right)$
(c) $(-1, -2)$

5. $y > 2$
(a) $(0, 0)$
(b) $(6, 2)$
(c) $(6, -1)$

6. $y > 7$
(a) $(0, 0)$
(b) $(0, 7)$
(c) $(5, 9)$

Problems 7–31: Graph the solution set of each inequality. It may be efficient to use the Intercept Method to graph each equation.

7. $x + 3y < 2$

x	y

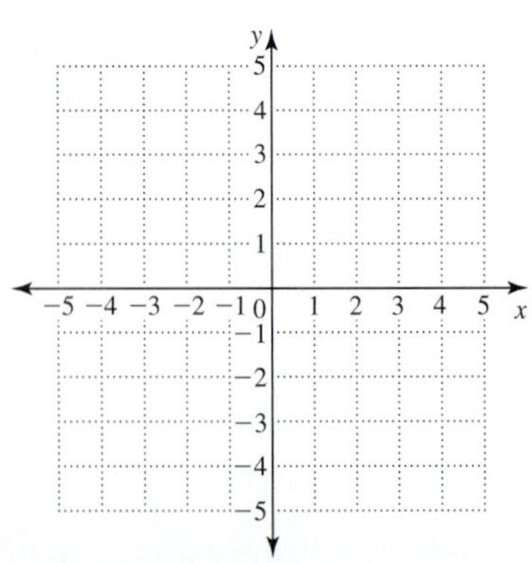

8. $x - y < -2$

x	y

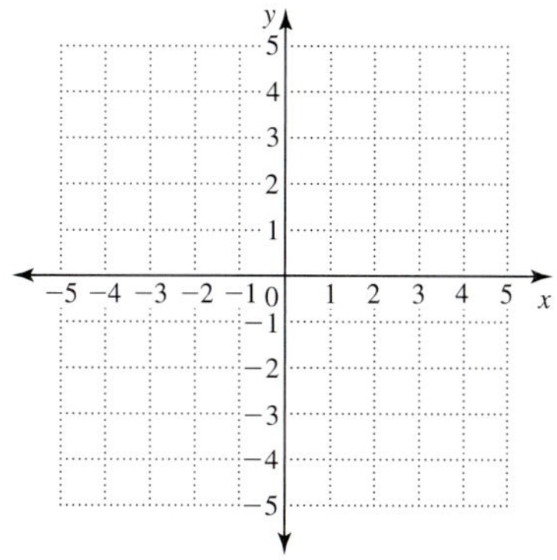

9. $x + y > 4$

x	y

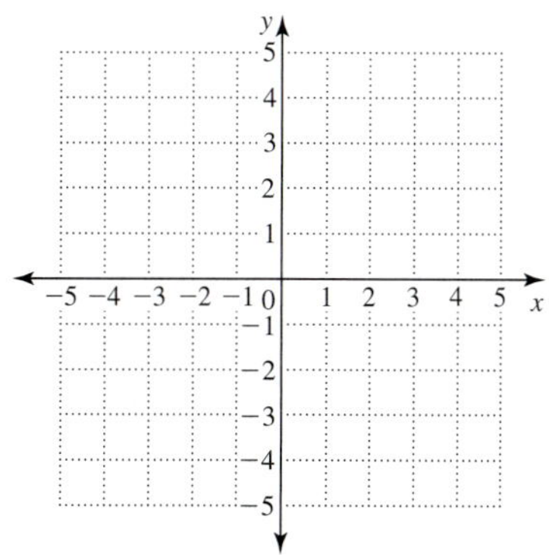

10. $3x - y > -1$

x	y

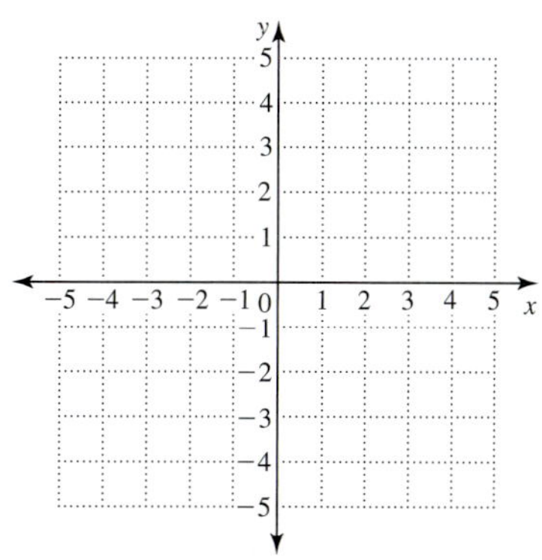

11. $x + 2y \geq -4$

x	y

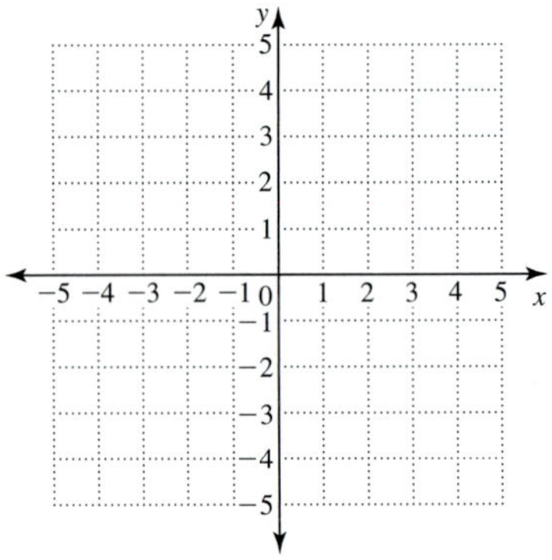

12. $x + y \geq -1$

x	y

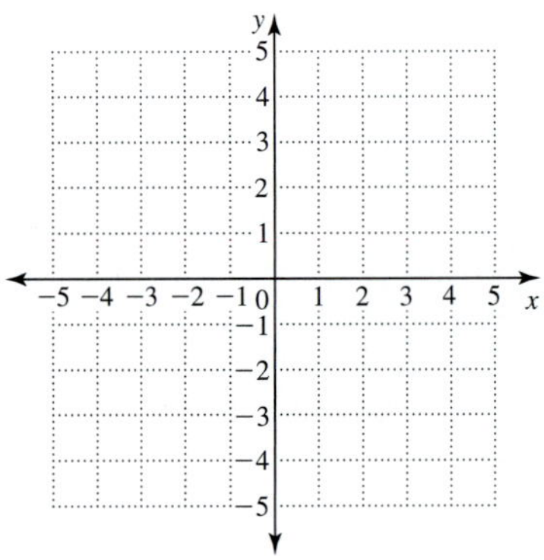

13. $3x - 4y \leq 12$

x	y

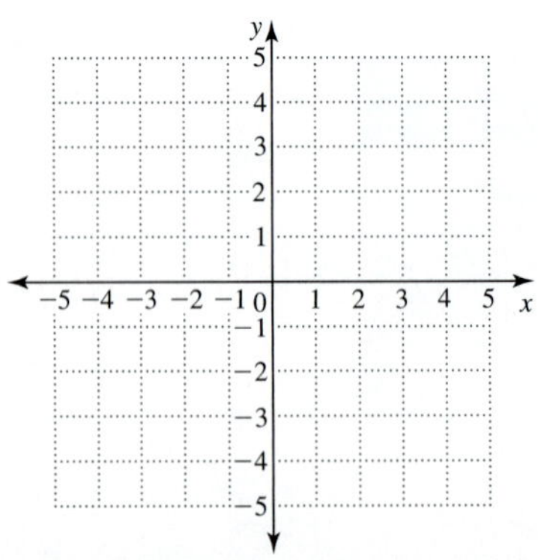

14. $2x - 3y \le 6$

x	y

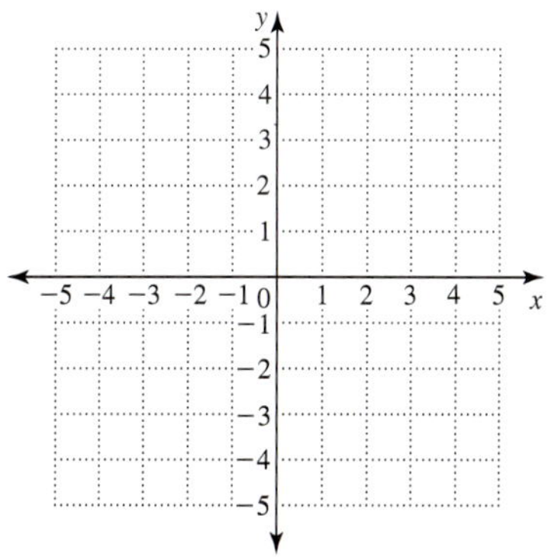

15. $y > -1$

x	y

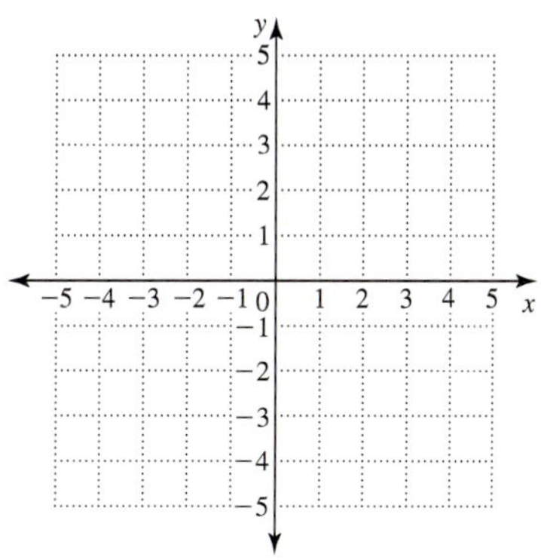

16. $y > 2$

x	y

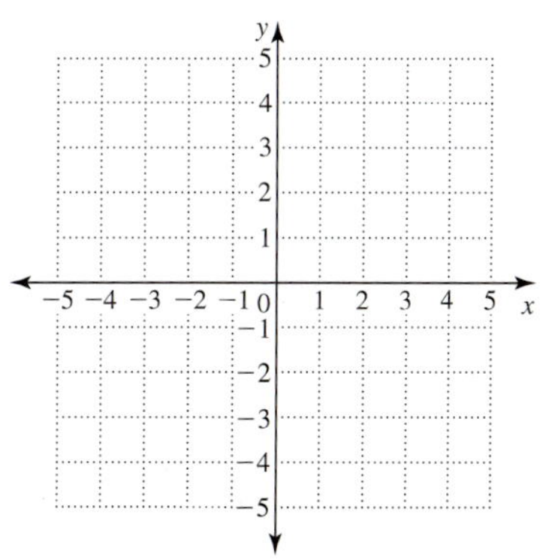

17. $y \leq 0$

x	y

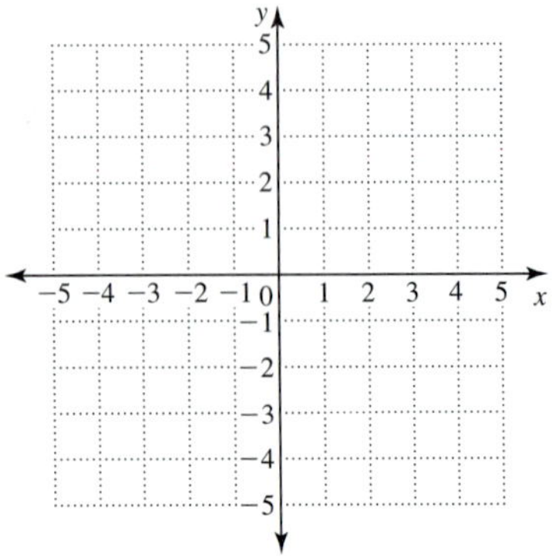

18. $y \leq -1$

x	y

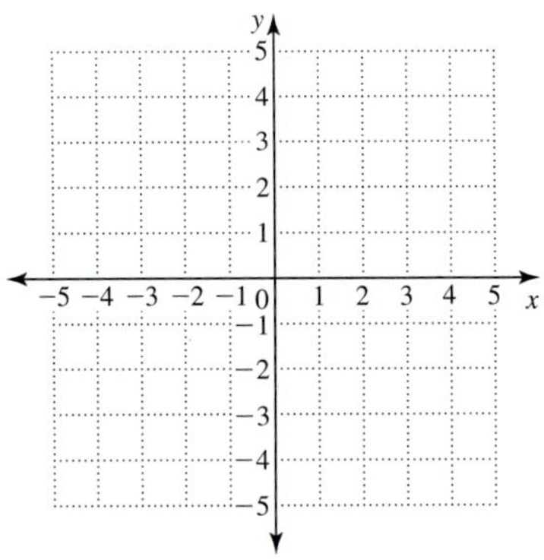

19. $y \geq -3$

x	y

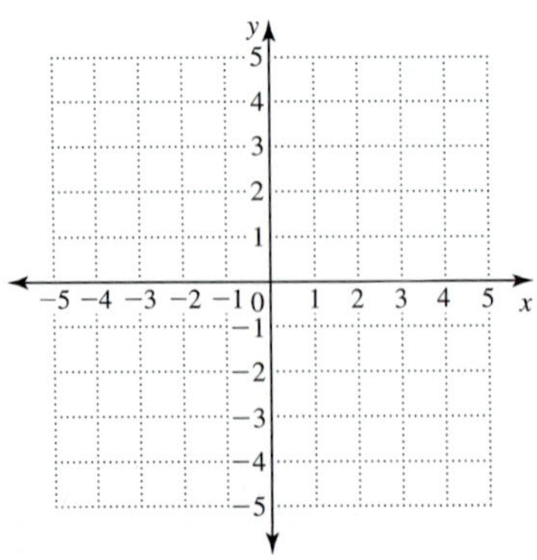

20. $y \geq 4$

x	y

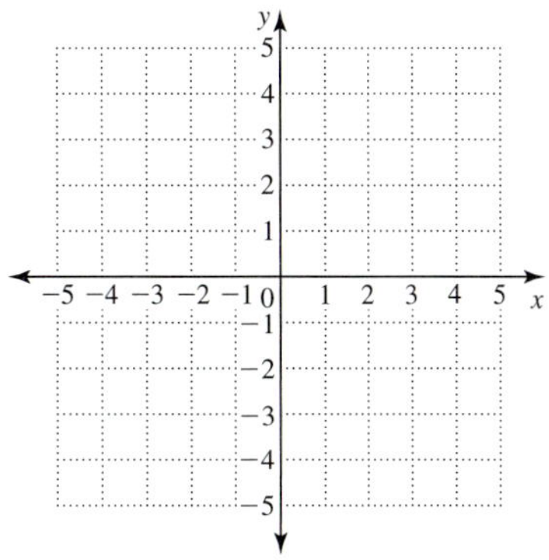

21. $x > 2$

x	y

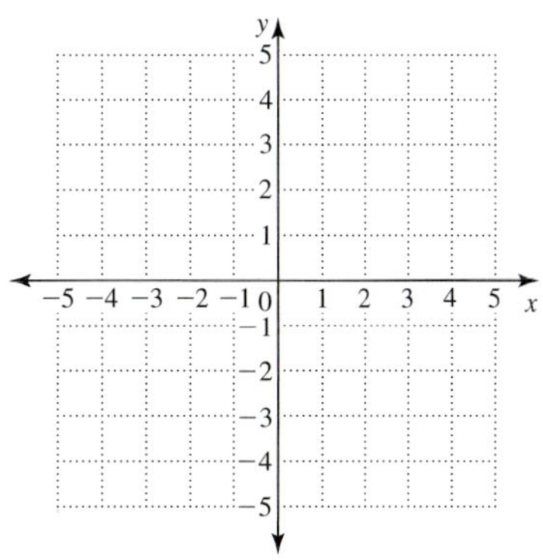

22. $x > 1$

x	y

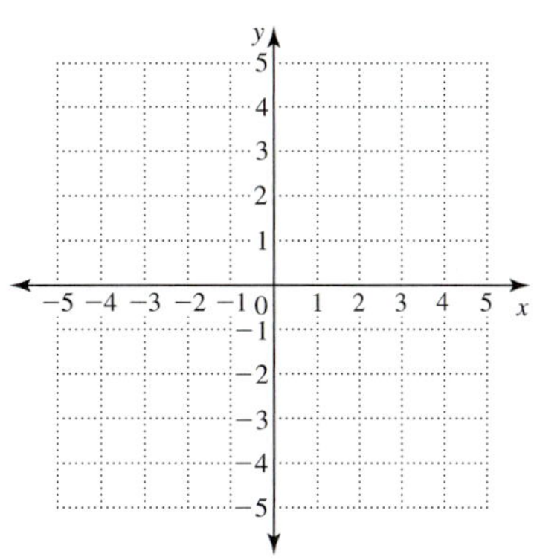

23. $x \geq 3$

x	y

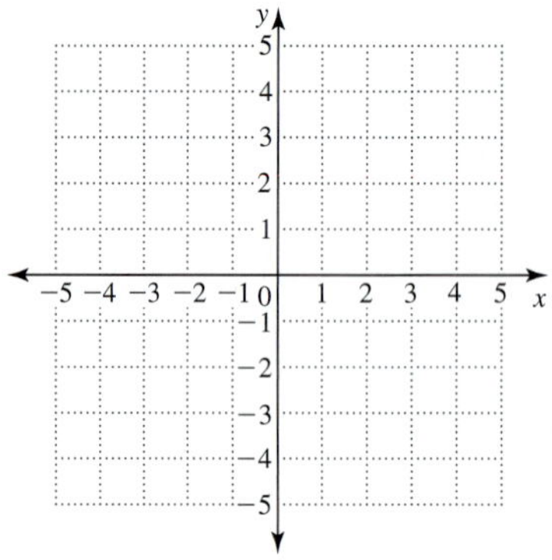

24. $x \geq -4$

x	y

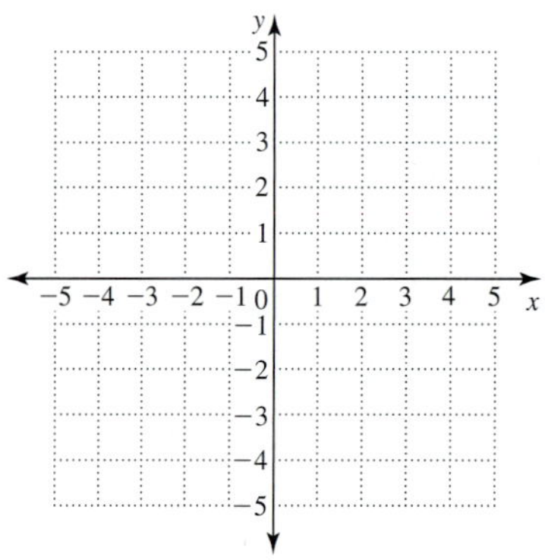

25. $x \leq -1.5$

x	y

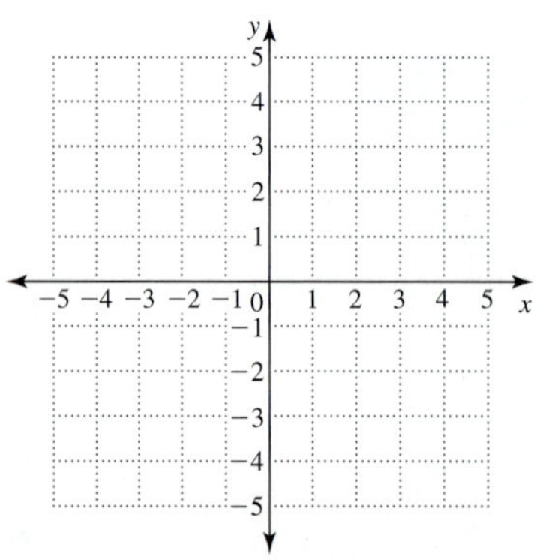

26. $x \leq -2$

x	y

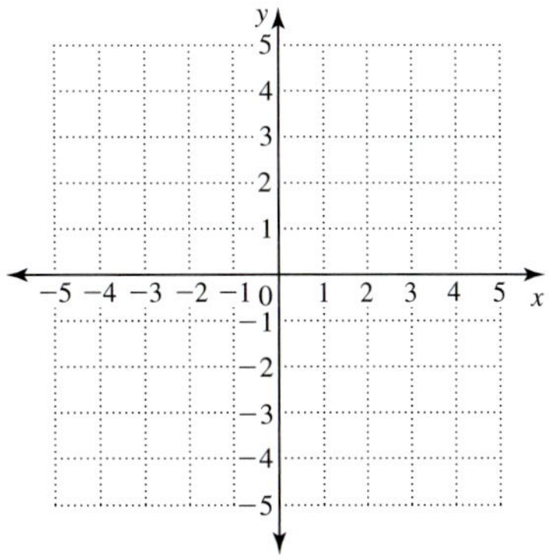

27. $y \leq -\frac{1}{2}x + 1$

x	y

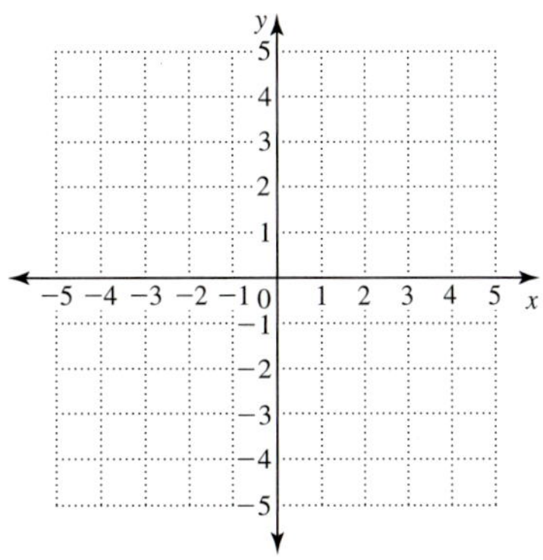

28. $y \leq -\frac{2}{3}x + 1$

x	y

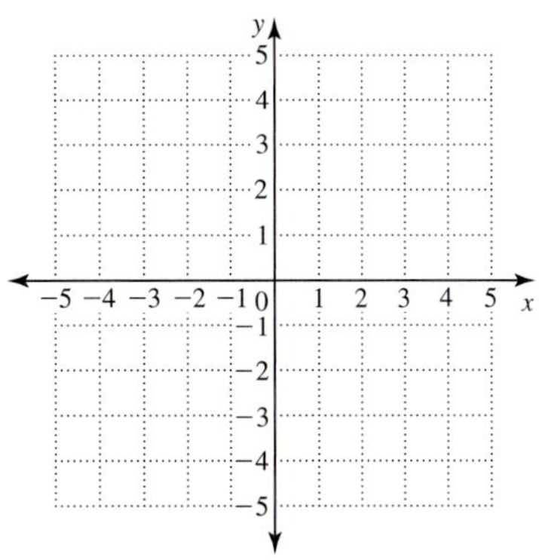

29. $y \geq -\frac{5}{3}x - 1$

x	y

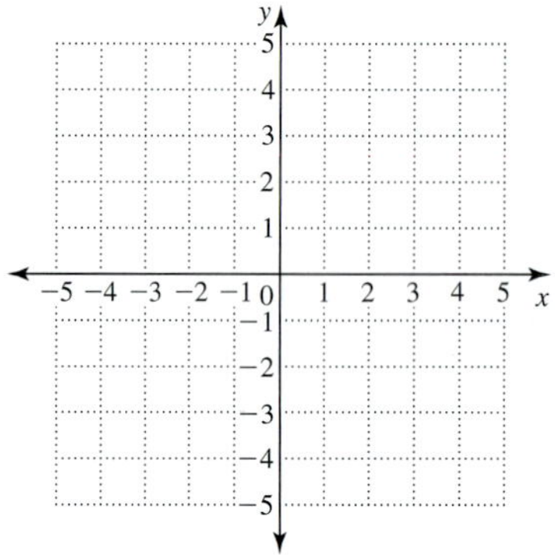

30. $y > -\frac{2}{3}x + 2$

x	y

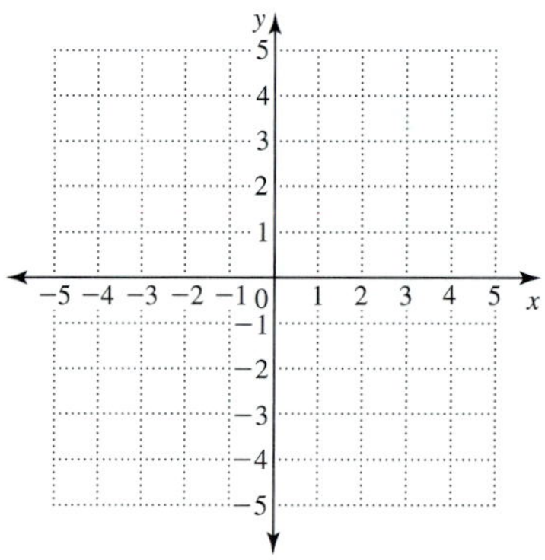

31. $y \geq \frac{1}{3}x$

x	y

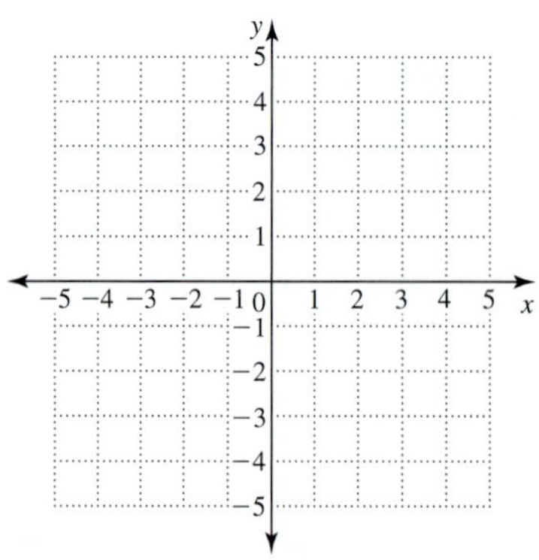

Chapter 4 Summary

SECTION 4.1:

The *X*-Intercept Graphical Method of Solving an Equation

1. Rewrite the equation with all nonzero terms on one side of the equation equal to zero.
2. Substitute y for zero in the transformed equation to form the associated function.
3. Graph the function by making a table of values, plotting and connecting the points with a line.
4. Identify the x-intercept of the graph. The solution of the equation is the x-coordinate of the x-intercept.
5. Verify the answer by substituting the solution into the original equation.

The Intersection-of-Graphs Method for Solving an Equation

1. Label the left side of the equation y_1. Label the right side of the equation y_2.
2. Make tables of values for y_1 and y_2.
3. Graph equations y_1 and y_2 on the same grid.
4. Identify the point of intersection of the two graphs where $y_1 = y_2$. The x-coordinate of the point of intersection is the solution of the original equation.
5. Verify the answer by substituting the solution into the original equation.

SECTION 4.2:

Visualizing the Graphs of a Linear System

1. A system of equations having at least one solution and one point of intersection is called **consistent**.

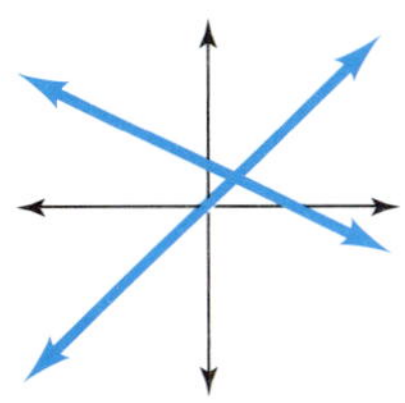

ONE point of intersection.

2. A consistent system of equations with an infinite number of solutions, where the graphs have an infinite number of points of intersection, is called **dependent**.

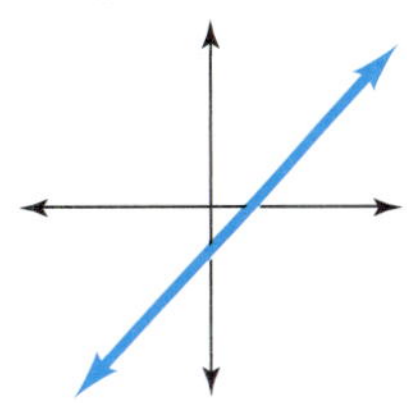

The two lines **COINCIDE**. They intersect at all points along the line.

3. A system of equations with no solution, where the graphs do not intersect, is called **inconsistent**.

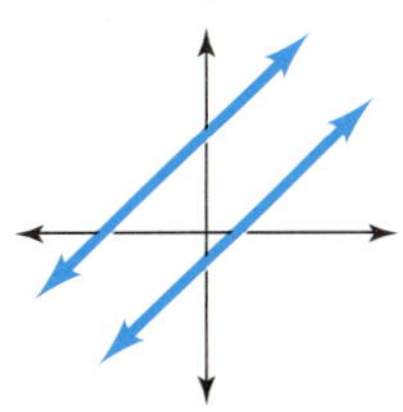

The two lines are parallel and **NEVER** intersect.

The Intersection-of-Graphs Method for Solving a Linear System

1. If necessary, assign variables as the unknowns in the problem to create equations describing the problem situation. Use a picture or table, as needed. Label the equations of the system Equation 1 and Equation 2.
2. Rewrite each equation in function form, solving for y. Label the first equation y_1 and the second equation y_2.
3. Complete tables of values for y_1 and y_2. Graph both equations on the same grid.
4. Identify the point(s) of intersection of the two graphs, where $y_1 = y_2$. The ordered pairs denoting the point(s) of intersection of the graphs is the solution of the system.
5. Verify the solution by substituting the coordinates of the point(s) of intersection into the original equations of the system. If the system describes a problem situation, label the answer(s).

SECTION 4.3:

Method of Substitution for Solving a Linear System in Two Variables

1. Assign variables as the unknowns in the problem to create the equations describing the problem situation. Use a picture or table, as needed. Label the equations of the system Equation 1 and Equation 2.
2. Solve one of the equations of the system for one of the variables. If one variable has a numerical coefficient of $+1$ or -1, solve for that variable. Otherwise, solve for any one of the variables in either equation. Label the result Equation 3.
3. Substitute the expression for the variable obtained in Step 2 into the other equation. Solve the resulting equation for the remaining variable.
4. Substitute the result of Step 3 into Equation 3 and solve for the second variable. This process is called back-substitution. Write the solution as an ordered pair.
5. Verify the answer by substituting the solution into each equation of the original system. If the system describes a problem situation, label the answer(s) with the appropriate units of measure.

Method of Elimination for Solving a Linear System in Two Variables

1. Create the equations describing the problem situation. Label the first equation Equation 1. Label the second equation Equation 2. Write each equation in standard form: $Ax + By = C$.
2. Look at the equations in the system and determine which variable can most easily be eliminated when the two equations are added together. It may be necessary to multiply one or both equations by a nonzero constant. Multiply EVERY term in the equation(s) by the constant(s).
3. Rewrite the system and add the equations together. The sum of the equations is a third equation (Equation 3) with one variable. Solve Equation 3 for the remaining variable.
4. Substitute the result of Step 4 into either of the original equations of the system and solve for the second variable. This process is called back-substitution. Write the solution as an ordered pair.
5. Verify the answer by substituting the solution into each equation of the original system. If the system describes a problem situation, label the answer(s) with the appropriate units of measure.

SECTION 4.4:

Shading the Solution of Inequality Statements in Slope-Intercept Form		
Shade ABOVE the line.	$y > mx + b$	The line is ***dotted*** for $y > mx + b$.
	$y \geq mx + b$	The line is ***solid*** for $y \geq mx + b$.
Shade BELOW the line.	$y < mx + b$	The line is ***dotted*** for $y < mx + b$.
	$y \leq mx + b$	The line is ***solid*** for $y \leq mx + b$.

Chapter 4 Review

- Review the definitions you have written for this chapter in the Dictionary.
- Review any properties or processes you have written on 3 × 5-inch cards.

SECTION 4.1

Problems 1–2: Solve each equation for x analytically. Confirm the solution graphically, using the X-Intercept Method. Verify each solution by substituting into the original equation.

1. Given $-\frac{1}{2}x + 3 = 0$.

(a) Solve the equation analytically.

(b) Confirm the solution graphically.

x	y
0	
2	
4	
6	
8	

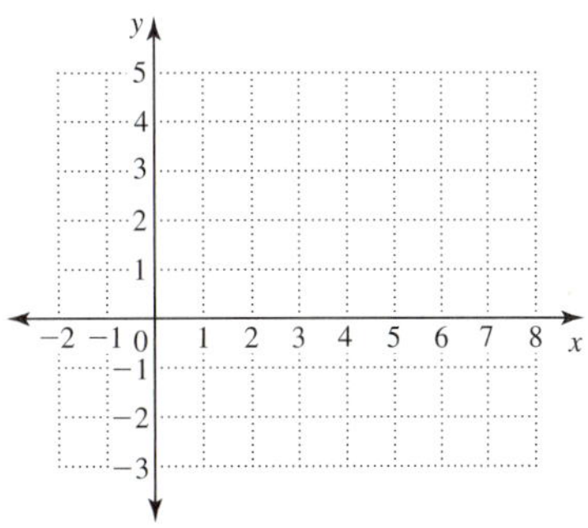

(c) What is the x-coordinate of the x-intercept of the graph?

(d) Does the graphical solution confirm the analytical solution? Why or why not?

2. Given $2x + 2(2x - 1) - 4 = 0$.

(a) Solve the equation for x analytically.

(b) Confirm the solution graphically.

x	y
−1	
0	
1	
2	
3	

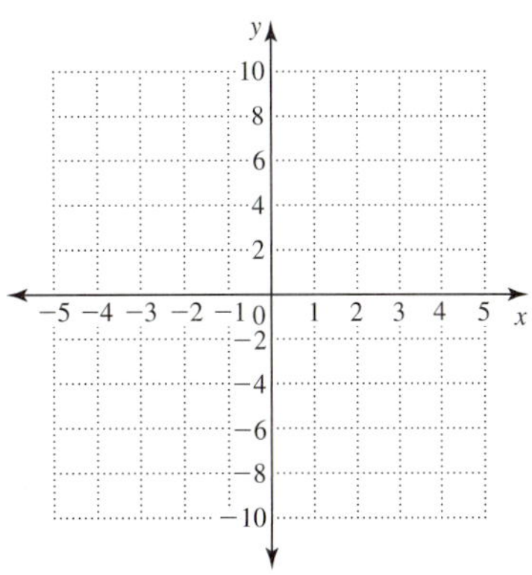

(c) What is the x-coordinate of the x-intercept of the graph?

(d) Does the graphical solution confirm the analytical solution? Why or why not?

Problems 3–5: Solve each equation for x analytically. Confirm the solution graphically, using the Intersection-of-Graphs Method (#3–4) and the X-Intercept Method (#5). Verify each solution by substituting into the original equation.

3. Given $9.3x - (1.3x + 2) = 4x + 6$.

(a) Solve the equation for x analytically.

(b) Confirm the solution graphically.

x	y_1	y_2
−1		
0		
1		
2		
3		

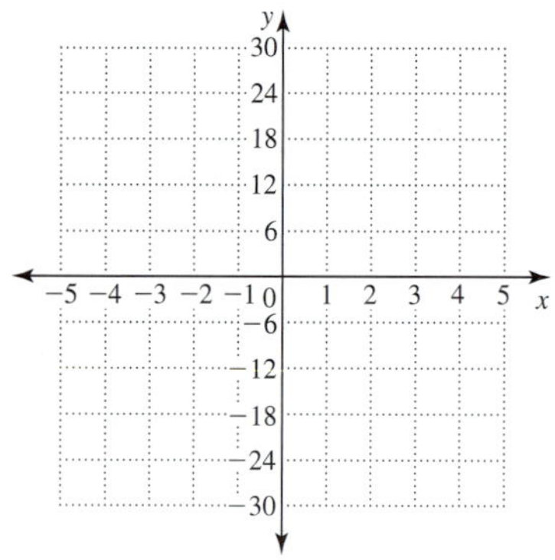

(c) What is the point of intersection of the graphs?

(d) Does the graphical solution confirm the analytical solution? Why or why not?

4. Given $5.7x - 3.6(x - 1) = 7.8$.

(a) Solve the equation for x analytically.

(b) Confirm the solution graphically.

x	y_1	y_2
−1		
0		
1		
2		
3		

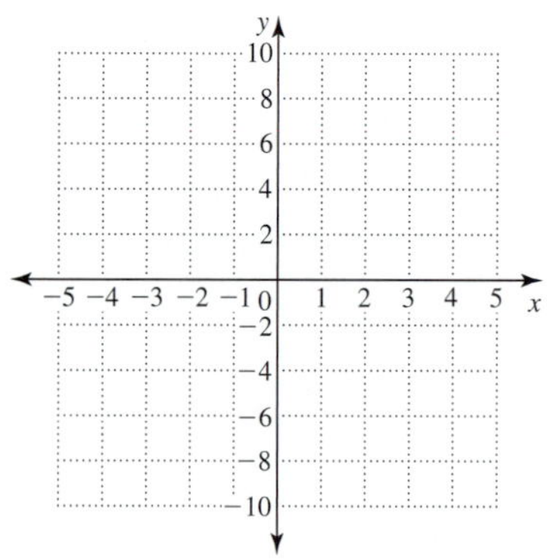

(c) What is the point of intersection of the graphs?

(d) Does the graphical solution confirm the analytical solution? Why or why not?

5. Given $(2x + 3) - (x - 6) = 3x - 5$.

(a) Solve the equation analytically.

(b) Confirm the solution graphically by the X-Intercept Method. Create your own table(s) of values.

x	y_1
3	
5	
7	
9	

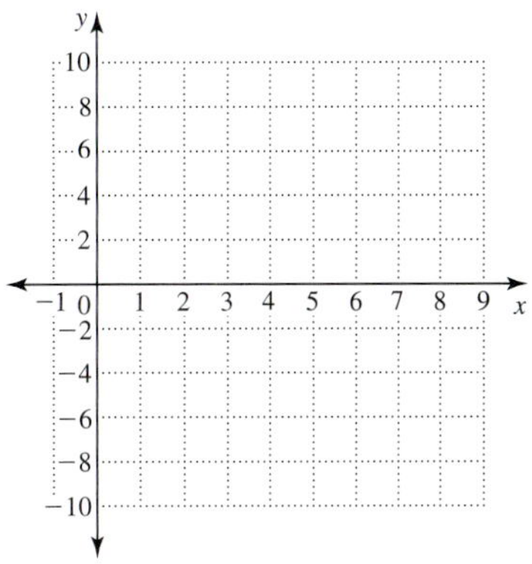

(c) What is the graphical solution of the equation?

(d) Does the graphical solution confirm the analytical solution? Why or why not?

6. Ace Tennis Racquets, Inc. has purchased 55 Top-Spin Premium racquets at a cost of \$75 per racquet. David, the CEO, has decided to sell the racquets for \$125 each.

(a) Complete the table of values. Let x represent the number of racquets sold and generalize the problem situation into a profit function, $P(x)$.

x = Number of racquets sold	Computations	$P(x)$ = Profit function
30		
40		
50		
x		$P(x) =$

(b) Substitute the number zero for $P(x)$ and solve the equation analytically.

(c) How many racquets must be sold to break even?

(d) Confirm the answer graphically. Determine a scale for the horizontal axis: Xscl = ______ . Determine a scale for the vertical axis: Yscl = ______ .

(e) Label each axis. Plot the points from the table on a separate piece of graph paper and connect them.

(f) Identify the x-intercept of the graph.

(g) How many racquets must be sold to make a profit of \$2500?

(h) Write a statement describing the relationship of the x-intercept of the graph and the breakeven point of the profit function.

7. The Smart College student choir is holding a spring concert. Tickets for the concert sell for \$5 and the choir spent \$500 in advertising.

   (a) Complete the table of values. Let x represent the number of tickets sold and generalize the problem situation into a profit function, $P(x)$.

x = Number of tickets sold	Computations	$P(x)$ = Profit function
0		
50		
100		
150		
200		
300		
400		
x		$P(x) =$

   (b) Substitute the number zero for $P(x)$ in the function and solve the equation for x analytically.

   (c) How many tickets must be sold to break even?

   (d) Confirm the answer graphically. Determine a scale for the horizontal axis: Xscl = ______ . Determine a scale for the vertical axis: Yscl = ______ .

   (e) Label the horizontal and vertical axes. Plot the points from the table on a separate piece of graph paper and connect them.

   (f) Identify the x-intercept of the graph.

   (g) How many tickets must be sold to make a profit of \$2000?

8. Karen and Stephanie went shopping. Karen spent \$50 more than Stephanie. Together they spent \$530.

   (a) Let x represent the amount spent by Stephanie. Write an expression in terms of x that represents the amount spent by Karen.

   (b) Write an equation in terms of x that represents the amount spent by Karen and Stephanie.

   (c) Solve the equation for x analytically.

   (d) How much did Stephanie spend?

   (e) How much did Karen spend?

   (f) Confirm the answer graphically on a separate piece of graph paper using the X-Intercept or Intersection-of Graphs Method.

SECTION 4.2

9. Pete was responsible for purchasing prizes for participants in a golf tournament. He decided to purchase gloves and towels as the prizes. Each golf glove costs \$10 and each golf towel costs \$8. He spent a total of \$256 on 28 prizes. How many golf gloves did he purchase? How many golf towels did he purchase?

(a) Let the variable x represent the number of golf gloves purchased. Let the variable y represent the number of golf towels purchased. The total number of prizes purchased is 28. Draw a picture representing the problem situation.

Equation 1:

	+		=	

Equation 2:

	+		=	

(b) Solve the system graphically. Rewrite Equation 1 in function form, solving for y. Substitute y_1 for y.

(c) Rewrite Equation 2 in function form, solving for y. Substitute y_2 for y.

(d) Complete the tables of values for the equations y_1 and y_2. Graph both equations on the same grid.

x	y_1	y_2
0		
8		
12		
16		
18		

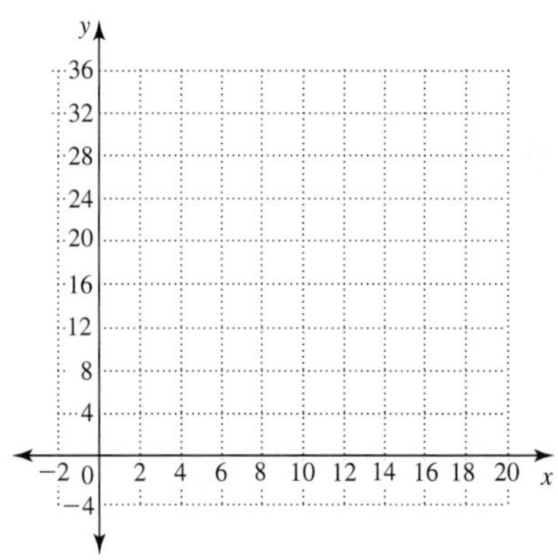

(e) Identify the point of intersection of the two graphs, where $y_1 = y_2$.

(f) How many golf gloves should be purchased?

(g) How many golf towels should be purchased?

(h) Verify the solution by substituting the coordinates of the point of intersection into the original equations of the system.

10. Tickets to the college play cost \$4 for students and \$6 for nonstudents. 600 people attended the play and \$2700 was collected at the door in ticket sales. How many students and how many nonstudents attended the play?

(a) Let x represent the number of student tickets sold. Let y represent the number of nonstudent tickets sold. Draw a picture of two equations representing the problem situation.

(b) Write an equation representing the total number of tickets sold.

(c) Solve the system graphically. Rewrite the equation in part (b) in function form, solving for y. Substitute y_1 for y.

(d) Write a second equation that represents the total amount collected from the sale of student and nonstudent tickets.

(e) Write the second equation in part (d) in function form, solving for y. Substitute y_2 for y.

(f) Complete the tables of values for the equations y_1 and y_2. Graph both equations on the same grid.

x	y_1	y_2
100		
200		
300		
400		
450		
500		

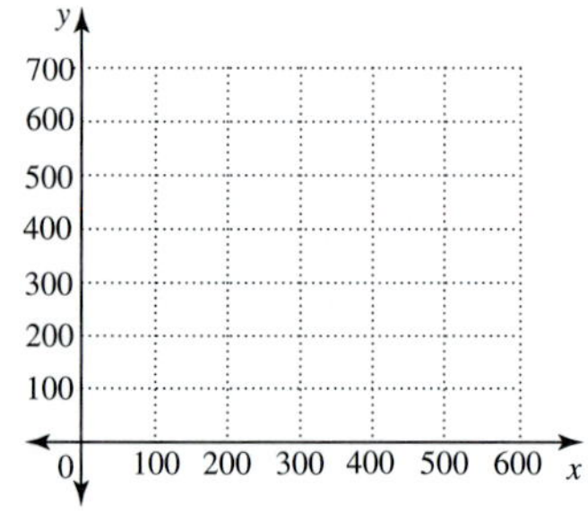

(g) Identify the point of intersection of the two graphs, where $y_1 = y_2$.

(h) How many student tickets were sold?

(i) How many nonstudent tickets were sold?

(j) Verify the solution by substituting the coordinates of the point of intersection into the original equations of the system.

Problems 11–17: Solve each system graphically and determine whether the system is consistent, dependent, or inconsistent. If the system is consistent, write the coordinates of the point of intersection. Otherwise, write dependent or inconsistent.

11. $y = 7 - 2x$
$3x - y = -2$

12. $x - 2y = -1$
$-3x + 6y = 8$

13. $-x + 6y = 10$
$x = y$

14. $2x = y + 5$
$x - y = 5$

15. $3x + 4y = 2$
$3x + 4y = 7$

16. $\frac{1}{3}x + \frac{1}{4}y = 1$
$4x + 3y = 12$

17. $\frac{1}{2}x + y = 3$
$x + 2y = 6$

SECTION 4.3

18. You have \$6000 to invest for one year. You invest in two different accounts because it may be too risky to invest all of the money in a single account. One account earns 6% annual interest and the other risky account earns 15% annual interest. Determine how much money to invest in each account in order to have earned \$585 by the end of the year to purchase a printer for the computer.

(a) Write two equations to form a system describing the problem situation. Let x represent the amount of money invested at 6%. Let y represent the amount of money invested at 15%.

Equation 1:

☐ + ☐ = ☐

Equation 2:

☐ + ☐ = ☐

(b) Solve the system using either substitution or elimination.

(c) How much money should be invested at 6%?

(d) How much money should be invested at 15%?

(e) Verify the solution by substituting into the original equations of the system. Confirm the answer graphically.

19. Brian has 10 bills in his wallet in \$5 and \$20 bills for a total of \$155. How many bills of each kind does he have?

(a) Write two equations to form a system describing the problem. Let x represent the number of \$5 bills. Let y represent the number of \$20 bills in Brian's wallet.

Equation 1:

$\square + \square = \square$

Equation 2:

$\square + \square = \square$

(b) Solve the system using either substitution or elimination.

(c) How many \$5 bills are in Brian's wallet?

(d) How many \$20 bills are in Brian's wallet?

(e) Verify the solution by substituting into the original equations of the system. Confirm the answer graphically.

20. Shirley is a nurse who has a solution containing 50% dextrose and needs to give a patient a solution that contains only 30% dextrose. How much solution containing no dextrose (therefore 0% dextrose) should she add to the 50% dextrose solution to obtain 100 cubic centimeters of solution containing 30% dextrose?

(a) Write two equations to form a system describing the problem. Let x represent the number of cubic centimeters of the 50% solution. Let y represent the number of cubic centimeters of the 0% solution.

Equation 1:

$\square + \square = \square$

Equation 2:

$\square + \square = \square$

(b) Solve the system using either substitution or elimination.

(c) How many cubic centimeters of 50% dextrose solution should be used?

(d) How many cubic centimeters of 0% dextrose solution should be used?

(e) Verify the solution by substituting into the original equations in the system. Confirm the answer graphically.

Problems 21–32: Solve each system using either the Method of Substitution or Elimination and determine whether the system is consistent, dependent, or inconsistent. If the system is consistent, write the coordinates of the point of intersection. Otherwise, write dependent or inconsistent.

21. $x = 4y - 2$
$5y = 2x + 1$

22. $2x - y = 5$
$x = 2 - y$

23. $-x + 6y = 10$
$0.5x - 3y = -5$

24. $x + y = 2$
$0.5x + 0.5y = 1$

25. $\frac{1}{2}x + y = 1$
$x + 2y = 3$

26. $x + y = 3$
$x + y = 5$

27. $x - 2y = 0$
$x + 2y = 8$

28. $x + 3y = 0$
$2x - y = 7$

29. $2x - y = 4$
$6x - 3y = 4$

30. $2x + 6y = 11$
$x + 3y = 3$

31. $x = y + \frac{4}{3}$
$\frac{3}{4}x - \frac{3}{4}y = 1$

32. $y = \frac{2}{3}x + 1$
$2x - 3y = -3$

SECTION 4.4

Problems 33–36: Determine whether the given ordered pairs are solutions to the given inequality.

33. $x + y \leq -2$

(a) $(0, 0)$

(b) $(-1, -1)$

(c) $\left(-\frac{5}{3}, -\frac{2}{3}\right)$

34. $3x - 2y \leq 0$

(a) $(0, 0)$

(b) $\left(\frac{1}{3}, 1\right)$

(c) $(1, -1)$

35. $x - y > 3$

(a) $(0, 0)$

(b) $(1, -2)$

(c) $(2, -1)$

36. $x - 2y > 5$

(a) $(0, 0)$

(b) $\left(7, \frac{1}{2}\right)$

(c) $(-2, 1)$

Problems 37–44: Graph the solution set of each inequality.

37. $3x + y > 4$

x	*y*

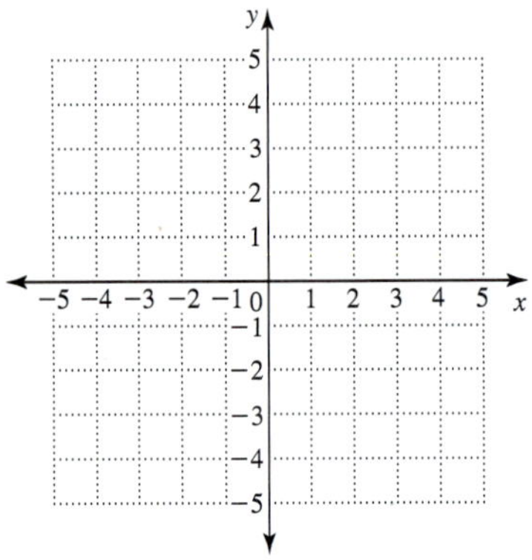

38. $2x + 3y \geq 0$

x	*y*

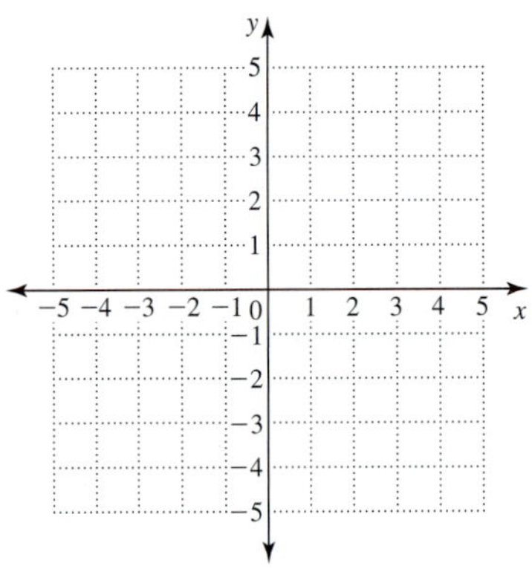

39. $x + 2y < 1$

x	y

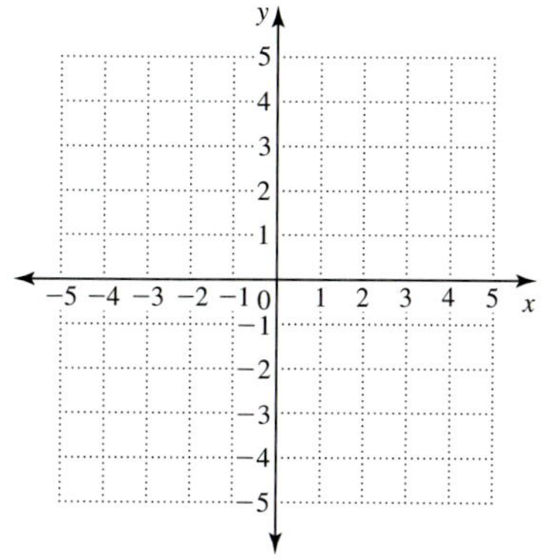

40. $3x + y < 2$

x	y

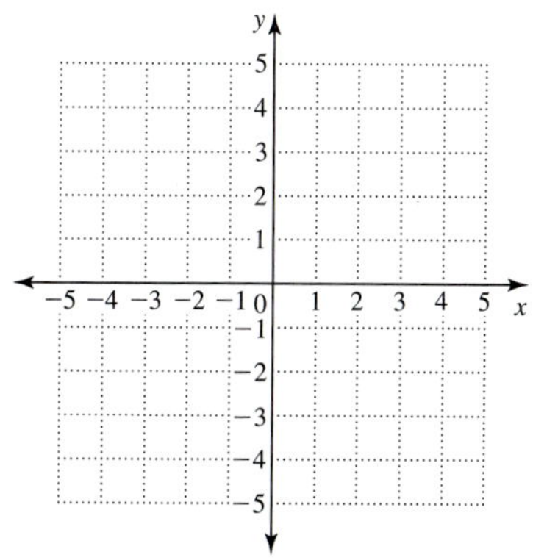

41. $x + y \leq -2$

x	y

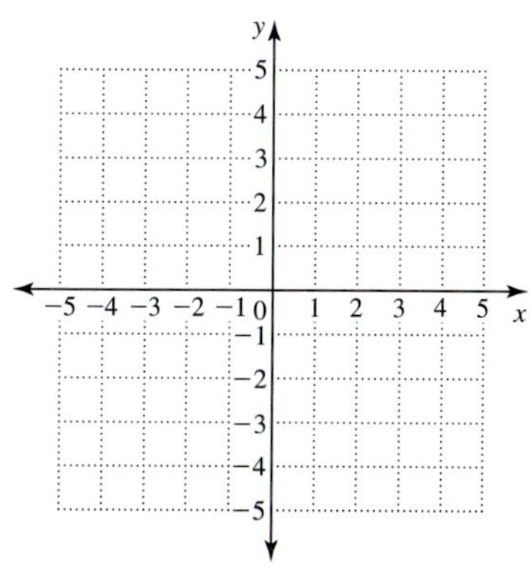

42. $x > 3$

x	*y*

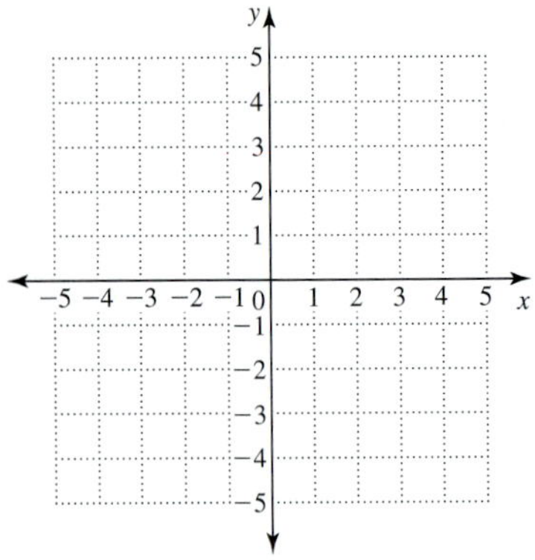

43. $y \geq -3$

x	*y*

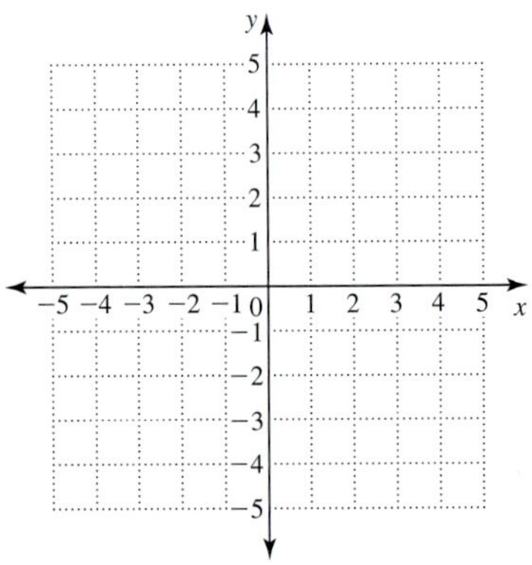

44. $y \leq 5$

x	*y*

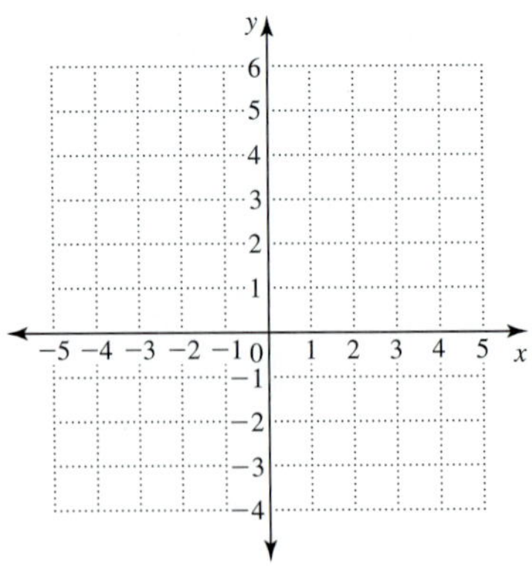

Practice Test—Chapter 4

SECTION 4.1

1. The College Student Government is holding a film festival. Tickets to see the films will sell for \$4 and the charge for renting the films will be \$700.

(a) Complete the table of values. Let x represent the number of tickets sold and generalize the problem situation into a profit function, $P(x)$.

x = Tickets sold	Computations	$P(x)$ = Profit function
100		
150		
200		
300		
Generalize x		$P(x) =$

(b) How many tickets must be sold to break even?

(c) Confirm the answer graphically. Determine a scale for the horizontal axis: Xscl = ______. Determine a scale for the vertical axis: Yscl = ______.

(d) Label each axis. Plot the points from the table on a separate piece of graph paper.

(e) Connect the points. Approximate the x-intercept of the graph.

(f) Does the graphical solution confirm the analytical solution? Why or why not?

(g) Use the profit function to determine how many tickets must be sold to make a profit of \$580.

2. Given $2x - 8 = 0$.

(a) Solve the equation analytically.

(b) Confirm the solution graphically using the X-Intercept Graphical Method.

x	y
2	
3	
4	
5	

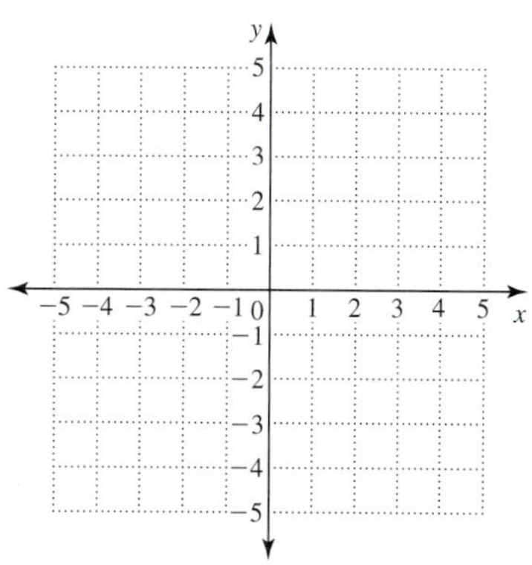

(c) What is the x-coordinate of the x-intercept of the graph?

(d) Does the graphical solution confirm the analytical solution? Why or why not?

(e) Verify the solution by substituting into the original equation.

3. Given $7x - 2 = 4x + 10$.

(a) Solve the equation analytically.

(b) Confirm the solution graphically using the Intersection-of-Graphs Method.

x	y_1	y_2
2		
3		
4		
5		

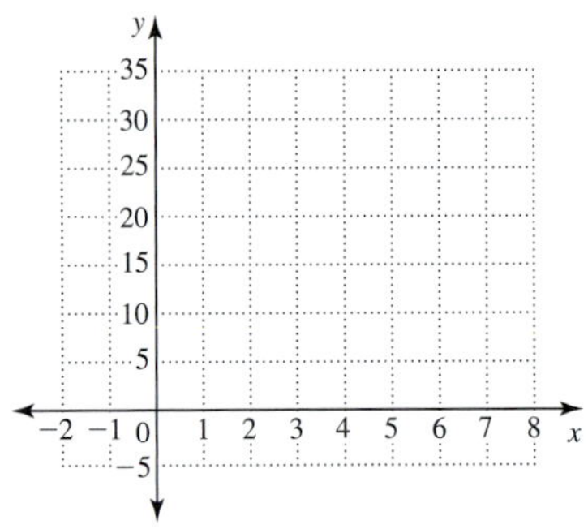

(c) What is the x-coordinate of the point of intersection?

(d) Does the graphical solution confirm the analytical solution? Why or why not?

(e) Verify the solution by substituting into the original equation.

4. Given $4(1.56x - 1) = 2.43x$.

(a) Solve the equation analytically. Round the answer to the nearest hundredth.

(b) Confirm the solution graphically using the Intersection-of-Graphs Method.

x	y_1	y_2
0		
1		
2		
3		

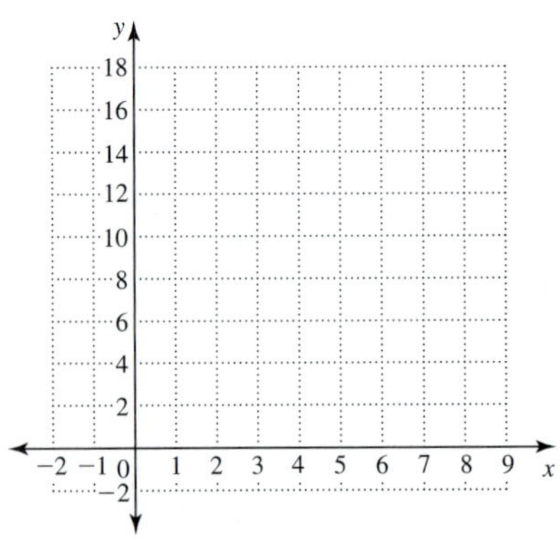

(c) Approximate the x-coordinate of the point of intersection.

(d) Does the graphical solution confirm the analytical solution? Why or why not?

(e) Verify the solution by substituting into the original equation.

SECTION 4.2

5. The bookstore had a special sale on notebooks and dividers. Notebooks sold for \$4 each and dividers sold for \$1.50 each. The total number of notebooks and dividers sold was 65 (notebooks and dividers together). The revenue collected was \$185. How many notebooks and how many dividers were sold?

(a) Let x represent the number of notebooks sold. Let y represent the number of dividers sold. Draw a picture of two equations representing the problem situation.

(b) Write an equation representing the total number of notebooks and dividers sold.

(c) Rewrite the equation in part (b) in function form, solving for y. Substitute y_1 for y.

(d) Write an equation representing the total revenue collected from the sale of all notebooks and dividers.

(e) Rewrite the equation in part (d) in function form, solving for y. Substitute y_2 for y.

(f) Complete the tables of values for equations y_1 and y_2 and graph both equations on the same grid.

x	y_1	y_2
15		
30		
35		
45		
60		

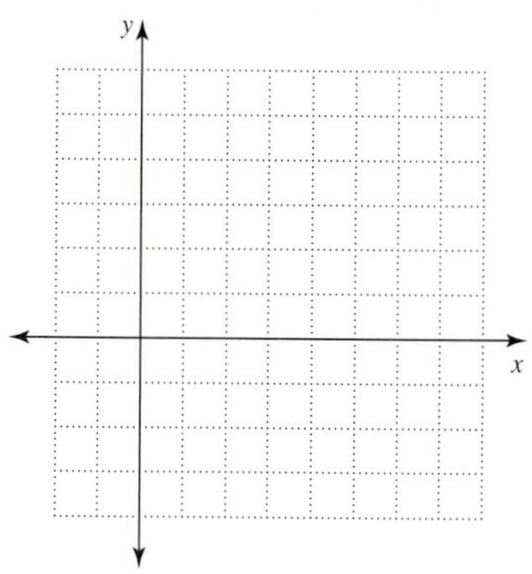

(g) Write the coordinates of the point of intersection.

(h) How many notebooks were sold?

(i) How many dividers were sold?

(j) Verify the solution by substituting the coordinates of the point of intersection into the original equations.

Problems 6–9: Solve each system graphically and determine whether the system is consistent, dependent, or inconsistent. If the system is consistent, write the coordinates of the point of intersection. Otherwise, write dependent or inconsistent.

6. $x + y = 6$
$3x - y = 2$

7. $x + 2y = 3$
$2x + 4y = 1$

8. $x + y = 5$
$y = 5 - x$

9. $4x - 3y = 2$
$5x + y = -7$

SECTION 4.3

10. You have \$10,000 to invest for one year. You invest in two different accounts because it is too risky to invest all of the money in a single account. One account earns 5% annual interest and the risky account earns 12% annual interest. Determine how much money to invest in each account in order to have earned \$815 by the end of the year.

(a) Write two equations to form a system describing the problem situation. Let x represent the amount of money invested at 5%. Let y represent the amount of money invested at 12%.

(b) Solve the system using either the Method of Substitution or Elimination.

(c) How much money should be invested at 5%?

(d) How much money should be invested at 12%?

(e) Verify the solution by substituting into the original equations. Confirm the answer graphically, if desired.

11. Mixture problem: Tracy is a chemist who has two acid solutions. One solution contains 50% acid and the second solution contains 10% acid. She needs to mix them together to create a solution that contains 35% acid. How much solution containing 10% acid should she add to the 50% acid solution to obtain 2 liters of solution containing 35% acid?

(a) Write two equations to form a system describing the problem. Let x represent the number of liters of the 50% solution. Let y represent the number of liters of the 10% solution.

(b) Solve the system using either the method of substitution or elimination.

(c) How many liters of 50% acid solution should be used?

(d) How many liters of 10% acid solution should be used?

(e) Verify the solution by substituting into the original equations. Confirm the answer graphically, if desired.

Problems 12–16: Solve each system using either the Method of Substitution or Elimination and determine whether the system is consistent, dependent, or inconsistent. If the system is consistent, write the coordinates of the point of intersection. Otherwise, write dependent or inconsistent.

12. $3x + y = 8$
$3x - y = -2$

13. $x + 1 = 2y$
$-3x + 6y = 15$

14. $2x + 3y = 8$
$4x + 3y = 4$

15. $y = 3x - 7$
$-3x + y = -7$

16. $x - 2y = 3$
$3x - 4y = 6$

Problems 17–20: Graph the solution set of each inequality.

17. $x + y > 1$

x	y

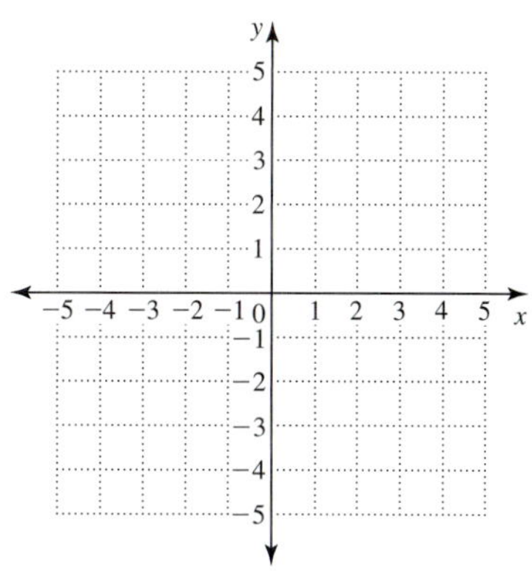

18. $y \leq 2$

x	y

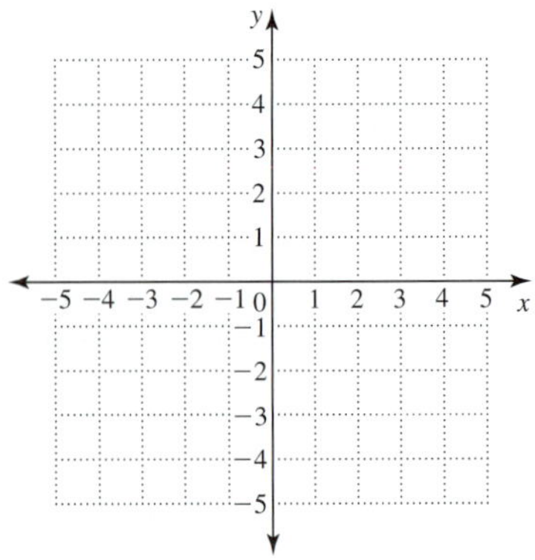

19. $3x - y \geq 4$

x	y

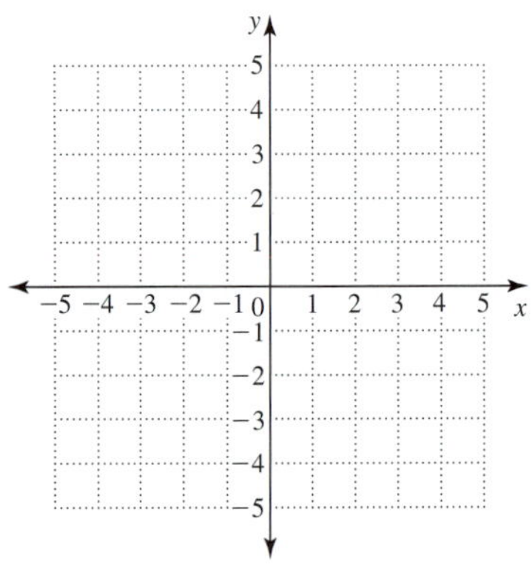

20. $x > 2.5$

x	y

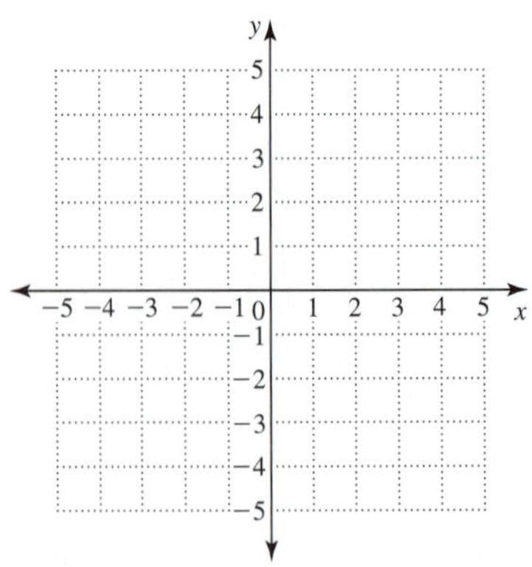

Answers to Getting Started—Chapter 4

SECTION 4.1

1. (a)

x	$y = 2x + 2$
−3	−4
−2	−2
−1	0
0	2
1	4
2	6
3	8

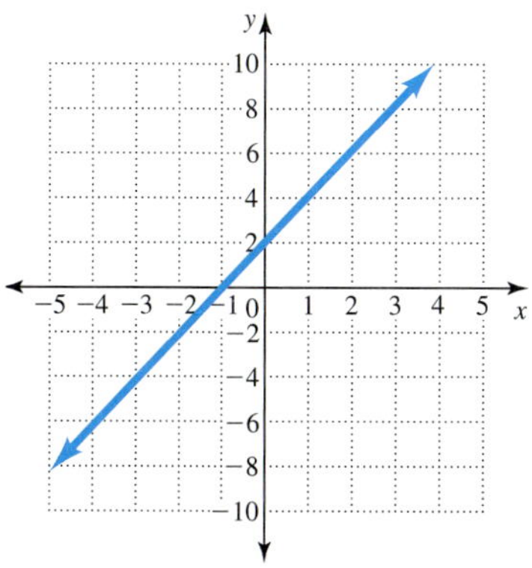

(b) $(-1, 0)$

(c) $x = -1$

(d) $0 = 2x + 2$

$-2 = 2x$

$-1 = x$

(e) Both solutions are $x = -1$. Verify the solution by substitution.

$0 = 2x + 2$

$0 \stackrel{?}{=} 2(-1) + 2$ ✓ It checks.

$0 = 0$

2. (a)

x	$y_1 = 4x - 1$	$y_2 = 2x + 3$
−2	−9	−1
−1	−5	1
0	−1	3
1	3	5
2	7	7
3	11	9

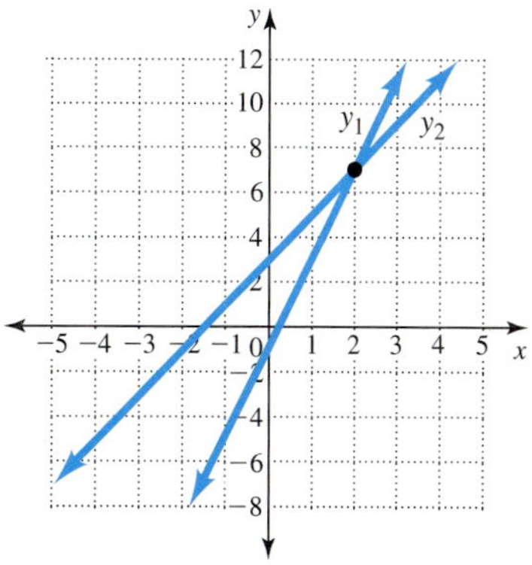

(b) $(2, 7)$

(c) $x = 2$

(d) $4x - 1 = 2x + 3$

$4x - 2x - 1 = 2x - 2x + 3$

$2x - 1 = 3$

$2x - 1 + 1 = 3 + 1$

$2x = 4$

$x = 2$

(e) Both solutions are $x = 2$. Verify the solution by substitution.

$4x - 1 = 2x + 3$

$4(2) - 1 \stackrel{?}{=} 2(2) + 3$

$8 - 1 \stackrel{?}{=} 4 + 3$ ✓ It checks.

$7 = 7$

SECTION 4.2

1. (a) Yes

(b) No

2. (a)

x	+	y	=	40

(b) Total amount of money spent on cans of tennis balls: $2x$

(c) Total amount of money spent on t-shirts: $5y$

(d)

$2x$	+	$5y$	=	\$125

(e) $y_1 = 40 - x$

(f) $y_2 = \dfrac{125 - 2x}{5}$

(g)

x	$y_1 = 40 - x$	$y_2 = \dfrac{125 - 2x}{5}$
10	30	21
15	25	19
20	20	17
25	15	15
30	10	13

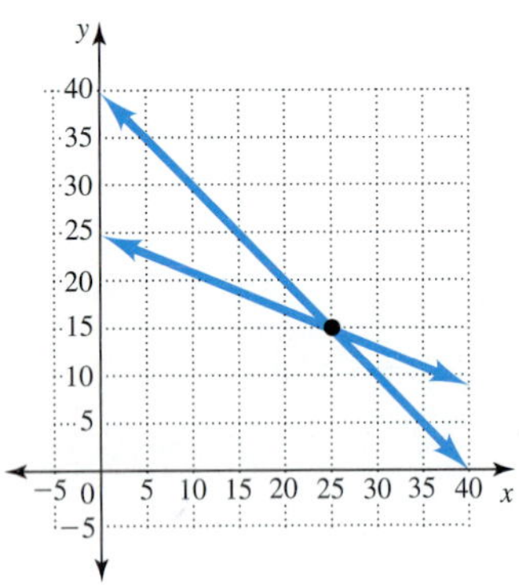

(h) (25, 15)

(i) 25 cans of tennis balls

(j) 15 t-shirts

(k) Verify the solution.

Equation 1: $25 + 15 \stackrel{?}{=} 40$

$40 = 40$

Equation 2: $2(25) + 5(15) \stackrel{?}{=} 125$

$50 + 75 \stackrel{?}{=} 125$

$125 = 125$

3. (a) $y_1 = 5 - x = -x + 5$

(b) Slope is -1, y-intercept is (0, 5)

(c) $y_2 = \dfrac{10 - 2x}{2}$ or $y_2 = 5 - x = -x + 5$

(d) Slope is -1, y-intercept is (0, 5)

(e)

x	$y_1 = 5 - x$	$y_2 = 5 - x$
−2	7	7
−1	6	6
0	5	5
1	4	4
2	3	3

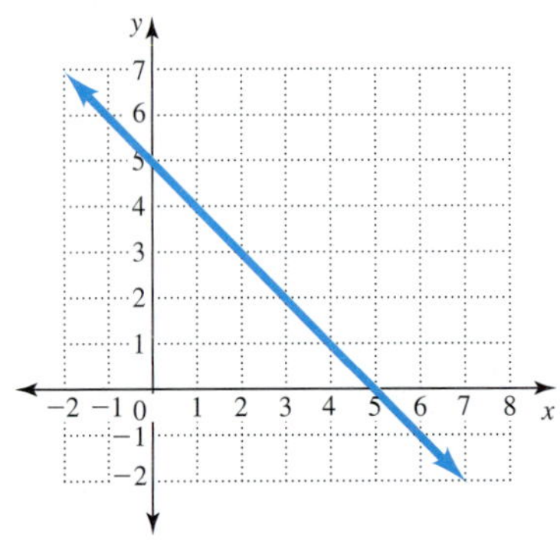

(f) The slopes of the two lines are equal. The y-intercepts are equal. The lines are identical.

(g) An infinite number of points exist.

(h) Infinite number of solutions.

4. (a) $y_1 = 5 - 2x = -2x + 5$

(b) Slope of y_1 is -2; the y-intercept is (0, 5)

(c) $y_2 = 1 - 2x = -2x + 1$

(d) Slope of y_2 is -2; the y-intercept is (0, 1)

(e)

x	y_1	y_2
−1	7	3
0	5	1
1	3	−1
2	1	−3

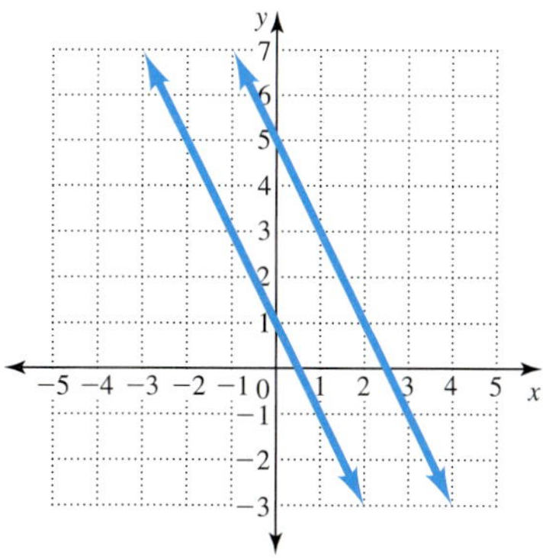

(f) The slopes of the two lines are equal. The y-intercepts are different. There are no points of intersection. The lines are parallel and do not intersect.

(g) There are no points of intersection.

(h) None, there is no solution.

SECTION 4.3

1. (a) Equation 1:

x	+	y	=	2000

(b) $I = 0.10y$

(c) Equation 2:

$0.06x$	+	$0.10y$	=	\$152

(d) Equation 1: $x + y = 2000$

Equation 2: $0.06x + 0.10y = 152$

(e) Equation 3: $y = 2000 - x$

(f) $0.06x + 0.10(2000 - x) = 152$

(g) $0.06x + 0.10(2000 - x) = 152$

$$0.06x + 200 - 0.10x = 152$$
$$-0.04x + 200 = 152$$
$$-0.04x = -48$$
$$x = 1200$$

(h) $y = 2000 - x$

$y = 2000 - 1200$

$y = 800$

(i) $(x, y) = (1200, 800)$

(j) \$1200 is invested at 6%.

(k) \$800 is invested at 10%.

(l) Equation 1:

$$x + y = 2000$$
$$1200 + 800 \stackrel{?}{=} 2000$$
$$2000 = 2000$$

Equation 2:

$$0.06x + 0.10y = 152$$
$$0.06(1200) + 0.10(800) \stackrel{?}{=} 152$$
$$72 + 80 \stackrel{?}{=} 152$$
$$152 = 152$$

SECTION 4.4

1. (a) No

(b) Yes

(c) Answers will vary: (0, 0), (−1, 1), (−2, −1)

2. (a) Compute the x-and y-intercepts of $2x + 3y = 6$.

x	y
0	2
3	0

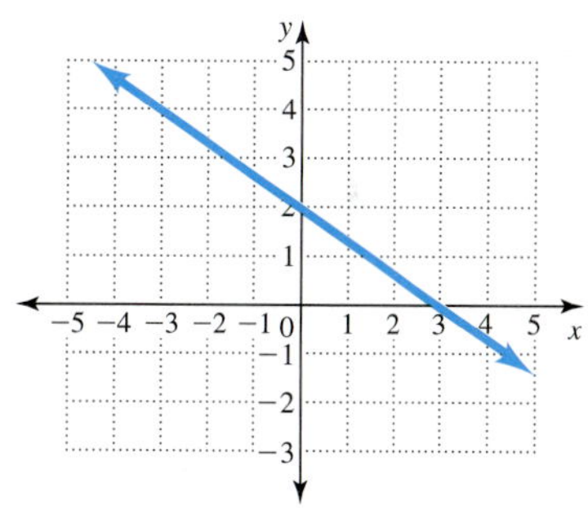

(b) Yes. Yes. All the points on the line satisfy the nonstrict inequality (≤). Points on the line where $2x + 3y = 6$ are included in the solutions set of the nonstrict inequality $2x + 3y \leq 6$.

(c) The point (−3, 2) is below the line. Shade below the line.

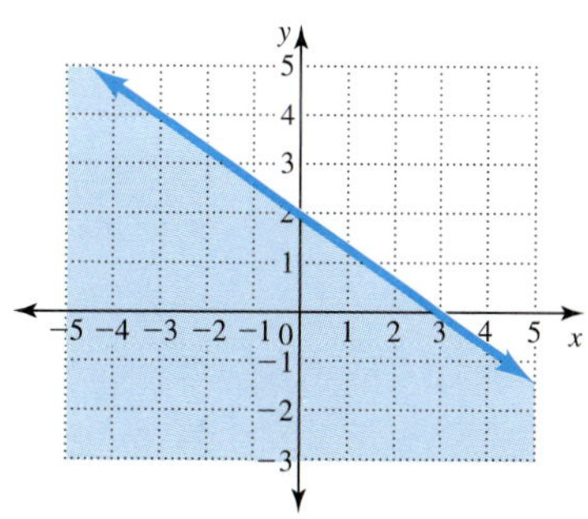

(d) Answers will vary: $(-3, -4)$ satisfies the inequality.

(e) The half-plane below the line, along with the points on the line satisfy the inequality $2x + 3y \leq 6$.

3. (a) Graph $3x - 4y = 12$.

x	y
0	−3
4	0

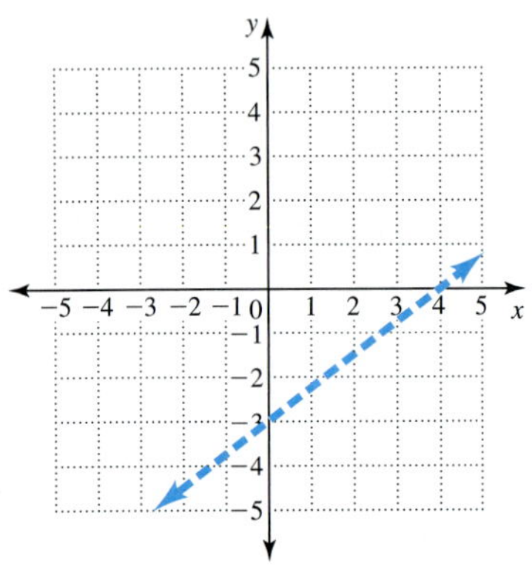

(b) No

(c) Points above the line do not satisfy the inequality. The test-point (0, 0) is above the line and does not satisfy the inequality:

$$3(0) - 4(0) \overset{?}{>} 12$$
$$0 - 0 \overset{?}{>} 12$$
$$0 \not> 12$$

(d) Points below the line satisfy the inequality. The second test-point $(3, -3)$ is below the line and satisfies the inequality:

$$3(3) - 4(-3) \overset{?}{>} 12$$
$$9 + 12 \overset{?}{>} 12$$
$$21 > 12$$

(e) Half-plane below the line.

(f) Shade the half-plane below the line.

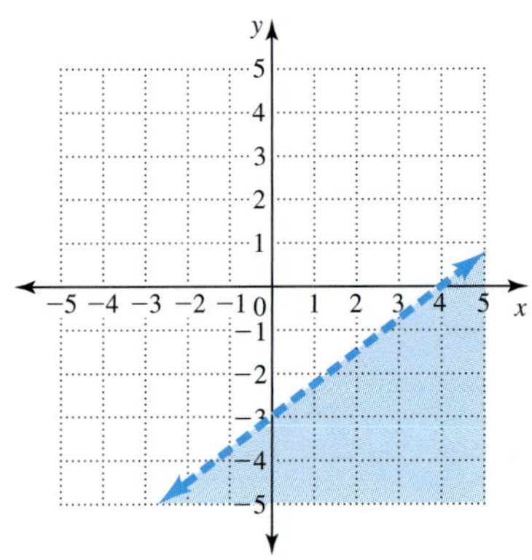

CHAPTER

5

Multiplication and Division of Polynomials

One of the most important contributions of the Greek mathematician Diophantus (mid-third century A.D.) is his introduction of symbolic abbreviations for squares, cubes, etc. "All numbers are made up of some multitude of units . . . the square is called dynamis *and its sign* Δ *with the index Y, that is* Δ^Y*; the cube is called* kubos *and has for its sign K with the index Y, that is* K^Y*; . . ."*

5.1 Properties of Exponents

OBJECTIVE

- Perform operations involving nonnegative real number exponents.

1 2 3 Getting Started

In this section, we develop a set of algebraic rules, called the Properties of Exponents that govern operations involving exponents and techniques for writing equivalent forms of expressions containing exponents. Exponential notation is shorthand notation for multiplication where exponents communicate the operation of multiplication.

2^3 is shorthand for $2 \cdot 2 \cdot 2 = 8$.

2 is the base. 3 is the exponent.

The base is the factor being multiplied. The exponent indicates how many times the base is multiplied. The exponent is written as a smaller number to the right and above the base.

2^3 is equivalent to the product of three factors of 2.

$$2^3 = 2 \cdot 2 \cdot 2 = 8$$

1. An expression is in **expanded form** when the individual factors and multiplication are written without exponents. The equal symbol denotes an equivalent form of the algebraic expression. Rewrite the following expressions in expanded form. Do not perform the operations.

EXAMPLE: $\underbrace{5x^2y^3}_{\text{Exponential form}} = \underbrace{5\cdot x\cdot x\cdot y\cdot y\cdot y}_{\text{Expanded form}}$

(a) $5^4 =$

(b) $\left(\frac{2}{3}\right)^3 =$

(c) $-7abc^3 =$

(d) $(x^2)^3 =$

2. When exponents are written, the expression is in **exponential form**. Rewrite the following expressions in exponential form. Do not perform the operations.

EXAMPLE: $\underbrace{4\cdot x\cdot y\cdot y}_{\text{Expanded form}} = \underbrace{4xy^2}_{\text{Exponential form}}$

(a) $(-2)(-2)(-2) =$

(b) $(-3)\cdot x\cdot y\cdot y\cdot y =$

(c) $\frac{6\cdot 6\cdot x\cdot x\cdot x}{y\cdot y} =$

(d) $(x-1)(x-1)(x-1) =$

3. Rewrite each expression in expanded form and then write the product. Look for patterns in the exponents of these multiplication problems.

		Expanded form	Product
(a)	$b^4\cdot b^2$		
(b)	$a^3\cdot a^2$		

(c) Write a shortcut method for writing exponents when multiplying bases that are ALIKE.

4. Exponents may appear inside and outside of parentheses. Rewrite each expression in expanded form and then write the product. Look for patterns in the exponents of these multiplication problems.

		Expanded form	Product
EXAMPLE	$(x^2)^3 =$	$(x^2)\cdot(x^2)\cdot(x^2) = (x\cdot x)\cdot(x\cdot x)\cdot(x\cdot x)$ $= x\cdot x\cdot x\cdot x\cdot x\cdot x$	$= x^6$
(a)	$(a^2)^4$		
(b)	$(b^3)^2$		

(c) Write a shortcut method for multiplying expressions when exponents appear inside and outside parentheses.

5. Rewrite each expression in expanded form and then write the product. Look for patterns in the exponents.

		Expanded form	Product
EXAMPLE	$(-2y)^3 =$	$(-2y)(-2y)(-2y) = (-2)\cdot y\cdot(-2)\cdot y\cdot(-2)\cdot y$ $= (-2)\cdot(-2)\cdot(-2)\cdot y\cdot y\cdot y$ $= (-2)^3y^3$	$= -8y^3$
(a)	$(-5y)^3$		
(b)	$(4a)^4$		

(c) Write a shortcut method for writing the product of an expression with more than one factor raised to a power.

6. Simplifying an expression may involve reducing a fraction to lowest terms and dividing out common number and variable factors in the numerator and denominator. This process is sometimes called "canceling." Write each expression in expanded form and identify common factors. Reduce the fraction and look for a pattern in the exponents.

		Expanded form with common factors divided out	Simplified or reduced form
EXAMPLE	$\frac{6x^5}{2x^3} =$	$\frac{\overset{1}{\cancel{2}}\cdot 3\cdot \overset{1}{\cancel{x}}\cdot\overset{1}{\cancel{x}}\cdot\overset{1}{\cancel{x}}\cdot x\cdot x}{\underset{1}{\cancel{2}}\cdot\underset{1}{\cancel{x}}\cdot\underset{1}{\cancel{x}}\cdot\underset{1}{\cancel{x}}} = \frac{3\cdot x\cdot x}{1}$	$= \frac{3x^2}{1} = 3x^2$
(a)	$\frac{3a^6}{12a^3}$		
(b)	$\frac{-8y^3}{10y^5}$		

(c) Write a shortcut method for writing exponents when dividing bases that are *alike*.

7. Rewrite each expression in expanded form and then in simplified form. Look for patterns in the exponents.

		Expanded form	Simplified form
EXAMPLE	$\left(\frac{a}{b}\right)^3 =$	$\left(\frac{a}{b}\right)\left(\frac{a}{b}\right)\left(\frac{a}{b}\right) = \frac{a\cdot a\cdot a}{b\cdot b\cdot b}$	$= \frac{a^3}{b^3}$
(a)	$\left(\frac{x}{y}\right)^4$		
(b)	$\left(\frac{2b}{a}\right)^3$		

(c) Write a shortcut method for simplifying a quotient raised to a power.

1 2 3 Discussing Mathematics

PROPERTIES OF EXPONENTS

The patterns observed in Getting Started can be generalized to form the first three **Properties of Exponents**. A total of six properties are presented in this section. The seventh property is presented in Section 5.2. Remember: division by zero is always undefined. A variable in the denominator of a fraction can never be assigned a value that causes the denominator to equal zero.

PROPERTIES OF EXPONENTS
GIVEN ANY REAL NUMBERS a, m, AND n:

Property I: $a^m \cdot a^n = a^{m+n}$

When multiplying factors where bases are ALIKE, the base stays the same and exponents are added.

EXAMPLE 1: Multiply $z^3 \cdot z^4$.

Expanded form:

$$z^3 \cdot z^4 = \underbrace{z \cdot z \cdot z \cdot z \cdot z \cdot z \cdot z}_{\text{(7 factors of } z)}$$
$$= z^7$$

Shortcut:

$$z^3 \cdot z^4 = z^{3+4}$$
$$= z^7$$

Add exponents.

Property II: $(a^m)^n = a^{m \cdot n}$

When a base raised to a power is raised to another power, the base stays the same and exponents are multiplied.

EXAMPLE 2: Multiply $(x^2)^4$.

Expanded form:

$$(x^2)^4 = (x^2)(x^2)(x^2)(x^2)$$
$$= \underbrace{x \cdot x \cdot x \cdot x \cdot x \cdot x \cdot x \cdot x}_{\text{(8 factors of } x)}$$
$$= x^8$$

Shortcut:

$$(x^2)^4 = x^{2 \cdot 4}$$
$$= x^8$$

Multiply exponents.

Property III: $(a \cdot b)^m = a^m \cdot b^m$

When more than one factor enclosed in parentheses is raised to a power of the exponent outside of the parentheses, "spread out" the exponent across the factors inside the parentheses.

EXAMPLE 3: Multiply $(xy)^4$.

Expanded form:

$$(xy)^4 = (xy)(xy)(xy)(xy)$$
$$= x \cdot x \cdot x \cdot x \cdot y \cdot y \cdot y \cdot y$$
(4 factors of x, 4 factors of y)
$$= x^4y^4$$

Shortcut:

$$(xy)^4 = (x)^4(y)^4$$
$$= x^4y^4$$
"Spread out" the exponent.

➢ *Do not apply Property III when addition or subtraction is indicated in parentheses raised to a power. The multiplication process for binomials and polynomials will be discussed in a later section.* $(x + 5)^2 \neq x^2 + 5^2$

Property IV (A): $\dfrac{a^m}{a^n} = \dfrac{a^{m-n}}{1}$ *when* $m \geq n$ *and* $a \neq 0$

When a fraction involves the same base as factors in the numerator and denominator, the base stays the same and exponents are subtracted.

Property IV (B): $\dfrac{a^m}{a^n} = \dfrac{1}{a^{n-m}}$ *when* $m \leq n$ *and* $a \neq 0$

When a fraction involves the same base as factors in the numerator and denominator, the base stays the same and exponents are subtracted.

EXAMPLE 4: Divide $\dfrac{x^4}{x^2}$, $x \neq 0$.

Expanded form:

$$\frac{x^4}{x^2} = \frac{\overset{1}{\cancel{x}} \cdot \overset{1}{\cancel{x}} \cdot x \cdot x}{\underset{1}{\cancel{x}} \cdot \underset{1}{\cancel{x}}} = x^2$$
Common factors are divided out.

Shortcut:

$$\frac{x^4}{x^2} = x^{4-2} = x^2$$
Subtract the exponents. The variable appears in the numerator.

EXAMPLE 5: Divide $\dfrac{y^3}{y^7}$, $y \neq 0$.

Expanded form:

$$\frac{y^3}{y^7} = \frac{\overset{1}{\cancel{y}} \cdot \overset{1}{\cancel{y}} \cdot \overset{1}{\cancel{y}}}{\underset{1}{\cancel{y}} \cdot \underset{1}{\cancel{y}} \cdot \underset{1}{\cancel{y}} \cdot y \cdot y \cdot y \cdot y} = \frac{1}{y^4}$$
Common factors are divided out.

Shortcut:

$$\frac{y^3}{y^7} = \frac{1}{y^{7-3}} = \frac{1}{y^4}$$
Subtract the exponents. The variable appears in the denominator.

EXAMPLE 6: Property IV (A) and (B) may occur in the same problem.

$$\frac{2x^4y^3}{6x^2y^5} = \frac{x^{4-2}}{3y^{5-3}} = \frac{x^2}{3y^2} \quad (x, y \neq 0)$$

➢ *Use Property IV (A) when bases are ALIKE and the exponent in the numerator is greater than the exponent in the denominator.*

➢ *Applying Property IV (A) when the degree in the denominator is less than the degree in the numerator will result in a negative exponent. Negative exponents will be discussed in the next section. When* $m \leq n$, *apply Property IV (B) to avoid negative exponents.*

➢ *Use Property IV (B) when bases are ALIKE and the exponent in the numerator is less than the exponent in the denominator.*

Property V: $\left(\frac{a}{b}\right)^m = \frac{a^m}{b^m}$, $(b \neq 0)$

When a fraction enclosed in parentheses is raised to an exponent, "spread out" the exponent across all factors inside the parentheses.

EXAMPLE 7: Simplify $\left(\frac{xy}{3}\right)^4$.

Expanded form:

$$\left(\frac{xy}{3}\right)^4 = \left(\frac{xy}{3}\right)\left(\frac{xy}{3}\right)\left(\frac{xy}{3}\right)\left(\frac{xy}{3}\right)$$

$$= \frac{x \cdot x \cdot x \cdot x \cdot y \cdot y \cdot y \cdot y}{3 \cdot 3 \cdot 3 \cdot 3} = \frac{x^4y^4}{81}$$

(4 factors of x, 4 factors of y, 4 factors of 3)

Shortcut:

$$\left(\frac{xy}{3}\right)^4 = \frac{x^4y^4}{3^4} = \frac{x^4y^4}{81}$$

"Spread out" exponent in the numerator and denominator.

Property VI: $a^0 = 1$, $(a \neq 0)$

Dividing a fraction where the numerator and denominator are identical results in a quotient equal to 1. Subtracting exponents results in an exponent of zero.

EXAMPLE 8: ANY real number base (except zero) raised to the zero power is equivalent to 1.

$$(5x^2y)^0 = 1 \ (x, y \neq 0)$$

➢ *Beware! Division by 0 is undefined.* $\frac{0^1}{0^1} = 0^{1-1} = 0^0$ *is undefined.*

EXAMPLE 9: Simplify $\frac{x^3}{x^3}$, $x \neq 0$.

Expanded form:

$$\frac{x^3}{x^3} = \frac{\overset{1}{\cancel{x}} \cdot \overset{1}{\cancel{x}} \cdot \overset{1}{\cancel{x}}}{\underset{1}{\cancel{x}} \cdot \underset{1}{\cancel{x}} \cdot \underset{1}{\cancel{x}}} = 1$$

Common factors are divided out.

Shortcut:

$$\frac{x^3}{x^3} = x^{3-3} = x^0 = 1$$

Any base raised to the zero power is equal to 1.

USING A GRAPHING CALCULATOR TO CONFIRM AND VISUALIZE EQUIVALENCE OF EXPRESSIONS

A graphing calculator can be a useful learning tool and used to show that two algebraic expressions are equivalent. Algebraic techniques should always be understood and sound mathematical concepts must be applied when drawing conclusions.

EXAMPLE 10: Given $x^2 \cdot x^3$ and x^5. To use a graphing calculator to show the equivalence of the two expressions, label each expression and compare the graphs and coordinates.

$y_1 = x^2 \cdot x^3$

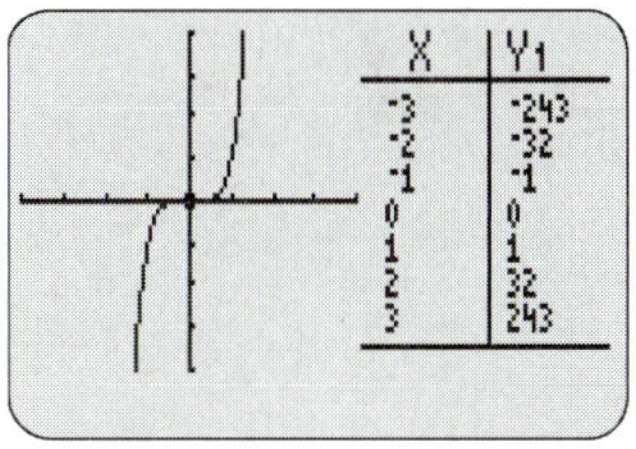

$y_2 = x^5$

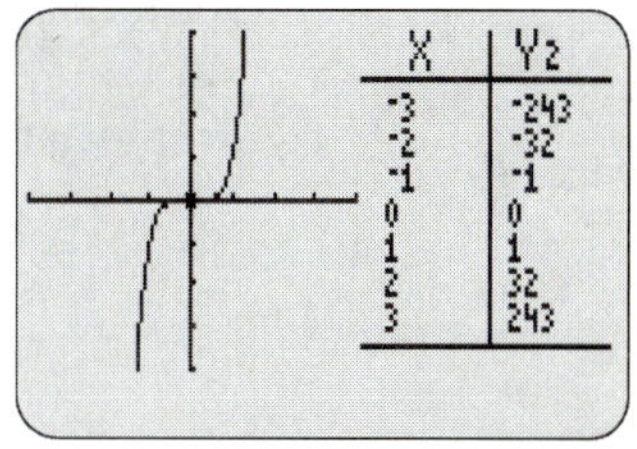

The graphs and tables are identical, therefore $x^2 \cdot x^3 = x^5$. The two expressions are equivalent.

➢ *When using graphical methods to determine equivalency of expressions, beware when two graphs appear identical, but are not.*

EXAMPLE 11: From Property of Exponent IV, we know that $\dfrac{x^5}{x^2} = x^{5-2} = x^3$.

However, the graph of $y_1 = \dfrac{x^5}{x^2}$ is not identical to $y_2 = x^3$ at $x = 0$.

$y_1 = \dfrac{x^5}{x^2}$

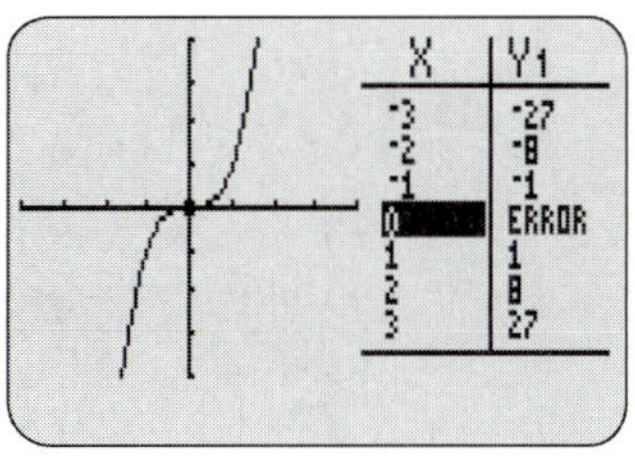

$y_2 = x^3$

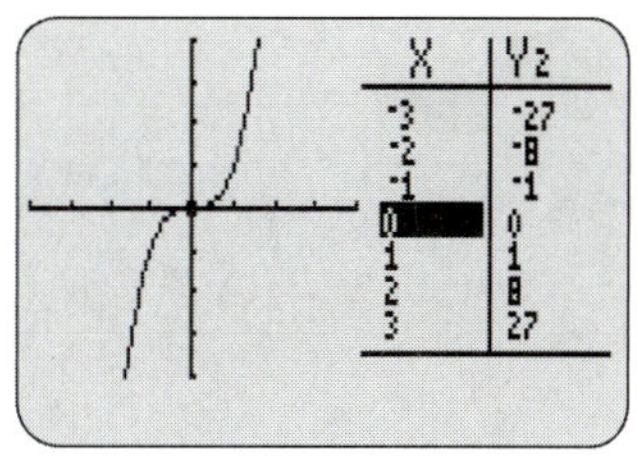

The domain of $y_1 = \dfrac{x^5}{x^2}$ does not include $x = 0$, the table indicates ERROR, and no point exists when $x = 0$. The graph has a "hole" at $x = 0$, although it is not easily seen on the graph. In contrast, the graph of $y_2 = x^3$ includes the point (0, 0).

1 2 3 Doing Mathematics

- Write the definitions of the words and symbols listed for this section in the Dictionary in Appendix A.
- On a 3 × 5-inch card, write the Properties of Exponents for future reference.

Problems 1–48: Simplify each expression using the indicated Property of Exponents. Assume variables represent nonzero quantities.

Property I

1. $x \cdot x^9$

2. $x^5 \cdot x^7$

3. $4^2 \cdot 4^3$

4. $5^3 \cdot 5^2$

5. $z^5 \cdot z^2$

6. $y^9 \cdot y^4$

7. $a^7 \cdot a^0$

8. $x^0 \cdot x^3$

Property II

9. $(n^3)^5$

10. $(y^4)^5$

11. $(y^7)^2$

12. $(x^2)^3$

13. $(2^5)^2$

14. $(10^3)^2$

15. $(b^6)^6$

16. $(z^{10})^3$

Property III

17. $(ab)^4$

18. $(xy)^3$

19. $(-3cd)^2$

20. $(-2xy)^2$

21. $(2cd)^4$

22. $(10xy)^2$

23. $(-3a)^3$

24. $(-2x)^3$

Property IV

25. $\frac{6^4}{6}$

26. $\frac{5^7}{5^4}$

27. $\frac{p^8}{p^9}$

28. $\frac{y^{10}}{y^{25}}$

29. $\frac{-12x^4y^3}{2x^5y^2}$

30. $\frac{-15x^5y^2}{5xy^3}$

31. $\frac{-36p^5q^4r^2}{-9p^7q^3r^6}$

32. $\frac{-20a^5b^2c^4}{-5a^7bc}$

Property V

33. $\left(\frac{-2}{5}\right)^2$

34. $\left(\frac{-3}{4}\right)^2$

35. $\left(\frac{x}{y}\right)^3$

36. $\left(\frac{x}{z}\right)^2$

37. $\left(\frac{-3a}{4b}\right)^3$

38. $\left(\frac{-4x}{3y}\right)^3$

39. $\left(\frac{-2xy}{z}\right)^4$

40. $\left(\frac{-3x}{yz}\right)^4$

Property VI

41. 7^0

42. 15^0

43. a^0

44. x^0

45. $-7x^2y^0$

46. $2a^0$

47. $(a-b)^0$

48. $(x+y)^0$

Problems 49–66: Simplify each expression using the appropriate Properties of Exponents. Assume variables represent nonzero quantities.

49. $b^3 \cdot b$

50. $x \cdot x^5$

51. $\frac{c^{10}}{2c^2}$

52. $\frac{2x^5}{x^3}$

53. $(-5x^{10}y^5)^2$

54. $(-6x^8y^3)^2$

55. $\frac{(-7x)^3}{(-7x)^2}$

56. $\frac{(-2x)^6}{(-2x)^5}$

57. $2a^{11} \cdot a^{12} \cdot a^{13}$

58. $5y^{23} \cdot y \cdot y^5$

59. $(-2x^3y^2)^4$

60. $(-3x^2y^3)^4$

61. $\frac{x^0}{x^{12}}$

62. $\frac{y^{10}}{y^0}$

63. $\left(\frac{3a^3}{b^2}\right)^4$

64. $\left(\frac{5x^2}{y^7}\right)^3$

65. $(2x^5y^0z^6)^3$

66. $(9a^0b^2c^3)^2$

Problems 67–68: Draw an Evaluating Machine and evaluate each function.

67. (a) $f(x) = (-3x)^2$

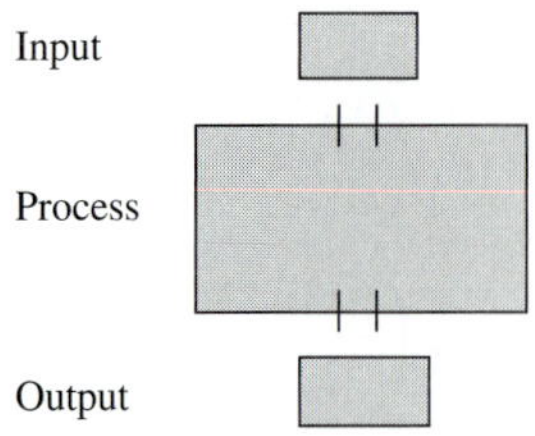

Find $f(2)$.

(b) $g(x) = 9x^2$

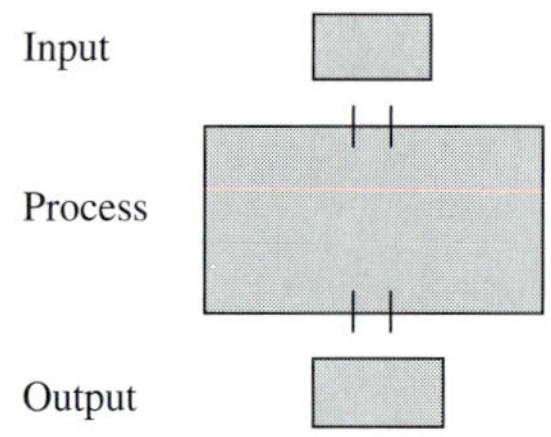

Find $g(2)$.

(c) Does $f(x) = g(x)$?

68. (a) $f(x) = (5x)^2$

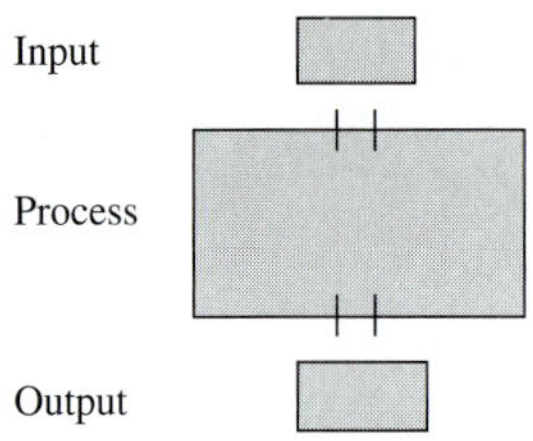

Find $f(2)$.

(b) $g(x) = 25x^2$

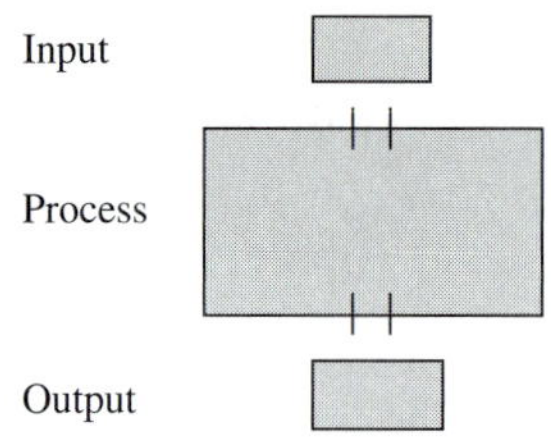

Find $g(2)$.

(c) Does $f(x) = g(x)$?

Problems 69–76: Insert the appropriate symbol, either = or ≠, between the expressions.

69. $x^3 + x^5$ ______ x^8

70. $x^2 + x^3$ ______ x^5

71. $y^2 \cdot y^3$ ______ y^6

72. $y^5 \cdot y^4$ ______ y^9

73. $a^5 \cdot a^7$ ______ a^{12}

74. $x^5 \cdot x^4$ ______ x^{20}

75. $4x^2 + 4x^2$ ______ $8x^2$

76. $2x^2 + 3x^2$ ______ $5x^4$

Problems 77–80: Use a graphing calculator to determine whether the two expressions are equivalent. Graph both expressions using a window $[-10, 10]$ Xscl 1 $[-10, 10]$ Yscl 1. Compare the graphs and the coordinates of the points. Insert the appropriate symbol, either = or ≠, between the expressions.

77. $x^2 + 2x^2$ ______ $3x^4$

78. $2x^2 + 3x^2$ ______ $5x^4$

79. $(-2x)^2$ ______ $4x^2$

80. $(2x^2)^2$ ______ $4x^4$

5.2 Integer Exponents

OBJECTIVE

- Perform operations involving positive and negative integer exponents.

Getting Started

In this section, we extend the Properties of Exponents to exponents that are integers that may be negative. Let's begin by reviewing Property of Exponents IV (A) and (B).

EXAMPLE 1: (a) $\dfrac{x^9}{x^2} = x^{9-2} = x^7$ Property of Exponents IV (A).

(b) $\dfrac{x^2}{x^5} = \dfrac{1}{x^{5-2}} = \dfrac{1}{x^3}$ Property of Exponents IV (B).

We will use our knowledge of Property IV to develop the seventh Property of Exponents.

1. Apply Property IV (A) to perform the division $\dfrac{x^2}{x^5}$, subtracting the exponent in the denominator from the exponent in the numerator.

(a) $\dfrac{x^2}{x^5}$

(b) Describe the circumstances when subtracting the exponent of the denominator from the exponent in the numerator will result in a negative number.

Two expressions represent the quotient of $\dfrac{x^2}{x^5}$:

$$\frac{x^2}{x^5} = x^{2-5} = x^{-3} \quad \text{and} \quad \frac{x^2}{x^5} = \frac{1}{x^{5-2}} = \frac{1}{x^3}.$$

Equivalent expressions

The expression x^{-3} can be rewritten in the equivalent form $\dfrac{1}{x^3}$. The equivalency shown above can be generalized to form Property of Exponents VII (A).

> **Property VII (A):** $a^{-n} = \dfrac{1}{a^n}$, where a and n are real numbers and $a \neq 0$.
>
> An algebraic expression with a negative exponent can be rewritten in an equivalent form by writing the reciprocal of the expression with a positive exponent.

2. Use Property VII (A) to rewrite each expression in an equivalent form with positive exponents.

EXAMPLE: $3^{-2} = \frac{1}{3^{+2}} = \frac{1}{3^2} = \frac{1}{9}$

(a) 5^{-2} (b) x^{-1}

(c) $x^2y^{-3}z$

Be careful! Each variable factor has its own exponent. Be sure to reciprocate only the factor with a negative exponent. All other variables remain in the numerator.

An algebraic expression with a negative exponent in the denominator can also be rewritten in an equivalent form, as generalized in Property of Exponents VII (B).

Property VII (B): $\frac{1}{a^{-n}} = a^n$, where a and n are real numbers and $a \neq 0$.

An algebraic expression with a negative exponent can be rewritten in an equivalent form by writing the reciprocal of the expression with a positive exponent.

3. Use Property VII (B) to rewrite each expression in an equivalent form with positive exponents.

EXAMPLE: $\frac{1}{4^{-2}} = \frac{4^{+2}}{1} = 16$

(a) $\frac{1}{2^{-2}}$ (b) $\frac{1}{x^{-2}}$ (c) $\frac{5a^2}{b^{-3}}$

Properties of Exponents VII (A) and (B) are used along with the other Properties of Exponents to simplify expressions.

EXAMPLE 2: Simplify.

(a) $(x^2)^{-3} = x^{2(-3)} = x^{-6} = \frac{1}{x^6}$ Apply Properties II and VII (A).

(b) $(2x)^{-3} = \frac{1}{(2x)^{+3}} = \frac{1}{2^3x^3} = \frac{1}{8x^3}$ Apply Properties VII (A) and III.

Note: Applying Property III first creates the same result: $(2x)^{-3} = 2^{-3} \cdot x^{-3} = \frac{1}{2^3x^3} = \frac{1}{8x^3}$.

(c) $\left(\frac{3}{2}\right)^{-1} = \frac{3^{-1}}{2^{-1}} = \frac{2^1}{3^1} = \frac{2}{3}$ Apply Properties V and VII (A) and (B).

4. Perform the operations and simplify.

(a) $(2^3)^{-4}$ (b) $(y^4)^{-2}$

(c) $(-3y)^{-2}$ (d) $(4x^2)^{-3}$

(e) $\left(\frac{x}{4}\right)^{-1}$ (f) $\left(\frac{x}{y}\right)^{-3}$

1 2 3

Discussing Mathematics

PROPERTY OF EXPONENTS VII (A) AND (B)

Properties of Exponents VII (A) and (B) define a process for rewriting an expression with a negative exponent in an equivalent form with a positive exponent. A negative exponent is an instruction to *reciprocate*. Properties of Exponents VII (A) and (B) are added to the Properties of Exponents discussed in Section 5.1.

PROPERTIES OF EXPONENTS

Given any real numbers a, b, m, and n:

I. $a^m \cdot a^n = a^{m+n}$

II. $(a^m)^n = a^{m \cdot n}$

III. $(a \cdot b)^m = a^m \cdot b^m$

IV (A). $\dfrac{a^m}{a^n} = \dfrac{a^{m-n}}{1}$, $(a \neq 0)$ IV (B). $\dfrac{a^m}{a^n} = \dfrac{1}{a^{n-m}}$, $(a \neq 0)$

V. $\left(\dfrac{a}{b}\right)^m = \dfrac{a^m}{b^m}$, $(b \neq 0)$

VI. $a^0 = 1$, $(a \neq 0)$

VII (A). $a^{-n} = \dfrac{1}{a^n}$, $(a \neq 0)$ VII (B). $\dfrac{1}{a^{-n}} = a^n$, $(a \neq 0)$

➢ *Working with negative exponents can be challenging. In general, the first step is to reciprocate and the second step is to simplify. A negative exponent is an instruction to reciprocate. After the base is reciprocated, the exponent is written as a positive number. A negative exponent does NOT mean that the fraction or base is negative.*

EXAMPLE 3: Simplify.

(a) $(5)^{-3} = \dfrac{1}{(5)^3} = \dfrac{1}{5 \cdot 5 \cdot 5} = \dfrac{1}{125}$ Apply Property of Exponents VII (A).

(b) $\dfrac{-8}{x^{-3}} = -8x^3$ Apply Property of Exponents VII (A).

(c) $6xy^{-2} = \dfrac{6x}{y^2}$ Apply Property of Exponents VII (A) only to the variable factor with a negative exponent.

In performing operations involving exponents, one or more Properties of Exponents may be applied according to the Order of Operations.

EXAMPLE 4: Simplify.

(a) $\dfrac{5}{(2x)^{-3}} = \dfrac{5}{1} \cdot \dfrac{(2x)^3}{1}$ Apply Property of Exponents VII (B).

$= 5 \cdot 2^3 \cdot x^3$ Apply Property of Exponents III.

$= 5 \cdot 8 \cdot x^3 = 40x^3$

(b) $y^{-9} \cdot y^5 = y^{-9+5} = y^{-4} = \dfrac{1}{y^4}$ Property of Exponents I and VII (A)

(c) $(-3x^{-2})^2 = (-3)^2(x^{-2})^2$ — Property of Exponents III

$= 9x^{-2(2)} = 9x^{-4} = \dfrac{9}{x^4}$ — Property of Exponents II and VII (A)

(d) $\left(\dfrac{3x^3}{2}\right)^{-2} = \dfrac{(3x^3)^{-2}}{2^{-2}}$ — Property of Exponents V

$= \dfrac{3^{-2}(x^3)^{-2}}{2^{-2}} = \dfrac{3^{-2}(x^{-6})}{2^{-2}}$ — Property of Exponents III

$= \dfrac{2^2}{3^2x^6} = \dfrac{4}{9x^6}$ — Property of Exponents VII (A) and (B)

(e) $\dfrac{(4x)^{-3}}{x^2} = \dfrac{1}{(4x)^3 \cdot x^2} = \dfrac{1}{4^3 \cdot x^3 \cdot x^2}$ — Property of Exponents VII (A) and III

$= \dfrac{1}{4^3 \cdot x^{3+2}} = \dfrac{1}{64x^5}$ — Property of Exponents I

EVALUATING FUNCTIONS INVOLVING NEGATIVE EXPONENTS

Apply the appropriate Property of Exponents and the Order of Operations when evaluating functions.

EXAMPLE 5: Given $f(x) = 5x^{-2}$, find $f(-3)$.

$$f(-3) = 5(-3)^{-2} = 5 \cdot \frac{1}{(-3)^2} = 5 \cdot \frac{1}{9} = \frac{5}{9}$$

USING THE GRAPHING CALCULATOR TO RECIPROCATE NUMBERS

The mathematics symbol for reciprocating is a -1 exponent. The x^{-1} key on some calculators is a reciprocating key. When writing the reciprocal of a number, the numerator becomes the denominator and the denominator becomes the numerator. In other words, the fraction is inverted. A number multiplied by its **reciprocal** results in a product of "1." The reciprocal of $\frac{1}{2}$ is $\frac{2}{1}$, as $\frac{1}{2} \cdot \frac{2}{1} = \frac{2}{2} = 1$. The reciprocal of $\frac{-3}{2}$ is $\frac{2}{-3}$. Some calculators use the 1/x key as the reciprocating key. Your instructor will discuss different keys on different calculators.

On a Texas Instrument model TI-83 calculator, when a reciprocating key is pressed, the calculator inverts the fraction (finding the reciprocal), performs the division, and returns a decimal value. An equivalent fractional answer can be computed by pressing the FRAC command on the MATH menu.

EXAMPLE 6: (a) $2^{-1} = \dfrac{1}{2^1} = 0.5$ (b) $(-5)^{-1} = \dfrac{1}{(-5)^{+1}} = -0.2$

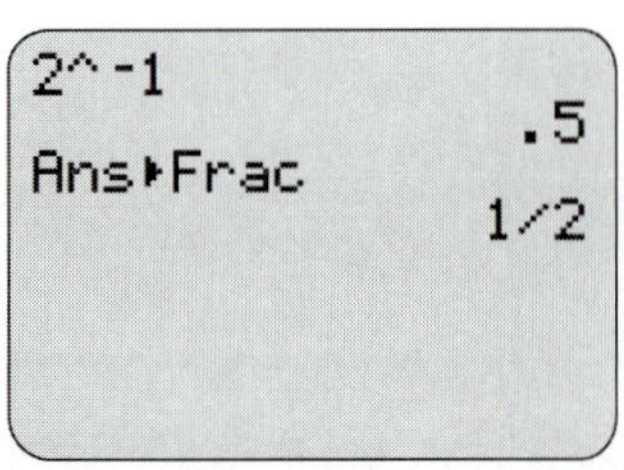

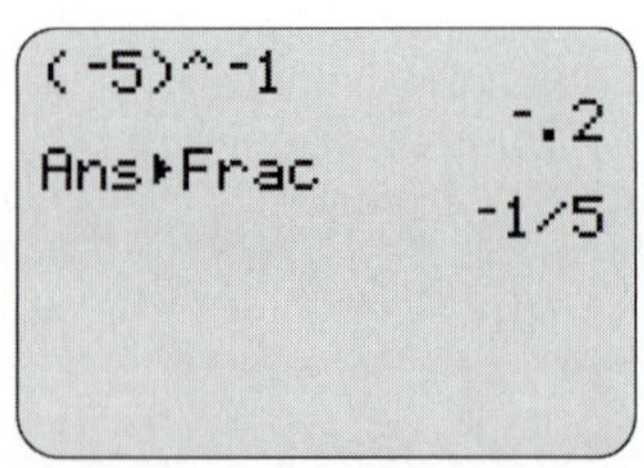

1 2 3
Doing Mathematics

- Write the definitions of the words and symbols listed for this section in the Dictionary in Appendix A.
- On a 3 × 5-inch card, write the Property of Exponents VII (A) and (B) for future reference.

Problems 1–20: Simplify each expression using Property of Exponents VII (A) or (B). Write all answers with positive exponents. Assume variables represent nonzero quantities.

Property VII (A)

1. a^{-5}

2. x^{-2}

3. $(-5)^{-2}$

4. $(-3)^{-4}$

5. $3x^4y^{-2}$

6. $4ab^{-3}$

7. $-2^2x^3y^{-2}$

8. -5^2xy^{-3}

9. $15c^{-3}d^{-6}$

10. $20a^{-4}b^{-2}$

Property VII (B)

11. $\dfrac{1}{3^{-3}}$

12. $\dfrac{1}{2^{-5}}$

13. $\dfrac{-1}{m^{-3}}$

14. $\dfrac{-1}{x^{-2}}$

15. $\dfrac{1}{p^{-1}q^{-3}}$

16. $\dfrac{1}{a^{-2}b^{-3}}$

17. $\dfrac{-2}{7x^5y^{-3}}$

18. $\dfrac{z}{15x^{-3}y^2}$

19. $\dfrac{w}{-5x^2y^{-1}z^{-4}}$

20. $\dfrac{2}{-7x^3y^{-1}z^{-4}}$

Problems 21–40: Simplify each expression using the appropriate Properties of Exponents. Write all answers with positive exponents. Assume variables represent nonzero quantities.

21. $\left(-\frac{1}{3}\right)^{-2}$

22. $\left(-\frac{1}{4}\right)^{-2}$

23. $(n^{-2})^{-3}$

24. $(a^{-4})^{-2}$

25. $(3x^4)^{-3}$

26. $(2x^4)^{-2}$

27. $(-8c^{-4}d^{-1})^{-2}$

28. $(-9a^2b^{-3})^{-2}$

29. $\left(\frac{2p^2}{3q^5}\right)^{-3}$

30. $\left(\frac{3x^2}{4y}\right)^{-2}$

31. $\left(\frac{-x^0}{2}\right)^{-5}$

32. $\left(\frac{-a^0}{5}\right)^{-3}$

33. $\frac{8x^6}{2x^{-1}}$

34. $\frac{6x^2}{3x^{-1}}$

35. $\frac{42y^{-3}z}{7y^{-8}z^2}$

36. $\frac{25x^{-5}}{5x^{-3}y}$

37. $(-2b^{-5})^0$

38. $(6a^{-4})^0$

39. $z^3 \cdot z^{-4} \cdot z^{-5}$

40. $x^{-6} \cdot x^4 \cdot x^{-1}$

Problems 41–42: Draw an Evaluating Machine and evaluate each function.

41. (a) $f(x) = -4x^{-2}$

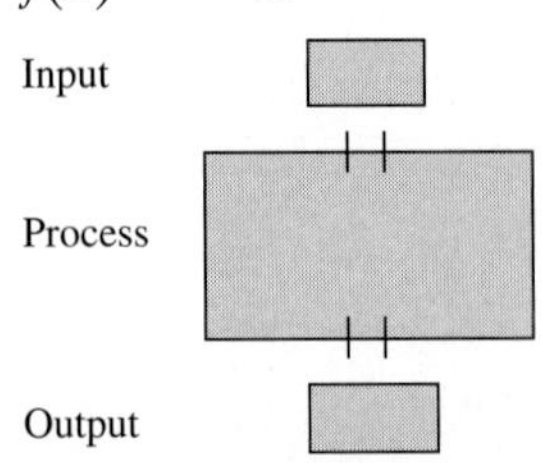

Find $f(2)$.

(b) $g(x) = \frac{-4}{x^2}$

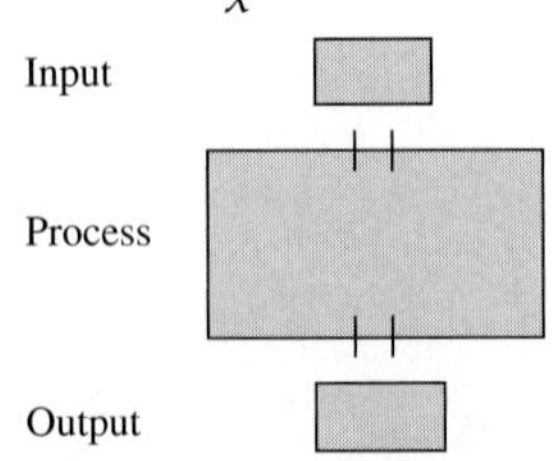

Find $g(2)$.

(c) Does $f(x) = g(x)$?

42. (a) $f(x) = 16x^{-3}$

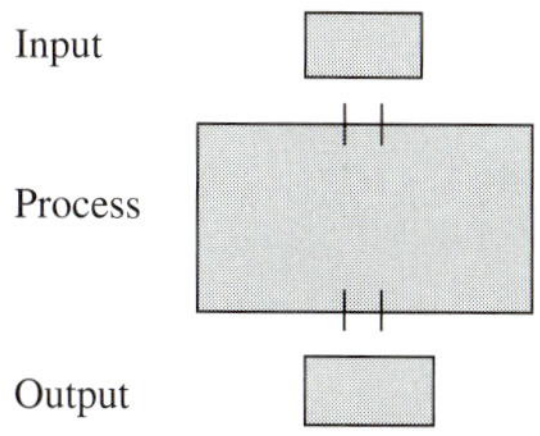

Find $f(2)$.

(b) $g(x) = \dfrac{16}{x^3}$

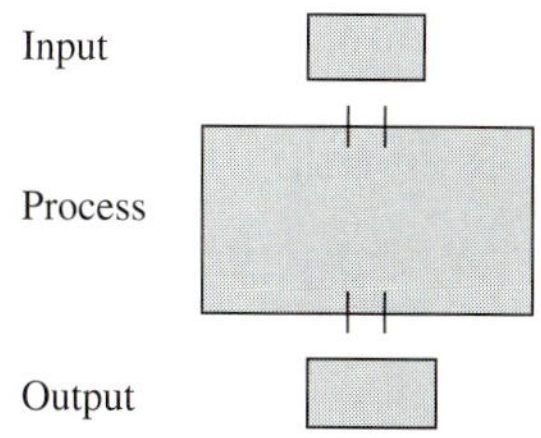

Find $g(2)$.

(c) Does $f(x) = g(x)$?

Problems 43–48: Evaluate each function.

43. Given $f(x) = 3x^{-3}$, find $f(2)$.

44. Given $f(x) = 3x^{-3}$, find $f(4)$.

45. Given $g(x) = \dfrac{4}{x^{-2}}$, find $g(-2)$.

46. Given $g(x) = \dfrac{4}{x^{-2}}$, find $g(-3)$.

47. Given $h(x) = (2x)^{-2}$, find $h(-3)$.

48. Given $h(x) = (2x)^{-2}$, find $h(-4)$.

Problems 49–56: Insert the appropriate symbol, either = or ≠, between the expressions.

49. $3x^{-2}$ ______ $\dfrac{1}{9x^2}$

50. $2x^{-2}$ ______ $\dfrac{1}{4x^2}$

51. $(-2)^{-3}$ ______ -8

52. 2^{-3} ______ -8

53. $(4x)^{-2}$ ______ $\dfrac{1}{16x^2}$

54. $(3x)^{-2}$ ______ $\dfrac{1}{9x^2}$

55. $-b^{-3}$ ______ b^3

56. x^{-2} ______ $-x^2$

Problems 57–60: Use a graphing calculator to determine whether the two expressions are equivalent. Graph both expressions using a window $[-10, 10]$ Xscl 1 $[-10, 10]$ Yscl 1. Compare the graphs and the coordinates of the points. Insert the appropriate symbol, either = or ≠, between the expressions.

57. $\left(-\dfrac{1}{3}\right)^{-2} x$ ______ $9x$

58. $\left(\dfrac{1}{2}\right)^{-2} x$ ______ $4x$

59. $(-3x^2)^{-2}$ ______ $-9x^4$

60. $(2x^2)^{-2}$ ______ $-4x^4$

5.3 Multiplication of Polynomials

OBJECTIVES

- Recognize different types of polynomial factors.
- Multiply polynomials.

1 2 3 Getting Started

Polynomials were presented in Section 2.1. In this section, we explore multiplication problems involving polynomials. Multiplication problems involving polynomials can be categorized into four types according to the type of polynomials being multiplied. It is important to recognize the operation of multiplication and type of factors in order to perform the appropriate operation and rewrite the expression in an equivalent form. The method used to perform the operation of multiplication depends upon the type of polynomial factors in the problem as listed in the following table.

	Type of polynomial factors	Example
Type 1	(monomial)(monomial)	$(2x^2)(5x^3)$
Type 2	(monomial)(polynomial)	$2x^2(x^3 + 9x^2 - x + 5)$
Type 3	(binomial)(binomial)	$(x + 3)(x + 1)$
Type 4	(polynomial)(polynomial)	$(3x^2 + 7x - 2)(x^3 + 5x^2 - x - 6)$

1. Using the categories above, determine the type of factors being multiplied in each problem.

(a) $(2x + 7)(x - 5)$ (b) $3x(2x^3 - 5y + 4xy - 9z)$

(c) $7x^3(-8x^2)$ (d) $2xyz(-3x^2y - 6y^2 + 4y^3z^4)$

(e) $(5x^2 - x - 3)(x^2 + 2x + 7)$

2. When multiplying Type 1 factors (monomial)(monomial), use the Commutative and Associative Properties to reorder and group numerical factors and variable factors with LIKE bases.

EXAMPLE: $4x^3 \cdot 5x^3 = (4 \cdot 5)(x^3 \cdot x^3)$ Rearrange factors.

$= 20x^{3+3}$ Multiply.

$= 20x^6$

(a) $(10z^4)(4z^4)$ (b) $(9x^3)(-2x^3)$

➢ ***Beware!*** *Do not confuse multiplication of monomials with adding monomials. When adding algebraic terms, the Distributive Property of Real Numbers is used to add the numerical coefficients of LIKE terms:* $4x^3 + 5x^3 = (4 + 5)x^3 = 9x^3$.

3. When multiplying Type 2 factors (monomial)(polynomial), use the Distributive Property to multiply each term of the polynomial by the factors in front of the parentheses.

EXAMPLE: $-2x^2(x^3 + 9x^2 - x + 5)$ Multiply $-2x^2$ by each term of the polynomial.

$$= (-2x^2)(x^3) + (-2x^2)(9x^2) - (-2x^2)(x) + (-2x^2)(5)$$
$$= -2x^{2+3} - 18x^{2+2} + 2x^{2+1} - 10x^2$$
$$= -2x^5 - 18x^4 + 2x^3 - 10x^2$$

(a) $6x(x^2 + 4x - 1)$

(b) $-5y^2(3y^3 - 4y^2 - 2y + 9)$

4. When multiplying Type 3 factors (binomial)(binomial), use the Distributive Property to multiply each term in the first binomial by each term in the second binomial. LIKE terms can be written in vertical columns in order to add LIKE terms easily. (Some call this process the FOIL method. The FOIL method and Types 3 and 4 factors will be discussed in detail in Discussing Mathematics.)

EXAMPLE: Multiply $(2x + 6)(x + 3)$.

Step 1: Multiply $2x$ by each term in $(x + 3)$.

$(2x + 6)(x + 3)$

Step 2: Multiply $+6$ by each term in $(x + 3)$.

$(2x + 6)(x + 3)$

$$\begin{aligned} &2x^2 + 6x \\ = &\qquad + 6x + 18 \end{aligned}$$

Step 3: Write the products, aligning LIKE terms in vertical columns.

$$= 2x^2 + 12x + 18$$

Step 4: Collect LIKE terms, adding vertically.

(a) Multiply $(x + 7)(x + 3)$. Complete Steps 1–4.

Step 1: Multiply x by $(x + 3)$.

$(x + 7)(x + 3)$

Step 2: Multiply $+7$ by $(x + 3)$.

$(x + 7)(x + 3)$

Step 3: Write the products, aligning LIKE terms in vertical columns.

Step 4: Collect LIKE terms, adding vertically.

(b) Multiply $(5x + 2)(3x + 1)$. Complete Steps 1–4.

Step 1: Multiply $5x$ by $(3x + 1)$.

$$(5x + 2)(3x + 1)$$

Step 2: Multiply $+2$ by $(3x + 1)$.

$$(5x + 2)(3x + 1)$$

Step 3: Write the products, aligning LIKE terms in vertical columns.

Step 4: Collect LIKE terms, adding vertically.

1 2 3 Discussing Mathematics

Multiplication of polynomials appears in many forms. Identify the type of factors in the problem and then apply the methods discussed in Getting Started.

MULTIPLYING POLYNOMIAL FACTORS

PRODUCT OF MONOMIALS: (MONOMIAL)(MONOMIAL)

1. Apply the Commutative Property to rearrange factors.
2. Apply the Associative Property to group LIKE factors.
3. Multiply numerical factors.
4. Multiply variables with LIKE bases.

EXAMPLE 1: Multiply.

(a)
$$\begin{aligned}(-7y^3)(-3y^2) &= \underbrace{(-7)\cdot(-3)}\cdot\underbrace{y^3\cdot y^2} && \text{Commutative and Associative Properties.}\\ &= 21y^{3+2} && \text{Multiply.}\\ &= 21y^5\end{aligned}$$

(b)
$$\begin{aligned}(3x^2y)(-4x)(x^3y^2) &= \underbrace{(3)\cdot(-4)}\cdot\underbrace{x^2\cdot x\cdot x^3}\cdot\underbrace{y\cdot y^2} && \text{Commutative and Associative Properties.}\\ &= -12x^{2+1+3}y^{1+2} && \text{Multiply.}\\ &= -12x^6y^3\end{aligned}$$

PRODUCT OF A MONOMIAL AND A POLYNOMIAL: (monomial)(polynomial)

1. Apply the Distributive Property to multiply the monomial by each term of the polynomial.
2. Multiply numerical coefficients and use the Property of Exponents I to rewrite variables with the same base with a single exponent.

EXAMPLE 2: Multiply $(-4x^2)(x^3 + 2x^2 - x - 7)$.

$(-4x^2)(x^3 + 2x^2 - x - 7)$ Distributive Property.

$$= \underbrace{(-4x^2)(x^3)} + \underbrace{(-4x^2)(2x^2)} - \underbrace{(-4x^2)(x)} - \underbrace{(-4x^2)(7)}$$

$$= \quad -4x^{2+3} \quad - 8x^{2+2} \quad + 4x^{2+1} \quad + 28x^2 \qquad \text{Multiply.}$$

$$= \quad -4x^5 - 8x^4 + 4x^3 + 28x^2$$

PRODUCT OF TWO BINOMIALS: (binomial)(binomial)

1. Apply the Distributive Property and multiply each term in the first binomial by each term in the second binomial. The algebraic signs preceding each term are part of each term. Be sure to include the algebraic signs when multiplying.
2. Write all four terms, aligning LIKE terms in vertical columns
3. Collect LIKE terms and write the final product.

EXAMPLE 3: Multiply $(2x + 3)(x - 5)$.

Multiply $2x$ by $(x - 5)$. Multiply $+3$ by $(x - 5)$.

$(2x + 3)(x - 5)$

$$(2x+3)(x-5) = \begin{array}{r} 2x^2 - 10x \\ +3x - 15 \end{array} \qquad \text{Write the products, aligning LIKE terms in vertical columns.}$$

$$(2x+3)(x-5) = 2x^2 - 7x - 15 \qquad \text{Collect LIKE Terms.}$$

To make it easier to remember the process of distributing terms in two binomials, it is often called the **FOIL method**. The terms in each binomial are labeled by their location: first term, inside term, outside term, last term. Each letter in the word FOIL represents the products of the respective terms.

USING THE FOIL METHOD TO MULTIPLY TWO BINOMIALS

The letters in FOIL represent the products of the terms in the binomials:

- **F** the product of the **first** terms
- **O** the product of the **outside** terms
- **I** the product of the **inside** terms
- **L** the product of the **last** terms

EXAMPLE 4: Multiply using the FOIL method.

(a)

$$\underset{}{(x+3)(x-5)} = \overset{\mathbf{F}}{x(x)} + \overset{\mathbf{O}}{x(-5)} + \overset{\mathbf{I}}{3(x)} + \overset{\mathbf{L}}{3(-5)}$$

$$= x^2 \quad \begin{matrix}-5x\\+3x\end{matrix} \quad -15$$

$$(x+3)(x-5) = x^2 \quad -2x \quad -15$$

Align LIKE terms in a vertical column.

The middle term, $-2x$, is the sum of the LIKE terms in O and I.

$$(x+3)(x-5) = x^2 - 2x - 15$$

(b)

$$(4x-3)(2x-1) = \overset{\mathbf{F}}{4x(2x)} + \overset{\mathbf{O}}{4x(-1)} + \overset{\mathbf{I}}{(-3)(2x)} + \overset{\mathbf{L}}{(-3)(-1)}$$

$$= 8x^2 \quad \begin{matrix}-4x\\-6x\end{matrix} \quad +3$$

$$(4x-3)(2x-1) = 8x^2 \quad -10x \quad +3$$

Align LIKE terms in a vertical column.

The middle term, $-10x$, is the sum of the LIKE terms in O and I.

$$(4x-3)(2x-1) = 8x^2 - 10x + 3$$

Some students have humorously drawn the FOIL method as a face, as an easy way to remember the process. The F and L steps of FOIL are drawn to look like eyebrows. The I step is drawn as a nose on the face. The O step is drawn as a mouth. The "Face of FOIL" looks like this:

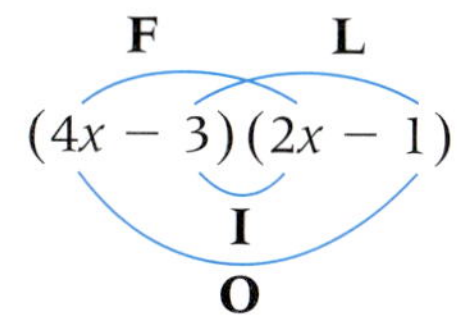

PRODUCT OF TWO POLYNOMIALS: (polynomial)(polynomial)

1. Apply the Distributive Property and multiply each term in the first polynomial across each term in the second polynomial. Be sure to multiply carefully, including the algebraic signs. Polynomials may be multiplied horizontally or vertically.
2. Write all terms, aligning LIKE terms in vertical columns.
3. Collect LIKE terms and write the final product.

EXAMPLE 5: Multiply $(x + 2)(5x^2 - 3x - 6)$.

Multiply x by $(5x^2 - 3x - 6)$. Multiply $+2$ by $(5x^2 - 3x - 6)$.

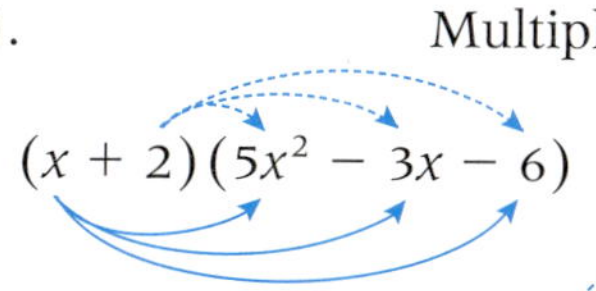

$$
\begin{aligned}
(x + 2)(5x^2 - 3x - 6) &= 5x^3 - 3x^2 - 6x \\
& + 10x^2 - 6x - 12 \\
(x + 2)(5x^2 - 3x - 6) &= 5x^3 + 7x^2 - 12x - 12
\end{aligned}
$$

Write the products, aligning LIKE terms in vertical columns.

Combine LIKE terms.

EXAMPLE 6: Multiply $(x^2 + 7x - 4)(2x^3 - 2x^2 - 5x + 7)$.

Multiply x^2 by $(2x^3 - 2x^2 - 5x + 7)$. Multiply $+7x$ by $(2x^3 - 2x^2 - 5x + 7)$. Multiply -4 by $(2x^3 - 2x^2 - 5x + 7)$.

$$
\begin{aligned}
&2x^5 - 2x^4 - 5x^3 + 7x^2 \\
= & + 14x^4 - 14x^3 - 35x^2 + 49x \\
& - 8x^3 + 8x^2 + 20x - 28 \\
= &\ 2x^5 + 12x^4 - 27x^3 - 20x^2 + 69x - 28
\end{aligned}
$$

Align LIKE terms in vertical columns.

Combine LIKE terms.

Multiplication of integers employs the Distributive Property, but is usually performed vertically. Vertical multiplication is appropriate for multiplying polynomials too! Apply the Distributive Property, align LIKE terms in vertical columns, and add each column vertically.

EXAMPLE 7: Multiply the polynomials in Example 10 *vertically*.

$$(x^2 + 7x - 4)(2x^3 - 2x^2 - 5x + 7)$$

$$\begin{array}{rrrrrrl}
 & & 2x^3 & -\ 2x^2 & -\ 5x & +\ 7 & \text{Write the polynomial with the most terms on the top.} \\
\times & & & x^2 & +\ 7x & -\ 4 & \\
\hline
 & & -\ 8x^3 & +\ 8x^2 & +\ 20x & -\ 28 & \text{Multiply } -4 \text{ by } 2x^3 - 2x^2 - 5x + 7. \\
 & +\ 14x^4 & -\ 14x^3 & -\ 35x^2 & +\ 49x & & \text{Multiply } +7x \text{ by } 2x^3 - 2x^2 - 5x + 7. \\
2x^5 & -\ 2x^4 & -\ 5x^3 & +\ 7x^2 & & & \text{Multiply } x^2 \text{ by } 2x^3 - 2x^2 - 5x + 7. \\
\downarrow & \downarrow & \downarrow & \downarrow & \downarrow & \downarrow & \text{Align LIKE terms in vertical columns.} \\
\hline
2x^5 & +\ 12x^4 & -\ 27x^3 & -\ 20x^2 & +\ 69x & -\ 28 & \text{Add each column vertically.}
\end{array}$$

USING A GRAPHING CALCULATOR TO CONFIRM AND VISUALIZE EQUIVALENCE OF EXPRESSIONS

Algebraic techniques should always be understood and used to perform operations, but a graphing calculator can be a useful tool to show that two algebraic expressions are equivalent.

EXAMPLE 8: Use a graphing calculator to show the equivalence of the two expressions $(x + 3)(x - 5)$ and $x^2 - 2x - 15$ from Example 7.

Label the first expression y_1: $y_1 = (x + 3)(x - 5)$

Label the first expression y_2: $y_2 = x^2 - 2x - 15$

Graph both y_1 and y_2. Compare the two graphs. The two expressions are equivalent if the graphs and values of the coordinates are *identical*.

$y_1 = (x + 3)(x - 5)$

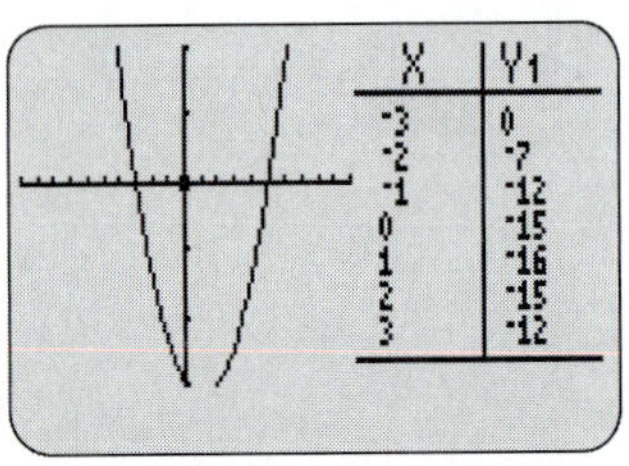

$y_2 = x^2 - 2x - 15$

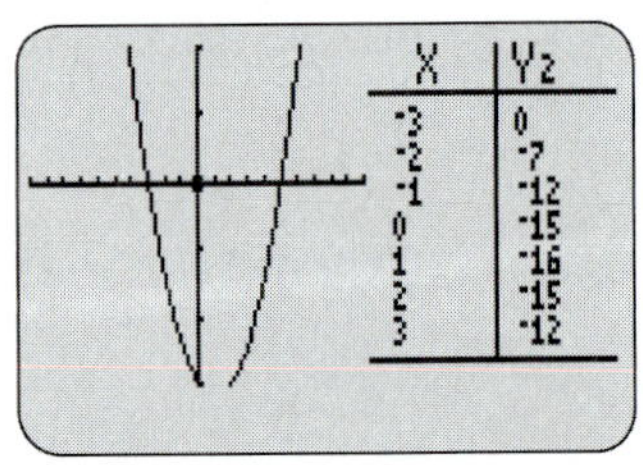

The graphs and tables are identical, therefore $(x + 3)(x - 5) = x^2 - 2x - 15$.

The two expressions are equivalent.

1 2 3

Doing Mathematics

- Write the definitions of the words and symbols listed for this section in the Dictionary in Appendix A.
- On a 3 × 5-inch card, write the following processes for future reference:
  (a) The process of multiplying two monomials
  (b) The process of multiplying a monomial and a polynomial
  (c) The process of multiplying two binomials
  (d) The process of multiplying any two polynomials

Problems 1–2: Determine the type of polynomial factors being multiplied in each problem.

Type 1: (monomial)(monomial)

Type 2: (monomial)(polynomial)

Type 3: (binomial)(binomial)

Type 4: (polynomial)(polynomial)

1. (a) $(-10t^2 + 5t - 2)(-10t)$ (b) $(x^2 + y^2)(x^2 - y^2)$
(c) $-4x^2(-5x^2)$ (d) $2p^4(3p^2 - 5p + 1)$
(e) $(a + b)(c + d)$ (f) $(3x^2 + 4x - 1)(x^2 - 5x + 7)$
(g) $(-18r^3)(-10r^4t)$ (h) $(x - 1)(x^2 + x + 1)$

2. (a) $(x + 7)(x + 5)$ (b) $(5x^2y)(-2x^3)$
(c) $(2x^2)(3x^2 - 5x - 10)$ (d) $(-9x^2y)2xyz$
(e) $(x^2 - x + 8)(x^2 - 9x + 7)$ (f) $(x - 3)(x + 4)$
(g) $(3x + 2)(x^3 - x^2 + 5x - 1)$ (h) $-a^2(7 + 3b + 4b^2)$

Problems 3–56: Multiply.

3. $(-5z^3)(-10z^3)$

4. $(2x^2)(-3x^4)$

5. $(1.3a^3bc^2)(-4.21ab^3c)$

6. $(4.6xy^3)(2.5x^2y^2)$

7. $(2y^3)(-4y)(6y^2)$

8. $(-2x)(-3x)(-4x^2)$

9. $\left(-\frac{1}{4}p^2q\right)\left(\frac{1}{12}pq^2r^2\right)$

10. $\left(\frac{1}{2}abc\right)\left(-\frac{1}{3}a^2b^2c^2\right)$

11. $(7.5z^2)(-0.5z^2)$

12. $(7y^3)(-5y^4)$

13. $(3x^2 + x - 2)(5x^2)$

14. $-5x(x^2 - x - 1)$

15. $(-16t^2)(4t^3 - 3t + 1)$

16. $(3x^2 - 5x - 10)(-2x)$

17. $lw(2l^2 + lw + w^2)$

18. $2xy(3x^2 + xy - y^2)$

19. $(-6c^3d^2)(-3b^2 + 8bd)$

20. $-9a^2(2b^2 + 5)$

21. $(-2k^3 + 3k^2 + 7k + 5)(5k)$

22. $-2x(4x^3 - 5x^2 + 10x - 8)$

23. $(v + 2)(v + 8)$

24. $(x + 5)(x + 9)$

25. $(t - 1)(t - 7)$

26. $(x - 5)(x - 9)$

27. $(y - 4)(y + 3)$

28. $(x + 1)(x - 5)$

29. $(7s + 2)(3s + 5)$

30. $(2x + 3)(5x + 4)$

31. $(3a - b)(a - 3b)$

32. $(x - 3y)(5x - y)$

33. $(u - 10)(u + 10)$

34. $(x + 5)(x - 5)$

35. $(2a - b)(a + 3b)$

36. $(x + y)(x - 5y)$

37. $(w + 4)(w - 9)$

38. $(y - 6)(y + 7)$

39. $(3x + 2)(7x - 1)$

40. $(2x - 3)(5x + 4)$

41. $(3y^2 + y - 7)(y^2 + 2y - 4)$

42. $(x^2 + 2x + 3)(2x^2 - 4x - 5)$

43. $(x - 1)(2x^3 + x^2 - 3x + 5)$

44. $(5x + 2)(4x^3 - 5x^2 + 10x - 8)$

45. $(4x^2 + x - 1)(2x^3 + 5x^2 - 3x + 1)$

46. $(5x^2 - 3x + 2)(x^3 - 7x^2 + 10x + 1)$

47. $(x^3 + x + 1)(-2x^3 + x^2 - 3)$

48. $(x^3 - x + 2)(4x^3 + 2x^2 - 9)$

49. $(x + 1)(x^3 - x + 1)$

50. $(3x - 1)(-x^2 + 15x - 2)$

51. $\begin{array}{r} 2x^2 + 3x + 1 \\ \times \quad x^2 - 5x - 2 \\ \hline \end{array}$

52. $\begin{array}{r} 2x^2 - 4x - 5 \\ \times \quad x^2 + 2x + 1 \\ \hline \end{array}$

53. $\begin{array}{r} 2x^3 - x^2 - 7x - 10 \\ \times \qquad 3x^2 + 2x + 1 \\ \hline \end{array}$

54. $\begin{array}{r} x^3 - 7x^2 + 10x + 1 \\ \times \qquad 5x^2 - 3x + 2 \\ \hline \end{array}$

55. $\begin{array}{r} -2x^3 - 3x^2 + 5 \\ \times \quad 4x^3 + 2x + 1 \\ \hline \end{array}$

56. $\begin{array}{r} 4x^3 + 2x^2 - 9 \\ \times \quad x^3 - x + 2 \\ \hline \end{array}$

Problems 57–60: The dimensions of a rectangle or square are given below.

(a) Express the area of each as a product of polynomials. The area of a rectangle is $A = l \cdot w$, where l represents the length and w represents the width. The area of a square is $A = s^2$, where s represents the length of each side of the square.

(b) Find the area when $x = 5$ inches.

57. Rectangle:
length $= 2x + 9$; width $= 3x - 2$

58. Rectangle:
length $= 4x + 5$; width $= 2x - 1$

59. Square: side $= 6x$

60. Square: side $= 4x$

Problems 61–62: (a) Express the area of the shaded region as a polynomial. (b) Find the area when $x = 5$ inches. (*Hint:* Find the area of the large rectangle and subtract the area of the smaller rectangle.)

61. Represent the area of the shaded region:

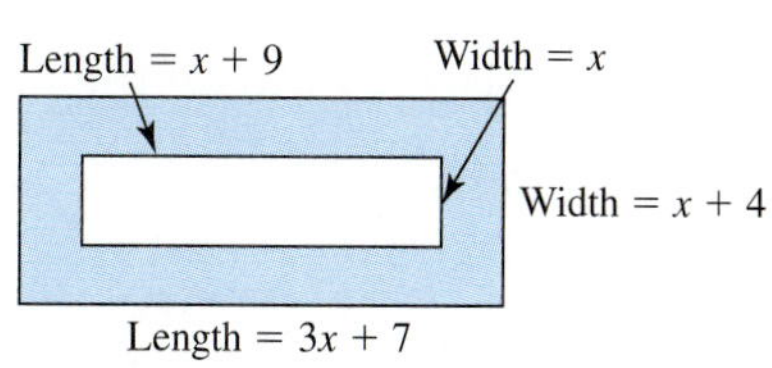

62. Represent the area of the shaded region:

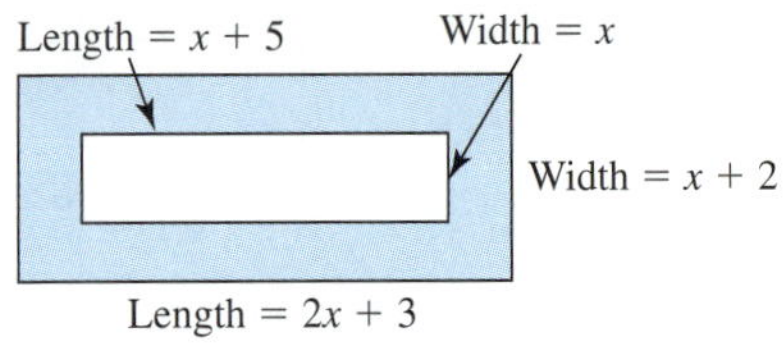

Problems 63–64: Draw an Evaluating Machine and evaluate each function.

63. (a) $f(x) = (x - 9)(x + 4)$

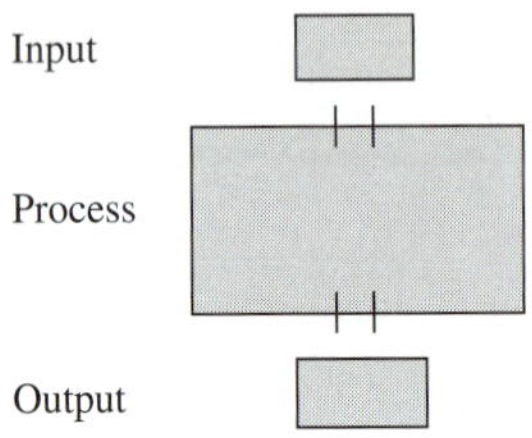

Find $f(-5)$.

(b) $g(x) = x^2 - 5x - 36$

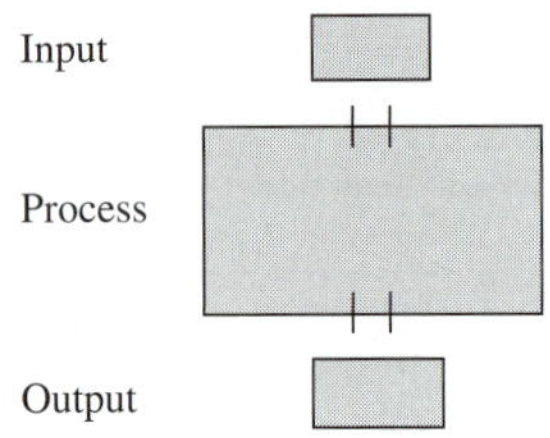

Find $g(-5)$.

(c) Does $f(x) = g(x)$?

64. (a) $f(x) = (x + 5)(x + 9)$

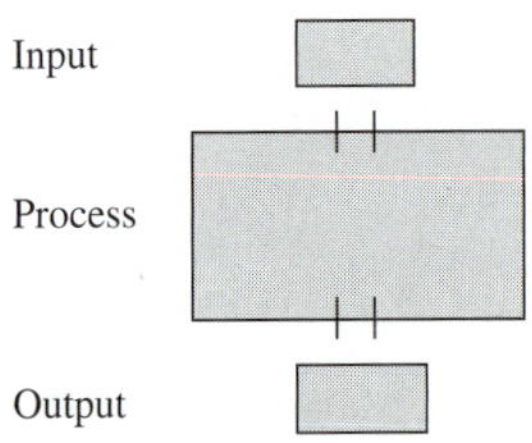

Find $f(-3)$.

(b) $g(x) = x^2 + 14x + 45$

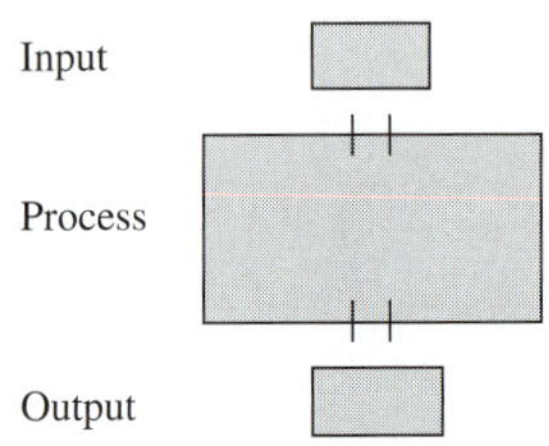

Find $g(-3)$.

(c) Does $f(x) = g(x)$?

Problems 65–74: Insert the appropriate symbol, either $=$ or $\neq$, between the expressions.

65. $(x - 3)^2$ ______ $x^2 + 9$

66. $(x + 2)^2$ ______ $x^2 + 4$

67. $(x - 5)^2$ ______ $x^2 - 10x + 25$

68. $(x + 3)^2$ ______ $x^2 + 6x + 9$

69. $-4a^5 \cdot 5a^2$ ______ $-20a^{10}$

70. $2x^5 \cdot 3x^2$ ______ $6x^{10}$

71. $(x + y)^2$ ______ $x^2 + y^2$

72. $(x - 1)^2$ ______ $x^2 + 2x + 1$

73. $3x(2x^2 + 5x - 1)$ ______ $6x^3 + 15x^2 - 3x$

74. $-2x(x^2 - x + 4)$ ______ $-2x^3 + 2x^2 - 8x$

Problems 75–78: Use a graphing calculator to determine whether the two expressions are equivalent. Graph both expressions using a window $[-10, 10]$ Xscl 1 $[-10, 10]$ Yscl 1. Compare the graphs and the coordinates of the points on the graphs. Insert the appropriate symbol, either $=$ or $\neq$, between the expressions.

75. $-3x(x + 2)$ ______ $-3x^2 + 6x$

76. $-2x(x + 3)$ ______ $-2x^2 + 6x$

77. $(x + 2)(x - 2)$ ______ $x^2 - 4$

78. $(x - 1)^2$ ______ $x^2 - 2x + 1$

5.4 Multiplication of Binomials—Special Products

OBJECTIVES

- Multiply two binomials, computing the middle term of the resulting trinomial mentally.
- Identify and multiply binomials that are the sum and difference of two terms.
- Identify and multiply the square of a binomial.

1 2 3 Getting Started

In this section, we continue our exploration of the multiplication of binomials, focusing on the special products called the sum and difference of two terms and the square of a binomial, and on computing the middle term of the product mentally.

EXAMPLE 1: Multiply $(3x + 1)(x - 7)$ using the FOIL method.

		F	O + I	L
$(3x + 1)(x - 7)$	=	$3x^2$	$-21x$	
			$+x$	-7
			↓	
$(3x + 1)(x - 7)$	=	$3x^2$	$-20x$	-7

➢ *The second term in the resulting trinomial is called the* **middle term**. *The middle term* $(-20x)$ *is the sum of the LIKE terms in the O and I steps.*

It will be useful to multiply binomials by computing the middle term of the trinomial mentally, finding the sum of the O and I steps without writing the individual products.

USING THE FOIL METHOD AND COMPUTING THE MIDDLE TERM OF THE TRINOMIAL MENTALLY

1. Find the product of the first terms in the binomials and write the product.
2. Find the product of the outside and inside terms and add them together. Perform these operations in your head, writing only the final sum.
3. Find the product of the last terms in the binomials and write the product.

EXAMPLE 2: Multiply $(x + 5)(x - 9)$, computing the middle term of the trinomial mentally.

	F	O + I	L
	↓	$(-9x + 5x)$	↓
$(x + 5)(x - 9) =$	x^2	$-4x$	-45

1. Multiply, computing the middle term of the trinomial mentally.

(a) $(x + 3)(x + 5)$

(b) $(2y - 1)(y + 3)$

(c) $(3x - 2)(2x + 5)$

2. Multiply, computing the middle term of the trinomial mentally.

(a) $(x + 2)^2 = (x + 2)(x + 2)$

(b) $(y - 5)^2 = (y - 5)(y - 5)$

(c) $(5x + 1)^2 = (5x + 1)(5x + 1)$

(d) $(3x - 4)^2 = (3x - 4)(3x - 4)$

(e) What pattern(s) do you observe in the products?

3. Multiply, computing the middle term of the trinomial mentally.

(a) $(y + 5)(y - 5)$

(b) $(x - 2)(x + 2)$

(c) $(2x + 3)(2x - 3)$

(d) What pattern(s) do you observe in the products?

4. Binomial factors with the following characteristics will be labeled as Type 1 or Type 2 factors:

		EXAMPLE
Type 1	Product of the Sum and Difference of Two Terms	$(y + 5)(y - 5)$
Type 2	Square of a Binomial	$(x + 2)^2 = (x + 2)(x + 2)$

Determine the type of binomials in each problem.

(a) $(2x + 3)^2$

(b) $(x - 4)(x + 4)$

(c) $(5x + 1)(5x - 1)$

(d) $(5x - 1)^2$

1 2 3 Discussing Mathematics

It is important to become proficient at multiplying binomials and computing the middle term of the resulting trinomial mentally. It is also important to be able to identify the type of binomials being multiplied. These skills will make it easier to perform the process called factoring which is presented in Chapter 6.

SQUARE OF A BINOMIAL

The patterns observed in Getting Started lead us to generalizations about the square of a binomial.

SQUARE OF A BINOMIAL

For any real number or algebraic expression X and Y:

Binomial Sum: $(X + Y)^2 = X^2 + 2XY + Y^2$

Binomial Difference: $(X - Y)^2 = X^2 - 2XY + Y^2$

EXAMPLE 3: Multiply.

(a) $(x + 5)^2 = (x + 5)(x + 5)$

	F	O + I	L
$=$	x^2	$+5x$	
$=$		$+5x$	$+5^2$
$=$	x^2	$+2(5x)$	$+25$
$(x + 5)^2 =$	x^2	$+10x$	$+25$

F is the square of first term in the binomial.

O and I are identical and result in a sum that is O or I doubled.

L is the square of last term in the binomial.

(b) $(x - 6)^2 = (x - 6)(x - 6)$

	F	O + I	L
$=$	x^2	$-6x$	
$=$		$-6x$	$+6^2$
$=$	x^2	$-2(6x)$	$+36$
$(x - 6)^2 =$	x^2	$-12x$	$+36$

(c) $(2x + 3y)^2 = (2x + 3y)(2x + 3y)$

	F	O + I	L
$=$	$(2x)^2$	$+6xy$	
$=$		$+6xy$	$+(3y)^2$
$=$	$4x^2$	$+2(6xy)$	$+9y^2$
$(2x + 3)^2 =$	$4x^2$	$+12xy$	$+9y^2$

SUM AND DIFFERENCE OF TWO TERMS

A generalization can also be made about the sum and difference of the same two terms.

> **PRODUCT OF THE SUM AND DIFFERENCE OF TWO TERMS**
> For any real number or algebraic expression X and Y: $(X + Y)(X - Y) = X^2 - Y^2$

EXAMPLE 4: Multiply.

(a) $(x + 3)(x - 3)$

	F	O + I	L
$=$	x^2	$-3x$	
		$+3x$	-3^2
$(x + 3)(x - 3) =$	x^2		-9

F is the square of first term in the binomial.

The sum of O and I is zero.

L is the square of last term in the binomial.

(b) $(4x + 3y)(4x - 3y)$

	F	O + I	L
$=$	$4(x)^2$	$-12xy$	
		$+12xy$	$-(3y)^2$
$(4x + 3y)(4x - 3y) =$	$16x^2$		$-9y^2$

REMOVING PARENTHESES AND COMBINING LIKE TERMS

Use the Order of Operations and the Commutative and Associative Properties to multiply expressions with more than two polynomial factors.

EXAMPLE 5: Multiply.

(a) $2x^2\underbrace{(x+5)(x-5)}$

$= 2x^2(x^2 - 25)$ — It is generally easier to multiply binomials first.

$= 2x^4 - 50x^2$ — Distribute $2x^2$.

(b) $-5a^2b(2a-1)^2$

$-5a^2b\underbrace{(2a-1)(2a-1)}$ — Rewrite the square of the binomial as two separate factors.

$= -5a^2b(4a^2 - 4a + 1)$ — Multiply binomials first.

$= -20a^4b + 20a^3b - 5a^2b$ — Distribute $-5ab^2$.

1 2 3 Doing Mathematics

- Write the definitions of the words and symbols listed for this section in the Dictionary in Appendix A.
- On a 3 × 5-inch card, write the following processes for future reference:
  (a) The process of multiplying two binomials that are the sum and difference of two terms
  (b) The process of squaring a binomial that is the sum of two terms
  (c) The process of squaring a binomial that is the difference of two terms

Problems 1–2: Determine the type of special product in each problem.

Type 1: Sum and Difference of Two Terms

Type 2: Square of a Binomial

1. (a) $(y-5)^2$ (b) $(3x-2y)(3x+2y)$ (c) $(x+7)(y-7)$
(d) $(a+b)(a+b)$ (e) $(a+b)(a-b)$

2. (a) $(x+4)^2$ (b) $(x-10)(x+10)$ (c) $(5x+9)(5x-9)$
(d) $(x-3)(x+3)$ (e) $(a-b)(a-b)$

Problems 3–28: Multiply.

3. $(t+5)(t-5)$

4. $(x+2)(x-2)$

5. $(6x-1)(6x+1)$

6. $(2a-3)(2a+3)$

7. $(x+8y)(x-8y)$

8. $(a-3b)(a+3b)$

9. $(2 + x)(2 - x)$

10. $(6 - y)(6 + y)$

11. $(3x + 4y)(3x - 4y)$

12. $(2x + 7y)(2x - 7y)$

13. $(w + 4)^2$

14. $(x + 1)^2$

15. $(y + 5)(y + 5)$

16. $(x + 3)(x + 3)$

17. $(3x + 4)^2$

18. $(2x + 5)^2$

19. $(3p + q)^2$

20. $(2x + y)^2$

21. $(r - 1)^2$

22. $(x - 2)^2$

23. $(b - 9)(b - 9)$

24. $(x - 5)(x - 5)$

25. $(4 - x)^2$

26. $(8 - y)^2$

27. $(5x - 4y)^2$

28. $(2a - 3b)^2$

Problems 29–42: Perform the operations and simplify the expressions applying the Properties of Real Numbers, the Properties of Exponents, and the Order of Operations.

29. $-5x(x + 5)^2$

30. $-4x(x - 7)^2$

31. $2x^3(x - 5)^2$

32. $7x^2(x + 2)^2$

33. $(2x - 3)^{-2}$

34. $(x + 2)^{-2}$

35. $(x + 3)(x - 3)^2$

36. $(x - 1)(x + 4)^2$

37. $-3x^3y^2(5x + 4)(5x - 4)$

38. $-5x^2y(2x - 1)(2x + 1)$

39. $-(3x - 7y)^2$

40. $-3(2x + 5y)^2$

41. $(x - 3)^2(x + 3)^2$

42. $(x + 1)^2(x + 2)^2$

Problems 43–46: The dimensions of a parallelepiped are given below.

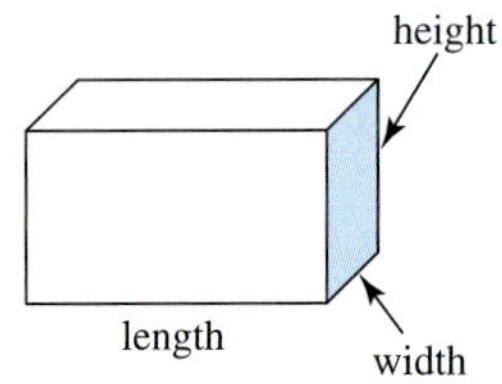

(a) Express the volume of each figure as a product of polynomials. The formula for the volume of a parallelepiped is $V = l \cdot w \cdot h$, where l represents the length, w represents the width, and h represents the height.

(b) Find the volume when $x = 6$ inches.

43. Length $= x + 4$; width $= x - 4$; height $= x$

44. Length $= x + 5$; width $= x - 5$ height $= x$

45. Length $= x + 6$; width $= x + 6$; height $= 2x$

46. Length $= x + 1$; width $= x + 1$; height $= 2x$

Problems 47–48: (a) Express the area of the shaded region as a polynomial. (b) Find the area of the shaded region when $x = 5$ feet. (*Hint:* Find the area of the large square and subtract the area of the smaller square.)

47. Represent the area of the shaded region:

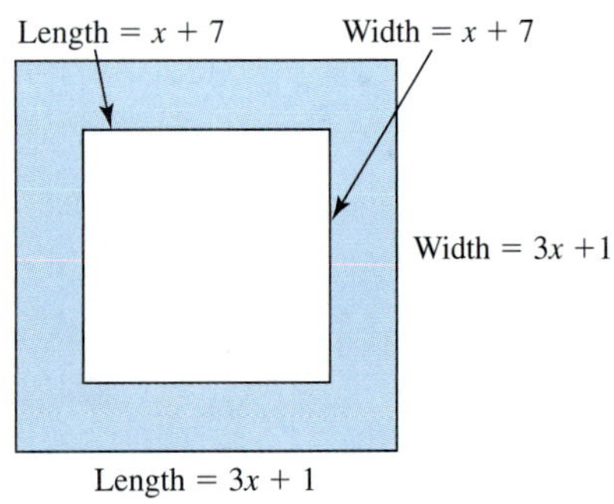

48. Represent the area of the shaded region:

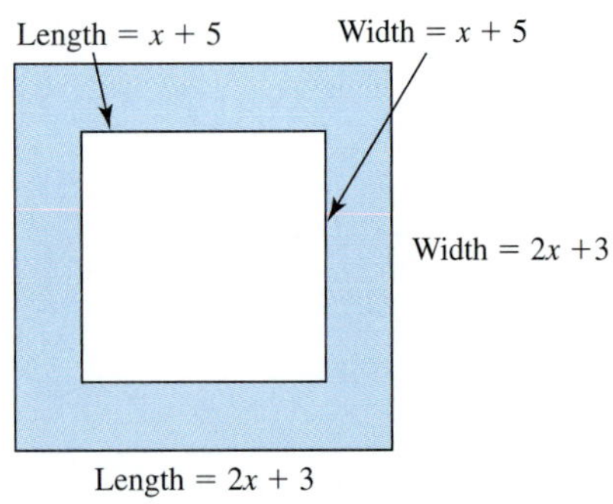

49. Problems 49–50: Draw an Evaluating Machine and evaluate each function.

(a) $f(x) = (x - 5)^2$

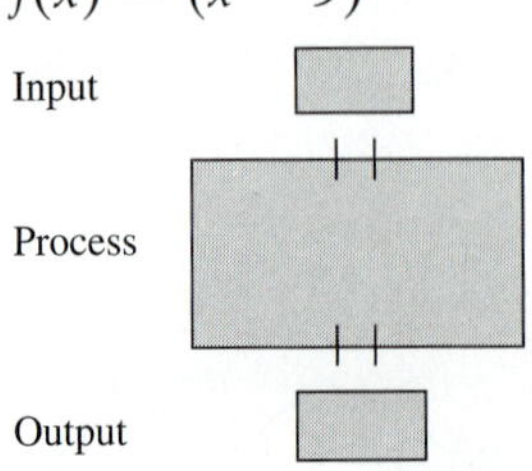

Find $f(-2)$.

(b) $g(x) = x^2 - 10x + 25$

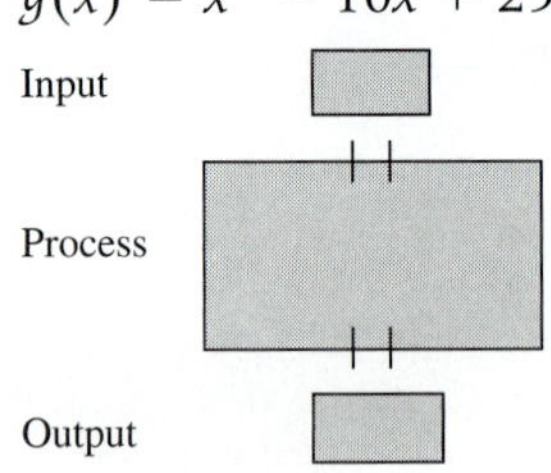

Find $g(-2)$.

(c) Does $f(x) = g(x)$?

50. (a) $f(x) = (x + 3)^2$

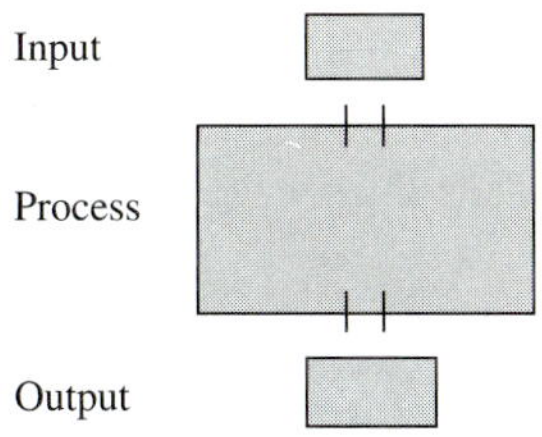

Find $f(-3)$.

(b) $g(x) = x^2 + 6x + 9$

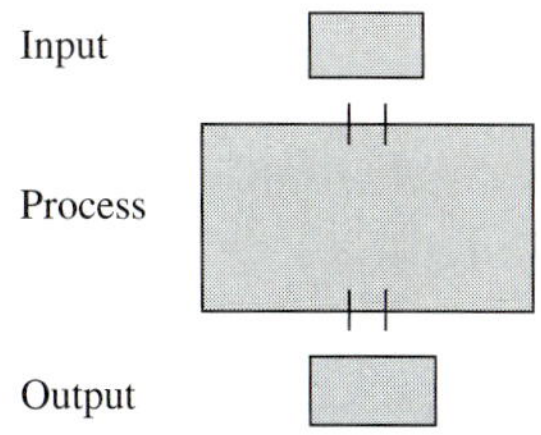

Find $g(-3)$.

(c) Does $f(x) = g(x)$?

Problems 51–60: Insert the appropriate symbol, either = or ≠, between the expressions.

51. $(x + 3)^2$ ______ $x^2 + 9$

52. $(x - 2)^2$ ______ $x^2 + 4$

53. $(x + 6)(x - 6)$ ______ $x^2 - 12$

54. $(x + 1)(x - 1)$ ______ $x^2 - 1$

55. $(p + q)^2$ ______ $p^2 + 2pq + q^2$

56. $(x - 3)^2$ ______ $x^2 - 6x + 9$

57. $(2x - 3)^2$ ______ $4x^2 - 12x + 9$

58. $(x - 1)^2$ ______ $x^2 + 2x + 1$

59. $x^2 + 4y^2$ ______ $(x + 2y)^2$

60. $-2x(x^2 + 5)$ ______ $-2x^3 + 10x$

Problems 61–64: Use a graphing calculator to determine whether the two expressions are equivalent. Graph both expressions using a window [−10, 10] Xscl 1 [−10, 10] Yscl 1. Compare the graphs and the coordinates of the points on the graphs. Insert the appropriate symbol, either = or ≠, between the expressions.

61. $\left(\frac{1}{2}x + 1\right)\left(\frac{1}{2}x - 1\right)$ ______ $\frac{1}{4}x^2 - 1$

62. $(x - 2)(x + 2)$ ______ $x^2 - 4$

63. $(3x - 2)^2$ ______ $(3x)^2 - 2^2$

64. $(2x + 3)^2$ ______ $4x^2 + 9$

5.5 Division of Polynomials

OBJECTIVES

- Recognize different types of polynomials in division problems.
- Divide polynomials.

1 2 3 Getting Started

In this section, we will review techniques for division of integers and extend those methods to division of polynomials. Remember, division by zero is undefined.

1. Divide each problem.

(a) $\frac{20}{15}$

(b) $\frac{20 + 3}{4} = \frac{20}{4} + \frac{3}{4}$

(c) $11\overline{)138}$

The methods used to divide integers are also used to divide polynomials. The decision of whether to divide out common factors, separate a fraction into individual fractions, or use long division depends on the specific problem. Division of polynomials can be categorized into three types according to the kinds of polynomials in the dividend and divisor. Identifying these types will help us determine which method to use.

	Type of polynomial in the numerator and denominator	Example	Division Method
Type 1	$\left(\dfrac{\text{monomial}}{\text{monomial}}\right)$	$\dfrac{12x^2y}{2xy}$	Divide out common factors.
Type 2	$\left(\dfrac{\text{polynomial (two or more terms)}}{\text{monomial}}\right)$	$\dfrac{12x^2 - x + 5}{3x}$	Separate the fraction into individual fractions.
Type 3	$\left(\dfrac{\text{polynomial (two or more terms)}}{\text{polynomial (two or more terms)}}\right)$	$\dfrac{x^2 + 3x - 5}{x + 2}$	Long division

2. Determine the type of polynomials being divided in each problem.

(a) $\dfrac{-15x}{xy}$ (b) $\dfrac{2x^3 - 5x^2 + 4x - 9}{x - 7}$

(c) $\dfrac{-4x^2 - 6x}{3x}$

3. Type 1 polynomials $\left(\dfrac{\text{monomial}}{\text{monomial}}\right)$ are divided by dividing out common integer and variable factors as discussed in Section 5.1. Perform the division.

EXAMPLE: $\dfrac{30x^2}{6x} = \dfrac{\overset{1}{\cancel{2}} \cdot \overset{1}{\cancel{3}} \cdot 5 \cdot x^{2-1}}{\underset{1}{\cancel{2}} \cdot \underset{1}{\cancel{3}}} = \dfrac{5x}{1} = 5x$

(a) $\dfrac{21x^3y}{3xy^2}$ (b) $\dfrac{-35a^2b}{14ab}$

4. Type 2 polynomials $\left(\dfrac{\text{polynomial (two or more terms)}}{\text{monomial}}\right)$ are divided by first writing individual fractions and then dividing out common factors. Perform the division.

EXAMPLE: $\dfrac{12x^2 + 5y}{3xy} = \dfrac{12x^2}{3xy} + \dfrac{5y}{3xy} = \dfrac{\overset{4}{\cancel{12}}x^{2-1}}{\underset{1}{\cancel{3}}y} + \dfrac{5\overset{1}{\cancel{y}}}{3x\underset{1}{\cancel{y}}} = \dfrac{4x}{y} + \dfrac{5}{3x}$

(a) $\dfrac{15b^2 - 4a}{6ab}$ (b) $\dfrac{-35a^2 - 2b}{14ab}$

BEWARE!	
Be sure to divide out only common factors. $\frac{6x}{3} = \frac{\overset{2}{\cancel{6}}x}{\underset{1}{\cancel{3}}} = \frac{2x}{1} = 2x$ Divide out the common factor of 3.	Write terms as separate fractions and reduce. $\frac{x+3}{3} = \frac{x}{3} + \frac{3}{3} = \frac{x}{3} + 1$ 3 is a term in the numerator.

Type 3 polynomials $\left(\frac{\text{polynomial (two or more terms)}}{\text{polynomial (two or more terms)}}\right)$ use long division. Look carefully at the examples in Discussing Mathematics.

1 2 3

Discussing Mathematics

Polynomial division can involve monomials, binomials, trinomials, or polynomials with more than three terms. Polynomial division will be discussed again in later chapters, when additional factoring techniques are discussed.

DIVIDING POLYNOMIALS

QUOTIENT OF MONOMIALS: $\left(\frac{\text{monomial}}{\text{monomial}}\right)$

1. Divide out common integer factors.
2. Apply Property of Exponents IV to variables with LIKE bases.
3. Multiply the remaining factors and write the fraction in reduced form.

EXAMPLE 1: Divide $\frac{-30x^3y}{5xy^2}$.

$$\frac{-30x^3y}{5xy^2} = \frac{\overset{-6}{\cancel{-30}}x^{3-1}}{\underset{1}{\cancel{5}}y^{2-1}}$$

Divide out common factors. Apply Property of Exponents IV to variables with LIKE bases.

$$= \frac{-6x^2}{y^1} = \frac{-6x^2}{y}$$

QUOTIENT OF A POLYNOMIAL AND A MONOMIAL: $\left(\frac{\text{polynomial (two or more terms)}}{\text{monomial}}\right)$

1. Separate the original fraction into individual fractions, each having the denominator of the original fraction.
2. Divide out common factors.
3. Multiply remaining factors and write the quotient.

EXAMPLE 2: Divide $\dfrac{45x^2y - 4x}{5xy^2}$.

$$\frac{45x^2y - 4x}{5xy^2} = \frac{45x^2y}{5xy^2} - \frac{4x}{5xy^2}$$

Separate the fraction into individual fractions, each with the same denominator.

$$= \frac{\overset{9}{\cancel{45}}x^{2-1}}{\underset{1}{\cancel{5}}y^{2-1}} - \frac{4x^{1-1}}{5y^2}$$

Divide out common factors.

$$= \frac{9x}{y} - \frac{4x^0}{5y^2} = \frac{9x}{y} - \frac{4}{5y^2}$$

QUOTIENT OF TWO POLYNOMIALS:
$\left(\dfrac{\text{polynomial (two or more terms)}}{\text{polynomial (two or more terms)}}\right)$

1. Use long division. Be careful to "change the signs and add" in each subtraction step.
2. Check using multiplication: Divisor · Quotient + Remainder = Dividend

Long division of polynomials uses the same process as long division of integers. Be sure to "change the signs and add" in each subtraction step.

EXAMPLE 3: Perform the division $\dfrac{6x^2 - x - 3}{2x + 7} = 2x + 7\overline{)6x^2 - x - 3}$.

Look at the first term in the divisor $(2x)$ and the first term in the dividend $(6x^2)$.

$$\begin{array}{rrrr}
 & & 3x & -\ 11 \\
2x + 7) & 6x^2 & -\ x & -\ 3 \\
 & + & - & - \\
- & 6x^2 & +\ 21x & \\
\hline
 & & -22x & -\ 3 \\
 & & + & + \\
 & - & -22x & -\ 77 \\
\hline
 & & & +74
\end{array}$$

Divide $2x$ into $6x^2$. Write $3x$ above $6x^2$.

Multiply $3x$ by each term in the divisor. Write each product so that LIKE terms are aligned vertically.

Bring down -3.

Divide $2x$ into $-22x$. Write -11 in the quotient above $-x$.

Multiply -11 by each term in the divisor. Write each product so that LIKE terms are aligned vertically.

The process stops because $2x$ cannot divide evenly into 74. The remainder is $+74$.

To subtract, change each sign of $6x^2 + 21x$ to $-6x^2 - 21x$ and add.

To subtract, change each sign of $-22x - 77$ to $+22x + 77$ and add.

Quotient: $\dfrac{6x^2 - x - 3}{2x + 7} = 3x - 11$ Remainder 74

$$= 3x - 11 + \frac{74}{2x + 7}$$

Check by multiplying: Divisor · Quotient + Remainder = Dividend

$$(2x + 7)\cdot(3x - 11) + 74 \stackrel{?}{=} 6x^2 - x - 3$$

$$6x^2 - x - 77 + 74 \stackrel{?}{=} 6x^2 - x - 3$$

$$6x^2 - x - 3 = 6x^2 - x - 3$$ ✓ It checks.

EXAMPLE 4: Perform the division and find the quotient:

$$\frac{x^3 - 10}{x - 1} = x - 1\overline{)x^3 - 10}$$

The x^2 and x terms do not appear in the dividend. The coefficients of the x^2 and x terms are zero. Before beginning the division process, rewrite the dividend with the x^2 and x terms written explicitly with a numerical coefficient of zero:

$$x - 1\overline{)x^3 + 0x^2 + 0x - 10}$$

$$\begin{array}{r} x^2 + x + 1 \\ x - 1\overline{)x^3 + 0x^2 + 0x - 10} \\ \underline{-\;\; x^3 - x^2 \qquad\qquad} \\ +x^2 + 0x \qquad \\ \underline{-\;\; x^2 - x \qquad} \\ +x - 10 \\ \underline{-\;\; x - 1} \\ -9 \end{array}$$

Divide x into x^3. Write x^2 above x^3.

Multiply x^2 by each term in the divisor. Write each product so that LIKE terms are aligned vertically.

To subtract, change each sign of $x^3 - x^2$ to $-x^3 + x^2$ and add. (+ − +)

Bring down $+0x$.
Divide x into $+x^2$. Write $+x$ in the quotient above $+0x^2$.
Multiply $+x$ by each term in the divisor. Write each product so that vertical columns form with LIKE terms.

To subtract, change each sign of $x^2 - x$ to $-x^2 + x$ and add. (+ − +)

Bring down -10.
Divide x into $+x$. Write $+1$ in the quotient above $+0x$.
Multiply $+1$ by each term in the divisor.

To subtract, change each sign of $x - 1$ to $-x + 1$ and add. (+ − +)

The process stops because x cannot divide evenly into -9. The remainder is -9.

Quotient: $\dfrac{x^3 - 10}{x - 1} = x^2 + x + 1$ Remainder -9

$$= x^2 + x + 1 + \frac{-9}{x - 1}$$

Check by multiplying: Divisor · Quotient + Remainder = Dividend

$$(x - 1)\cdot(x^2 + x + 1) + (-9) \stackrel{?}{=} x^3 - 10$$

$$x^3 + x^2 + x - x^2 - x - 1 + (-9) \stackrel{?}{=} x^3 - 10 \qquad \checkmark \text{ It checks.}$$

$$x^3 - 10 = x^3 - 10$$

USING A GRAPHING CALCULATOR TO CONFIRM AND VISUALIZE EQUIVALENCE OF EXPRESSIONS

We used a graphing calculator to confirm the equivalence of two algebraic expressions in previous sections. A similar process confirms the quotient of a polynomial and a monomial.

EXAMPLE 5: Use the graphing calculator to show the equivalence of the two expressions.

$$\frac{2x + 5}{4x} \quad \text{and} \quad \frac{1}{2} + \frac{5}{4x}$$

Label $y_1 = \dfrac{2x + 5}{4x}$ and $y_2 = \dfrac{1}{2} + \dfrac{5}{4x}$, $x \neq 0$, division by zero is undefined. Graph both y_1 and y_2. Compare the two graphs and the coordinates of points on the graphs. The graphs and the values of the coordinates are identical. The two expressions are equivalent.

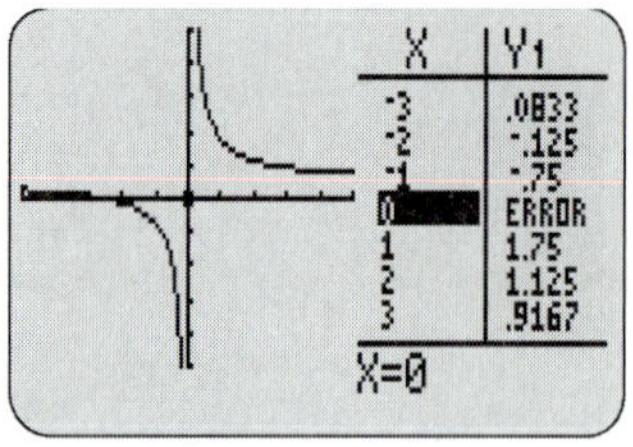

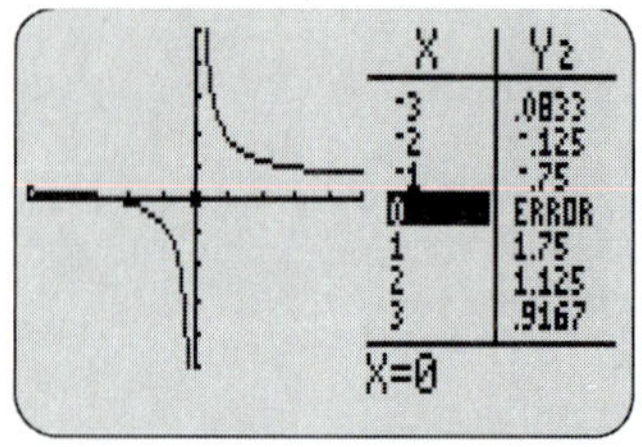

1 2 3 Doing Mathematics

- Write the definitions of the words and symbols listed in this section in the Dictionary in Appendix A.
- On a 3 × 5-inch card, write the following processes for future reference:
  (a) The process of dividing two monomials
  (b) The process of dividing a monomial and a polynomial
  (c) The process of dividing any two polynomials

Problems 1–2: Determine the type of polynomials being divided in each problem.

Type 1: $\left(\dfrac{\text{monomial}}{\text{monomial}}\right)$

Type 2: $\left(\dfrac{\text{polynomial (two or more terms)}}{\text{monomial}}\right)$

Type 3: $\left(\dfrac{\text{polynomial (two or more terms)}}{\text{polynomial (two or more terms)}}\right)$

1. (a) $\dfrac{-12a^3 + 4a^2}{2a}$ (b) $\dfrac{8pq^2r}{4pq}$ (c) $\dfrac{x^2 + 5x + 6}{x + 2}$

(d) $\dfrac{x^2 + 2xy + y^2}{xy}$ (e) $\dfrac{x^3 - 1}{x - 1}$ (f) $\dfrac{36t^3}{-8t^5}$

2. (a) $\dfrac{8y^2 - 6x}{2y}$ (b) $\dfrac{12x^3yz}{6xy}$ (c) $\dfrac{4x^2 - 6x + 5}{3x + 1}$

(d) $\dfrac{12a}{36ab^2}$ (e) $\dfrac{x^2 - 6x + 14}{x - 2}$ (f) $\dfrac{-x^2 + 10x - 5}{5x}$

Problems 3–26: Divide. Assume that variables represent nonzero quantities.

3. $\dfrac{56x^3y^2}{7xy}$

4. $\dfrac{25x^2y}{5y}$

5. $\dfrac{-121mn^2}{11m^2n^3}$

6. $\dfrac{-36ab}{6a^3b^4}$

7. $\dfrac{52p^4q^3r^2}{-4p^3q^4}$

8. $\dfrac{100x^5y^3z}{-25x^3z^4}$

9. $\dfrac{-54xy^3z^2}{-6x^2yz}$

10. $\dfrac{-55a^2b^5c^7}{-5a^3b^4c^5}$

11. $\dfrac{21w^3 - 14z^2}{7z^2}$

12. $\dfrac{25x^2 + 15y}{5y}$

13. $\dfrac{w^2 - 3w + 12}{w^2}$

14. $\dfrac{x^2 + 3x - 7}{x}$

15. $\dfrac{y^2 - 4y - 28}{2y}$

16. $\dfrac{x^2 + 6x + 15}{-3x}$

17. $\dfrac{-12m^3 - 9mn + 3n^2}{-6m^2n^2}$

18. $\dfrac{-15x^2 + 3xy - 6y^2}{3xy}$

19. $\dfrac{x^2 - 5x - 8}{x - 6}$

20. $\dfrac{x^2 + 2x - 10}{x - 1}$

21. $\dfrac{6x^2 + 13x + 7}{3x + 5}$

22. $\dfrac{2x^2 - 5x - 30}{2x + 3}$

23. $\dfrac{2x^3 + 16}{x + 2}$

24. $\dfrac{2x^3 - 8}{x - 1}$

25. $\dfrac{y^3 + 3y^2 - 3y - 9}{y + 3}$

26. $\dfrac{6x^4 + 2x^3 + 5}{x - 1}$

Problems 27–38: Divide. Assume variables represent nonzero quantities.

27. $\dfrac{15j^3k}{-5jk^3}$

28. $\dfrac{9x^2y^3z}{-3y^5z}$

29. $\dfrac{2.2t^2 + 4.4t - 11}{-1.1t}$

30. $\dfrac{-6y^2 - 5y + 9}{-2y}$

31. $\dfrac{2x^2 - 3x + 10}{x - 2}$

32. $\dfrac{x^2 + 7x - 30}{x + 3}$

33. $\dfrac{x^4 + 1}{x^2 + 1}$

34. $\dfrac{x^4 + 9}{x^2 - 3}$

35. $\dfrac{18xy^2z + 54x^2yz - 27xyz^2}{-9xyz}$

36. $\dfrac{-36a^3b^4 - 42a^2b^2c^2 + 24a^2b^3}{6a^2b^2}$

37. $\dfrac{-123rs^4t}{-41r^3t}$

38. $\dfrac{-68a^2bc^3}{-2ab^2c}$

Problems 39–42: The formula for the area of a rectangle is $A = l \cdot w$, where l represents the length and w represents the width. Solving the area formula of a rectangle for the length results in another formula: $l = \dfrac{A}{w}$. Given an expression representing the area and width, do the following:

(a) Write a polynomial expression for the length (reduce, if possible)

(b) Find the length if $x = 5$ feet

39. Area $= x^2 + 2x - 8$; width $= x - 2$

40. Area $= x^2 - x - 6$; width $= x - 3$

41. Area $= 9x^2 - 4$; width $= 3x - 2$

42. Area $= 4x^2 - 1$; width $= 2x + 1$

Problems 43–54: Insert the appropriate symbol, either = or ≠, between the expressions.

43. $\dfrac{-36x^{14}}{4x^7}$ ______ $-9x^2$

44. $\dfrac{20x^{15}}{5x^3}$ ______ $4x^5$

45. $\dfrac{12x^2 - 4x}{4x}$ ______ $3x - 1$

46. $\dfrac{4x^2 + 2x}{2x}$ ______ $2x + 1$

47. $\dfrac{4x - 4}{2}$ ______ $2x - 2$

48. $\dfrac{2x + 2}{2}$ ______ $x + 1$

49. $\dfrac{x^3 + 1}{x + 1}$ ______ $x^2 + 1$

50. $\dfrac{x^3 - 1}{x - 1}$ ______ $x^2 - 1$

51. $\dfrac{x^2 + y^2}{x + y}$ ______ $x + y$

52. $\dfrac{2x^2 + 8}{x + 2}$ ______ $2x + 4$

53. $\dfrac{x^4 - 81}{x - 3}$ ______ $x^3 + 27$

54. $\dfrac{x^3 - 8}{x - 2}$ ______ $x^2 + 4$

Problems 55–58: Use a graphing calculator to determine whether the two expressions are equivalent. Graph both expressions using a window $[-10, 10]$ Xscl 1 $[-10, 10]$ Yscl 1. Compare the graphs and the coordinates of the points on the graphs. Insert the appropriate symbol, either = or ≠, between the expressions.

55. $\dfrac{6x + 3}{3}$ ______ $2x + 1$

56. $\dfrac{4x^2 + 2x}{2}$ ______ $2x^2 + x$

57. $\dfrac{21x^2}{-7} + 1$ ______ $-3x^2 + 1$

58. $\dfrac{5}{5} + 10x$ ______ $1 + 10x$

Problems 59–60: Writing Exercises.

59. Write a fraction involving polynomials that has a common factor in the numerator and denominator.

60. Write a fraction involving polynomials that would require long division to perform the division.

Chapter 5 Summary

SECTIONS 5.1 AND 5.2:

Properties of Exponents

Given any real numbers a, b, m, and n:

I. $a^m \cdot a^n = a^{m+n}$

II. $(a^m)^n = a^{m \cdot n}$

III. $(a \cdot b)^m = a^m \cdot b^m$

IV (A). $\dfrac{a^m}{a^n} = \dfrac{a^{m-n}}{1}$, $(a \neq 0)$

IV (B). $\dfrac{a^m}{a^n} = \dfrac{1}{a^{n-m}}$, $(a \neq 0)$

V. $\left(\dfrac{a}{b}\right)^m = \dfrac{a^m}{b^m}$, $(b \neq 0)$

VI. $a^0 = 1$, $(a \neq 0)$

VII (A). $a^{-n} = \dfrac{1}{a^n}$, $(a \neq 0)$

VII (B). $\dfrac{1}{a^{-n}} = a^n$, $(a \neq 0)$

SECTION 5.3:

Product of Monomials: (monomial)(monomial)

1. Apply the Commutative Property to rearrange factors.
2. Apply the Associative Property to group LIKE factors.
3. Multiply numerical factors.
4. Multiply variables with LIKE bases.

Product of a Monomial and a Polynomial: (monomial)(polynomial)

1. Apply the Distributive Property to multiply the monomial by each term of the polynomial.
2. Multiply numerical coefficients and use the Property of Exponents I to rewrite variables with the same base with a single exponent.

The FOIL Method to Multiply Two Binomials:

The letters in FOIL represent the products of the terms in the binomials.

F the product of the **first** terms

O the product of the **outside** terms

I the product of the **inside** terms

L the product of the **last** terms

Product of Two Polynomials: (polynomial)(polynomial)

1. Distribute each term in the first polynomial across each term in the second polynomial. Be sure to multiply carefully, including the algebraic signs. Polynomials may be multiplied horizontally or vertically.
2. Write all terms, aligning LIKE terms in vertical columns.
3. Collect LIKE terms and write the final product.

SECTION 5.4:

Product of the Sum and Difference of Two Terms

For any real number or algebraic expression X and Y: $(X + Y)(X - Y) = X^2 - Y^2$

Square of a Binomial

For any real number or algebraic expression X and Y:

Binomial Sum: $(X + Y)^2 = X^2 + 2XY + Y^2$

Binomial Difference: $(X - Y)^2 = X^2 - 2XY + Y^2$

SECTION 5.5:

Quotient of Monomials: $\left(\dfrac{\text{monomial}}{\text{monomial}}\right)$

1. Divide out common integer factors.
2. Apply Property of Exponents IV to variables with LIKE bases.
3. Multiply the remaining factors and write the fraction in reduced form.

Quotient of a Polynomial and a Monomial: $\left(\dfrac{\text{polynomial (two or more terms)}}{\text{monomial}}\right)$

1. Separate the fraction into individual fractions, each having the denominator of the original fraction.
2. Factor the numerators and denominators. Divide out common factors.
3. Multiply remaining factors and write the quotient.

Quotient of Two Polynomials: $\left(\dfrac{\text{polynomial (two or more terms)}}{\text{polynomial (two or more terms)}}\right)$

1. Use long division. Be sure to change the sign and add in each subtraction.
2. Check using multiplication: Divisor·Quotient + Remainder = Dividend

Chapter 5 Review

- Review the definitions you have written for this chapter in the Dictionary.
- Review any properties and/or processes you have written on 3 × 5-inch cards.

SECTIONS 5.1 AND 5.2

Problems 1–26: Simplify each expression using the appropriate Properties of Exponents. Assume variables represent nonzero quantities. Write all answers with positive exponents.

1. $x^4 \cdot x$ **2.** $5^7 \cdot 5^{-10}$ **3.** $(-x^2)^5$

4. $(-4^3)^2$ **5.** $(-1.5xy)^3$ **6.** $(-x^4y^3)^2$

7. $\dfrac{3^{12}}{3^4}$ **8.** $\dfrac{y^{10}}{y^{15}}$ **9.** $\left(\dfrac{x}{y}\right)^2$

10. $\left(\dfrac{-3}{5}\right)^2$ **11.** x^0 **12.** $(-3)^0$

13. y^{-3} **14.** $\dfrac{1}{(2x)^{-3}}$ **15.** $5x^{-4}$

16. $3x^2 \cdot x^5$ **17.** $\dfrac{2x^9}{4x^3}$ **18.** $(3x^2y^{-3})^2$

19. $\dfrac{(4x)^2}{(4x)^{-3}}$ **20.** $5y^{-5} \cdot y \cdot y^5$ **21.** $\dfrac{7^3}{7^0}$

22. $(-3x^2y^3)^{-2}$ **23.** $\dfrac{6y^{10}}{y^{-5}}$ **24.** $\left(\dfrac{2x^2}{y^7}\right)^{-3}$

25. $(5a^0b^{-2}c^3)^{-2}$ **26.** $\left(\dfrac{2x}{5}\right)^3$

Problems 27–30: Insert the appropriate symbol, either = or ≠, between the expressions.

27. $x^2 \cdot x^3$ ______ x^6 **28.** $y^5 \cdot y^{-4}$ ______ y^{+1}

29. $(2y)^{-2}$ ______ $-4y^2$ **30.** $6x^2 - 3x^2$ ______ 3

SECTIONS 5.3 AND 5.4

Problems 31–49: Multiply.

31. $(-0.5x^2)(-4x^5)$ **32.** $(-x)(-3x)(-4x)^2$

33. $\left(\frac{2}{3}abc\right)\left(\frac{1}{4}a^2b^2c^2\right)$

34. $-5x(x^2 - 2x + 3)$

35. $(4x^2 + 5x - 10)(-2x)$

36. $-a^2(3ab - 10ab^2)$

37. $(x + 6)(x - 5)$

38. $(x - y)(x + y)$

39. $(x + 7)^2$

40. $(2x - 5)(9x + 4)$

41. $(2x - 1)(2x + 1)$

42. $(5x - 3y)(x + y)$

43. $(y - 4)^2$

44. $(3x + 2)^2$

45. $(a - b)^2$

46. $(x - 5y)^2$

47. $(2x + 3)(5x + 4)$

48. $(3x^2 - 2x + 1)(x^2 - 4x - 5)$

49. $(5x - 1)(x^3 - 6x^2 - 9x - 8)$

Problems 50–53: Multiply, vertically.

50. $\begin{array}{r} 8x^2 - 4x - 5 \\ \times \quad x^2 - 3x + 1 \\ \hline \end{array}$

51. $\begin{array}{r} x^3 - 2x^2 + 10x + 1 \\ \times \qquad 7x^2 - 3x - 2 \\ \hline \end{array}$

52. $\begin{array}{r} x^3 + x^2 + 10x - 8 \\ \times \qquad\qquad 3x + 2 \\ \hline \end{array}$

53. $\begin{array}{r} 5x^3 - 7x^2 + 10x - 1 \\ \times \qquad 5x^2 - 7x + 2 \\ \hline \end{array}$

Problems 54–55: The dimensions of a rectangle and square are given below.

(a) Express the area of each as a product of polynomials. The formula for the area of a rectangle is $A = l \cdot w$, where l represents the length and w represents the width. The area of a square is $A = s^2$, where s represents the length of each side.

(b) Find the area when $x = 5$ inches.

54. Rectangle: length $= 6x + 2$; width $= 3x - 5$

55. Square: side $= 2x - 5$

56. (a) Express the area of the shaded region as a polynomial:

(Hint: Find the area of the circle and subtract the area of the square. The formula for the area of a circle is $A = \pi r^2$. The formula for the area of a square is $A = s^2$.)

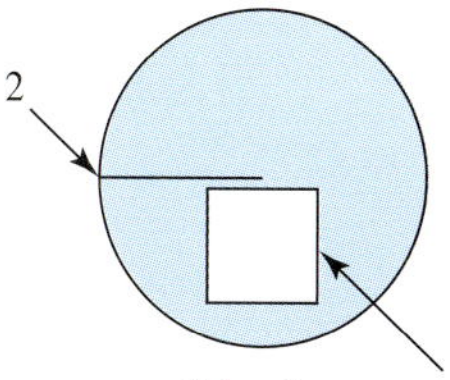

(b) Area when $x = 5$ inches. Use $\pi \approx 3.14$.

Problems 57–60: Insert the appropriate symbol, either = or ≠, between the expressions.

57. $(x - 5)^2$ ______ $x^2 - 25$

58. $(x - 3)^2$ ______ $x^2 + 6x + 9$

59. $-3x(x^2 - 4x + 7)$ ______ $-3x^3 + 12x^2 - 21x$

60. $(x + 1)^2$ ______ $x^2 + 2x + 1$

Problems 61–62: Use a graphing calculator to determine the equivalence of the given expressions. Graph both expressions using a window of $[-10, 10]$ Xscl 1 $[-10, 10]$ Yscl 1. Compare the graphs and the coordinates of the points. Insert the appropriate symbol, either = or ≠, between the expressions.

61. $-x(x + 1)^2$ ______ $-x^3 - 2x^2 - x$

62. $(x + 3)^2$ ______ $x^2 + 9$

SECTION 5.5

Problems 63–73: Divide. Assume variables represent nonzero quantities.

63. $\dfrac{-30a^5b^6}{5a^2b^4}$

64. $\dfrac{-5x^2y^5z}{-5x^2y^3z^5}$

65. $\dfrac{x^3 + 3x^2 - 7}{x^2}$

66. $\dfrac{-15x^2 + 6xy + 5y^2}{-3xy}$

67. $\dfrac{x^2 + 5x - 10}{x + 1}$

68. $\dfrac{x^3 - 8}{x - 1}$

69. $\dfrac{-24x^2y^{-2}z}{3y^5z}$

70. $\dfrac{-6y^2 - 5y + 8}{-2y}$

71. $\dfrac{x^2 + 4x - 12}{x + 3}$

72. $\dfrac{2x^2 + 13x - 7}{2x - 1}$

73. $\dfrac{-36a^3b - 44ab^2c + 24a^2b^3}{-4a^3b^4}$

Problems 74–75: The dimensions of a parallelepiped are given below. Express the missing quantity as a product or quotient of polynomials. The formula for the volume of a parallelepiped is $V = l \cdot w \cdot h$, where l represents the length, w represents the width, and h represents the height.

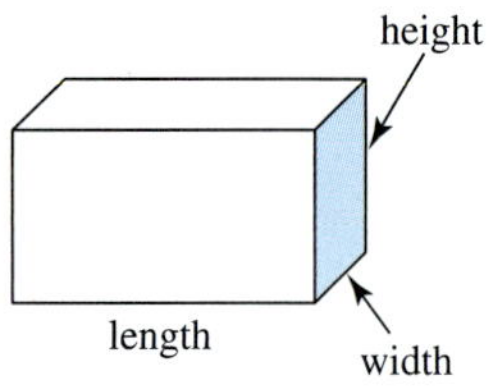

74. Length $= x + 2$; width $= x + 2$; height $= x$. Find the volume when $x = 9$ feet.

75. Volume $= 3x^3 - 75x$; length $= x + 5$; width $= x - 5$. Find an expression for the height.

Problems 76–79: Insert the appropriate symbol, either $=$ or $\neq$, between the expressions.

76. $\dfrac{y^{10}}{y^5}$ ______ y^2

77. $\dfrac{3x + 6x}{3x}$ ______ 3

78. $\dfrac{3 + 2}{2}$ ______ 3

79. $x^2 + 5$ ______ $x^2 + 0x + 5$

Practice Test—Chapter 5

SECTIONS 5.1 AND 5.2

Problems 1–10: Simplify each expression using the appropriate Properties of Exponents. Assume variables represent nonzero quantities. Write all answers with positive exponents.

1. $7x^3 \cdot x^5$

2. $5x^{-6}$

3. $(-3xy^2)^3$

4. $\left(\dfrac{-15x^2y}{3y^3}\right)^3$

5. $\dfrac{-9a^8b^2c^4}{81a^{10}bc^2}$

6. $\dfrac{(-5x^2y^3)^2}{5x^3}$

7. $(8x^0)^2$

8. $\dfrac{a^{10}}{a^{-2}}$

9. $\left(\dfrac{-7xz}{y}\right)^{-2}$

10. $(-2x^{-3}yz)^{-4}$

SECTIONS 5.3 AND 5.4

Problems 11–19: Multiply.

11. $(6x^2)(-3x)^2$

12. $-2x^2(x-5)(x+6)$

13. $(2x+1)(7x-5)$

14. $-3x^2y(x-2y)$

15. $(2x-3)^2$

16. $-3a^2b^2c^2(5ab^2-4c)$

17. $(x+2)(x^2-5x+4)$

18. $-5xy(x+y)(x-y)$

19. $(2x^2-x+5)(x^2+3x-7)$

SECTION 5.5

Problems 20–24: Divide. Assume variables represent nonzero quantities.

20. $\dfrac{-24a^2b^2}{8a^2b}$

21. $\dfrac{36x^3 - 18x^2}{6x^2}$

22. $\dfrac{4ab - 12a^2b^2 - 16ab^2}{4ab}$

23. $\dfrac{x^2 - x - 2}{x + 1}$

24. $\dfrac{x^3 - 27}{x - 3}$

25.

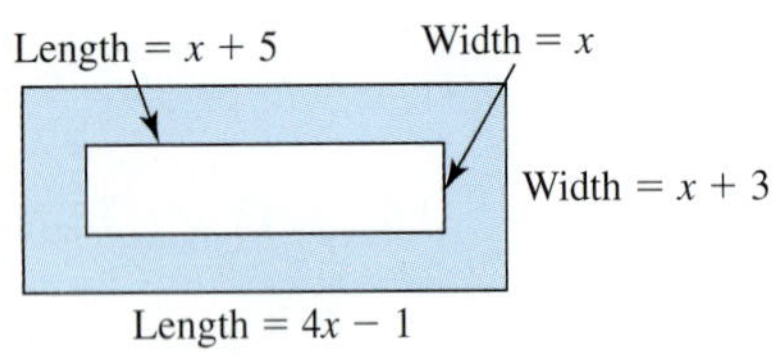

(a) Express the area of the shaded region as a polynomial.

(b) Find the area when $x = 6$ inches.

Answers to Getting Started—Chapter 5

SECTION 5.1

1. (a) $5 \cdot 5 \cdot 5 \cdot 5$

(b) $\left(\frac{2}{3}\right) \cdot \left(\frac{2}{3}\right) \cdot \left(\frac{2}{3}\right)$ or $\frac{2 \cdot 2 \cdot 2}{3 \cdot 3 \cdot 3}$

(c) $-7 \cdot a \cdot b \cdot c \cdot c \cdot c$

(d) $(x^2)(x^2)(x^2)$ or $x \cdot x \cdot x \cdot x \cdot x \cdot x$

2. (a) $(-2)^3$

(b) $-3xy^3$

(c) $\frac{6^2x^3}{y^2}$

(d) $(x - 1)^3$

3. (a) $(b \cdot b \cdot b \cdot b)(b \cdot b) = b \cdot b \cdot b \cdot b \cdot b \cdot b = b^6$

(b) $(a \cdot a \cdot a)(a \cdot a) = a \cdot a \cdot a \cdot a \cdot a = a^5$

(c) Add the exponents when the bases are ALIKE and keep the base the same.

4. (a) $(a^2)(a^2)(a^2)(a^2) = a \cdot a \cdot a \cdot a \cdot a \cdot a \cdot a \cdot a = a^8$

(b) $(b^3)(b^3) = b \cdot b \cdot b \cdot b \cdot b \cdot b = b^6$

(c) Multiply the exponents and keep the base the same.

5. (a) $(-5y)(-5y)(-5y)$
$= (-5) \cdot (-5) \cdot (-5) \cdot y \cdot y \cdot y$
$= (-5)^3y^3 = -125y^3$

(b) $(4a)(4a)(4a)(4a)$
$= 4 \cdot 4 \cdot 4 \cdot 4 \cdot a \cdot a \cdot a \cdot a = 4^4a^4 = 256a^4$

(c) "Spread out the exponent." Raise each factor in the parentheses to the exponent outside the parentheses.

6. (a) $\frac{\overset{1}{\cancel{3}} \cdot \overset{1}{\cancel{a}} \cdot \overset{1}{\cancel{a}} \cdot \overset{1}{\cancel{a}} \cdot a \cdot a \cdot a}{2 \cdot 2 \cdot \underset{1}{\cancel{3}} \cdot \underset{1}{\cancel{a}} \cdot \underset{1}{\cancel{a}} \cdot \underset{1}{\cancel{a}}} = \frac{a^3}{4}$

(b) $\frac{-1 \cdot 2 \cdot 2 \cdot \overset{1}{\cancel{2}} \cdot \overset{1}{\cancel{y}} \cdot \overset{1}{\cancel{y}} \cdot \overset{1}{\cancel{y}}}{\underset{1}{\cancel{2}} \cdot 5 \cdot \underset{1}{\cancel{y}} \cdot \underset{1}{\cancel{y}} \cdot \underset{1}{\cancel{y}} \cdot y \cdot y} = \frac{-4}{5y^2}$

(c) Divide out common numerical factors. Subtract exponents for bases that are ALIKE in the numerator and denominator and keep the same base.

7. (a) $\left(\frac{x}{y}\right)\left(\frac{x}{y}\right)\left(\frac{x}{y}\right)\left(\frac{x}{y}\right) = \frac{x \cdot x \cdot x \cdot x}{y \cdot y \cdot y \cdot y} = \frac{x^4}{y^4}$

(b) $\left(\frac{2b}{a}\right)\left(\frac{2b}{a}\right)\left(\frac{2b}{a}\right) = \frac{2 \cdot 2 \cdot 2 \cdot b \cdot b \cdot b}{a \cdot a \cdot a} = \frac{2^3b^3}{a^3} = \frac{8b^3}{a^3}$

(c) Raise the numerator and the denominator to the power of the exponent outside the parentheses, keeping the base the same.

SECTION 5.2

1. (a) $\frac{x^2}{x^5} = x^{2-5} = x^{-3}$

(b) When the exponent in the denominator is a larger number than the exponent in the numerator.

2. (a) $5^{-2} = \frac{1}{5^2} = \frac{1}{25}$

(b) $x^{-1} = \frac{1}{x^1} = \frac{1}{x}$

(c) $x^2y^{-3}z = \frac{x^2z}{y^3}$

3. (a) $\frac{1}{2^{-2}} = 2^2 = 4$

(b) $\frac{1}{x^{-2}} = x^2$

(c) $\frac{5a^2}{b^{-3}} = \frac{5a^2b^3}{1} = 5a^2b^3$

4. (a) $(2^3)^{-4} = 2^{3(-4)} = 2^{-12} = \frac{1}{2^{12}} = \frac{1}{4096}$

(b) $(y^4)^{-2} = y^{4(-2)} = y^{-8} = \frac{1}{y^8}$

(c) $(-3y)^{-2} = \frac{1}{(-3y)^2} = \frac{1}{(-3)^2y^2} = \frac{1}{9y^2}$

(d) $(4x^2)^{-3} = \frac{1}{(4x^2)^3} = \frac{1}{4^3 \cdot (x^2)^3} = \frac{1}{64x^6}$

(e) $\left(\frac{x}{4}\right)^{-1} = \frac{x^{-1}}{4^{-1}} = \frac{4}{x}$

(f) $\left(\frac{x}{y}\right)^{-3} = \frac{x^{-3}}{y^{-3}} = \frac{y^3}{x^3}$

SECTION 5.3

1. (a) Type 3
   (b) Type 2
   (c) Type 1
   (d) Type 2
   (e) Type 4
2. (a) $40z^8$
   (b) $-18x^6$
3. (a) $6x^3 + 24x^2 - 6x$
   (b) $-15y^5 + 20y^4 + 10y^3 - 45y^2$
4. (a)
$$(x + 7)(x + 3) = x^2 + 3x + 7x + 21$$
$$(x + 7)(x + 3) = x^2 + 10x + 21$$
   (b)
$$(5x + 2)(3x + 1) = 15x^2 + 5x + 6x + 2$$
$$(5x + 2)(3x + 1) = 15x^2 + 11x + 2$$

SECTION 5.4

1. (a) $x^2 + 8x + 15$
   (b) $2y^2 + 5y - 3$
   (c) $6x^2 + 11x - 10$
2. (a) $x^2 + 4x + 4$
   (b) $y^2 - 10y + 25$
   (c) $25x^2 + 10x + 1$
   (d) $9x^2 - 24x + 16$
   (e) The first term of the trinomial is a square. The second term of the trinomial is twice the product of the square root of the first and last terms. The last term of the trinomial is a square.
3. (a) $y^2 - 25$
   (b) $x^2 - 4$
   (c) $4x^2 - 9$
   (d) The first term is a square. The last term is a square. The operation is subtraction.
4. (a) Type 2
   (b) Type 1
   (c) Type 1
   (d) Type 2

SECTION 5.5

1. (a)
$$\frac{20}{15} = \frac{\overset{4}{\cancel{20}}}{\underset{3}{\cancel{15}}} = \frac{4}{3} = 1\frac{1}{3} \approx 1.33$$
   (b)
$$\frac{20 + 3}{4} = \frac{20}{4} + \frac{3}{4} = 5 + \frac{3}{4} = 5\frac{3}{4}$$
   (c)
$$\begin{array}{r} 12 \\ 11\overline{)138} \\ -11 \\ \hline 28 \\ -22 \\ \hline 6 \end{array} \qquad \frac{138}{11} = 12 + \frac{6}{11} = 12\frac{6}{11}$$
2. (a) Type 1
   (b) Type 3
   (c) Type 2
3. (a) $\frac{7x^2}{y}$
   (b) $\frac{-5a}{2}$
4. (a)
$$\frac{15b^2 - 4a}{6ab} = \frac{15b^2}{6ab} - \frac{4a}{6ab} = \frac{5b}{2a} - \frac{2}{3b}$$
   (b)
$$\frac{-35a^2 - 2b}{14ab} = \frac{-35a^2}{14ab} - \frac{2b}{14ab} = \frac{-5a}{2b} - \frac{1}{7a}$$

CHAPTER

6

Factoring and Solving Equations by Factoring

French mathematician Vieté (1540–1603) introduced a new form of symbolism using letters to represent numbers. In his Prior Notes on Symbolic Logistic, *published after his death in 1631, he shows how to operate on symbolic quantities such as $A - B$ times $A + B$ equals $A^2 - B^2$.*

6.1 Factoring Common Monomials and Factoring by Grouping

OBJECTIVES

- Identify a common monomial and/or binomial factor in a polynomial.
- Factor a common monomial from a polynomial.
- Factoring a common binomial from a polynomial.
- Factor by grouping.
- Reduce rational expressions using factoring.

1 2 3 Getting Started

In Chapter 5, we multiplied polynomials. In some mathematical situations we need to "undo" the multiplication with a process called **factoring**. An expression written as the product of its factors is in **factored form** and can be used to reduce fractions and solve equations. In this section, factoring polynomials with a common monomial factor and factoring polynomials with a common binomial factor are discussed.

The goal of factoring is to write an expression as a product of factors, "undoing" multiplication, and reversing the distributive process. The product of the individual common monomial factors is called the **Greatest Common Factor (GCF)**. Identifying common monomial factor(s) and the Greatest Common Factor (GCF) is the first step in factoring a polynomial.

EXAMPLE 1: Factor the greatest common monomial factors from the expression $3a^2b - 6b$.

$3a^2b - 6b$

$= 3 \cdot a \cdot a \cdot b - 3 \cdot 2 \cdot b$ — Write each term of the polynomial as a product of prime factors. Look for common factors.

3 and b are common factors.

$\boxed{3} \cdot a \cdot a \cdot \boxed{b} - \boxed{3} \cdot 2 \cdot \boxed{b}$ — Draw a box around individual common factors.

$3b$ is the GCF. — The product of the individual common factors is the Greatest Common Factor (GCF).

$= 3b(\quad)$ — Write the GCF in front of parentheses.

$= 3b\left(\frac{3a^2b}{3b} - \frac{6b}{3b}\right)$ — Divide the GCF from each term, reversing the Distributive Property.

$= 3b(a^2 - 2)$ — Factored form.

$3b(a^2 - 2)$ — Confirm the factorization, multiplying the factors using the Distributive Property. The resulting product should be equivalent to the original polynomial.

$= 3a^2b - 6b$ ✓ It checks.

1. Factor all common monomial factors from each expression.

(a) $5x^2 + 10x$ (b) $4y^3 + 20y^2$

If a common monomial does not appear in all terms, it may be possible to factor a *common binomial* factor.

EXAMPLE 2: Factor the common binomial factor from the expression $x(y + 3) + 5(y + 3)$.

$x\boxed{(y + 3)} + 5\boxed{(y + 3)}$ — Identify a common binomial factor in each group. Draw a box around the common binomial in each term.

$(y + 3)$ is a common binomial in both terms

$= (y + 3)(\quad)$ — Write the common binomial factor in front of another set of parentheses.

$= (y + 3)(x + 5)$ — Divide the common binomial factor from each group, reversing the Distributive Property. Factored form.

$(y + 3)(x + 5)$ — Confirm the factorization, multiplying the factors using the FOIL method. The resulting product should be equivalent to the original polynomial. It may be necessary to change the order of the resulting terms and factor to match the original polynomial.

F O I L

$= xy + 5y + 3x + 15$

$= xy + 3x + 5y + 15$

$= x(y + 3) + 5(y + 3)$

2. Factor the common binomial factor from each expression.

(a) $2(x + 4) + y(x + 4)$ (b) $x(y - 1) + 5(y - 1)$

1 2 3

Discussing Mathematics

There are several techniques used to factor polynomials. Factoring common monomial factors and binomial factors from a polynomial are presented in this section. Other types of factoring methods will be discussed later in this chapter.

FACTORING COMMON MONOMIAL FACTORS FROM A POLYNOMIAL

The first step of factoring involves identifying common monomial factors (numbers or variables common to each term in the polynomial) and the GCF.

FACTORING COMMON MONOMIAL FACTORS FROM A POLYNOMIAL

1. Identify all individual common monomial factors in each term. Draw boxes around common factors. The product of the individual common factors is the Greatest Common Factor (GCF).
2. Write the GCF in front of parentheses.
3. Divide the GCF from each term, reversing the Distributive Property. The factors not enclosed in the boxes in Step 2 are written inside the parentheses.
4. Confirm the factorization, multiplying the factors using the Distributive Property. The resulting product should be equivalent to the original polynomial.

EXAMPLE 3: Factor.

$$-2x^3 + 10x^2 - 4x$$

$$= -2 \cdot x \cdot x \cdot x + 2 \cdot 5 \cdot x \cdot x - 2 \cdot 2 \cdot x$$

To help identify common factors, factor each term into prime factors.

$$= (-2) \cdot x \cdot x \cdot x - (-2) \cdot 5 \cdot x \cdot x + (-2) \cdot 2 \cdot x$$

Because the leading term is negative, rewrite addition and subtraction so that -2 is shown as a common negative factor.

$$= (\boxed{-2}) \cdot \boxed{x} \cdot x \cdot x - (\boxed{-2}) \cdot 5 \cdot \boxed{x} \cdot x + (\boxed{-2}) \cdot 2 \cdot \boxed{x}$$

Draw boxes around common individual factors: -2 and x. $-2x$ is the GCF.

$$= -2x(\qquad)$$

Write the GCF in front of the parentheses.

$$= -2x(x^2 - 5x + 2)$$

Divide $-2x$ from each term, reversing the Distributive Property. Enclose remaining factors in the parentheses. Factored form.

$$-2x(x^2 - 5x + 2)$$

$$= -2x^3 + 10x^2 - 4x$$ ✓ It checks.

Confirm the factorization, multiplying the factors using the Distributive Property.

FACTORING BY GROUPING

When a common factor does not appear in each term of a polynomial, a common binomial factor may be found by grouping terms of a polynomial.

FACTORING BY GROUPING

1. Look at all terms of the polynomial and identify common monomial factors (GCF) in groups of two terms. Rewrite the polynomial, grouping terms with common monomial factors together.
2. Write each GCF in front of a parenthesis and factor the GCF from each group.
3. Identify any common binomial factor in each group. Draw a box around the common binomial in each group. Write the common binomial factor in front of another set of parentheses.
4. Divide the common binomial factor from each group, reversing the Distributive Property. The factors not enclosed in the boxes remain in the parentheses.
5. Confirm the factorization, multiplying the factors using the FOIL method. The resulting product should be equivalent to the original polynomial. It may be necessary to change the order of the resulting terms to match the original polynomial.

EXAMPLE 4: Factor by grouping.

$2xy - 10x + 7y - 35$	Note: $2x$ is a common monomial factor in the first two terms and 7 is a common monomial factor in the third and fourth terms.
$= (2xy - 10x) + (7y - 35)$ First group: $2x$ is a common monomial factor in the first group. Second group: 7 is a common monomial factor in the second group.	Group the first two terms and last two terms together.
$= 2x\boxed{(y - 5)} + 7\boxed{(y - 5)}$	Write each GCF in front of parentheses and factor the GCF from each group. $(y - 5)$ is a common binomial. Draw a box around $(y - 5)$ in each group.
$= (y - 5)(\quad)$	Write $(y - 5)$ in front of another set of parentheses.
$= (y - 5)(2x + 7)$	Divide $(y - 5)$ from each group. The factors $2x$ and 7 remain in the parentheses. Factored form.
F O I L $2xy + 7y - 10x - 35$	Confirm the factorization, multiplying the factors using the FOIL method.
$= 2xy - 10x + 7y - 35$	Rearrange the resulting terms to match the original polynomial.

EXAMPLE 5: Factor by grouping.

$xy + x + y + 1$	x is a common monomial factor in the first two terms and 1 is the largest common monomial factor in the third and fourth term.
$= (xy + x) + (y + 1)$ First group: x is a common monomial factor in the first group. Second group: 1 is the largest common monomial factor in the second group.	Group the first two terms and last two terms together.
$= x(y + 1) + 1(y + 1)$	Write each GCF in front of parentheses and factor the GCF from each group.
$= (y + 1)(\quad)$	$(y + 1)$ is a common binomial. Draw a box around $(y + 1)$ in each group. Write $(y + 1)$ in front of another set of parentheses.
$= (y + 1)(x + 1)$	Divide $(y + 1)$ from each group. Factored form.
F O I L $xy + y + x + 1$	Confirm the factorization, multiplying the factors using the FOIL method.
$= xy + x + y + 1$	Rearrange the resulting terms to match the original polynomial.

EXAMPLE 6: Factor by grouping $6x^2 - 9x - 10x + 15$.

$(6x^2 - 9x) + (-10x + 15)$ First group: $3x$ is a common monomial factor in the first group. Second group: -5 is a common monomial factor in the second group.	Group the first two terms and last two terms together.
$= 3x(2x - 3) - 5(2x - 3)$	Write each GCF in front of parentheses and factor the GCF from each group. $(2x - 3)$ is a common binomial. Draw a box around $(2x - 3)$ in each group.
$= (2x - 3)(3x - 5)$	Divide $(2x - 3)$ from each group. Factored form.
$6x^2 - 10x - 9x + 15$	Confirm the factorization, multiplying the factors using the FOIL method and rearranging terms.
$= 6x^2 - 9x - 10x + 15$	

USING FACTORING TO REDUCE FRACTIONS

Factoring is used to reduce fractions. Common factors are divided from the numerator and denominator. Be careful to divide out factors, not terms or parts of terms.

EXAMPLE 7: Use factoring to reduce the fraction $\dfrac{2x+6}{x+3}$.

$$\frac{2x+6}{x+3} = \frac{2(x+3)}{(x+3)}$$

Factor the numerator and denominator completely. Enclose the denominator in parentheses.

$$\frac{2\,(x+3)}{(x+3)}$$

The factor $(x + 3)$ is common to the numerator and denominator.

$$= \frac{2\overset{1}{\cancel{(x+3)}}}{\underset{1}{\cancel{(x+3)}}}$$

Divide $(x + 3)$ from the numerator and denominator.

$$= \frac{2}{1} = 2$$

USING A GRAPHING CALCULATOR TO CONFIRM AND VISUALIZE EQUIVALENCE OF EXPRESSIONS

A graphing calculator is used in this chapter to show that two algebraic expressions are equivalent using a similar process presented in earlier chapters.

EXAMPLE 8: Use a graphing calculator to show the equivalence of the two expressions in Example 6: $6x^2 - 9x - 10x + 15$ and $(2x - 3)(3x - 5)$.

Label $y_1 = 6x^2 - 9x - 10x + 15$. Label $y_2 = (2x - 3)(3x - 5)$.

Graph both y_1 and y_2. Compare the two graphs.

$y_1 = 6x^2 - 9x - 10x + 15$

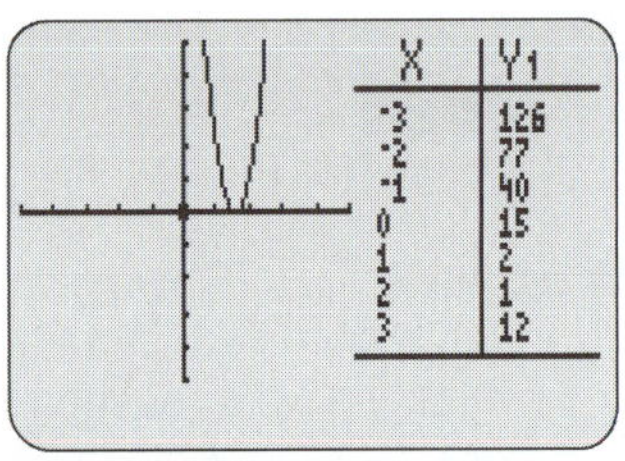

$y_2 = (2x - 3)(3x - 5)$

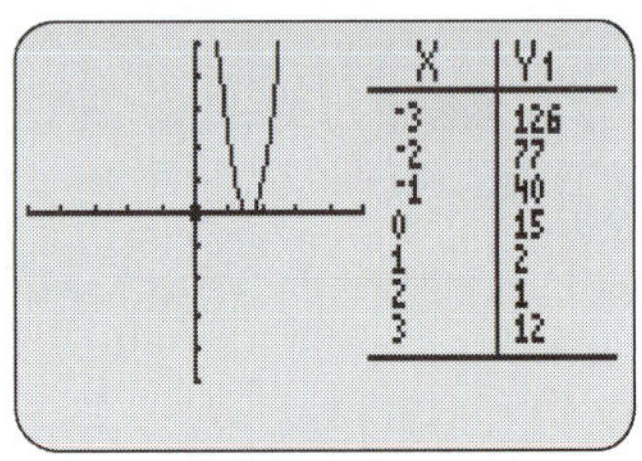

The graphs and the values of the coordinates on the graphs are identical. $6x^2 - 9x - 10x + 15 = (2x - 3)(3x - 5)$. The two expressions are equivalent.

1 2 3 Doing Mathematics

- Write the definitions of the words and symbols listed for this section in the Dictionary in Appendix A.
- On a 3×5-inch card, write the following processes for future reference:
  (a) Factoring a common monomial factor from a polynomial
  (b) Factoring a common binomial factor from a polynomial

Problems 1–2: In each expression, identify any common monomial or binomial factor(s) of the polynomial. (*Hint:* The answer may be "none.")

1. (a) $-7x + 14$ (b) $4x(2x - 3) - 5(2x - 3)$
(c) $3y^2 - 6y + 12$ (d) $2a(5c + 1) + 3b(5c + 1)$
(e) $3x^2 + 6x + 1$ (f) $3y^4 - 9y^3 - 12y$

2. (a) $3x - 6$ (b) $x(x + 1) + 2(x + 1)$
(c) $4x^2 + 4x - 10$ (d) $3x(2x - 3) + 1(2x - 3)$
(e) $x^2 + 3x + 2$ (f) $2y^3 - 4y^2 - 10y$

Problems 3–18: Factor the GCF from each polynomial.

3. $3x - 6$

4. $5x + 30$

5. $-3x + 6$

6. $-12y - 9$

7. $-5x^2 - 5x$

8. $-4x - 4x^2$

9. $-3x^5 + 6x^3$

10. $-2m^4 + 8m^3$

11. $2x^3 + 4x^2 + 10x$

12. $3x^4 + 9x^3 + 24x^2 + 36x$

13. $-2x^3 - x^2 - x$

14. $-2y^3 + 3y^2 - 5y$

15. $15a^2b^3c - 12abc^2$

16. $7p^2qr^3 + 28pq^2r^3$

17. $-4x^4 + 2x^3 - 8x^2$

18. $-6x^3 - 3x^2 - 9x$

Problems 19–34: Factor by grouping.

19. $xy + 7x + 2y + 14$

20. $4ab + 3b + 12a + 9$

21. $3x - 6 - xy + 2y$

22. $5mn - 15m - 2n + 6$

23. $5x - 5 - xy + y$

24. $3u + 6 - uv - 2v$

25. $abc + 4ab + c + 4$

26. $xyz - 3xy + z - 3$

27. $10x^2 - 15x - 8x + 12$

28. $6x^2 + 9x - 4x - 6$

29. $6a^2 + 2a + 15a + 5$

30. $15y^2 + 20y + 6y + 8$

31. $x^2 + 6x - 5x - 30$

32. $x^2 + 2x - 3x - 6$

33. $2x^2 - 10x + x - 5$

34. $21x^2 - 35x + 6x - 10$

Problems 35–46: Reduce each fraction by factoring.

35. $\dfrac{10x + 30}{5x + 15}$

36. $\dfrac{6x - 24}{2x - 8}$

37. $\dfrac{20x + 15}{8x + 6}$

38. $\dfrac{7x + 21}{14x + 42}$

39. $\dfrac{x^2 + 2x - 3x - 6}{4x - 12}$

40. $\dfrac{3x^2 - 21x + x - 7}{6x + 2}$

41. $\dfrac{2a^2 - 10a + a - 5}{3a - 15}$

42. $\dfrac{15u^2 + 20u - 6u - 8}{10u - 4}$

43. $\dfrac{x^2 + 2x - 3x - 6}{2x^2 + 4x + x + 2}$

44. $\dfrac{8x^2 - 10x + 12x - 15}{2x^2 + 3x - 8x - 12}$

45. $\dfrac{-3x^2 - 6x}{5x + 10}$

46. $\dfrac{-9y^2 + 18y}{3y - 6}$

Problems 47–50: Use a graphing calculator to determine the equivalence of the two expressions. Graph both equations. Compare the graphs and the coordinates of the points on the graphs. Insert the appropriate symbol, either = or ≠, between the expressions.

47. $3x^2 - 3x$ ______ $3x(x - 1)$

48. $2x^2 - 6x$ ______ $2x(x - 3)$

49. $x^2 - 6x + 4x - 24$ ______ $(x + 6)(x - 4)$

50. $x^2 - 3x + 2x - 6$ ______ $(x - 2)(x + 3)$

6.2 Factoring Trinomials of the Form $ax^2 + bx + c$ $(a = 1)$, Perfect Square Trinomials, Difference of Squares

OBJECTIVES

- Factor trinomials of the form $ax^2 + bx + c$ $(a = 1)$.
- Factor perfect square trinomials.
- Factor the difference of squares.
- Reduce rational expressions using factoring.

1 2 3 Getting Started

In this section, we discuss factoring three kinds of polynomial expressions:

- Trinomials of the form $ax^2 + bx + c$ $(a = 1)$
- Perfect square trinomials
- The difference of squares.

Some binomials and trinomials have particular characteristics and are given specific names. Our goal in this section is to be able to identify these expressions and write their factors.

We begin by looking for patterns in the products of binomials.

1. Use the FOIL method to multiply the binomial factors.

(a) $(x + 3)(x + 5)$ (b) $(x + 6)(x + 4)$

(c) Look carefully at the factors and the resulting trinomial. Describe the relationship between the original binomials and their products.

(d) Outline a method for finding the factors of similar trinomials.

2. Use your method from problem 1 to factor the following trinomials. Check the factors using the FOIL method to confirm that the product is equivalent to the original expression.

(a) $x^2 + 7x + 12$ (b) $x^2 + 12x + 35$

3. Use the FOIL method to multiply the binomials.

(a) $(x - 7)(x + 7)$ (b) $(x + 5)(x - 5)$

(c) Describe the relationship between the original binomials and their products.

(d) A binomial where one perfect square is subtracted from another perfect square is called the *difference of squares*: $x^2 - 4 = x^2 - 2^2$. Outline a method for finding the factors of expressions that are the difference of squares.

4. Use your method from problem 3 to factor the following expressions. Check the factors using the FOIL method to confirm that the product is equivalent to the original expression.

(a) $x^2 - 36$ (b) $x^2 - 81$

5. Use the FOIL method to multiply the binomials.

(a) $(x + 6)(x + 6) = (x + 6)^2$ (b) $(x - 5)(x - 5) = (x - 5)^2$

(c) Describe the relationship between the original binomials and their products.

(d) A trinomial whose factors are two identical binomials is called a *perfect square trinomial*: $x^2 + 6x + 9 = (x + 3)^2$ or $x^2 - 6x + 9 = (x - 3)^2$. Outline a method for finding the factors of expressions that are perfect square trinomials.

6. Use your method from problem 5 to factor the following trinomials. Check the factors using the FOIL method to confirm that the product is equivalent to the original expression.

(a) $x^2 + 8x + 16$ (b) $x^2 - 16x + 64$

Discussing Mathematics

It is essential to be proficient in multiplying binomials and understand the FOIL method in order to understand factoring of polynomials. The process of factoring a trinomial of the form $x^2 + bx + c$ "undoes" the operation of multiplication and reverses the FOIL process.

When multiplying, factors are multiplied: $(x + 6)(x + 8) = x^2 + 14x + 48$.

When factoring, factors are identified: $x^2 + 14x + 48 = (x + 6)(x + 8)$.

Note the direct relationship between the constants in the factors and the coefficient of the middle term (b) and the constant (c) of the related trinomial $ax^2 + bx + c$, when $a = 1$:

$$(x + 6)(x + 8) = \quad x^2 + 14x + 48 \quad a = 1, b = 14, c = 48$$

The sum of 6 and 8 is 14. Notice $b = 14$.

The product of 6 and 8 is 48. Notice $c = 48$.

When the constants in the binomial factors are labeled r and s, this generalization is formed:

FACTORING A TRINOMIAL OF THE FORM $x^2 + bx + c$ WHERE b, c, r, AND s ARE CONSTANTS

$x^2 + bx + c = (x + r)(x + s)$, then $r \cdot s = c$ and $r + s = b$.

It is also important to observe the patterns in the algebraic signs in a trinomial $x^2 + bx + c$ and the associated binomial factors $(x + r)(x + s)$.

Signs of				Example
c	***b***	***r***	***s***	$x^2 + bx + c = (x + r)(x + s)$
Positive	Positive	Both r and s are positive.		$x^2 + 14x + 48 = (x + 6)(x + 8)$
	Negative	Both r and s are negative.		$x^2 - 14x + 48 = (x - 6)(x - 8)$
Negative	Positive	One is negative, one is positive. The larger of $\lvert r \rvert$ or $\lvert s \rvert$ is positive.		$x^2 + 2x - 48 = (x - 6)(x + 8)$
	Negative	One is negative, one is positive. The larger of $\lvert r \rvert$ or $\lvert s \rvert$ is negative.		$x^2 - 2x - 48 = (x + 6)(x - 8)$

➢ *With practice, students often discover that they can find the pair of numbers of whose product is c and sum is b in their heads. As a result, this process might be called "factoring by inspection" or "trial and error."*

FACTORING TRINOMIALS USING A TABLE OF VALUES

EXAMPLE 1: Factor $x^2 - 5x - 14$ using a table of values.

Write two sets of parentheses. Write the factors of the first term of the trinomial in the parentheses.

$(x \quad)(x \quad)$

List all pairs of integer factors of c and their sums.

Begin the list with $+1$, then -1, then $+2$, then -2 and so forth.

Find a pair of factors whose sum is $b = -5$, the numerical coefficient of the middle term of the trinomial (shaded row).

Factors of $c = -14$		Sum of factors
+1	−14	−13
−1	+14	+13
+2	−7	−5
−2	+7	+5
3, 4, 5, 6	Not factors	
+7	The pairs are repeating.	

Write the factors chosen as the last terms of each binomial.

Factors of −14		Sum	
+2	−7	−5	There is the only pair of factors whose sum is $b = -5$: $(+2)$ and (-7).

Factored form. $(x + 2)(x - 7)$

Check the factors using the FOIL method.

$$(x + 2)(x - 7) = x^2 - 7x + 2x - 14$$
$$= x^2 - 5x - 14 \quad ✓ \text{ It checks.}$$

EXAMPLE 2: Factor $x^2 + 6x - 27$ using a table of values.

Write two sets of parentheses and the first term in each binomial.

$(x \quad)(x \quad)$

List pairs of integer factors of c and their sums.

Stop when a pair of factors has a sum equal to $b = +6$ (shaded row).

Factors of $c = -27$		Sum of factors
+1	−27	−26
−1	+27	+26
+2	Not a factor	
+3	−9	−6
−3	+9	+6

Write the factors chosen as the last terms of each binomial.

Factors of −27		Sum
−3	+9	+6

There is only one pair of factors whose sum is $b = +6$: (-3) and $(+9)$.

Factored form. $(x - 3)(x + 9)$

Check the factors using the FOIL method.

$$(x - 3)(x + 9) = x^2 + 9x - 3x - 27$$
$$= x^2 + 6x - 27 \quad ✓ \text{ It checks.}$$

FACTORING THE DIFFERENCE OF SQUARES, $X^2 - Y^2$

The process of multiplying two binomials containing the sum and difference of two terms was generalized in Section 5.4: $(X + Y)(X - Y) = X^2 - Y^2$. Now we want to take the product $X^2 - Y^2$, called the *difference of squares,* and find the factors.

> **FACTORING THE DIFFERENCE OF SQUARES**
> For any real number or algebraic expression X and Y: $X^2 - Y^2 = (X + Y)(X - Y)$

EXAMPLE 3: Factor the difference of squares $x^2 - 9$.

Rewrite each term as a square.

$$x^2 - 9 = (x)^2 - (3)^2$$

Write two sets of parentheses.

Write the square root of $x^2 = x$ and the square root of $9 = 3$, as the first and second terms in each binomial. Write the operations inside the parentheses.

$$= (x + 3)(x - 3)$$

One factor with addition — The other factor with subtraction.

Check the factors using the FOIL method.

$$(x + 3)(x - 3) = x^2 - 3x + 3x - 9$$
$$= x^2 - 9 \quad ✓ \text{ It checks.}$$

➤ *Beware! The sum of two squares, $X^2 + Y^2$, is prime and does not factor over the set of real numbers.*

EXAMPLE 4: Factor the difference of squares $16x^2 - 25$.

Rewrite each term as a square.	$16x^2 - 25 = (4x)^2 - (5)^2$
Write the first and second terms in each binomial, along with the operations inside the parentheses.	$= (4x + 5)(4x - 5)$
	One factor with addition — The other factor with subtraction
Check the factors using the FOIL method.	$(4x + 5)(4x - 5) = 16x^2 - 5x + 5x - 25$
	$= 16x^2 - 25$ ✓ It checks.

FACTORING PERFECT SQUARE TRINOMIALS

Trinomials of the form $X^2 + 2XY + Y^2$ and $X^2 - 2XY + Y^2$ are called *perfect square trinomials* because they factor into two identical binomials. In Section 5.4, we generalized the product of binomials squared as $(X + Y)^2 = X^2 + 2XY + Y^2$ and $(X - Y)^2 = X^2 - 2XY + Y^2$. Now we want to factor the resulting trinomial.

Perfect square trinomials are easy to factor, once we recognize them. If the first and last terms are perfect squares and the coefficient of the middle term is two times the product of the square root of the first term and the square root of the last term, the trinomial is a perfect square trinomial and factors as follows.

FACTORING A PERFECT SQUARE TRINOMIAL

For any real number or algebraic expression X and Y:

$$X^2 + 2XY + Y^2 = (X + Y)(X + Y) = (X + Y)^2$$
$$X^2 - 2XY + Y^2 = (X - Y)(X - Y) = (X - Y)^2$$

EXAMPLE 5: Factor the perfect square trinomial $x^2 + 6x + 9$.

Rewrite $6x$ as $2(3x)$ and 3 as $(3)^2$.	$x^2 + 6x + 9 = x^2 + 2(3x) + (3)^2$
Write the square root of the first and last term of the trinomial as the first and last term in each binomial.	$= (x \quad 3)(x \quad 3)$
Write the operation of addition in each binomial (the same operation as the middle term).	$= (x + 3)(x + 3)$
	$= (x + 3)^2$
Check the factors using the FOIL method.	$(x + 3)(x + 3) = x^2 + 3x + 3x + 9$
	$= x^2 + 6x + 9$ ✓ It checks.

EXAMPLE 6: Factor the perfect square trinomial $x^2 - 10x + 25$.

Rewrite $10x$ as $2(5x)$ and 25 as $(5)^2$.

$$x^2 - 10x + 25 = x^2 - 2(5x) + (5)^2$$

Write the square root of the first and last term of the trinomial as the first and last term in each binomial.

$$= (x \quad 5)(x \quad 5)$$

Write the operation of subtraction in each binomial.

$$= (x - 5)(x - 5)$$
$$= (x - 5)^2$$

Check the factors using the FOIL method.

$$(x - 5)(x - 5) = x^2 - 5x - 5x + 25$$
$$= x^2 - 10x + 25$$ ✓ It checks.

FACTORING POLYNOMIALS INVOLVING TWO VARIABLES

It is often necessary to factor polynomials containing two or more variables using techniques discussed in this section.

EXAMPLE 7: Factor $x^2 - 8xy + 15y^2$ using a table of values.

Write the variable factors as the first and last terms of each binomial.

$$(x \quad y)(x \quad y)$$

List all pairs of integer factors of c and their sums.

Factors of $c = +15$		Sum of factors
+1	+15	+16
−1	−15	−16
+3	+5	+8
−3	−5	−8

Stop when a pair of factors has a sum equal to $b = -8$.

Write the factors chosen as coefficients of the y variable. Factored form.

$$(x - 3y)(x - 5y)$$

Check the factors using the FOIL method.

$$(x - 3y)(x - 5y) = x^2 - 5xy - 3xy + 15y^2$$
$$= x^2 - 8xy + 15y^2$$ ✓ It checks.

EXAMPLE 8: Factor the perfect square trinomial $x^2 + 4xy + 4y^2$.

Rewrite the first and last terms as a base squared and the middle term as $2(2xy)$.

$$x^2 + 4xy + 4y^2 = (x)^2 + 2(2xy) + (2y)^2$$

Apply the formula for factoring a perfect square trinomial.

$$= (x + 2y)(x + 2y)$$
$$= (x + 2y)^2$$

Check the factors using the FOIL method.

$$(x + 2y)(x + 2y) = x^2 + 2xy + 2xy + 4y^2$$
$$= x^2 + 4xy + 4y^2$$ ✓ It checks.

USING FACTORING TO REDUCE FRACTIONS

Factorizations of the numerators and denominators of fractions are used to reduce the fractions. After the numerator and denominator of a fraction are written in factored form, common factors are divided out.

EXAMPLE 9: Reduce to lowest terms $\dfrac{x^2 + 2x - 48}{x^2 - 14x + 48}$.

Factor the numerator and denominator completely. $\dfrac{(x + 8)(x - 6)}{(x - 6)(x - 8)}$

$(x - 6)$ is a common factor. $= \dfrac{(x + 8)(x - 6)}{(x - 6)(x - 8)}$

Divide the common factor in the numerator and denominator. $= \dfrac{(x + 8)\overset{1}{\cancel{(x - 6)}}}{\underset{1}{\cancel{(x - 6)}}(x - 8)}$

Rewrite the fraction listing the remaining factors. $= \dfrac{x + 8}{x - 8}$

➢ *Beware! Only common factors can be divided out. Do not attempt to divide out variables or numbers that are parts of terms. Factor the numerator and denominator and look for common factors.*

1 2 3 Doing Mathematics

- Write the definitions of the words and symbols listed for this section in the Dictionary in Appendix A.
- On a 3 × 5-inch card, write the following processes for future reference:
  - (a) Factoring a trinomial of the form $ax^2 + bx + c$, where $a = 1$
  - (b) Factoring the difference of squares
  - (c) Factoring a perfect square trinomial

Problems 1–2: Identify each expression as one or more types of polynomials:

Type 1: A trinomial of the form $ax^2 + bx + c$, where $a = 1$ and $b \neq 0$.

Type 2: A binomial that is the difference of squares.

Type 3: A perfect square trinomial.

Type 4: None of the above.

1. (a) $x^2 + 14x + 49$ (b) $u^2 + 6u + 8$
(c) $x^2 + 36$ (d) $t^2 - 121$
(e) $x^2 - x - 30$ (f) $z^2 - 6z + 9$
(g) $x^2 + y$ (h) $81 - c^2$
(i) $a^2 + b + 1$ (j) $x^2 + 5x - 14$

2. (a) $x^2 - 2x + 1$ (b) $x^2 - 100$
(c) $x^2 + 12x + 36$ (d) $x^2 - 10x + 25$
(e) $x^2 - 2x - 10$ (f) $5ab^2 - 4c$
(g) $y^2 - 5$ (h) $a^2 - b^2$
(i) $y^2 + 7x + 12$ (j) $y^2 - 2y + 1$

Problems 3–18: Factor each trinomial of the form $ax^2 + bx + c$, $(a = 1)$.

3. $y^2 + 7y + 10$

4. $x^2 + 10x + 21$

5. $m^2 + 13m + 30$

6. $x^2 + 9x + 20$

7. $n^2 - 4n - 32$

8. $x^2 - 4x - 5$

9. $y^2 + y - 42$

10. $x^2 - x - 12$

11. $x^2 + 5x + 6$

12. $x^2 + 6x + 5$

13. $x^2 - 10x + 24$

14. $x^2 - 11x + 24$

15. $c^2 - 3c - 40$

16. $x^2 + 4x - 21$

17. $x^2 + x - 56$

18. $x^2 + x - 20$

Problems 19–34: Factor the difference of squares.

19. $x^2 - 81$

20. $x^2 - 4$

21. $m^2 - 1$

22. $y^2 - 9$

23. $z^2 - 144$

24. $x^2 - 100$

25. $b^2 - 16c^2$

26. $a^2 - 36b^2$

27. $x^2 - 49y^2$

28. $x^2 - 16y^2$

29. $1 - 100a^2$

30. $1 - 36y^2$

31. $a^2 - 169b^2$

32. $y^2 - 49z^2$

33. $x^2 - 4y^2$

34. $x^2 - 81y^2$

Problems 35–50: Factor each perfect square trinomial.

35. $y^2 - 18y + 81$

36. $x^2 - 4x + 4$

37. $x^2 + 10x + 25$

38. $x^2 - 10x + 25$

39. $a^2 + 20a + 100$

40. $x^2 + 8x + 16$

41. $n^2 + 12n + 36$

42. $y^2 + 14y + 49$

43. $p^2 - 4p + 4$

44. $x^2 + 2x + 1$

45. $x^2 + 22xy + 121y^2$

46. $x^2 + 10xy + 25y^2$

47. $x^2 - 24x + 144$

48. $x^2 - 18x + 81$

49. $c^2 - 8cd + 16d^2$

50. $x^2 - 12xy + 36y^2$

Problems 51–66: Factor each expression. (*Beware:* some expressions may be prime.)

51. $x^2 - 5x - 24$

52. $x^2 - 3x - 4$

53. $k^2 - 12$

54. $x^2 - 5$

55. $x^2 + 2xy + y^2$

56. $x^2 + 6xy + 9y^2$

57. $s^2 + 18s + 77$

58. $x^2 - 14x + 49$

59. $z^2 + 100$

60. $a^2 + 49$

61. $a^2 - 1$

62. $y^2 + 27y + 50$

63. $x^2 - 8x + 16$

64. $a^2 - 6a + 9$

65. $x^2 + 2x + 4$

66. $x^2 - 81y^2$

Problems 67–82: Reduce each rational expression to lowest terms.

67. $\dfrac{x^2 - 3x - 10}{x^2 - 8x - 20}$

68. $\dfrac{x^2 - 2x - 3}{x^2 + 6x + 5}$

69. $\dfrac{r^2 + 8r + 7}{r^2 + 4r + 3}$

70. $\dfrac{y^2 - 3y - 4}{y^2 - 7y + 12}$

71. $\dfrac{z^2 - 6z - 16}{z^2 - 12z + 32}$

72. $\dfrac{a^2 - 7a + 10}{a^2 - 9a + 14}$

73. $\dfrac{x^2 + 3x - 54}{x^2 - 81}$

74. $\dfrac{y^2 - 25}{y^2 - 11y + 30}$

75. $\dfrac{x^2 + 13x + 42}{x^2 + 7x + 6}$

76. $\dfrac{x^2 + 5x + 6}{x^2 + x - 2}$

77. $\dfrac{m^2 + 16m + 64}{m^2 - 64}$

78. $\dfrac{a^2 + 6a + 9}{a^2 - 9}$

79. $\dfrac{x^2 - 36}{x^2 + 12x + 36}$

80. $\dfrac{y^2 - 1}{y^2 - 2y + 1}$

81. $\dfrac{x^2 + 7x + 12}{x^2 - 9}$

82. $\dfrac{x^2 - 10x + 16}{x^2 - 4}$

Problems 83–86: Use a graphing calculator to determine the equivalence of the two expressions. Graph both equations. Compare the graphs and the coordinates of the points on the graphs. Insert the appropriate symbol, either = or ≠, between the expressions.

83. $x^2 + 6x + 9$ ______ $(x + 3)^2$

84. $x^2 - 4x + 4$ ______ $(x - 2)^2$

85. $x^2 + 4$ ______ $(x - 2)(x - 2)$

86. $x^2 + 36$ ______ $(x - 6)(x + 6)$

6.3 Factoring Trinomials of the Form $ax^2 + bx + c$ $(a \neq 1)$

OBJECTIVES

- Factor trinomials of the form $ax^2 + bx + c$ $(a \neq 1)$.
- Factor the difference of squares and perfect square trinomials.
- Reduce rational expressions using factoring.

1 2 3 Getting Started

In this section, we focus on trinomials of the form $ax^2 + bx + c$ $(a \neq 1)$ and the difference of squares $(a \neq 1)$, extending the processes outlined in the last section.

1. Use the FOIL method to multiply the binomial factors.

(a) $(2x + 3)(x + 1)$ (b) $(5x + 1)(x + 3)$

(c) $(3x - 4)(x - 5)$ (d) $(2x + 5)(3x - 4)$

(e) Look carefully at the factors and the resulting trinomial. Describe the relationship between each trinomial and its equivalent factored form.

(f) Outline a method for finding the factors of the trinomials.

2. Use your method to factor each trinomial. Check the answer by using the FOIL method to confirm that the product is equivalent to the original trinomial.

(a) $6x^2 + 7x - 3$ (b) $3x^2 - 7x + 2$

(c) $2x^2 + 5x - 12$ (d) $3x^2 + 14x + 8$

3. Factor each perfect square trinomial. Check the answer by using the FOIL method to confirm that the product is equivalent to the original trinomial. (*Hint:* Rewrite the first and last terms of the trinomial as a perfect square.)

(a) $4x^2 + 4x + 1$ (b) $9x^2 - 12x + 4$

(c) $25x^2 + 70x + 49$ (d) $4x^2 - 12x + 9$

4. Factor the difference of squares. (*Hint:* Rewrite the terms as perfect squares.) Check the answer by using the FOIL method to confirm that the product is equivalent to the original trinomial.

(a) $25x^2 - 49$ (b) $4x^2 - 9$

(c) $36x^2 - 16y^2$ (d) $64x^2 - y^2$

1 2 3 Discussing Mathematics

FACTORING BY GROUPING—TRINOMIALS OF THE FORM $ax^2 + bx + c, (a \neq 1)$

When the leading coefficient of a trinomial is a number other than 1 and factoring by inspection is unsuccessful, a table of values and factoring by grouping can be helpful. An extra step of finding the product of the leading coefficient and the constant (or coefficient of the third term) of the trinomial is added to the process presented in Section 6.1. This product is labeled $p = a \cdot c$. The factors of p are used to rewrite the middle term of the trinomial. Factoring by grouping of trinomials of the form $ax^2 + bx + c$ $(a \neq 1)$ can be challenging. Don't get discouraged.

EXAMPLE 1: Factor by grouping $2x^2 + 7x + 3$.

Given $a = 2$, $b = +7$, $c = +3$. Multiply $a = 2$ and $c = +3$. Label the product $p = a \cdot c$.

$$p = a \cdot c = 2(+3) = +6$$

List pairs of integer factors of p. Begin the list with $+1$. Stop when a pair of factors has a sum equal to $b = 7$. →

Factors of $p = +6$		Sum of factors
+1 Stop here.	+6	+7

Rewrite the middle term of the original trinomial as two terms using $+1$ and $+6$ as coefficients.

Original trinomial: $2x^2 + 7x + 3 =$

Use factors from table: $2x^2 + x + 6x + 3 =$

Factor the four-term polynomial by grouping.

$$(2x^2 + x) + (6x + 3) =$$
$$x(2x + 1) + 3(2x + 1) =$$

Factored form.

$$(2x + 1)(x + 3)$$

Check the factors using the FOIL method.

$$(2x + 1)(x + 3) = 2x^2 + 6x + x + 3$$
$$= 2x^2 + 7x + 3$$

✓ It checks.

EXAMPLE 2: Factor by grouping $6x^2 + 11x - 10$.

Find the product $p = a \cdot c$, where $a = 6$ and $b = -10$.

$p = a \cdot c = 6(-10) = -60$

List pairs of integer factors of p and their sums. Begin the list with $+1$, then -1, then $+2$, then -2 and so forth.*

Factors of $p = -60$		Sum of factors
+1	−60	−59
−1	+60	+59
+2	−30	−28
−2	+30	+28
+3	−20	−17
−3	+20	+17
+4	−15	−11
−4	+15	+11

Stop when a pair of factors has a sum equal to $b = +11$ (shaded row). → Stop here.

Rewrite the middle term of the original trinomial as two terms using -4 and $+15$ as coefficients.

$$6x^2 + 11x - 10 =$$
$$6x^2 - 4x + 15x - 10 =$$

Factor the four-term polynomial by grouping.

$$(6x^2 - 4x) + (15x - 10) =$$
$$2x(3x - 2) + 5(3x - 2) =$$

Factored form.

$$(3x - 2)(2x + 5)$$

Check the factors using the FOIL method.

$(3x - 2)(2x + 5) = 6x^2 + 11x - 10$ ✓ It checks.

* *The list of factors may include many pairs of factors. If you don't find the right pair of factors, keep going. Don't get discouraged. You will eventually find the pair of factors whose sum is the numerical coefficient of the middle term of the trinomial.*

EXAMPLE 3: Factor by grouping $10x^2 - 3x - 4$.

Find the product $p = a \cdot c$, where $a = 10$ and $c = -4$.

$p = a \cdot c = 10(-4) = -40$

List pairs of integer factors of p and their sums.

Factors of $p = -40$		Sum of factors
+1	−40	−39
−1	+40	+39
+2	−20	−18
−2	+20	+18
±3	Not a factor	
+4	−10	−6
−4	+10	+6
+5	−8	−3

Stop when a pair of factors has a sum equal to $b = -3$ (shaded row). → Stop here.

Rewrite the middle term of the original trinomial as two terms using $+5$ and -8 as coefficients.

$$10x^2 - 3x - 4 =$$
$$10x^2 + 5x - 8x - 4 =$$

Factor the four-term polynomial by grouping.

$$(10x^2 + 5x) + (-8x - 4) =$$
$$5x(2x + 1) - 4(2x + 1) =$$

Factored form.

$$(2x + 1)(5x - 4)$$

Check the factors using the FOIL method.

$(2x + 1)(5x - 4) = 10x^2 - 3x - 4$ ✓ It checks.

FACTORING THE DIFFERENCE OF TWO SQUARES AND PERFECT SQUARE TRINOMIALS

To factor the difference of two squares, use the formula discussed in Section 6.2, page 392: $X^2 - Y^2 = (X + Y)(X - Y)$, where X and Y are algebraic expressions.

EXAMPLE 4: Factor the difference of squares $9x^2 - 49y^2$.

Rewrite each term as a base squared.

$$9x^2 - 49y^2 = (3x)^2 - (7y)^2$$

Write the square root of each term as the first and second terms in each binomial. Write the operations of + and − inside the parentheses.

$$= (3x + 7y)(3x - 7y)$$

One factor with addition — The other factor with subtraction

Check the factors using the FOIL method. $(3x + 7y)(3x - 7y) = 9x^2 - 49y^2$ ✓ It checks.

To factor a perfect trinomial square when the leading coefficient is not 1, use the formula discussed in Section 6.2, page 393: $X^2 + 2XY + Y^2 = (X + Y)(X + Y) = (X + Y)^2$ or $X^2 - 2XY + Y^2 = (X - Y)(X - Y) = (X - Y)^2$, where X and Y are algebraic expressions.

EXAMPLE 5: Factor the perfect square trinomial $25x^2 - 20xy + 4y^2$.

Rewrite the first and last terms as a base squared. Write the middle term as 2 times the product of the bases.

$$25x^2 - 20xy + 4y^2 = (5x)^2 - 2(5x \cdot 2y) + (2y)^2$$

Write the square root of the first and last terms and the operation of subtraction in each binomial.

$$= (5x - 2y)(5x - 2y)$$
$$= (5x - 2y)^2$$

Check the factors using the FOIL method. $(5x - 2y)(5x - 2y) = 25x^2 - 20xy + 4y^2$ ✓ It checks.

USING FACTORING TO REDUCE FRACTIONS

Factoring is used to reduce fractions in this section as discussed in previous sections. After the numerator and denominator of the fraction are rewritten in factored form, common factors are divided out.

EXAMPLE 6: Reduce to lowest terms $\frac{2x^2 + 5x - 12}{6x^2 - 7x - 3}$.

Factor the numerator and denominator completely.

$$\frac{2x^2 + 5x - 12}{6x^2 - 7x - 3} = \frac{(2x - 3)(x + 4)}{(3x + 1)(2x - 3)}$$

$(2x - 3)$ is a common factor in the numerator and denominator.

$$= \frac{(2x - 3)(x + 4)}{(3x + 1)(2x - 3)}$$

Divide out the common factor in the numerator and denominator.

$$= \frac{\overset{1}{\cancel{(2x - 3)}}(x + 4)}{(3x + 1)\underset{1}{\cancel{(2x - 3)}}}$$

$$= \frac{x + 4}{3x + 1}$$

➢ *Remember: Only common factors can be divided out. Do not attempt to divide out variables or numbers that are parts of terms.*

1 2 3

Doing Mathematics

- Write the definitions of the words and symbols listed for this section in the Dictionary in Appendix A.
- On a 3 × 5-inch card, write the steps for factoring a trinomial of the form $ax^2 + bx + c$, where $a \neq 1$, for future reference.

Problems 1–2: Identify each expression as one or more of the following types of polynomials:

Type 1: A trinomial of the form $ax^2 + bx + c$, where $a \neq 1$ and $b \neq 0$.

Type 2: A trinomial of the form $ax^2 + bx + c$, where $a = 1$ and $b \neq 0$.

Type 3: A binomial that is the difference of squares.

Type 4: A perfect square trinomial.

Type 5: None of the above.

1. (a) $b^2 - 2b + 1$ (b) $t^2 - 16t + 60$
(c) $n^2 - 49$ (d) $12x^2 + 7x + 1$
(e) $k^2 + 100$ (f) $16a^2 - 25$
(g) $u^2 + 22u + 121$ (h) $p^2 - 10p + 21$

2. (a) $3x^2 - 2x + 1$ (b) $x^2 - 20x + 100$
(c) $x^2 + 36$ (d) $x^2 + 10x + 25$
(e) $x^2 - 2x - 10$ (f) $25b^2 - 4$
(g) $y^2 - 5$ (h) $a^2 - 12x + 36$

Problems 3–28: Factor.

3. $3y^2 - 8y - 3$

4. $2x^2 - 9x - 5$

5. $5z^2 - 21z + 4$

6. $3x^2 - 14x + 8$

7. $12a^2 + 25a + 12$

8. $6y^2 - 5y - 6$

9. $4a^2 - 8a - 5$

10. $6x^2 + 17x - 3$

11. $6p^2 + p - 12$

12. $4x^2 - x - 14$

13. $12m^2 + 4m - 1$

14. $12m^2 - m - 1$

15. $4x^2 - 20x + 9$

16. $6x^2 + 11x + 4$

17. $14x^2 - 31x - 10$

18. $4x^2 + 9x - 9$

19. $9x^2 - 4$

20. $25x^2 - 49$

21. $16x^2 - 25y^2$

22. $9a^2 - 25b^2$

23. $4x^2 + 12x + 9$

24. $4x^2 - 4x + 1$

25. $16x^2 - 8xy + y^2$

26. $4x^2 + 20xy + 25y^2$

27. $6x^2 + 5xy + y^2$

28. $6x^2 + 7xy - 5y^2$

Challenge problems 29–36:

29. $24x^2 + 59x + 36$

30. $20a^2 - 13a + 2$

31. $12x^2 - 16x - 35$

32. $21y^2 + 17y + 2$

33. $16x^2 + 88x + 121$

34. $2x^2 - 15x + 25$

35. $27x^2 + 39x - 10$

36. $7x^2 + 25x - 12$

Problems 37–52: Reduce each rational expression to lowest terms.

37. $\dfrac{6v^2 + 5v + 1}{2v^2 + 3v + 1}$

38. $\dfrac{6x^2 + 7x - 5}{2x^2 + x - 1}$

39. $\dfrac{3x^2 + 14x - 80}{x^2 - 64}$

40. $\dfrac{x^2 - 16}{x^2 + 5x + 4}$

41. $\dfrac{8x^2 + 6x + 1}{4x^2 - 7x - 2}$

42. $\dfrac{15x^2 + 7x - 2}{5x^2 + 14x - 3}$

43. $\dfrac{5z^2 - 13z - 6}{15z^2 + 11z + 2}$

44. $\dfrac{3y^2 - 14y + 8}{7y^2 - 31y + 12}$

45. $\dfrac{3x^2 + 2x - 1}{2x^2 + x - 1}$

46. $\dfrac{2x^2 - x - 1}{x^2 - 2x + 1}$

47. $\dfrac{7x^2 + 34x - 5}{3x^2 + 13x - 10}$

48. $\dfrac{2x^2 - 9x - 5}{2x^2 - 15x + 25}$

49. $\dfrac{6x^2 + 19x + 15}{6x^2 + 11x + 3}$

50. $\dfrac{2x^2 - 17x + 36}{4x^2 - 19x + 12}$

51. $\dfrac{x^2 - 25y^2}{4x^2 - 23xy + 15y^2}$

52. $\dfrac{5x^2 - 9xy - 2y^2}{x^2 - 4y^2}$

Problems 53–56: Use a graphing calculator to determine the equivalence of the two expressions. Graph both equations. Compare the graphs and the coordinates of the points on the graphs. Insert the appropriate symbol, either = or ≠, between the expressions.

53. $6x^2 - 5x - 6$ ______ $(3x + 2)(2x - 3)$

54. $2x^2 - x - 1$ ______ $(2x + 1)(x - 1)$

55. $6x^2 - 14x + 3$ ______ $(4x - 1)(2x - 3)$

56. $2x^2 - 15x + 25$ ______ $(2x - 5)(x - 5)$

6.4 Solving Equations Using the Zero-Product Property

OBJECTIVE

- Solve second-degree equations using factoring.

Getting Started

In this section, factoring and the Zero-Product Property are presented to solve second-degree polynomial equations. The **degree of a polynomial equation** in one variable is determined by the largest exponent of the variable. An example of a second-degree equation in one variable is $x^2 + 14x + 48 = 0$.

1. Denise is planting a garden and has enough topsoil to cover 96 square feet. She would like a rectangular garden with the length 4 feet more than the width.

(a) Determine the length and width of the garden by drawing a picture and using the guess and check method.

Width

Length

(b) Let x represent the width of the garden. Write an expression for the length of the garden.

(c) Use the formula of the area of a rectangle, $Area = length \cdot width$, to write an equation of the problem situation.

(d) How could you solve the equation analytically, instead of drawing a picture and using guess and check?

Some second-degree polynomial equations can be solved using factoring and applying the Zero-Product Property.

> **ZERO-PRODUCT PROPERTY**
> Given a polynomial equation with real number or algebraic expression factors a and b, where $a \cdot b = 0$, then either $a = 0$ and/or $b = 0$.

The factors a and b are often binomials when solving higher-degree polynomials. The Zero-Product Property could be written as $(x + r)(x + s) = 0$, then $x + r = 0$ or $x + s = 0$.

EXAMPLE 1: Solve $x^2 + 14x = -48$ using the Zero-Product Property.

Add 48 on each side of the equation. $x^2 + 14x + 48 = -48 + 48$

Factor the nonzero side of the equation. $x^2 + 14x + 48 = 0$

$(x + 6)(x + 8) = 0$

Apply the Zero-Product Property and set each factor equal to zero. $x + 6 = 0$ or $x + 8 = 0$

Solve each new equation separately. $x = -6$ or $x = -8$

Check the answers by substituting into the original equation $x^2 + 14x = -48$ to make sure that each yields a true statement.

Check $x = -6$:

$$(-6)^2 + 14(-6) \stackrel{?}{=} -48$$
$$36 - 84 \stackrel{?}{=} -48$$
$$-48 = -48$$

✓ True statement.

Check $x = -8$:

$$(-8)^2 + 14(-8) \stackrel{?}{=} -48$$
$$64 - 112 \stackrel{?}{=} -48$$
$$-48 = -48$$

✓ True statement.

2. Solve each equation by factoring and applying the Zero-Product Property. Check each answer by substituting into the original equation.

(a) $x^2 + 7x + 12 = 0$

(b) $x^2 + 2x - 15 = 0$

(c) $2x^2 + 11x = -5$

(d) Describe the relationship between the factors in the equations above and the solutions of the equations.

1 **2** 3

Discussing Mathematics

SOLVING EQUATIONS USING THE ZERO-PRODUCT PROPERTY

The Equality-Principles of Equations discussed in Section 2.3 are effective for solving first-degree polynomial equations in one variable. The Zero-Product Property is needed to solve second-degree and higher degree polynomial equations.

SOLVING A POLYNOMIAL EQUATION USING THE ZERO-PRODUCT PROPERTY

1. Add or subtract terms from each side of the equation so that all nonzero terms are on one side of the equation. The other side of the equation is equal to zero.
2. Factor the nonzero side of the equation using methods presented in earlier sections. Constant factors may be divided out.
3. Apply the Zero-Product Property and set each factor equal to zero.
4. Solve each equation separately. The answers are the solutions of the original equation.
5. Check by substituting the answers into the original equation to make sure that each yields a true statement.

EXAMPLE 2: Solve the second-degree equation $6x^2 = 3 - 7x$ using the Zero-Product Property.

Add $7x$ to each side of the equation.

$$6x^2 + 7x = 3 - 7x + 7x$$
$$6x^2 + 7x = 3$$

Subtract 3 from each side of the equation.

$$6x^2 + 7x - 3 = 3 - 3$$
$$6x^2 + 7x - 3 = 0$$

Factor the nonzero side of the equation.

$$6x^2 + 7x - 3 = 0$$
$$(2x + 3)(3x - 1) = 0$$

Apply the Zero-Product Property and set each factor equal to zero.

$$2x + 3 = 0 \quad \text{or} \quad 3x - 1 = 0$$
$$2x = -3 \qquad 3x = 1$$

Solve each new equation separately.

$$x = -\frac{3}{2} \quad \text{or} \quad x = \frac{1}{3}$$

Check by substituting the answers into the original equation $6x^2 = 3 - 7x$ to make sure that each yields a true statement.

Check $x = -\frac{3}{2}$:

$$6\left(-\frac{3}{2}\right)^2 \stackrel{?}{=} 3 - 7\left(-\frac{3}{2}\right)$$
$$6\left(\frac{9}{4}\right) \stackrel{?}{=} 3 - \left(\frac{-21}{2}\right)$$
$$\frac{27}{2} \stackrel{?}{=} 3 + \frac{21}{2}$$
$$\frac{27}{2} \stackrel{?}{=} \frac{6}{2} + \frac{21}{2}$$
$$\frac{27}{2} = \frac{27}{2}$$

✓ It checks.

Check $x = \frac{1}{3}$:

$$6\left(\frac{1}{3}\right)^2 \stackrel{?}{=} 3 - 7\left(\frac{1}{3}\right)$$
$$6\left(\frac{1}{9}\right) \stackrel{?}{=} 3 - 7\left(\frac{1}{3}\right)$$
$$\frac{2}{3} \stackrel{?}{=} 3 - \frac{7}{3}$$
$$\frac{2}{3} \stackrel{?}{=} \frac{9}{3} - \frac{7}{3}$$
$$\frac{2}{3} = \frac{2}{3}$$

✓ It checks.

EXAMPLE 3: Solve $4x^2 = 9$ using the Zero-Product Property.

Subtract 9 from each side of the equation. $4x^2 - 9 = 9 - 9$

The right side of the equation is equal to zero. $4x^2 - 9 = 0$

Factor the nonzero side of the equation. $(2x - 3)(2x + 3) = 0$

Apply the Zero-Product Property and set each factor equal to zero. $2x - 3 = 0$ or $2x + 3 = 0$

Solve each new equation separately. $x = \frac{3}{2}$ or $x = -\frac{3}{2}$

Check by substituting the answers into the original equation. ✓ Both answers check.

EXAMPLE 4: Solve $x^2 + x = 2x + 2$ using the Zero-Product Property.

Subtract $2x$ from each side of the equation. $x^2 + x - 2x = 2x - 2x + 2$

$x^2 - x = 2$

Subtract 2 from each side of the equation. $x^2 - x - 2 = 2 - 2$

$x^2 - x - 2 = 0$

Factor the nonzero side of the equation. $(x + 1)(x - 2) = 0$

Apply the Zero-Product Property and set each factor equal to zero. $x + 1 = 0$ or $x - 2 = 0$

Solve each new equation separately. $x = -1$ or $x = 2$

Check by substituting the answers into the original equation. ✓ Both answers check.

CONFIRMING THE SOLUTION OF AN EQUATION GRAPHICALLY

The **zeros of a polynomial** are the x-coordinates of the points where the graph of the related polynomial equation crosses the x-axis (see Section 3.3), the x-intercepts of the graph, where $y = 0$. We can use this definition and a graph of the associated function to confirm solutions of the second-degree equations with the X-Intercept Graphical Method presented in Section 4.1.

EXAMPLE 5: Use the X-Intercept Graphical Method to confirm the solutions $x = -1$ and $x = 2$ of the equation $x^2 + x = 2x + 2$ in Example 5.

$x^2 - x - 2 = 0$

With the equation written with all nonzero terms on one side, substitute 0 for y in the equation to form the associated function.

$x^2 - x - 2 = y$

$y = x^2 - x - 2$

Graph $y = x^2 - x - 2$ in the window $[-5, 5]$ Xscl 1 $[-10, 10]$ Yscl 2.

The x-intercepts of the graph are $x = -1$ and $x = 2$. The zeros of the equation are $x = -1$ and $x = 2$, confirming the analytical solutions.

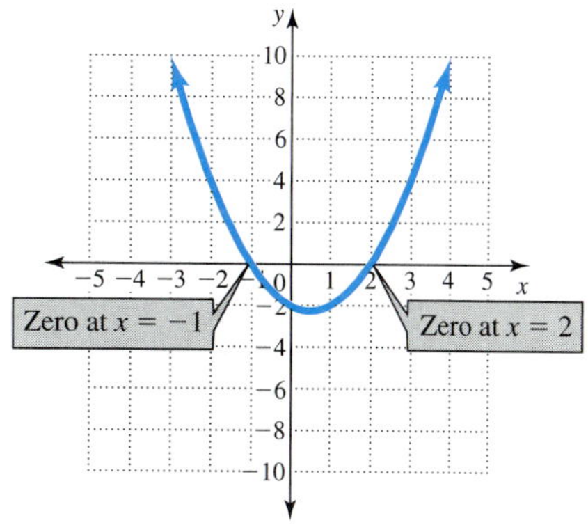

X	Y_1
−3	10
−2	4
−1	0
0	−2
1	−2
2	0
3	4

$x = -1$

SOLVING REAL-WORLD PROBLEMS

Some real-world problem situations can be modeled with second-degree equations and solved using the factoring and the Zero-Product Property.

EXAMPLE 6: Janice can afford to buy 300 square feet of carpet for her living room. She would like the width of the carpet to be 5 feet less than the length. Find the dimensions of the carpet. Let x represent the length of the carpet.

Carpeted portion of the living room

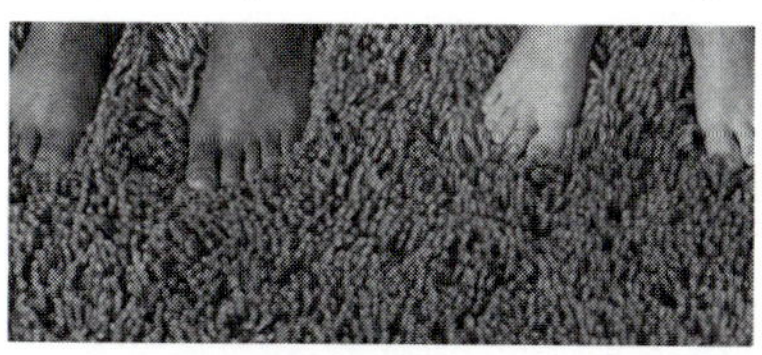

Width: $x - 5$

Length: x

Write the equation describing the problem situation. Solve for x.

Area of a rectangle = length × width

$$300 = x(x - 5)$$
$$300 = x^2 - 5x$$
$$0 = x^2 - 5x - 300$$
$$0 = (x - 20)(x + 15)$$
$$x - 20 = 0 \qquad x + 15 = 0$$
$$x = 20 \qquad x = -15$$

The length (x) of a rectangle cannot be a negative number.

$x = -15$ is called an extraneous solution and is discarded.

Carpet length: $x = 20$ feet; width: $x - 5 = 20 - 5 = 15$ feet

1 2 3 Doing Mathematics

- Write the definitions of the words and symbols listed for this section in the Dictionary in Appendix A.
- On a 3 × 5-inch card, write the process of solving a polynomial equation using the Zero-Product Property for future reference.

Problems 1–40: Solve each equation using factoring and the Zero-Product Property. Check by substituting. Confirm solutions graphically, if desired.

1. $x^2 - x - 30 = 0$

2. $x^2 - 2x - 3 = 0$

3. $y^2 + 6x = -9$

4. $y^2 - 3y = 4$

5. $x^2 + 8x + 7 = 0$

6. $x^2 + 6x + 5 = 0$

7. $y^2 + 20 = 9y$

8. $y^2 + 12 = 7y$

9. $x^2 - 14x = -45$

10. $x^2 - 9x = -14$

11. $a^2 + 1 = 2a$

12. $a^2 + 2a = 3$

13. $x^2 - 2x - 12 = 2x$

14. $x^2 - 3x = 3x - 9$

15. $4x^2 + 4x = 0$

16. $3x^2 - 3x = 0$

17. $2x^2 = -10x$

18. $5x^2 = 15x$

19. $-9x^2 + 36x = 0$

20. $-7x^2 - 28x = 0$

21. $y^2 = y$

22. $6x^2 = -12x$

23. $3x^2 - 15x = 5 - x$

24. $2x^2 + 10x = x + 5$

25. $x^2 + 1 = 2x$

26. $x^2 + 25 = 10x$

27. $6x = 8 + 28x - 21x^2$

28. $12 = 15x + 8x - 10x^2$

29. $3x^2 = 8 - 10x$

30. $2x^2 = 9x + 5$

31. $2y^2 + 36 = 17y$

32. $3y^2 + 8 = 14y$

33. $4x^2 + 9 = 12x$

34. $9x^2 + 1 = 6x$

35. $19y + 36 = 6y^2$

36. $7y = 5 - 6y^2$

37. $6x^2 + 7x + 2 = 0$

38. $12x^2 = 19x + 18$

39. $25a^2 = 81$

40. $9a^2 = 49$

41. Laurie is planning to carpet a portion of her living room. She can afford to buy 630 square feet of carpet. She would like the width of the carpeted portion to be 9 feet less than the length. Find the dimensions of the carpet. Let x represent the length of the carpeted portion.

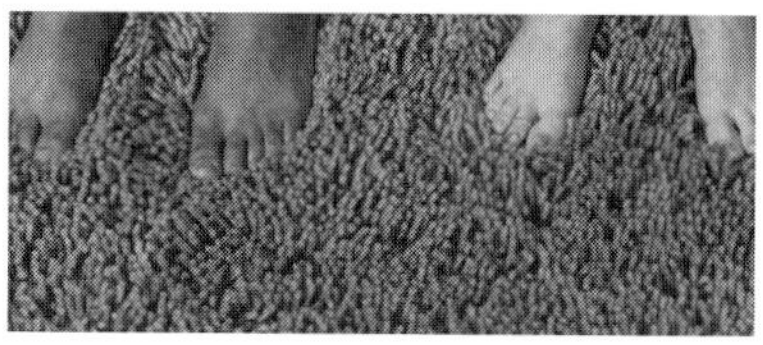

Width: $x - 9$

Length: x

42. David plans to fertilize his lawn. The lawn requires 20,000 square feet of fertilizer. The lawn is twice as long as it is wide. What are the dimensions of his lawn?

Width: x

Length: $2x$

43. The first Mars colonists will experience limited space for their living accommodations. Modular housing units will be constructed so that the length of each room is 5 feet more than the width. The floor area of each room is to be 500 square feet. Determine the dimensions of each room.

44. The position of a dropped object from an initial height of s_0 meters above the surface of Mars is given by the function $h(t) = -1.9t^2 + s_0$, where t is in seconds. Mars colonists Mack and Julie are repairing a sensor on a tower 47.5 meters tall. Julie climbs the tower to perform the repair and Mack assists her at the foot of tower. Julie accidentally drops a wrench from the top of the tower. How long does Mack have to get out of the way before the wrench hits the surface of Mars? (*Hint:* Let $s_0 = 47.5$ and $h(t) = 0$.)

45. Jamal wants to build a ramp for his remote control 4×4 truck. His design calls for the base to be 1 foot longer than the height. The length of the ramp is to be 5 feet.

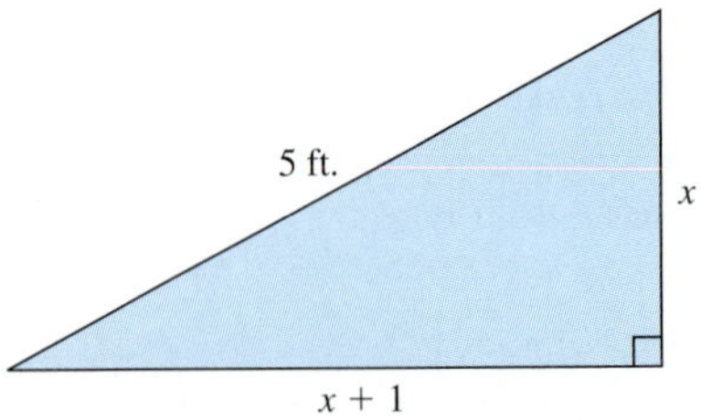

Apply the Pythagorean Theorem (see Section 2.5) to determine the length of the base and the height of Jamal's ramp.

46. Hank wants to build a different ramp for his remote control truck. He designed the base of the ramp to be 7 feet longer than its height. The length of the ramp is to be 13 feet.

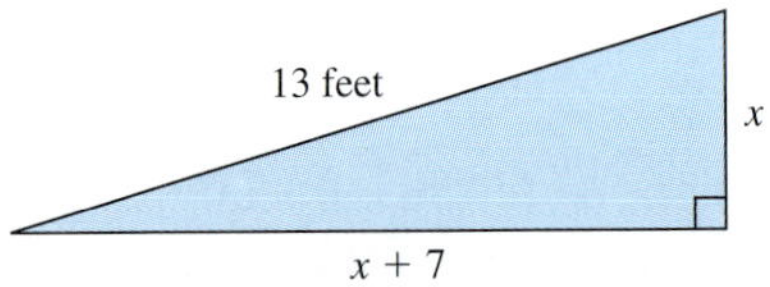

Apply the Pythagorean Theorem (see Section 2.5) to determine the length of the base and the height of Hank's ramp.

47. The graph of a quadratic function $f(x) = ax^2 + bx + c$, where a, b, and c are real numbers and $a \neq 0$, is a vertical parabola. If the parabola graph has two x-intercepts, then the x-coordinate of its vertex (high point or low point) can be found by computing the average of the x-intercepts.

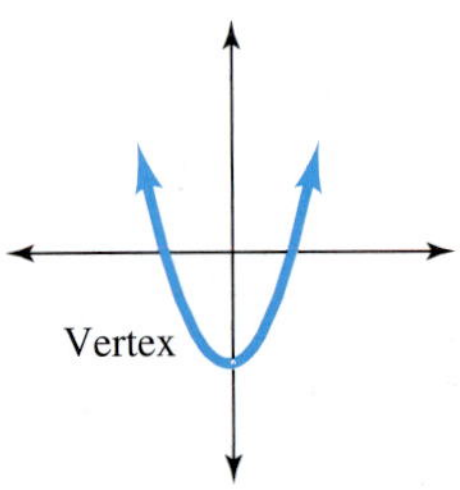

(a) Given $f(x) = x^2 + 5x + 6$, find the x-coordinate of the vertex. (*Hint:* Let $f(x) = 0$, solve the equation for x and compute the average of the solutions.)

(b) Find the y-coordinate of the vertex by evaluating the function at the calculated x-coordinate of the vertex. Write the coordinates of the vertex.

6.5 Multistep Factoring Strategy

OBJECTIVES

- Combine different factoring techniques to completely factor an algebraic expression, including expressions involving two variables.
- Factor the sum and difference of cubes.
- Solve equations using factoring.
- Reduce rational expressions using factoring.

1 2 3 Getting Started

In this section, the four factoring strategies discussed earlier will be used as part of a Multistep Factoring Strategy for polynomials with more than two factors, for the sum and difference of cubes, and for reducing fractions and solving equations by factoring.

1. Factors can be multiplied in any order applying the Commutative and Associative Properties. Find each product by first multiplying the binomials and then distributing the monomial factor.

EXAMPLE: $2\underbrace{(x + 5)(x - 4)}$ First, multiply binomials.

$= 2(x^2 + x - 20)$ Second, distribute monomial.

$= 2x^2 + 2x - 40$

(a) $3(2x - 5)(x + 4)$ (b) $y(y + 3)^2$

Multiplying the polynomials above involved more than one step. Reversing multiplication and factoring some polynomials employs a multistep process. The first step is to identify and factor the greatest common factor (GCF) from each term. Then factor the remaining polynomial again, if possible.

2. Factor the greatest common factor from each polynomial and then factor the remaining trinomial. Check the factors by multiplying the factors.

EXAMPLE: Factor $2x^3 + 2x^2 - 40x$.

$= 2x(x^2 + x - 20)$ Factor the GCF $2x$.

$= 2x(x + 5)(x - 4)$ Then factor the remaining trinomial.

Multiply to check:

$2x(x + 5)(x - 4)$

$= 2x(x^2 + x - 20)$

$= 2x^3 + 2x^2 - 40x$

(a) $12x^3 + 34x^2 + 10x$ (b) $5y^3 - 20y^2 + 20y$

3. Raise each integer to the third power. Remembering these cubes will help us to factor polynomials that are the sum or difference of cubes.

	Integer	Raise the integer to the third power	Cube
(a)	1	$(1)^3 = 1 \cdot 1 \cdot 1$	1
(b)	2	$(2)^3 = 2 \cdot 2 \cdot 2$	8
(c)	3		
(d)	4		
(e)	5		

4. Rewrite each expression as the sum or difference of perfect cubes.

EXAMPLE: $x^3 + 64 = (x)^3 + (4)^3$ is the sum of perfect cubes.

(a) $x^3 + 8$

(b) $x^3 + 27$

(c) $x^3 - 125$

(d) $x^3 - 216$

5. Use the Distributive Property to multiply each binomial and trinomial. Combine LIKE terms. Look for patterns in the products.

(a) $(x + 2)(x^2 - 2x + 4)$

(b) $(x - 5)(x^2 + 5x + 25)$

(c) Describe any pattern you observe in the products above.

➢ *Factoring the sum or difference of cubes involves several steps. The process is presented in Discussing Mathematics.*

1 2 3 Discussing Mathematics

MULTISTEP FACTORING STRATEGY

It may be necessary to use one or more factoring methods to completely factor some polynomials. It is most efficient to look for common monomial factors (GCF) as a first step. After the greatest common factor is factored, the remaining expression may factor again. All the factoring strategies discussed in this chapter are combined to form the **Multistep Factoring Strategy**.

MULTISTEP FACTORING STRATEGY

1. When common monomial factors appear, divide the greatest common factor (GCF) from each term. Write the GCF in front of parentheses and enclose the remaining terms inside the parentheses.
2. Determine whether the expression inside the parentheses factors again:
   (a) If the parentheses contain a binomial factor, determine if the factor is the difference of two squares or the sum or difference of two cubes and factor appropriately.
   (b) If the parentheses contain a trinomial, determine if the factor is a perfect square trinomial and apply the appropriate formula.
   (c) If neither (a) nor (b) applies and the factor is a trinomial, try to factor by inspection or by grouping.
   (d) If the remaining factor is a four-term polynomial, try to factor by grouping.
   (e) If none of (a)–(d) apply, the polynomial inside the parentheses is prime.
3. Be sure each factor is a prime factor. If not, factor again.
4. Check the factors using the Distributive Property. The resulting product should be the original expression.

EXAMPLE 1: Factor $3x^2 - 9x - 30$ using the Multistep Factoring Strategy.

Identify the GCF = 3. Divide the GCF from each term.

$3(x^2 - 3x - 10)$

Determine whether the trinomial $(x^2 - 3x - 10)$ factors again.

(a) $(x^2 - 3x - 10)$ is not the difference of two squares or the sum or difference of two cubes.

(b) $(x^2 - 3x - 10)$ is not a perfect square trinomial.

(c) $(x^2 - 3x - 10)$ is a trinomial that factors by inspection or by making a table of values.

Find a pair of factors whose sum is -3.

$c = -10$ Factors of -10		Sum of factors
+1	−10	−9
−1	+10	+9
+2	−5	−3

Write the binomial factors, along with the GCF. Be sure each is a prime factor. The factors 3, $(x + 2)$, and $(x - 5)$ are prime.

Factored form.

$3(x + 2)(x - 5)$

Check the factors using the Distributive Property. Multiply binomial factors first.

$$3(x + 2)(x - 5) = 3(x^2 - 3x - 10)$$
$$= 3x^2 - 9x - 30$$ ✓ It checks.

EXAMPLE 2: Use the Multistep Factoring Strategy to factor $2x^4 - 2x^2y^2$.

Divide the GCF $= 2x^2$ from each term.	$2x^4 - 2x^2y^2 = 2x^2(x^2 - y^2)$
Factor the difference of squares. Write the factors along with the GCF.	$= 2x^2(x - y)(x + y)$
Check the factors using the Distributive Property. Multiply binomial factors first.	$2x^2(x - y)(x + y) = 2x^2(x^2 - y^2)$ $= 2x^4 - 2x^2y^2$ ✓ It checks.

FACTORING THE SUM OR DIFFERENCE OF CUBES

The factors of the sum (or difference) of perfect cubes are always a binomial factor and a trinomial factor with the following patterns:

$$x^3 + 8 = (x)^3 + (2)^3 = (x + 2)(x^2 - 2x + 4)$$

Patterns in the binomial factor			**Patterns in the trinomial factor**		
$(x)^3 + (2)^3$ → $(x + 2)(x^2 - 2x + 4)$			$(x)^3 + (2)^3$ → $(x + 2)(x^2 - 2x + 4)$		
The cube root of x^3 is x.	The arithmetic operation in the binomial factor is the same as in the original expression.	The cube root of 8 is 2.	The first term in the trinomial factor is $(x)^2$.	The middle term of the trinomial factor is $-(x \cdot 2) = -2x$	The last term in the trinomial factor is $2^2 = 4$.

These observations can be generalized into the formulas for factoring the sum or difference of cubes. The two generalizations are similar, except for the location of the subtraction sign.

FACTORING THE SUM OR DIFFERENCE OF CUBES

For any real number or algebraic expression X and Y:

The Sum of Cubes: $X^3 + Y^3 = (X + Y)(X^2 - XY + Y^2)$

The Difference of Cubes: $X^3 - Y^3 = (X - Y)(X^2 + XY + Y^2)$

EXAMPLE 3: Factor $a^3 + 27$ using the formula for factoring the sum of cubes.

Rewrite each term as a base cubed.	$X^3 + Y^3 = (X + Y)(X^2 - X \cdot Y + Y^2)$
Substitute a for X and 3 for Y in the formula.	$(a)^3 + (3)^3 = (a + 3)(a^2 - (a \cdot 3) + 3^2)$ $= (a + 3)(a^2 - 3a + 9)$
Check the factors using the Distributive Property.	$(a + 3)(a^2 - 3a + 9)$ $= a(a^2 - 3a + 9) + 3(a^2 - 3a + 9)$ $= a^3 - 3a^2 + 9a + 3a^2 - 9a + 27$ $= a^3 + 27$ ✓ It checks.

EXAMPLE 4: Factor $27x^3 - 8y^3$ using the formula for factoring the difference of cubes.

Rewrite each term as a base cubed.

$$X^3 - Y^3 = (X - Y)(X^2 + X \cdot Y + Y^2)$$

Substitute $3x$ for X and $2y$ for Y in the formula.

$$(3x)^3 - (2y)^3 = (3x - 2y)((3x)^2 + (3x \cdot 2y) + (2y)^2)$$
$$= (3x - 2y)(9x^2 + 6xy + 4y^2)$$

Check the factors using the Distributive Property.

$$(3x - 2y)(9x^2 + 6xy + 4y^2)$$
$$= 3x(9x^2 + 6xy + 4y^2) - 2y(9x^2 + 6xy + 4y^2)$$
$$= 27x^3 + 18x^2y + 12xy^2 - 18x^2y - 12xy^2 - 8y^3$$
$$= 27x^3 - 8y^3 \quad \checkmark \text{ It checks.}$$

USING FACTORING TO REDUCE FRACTIONS

Use the Multistep Factoring Strategy, when necessary, to factor the numerator and denominator and reduce fractions.

EXAMPLE 5: Reduce to lowest terms $\dfrac{5x^3 + 40}{10x^2 - 50x - 140}$.

Factor the GCF = 5 from the numerator and the GCF = 10 in the denominator.

$$\frac{5x^3 + 40}{10x^2 - 50x - 140} = \frac{5(x^3 + 8)}{10(x^2 - 5x - 14)}$$

Factor $(x^3 + 8) = ((x)^3 + (2)^3)$.

Factor $(x^2 - 5x - 14)$.

$$= \frac{5(x + 2)(x^2 - 2x + 4)}{10(x + 2)(x - 7)}$$

Divide common factors of 2 and $(x + 2)$ from the numerator and denominator.

$$= \frac{\overset{1}{\cancel{5}}\overset{1}{\cancel{(x + 2)}}(x^2 - 2x + 4)}{\underset{2}{\cancel{10}}\underset{1}{\cancel{(x + 2)}}(x - 7)}$$

Rewrite the fraction listing the remaining factors. Multiply factors.

$$= \frac{(x^2 - 2x + 4)}{2(x - 7)}$$
$$= \frac{x^2 - 2x + 4}{2x - 14}$$

Expressions such as $b - a$ and $a - b$ are called **additive inverses** because their sum is zero. When reducing rational expressions involving additive inverses, it may be useful to rewrite one of the expressions in an equivalent form. Note that the expression $b - a$ can be rewritten as follows:

$$b - a = (-1)(-b + a) \quad \text{Factor } -1 \text{ from each term.}$$
$$= (-1)(a - b)$$

$b - a$ is equivalent to $(-1)(a - b)$. This equivalence is used to reduce fractions in the next examples.

EXAMPLE 6: Reduce to lowest terms $\frac{b - a}{a - b}$.

Note: The numerator and denominator are additive inverses.
$$\frac{b-a}{a-b} = \frac{(-1)(-b+a)}{a-b}$$

Factor (-1) from each term in the numerator.
$$= \frac{(-1)(a-b)}{a-b}$$

Rearrange terms inside the parentheses in the numerator.

Enclose denominator in parentheses.
$$= \frac{(-1)(a-b)}{(a-b)}$$

Divide the common factor $(a - b)$ from the numerator and denominator.
$$= \frac{(-1)\cancelto{1}{(a-b)}}{\cancelto{1}{(a-b)}}$$

$$= \frac{-1}{1} = -1$$

EXAMPLE 7: Reduce to lowest terms $\frac{50 - 2x^2}{3x^2 - 17x + 10}$.

Factor the GCF = 2 from the numerator.
$$\frac{50 - 2x^2}{3x^2 - 17x + 10} = \frac{2(25 - x^2)}{3x^2 - 17x + 10}$$

Factor the difference of squares in the numerator. Factor the trinomial in the denominator.
$$= \frac{2(5-x)(5+x)}{(3x-2)(x-5)}$$

Factor (-1) from $(5 - x)$: $(5 - x) = (-1)(-5 + x) = (-1)(x - 5)$.
$$= \frac{2\cdot(-1)(x-5)(5+x)}{(3x-2)(x-5)}$$

$$= \frac{-2(x-5)(5+x)}{(3x-2)(x-5)}$$

Divide the common factor $(x - 5)$ from the numerator and denominator.
$$= \frac{-2\cancelto{1}{(x-5)}(5+x)}{(3x-2)\cancelto{1}{(x-5)}}$$

Rewrite the fraction listing the remaining factors. Multiply factors.
$$= \frac{-2(5+x)}{(3x-2)} = \frac{-10-2x}{3x-2}$$

$$= \frac{-2x-10}{3x-2}$$

EXTENDING THE ZERO-PRODUCT PROPERTY TO SOLVE EQUATIONS

Multistep Factoring and the Zero-Product Property are used to find solutions of polynomial equations.

EXAMPLE 8: Solve the equation using the Zero-Product Property $3x^3 = 12x$.

Rewrite the equation with all terms of the algebraic expression on one side of the equation.

$$3x^3 - 12x = 12x - 12x$$
$$3x^3 - 12x = 0$$

Factor the nonzero side of the equation.

$$3x^3 - 12x = 0$$
$$3x(x^2 - 4) = 0$$
$$3x(x + 2)(x - 2) = 0$$

Use the Zero-Product Property and set each factor equal to zero.

$$3x = 0 \quad \text{or} \quad x + 2 = 0 \quad \text{or} \quad x - 2 = 0$$

Solve each new equation separately.

$$x = 0 \quad \text{or} \quad x = -2 \quad \text{or} \quad x = 2$$

Check the answers by substituting the answers into the original equation $3x^3 = 12x$ to make sure that each yields a true statement.

Check $x = 0$:

$$3(0)^3 \stackrel{?}{=} 12(0)$$
$$0 = 0$$

Check $x = -2$:

$$3(-2)^3 \stackrel{?}{=} 12(-2)$$
$$3(-8) \stackrel{?}{=} 12(-2)$$
$$-24 = -24$$

Check $x = 2$:

$$3(2)^3 \stackrel{?}{=} 12(2)$$
$$3(8) \stackrel{?}{=} 12(2)$$
$$24 = 24$$

✓ All three solutions check.

CONFIRMING THE SOLUTIONS OF AN EQUATION GRAPHICALLY

As discussed earlier, the zeros of an equation are the x-coordinates of the points where the graph of the related equation crosses the x-axis. These points are the x-intercepts of the graph, where $y = 0$.

EXAMPLE 9: Use the X-Intercept Graphical Method to confirm the analytical solutions $x = 0$, $x = -2$, and $x = 2$ of $3x^3 = 12x$.

$$3x^3 - 12x = 0$$

With all nonzero terms on one side and zero on the other side, substitute 0 for y to form the associated function.

$$3x^3 - 12x = y$$

Graph $y = 3x^3 - 12x$. The zeros or x-intercepts of the graph should appear when $y = 0$ at the analytical solutions $x = 0$, $x = -2$, and $x = 2$.

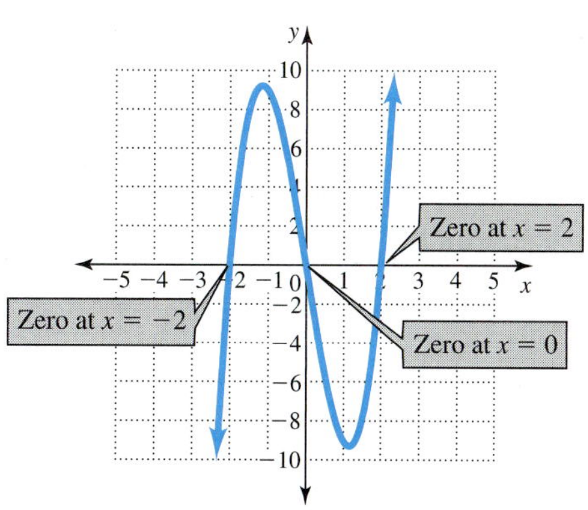

x	Y_1
−3	−45
−2	0
−1	5
0	0
1	−5
2	0
3	45

$x = -2$

The x-intercepts and zeros of the function are $x = 0$, $x = -2$, and $x = 2$, confirming the analytical solutions.

1 2 3

Doing Mathematics

- Write the definitions of the words and symbols listed for this section in the Dictionary in Appendix A.
- On a 3×5-inch card, write the processes discussed in this section for future reference:
  (a) Factoring the difference of cubes
  (b) Factoring the sum of cubes
  (c) Outline the Multistep Factoring Strategy

Problems 1–2: Identify each expression as one of the following:

Type 1: The difference of squares.

Type 2: The sum of cubes.

Type 3: The difference of cubes.

Type 4: A perfect square trinomial.

Type 5: None of the above.

1. (a) $x^3 + 1$ (b) $4x^2 - y^2$ (c) $x^3 + 8$ (d) $27x^3 - 8y^3$
(e) $x^2 + 12x + 9$ (f) $25b^2 - 16c^2$ (g) $y^2 + 2xy + x^2$ (h) $y^2 + x^2$

2. (a) $y^2 - x^2$ (b) $x^3 + 27$ (c) $9x^2 - 4$ (d) $x^3 - 1$
(e) $y^2 - 4y + 4$ (f) $a^3 - b^3$ (g) $x^2 - 12x + 36$ (h) $2x^2 + 7x + 1$

Problems 3–22: Factor the sum or difference of cubes.

3. $x^3 + 216$

4. $y^3 + 27$

5. $x^3 + 1$

6. $x^3 + 64$

7. $x^3 + y^3$

8. $27x^3 + 8y^3$

9. $8y^3 + 125$

10. $64x^3 + 27$

11. $8a^3 + 27b^3$

12. $a^3 - 27b^3$

13. $y^3 - 27$

14. $x^3 - 216$

15. $x^3 - 64$

16. $x^3 - 1$

17. $27x^3 - 8y^3$

18. $x^3 - y^3$

19. $64y^3 - z^3$

20. $125x^3 - 27$

21. $8(x + 1)^3 + (y - 2)^3$
(This problem is challenging.)

22. $8(x - 5)^3 - (y + 1)^3$
(This problem is challenging.)

Problems 23–38: Factor using the Multistep Factoring Strategy.

23. $3x^2y - 3xy$

24. $5x^3 + 10x^2 + 20x$

25. $5x^2 - 500y^2$

26. $12x^2 - 75z^2$

27. $-2x^3 + 20x^2y - 50xy^2$

28. $7x^3 - 9x^2y - 10xy^2$

29. $x^2y - 11xy - 60y$

30. $2x^3 - 50x$

31. $-3x^3 - 6x^2 - 3x$

32. $-5x^3 - 20x^2 - 20x$

33. $-6x^3 - 13x^2 + 5x$

34. $16x^3 + 48x^2y + 36xy^2$

35. $4x^3 + 108$

36. $-54y^4 + 16y$

37. $2x^3 - 2xy^2 + 2x^2 - 2y^2 =$
(Hint: This problem is challenging. Factor the GCF first and then factor by grouping.)

38. $-3x^4 - 3xy^3 + 3x^3y + 3y^4$
(Hint: This problem is challenging. Factor the GCF first and then factor by grouping.)

Problems 39–50: Reduce each rational expression to lowest terms by factoring.

39. $\dfrac{x^3 - 27}{2x^2 + 6x + 18}$

40. $\dfrac{x^3 + 64}{3x^2 - 12x + 48}$

41. $\dfrac{24x^2 - 6y^2}{18x^2 + 3xy - 3y^2}$

42. $\dfrac{18x^2 - 2y^2}{12x^2 + 8xy - 4y^2}$

43. $\dfrac{5x^3 - 15x^2 + 45x}{5x^4y + 135xy}$

44. $\dfrac{x^3 + 4x^2 + 4x}{x^4 + 8x}$

45. $\dfrac{-24x^3 - 3}{36x^2 - 9}$

46. $\dfrac{-6x^2 - 6x + 36}{3x^2 - 27}$

47. $\dfrac{x^2 - y^2}{y - x}$

48. $\dfrac{b - a}{a^2 - b^2}$

49. $\dfrac{8 - 2x^2}{x^2 + x - 6}$

50. $\dfrac{27 - 3x^2}{2x^2 - 5x - 3}$

Problems 51–62: Solve each equation using the Zero-Product Property.

51. $24x^2 = 3x^3$

52. $4x^3 = 500x^2$

53. $2x^3 - 8x + x^2 = 4$

54. $x^3 - 9x + x^2 = 9$

55. $75x^3 - 60x^2 + 12x = 0$

56. $2x^3 + 50x = 20x^2$

57. $30x^3 - 85x^2 = -60x$

58. $5x^3 = 14x^2 + 3x$

59. $x^2(3x + 20) = -25x$

60. $2x^2(4x + 1) = 3x$

61. $5x^2 = 2x^3 + 3x$

62. $12x^2 + 12x^3 + 3x = 0$

63. Graphing Calculator Exercises

(a) Use a graphing calculator to confirm the equivalence of the expressions and factors obtained in problems 3–6, 13–16, 30–33, and 35–36.

(b) Use a graphing calculator to confirm your solutions obtained analytically in problems 51–62.

Chapter 6 Summary

SECTION 6.1:

Factoring Common Monomial Factors from a Polynomial

1. Identify all individual common monomial factors in each term. Draw boxes around common factors. The product of the individual common factors is called the Greatest Common Factor (GCF).
2. Write the GCF in front of parentheses.
3. Divide the GCF from each term, reversing the Distributive Property.
4. Confirm the factorization, multiplying the factors using the Distributive Property.

Factoring by Grouping

1. Identify common monomial factors (GCF) in groups of two terms. Rewrite the polynomial, grouping terms with common monomial factors together.
2. Write each GCF in front of a parenthesis and factor the GCF from each group.
3. Identify any common binomial factors from each group. Draw a box around the common binomial in each group. Write the common binomial factor in front of another set of parentheses.
4. Divide the common binomial factor from each group, reversing the Distributive Property.
5. Confirm the factorization, multiplying the factors using the FOIL method.

SECTION 6.2:

Factoring Using a Table of Values (inspection) $ax^2 + bx + c = 0$, $(a = 1)$

1. Identify the values of a, b, and c. Write two sets of parentheses. Write the variable factors as the first terms of each binomial.
2. Make a list of pairs of integer factors of c and their sums.
3. Find a pair of factors whose sum is b.
4. Write the chosen factors as the last terms of each binomial.
5. Check the factors using the FOIL method.

Factoring the Difference of Squares

For any real number or algebraic expression X and Y: $X^2 - Y^2 = (X + Y)(X - Y)$

Factoring a Perfect Square Trinomial

For any real number or algebraic expression X and Y:

$$X^2 + 2XY + Y^2 = (X + Y)(X + Y) = (X + Y)^2$$
$$X^2 - 2XY + Y^2 = (X - Y)(X - Y) = (X - Y)^2$$

SECTION 6.3:

Factoring by Grouping $ax^2 + bx + c = 0, (a \neq 1)$

1. Identify the values of a, b, and c. Multiply $a \cdot c$. Label the product p.
2. Make a list of pairs of integer factors of p and their sums.
3. Find a pair of factors whose sum is equal to b.
4. Rewrite the original trinomial with the middle term expressed as two terms whose coefficients are the factors in Step 3.
5. Factor the four-term polynomial by grouping.
6. Check the factors using the FOIL method.

SECTION 6.4: Use factoring and the Zero-Product Property to solve equations of the form $ax^2 + bx + c = 0$.

Zero-Product Property

Given a polynomial equation with real number or algebraic expression factors a and b, where $a \cdot b = 0$, then either $a = 0$ and/or $b = 0$.

SECTION 6.5:

Factoring the Sum or Difference of Cubes

For any real number or algebraic expression X and Y:

The Sum of Cubes: $X^3 + Y^3 = (X + Y)(X^2 - XY + Y^2)$

The Difference of Cubes: $X^3 - Y^3 = (X - Y)(X^2 + XY + Y^2)$

Multistep Factoring Strategy

1. When common monomial factors appear, divide the greatest common factor (GCF) from each term. Write the GCF in front of parentheses and enclose the remaining terms inside the parentheses.
2. Determine whether the expression inside the parentheses factors again:
   (a) If the parentheses contain a binomial factor, determine if the factor is the difference of squares or the sum or difference of cubes. If so, apply the appropriate formula: $X^2 - Y^2 = (X + Y)(X - Y)$ or $X^3 + Y^3 = (X + Y)(X^2 - XY + Y^2)$ or $X^3 - Y^3 = (X - Y)(X^2 + XY + Y^2)$.
   (b) If the parentheses contain a trinomial, determine if the factor is a perfect trinomial square. If so, apply the appropriate formula:
   $$X^2 + 2XY + Y^2 = (X + Y)^2 \text{ or } X^2 - 2XY + Y^2 = (X - Y)^2.$$
   (c) If neither (a) nor (b) applies and the factor is a trinomial, try to factor by inspection or by grouping.
   (d) If the remaining factor is a four-term polynomial, try to factor by grouping.
   (e) If none of (a)–(d) apply, the polynomial inside the parentheses is prime.
3. Be sure each factor is a prime factor. If not, factor again.
4. Check the factors using the Distributive Property.

Chapter 6 Review

- Review the definitions you have written for this chapter in the Dictionary.
- Review any properties and/or processes you have written on 3 × 5-inch cards.

SECTION 6.1

1. Write an example of a polynomial with the given characteristics:

(a) A polynomial with common monomial factor.

(b) A polynomial with common binomial factor.

(c) A trinomial of the form $ax^2 + bx + c$, where $a = 1$.

(d) A trinomial of the form $ax^2 + bx + c$, where $a \neq 1$.

(e) A perfect square trinomial.

(f) The difference of two squares.

(g) The sum of two cubes.

(h) The difference of two cubes.

Problems 2–66: Factor completely.

Section 6.1: Factor the GCF.

2. $2x^2y - 4xy$

3. $-5y^2 - 30xy$

4. $-3y^3 - 9y^2 + 3y$

5. $-24x^7 - 48x^6$

6. $6x^3y - 12y^3$

7. $-2x^4y + 4x^3y^2 - 8x^2y^3 + 20xy^4$

Section 6.1: Factor by grouping.

8. $2a + 2b + a^2 + ab$

9. $xy - 2x + 3y - 6$

10. $5x - 5 - xy + y$

11. $a^2 - 7a + ab - 7b$

12. $12x^2 + 15x - 8x - 10$

13. $3x^2 - 15x - 4x + 20$

Section 6.2: Factor each trinomial of the form $ax^2 + bx + c$, $(a = 1)$.

14. $x^2 - x - 12$

15. $x^2 - 3x + 2$

16. $x^2 + 12x + 20$

17. $x^2 - 11x + 30$

18. $x^2 - 2x - 8$

19. $x^2 + 9x + 18$

Section 6.2: Factor the difference of squares.

20. $x^2 - 49$
21. $x^2 - 100$
22. $9x^2 - 49$
23. $16x^2 - 81y^2$
24. $x^2 - y^2$
25. $4x^2 - 25y^2$

Section 6.2: Factor each perfect square trinomial.

26. $y^2 - 14y + 49$
27. $x^2 + 10x + 25$
28. $x^2 - 8xy + 16y^2$
29. $25x^2 - 30x + 9$
30. $4x^2 + 20x + 25$
31. $9x^2 + 6xy + y^2$

Section 6.3: Factor each trinomial of the form $ax^2 + bx + c$, $(a \neq 1)$.

32. $2x^2 + 3x - 5$
33. $4x^2 - 19x - 5$
34. $6x^2 + 5x - 4$
35. $2y^2 - 15y + 25$
36. $6y^2 + 7y - 5$
37. $15x^2 + 11x + 2$

Section 6.4: Solve each equation using the Zero-Product Property.

38. $x^2 + 2x + 1 = 0$
39. $x^2 + 7x + 12 = 0$
40. $x^2 - 1 = 0$
41. $x^2 - 4 = 0$
42. $4x^2 = 4x + 3$
43. $3x^2 + 14x = 5$

Section 6.5: Factor the sum of cubes.

44. $x^3 + 8$
45. $8x^3 + 27y^3$
46. $1 + x^3$
47. $a^3 + b^3$
48. $216x^3 + 125$
49. $125x^3 + 64y^3$

Section 6.5: Factor the difference of cubes.

50. $a^3 - b^3$
51. $x^3 - 1$
52. $8 - 125x^3$
53. $27a^3 - b^3$
54. $216x^3 - 125y^3$
55. $27x^3 - y^3$

Section 6.5: Factor using the Multistep Factoring Strategy.

56. $2x^3 - 128y^3$

57. $2x^2 - 8y^2$

58. $12a^2 + 36ab + 27b^2$

59. $-8x^3 - 12x^2y + 20xy^2$

60. $-90x^4y^2 - 120x^3y^2 - 40x^2y^2$

61. $-3xy^4 - 3xy$

62. $12x^2y - 27y$

63. $-45x^5 - 120x^4 - 80x^3$

64. $4x^2y^3 - 32x^2 - 9y^3 + 72$

65. $-2x^4 + 16x + 2x^3y - 16y$

66. $-3x^4 - 3xy^3 + 3x^3y + 3y^4$

Problems 67–74: Reduce each rational expression to lowest terms using factoring.

67. $\dfrac{5x^4 - 135x}{40x^3 + 120x^2 + 360x}$

68. $\dfrac{2x^2 - 18}{20x^2 - 70x + 30}$

69. $\dfrac{18x^2y + 3xy^2 - 3y^3}{36x^3 - 9xy^2}$

70. $\dfrac{20x^2 - 90xy + 40y^2}{80x^3 - 10y^3}$

71. $\dfrac{8a^2 + 4ab - 12b^2}{8a^3 + 27b^3}$

72. $\dfrac{x^2y + 10xy + 21y}{2x^2y - 98y}$

73. $\dfrac{6x^2 + 12x - 90}{18 - 2x^2}$

74. $\dfrac{5a^2 - 20b^2}{40b^3 - 5a^3}$

Problems 75–82: Solve each equation using the Zero-Product Property and multistep factoring.

75. $5x^2 = 135x$

76. $48x^3 + 3x = 24x^2$

77. $3x^3 + 10x^2 = 25x$

78. $y^3 + 49y = 14y^2$

79. $(x + 5)^2 - 4 = 0$

80. $2x(x - 5) = 15 - 3x$

81. $2x^3 = 32x$

82. $3y^3 - 12y^2 = 18y^2 - 72y$

83. Use a graphing calculator to confirm solutions obtained analytically in problems 75–82.

84. The width of a rectangular family room is 6 feet shorter than the length. The area of the family room is 280 square feet. Write an equation for the area of the family room. Find the length and width of the family room.

85. The length of a rectangular pasture is 5 feet longer than the width. The area of the pasture is 300 square feet. Write an equation for the area of the pasture. Find the length and width of the pasture.

86. A square swimming pool has a 2-foot sidewalk around all four sides. Let x represent the length and width of the pool in feet. The area of the pool and the sidewalk is 256 square feet. Find the dimensions of the pool.

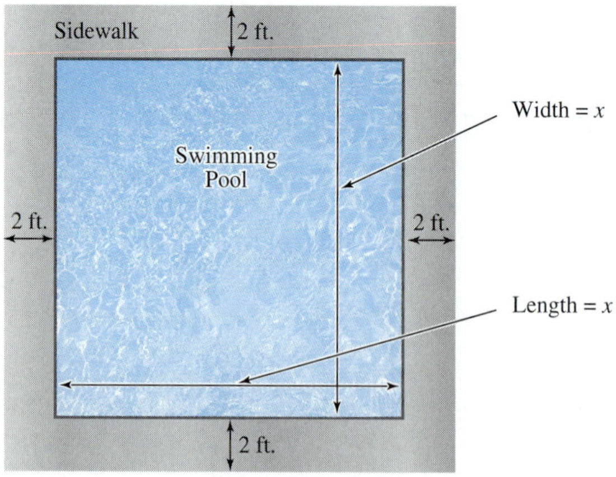

(a) Write a polynomial equation for the total area of the pool and sidewalk (large square).

(b) What is the length and width of the swimming pool?

(c) What is the area (in square feet) of the swimming pool?

(d) What is the area (in square feet) of the sidewalk?

Practice Test—Chapter 6

SECTIONS 6.1–6.5

1. Write an example of a polynomial with the given characteristics:

(a) A polynomial with a common monomial factor:

(b) A polynomial with a common binomial factor:

(c) A trinomial of the form $ax^2 + bx + c$, where $a = 1$:

(d) A trinomial of the form $ax^2 + bx + c$, where $a \neq 1$:

(e) A perfect trinomial square:

(f) The difference of squares:

(g) The sum of cubes:

(h) The difference of cubes:

(i) The sum of squares:

(j) A trinomial that is prime:

Problems 2–15: Factor completely.

2. $-2x^3y + 50xy$

3. $6x^3 + 54x^2 - 60x$

4. $-36a^2b^2c^2 - 54a^3b^3c^3$

5. $xz - yz + 3x - 3y$

6. $a^2 + 5a - ab - 5b$

7. $x^2 - 6x - 27$

8. $y^2 - 10y + 25$

9. $3x^2 - 20x + 12$

10. $4x^2 - 9y^2$

11. $6x^2 + 5xy - 4y^2$

12. $5x^3 - 80x$

13. $x^2(y + 7) - 4(y + 7)$

14. $-9a^2 - 3ab + 30b^2$

15. $x^3 - 27$

Problems 16–19: Reduce each rational expression to lowest terms using factoring.

16. $\dfrac{x^2 - 9}{x^2 + 2x - 15}$

17. $\dfrac{x^2y - 2xy - 3y}{2x^2y + 6xy + 4y}$

18. $\dfrac{2a^2 + 10a + 8}{4a^2 - 8a - 12}$

19. $\dfrac{4x^2 + 8x - 5}{6x^2 + 25x + 25}$

Problems 20–23: Use the Zero-Product Property to solve each equation.

20. $x^3 - 81x = 0$

21. $4x^2 + x = 3$

22. $4y^2 = 3y + 7$

23. $5y = 3y^2 + 2y^3$

24. A rectangular swimming pool has a 2-foot sidewalk around all four sides. The width of the swimming pool is 5 feet less than the length. Let x represent the length of the swimming pool in feet. The area of the swimming pool is 150 square feet. Find the length and width of the outer edge of the sidewalk.

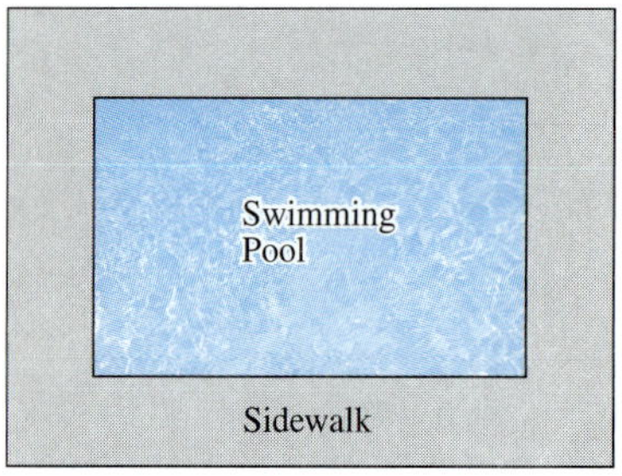

Answers to Getting Started—Chapter 6

SECTION 6.1

1. (a) $5x(x + 2)$ (b) $4y^2(y + 5)$

2. (a) $(x + 4)(2 + y)$ (b) $(y - 1)(x + 5)$

SECTION 6.2

1. (a) $x^2 + 8x + 15$ (b) $x^2 + 10x + 24$

(c) The first term of the trinomial is the product of the first terms in the binomials. The middle term of the trinomial is the sum of the "O" and "I" steps of the FOIL process. The last term of the trinomial (the constant) is the product of the last terms in the binomials.

(d) The integers in the binomials are numbers whose sum is the middle term of the trinomial and whose product is the last term of the trinomial.

2. (a) $(x + 3)(x + 4)$ (b) $(x + 7)(x + 5)$

3. (a) $x^2 - 49$ (b) $x^2 - 25$

(c) and (d) The first and last terms in the binomial factors equal the values of the square root of the first and last terms of the original algebraic expression. One binomial factor will have the operation of addition, the other subtraction.

4. (a) $(x + 6)(x - 6)$ (b) $(x + 9)(x - 9)$

5. (a) $x^2 + 12x + 36$ (b) $x^2 - 10x + 25$

(c) and (d) The constant term of each trinomial is a perfect square. The coefficient of the middle term of the trinomial is equal to two times the square root of the constant. The operation in the binomial factors is the same as the middle term of the trinomial.

6. (a) $(x + 4)^2$ (b) $(x - 8)^2$

SECTION 6.3

1. (a) $2x^2 + 5x + 3$ (b) $5x^2 + 16x + 3$

(c) $3x^2 - 19x + 20$ (d) $6x^2 + 7x - 20$

(e) The first term of the trinomial is the product of the first terms in the binomials. The middle term of the trinomial is the sum of the "O" and "I" steps of the FOIL process. The last term of the trinomial (the constant) is the product of the last terms in the binomials.

(f) Write factors of the first term of the trinomial in parentheses as the first terms of the binomials. Write factors of the constant (third term of the trinomial) in parentheses as the last terms of the binomials. Try different combinations of numbers and algebraic signs to produce a product equivalent to the original trinomial.

2. (a) $(2x + 3)(3x - 1)$ (b) $(3x - 1)(x - 2)$

(c) $(2x - 3)(x + 4)$ (d) $(3x + 2)(x + 4)$

3. (a) $(2x + 1)^2$ (b) $(3x - 2)^2$

(c) $(5x + 7)^2$ (d) $(2x - 3)^2$

4. (a) $(5x - 7)(5x + 7)$ (b) $(2x - 3)(2x + 3)$

(c) $(6x - 4y)(6x + 4y)$ (d) $(8x - y)(8x + y)$

SECTION 6.4

1. (a) Length: 12 feet; Width: 8 feet

(b) $(x + 4)$ (c) $96 = (x + 4)(x)$

(d) Multiply to remove parentheses. Subtract 96 from each side of equation. Factor the polynomial. Set each factor to zero and solve.

2. (a)
$$x^2 + 7x + 12 = 0$$
$$(x + 3)(x + 4) = 0$$
$$x + 3 = 0 \quad \text{or} \quad x + 4 = 0$$
$$x = -3 \quad \text{or} \quad x = -4$$

(b)
$$x^2 + 2x - 15 = 0$$
$$(x + 5)(x - 3) = 0$$
$$x + 5 = 0 \quad \text{or} \quad x - 3 = 0$$
$$x = -5 \quad \text{or} \quad x = 3$$

(c)
$$2x^2 + 11x = -5$$
$$2x^2 + 11x + 5 = -5 + 5$$
$$2x^2 + 11x + 5 = 0$$
$$(2x + 1)(x + 5) = 0$$
$$2x + 1 = 0 \quad \text{or} \quad x + 5 = 0$$
$$2x = -1$$
$$x = -\frac{1}{2} \quad \text{or} \quad x = -5$$

(d) The solutions of the individual equations are the values of the variable that cause each factor to equal zero. Each solution also yields a true statement in the original equation.

SECTION 6.5

1. (a) $6x^3 + 9x - 60$ (b) $y^3 + 6y^2 + 9y$

2. (a) $2x(6x^2 + 17x + 5) = 2x(3x + 1)(2x + 5)$

(b) $5y(y^2 - 4y + 4) = 5y(y - 2)^2$

3.

(c)	3	$(3)^3 = 3\cdot3\cdot3$	27
(d)	4	$(4)^3 = 4\cdot4\cdot4$	64
(e)	5	$(5)^3 = 5\cdot5\cdot5$	125

4. (a) $(x)^3 + (2)^3$ (b) $(x)^3 + (3)^3$

(c) $(x)^3 - (5)^3$ (d) $(x)^3 - (6)^3$

5. (a) $x^3 + 8$ (b) $x^3 - 125$

(c) When the product is written as the sum or difference of cubes, $x^3 + 8 = (x)^3 + (2)^3$, the x and 2 appear in the binomial factor. x^2 and $2^2 = 4$ are the first and third terms of the trinomial. The middle term of the trinomial is $2\cdot x$.

Cumulative Test Chapters 4–6

Problems 1–2: Solve each system graphically.

1. $3x - y = -6$
$x + 2y = 5$

2. $2x = 3y - 6$
$2x - 3y = 3$

3. Mixture problem: A chemist has two acid solutions, one containing 40% acid and a second solution containing 8% acid. She needs to mix them together to create a solution that contains 32% acid. How much solution containing 8% acid should she add to the 40% acid solution to obtain 3 liters of solution containing 32% acid?

(a) Let x represent the number of liters of the 40% solution. Let y represent the number of liters of the 8% solution. Write two equations to form a system describing the problem.

(b) Solve the system.

(c) How many liters of 40% acid solution should be used?

(d) How many liters of 8% acid solution should be used?

(e) Verify the solution by substituting into the original equations.

Problems 4–6: Solve each system using either substitution or elimination.

4. $4x - 3y = 17$
$3x - 5y = 21$

5. $3x = 3y + 12$
$x - y = 3$

6. $x - y = 0$
$5x = 5y$

7. Graph the solution set of the inequality $3y - 5x \geq 15$.

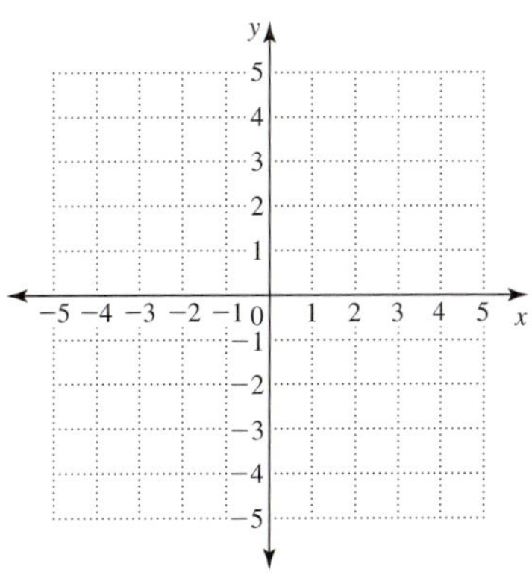

Problems 8–13: Simplify each expression using the Properties of Exponents. Assume all variables are non-negative quantities.

8. $(-5x^2y^2)^3$

9. $\dfrac{-5a^6b^3c^4}{30a^4bc^5}$

10. $\dfrac{(-10x^2y^3)^2}{10x^4}$

11. $(-x^2)^0$

12. $\left(\dfrac{-2x^3z}{4xyz}\right)^{-2}$

13. $(6x^2y^{-2}z)^{-4}$

Problems 14–20: Perform the operations.

14. $(-x^2)(-3x)^2(-4x^3)$

15. $-5x^2y(2x+1)^2$

16. $(2x-3)(5x-6)$

17. $-2x(x^2-x+5)(2x^2+x-7)$

18. $\dfrac{-4x^2y^3}{-32x^2y^4}$

19. $\dfrac{3ab-9a^2b^2+12ab^2}{6ab}$

20. $\dfrac{x^3-5}{x-1}$

21. Given the figure:

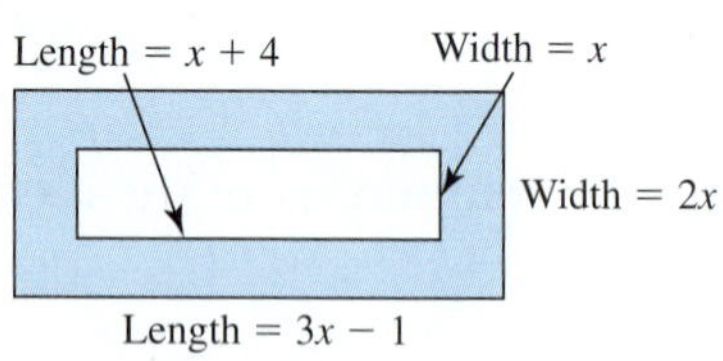

(a) Express the area of the shaded region as a polynomial.

(b) Find the area of the shaded region when $x = 6$ inches.

Problems 22–26: Factor completely.

22. $-50a^2b^3c^2 - 10a^3b^4c^5$

23. $a^2 - 7a - ab + 7b$

24. $3x^4 - 24x$

25. $2x^2 + 50$

26. $4a^2 + 4ab - 3b^2$

27. Reduce the rational expression to lowest terms using factoring.

$$\frac{5a^2 + 3a - 2}{4a^2 + 7a + 3}$$

Problems 28–31: Use the Zero-Product Property to solve each equation.

28. $x^2 = 144$

29. $x^3 = x^2 + 30x$

30. $4y^2 = 3 - 4y$

31. $3y^3 = 48y$

32. A rectangular swimming pool has a 2-foot sidewalk around all four sides. The width of the swimming pool is 5 feet less than the length. Let x represent the length of the swimming pool in feet. The area of the swimming pool is 126 square feet.

(a) Find the length of the sidewalk.

(b) Find the width of the sidewalk.

(c) What is the area of the sidewalk?

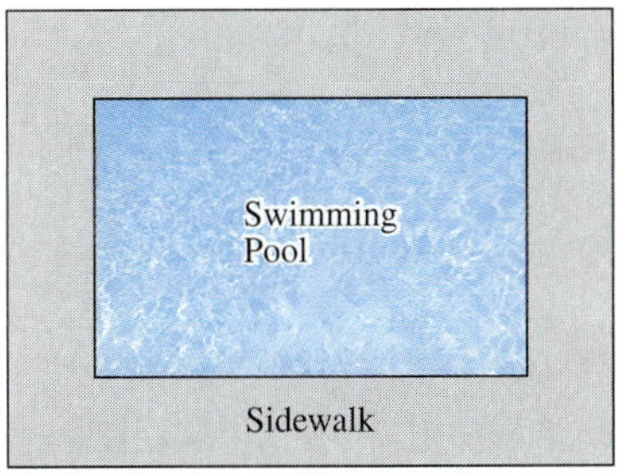

CHAPTER

7

Rational Expressions and Equations

Abu Bekr ibn Muhammad ibn al-Husayn Al-Karaji (953–1029) and other Arab algebraists hold an important place in the history of mathematics. Al-Karaji is best known for his methods of operating on unknowns, including rational expressions, using all the tools from arithmetic in the same way as the arithmetician operates on numbers.

7.1 Rational Expressions and Proportions

OBJECTIVES

- Define a rational expression.
- Reduce and build equivalent rational expressions.
- Identify restrictions in the domain of a rational expression.
- Solve a proportion.

1 2 3

Getting Started

In this chapter, we discuss operations on rational expressions involving variables. A rational number is the quotient of two integers. A **rational expression** is the quotient of two polynomials, where variables appear in the numerator and/or denominator. If you need more practice reducing, building, adding, subtracting, multiplying, or dividing rational numbers (fractions), or working with proportions, review Sections R.2 and R.3.

1. In Chapter 6, common factors were divided out to reduce rational expressions. To review, divide, and reduce the rational expression $\dfrac{12x^2y}{4xy^2}$.

2. Consult your Dictionary for Section 2.1. What is the difference between a term and a factor?

3. Given the rational expression $\frac{x + 5}{5}$, answer the following.

(a) Is 5 a common factor in the numerator and denominator of the fraction?

(b) Can 5 be divided out of the numerator and denominator? Why or why not?

4. Given the rational expression $\frac{2x + 5}{x - 3}$, answer the following.

(a) Evaluate the expression when $x = 3$.

(b) Is there any reason why x cannot equal 3 in the given rational expression?

5. It may be necessary to determine if two rational expressions are equivalent by reducing or building fractions. Determine whether or not the rational expressions are equivalent. Insert the appropriate symbol, either $=$ or $\neq$, between each expression.

(a) $\frac{5}{20}$ ______ $\frac{1}{4}$

(b) $\frac{2x^2}{2x}$ ______ x

(c) $\frac{x}{5}$ ______ $\frac{xy}{5y}$

(d) $\frac{3a^2}{bc}$ ______ $\frac{3a^2d}{bcd}$

6. To build a rational expression, multiply the numerator and denominator by the same nonzero number or expression. Build each rational expression into an equivalent fraction with the given denominator.

(a) $\frac{x}{y} = \frac{\quad}{2xy}$

Hint: Multiply the numerator and denominator of the first fraction by $2x$.

(b) $\frac{4ab}{3cd} = \frac{\quad}{15c^2d^2}$

Hint: Multiply the numerator and denominator of the first fraction by $5cd$.

Discussing Mathematics

Fractions are a mathematical notation representing parts of a whole and communicate relationships between numbers or quantities. Fractions with variables in the numerator or denominator are called rational expressions.

RATIONAL EXPRESSION

A rational expression is a fraction, which is the quotient of two polynomials:

$$\frac{p}{q}, \text{ where } p \text{ and } q \text{ are polynomials and } q \neq 0.$$

EQUIVALENT FRACTIONS

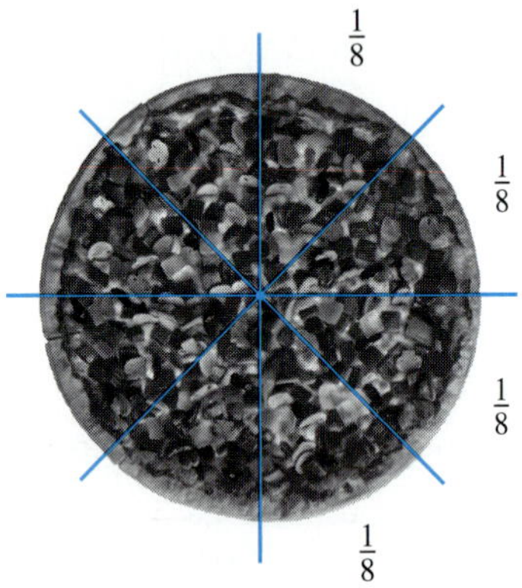

Two fractions are equivalent if they represent the same quantity. Half of a pizza is equivalent to 4 pieces if the pizza is cut into 8 pieces. The fraction $\frac{1}{2}$ can be represented in equivalent forms:

$$\frac{1}{2} = \frac{1}{8} + \frac{1}{8} + \frac{1}{8} + \frac{1}{8} = \frac{4}{8}.$$

The rational numbers $\frac{1}{2}$ and $\frac{4}{8}$ are equivalent.

Dividing (or multiplying) the numerator and denominator by the same nonzero quantity is equivalent to dividing (or multiplying) a fraction by 1 and produces an equivalent fraction.

Dividing the numerator and denominator by the same nonzero quantity is called **reducing a fraction**:

$$\frac{4}{8} = \frac{4 \div 4}{8 \div 4} = \frac{\overset{1}{\cancel{4}}}{\underset{2}{\cancel{8}}} = \frac{1}{2}.$$

Multiplying the numerator and denominator by the same nonzero quantity is called **building a fraction**:

$$\frac{1}{2} = \frac{1 \cdot 4}{2 \cdot 4} = \frac{4}{8}.$$

The process of reducing and building a fraction is generalized in the Fundamental Principle of Equivalent Fractions.

> **FUNDAMENTAL PRINCIPLE OF EQUIVALENT FRACTIONS**
>
> When the numerator (p) and the denominator (q) of a fraction are divided (or multiplied) by the same nonzero number (n), the result is an equivalent fraction.
>
> **Reducing a fraction:** $\frac{p}{q} = \frac{p \div n}{q \div n}$ **Building a fraction:** $\frac{p}{q} = \frac{p \cdot n}{q \cdot n}$

REDUCING RATIONAL EXPRESSIONS

To reduce a fraction, divide out factors common to both the numerator and denominator and/or apply the appropriate Property of Exponents. Common factors may be numbers and/or polynomials (monomials, binomials, trinomials, etc). The **Greatest Common Factor (GCF)** is the product of all the common factors of the numerator and denominator.

EXAMPLE 1: Reduce the rational expression $\frac{-20xy^3z^2}{5y^2z}$ to lowest terms.

$$\frac{-20xy^3z^2}{5y^2z} = \frac{-\overset{-4}{\cancel{20}}xy^3z^2}{\underset{1}{\cancel{5}}y^2z}$$

Divide the numerator and denominator by the GCF, $5y^2z$. Divide out common numerical factors.

$$= -4xy^{3-2}z^{2-1} = -4xyz$$

Apply Property of Exponents IV to divide out variable factors.

EXAMPLE 2: Reduce the rational expression $\dfrac{2x^2 + 2x - 12}{4x^2 + 2x - 20}$ to lowest terms.

$$\frac{2x^2 + 2x - 12}{4x^2 + 2x - 20} = \frac{2(x^2 + x - 6)}{2(2x^2 + x - 10)}$$ Factor the numerator and denominator.

$$= \frac{\overset{1}{\cancel{2}}(x + 3)\overset{1}{\cancel{(x - 2)}}}{\underset{1}{\cancel{2}}(2x + 5)\underset{1}{\cancel{(x - 2)}}}$$ Divide out the GCF, $2(x - 2)$.

$$= \frac{x + 3}{2x + 5}$$

➢ *Be sure to divide out only factors that are common in the numerator and denominator of a fraction. The rational expression $\frac{2x + 7}{7}$ has no common factors: 7 is a term in the numerator and cannot be divided out in this form. The fraction can be rewritten as two fractions and the second fraction reduced:* $\frac{2x + 7}{7} = \frac{2x}{7} + \frac{7}{7} = \frac{2x}{7} + 1.$

BUILDING EQUIVALENT RATIONAL EXPRESSIONS

To build an equivalent rational expression, multiply the numerator and denominator by the same nonzero number or expression.

EXAMPLE 3: Build each rational expression into an equivalent fraction with the given denominator.

(a) $$\frac{5x}{y} = \frac{?}{2 \cdot x \cdot y \cdot y \cdot z}$$ Compare the factors of the denominators. The factors 2, x, y, and z do not appear in the denominator of first fraction.

$$\frac{5x \cdot 2xyz}{y \cdot 2xyz} = \frac{10x^2yz}{2xy^2z}$$ Multiply the numerator and denominator of $\frac{5x}{y}$ by $2xyz$.

$$\frac{5x}{y} = \frac{10x^2yz}{2xy^2z}$$ Equivalent fractions.

(b) $$\frac{3}{(x - 2)} = \frac{?}{(x - 2)(x + 3)}$$ Factor the denominator of the second fraction and compare the factors. The factor $(x + 3)$ does not appear in the first fraction.

$$\frac{3 \cdot (x + 3)}{(x - 2) \cdot (x + 3)} = \frac{3(x + 3)}{(x - 2)(x + 3)}$$ Multiply the numerator and denominator of $\frac{3}{x - 2}$ by $(x + 3)$.

$$\frac{3}{x - 2} = \frac{3x + 9}{x^2 + x - 6}$$ Equivalent fractions.

(c) $\dfrac{9}{y} = \dfrac{?}{y(y-2)}$ Factor the denominator of the second fraction and compare the factors. The factor $(y - 2)$ does not appear in the first fraction.

$\dfrac{9 \cdot (y-2)}{y \cdot (y-2)} = \dfrac{9y - 18}{y^2 - 2y}$ Multiply the numerator and denominator of $\dfrac{9}{y}$ by $(y - 2)$.

$\dfrac{9}{y} = \dfrac{9y - 18}{y^2 - 2y}$ Equivalent fractions.

RESTRICTED VALUES OF THE VARIABLE

Remember that the denominator of a fraction can never be 0 because division by 0 is undefined. The value(s) of the variable that cause the denominator to be 0 are called **restricted values**. Never assign a restricted value to a variable.

EXAMPLE 4: Determine the restricted value(s) in each rational expression.

(a) The value $x = -1$ causes the denominator of the fraction $\dfrac{4x - 3}{x + 1}$ to equal zero. The restricted value is -1 and $x \neq -1$ in $\dfrac{4x - 3}{x + 1}$.

(b) Restricted value(s) may not be obvious. The restricted value of $\dfrac{x + 4}{3x - 7}$ can be identified by setting the denominator equal to zero and solving the equation: $3x - 7 = 0$

$$x = \frac{7}{3}$$

The number $\dfrac{7}{3}$ causes the denominator of the fraction to equal zero.

The restricted value is $\dfrac{7}{3}$ and $x \neq \dfrac{7}{3}$.

(c) A rational expression may have more than one restricted value.

When the denominator is a second-degree polynomial, as in $\dfrac{2x + 1}{x^2 + 2x - 15}$, use factoring to identify restricted values:

$$x^2 + 2x - 15 = 0$$

$$(x + 5)(x - 3) = 0$$

$$x + 5 = 0 \qquad x - 3 = 0$$

$$x = -5 \qquad x = 3$$

The restricted values are -5 and 3. $x \neq -5$ and $x \neq 3$ in $\dfrac{2x + 1}{x^2 + 2x - 15}$.

RATIOS, RATES, AND PROPORTIONS

Fractions express relationships of one quantity to another.

A **ratio** is the quotient of two quantities with the same unit of measure.	A **rate** is the quotient of two quantities with different units of measure in the numerator or denominator.
$\dfrac{\text{6 number of pieces of pizza eaten}}{\text{8 number of total pieces}}$	$\dfrac{\text{28 persons attending the meeting}}{\text{74 pieces of pizza eaten}}$
$\dfrac{\text{11 pizzas ordered}}{\text{17 pizzas ordered}}$	$\dfrac{\text{500 miles traveled}}{\text{10 hours}} = \text{50 miles per hour}$

An equation where two ratios or two rates are equal is called a **proportion**. Consult Section R.4 for more examples of solving proportions using the Extremes-Means Property of Proportions.

EXTREMES-MEANS PROPERTY OF PROPORTIONS

In any proportion, the product of the Extremes is equal to the product of the Means:

$$\text{Given } \frac{a}{b} = \frac{c}{d} \text{ then } \underbrace{a \cdot d}_{\text{Extremes}} = \underbrace{b \cdot c}_{\text{Means}}.$$

➢ *The product of the Extremes equal to the product of the Means is often called a "cross-product" and can be visualized as two crossed arrows:*

$$\frac{a}{b} \times \frac{c}{d} \text{ then } \underbrace{a \cdot d}_{\text{Extremes}} = \underbrace{b \cdot c}_{\text{Means}}.$$

EXAMPLE 5: Solve for x using the Extremes-Means Property of Proportions: $\dfrac{8}{x} = \dfrac{4}{13}$.

$\dfrac{8}{x} \times \dfrac{4}{13}$ Apply the Extremes-Means Property of Proportions.

$8 \cdot 13 = x \cdot 4$

$104 = 4x$ Solve for x.

$\dfrac{104}{4} = x$

$26 = x$

EXAMPLE 6: Equivalent fractions form a proportion. If a cross-product results in a true statement, the fractions are equivalent. Use a proportion to determine whether $\frac{5}{8}$ is equivalent to $\frac{15}{24}$.

$$\frac{5}{8} \stackrel{?}{=} \frac{15}{24}$$ Write a proportion. Find the products of the Extremes and the Means.

$$5 \cdot 24 \stackrel{?}{=} 8 \cdot 15$$

$$120 = 120$$ The products are equal. The fractions are equivalent.

EXAMPLE 7: One cell phone contract costs \$30 per month for 525 unlimited minutes, but offers additional minutes at the same rate. How much would it cost Jane per month if she used 700 minutes?

$$\frac{\$30}{525 \text{ minutes}} = \frac{x}{700 \text{ minutes}}$$ Write a proportion describing the problem situation.

$$30 \cdot 700 = 525x$$ Solve for x.

$$40 = x$$

It would cost Jane \$40 per month if she used 700 minutes.

1 2 3 Doing Mathematics

- Write the definitions of the words listed for this section in the Dictionary in Appendix A.
- On a 3 × 5-inch card, write the following processes for future reference:
  (a) Building a fraction
  (b) Reducing or simplifying a fraction
  (c) Solving a proportion

Problems 1–15: Reduce each rational expression to lowest terms.

1. $\frac{24}{48}$

2. $\frac{-52}{4}$

3. $\frac{3x}{12x}$

4. $\frac{16ab}{12a}$

5. $\frac{18a^2bc^3}{16a^2b^2c}$

6. $\frac{-36x^3yz^2}{9x^2y^2z^2}$

7. $\frac{2x + 4}{2}$

8. $\frac{5a - 20}{5}$

9. $\frac{5x^2 + 20x}{6x^2 + 24x}$

10. $\frac{4x^3 - 12x}{2x^2 - 6}$

11. $\dfrac{x^2 - x - 2}{x^2 + 8x + 7}$

12. $\dfrac{x^2 + 7x + 12}{x^2 - 16}$

13. $\dfrac{y^3 - 27}{y^3 + 3y^2 + 9y}$

14. $\dfrac{3x^2 - 6xy + 3y^2}{3x^2 + 6xy - 9y^2}$

15. $\dfrac{a^3 - a^2 + a - 1}{a^2 - 1}$

Problems 16–30: Build each rational expression to an equivalent fraction with the given denominator.

16. $\dfrac{1}{2} = \dfrac{?}{14}$

17. $\dfrac{5}{16} = \dfrac{?}{48}$

18. $\dfrac{2x}{5y} = \dfrac{?}{5xy}$

19. $\dfrac{-4a}{6b} = \dfrac{?}{12ab}$

20. $\dfrac{5x}{yz^2} = \dfrac{?}{2xy^2z^2}$

21. $\dfrac{-3ab}{xy^2} = \dfrac{?}{10x^2y^2}$

22. $\dfrac{6a}{b} = \dfrac{?}{b(b + 1)}$

23. $\dfrac{x}{3y} = \dfrac{?}{3y(y - 1)}$

24. $\dfrac{-4x}{x(x - 1)} = \dfrac{?}{x(x - 1)(x + 3)}$

25. $\dfrac{2x + 5}{3(x + 1)} = \dfrac{?}{3(x + 1)(x - 4)}$

26. $\dfrac{x}{x^2 - 1} = \dfrac{?}{9(x + 1)(x - 1)}$

27. $\dfrac{4a}{a^2 - 4} = \dfrac{?}{5(a + 2)(a - 2)}$

28. $\dfrac{4x - 9}{x^2 - x - 6} = \dfrac{?}{5(x + 2)(x - 3)}$

29. $\dfrac{7x}{x - 2} = \dfrac{?}{(x - 2)(x^2 + 2x + 4)}$

30. $\dfrac{-ab}{a - 1} = \dfrac{?}{a^2 - 2a + 1}$

Problems 31–38: Determine the restricted value(s) of the variables in each rational expression.

31. $\dfrac{5x}{x - 1}$

32. $\dfrac{2x - 5}{2x + 1}$

33. $\dfrac{20}{(x - 1)(x + 2)}$

34. $\dfrac{x + 3}{(x - 2)(x - 5)}$

35. $\dfrac{(x - 2)(x + 2)}{7x + 1}$

36. $\dfrac{(x + 1)(x - 5)}{3x - 5}$

37. $\dfrac{2x - 1}{x^2 + 7x + 12}$

38. $\dfrac{4x - 9}{x^2 - x - 6}$

Problems 39–40: Write a ratio expressing each statement.

39. (a) The number of females in your immediate family to the total number of persons in the family

(b) The number of males in your immediate family to the total number of persons in the family

(c) The number of females in this class to the number of men

40. (a) The number of DVD players in your home to the total number of televisions in your home

(b) The number of computers in your home to the number of persons in your family

(c) The number of cars in your home to the number of persons in your family

Problems 41–42: Write a rate expressing each relationship.

41. (a) 60 meters to 2.5 seconds

(b) 3 cups of flour to 1 batch of cookies

(c) \$1.98 for 2 cantaloupes

42. (a) 100 miles to 2 hours

(b) 240 words typed in 3 minutes

(c) 250 CDs produced for every 10 workers

Problems 43–52: Use the Extremes-Means Property of Proportions to determine whether or not the following fractions are equivalent. Write the appropriate symbol, either $=$ or $\neq$, on the line between the expressions.

43. $\dfrac{3}{7}$ ______ $\dfrac{9}{21}$

44. $\dfrac{6}{21}$ ______ $\dfrac{12}{42}$

45. $\dfrac{2x}{y}$ ______ $\dfrac{8x^2}{xy}$

46. $\dfrac{2x}{12}$ ______ $\dfrac{-x}{6}$

47. $\dfrac{(x+2)(x-1)}{(x-1)(x+3)}$ ______ $\dfrac{(x+2)}{(x+3)}$

48. $\dfrac{5+7}{7}$ ______ $\dfrac{12}{7}$

49. $\dfrac{x+5}{5}$ ______ $\dfrac{x}{1}$

50. $\dfrac{x+1}{-x}$ ______ $\dfrac{-x-1}{x}$

51. $\dfrac{a^2-4}{a-2}$ ______ $\dfrac{a+2}{1}$

52. $\dfrac{-(x-1)}{5xy}$ ______ $\dfrac{(1-x)y}{5xy^2}$

Problems 53–58: Solve each proportion.

53. $\dfrac{2}{5}=\dfrac{x}{10}$

54. $\dfrac{5}{11}=\dfrac{x}{66}$

55. $\dfrac{4}{x}=\dfrac{3}{66}$

56. $\dfrac{7}{x}=\dfrac{3}{60}$

57. $\dfrac{9}{11}=\dfrac{198}{x}$

58. $\dfrac{1}{12}=\dfrac{7}{x}$

59. Marat needs to purchase 12 coconuts for a party. He wants the best bargain for his purchase. Friendly Market advertises 3 coconuts for \$4.99. Biggie Mart sells 4 coconuts for \$6.60.

(a) Where should Marat purchase the coconuts considering price per coconut only?

(b) Marat found that the coconuts at Biggie Mart were much smaller than those at the Friendly Market. Which store really offers the best buy?

60. Bob has gone away to college and must do his laundry for the first time. He is searching for the best bargain in detergents. Brand X is priced at \$5.99 for 64 ounces. Brand Y is priced at \$6.99 for 84 ounces.

(a) Which brand should Bob purchase, considering only price per ounce? Why?

(b) Bob read the directions and found that 1 ounce of Brand X detergent is used per load, while 2 ounces of Brand Y detergent are used per load. Which brand is the most economical?

61. One cell phone contract costs \$25 per month for 400 unlimited minutes, but offers additional minutes at the same rate. How much would it cost Bill per month if he uses 600 minutes?

62. It takes Cindy 14 minutes to type 5 pages. How long will it take her to type 15 pages?

63. In the fall semester, Mr. Roskos mixed a solution in a ratio of 6 liters of acid to 1.5 liters of water for a chemistry experiment. In the spring semester, he has more students and plans to use 13 liters of acid. How many liters of water should he add to the 13 liters of acid to create a solution that is in the same ratio as the solution used in fall semester?

(a) Write a ratio of the number of liters of acid to the number of liters of water used in the solution fall semester: $\dfrac{\text{number of liters of acid}}{\text{number of liters of water}} =$ ______ .

(b) Let x represent the number of liters of water to be mixed with acid for the spring semester. Write a ratio of the number of liters of acid to the number of liters of water used in the solution for the spring semester:
$\dfrac{\text{total liters of acid}}{\text{total liters of water}} =$ ______ .

(c) Use the ratios in (a) and (b) to write a proportion. Solve the proportion for x.

______ = ______

(d) How many liters of water should Mr. Roskos add to 13 liters of acid to create a solution that is in the same ratio as the solution used fall semester?

Writing Exercise

64. Your calculator displays this table of values for the function $y_1 = \dfrac{4x + 2}{x - 1}$. Why did the calculator display "ERROR" when the function was evaluated at $x = 1$?

X	Y1
-1	1
0	-2
1	ERROR
2	10
3	7
4	6
5	5.5

X=1

7.2 Multiplication and Division of Rational Expressions

OBJECTIVE

- Multiply and divide rational expressions.

1 2 3 Getting Started

In this section, we review the processes of multiplication and division of rational numbers and extend those concepts to rational expressions.

1. Mary Lou wants to make her famous pound cake this weekend. The recipe makes 6 small cakes using these ingredients:

3 cups flour
6 eggs
$\frac{3}{4}$ cup butter
$1\frac{1}{2}$ cups sugar
2 teaspoons vanilla
$\frac{1}{2}$ teaspoon salt
$\frac{3}{4}$ teaspoon baking powder
$1\frac{1}{4}$ cups milk

(a) What amount of each ingredient should Mary Lou use to make 12 cakes?

____ cup(s) flour	____ egg(s)	____ cup(s) butter
____ cup(s) sugar	____ teaspoon(s) vanilla	____ teaspoon(s) salt
____ teaspoon(s) baking powder		____ cup(s) milk

(b) Mary Lou wants to make 2 cakes. What fraction of the recipe is she making?

(c) What amount of each ingredient should Mary Lou use to make 2 cakes?

____ cup(s) flour	____ egg(s)	____ cup(s) butter
____ cup(s) sugar	____ teaspoon(s) vanilla	____ teaspoon(s) salt
____ teaspoon(s) baking powder		____ cup(s) milk

(d) Describe the methods you used to find your answers.

2. Paul is remodeling his kitchen and needs to purchase cove molding to install under the cabinets. He needs 4 pieces of molding, each $3\frac{1}{2}$ feet long.

(a) What is the total length of molding that he needs?

(b) If the molding comes in lengths of 15 feet, how many lengths should he purchase?

(c) Describe the method you used to arrive at the answer.

3. Suppose you have a pizza that is cut into 8 pieces.

(a) If you divide the 8 pieces of pizza into groups of two pieces, how many groups would be formed?

(b) Write a mathematical statement for the process of dividing by 2 in part (a).

(c) If you divide each of the 8 pieces of pizza into halves, how many pieces would be formed?

(d) Write a mathematical statement for the process of dividing by $\frac{1}{2}$ in part (c).

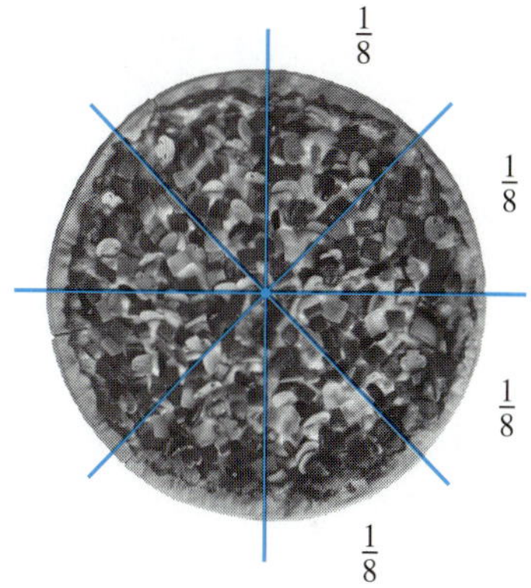

(e) Does dividing 8 by 2 result in a larger or smaller number?

(f) Does dividing 8 by $\frac{1}{2}$ result in a larger or smaller number?

4. Don is the statistician for a major league baseball team and earns an annual salary of \$80,000. His contract ends next month. He met with his boss to negotiate his salary for next year. Don told his boss that he liked working for the organization so much that he was willing to have his salary divided by $\frac{1}{2}$ for next year. His boss was delighted and agreed immediately.

(a) What would Don's salary be for next year?

(b) Did the boss make a wise decision?

Discussing Mathematics

The processes used to multiply and divide rational numbers are the same used to multiply or divide rational expressions. If you need more practice in multiplying or dividing fractions, additional exercises can be found in Section R.2.

MULTIPLYING RATIONAL EXPRESSIONS

MULTIPLYING RATIONAL EXPRESSIONS

When multiplying rational expressions, multiply the numerators of the fractions and multiply the denominators of the fractions:

$$\frac{A}{B}\cdot\frac{C}{D}=\frac{A\cdot C}{B\cdot D}, \text{ where } B \neq 0 \text{ and } D \neq 0.$$

Reduce the resulting product by dividing out common factors, if possible.

Common factors may be divided out in the first step and then multiply *or* multiply first and then reduce. Final products should always be reduced to lowest terms.

1. Divide out common factors.	*or*	1. Multiply numerators and denominators.
2. Multiply numerators and denominators.		2. Divide out common factors.

EXAMPLE 1: Multiply. $\frac{2x^2}{7}\cdot\frac{5}{9y}$

$$=\frac{2x^2\cdot 5}{7\cdot 9y}=\frac{10x^2}{63y}$$

Multiply the numerators and denominators.

EXAMPLE 2: Multiply. $\frac{5xy}{-3a^2b}\cdot\frac{12ab}{-25xy^2}$

$$=\frac{\overset{1}{\cancel{5}}x^{1-1}}{\underset{-1}{\cancel{-3}}a^{2-1}}\cdot\frac{\overset{4}{\cancel{12}}b^{1-1}}{\underset{-5}{\cancel{-25}}y^{2-1}}$$

Divide out common factors numerical factors. Apply Property of Exponents IV.

$$=\frac{1}{-a}\cdot\frac{4}{-5\cdot y}=\frac{4}{5ay}$$

Multiply remaining factors.

EXAMPLE 3: Multiply. $\dfrac{x+1}{3}\cdot\dfrac{-18x}{x^2+3x+2}$

$$=\frac{(x+1)}{3}\cdot\frac{-18x}{(x+2)(x+1)}$$

Factor the numerators and denominators. The expression $x+1$ cannot be factored. Enclose $x+1$ in parentheses.

$$=\frac{\overset{1}{\cancel{(x+1)}}}{\underset{1}{\cancel{3}}}\cdot\frac{\overset{-6}{\cancel{-18}}\cdot x}{(x+2)\underset{1}{\cancel{(x+1)}}}$$

Divide out common factors.

$$=\frac{1}{1}\cdot\frac{-6x}{(x+2)}=\frac{-6x}{x+2}$$

Multiply remaining factors.

EXAMPLE 4: Multiply. $\dfrac{1-x}{15}\cdot\dfrac{5x+5}{x^2-1}$

$$=\frac{-(x-1)}{15}\cdot\frac{5(x+1)}{(x-1)(x+1)}$$

Rewrite the numerator of the first fraction in its equivalent form $1-x=-(x-1)$. Factor the numerator and denominator of the second fraction.

$$=\frac{-\overset{1}{\cancel{(x-1)}}}{\underset{3}{\cancel{15}}}\cdot\frac{\overset{1}{\cancel{5}}\overset{1}{\cancel{(x+1)}}}{\underset{1}{\cancel{(x-1)}}\underset{1}{\cancel{(x+1)}}}$$

Divide out common factors.

$$=\frac{-1}{3}\cdot\frac{1\cdot 1}{1\cdot 1}=-\frac{1}{3}$$

Multiply remaining factors.

DIVIDING RATIONAL EXPRESSIONS

Division of rational expressions is equivalent to multiplying the dividend by the reciprocal of the divisor.

EXAMPLE 5: Divide $10 \div 2$.

$$10\div 2=10\div\frac{2}{1}=\frac{10}{1}\cdot\frac{1}{2}=\frac{10}{2}=5$$

Multiply by the reciprocal of the divisor.

DIVIDING RATIONAL EXPRESSIONS

The operation of division is equivalent to multiplying the dividend by the reciprocal of the divisor:

$\dfrac{A}{B}\div\dfrac{C}{D}=\dfrac{A}{B}\cdot\dfrac{D}{C}$, where $B\neq 0$, $D\neq 0$, and $C\neq 0$.

Reduce the resulting product by dividing out common factors, if possible.

1. Rewrite the division problem as a multiplication problem (multiply the dividend by the reciprocal of the divisor).
2. Factor the numerators and denominators.
3. Divide out any common factors.
4. Multiply remaining factors in the numerators and denominators.
5. Check the resulting product for any common factors. If any common factors are present, repeat steps 2–4.

EXAMPLE 6: Divide. $\dfrac{-10xyz}{9ab} \div \dfrac{5xz^3}{-3ab}$

$= \dfrac{-10xyz}{9ab} \cdot \dfrac{-3ab}{5xz^3}$ Rewrite division as multiplication.

$= \dfrac{\overset{-2}{\cancel{-10}}x^{1-1}y}{\underset{-3}{\cancel{9}}} \cdot \dfrac{\overset{1}{\cancel{-3}}a^{1-1}b^{1-1}}{\underset{1}{\cancel{5}}z^{3-1}}$ Divide out common factors numerical factors. Apply Property of Exponents IV.

$= \dfrac{-2y}{-3} \cdot \dfrac{1}{z^2} = \dfrac{-2y}{-3z^2}$ Multiply remaining factors.

$= \dfrac{2y}{3z^2}$ Reduce the fraction to lowest terms.

EXAMPLE 7: Divide. $\dfrac{x^2 - y^2}{y + 1} \div \dfrac{x - y}{5y + 5}$

$= \dfrac{x^2 - y^2}{y + 1} \cdot \dfrac{5y + 5}{x - y}$ Rewrite division as multiplication.

$= \dfrac{\overset{1}{\cancel{(x - y)}}(x + y)}{\underset{1}{\cancel{(y + 1)}}} \cdot \dfrac{5\overset{1}{\cancel{(y + 1)}}}{\underset{1}{\cancel{(x - y)}}}$ Factor the numerators and the denominators. Divide out common factors.

$= \dfrac{(x + y)}{1} \cdot \dfrac{5}{1} = 5(x + y)$ Multiply remaining factors.

$= 5x + 5y$ The final answer may be written in factored form or expanded form.

➢ *Be sure to follow the Order of Operations when problems involve more than one operation.*

EXAMPLE 8: Perform the operations. $\dfrac{x^2 + x - 6}{5x^2y} \div \dfrac{2x^2 - 18}{5x^2 + 30x + 45} \div \dfrac{x - 2}{4x}$

$= \dfrac{x^2 + x - 6}{5x^2y} \cdot \dfrac{5x^2 + 30x + 45}{2x^2 - 18} \cdot \dfrac{4x}{x - 2}$ Rewrite division as multiplication.

$= \dfrac{(x + 3)\overset{1}{\cancel{(x - 2)}}}{\underset{1}{\cancel{5}}x^{2-1}y} \cdot \dfrac{\overset{1}{\cancel{5}}\overset{1}{\cancel{(x + 3)}}(x + 3)}{\underset{1}{\cancel{2}}\underset{1}{\cancel{(x + 3)}}(x - 3)} \cdot \dfrac{\overset{2}{\cancel{4}}}{\underset{1}{\cancel{(x - 2)}}}$ Factor the numerators and the denominators. Divide out common factors.

$= \dfrac{(x + 3)}{xy} \cdot \dfrac{(x + 3)}{(x - 3)} \cdot \dfrac{2}{1} = \dfrac{2(x + 3)(x + 3)}{xy(x - 3)}$ Multiply remaining factors.

$= \dfrac{2x^2 + 12x + 18}{x^2y - 3y}$ The final answer may be written in factored form or expanded form.

1 2 3 Doing Mathematics

- On a 3 × 5-inch card, write the following processes for future reference:
  - (a) Multiplying rational expressions
  - (b) Dividing rational expressions

Problems 1–2: Claude is a chef and often makes crepes. The recipe he uses makes 32 crepes and includes these ingredients:

4 eggs	$\frac{1}{2}$ cup milk	2 tablespoons clarified butter
$\frac{1}{2}$ cup water	1 cup flour	2 tablespoons brandy (optional)
$\frac{3}{4}$ teaspoon salt	1 teaspoon vanilla extract	1 to 2 teaspoons sugar (for dessert crepes)

1. What amount of each ingredient should Claude use to make 96 crepes?

___ eggs	___ cup milk	___ tablespoons clarified butter
___ cup water	___ cup flour	___ tablespoons brandy (optional)
___ teaspoon salt	___ teaspoon vanilla extract	___ teaspoons sugar (for dessert crepes)

2. What amount of each ingredient should Claude use to make 24 crepes?

___ eggs	___ cup milk	___ tablespoons clarified butter
___ cup water	___ cup flour	___ tablespoons brandy (optional)
___ teaspoon salt	___ teaspoon vanilla extract	___ teaspoons sugar (for dessert crepes)

3. Ron drove 120 miles from Cleveland to Pittsburgh in 2 hours.

(a) Using the formula $d = r \cdot t$, distance = (rate)(time), find Ron's average rate of speed on this trip. Write your answer with the proper units to describe his speed.

(b) If Ron averages the same speed driving 40 miles from Pittsburgh to Youngstown, how long should he plan for the trip?

(c) Ron plans to leave Pittsburgh at 4:00 P.M. What time will he arrive in Youngstown?

Problems 4–17: Multiply the rational expressions. Reduce each answer to lowest terms.

4. $1\frac{1}{3} \cdot 5\frac{5}{8}$

5. $2\frac{1}{2} \cdot 4\frac{3}{5}$

6. $\frac{16}{15} \cdot \frac{-25}{8} \cdot \frac{-3}{2}$

7. $\frac{-32}{85} \cdot \frac{-25}{8} \cdot \frac{1}{-2}$

8. $\frac{18}{(-3)^3} \cdot \frac{9}{-2^2}$

9. $\frac{(-3)^3}{5} \cdot \frac{20}{-3^2}$

10. $\frac{-6ab}{5} \cdot \frac{25}{12a^2b}$

11. $\frac{-2x}{5y} \cdot \frac{3}{4x^2}$

12. $\dfrac{(2a)^2}{5xyz}\cdot\dfrac{15xz^2}{8a}$

13. $\dfrac{(2a)^2}{12ab}\cdot\dfrac{(-3b)^3}{a^2}$

14. $\dfrac{5r-10}{r^2-r-2}\cdot\dfrac{r+1}{10r}$

15. $\dfrac{3x+6}{x^2+x-2}\cdot\dfrac{x-1}{18x}$

16. $\dfrac{-2x+6}{x^2-4}\cdot\dfrac{15x^2}{x^2-9}\cdot\dfrac{x^2+x-6}{6x}$

17. $\dfrac{2x^2+3x-2}{-3x-6}\cdot\dfrac{6x^2-6x-72}{x^2-2x-8}\cdot\dfrac{-x-2}{10}$

Problems 18–31: Divide the rational expressions. Reduce each answer to lowest terms.

18. $-1\dfrac{1}{3}\div\dfrac{-2}{21}$

19. $5\dfrac{1}{2}\div\dfrac{-22}{10}$

20. $\dfrac{(5c)^2}{6a^2}\div\dfrac{25c}{12ab}$

21. $\dfrac{(2a)^2}{15b^2c}\div\dfrac{8a^2b^2}{25c}$

22. $\dfrac{-2z}{3xy^2}\div\dfrac{-4z}{9y}$

23. $\dfrac{3xy}{10a^3b^3}\div\dfrac{3x^2y^2z^2}{10(ab)^2}$

24. $\dfrac{a^2-b^2}{-6}\div\dfrac{a+b}{60}$

25. $\dfrac{x^2-y^2}{24}\div\dfrac{x-y}{-48}$

26. $\dfrac{x^2-y^2}{-3x-3z}\div\dfrac{5x+5y}{6x^2-6z^2}$

27. $\dfrac{3x^2-3z^2}{6x+6y}\div\dfrac{6x^2-6z^2}{-2x-2y}$

28. $\dfrac{4x^2-9x-9}{4x^2-21x-18}\div\dfrac{2x^2-3x-9}{2x^2-9x-18}$

29. $\dfrac{3x^2-18x-21}{-12x^2-20x+8}\div\dfrac{x^2-49}{9x^2+15x-6}$

30. $\dfrac{x^3-8}{x^2-4}\div\dfrac{3x^2+6x+12}{15x+30}$

31. $\dfrac{x^2-9}{x^2-16}\div\dfrac{x^2+6x+9}{x^2+5x+4}$

Problems 32–38: Perform the operations. Leave answers in factored form.

32. $\dfrac{x^2 + 2x - 3}{x - 1} \cdot \dfrac{x^2 - 2x + 1}{-2x^2 - 6x} \div \dfrac{2x^2 - x - 1}{2x}$

33. $\dfrac{x^2 + 2xy + y^2}{x + y} \div \dfrac{x^2 - y^2}{-3x - 3y} \cdot \dfrac{x - y}{x + y}$

34. $\dfrac{2x + 6}{2x^2 - 3x + 1} \cdot \dfrac{18x^2 + 81x - 45}{x^2 + 2x - 3} \div \dfrac{3x + 3y}{x^2 - y^2}$

35. $\dfrac{15x^2 - 15y^2}{x^2 - 1} \cdot \dfrac{2x^2 - 3x + 1}{3x + 3y} \div \dfrac{5x - 5y}{3x + 3y}$

36. $\dfrac{6x^2 + x - 1}{4x^2 - 4x - 35} \cdot \dfrac{6x^2 + 11x - 10}{3x^2 + 2x - 1} \div \dfrac{6x^2 - x - 2}{6x^2 - 17x - 14}$

37. $\dfrac{2x + 2}{4x + 2} \cdot \dfrac{2x^2 + 7x + 3}{x^3 + 1} \div \dfrac{8x}{4x^2 - 4x + 1} \cdot \dfrac{x^2 - x + 1}{2x^2 + 5x - 3}$

38. $\dfrac{x^3 - 1}{x^2 - 1} \cdot \dfrac{2x - 2}{x^2 + x + 1} \div \dfrac{10x^2 - 10}{x^2 + 2x + 1} \cdot \dfrac{5x + 15}{x + 3}$

Problems 39–42: Given the dimensions for each rectangle, represent the area of each rectangle as a single expression in terms of x. Find the area when $x = 5$ inches.

39. Length $= \dfrac{x^2}{2}$, Width $= \dfrac{8}{x + 5}$, $x \neq -5$

40. Length $= x + 3$, Width $= \dfrac{x}{2}$

41. Length $= \dfrac{x^2 - x - 6}{x}$, $x \neq 0$

Width $= \dfrac{x}{x - 3}$, $x \neq 3$

42. Length $= \dfrac{x}{x - 2}$, $x \neq 2$,

Width $= \dfrac{x^2 - 4}{2x}$, $x \neq 0$

7.3 Addition and Subtraction of Rational Expressions

OBJECTIVES

- Identify the LCM of two or more numbers.
- Use the LCM as the LCD and write equivalent rational expressions.
- Add and subtract rational expressions.

1 2 3 Getting Started

In this section, the operations of addition and subtraction of rational expressions are presented, along with a review of adding and subtracting rational numbers. If you find you need more practice with operations on rational numbers, consult Section R.2.

1. (a) Add the rational numbers. $\frac{5}{7} + \frac{2}{7}$

(b) List the steps involved in adding (or subtracting) fractions that have the same denominator.

2. Adding and/or subtracting rational expressions with LIKE denominators uses the same process as adding or subtracting rational numbers.

EXAMPLE: $\frac{1}{x} + \frac{1}{x} = \frac{1+1}{x} = \frac{2}{x}$ Add the numerators and keep the denominator the same.

Add or subtract.

(a) $\frac{10}{y} + \frac{5}{y}$ (b) $\frac{12}{x+3} - \frac{4}{x+3}$

(c) $\frac{7}{x^2} + \frac{3}{x^2}$

3. What if the denominators of the fractions are not the same? Suppose you wanted to add 5 pieces of a 12-cut pizza and 3 pieces of an 8-cut pizza to make a fraction of one pizza: $\frac{5}{12} + \frac{3}{8}$. To add or subtract fractions, the denominators of the fractions must be the same. To add $\frac{5}{12} + \frac{3}{8}$, a common multiple of 12 and 8 must be identified and used as the common denominator. A common multiple is a number that is a multiple of all denominators under consideration. The smallest common multiple is called the least common multiple (LCM) and used as the **least common denominator (LCD)**. Use the following steps to find the LCD of 12 and 8, and use the LCD to add the fractions $\frac{5}{12}$ and $\frac{3}{8}$.

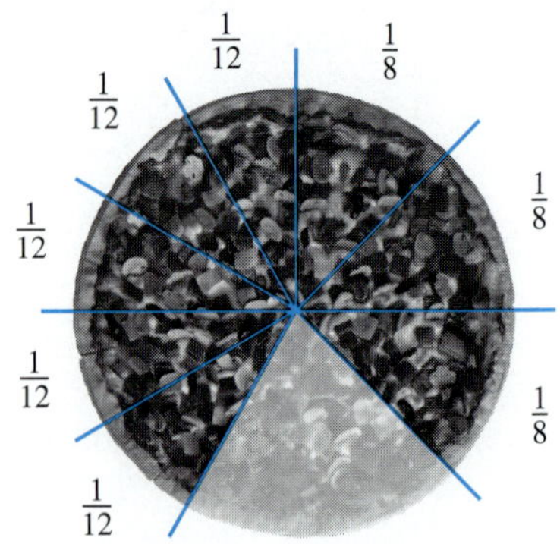

(a) The factors of 12 and 8 are written in the table with LIKE factors in vertical columns. Draw a vertical arrow through each vertical column (even if a factor appears only in one factorization). Write the factor in each column in the last row, as a factor of the LCD.

Factors of 12:		2	2	3
Factors of 8:	2	2	2	
LCD:				

(b) The product of the factors in the last row is the Least Common Denominator (LCD). Multiply the factors in the last row and write the LCD.

(c) Use the LCD and build each fraction to an equivalent fraction.

$$\frac{5}{12} = \frac{\quad}{24} \qquad \text{and} \qquad \frac{3}{8} = \frac{\quad}{24}$$

(d) Rewrite the problem using equivalent fractions and add.

$$\frac{5}{12} + \frac{3}{8}$$

➢ *When adding or subtracting rational expressions with denominators that are* not the same, *the least common multiple is identified and used as the least common denominator (LCD).*

EXAMPLE 1: Find the LCD and add $\frac{3}{xy} + \frac{1}{5y}$.

Factors of xy: $x \cdot y$
Factors of $5y$: $5 \cdot y$
LCD: $5\ x\ y$

$$\frac{3}{xy} + \frac{1}{5y} = \frac{3 \cdot 5}{xy \cdot 5} + \frac{1 \cdot x}{5y \cdot x}$$

Multiply the numerator and denominator of the first fraction by 5 and of the second fraction by x.

$$= \frac{15}{5xy} + \frac{x}{5xy}$$

$$= \frac{15 + x}{5xy}$$

Add the numerators. The denominator remains the same.

4. Add or subtract.

(a) $\frac{5}{xy} + \frac{7}{2x}$

(*Hint:* The LCD is $2xy$. Multiply the numerator and denominator of the first fraction by 2 and of the second fraction by y to build an equivalent fraction with a denominator of $2xy$. Then add the two fractions.)

(b) $\frac{2x}{4y} - \frac{3xy}{5y^2}$

(*Hint:* The LCD is $20y^2$. Multiply the numerator and denominator of the first fraction by $5y$. Multiply the numerator and denominator of the second fraction by 4. Then subtract the two fractions.)

1 2 3

Discussing Mathematics

ADDING OR SUBTRACTING RATIONAL EXPRESSIONS

The same process used to add or subtract rational numbers is used to add or subtract rational expressions with denominators that are the *same:* find the sum or difference of the numerators and keep the denominator the same; reduce final answers to lowest terms.

> **ADDING OR SUBTRACTING RATIONAL EXPRESSIONS WITH DENOMINATORS THAT ARE THE *SAME***
>
> Given the rational expressions $\frac{A}{B}$ and $\frac{C}{B}$, then
>
> **Sum:** $\frac{A}{B} + \frac{C}{B} = \frac{A + C}{B}, (B \neq 0)$; **Difference:** $\frac{A}{B} - \frac{C}{B} = \frac{A - C}{B}, (B \neq 0)$

EXAMPLE 2: Subtract.

$$\frac{12}{54} - \frac{2}{54} = \frac{12 - 2}{5y} = \frac{10}{5y}$$

Subtract the numerators. The denominator remains the same.

$$= \frac{2 \cdot \overset{1}{\cancel{5}}}{\underset{1}{\cancel{5}}y} = \frac{2}{y}$$

Reduce fraction to lowest terms.

EXAMPLE 3: Add.

$$\frac{2}{x + 3} + \frac{4}{x + 3} = \frac{2 + 4}{x + 3}$$

Add the numerators. The denominator remains the same.

$$= \frac{6}{x + 3}$$

Final answer.

➢ *3 is not a common factor of the numerator and denominator and cannot be divided out. The fraction is in lowest terms.*

When adding or subtracting fractions when the denominators of the fractions *are not the same,* the least common denominator (LCD) must be identified.

> **FINDING THE LCD OF THE DENOMINATORS OF FRACTIONS**
>
> 1. Write the prime factorization of each denominator. Write LIKE factors in vertical columns.
> 2. Draw an arrow through each factor and vertical column. Write the factors of the LCD. (Each factor should be included the greatest number of times it appears in any one factorization.)
> 3. Multiply the factors to form the LCD.

The process of identifying the LCM for rational expressions is the same as with rational numbers. However, the factors may be polynomials.

ADDING OR SUBTRACTING RATIONAL EXPRESSIONS WITH DENOMINATORS THAT ARE *DIFFERENT*

1. Write the factors of all denominators. Find the LCM of the denominators.
2. Use the LCM as the LCD and build the original fractions to equivalent fractions with the LCD as the new denominators.
3. Add or subtract the numerators. The denominator is the LCD.
4. Factor the resulting numerator, if possible. Identify any common factors and reduce the fraction to lowest terms, if possible.

EXAMPLE 4: Add $\frac{7}{x} + \frac{3}{x^2}$.

$$x = x$$ Write the factors of each denominator.

$$x^2 = x \cdot x$$ Draw an arrow through each factor and vertical column.

$$\textbf{LCD} = x \cdot x = x^2$$

$$\frac{7}{x} + \frac{3}{x^2} = \frac{7 \cdot x}{x \cdot x} + \frac{3}{x^2} = \frac{7x}{x^2} + \frac{3}{x^2}$$ Build the first fraction to a new fraction with a denominator of LCD = x^2.

$$= \frac{7x + 3}{x^2}$$ Add the numerators. The denominator remains same. The fraction cannot be reduced.

EXAMPLE 5: Perform the operations $\frac{4}{xy} + \frac{1}{2y} - \frac{2}{5xy^2}$.

$$x = x \cdot y$$ Write the factors of each denominator.

$$2y = 2 \cdot y$$

$$5xy^2 = 5 \cdot x \cdot y \cdot y$$ Draw an arrow through each factor and vertical column.

$$\textbf{LCD} = 2 \cdot 5 \cdot x \cdot y \cdot y = 10xy^2$$

$$\frac{4}{xy} + \frac{1}{2y} - \frac{2}{5xy^2} = \frac{4 \cdot 10y^2}{x \cdot 10y^2} + \frac{1 \cdot 5xy}{2y \cdot 5xy} - \frac{2 \cdot 2}{5xy^2 \cdot 2}$$ Build each fraction to a new fraction with a denominator of LCD = $10xy^2$.

$$= \frac{40y^2}{10xy^2} + \frac{5xy}{10xy^2} - \frac{4}{10xy^2}$$

$$= \frac{40y^2 + 5xy - 4}{10xy^2}$$ Add and subtract the numerators. The denominator remains the same.

EXAMPLE 6: Subtract $\dfrac{x+6}{9x+81} - \dfrac{3}{x^2+9x}$.

$$9x + 81 = 9 \quad (x+9)$$
$$x^2 + 9x = \quad x(x+9)$$

Write the factors of each denominator.

Draw an arrow through each factor and vertical column.

$$\textbf{LCD} = 9 \cdot x(x+9)$$

$$\frac{x+6}{9x+81} - \frac{3}{x^2+9x} = \frac{(x+6)\cdot x}{9(x+9)\cdot x} - \frac{3 \cdot 9}{x(x+9)\cdot 9}$$

Build each fraction to a new fraction with a denominator of LCD = $9x(x+9)$.

$$= \frac{x^2+6x}{9x(x+9)} - \frac{27}{9x(x+9)}$$

Multiply each numerator. Keep the denominators in factored form.

$$= \frac{x^2+6x-27}{9x(x+9)}$$

Subtract the numerators. The denominator remains same.

$$= \frac{(x+9)(x-3)}{9x(x+9)}$$

Factor the numerator and look for common factors.

$$= \frac{\overset{1}{\cancel{(x+9)}}(x-3)}{9x\underset{1}{\cancel{(x+9)}}} = \frac{x-3}{9x}$$

Divide out common factors.

EXAMPLE 7: Subtract $\dfrac{5y}{y+1} - \dfrac{5y^2-6y+1}{y^2-2y-3}$.

$$y + 1 = (y+1)$$
$$y^2 - 2y - 3 = (y+1)(y-3)$$

Write the factors of each denominator.

Draw an arrow through each factor and vertical column.

$$\textbf{LCD} = (y+1)(y-3)$$

$$\frac{5y}{y+1} - \frac{5y^2-6y+1}{y^2-2y-3} = \frac{5y\cdot(y-3)}{(y+1)\cdot(y-3)} - \frac{5y^2-6y+1}{(y+1)(y-3)}$$

Build the first fraction to a new fraction with a denominator of LCD = $(y+1)(y-3)$.

$$= \frac{5y^2-15y}{(y+1)(y-3)} - \frac{5y^2-6y+1}{(y+1)(y-3)}$$

Multiply the numerator in the first fraction.

$$= \frac{5y^2-15y-5y^2+6y-1}{(y+1)(y-3)}$$

Subtract the numerators. The denominator remains same.

$$= \frac{-9y-1}{(y+1)(y-3)} = \frac{-1(9y+1)}{(y+1)(y-3)}$$

Combine LIKE terms. Factor the numerator and look for common factors.

$$= -\frac{9y+1}{(y+1)(y-3)}$$

There are no common factors.

1 2 3 Doing Mathematics

- Write the definitions of the words and symbols listed for this section in the Dictionary in Appendix A.
- On a 3 × 5-inch card, write the following processes for future reference
  (a) Finding the least common denominator
  (b) Adding or subtracting rational expressions with denominators that are the same
  (c) Adding or subtracting rational expressions with denominators that are different

Problems 1–12: Factor each denominator and find the Least Common Denominator (LCD).

1. $\frac{2}{9}, \frac{4}{15}$

Factors of 9:
Factors of 15:
LCD:

2. $\frac{7}{12}, \frac{11}{15}$

Factors of 12:
Factors of 15:
LCD:

3. $\frac{2}{xy}, \frac{1}{2x^2}$

Factors of xy:
Factors of $2x^2$:
LCD:

4. $\frac{a}{3b^2}, \frac{4}{ab}$

Factors of $3b^2$:
Factors of ab:
LCD:

5. $\frac{5}{3xy}, \frac{3z}{14x^2y}$

Factors of $3xy$:
Factors of $14x^2y$:
LCD:

6. $\frac{10}{30xy}, \frac{1}{7xy^3}$

Factors of $30xy$:
Factors of $7xy^3$:
LCD:

7. $\frac{5}{x}, \frac{6}{x(x+1)}$

Factors of x:
Factors of $x(x+1)$:
LCD:

8. $\frac{7}{2y}, \frac{x}{y(y-1)}$

Factors of $2y$:
Factors of $y(y-1)$:
LCD:

9. $\frac{x}{(y-2)^2}, \frac{4x}{3(y-2)}$

Factors of $(y-2)^2$:
Factors of $3(y-2)$:
LCD:

10. $\frac{z}{(y+3)^2}, \frac{2z}{5(y+3)}$

Factors of $(y+3)^2$:
Factors of $5(y+3)$:
LCD:

11. $\frac{2x}{y^2-1}, \frac{x}{y^2-y-2}$

Factors of y^2-1:
Factors of y^2-y-2:
LCD:

12. $\frac{5x}{y^2+5y+6}, \frac{2x}{y^2-4}$

Factors of y^2+5y+6:
Factors of y^2-4:
LCD:

Problems 13–40: Perform the operations.

13. $\frac{9}{5} - \frac{4}{5}$

14. $\frac{143}{16} - \frac{15}{16}$

15. $\frac{19}{3} - \frac{11}{7}$

16. $\frac{3}{8} - \frac{5}{12}$

17. $\frac{3x}{4} + \frac{7x}{20}$

18. $\frac{x}{15} + \frac{7x}{25}$

19. $\frac{11}{y} + \frac{4}{y}$

20. $\frac{5}{x} + \frac{9}{x}$

21. $\frac{11}{3x} + \frac{7}{3x}$

22. $\frac{25}{2x} + \frac{35}{2x}$

23. $\frac{15}{y-5} - \frac{2}{y-5} - \frac{7}{y-5}$

24. $\frac{5}{x+3} - \frac{2}{x+3} + \frac{6}{x+3}$

25. $\frac{8}{3x} - \frac{5}{2x}$

26. $\frac{9}{5x} - \frac{1}{2x}$

27. $\frac{2}{ab} + \frac{8}{c}$

28. $\frac{3}{x} + \frac{4}{y}$

29. $\frac{2x}{3xy} + \frac{5}{4x^2y^2}$

30. $\frac{3x}{2y} + \frac{23}{5x^2y^2}$

31. $\frac{2}{(x-2)(x+3)} - \frac{10}{(x-2)}$

32. $\frac{5}{x+1} - \frac{3}{(x+1)(x+2)}$

33. $\frac{3}{x-y} + \frac{2xy}{x^2-y^2}$

34. $\frac{a}{a^2-b^2} + \frac{b}{a+b}$

35. $\frac{2x}{x^2+x} + \frac{3x}{x^2-1}$

36. $\frac{5y}{y^2-y} + \frac{3y}{y^2-1}$

37. $\dfrac{x}{x^2 - 4} - \dfrac{5x}{x + 2} - \dfrac{1}{x - 2}$

38. $\dfrac{6a}{a^2 - 9} - \dfrac{5}{a + 3} - \dfrac{2}{a - 3}$

39. $\dfrac{1}{3x^2 + 2x - 1} + \dfrac{1}{3x^2 - 4x + 1}$

40. $\dfrac{3}{2x^2 + 3x + 1} + \dfrac{5}{2x^2 - 5x - 3}$

41. Linda and Claude have a construction business and have been hired to replace the baseboards in a living room that has an unusual shape. The dimensions in feet of the room are shown in the figure.

What is the exact amount of baseboard required for the living room? (*Hint:* The distance around the outside of the room is called the perimeter.)

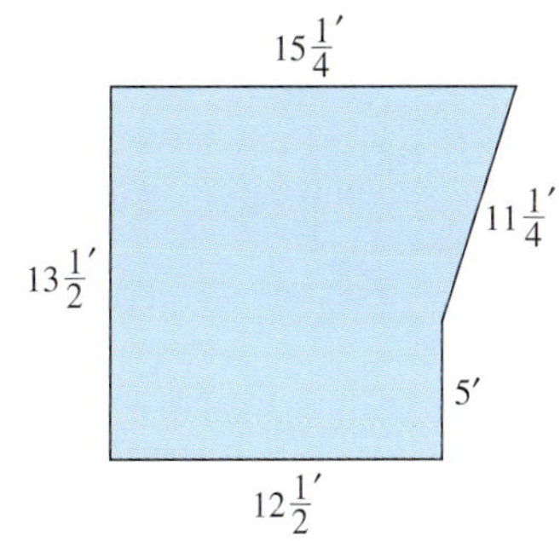

42. Jan is making stained glass jewelry boxes for gifts. Each box has decorative lead trim around the top. The dimensions of each box are $6\frac{3}{4}$ inches long by $4\frac{1}{2}$ inches wide. Jan plans to make 5 boxes for gifts this year. How much decorative lead trim should she purchase?

43. Tony is making music CDs as favors for the guests at his wedding. Each CD takes him $\frac{1}{3}$ of an hour to copy on his CD writer. His CD writer also requires an additional 4 minutes per CD to finalize and complete the copy process. How much time will it take Tony to copy and finalize 150 CDs for his guests?

Problems 44–47: Given the dimensions of each rectangle, represent the perimeter as a single expression in terms of x: Perimeter $= 2 \cdot \text{Length} + 2 \cdot \text{Width}$. Find the perimeter when $x = 4$ feet.

44. Length $= \dfrac{x}{2}$; Width $= \dfrac{x}{3}$

45. Length $= \dfrac{x}{4}$; Width $= \dfrac{x}{5}$

46. Length $= \dfrac{7}{x + 1}$ $(x \neq -1)$;
Width $= \dfrac{3}{x - 1}$ $(x \neq 1)$

47. Length $= \dfrac{9}{x + 2}$ $(x \neq -2)$;
Width $= \dfrac{5}{x - 2}$ $(x \neq 2)$

7.4 Introduction to Graphing Rational Functions

OBJECTIVES

- Identify restrictions of the independent and dependent variables.
- Graph a rational function.

1 2 3 Getting Started

Rational functions comprise a "family" of functions whose graphs are often curves with special characteristics. A function of the form $f(x) = \dfrac{p(x)}{q(x)}$ is called a **rational function**. The numerator and the denominator are polynomials and $q(x) \neq 0$.

1. The rational function $f(x) = \dfrac{1}{x}$ is called the reciprocal function.

(a) What value of the independent variable x is restricted in $f(x) = \dfrac{1}{x}$?

(b) Complete the table of values and look carefully at the behavior of $f(x)$ values as x gets closer and closer to the restricted value of $x \neq 0$ from the negative (or left) side and from the positive (or right) side. Write $f(x)$ values as decimals.

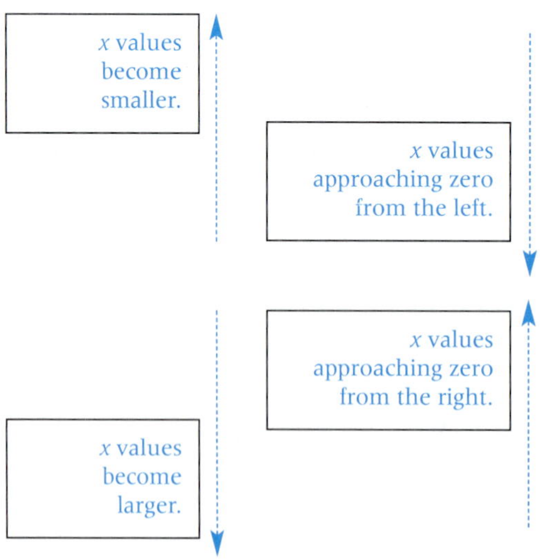

x	$f(x) = \dfrac{1}{x}$
−100	
−10	
−2	
−1	
−0.4	
−0.1	
−0.01	
0	**Undefined**
0.01	
0.4	
0.1	
1	
2	
10	
100	

(c) Are the $f(x)$ values increasing or decreasing as x approaches zero from the negative (or left) side?

(d) Are the $f(x)$ values increasing or decreasing as x approaches zero from the positive (or right) side?

(e) What constant value is $f(x)$ approaching as x becomes smaller and smaller and larger and larger?

(f) Will $f(x)$ ever equal zero?

2. Draw the graph of the curve of $f(x) = \frac{1}{x}$ using these steps.

Since $x \neq 0$, a darkened dashed line is drawn through $x = 0$ (on top of the y-axis). The line $x = 0$ is called a vertical asymptote and the graph will not cross it.

Since $y \neq 0$, a darkened dashed line is drawn through $y = 0$ (on top of the x-axis). The line $y = 0$ is called a horizontal asymptote and the graph will not cross it.

Plot the ordered pairs from the table of values in problem 1(b). It will be necessary to estimate the placement of the points with non-integer values. Connect the points with two curves and draw arrows on the ends of each. One curve should appear in Quadrant I, approaching the dashed asymptotes, and another curve should appear in Quadrant III, with a similar shape.

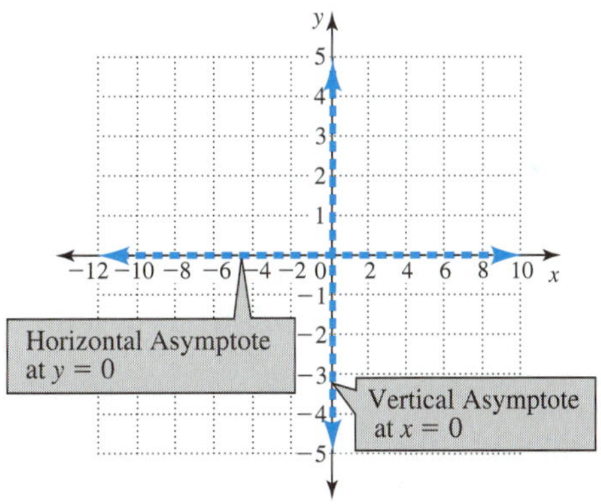

➢ *The curves should not touch nor cross the vertical or horizontal asymptotes. The two distinct curves form the graph of the function* $f(x) = \frac{1}{x}$.

3. Rational functions can be graphed on a graphing calculator. Your instructor will show you how to enter the function, $f(x) = \frac{1}{x}$ or $y_1 = \frac{1}{x}$, view a table of values, and graph the function to obtain these screens.

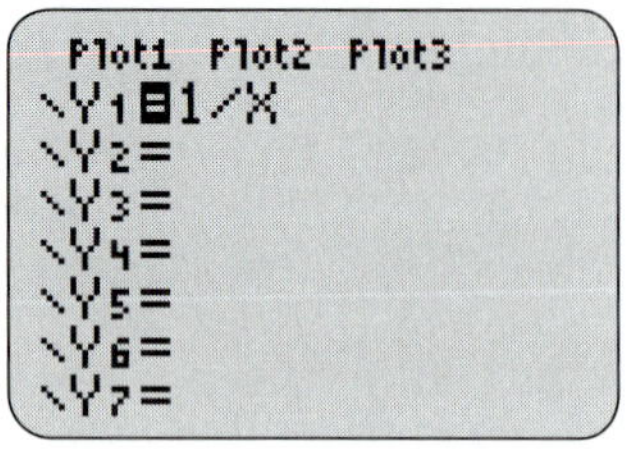

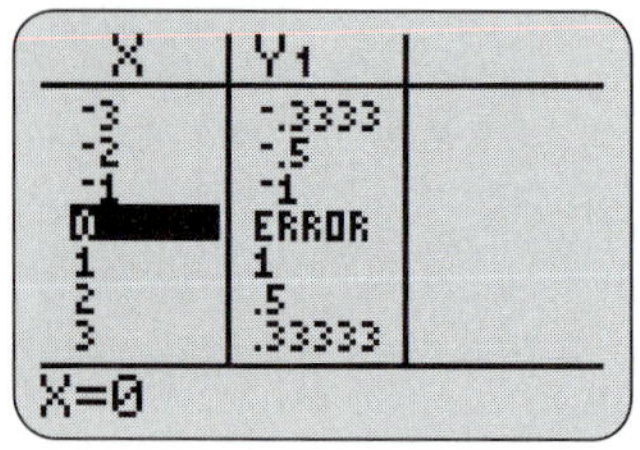

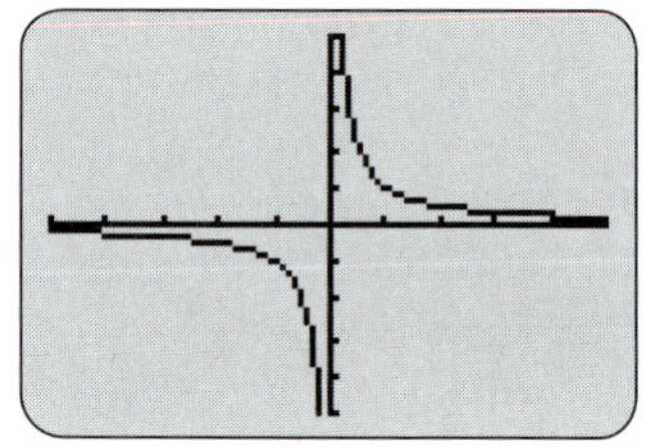

Why does the table of values shown above, generated by a graphing calculator, show ERROR for y_1 when $x = 0$?

Discussing Mathematics

A rational function is the quotient of two polynomials.

> **RATIONAL FUNCTION**
>
> A rational function is a function in the form of
>
> $$f(x) = \frac{p(x)}{q(x)}, \text{ where } p(x) \text{ and } q(x) \text{ are polynomials and } q(x) \neq 0.$$

Restricted values in the independent variable may be identified by setting the denominator of the rational function equal to zero and solving the equation(s). Restricted value(s) of the independent and dependent variables may define asymptote(s) of the graph of the function.

EXAMPLE 1: Identify the restricted value of x in $f(x) = \frac{5}{2x - 1}$.

Let $2x - 1 = 0$ and solve for x: $\quad 2x - 1 = 0$

$$x = \frac{1}{2}$$

$x \neq \frac{1}{2}$. $\frac{1}{2}$ is a restricted value of the function $f(x) = \frac{5}{2x - 1}$.

VERTICAL ASYMPTOTE

If $f(x)$ increases or decreases infinitely as x approaches the restricted value r, the restricted value defines a vertical asymptote. This behavior is written symbolically: $f(x) \to +\infty$ or $f(x) \to -\infty$ as $x \to r$ from the left or the right.

This means that "$f(x)$ approaches negative or positive infinity as x approaches a restricted value r from the left or the right." A vertical asymptote is drawn as a dashed vertical line on the graph at $x = r$.

➢ *Vertical asymptotes are identified from the restrictions in x and by looking carefully at the behavior of f(x) values on either side of the restricted values of x.*

HORIZONTAL ASYMPTOTE

If x increases or decreases infinitely as $f(x)$ approaches a constant c, a horizontal asymptote occurs at $f(x) = y = c$. This behavior is written symbolically: as $x \to -\infty$ or $x \to +\infty$, $f(x) \to c$.

This means that "as x approaches negative or positive infinity, $f(x)$ approaches a constant c." A horizontal asymptote is drawn as a dashed horizontal line on the graph.

➢ *Horizontal asymptotes are identified by looking at the behavior of f(x) as x becomes larger and smaller. The graph should not cross a horizontal asymptote for very large or very small values of x.*

GRAPHING A RATIONAL FUNCTION

1. Identify any restricted values (r) in the independent variable (x) by setting the denominator equal to zero and solving the equation.
2. Make a table of values:
   (a) Include at least three or four x values that are larger and smaller than the restricted value(s). These should include decimal values that are very close to the restricted value(s).
   (b) Include three or four x values that are very large and very small.
3. Identify any vertical asymptotes and draw dashed vertical lines on the grid. Look at the $f(x)$ values as x approaches the restricted value(s) (r) from the left and from the right. If $f(x)$ increases or decreases infinitely, a vertical asymptote is defined at $x = 4$.
4. Identify any horizontal asymptotes and draw dashed horizontal lines. Look at $f(x)$ values as x takes on very large and very small values. If the $f(x)$ values approach the same value (c), a horizontal asymptote is defined at $y = c$.
5. Plot the points.
6. Connect the points. Do not cross any vertical asymptotes.
7. Draw arrows on the ends of the curves to denote that the curve continues in that pattern.

EXAMPLE 2: Graph $f(x) = \dfrac{5}{x+2}$.

$$x + 2 = 0$$
$$x = -2$$
$$x \neq -2 \text{ in the function}$$
$$f(x) = \frac{5}{x+2}$$

To identify restricted values in the independent variable (x), set the denominator of the function equal to zero and solve the equation.

Make a table of values. Be sure to include x values that are smaller and larger than the restricted value. Look for vertical and horizontal asymptotes. Draw a vertical asymptote at $x = -2$. Draw a horizontal asymptote at $y = 0$. Plot the points. Connect the points with curves. Draw arrows on the ends of the curves.

As $x \to -\infty$, $f(x) \to 0$.

$f(x) \to -\infty$ as $x \to -2$ from the left.

$f(x) \to +\infty$ as $x \to -2$ from the right.

As $x \to +\infty$, $f(x) \to 0$.

x	$f(x) = \dfrac{5}{x+2}$
−1000	−0.01
−100	−0.05
−10	−0.10
−6	−1.25
−5	−1.67
−4	−2.5
−3	−5
−2.1	−50
−2	**Undefined**
−1.9	50
−1	5
0	2.5
1	1.67
2	1.25
10	0.42
100	0.05
1000	0.01

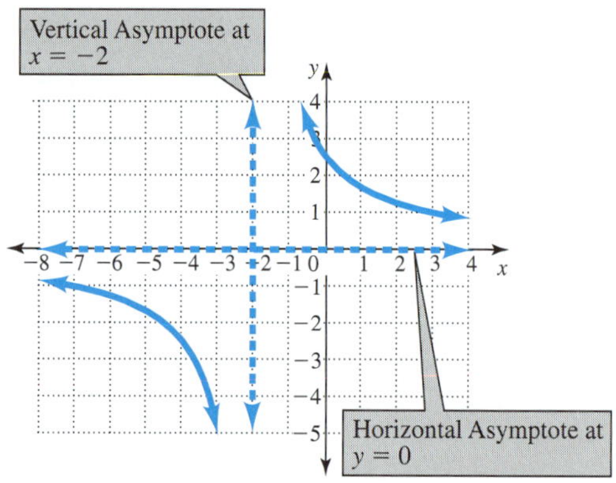

EXAMPLE 3: Graph $f(x) = \dfrac{5}{x^2 + 2x - 3}$.

$$x^2 + 2x - 3 = 0$$
$$(x - 1)(x + 3) = 0$$
$$x - 1 = 0 \qquad x + 3 = 0$$
$$x = 1 \qquad\qquad x = -3$$

To identify restricted values in the independent variable (x), factor the denominator of the function, apply the Zero-Product Property, and solve each equation. $x \neq 1$ and $x \neq -3$ are restricted values in the function.

Make a table of values. Be sure to include x values that are smaller than, larger than, and between the restricted values. Draw vertical asymptotes at $x = -3$ and $x = 1$. Draw a horizontal asymptote at $y = 0$. Plot the points. Connect the points with curves. Draw arrows on the ends of the curves.

As $x \to -\infty$, $f(x) \to 0$.

$f(x) \to +\infty$ as $x = -3$ from the left and $f(x) \to -\infty$ as $x = -3$ from the right.

$f(x) \to -\infty$ as $x = 1$ from the left and $f(x) \to +\infty$ as $x = 1$ from the right.

As $x \to +\infty$, $f(x) \to 0$.

x	$f(x)$
−100	0.0005
−50	0.0021
−10	0.06
−4	1
−3.5	2.22
−3.01	124.69
−3	**Undefined**
−2.99	−125.3
−2	−1.67
−1	−1.25
0	−1.67
0.99	−125.3
1	**Undefined**
1.01	124.69
1.5	2.22
2	1
10	0.04
50	0.0019
100	0.0005

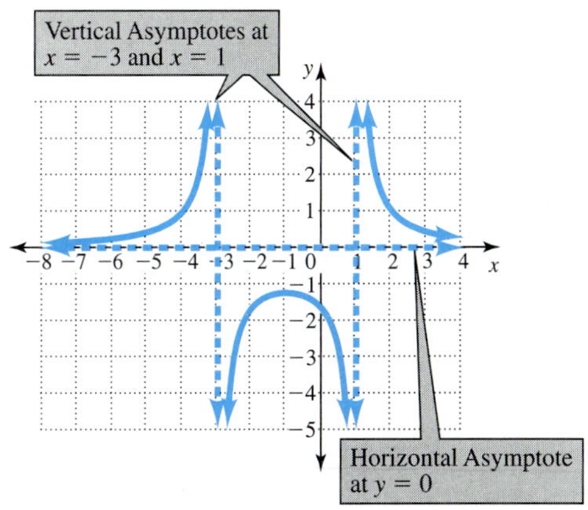

EXAMPLE 4: Graph $f(x) = \dfrac{3}{x^2 + 2x + 1}$.

$$x^2 + 2x + 1 = 0$$
$$(x + 1)(x + 1) = 0$$
$$x + 1 = 0$$
$$x = -1$$

Factor the denominator of the function, apply the Zero-Product Property, and solve each equation.
$x \neq -1$ is restricted in the function.

Make a table of values. Draw a vertical asymptote at $x = -1$. Draw a horizontal asymptote at $y = 0$. Plot the points. Connect the points with curves. Draw arrows on the ends of the curves.

As $x \to -\infty$, $f(x) \to 0$.

$f(x) \to \infty$ as $x = -1$ from the left and from the right.

As $x \to +\infty$, $f(x) \to 0$.

x	$f(x)$
-100	0.0003
-10	0.0370
-4	$0.\overline{3}$
-3	0.75
-2	3
-1.1	300
-1	**Undefined**
-0.9	300
0	3
1	0.75
2	$0.\overline{3}$
10	0.0248
100	0.00003

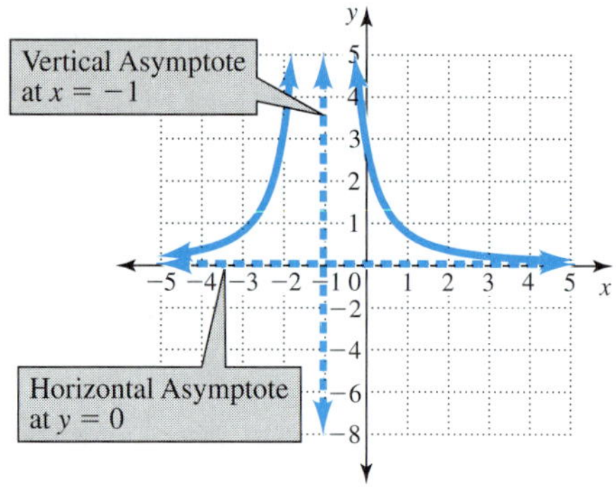

USING A GRAPHING CALCULATOR TO GRAPH RATIONAL FUNCTIONS

The graph of the rational function, $f(x) = \dfrac{5}{x^2 + 2x - 3}$, also can be generated on a graphing calculator.

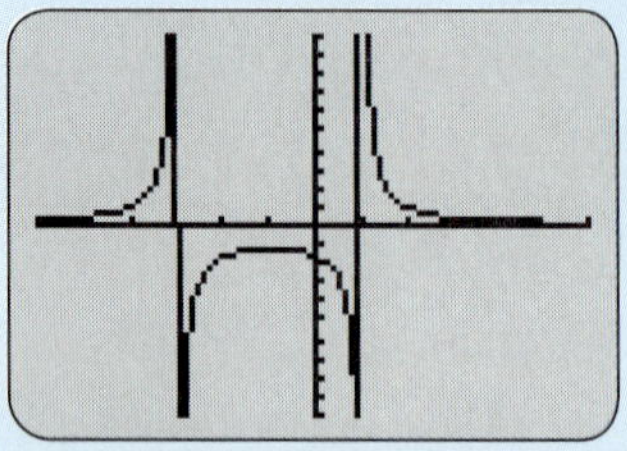
$[-6, 6]$ Xscl 1 $[-10, 10]$ Yscl 1

Some graphing calculators will appear to draw vertical asymptotes as *solid lines* as the calculator attempts to connect all the points in the table of values. It is essential that these lines on a graphing calculator be recognized as vertical asymptotes and not part of the graph of the function.

1 2 3

Doing Mathematics

- Write the definitions of the words and symbols listed for this section in the Dictionary in Appendix A.
- On a 3 × 5-inch card, write the steps to graph a rational function for future reference.

Problems 1–10: Determine the restricted value(s) of the independent variable. Complete the table of values rounding decimals to two decimal places. Draw any asymptotes, plot the points, and connect the points with curves.

1. Graph the function $f(x) = \dfrac{1}{x - 3}$.

x	$f(x)$

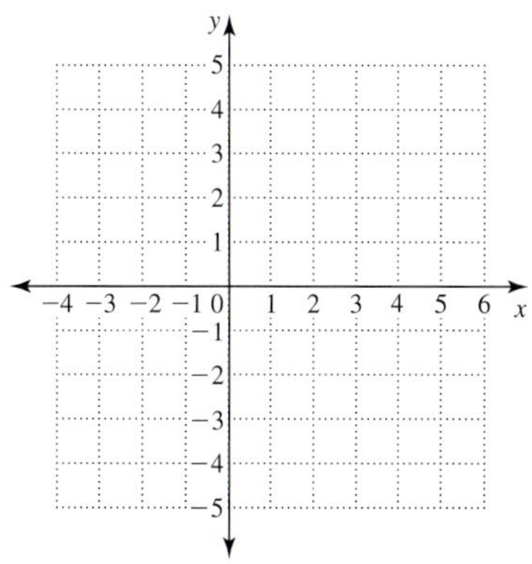

2. Graph the function $f(x) = \dfrac{1}{x - 2}$.

x	$f(x)$

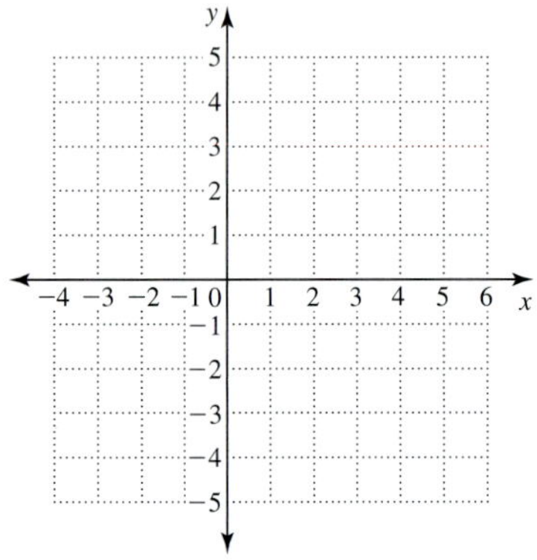

3. Graph the function $f(x) = \dfrac{3}{x - 1}$.

x	$f(x)$

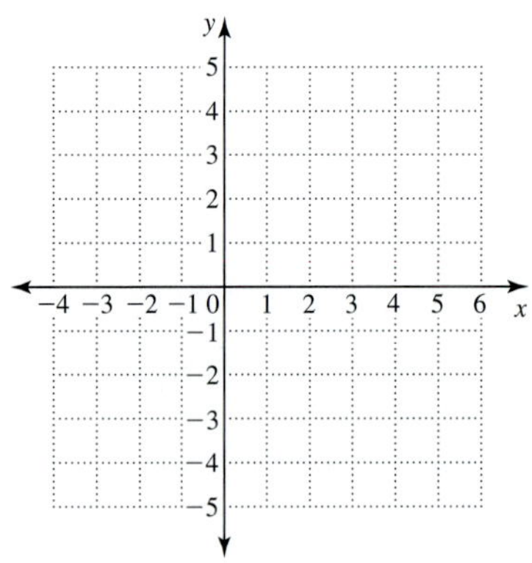

4. Graph the function $f(x) = \dfrac{2}{x + 3}$.

x	$f(x)$

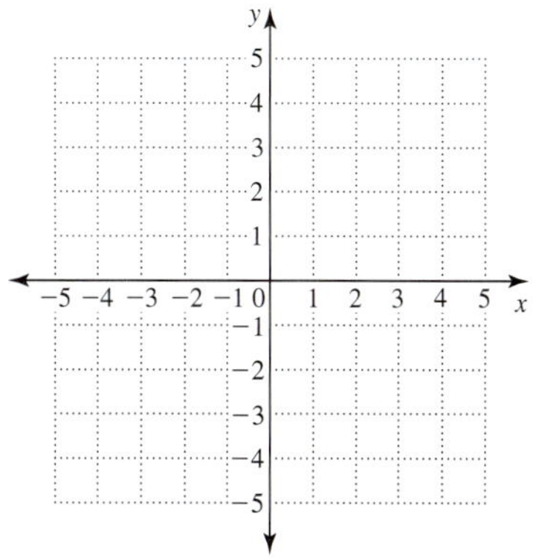

5. Graph the function $f(x) = \dfrac{1}{x^2 - x - 2}$.

x	$f(x)$

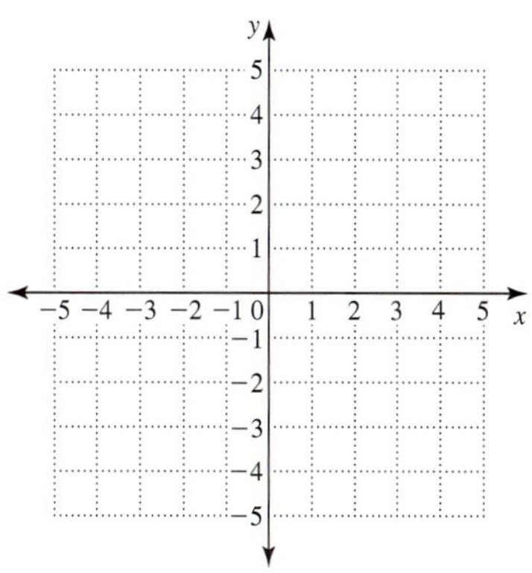

6. Graph the function $f(x) = \dfrac{2}{x^2 + 2x - 3}$.

x	$f(x)$

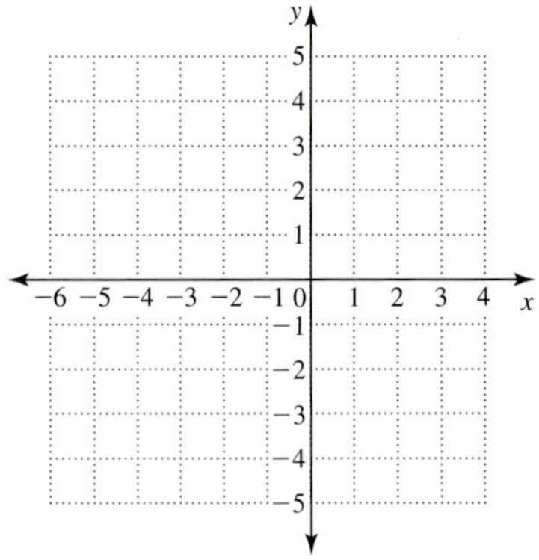

7. Graph the function $f(x) = \dfrac{1}{x^2 - x}$.

x	$f(x)$

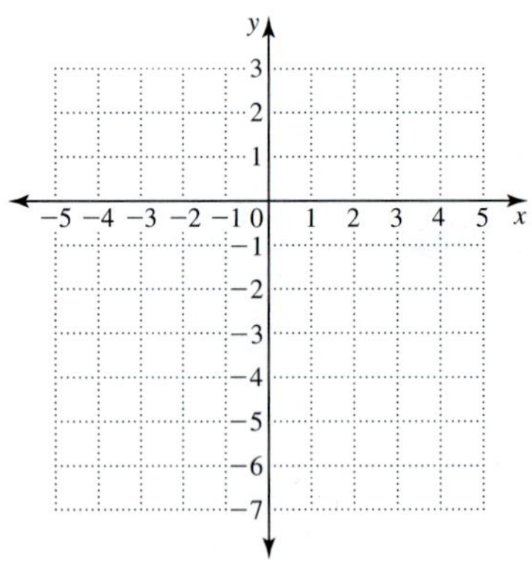

8. Graph the function $f(x) = \dfrac{1}{x^2 + x}$.

x	$f(x)$

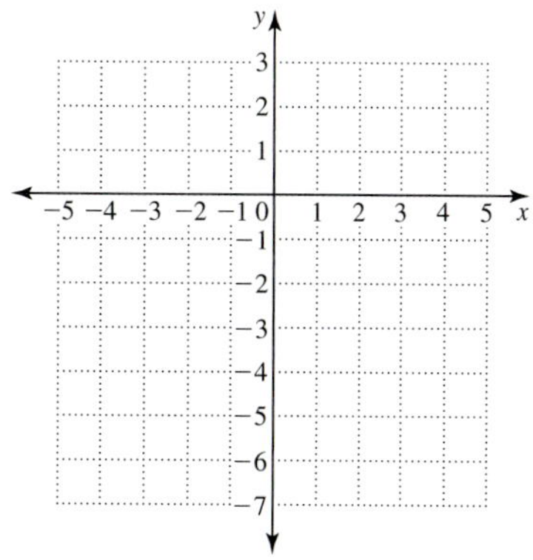

9. Graph the function $f(x) = \dfrac{1}{x^2 - 2x - 8}$.

x	$f(x)$

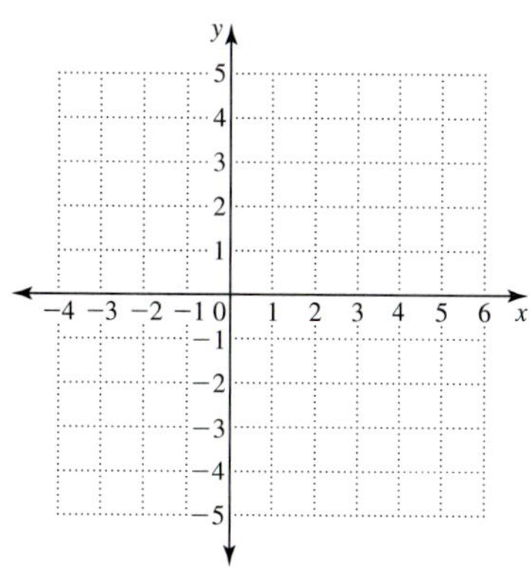

10. Graph the function $f(x) = \dfrac{1}{x^2 - 4}$.

x	$f(x)$

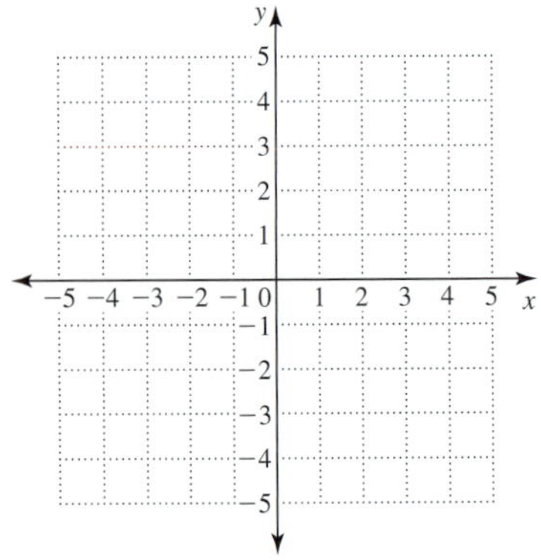

Problems 11–13: Writing Exercises

11. Why do restrictions exist in the independent variable and dependent variable of a rational function?

12. Describe in your own words why zero divided by a nonzero number is zero.

13. Describe in your own words why division by zero is undefined.

7.5 Solving Rational Equations

OBJECTIVE

- Solve rational equations analytically (and graphically, if desired).

Getting Started

A **rational equation** is an equation containing one or more rational expressions. Examples of rational equations are as follows:

$$\frac{1}{3}+\frac{1}{5}=\frac{1}{t} \qquad \frac{5}{6x}-\frac{1}{2}=\frac{1}{x}-\frac{1}{3} \qquad \frac{2}{x-2}-\frac{5}{x+2}=\frac{4x}{x^2-4}.$$

We will now use the techniques presented earlier in this chapter for solving rational equations.

1. Suppose you had a few pieces of pizza left from 8-cut pizzas and 15 pieces of pizza left from 12-cut pizzas. How many pieces of pizza from the 8-cut pizzas must be combined with the 15 pieces of 12-cut pizzas to form two whole pizzas?

(a) Draw a picture of two pizzas made from the 15 pieces of 12-cut pizza and the appropriate number of pieces of 8-cut pizza.

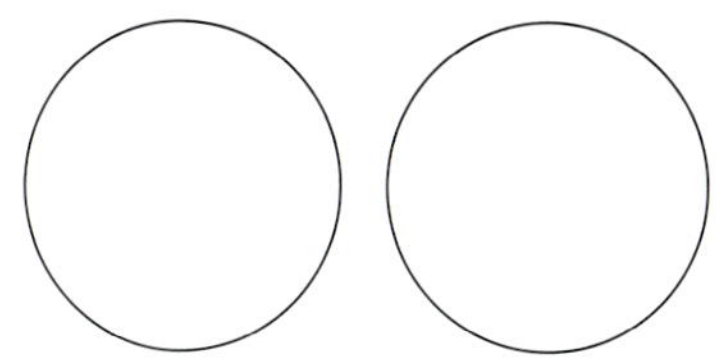

(b) How many pieces of 8-cut pizza must be combined with 15 pieces of 12-cut pizza to form two whole pizzas?

(c) Let x represent the unknown number of pieces of 8-cut pizza to be combined with 15 pieces of 12-cut pizza. To solve the equation describing the problem situation, $\frac{x}{8}+\frac{15}{12}=2$, build all fractions in the equation to a common denominator of 24 and then solve for x:

(d) The graphical solution of an equation is the x-coordinate of the x-intercept. Graph $y=\frac{x}{8}+\frac{15}{12}-2$ and use the X-Intercept Graphical Method to confirm the analytical answer.

x	$y=\frac{x}{8}+\frac{15}{12}-2$
−2	
0	
2	
4	
6	
8	

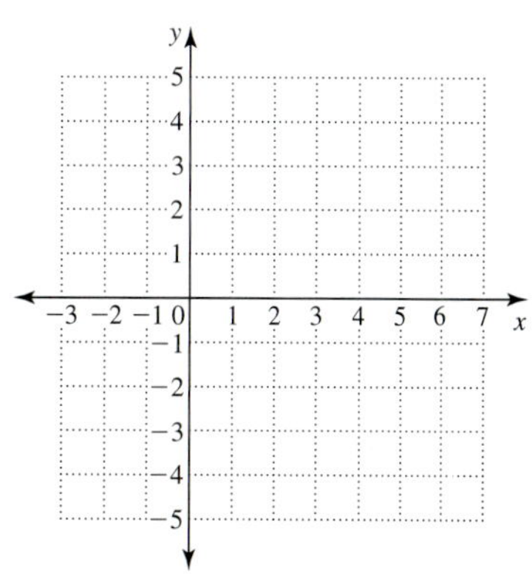

(e) Write the x-coordinate of the x-intercept. Does your graphical solution match your analytical solution? If not, what should you do?

1 2 3

Discussing Mathematics

Rational equations can be solved using either of two methods—Method 1: Building fractions or Method 2: Clearing fractions. Methods 1 and 2 are presented in Example 1. However, clearing fractions is usually the most efficient process for solving rational equations, therefore Method 2 is used exclusively in Examples 2–5. Notice that steps 2 and 3 are different, but steps 1, 4, 5, and 6 are the same.

Method 1—Solving a Rational Equation by Building All Fractions to the LCD	Method 2—Solving a Rational Equation by Clearing Fractions
This method uses the fact that when denominators are equivalent, the numerators are equivalent.	*Multiplying an equation by a common multiple is called "clearing" fractions because each term reduces to a quantity without fractions. Any common multiple will accomplish our goal of solving the equation, but we will choose the least common multiple (LCD) to make calculations easiest.*
1. Factor all denominators in the equation and identify the LCD.	
2. Build each fraction into an equivalent fraction with the LCD as the denominator.	2. Multiply every term in the equation by the LCD.
3. Write a new equation, setting the numerators equal.	3. Divide out common factors in each term. Multiply remaining factors.
4. Solve for the unknown.	
5. Check the answer(s) analytically.	
6. Confirm the solution graphically, if desired.	

EXAMPLE 1: Solve the equation $\frac{5}{6x} - \frac{1}{2} = \frac{1}{x} - \frac{1}{3}$.

Method 1—Building fractions | Method 2—Clearing fractions

1. Identify the LCD $= 2 \cdot 3 \cdot x = 6x$

Method 1

2. Build all fractions to a denominator of $6x$:

$$\frac{5}{6x} - \frac{3x \cdot 1}{3x \cdot 2} = \frac{6 \cdot 1}{6 \cdot x} - \frac{2x \cdot 1}{2x \cdot 3}$$

$$\frac{5}{6x} - \frac{3x}{6x} = \frac{6}{6x} - \frac{2x}{6x}$$

3. Write a new equation, setting the numerators equal:
$5 - 3x = 6 - 2x$

When the denominators of an equation are equivalent, the numerators are equivalent (the result of multiplying the equation by $6x$).

Method 2

2. Multiply each term in the equation by the LCD $= \frac{6x}{1}$:

$$\frac{6x}{1} \cdot \left(\frac{5}{6x}\right) - \frac{6x}{1} \cdot \left(\frac{1}{2}\right) = \frac{6x}{1} \cdot \left(\frac{1}{x}\right) - \frac{6x}{1} \cdot \left(\frac{1}{3}\right)$$

3. Divide out common factors in each term:

$$\frac{\overset{1}{\cancel{6x}}}{1} \cdot \left(\frac{5}{\underset{1}{\cancel{6x}}}\right) - \frac{\overset{3}{\cancel{6}}x}{1} \cdot \left(\frac{1}{\underset{1}{\cancel{2}}}\right) = \frac{6\overset{1}{\cancel{x}}}{1} \cdot \left(\frac{1}{\underset{1}{\cancel{x}}}\right) - \frac{\overset{2}{\cancel{6}}x}{1} \cdot \left(\frac{1}{\underset{1}{\cancel{3}}}\right)$$

$$1 \cdot 5 - 3 \cdot x = 6 \cdot 1 - 2x \cdot 1$$

$$5 - 3x = 6 - 2x$$

4. Solve for x: $5 - 3x = 6 - 2x$

$$5 = 6 + x$$

$$-1 = x$$

5. Verify the answer by substituting into the original equation $\frac{5}{6x} - \frac{1}{2} = \frac{1}{x} - \frac{1}{3}$.

$$\text{Let } x = -1: \quad \frac{5}{6(-1)} - \frac{1}{2} \stackrel{?}{=} \frac{1}{(-1)} - \frac{1}{3}$$

$$-\frac{5}{6} - \frac{3}{6} \stackrel{?}{=} -\frac{3}{3} - \frac{1}{3}$$

$$\frac{-5-3}{6} \stackrel{?}{=} \frac{-3-1}{3}$$

$$\frac{-4}{3} = \frac{-4}{3}$$

✓ It checks.

EXAMPLE 2: Solve the equation by clearing fractions (Method 2).

$$\frac{\overset{5}{\cancel{15}}}{1}\cdot\left(\frac{x-2}{\underset{1}{\cancel{3}}}\right) - \frac{\overset{3}{\cancel{15}}}{1}\cdot\left(\frac{x-3}{\underset{1}{\cancel{5}}}\right) = \frac{\overset{1}{\cancel{15}}}{1}\cdot\left(\frac{13}{\underset{1}{\cancel{15}}}\right)$$

Clear fractions by multiplying each term of the equation by the LCD $= \frac{15}{1}$. Divide out common factors in each term.

$$5(x-2) - 3(x-3) = 1\cdot 13$$

Multiply remaining factors.

$$5x - 10 - 3x + 9 = 13$$

Solve for x.

$$2x - 1 = 13$$

$$2x = 14$$

$$x = 7$$

Verify the answer by substituting into the original equation. ✓ It checks.

EXAMPLE 3: Solve the equation by clearing fractions (Method 2).

$$\frac{2}{(x-2)} - \frac{5}{(x+2)} = \frac{4x}{(x-2)(x+2)}$$

Factor the denominators and identify the LCD.

$$\frac{\overset{1}{\cancel{(x-2)}}(x+2)}{1}\cdot\left(\frac{2}{\underset{1}{\cancel{(x-2)}}}\right) - \frac{(x-2)\overset{1}{\cancel{(x+2)}}}{1}\cdot\left(\frac{5}{\underset{1}{\cancel{(x+2)}}}\right)$$

Clear fractions by multiplying each term of the equation by the LCD $= \frac{(x-2)(x+2)}{1}$.

$$= \frac{\overset{1}{\cancel{(x-2)}}\overset{1}{\cancel{(x+2)}}}{1}\cdot\left(\frac{4x}{\underset{1}{\cancel{(x-2)}}\underset{1}{\cancel{(x+2)}}}\right)$$

$$\frac{(x+2)}{1}\cdot\frac{2}{1} - \frac{(x-2)}{1}\cdot\frac{5}{1} = \frac{1}{1}\cdot\frac{4x}{1}$$

Divide out common factors in each term.

$$(x+2)\cdot 2 - (x-2)\cdot 5 = 1\cdot 4x$$

Multiply remaining factors.

$$2x + 4 - 5x + 10 = 4x$$

Solve for x.

$$-3x + 14 = 4x$$

$$14 = 7x$$

$$2 = x$$

$$\text{Let } x = 2: \quad \frac{2}{2-2} - \frac{5}{2+2} \stackrel{?}{=} \frac{4(2)}{2^2-4}$$

$$\frac{2}{0} - \frac{5}{4} = \frac{8}{0} \longleftarrow \boxed{\text{Undefined}}$$

Verify the answer by substituting into the original equation.

$\frac{2}{x-2} - \frac{5}{x+2} = \frac{4x}{x^2-4}$. Division by zero is undefined. **The equation has no solution.**

To confirm the solution of Example 3 graphically use the x-intercept method and graph

$$y = \frac{2}{x-2} - \frac{5}{x+2} - \frac{4x}{x^2-4}.$$

The graph has no x-intercept, confirming that there is no real number value of the unknown that satisfies the equation.

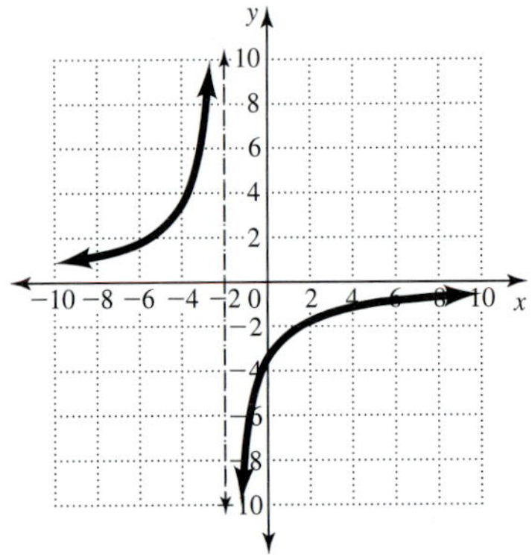

EXAMPLE 4: Solve the rational equation $x + \frac{1}{x} = \frac{5}{2}$ by clearing fractions.

$\frac{2x}{1}\cdot\left(\frac{x}{1}\right) + \frac{\overset{1}{2\cancel{x}}}{1}\cdot\left(\frac{1}{\underset{1}{\cancel{x}}}\right) = \frac{\overset{1}{\cancel{2}x}}{1}\cdot\left(\frac{5}{\underset{1}{\cancel{2}}}\right)$	Identify the LCD and multiply each term of the equation by the LCD $= \frac{2x}{1}$.
$2x\cdot x + 2\cdot 1 = x\cdot 5$	Divide out common factors in each term. Multiply remaining factors.
$2x^2 + 2 = 5x$	
$2x^2 - 5x + 2 = 0$	
$(2x-1)(x-2) = 0$	Solve the resulting quadratic equation by factoring.
$2x - 1 = 0$ or $x - 2 = 0$	
$x = \frac{1}{2}$ or $x = 2$	

Let $x = \frac{1}{2}$:	Let $x = 2$:	
$\frac{1}{2} + \frac{1}{\frac{1}{2}} \stackrel{?}{=} \frac{5}{2}$	$2 + \frac{1}{2} \stackrel{?}{=} \frac{5}{2}$	Verify the answers by substituting into the original equation $x + \frac{1}{x} = \frac{5}{2}$.
$\frac{1}{2} + \frac{2}{1} \stackrel{?}{=} \frac{5}{2}$	$\frac{4}{2} + \frac{1}{2} \stackrel{?}{=} \frac{5}{2}$	
$\frac{1}{2} + \frac{4}{2} \stackrel{?}{=} \frac{5}{2}$	$\frac{5}{2} = \frac{5}{2}$	
$\frac{5}{2} = \frac{5}{2}$		✓ Both solutions check.

To confirm the solution of Example 4 graphically use the x-intercept method and graph $y = x + \frac{1}{x} - \frac{5}{2}$. The graph has two x-intercepts, $x = \frac{1}{2}$ and $x = 2$.

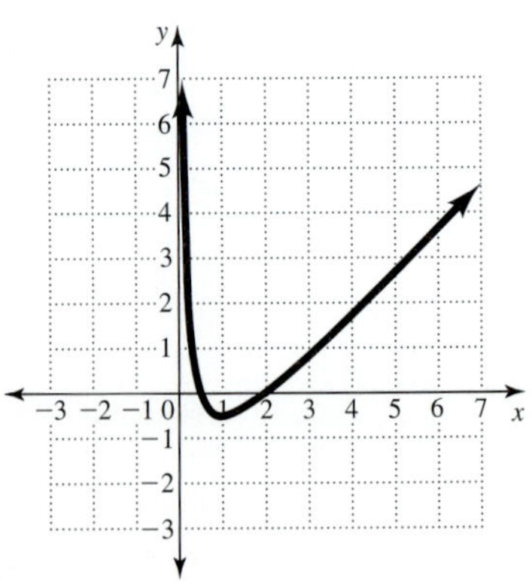

SOLVING REAL-WORLD PROBLEMS

EXAMPLE 5: Christopher and Jeremy have been hired to spread 10 cubic yards of mulch for a friend. Christopher estimates that he could spread all 10 cubic yards of mulch in 3 hours if he did it alone. Jeremy estimates that he could complete the job in 5 hours if he did it alone. It should take less time to complete the job if Christopher and Jeremy work together than if either person did the job alone. How long will it take them to spread the 10 cubic yards of mulch if they work together?

Let t represent the total number of hours to complete the job when Christopher and Jeremy are working together. The problem situation can be described in this way:

Christopher can complete $\frac{1}{3}$ of the job in 1 hour.	Jeremy can complete $\frac{1}{5}$ of the job in 1 hour.	Working together, they can complete $\frac{1}{t}$ of the job in 1 hour.

Amount of the job Christopher can complete in 1 hour	+	Amount of the job Jeremy can complete in 1 hour	=	Amount of the job completed in 1 hour working together
$\frac{1}{3}$		$\frac{1}{5}$		$\frac{1}{t}$

Multiply the equation by the LCD, $15t$, and solve for t:

$$\frac{15t}{1}\cdot\frac{1}{3} + \frac{15t}{1}\cdot\frac{1}{5} = \frac{15t}{1}\cdot\frac{1}{t}$$ Multiply each term by $15t$ and divide out common factors.

$$5t + 3t = 15$$

$$8t = 15$$ Combine LIKE terms.

$$t = \frac{15}{8} = 1.875$$ Number of hours to complete the job, working together.

Christopher and Jeremy rounded their answer above to the nearest hour, 2 hours, to compute the amount to charge their friend:
2 hours multiplied by \$15 per hour per person $= 2\cdot 15\cdot 2 =$ \$60.

COMPARING THE OPERATION OF ADDING/SUBTRACTING RATIONAL EXPRESSIONS TO SOLVING RATIONAL EQUATIONS

The process of clearing fractions is appropriate *only when solving equations*. Do not clear fractions when adding or subtracting fractions.

EXAMPLE 6: Comparing adding fractions to solving an equation involving similar fractions.

Adding fractions: $\frac{1}{3} + \frac{1}{5}$

$\frac{1}{3} + \frac{1}{5} = ?$	**The goal is to ADD the factions.**
$\frac{5}{15} + \frac{3}{15}$	Build fractions with a denominator of the LCD = 15
$\frac{8}{15}$	Add the numerators and keep the denominator.

When adding the fractions, $\frac{1}{3} + \frac{1}{5}$, the sum includes the denominator.

➢ *Remember: When adding or subtracting fractions, keep the common denominator (LCD) throughout the process and reduce your final answer.*

Solving for *t*: $\frac{1}{3} + \frac{1}{5} = \frac{1}{t}$

$\frac{1}{3} + \frac{1}{5} = \frac{1}{t}$	**The goal is to SOLVE for *t*.**
$\frac{15t}{1} \cdot \frac{1}{3} + \frac{15t}{1} \cdot \frac{1}{5} = \frac{15t}{1} \cdot \frac{1}{t}$	Clear fractions by multiplying each term of the equation by the LCD = 15*t*.
$5t + 3t = 15$	
$8t = 15$	Solve for *t*.
$t = \frac{15}{8}$	

When solving the equation, $\frac{1}{3} + \frac{1}{5} = \frac{1}{t}$, the process of solving a rational equation includes the process of clearing fractions.

➢ *Remember: When solving an equation containing fractions, clear the fractions by multiplying by the LCD, and solve for the unknown.*

1 2 3 Doing Mathematics

- On a 3 × 5-inch card, write the process of solving a rational equation by clearing fractions for future reference.

Problems 1–10: Solve each equation by any method. Verify the solution by substituting into the original equation. Confirm solutions graphically, if desired.

1. $\frac{6}{7} = \frac{x}{84}$

2. $\frac{3}{5} = \frac{x}{10}$

3. $\frac{x+5}{4} = \frac{6}{x}$

4. $\frac{x-3}{5} = \frac{8}{x}$

5. $\frac{x+5}{x-2} = \frac{x+1}{x+3}$

6. $\frac{x+3}{x+1} = \frac{x+1}{x-3}$

7. $\frac{2y-1}{3} = \frac{y+2}{4}$

8. $\frac{2x}{5} = \frac{36}{5}$

9. $\frac{3}{x-1} = \frac{3}{5}$

10. $\frac{10}{x-2} = \frac{5}{4}$

Problems 11–38: Solve each rational equation. Verify the answer(s) by substituting into the original equation. Confirm solutions graphically, if desired.

11. $\frac{7}{6} + x = \frac{5}{8}$

12. $\frac{5}{12} + x = \frac{3}{4}$

13. $\frac{1}{4}y - \frac{1}{5}y = 1$

14. $\frac{1}{5}y - \frac{1}{6}y = 1$

15. $\frac{7}{x} = \frac{6}{x} - 2$

16. $\frac{3}{x} = \frac{2}{x} - 1$

17. $x - \frac{6}{x} = 1$

18. $x + \frac{12}{x} = -7$

19. $\frac{x}{4} + \frac{2x}{3} = \frac{55}{12}$

20. $\frac{x}{5} + \frac{1}{6} = \frac{17}{30}$

21. $\frac{5}{6} - \frac{1}{3x} = \frac{3}{2}$

22. $\frac{1}{2x} + \frac{5}{8} = \frac{3}{4}$

23. $\frac{x - 2}{2x} - \frac{1}{x} = \frac{3}{4}$

24. $\frac{2}{y} + \frac{y - 1}{3y} = \frac{2}{5}$

25. $\frac{2}{3x} + \frac{5}{6x} = \frac{1}{4}$

26. $\frac{1}{2x} + \frac{1}{4x} = \frac{3}{8}$

27. $\frac{x}{x - 2} = \frac{2}{x - 2} - 2$

28. $\frac{x}{x - 1} = \frac{1}{x - 1} - 1$

29. $\frac{11}{3x + 2} = \frac{5}{3x + 2} + \frac{3}{4}$

30. $\frac{2}{5x - 1} = \frac{4}{5x - 1} + \frac{1}{3}$

31. $\frac{y}{y - 1} = \frac{3y}{y - 1} + 1$

32. $\frac{2x}{x + 3} + 1 = \frac{4x}{x + 3}$

33. $1 - \frac{x}{x + 1} = \frac{-1}{x - 1}$

34. $1 + \frac{3}{x - 2} = \frac{x}{x + 2}$

35. $\dfrac{2x}{x-3} + 5 = \dfrac{2x}{x-3}$

36. $\dfrac{5x}{x+1} = \dfrac{5x}{x+1} - 7$

37. $\dfrac{5}{3x+2} + \dfrac{13}{3x^2-x-2} = \dfrac{-7}{x-1}$

38. $\dfrac{6}{x+3} - \dfrac{9}{2x-7} = \dfrac{2x-59}{2x^2-x-21}$

Problems 39–45: Solve each rational equation. (Hint—Use factoring before and after clearing fractions in 39–42.)

39. $\dfrac{a}{a+1} = \dfrac{a-7}{a^2-1} + \dfrac{3}{a-1}$

40. $\dfrac{x+9}{x^2-4} = \dfrac{x}{x+2} + \dfrac{5}{x-2}$

41. $\dfrac{3}{3-x} + 1 = \dfrac{x}{2-x}$

42. $\dfrac{x+1}{5} + \dfrac{4}{x} = 2$

43. $\dfrac{x+2}{x+4} - \dfrac{2x^2-3x}{3x^2+14x+8} = \dfrac{x-3}{3x+2}$

44. $\dfrac{x-3}{x-4} = \dfrac{x^2}{2x^2-5x-12} + \dfrac{x}{2x+3}$

45. $\dfrac{x^2-2}{3x^2+10x-8} = \dfrac{x+2}{x+4} - \dfrac{2x-1}{3x-2}$

46. After a math club meeting, Bill had 9 pieces of pizza that were remaining from a 12-cut pizza and 11 pieces remaining from an 8-cut pizza. How many pieces of 8-cut pizza must be combined with the 9 pieces of 12-cut pizza to form 2 whole pizzas?

47. The Cheesecake Emporium sells cheesecakes cut in 12 pieces or 8 pieces. Customers may choose any flavor of cheesecake and any number of pieces to form a whole cheesecake to purchase. How many pieces of 12-cut chocolate cheesecake should be combined with 2 pieces of 8-cut pumpkin cheesecake to form one whole cheesecake?

48. Justin can usually fill his swimming pool in 7 hours using water pumped from his well. If he fills the pool using city water from a garden hose in addition to his well water, Justin can complete the task in 4 hours. The weather has been very dry at Justin's house and his well is empty. How long will it take to fill the pool using only the city water from a garden hose?

49. Wendy and her brother Rob have a lawn mowing business. Wendy is a runner and likes to walk very quickly behind the mower. She can mow an acre of lawn in 40 minutes. It takes Rob 60 minutes to mow an acre. How many minutes will it take if Wendy and Rob are mowing the acre of lawn together?

50. The length of a rectangular garden is 8 feet longer than its width. The ratio of the length to the width is $\frac{4}{3}$. Find the dimensions of the garden.

51. The width of a rectangular garden is 3 feet shorter than its length. The ratio of the width to the length is $\frac{2}{5}$. Find the dimensions of the garden.

52. Consider the following experiment:

(i) Wash and empty a 2-liter plastic bottle.

(ii) Make a small hole (about $\frac{1}{8}''$ in diameter) in the bottom of the bottle.

(iii) Cover the hole with your finger and completely fill the bottle with water.

(iv) Placing the bottle over a sink, carefully uncover the hole. Using a stopwatch determine how long it takes to drain the water from the bottle.

(v) Record this value as S.

(vi) Make a separate larger hole (about $\frac{3}{8}''$ in diameter) in the bottom of the bottle.

(vii) Cover both holes and completely fill the bottle with water.

(viii) Placing the bottle over a sink, carefully uncover both holes (at the same time). Using a stopwatch, determine how long it takes to drain the water from the bottle.

(ix) Record this value as T.

S represents the time it takes to empty the bottle with the small hole alone.

T represents the time it takes to empty the bottle with the small hole and larger hole together.

(a) Find the time it should take to empty the bottle with the larger hole alone.

(b) Actually perform the experiment of emptying the bottle uncovering only the larger hole. How does this result compare with your answer from part (a) above?

7.6 Complex Fractions

OBJECTIVES

- Identify a complex fraction.
- Simplify complex fractions.

Getting Started

The fraction bar is a symbol for the operation of division. The fraction $\frac{6}{2}$ represents the operation of dividing 6 by 2. What happens if the denominator is a fraction? What happens when both the numerator and denominator are fractions? In this section, we discuss **complex fractions**, where the numerator and/or denominator are fractions or the sum or difference of fractions.

1. Teresa is planning dessert and has 6 apples to use. She isn't sure how many people will attend dinner. Let's help Teresa determine how many people can be served if the apples are divided into different size portions.

(a) Draw a picture of the result of dividing 6 apples into servings of 3 apples. How many people can be served if each person receives 3 apples?

(b) Draw a picture of the result of dividing 6 apples into servings of 2 apples. How many people can be served if each person receives 2 apples?

(c) Let's continue decreasing the serving size. Draw a picture of the result of dividing 6 apples into servings of $\frac{1}{2}$ apple. How many people can be served if each person receives $\frac{1}{2}$ of an apple?

(d) Draw a picture of the result of dividing 6 apples into servings of $\frac{1}{3}$ apple. How many people can be served if each person receives $\frac{1}{3}$ of an apple?

(e) To divide fractions, the numerator (dividend) is multiplied by the reciprocal of the denominator (divisor). In each row, begin with 6 apples and divide by the serving size. Complete the table and look for a pattern in the quotients.

Number of apples	Serving size	Process written symbolically	Number of people served
6	$\frac{1}{2}$ apple	$6 \div \frac{1}{2} = 6 \cdot \frac{2}{1} = 6 \cdot 2$	12 people
6	$\frac{1}{3}$		
6	$\frac{1}{4}$		
6	$\frac{1}{6}$		

(f) Does the quotient (number of people served) get larger or smaller as the divisor (the size of each serving) gets smaller?

2. One method of dividing complex fractions involves multiplying the numerator (dividend) by the reciprocal of the denominator (divisor). Divide each complex fraction.

EXAMPLE: $$\frac{\frac{3}{8}}{\frac{1}{2}} = \frac{3}{8} \div \frac{1}{2} = \frac{3}{\cancel{8}_{4}} \cdot \frac{\cancel{2}^{1}}{1} = \frac{3 \cdot 1}{4 \cdot 1} = \frac{3}{4}$$

(a) $\dfrac{\frac{3}{4}}{\frac{9}{16}} =$

(b) $\dfrac{\frac{y^2}{6}}{\frac{y}{36x}} =$

3. A second method for dividing complex fractions involves multiplying the numerator and denominator of the complex fraction by the LCD of the denominators of the individual fractions.

(a) Multiply the numerator and denominator of $\dfrac{\frac{y^2}{6}}{\frac{y}{36x}}$ by the LCD $= 36x = \dfrac{36x}{1}$.

$$\frac{\frac{y^2}{6}}{\frac{y}{36x}} = \frac{\frac{36x}{1}\cdot\left(\frac{y^2}{6}\right)}{\frac{36x}{1}\cdot\left(\frac{y}{36x}\right)} =$$

(b) What happens to a complex fraction when the numerator and denominator are multiplied by the LCD of the denominators of the individual fractions?

1 **2** 3

Discussing Mathematics

Complex fractions, where the numerator and/or denominator include fractions, are divided using either of two methods.

DIVIDING COMPLEX FRACTIONS		
	Method 1—The numerator (dividend) is multiplied by the reciprocal of the denominator (divisor).	**Method 2—The numerator and denominator of the complex fraction are multiplied by the LCD of all denominators.**
	➢ *Use Method 1 when you have one fraction divided by one fraction or when the individual fractions are uncomplicated.*	➢ *Method 2 is efficient when both the numerator and denominator have more than one fraction.*
1.	Perform the operations of addition and/or subtraction in each of the numerator and denominator, if necessary.	Identify the LCD of all individual denominators of the complex fraction.
2.	Rewrite the problem using a division symbol.	Multiply the numerator and denominator of the complex fraction by $\dfrac{\text{LCD}}{1}$.
3.	Multiply the dividend by the reciprocal of the divisor. Factor the numerators and denominators and divide out any common factors, if possible.	Factor the numerator and denominator and divide out any common factors, if possible.
4.	Multiply remaining factors. Perform operations and reduce answer to lowest terms, if necessary.	Multiply remaining factors. Perform operations and reduce answer to lowest terms, if necessary.

EXAMPLE 1: Divide $\dfrac{\dfrac{x-y}{2xy}}{\dfrac{x^2-y^2}{4x^2+6x}}$ using Method 1.

$$\frac{x-y}{2xy} \div \frac{x^2-y^2}{4x^2+6x} = \frac{x-y}{2xy}\cdot\frac{4x^2+6x}{x^2-y^2}$$

Multiply the dividend by the reciprocal of the divisor.

$$= \frac{(x-y)}{2xy}\cdot\frac{2x(2x+3)}{(x-y)(x+y)}$$

Factor the numerators and denominators.

$$= \frac{\overset{1}{\cancel{(x-y)}}}{\underset{1}{\cancel{2x}}y}\cdot\frac{\overset{1}{\cancel{2x}}(2x+3)}{\underset{1}{\cancel{(x-y)}}(x+y)}$$

Divide out common factors. Multiply remaining factors.

$$= \frac{1}{y}\cdot\frac{(2x+3)}{(x+y)}$$

$$= \frac{2x+3}{y(x+y)} = \frac{2x+3}{xy+y^2}$$

The denominator can be written in either factored or expanded form.

EXAMPLE 2: Divide $\dfrac{\dfrac{1}{6x^2}-\dfrac{4}{3x}}{\dfrac{3}{2x}+\dfrac{5}{2x^3}}$ by clearing fractions (Method 2).

The denominators of the individual fractions are $6x^2$, $3x$, $2x$, and $2x^3$. The LCD of the denominators is $12x^3$.

$$\frac{\dfrac{1}{6x^2}-\dfrac{4}{3x}}{\dfrac{3}{2x}+\dfrac{5}{2x^2}} = \frac{\dfrac{12x^3}{1}\cdot\left(\dfrac{1}{6x^2}\right)-\dfrac{12x^3}{1}\cdot\left(\dfrac{4}{3x}\right)}{\dfrac{12x^3}{1}\cdot\left(\dfrac{3}{2x}\right)+\dfrac{12x^3}{1}\cdot\left(\dfrac{5}{2x^3}\right)}$$

Multiply each term by $\dfrac{\text{LCD}}{1} = \dfrac{12x^3}{1}$.

$$= \frac{\dfrac{\overset{2x}{\cancel{12x^3}}}{1}\cdot\dfrac{1}{\underset{1}{\cancel{6x^2}}}-\dfrac{\overset{4x^2}{\cancel{12x^3}}}{1}\cdot\dfrac{4}{\underset{1}{\cancel{3x}}}}{\dfrac{\overset{6x^2}{\cancel{12x^3}}}{1}\cdot\dfrac{3}{\underset{1}{\cancel{2x}}}+\dfrac{\overset{6}{\cancel{12x^3}}}{1}\cdot\dfrac{5}{\underset{1}{\cancel{2x^3}}}}$$

Divide out common factors.

$$= \frac{2x\cdot 1-4x^2\cdot 4}{6x^2\cdot 3+6\cdot 5} = \frac{2x-16x^2}{18x^2+30}$$

Multiply remaining factors.

$$= \frac{2x(1-8x)}{6(3x^2+5)} = \frac{\overset{1}{\cancel{2}}x(1-8x)}{\underset{3}{\cancel{6}}(3x^2+5)}$$

Factor and divide out common factors to reduce answer to lowest terms.

$$= \frac{x(1-8x)}{3(3x^2+5)} = \frac{x-8x^2}{9x^2+15}$$

The denominator can be written in either factored or expanded form.

EXAMPLE 3: Divide $\dfrac{x - 6 - \frac{7}{x}}{x - 3 - \frac{4}{x}}$ by clearing fractions (Method 2).

$$\frac{x - 6 - \frac{7}{x}}{x - 3 - \frac{4}{x}} = \frac{\frac{x}{1}\cdot(x) - \frac{x}{1}\cdot(6) - \frac{\cancel{x}}{1}\left(\frac{7}{\cancel{x}}\right)}{\frac{x}{1}\cdot(x) - \frac{x}{1}\cdot(3) - \frac{\cancel{x}}{1}\left(\frac{4}{\cancel{x}}\right)}$$

Multiply each term by $\frac{\text{LCD}}{1} = \frac{x}{1}$.

Divide out common factors.

$$= \frac{x^2 - 6x - 7}{x^2 - 3x - 4}$$

Multiply remaining factors.

$$= \frac{(x - 7)\cancel{(x + 1)}}{(x - 4)\cancel{(x + 1)}} = \frac{x - 7}{x - 4}$$

Factor and divide out common factors to reduce answer to lowest terms.

➢ *Either Method 1 or Method 2 is an appropriate method to simplify complex fractions. As you become more comfortable with Method 2, you will probably find it is easier than Method 1. Method 2 is especially useful when a sum or difference of rational expressions appears in the numerator or denominator of a complex fraction.*

1 2 3 Doing Mathematics

- Write the definitions of the words and symbols listed for this section in the Dictionary in Appendix A.
- On a 3 × 5-inch card, write the processes (Method 1 and Method 2) for simplifying complex fractions for future reference.

Problems 1-2: Circle each fraction that is a complex fraction.

1. (a) $\dfrac{5}{7}$ (b) $\dfrac{5}{\frac{1}{7}}$ (c) $\dfrac{\frac{x}{2}}{5}$ (d) $\dfrac{2}{1}$ (e) $\dfrac{\frac{1}{x} + y}{\frac{x}{z}}$

2. (a) $\dfrac{-x}{y}$ (b) $\dfrac{\frac{x}{y}}{2}$ (c) $\dfrac{1 + 5}{9}$ (d) $\dfrac{\frac{2ab}{x + y}}{\frac{-4ab^2}{x + y}}$ (e) $\dfrac{\frac{1}{x} + \frac{1}{y}}{\frac{x^2 - y^2}{xy}}$

3. Describe the difference between dividing the number 10 by 2 and dividing the number 10 by $\frac{1}{2}$.

4. Answer true or false.

(a) Dividing a counting number by a larger counting number produces a number that is smaller than the original counting number.

(b) Dividing a counting number by a number x, where $0 < x < 1$, produces a number that is smaller than the original counting number.

Problems 5–20: Divide each complex fraction using Method 1.

5. $\dfrac{\frac{5}{6}}{\frac{2}{3}}$

6. $\dfrac{\frac{5}{7}}{\frac{10}{21}}$

7. $\dfrac{\frac{x^2}{6}}{\frac{x}{3}}$

8. $\dfrac{\frac{2y^3}{5}}{\frac{4y}{10}}$

9. $\dfrac{\frac{4xy}{7ab}}{\frac{-24xy^2}{7a^2b^2}}$

10. $\dfrac{\frac{-6xy^2}{5z}}{\frac{42x^2y}{25wz^2}}$

11. $\dfrac{\frac{x^2 - x - 2}{5xy}}{\frac{x^2 - 1}{10x}}$

12. $\dfrac{\frac{x^2 + 2x + 1}{4x^2}}{\frac{x + 1}{8x}}$

13. $\dfrac{\frac{x^2}{x + y}}{\frac{x^3}{(x + y)^2}}$

14. $\dfrac{\frac{ab^2}{a - b}}{\frac{a^3b}{(a - b)^2}}$

15. $\dfrac{\frac{x^2 - 1}{x + 2}}{\frac{x^2 - x - 2}{x^2 - 4}}$

16. $\dfrac{\frac{x^2 - 9}{x + 1}}{\frac{x^2 + 4x + 3}{x^2 + 2x + 1}}$

17. $\dfrac{\dfrac{-10xy^2}{2x^2 - 5x - 3}}{\dfrac{15x^2y}{2x^2 + 3x + 1}}$

18. $\dfrac{\dfrac{30xyz}{2x^2 + 7x - 4}}{\dfrac{15x^2y^2z^2}{2x^2 + 9x - 5}}$

19. $\dfrac{\dfrac{x^2 - x - 2}{5xy}}{\dfrac{x^2 - 1}{10x}}$

20. $\dfrac{\dfrac{x^2 + 6x + 9}{-8x^2y}}{\dfrac{x + 3}{4xy}}$

Problems 21–38: Divide each complex fraction using Method 2.

21. $\dfrac{\dfrac{5}{6} - \dfrac{1}{2}}{\dfrac{2}{3}}$

22. $\dfrac{\dfrac{-18}{5} + \dfrac{4}{3}}{\dfrac{2}{15}}$

23. $\dfrac{\dfrac{4}{x} - \dfrac{8}{y}}{2}$

24. $\dfrac{\dfrac{3}{a} - \dfrac{5}{b}}{3}$

25. $\dfrac{\dfrac{3}{xy^2} + \dfrac{2}{x^2y}}{\dfrac{1}{x^2y} + \dfrac{2}{xy^2}}$

26. $\dfrac{\dfrac{5}{x} - \dfrac{3}{y}}{\dfrac{4}{x} - \dfrac{2}{y}}$

27. $\dfrac{\dfrac{2}{x^3} + \dfrac{3}{x^2}}{\dfrac{4x}{x^4} + \dfrac{5}{x^3}}$

28. $\dfrac{\dfrac{6}{x^2} - \dfrac{8}{x^3}}{\dfrac{10}{x^3} - \dfrac{2x}{x^4}}$

29. $\dfrac{x + \dfrac{1}{x}}{x + \dfrac{5}{x}}$

30. $\dfrac{\dfrac{2a}{b} - \dfrac{b}{2a}}{\dfrac{1}{b} + \dfrac{1}{2a}}$

31. $\dfrac{x - \dfrac{1}{x}}{1 - \dfrac{1}{x^2}}$

32. $\dfrac{\dfrac{1}{4x^2} - \dfrac{1}{25}}{\dfrac{1}{2x} - \dfrac{1}{5}}$

33. $\dfrac{\dfrac{2}{ab^2} + \dfrac{1}{a^2b}}{\dfrac{3}{ab} - \dfrac{5}{a^2b}}$

34. $\dfrac{3 + \dfrac{3}{x - 1}}{\dfrac{3}{x - 1}}$

35. $\dfrac{1 - \dfrac{1}{a - 2}}{\dfrac{1}{a - 2}}$

36. $\dfrac{6 - \dfrac{1}{x} - \dfrac{2}{x^2}}{3 + \dfrac{1}{x} - \dfrac{2}{x^2}}$

37. $\dfrac{\dfrac{7}{y} + \dfrac{4}{y^2} + 3}{3 - \dfrac{8}{y^2} - \dfrac{2}{y}}$

38. $\dfrac{\dfrac{3}{x + y} - \dfrac{3}{x - y}}{\dfrac{5}{x^2 - y^2}}$

39. Simplify $\dfrac{\dfrac{1}{xy^2} - \dfrac{3}{xy}}{\dfrac{3}{x^2y} + \dfrac{1}{xy^2}}$ using Method 1 and Method 2. Which method is more efficient? Why?

Method 1:	Method 2:

Chapter 7 Summary

SECTION 7.1:

Fundamental Principle of Equivalent Fractions

When the numerator (p) and the denominator (q) of a fraction are divided (or multiplied) by the same nonzero number (n), the result is an equivalent fraction.

Reducing a fraction: $\frac{p}{q} = \frac{p \div n}{q \div n}$ Building a fraction: $\frac{p}{q} = \frac{p \cdot n}{q \cdot n}$

SECTION 7.2:

Multiplying Rational Expressions

When multiplying rational expressions, multiply the numerators of the fractions and multiply the denominators of the fractions:

$$\frac{A}{B} \cdot \frac{C}{D} = \frac{A \cdot C}{B \cdot D}, \text{ where } B \neq 0 \text{ and } D \neq 0.$$

Reduce the resulting product by dividing out common factors, if possible.

Dividing Rational Expressions

The operation of division is equivalent to multiplying the dividend by the reciprocal of the divisor:

$$\frac{A}{B} \div \frac{C}{D} = \frac{A}{B} \cdot \frac{D}{C}, \text{ where } B \neq 0,\ D \neq 0, \text{ and } C \neq 0.$$

SECTION 7.3:

Adding or Subtracting Rational Expressions with Denominators that Are the *Same*

Given the rational expressions $\frac{A}{B}$ and $\frac{C}{B}$, then

Sum: $\frac{A}{B} + \frac{C}{B} = \frac{A + C}{B}, \ (B \neq 0)$; Difference: $\frac{A}{B} - \frac{C}{B} = \frac{A - C}{B}, \ (B \neq 0)$.

Process for Adding or Subtracting Rational Expressions with Denominators that Are *Different*

1. Write the factors of all denominators. Find the LCM of the denominators.
2. Use the LCM as the LCD and build the original fractions to equivalent fractions with the LCD as the new denominators.
3. Add or subtract the numerators. The denominator is the LCD.
4. Factor the resulting numerator, if possible. Identify any common factors and reduce the fraction to lowest terms, if possible.

SECTION 7.4:

Rational Function

$$f(x) = \frac{p(x)}{q(x)}, \text{ where } p(x) \text{ and } q(x) \text{ are polynomials and } q(x) \neq 0.$$

Vertical Asymptote

If $f(x)$ increases or decreases infinitely as x approaches the restricted value r, the restricted value defines a **vertical asymptote**. This behavior is written symbolically: $f(x) \to +\infty$ or $f(x) \to -\infty$ as $x \to r$ from the left or the right. A vertical asymptote is drawn as a dashed vertical line on the graph at $x = r$.

Horizontal Asymptote

If x increases or decreases infinitely as $f(x)$ approaches a constant c, a **horizontal asymptote** occurs at $f(x) = y = c$. This behavior is written symbolically: as $x \to -\infty$ or $x \to +\infty$, $f(x) \to c$. A horizontal asymptote is drawn as a dashed horizontal line on the graph.

Graphing a Rational Function

1. Identify any restricted values (r) in the independent variable (x) by setting the denominator equal to zero and solving the equation.
2. Make a table of values:
   (a) Include at least three or four x values that are larger and smaller than the restricted value(s). These should include decimal values that are very close to the restricted value(s).
   (b) Include three or four x values that are very large and very small.
3. Identify any vertical asymptotes and draw dashed vertical lines on the grid. Look at the $f(x)$ values as x approaches the restricted value(s) (r) from the left and from the right. If $f(x)$ increases or decreases infinitely, a vertical asymptote is defined at $x = 4$.
4. Identify any horizontal asymptotes and draw dashed horizontal lines. Look at $f(x)$ values as x takes on very large and very small values. If the $f(x)$ values approach the same value (c), a horizontal asymptote is defined at $y = c$.
5. Plot the points.
6. Connect the points. Do not cross any vertical asymptotes.
7. Draw arrows on the ends of the curves to denote that the curve continues in that pattern.

SECTION 7.5:

Solving a Rational Equation by Clearing Fractions

1. Factor all denominators in the equation and identify the LCD.
2. Multiply the $\frac{\text{LCD}}{1}$ by every term in the equation.
3. Divide out common factors in each term.
4. Solve for the unknown.
5. Check the answer(s) analytically.
6. Confirm the solution graphically, if desired.

SECTION 7.6:

Dividing Complex Fractions (Method 2)

1. Identify the LCD of all individual denominators of the complex fraction.
2. Multiply the numerator and denominator of the complex fraction by the $\frac{\text{LCD}}{1}$.
3. Factor the numerator and denominator and divide out any common factors, if possible.
4. Multiply remaining factors. Perform operations and reduce the answer to lowest terms, if necessary.

Chapter 7 Review

- Review the definitions you have written for this chapter in the Dictionary.
- Review any properties and/or processes you have written on 3×5-inch cards.

SECTION 7.1

Problems 1–6: Reduce each rational expression to lowest terms.

1. $\dfrac{15x}{40x}$

2. $\dfrac{3x + 6}{3}$

3. $\dfrac{21xyz^2}{3xz}$

4. $\dfrac{x^2 - 1}{x^2 + 3x - 4}$

5. $\dfrac{10x^3 - 25x^2 + 15x}{30x^2 - 45x}$

6. $\dfrac{x^2 - x^3 - 1 + x}{1 - x^2}$

Problems 7–10: Build each rational expression to an equivalent fraction with the given denominator.

7. $\dfrac{4y}{x} = \dfrac{?}{x(x + 2)}$

8. $\dfrac{-3(x - 1)}{3xy^2z} = \dfrac{?}{27x^2y^2z}$

9. $\dfrac{-x}{(x + 3)} = \dfrac{?}{x(x + 1)(x + 3)}$

10. $\dfrac{3x}{x - 2} = \dfrac{?}{x^2 + 3x - 10}$

Problems 11–14: Determine the restricted value(s) for the variable in each rational expression.

11. $\dfrac{x}{x + 1}$

12. $\dfrac{x + 3}{2x - 1}$

13. $\dfrac{2x}{(x + 4)(x - 3)}$

14. $\dfrac{x - 5}{3x^2 - 5x - 2}$

15. Dennis purchased sports drinks for 42 people participating in a tennis tournament. After the tournament, 27 bottles had been consumed. Dennis would like to order the same ratio of drinks for a second tournament. Let x represent the number of drinks to be purchased for the second tournament. Determine the number of drinks to be purchased for the second tournament if 84 people enter the tournament.

16. Anne can read 5 pages in 12 minutes. How long will it take her to read 28 pages?

17. Elaine needs to choose between two kinds of mulch for her garden. Pine mulch costs \$15.00 for 6 cubic yards and hardwood mulch costs \$18.99 for 8 cubic yards. Which mulch is less expensive? Why?

18. Why did the calculator display "ERROR" when the function $y_1 = \dfrac{x-2}{2x+1}$ was evaluated at $x = -\dfrac{1}{2} = -0.5$?

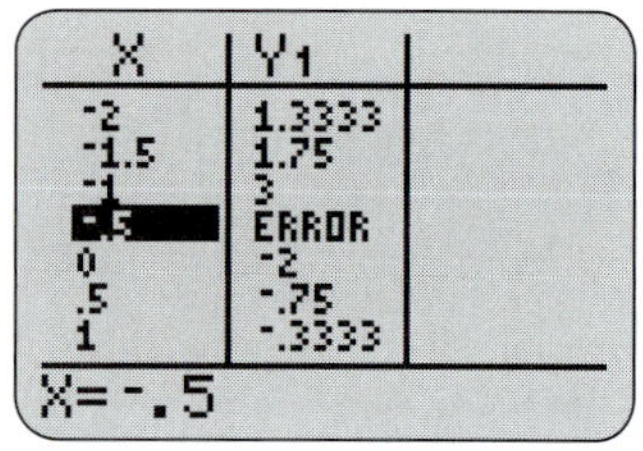

SECTION 7.2

Problems 19–30: Perform the operations. Reduce each answer to lowest terms.

19. $\dfrac{-8}{85} \cdot \dfrac{5}{-16} \cdot \dfrac{-1}{2}$

20. $\dfrac{-5^3}{3^2} \cdot \dfrac{9}{(-2)^2}$

21. $\dfrac{-2x}{15y^2} \cdot \dfrac{3y^2}{4x^2}$

22. $\dfrac{(2a)^2}{15xyz} \cdot \dfrac{5xz^2}{8a^2}$

23. $\dfrac{-x}{4} \cdot \dfrac{4x^2-4}{x-1} \cdot \dfrac{x^2+x-6}{x^2-x-2}$

24. $\dfrac{-1}{6} \div \dfrac{5}{12} \cdot (-7)$

25. $\dfrac{3xy}{20z} \div \dfrac{9xy^2}{-4z^2}$

26. $\dfrac{(5a)^3}{16b} \div \dfrac{50a}{(2b)^3}$

27. $\dfrac{2}{3} \cdot \dfrac{x+1}{x-1} \div \dfrac{x+1}{x-1}$

28. $\dfrac{x^2-4}{-2x-2y} \div \dfrac{x+2}{x^2-y^2} \cdot \dfrac{10}{x-2}$

29. $\dfrac{x^2-6x+8}{x^2-2x-24} \cdot \dfrac{x^2+7x+12}{x^2-7x+10} \div \dfrac{x^2-9}{x^2-36}$

30. $\dfrac{8x^2-30x-27}{8x^2-2x-6} \div \dfrac{2x^2+7x-72}{12x^2+11x-36} \cdot \dfrac{2x^2+14x-16}{3x^2-13x+12}$

SECTION 7.3

31. Pete is going to install anti-skid strips on the steps of his outdoor deck. Five steps measure $3\frac{1}{2}$ feet across, two steps measure $5\frac{1}{4}$ feet across, and the last step measures $10\frac{3}{8}$ feet across. What is the exact amount of anti-skid stripping Pete should buy?

Problems 32–43: Add or subtract.

32. $\dfrac{25}{x} - \dfrac{15}{x}$

33. $\dfrac{10}{x-1} - \dfrac{2}{x-1} + \dfrac{7}{x-1}$

34. $\dfrac{9x}{2x+3} - \dfrac{5x}{2x+3}$

35. $\dfrac{x}{15} + \dfrac{x}{20}$

36. $\dfrac{15}{x} - \dfrac{4}{5x}$

37. $\dfrac{13}{x} + \dfrac{14}{y}$

38. $\dfrac{2x}{5y} - \dfrac{x^2}{25xy^2}$

39. $\dfrac{2}{x+3} - \dfrac{4}{(x+3)(x-2)}$

40. $\dfrac{3}{x+y} + \dfrac{10xy}{x^2-y^2}$

41. $\dfrac{2y}{y^2+y} + \dfrac{4y}{y^2-1}$

SECTION 7.4

Problems 42–43: Graph each function. Complete the table of values rounding decimals to two decimal places. State any restricted value(s) of the independent variable.

42. $f(x) = \dfrac{1}{x + 3}$

x	$f(x)$

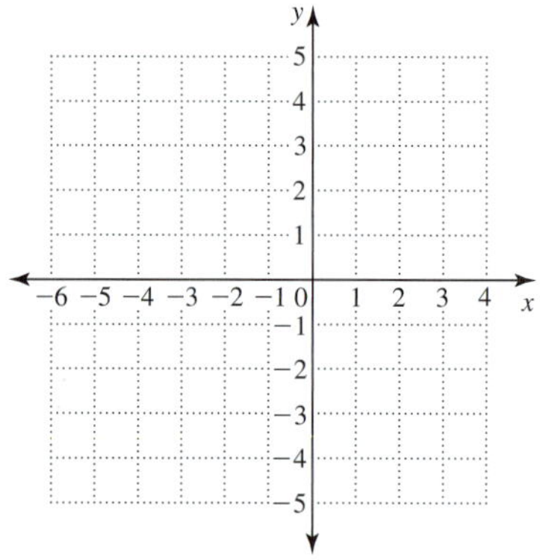

43. $f(x) = \dfrac{1}{x^2 - x - 6}$

x	$f(x)$

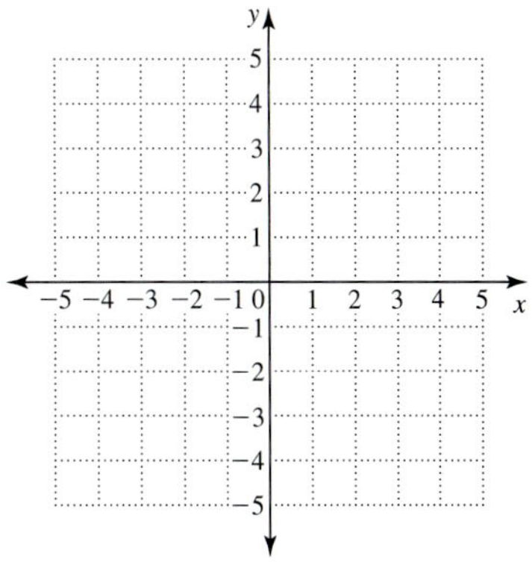

44. $f(x) = \dfrac{3}{2x^2 - 2x}$

x	$f(x)$

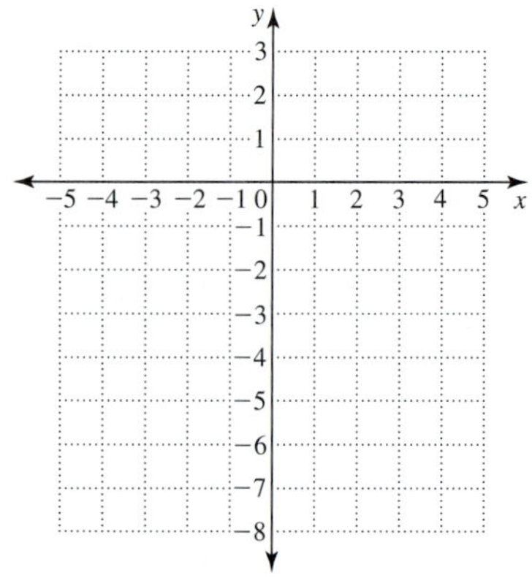

SECTION 7.5

45. Bill and Ken are window washers. Working alone, Bill can wash 60 windows in 4 hours. Ken can wash 60 windows in 6 hours. Let t represent the total number of hours required to complete the job when Bill and Ken work together. How many hours will it take to wash 60 windows if Bill and Ken work together?

46. The length of a rectangular garden is 2 feet longer than its width. The ratio of the length to the width is $\frac{5}{4}$. Let w represent the width of the garden. Find the dimensions of the garden.

Problems 47–50: Solve each proportion using the Extremes-Means Property of Proportions and Equivalent Fractions. Verify the solution by substituting into the original equation. Confirm solutions graphically, if desired.

47. $\frac{3}{5} = \frac{x}{20}$

48. $\frac{x+5}{x-2} = \frac{x+1}{x+3}$

49. $\frac{x}{7} = 11$

50. $\frac{5y}{12} = \frac{5}{3}$

Problems 51–60: Solve each rational equation. Verify the answer(s) by substituting into the original equation. Confirm solutions graphically, if desired.

51. $\frac{1}{15} + x = \frac{-3}{5}$

52. $\frac{16}{x} = \frac{1}{x} - 5$

53. $\frac{4}{5}y - \frac{5}{6}y = 1$

54. $\frac{2}{x} + \frac{3}{x^2} = \frac{7}{2x}$

55. $\frac{5}{x-2} = \frac{3}{x+1} - \frac{1}{x-2}$

56. $\frac{3}{x-2} - \frac{5}{2x-4} = \frac{2}{x+1}$

57. $\dfrac{2x}{x+2} - \dfrac{5}{x-1} = \dfrac{2x^2}{x^2+x-2}$

58. $\dfrac{x}{2x+1} - \dfrac{3x^2-11x-5}{6x^2+x-1} = \dfrac{4x}{3x-1}$

59. $\dfrac{3}{2x+1} - \dfrac{1}{3x-4} = \dfrac{1}{4x+2}$

60. $\dfrac{7}{2x-1} - \dfrac{2}{3x+4} = \dfrac{15x+31}{6x^2+5x-4}$

SECTION 7.6

61. Circle the fractions that are complex fractions.

(a) $\dfrac{1}{x+2}$ (b) $\dfrac{5+\frac{2}{x}}{\frac{1}{3}}$ (c) $\dfrac{2x-1}{3}$ (d) $\dfrac{\frac{5}{x}}{\frac{2}{x+1}}$ (e) $\dfrac{(3)(-4)}{-2}$ (f) $\dfrac{\frac{2x+3}{5x}}{\frac{4y}{3xz}}$

Problems 62–70: Divide each complex fraction.

62. $\dfrac{\frac{3}{8}}{\frac{2}{9}}$

63. $\dfrac{\frac{-19}{7}+1}{\frac{4}{3}}$

64. $\dfrac{\frac{9xy^2}{wz}}{\frac{-y^2}{12wz}}$

65. $\dfrac{\frac{x^2-x-2}{5xy}}{\frac{x^2-1}{10x}}$

66. $\dfrac{\frac{x^2-4x-5}{3xy}}{\frac{x^2-7x+10}{15x^2y^2}}$

67. $\dfrac{\frac{2a}{b}-\frac{b}{2a}}{\frac{1}{b}+\frac{1}{2a}}$

68. $\dfrac{5+\frac{5}{x-1}}{\frac{5}{x-1}}$

69. $\dfrac{\frac{4}{ab}+\frac{8}{a^2b}}{\frac{6}{ab}-\frac{2}{a^2b}}$

70. $\dfrac{\frac{5}{y+3}-\frac{1}{y-1}}{\frac{2}{y+3}+\frac{4}{y-1}}$

Practice Test—Chapter 7

SECTION 7.1

Problems 1–3: Reduce each rational expression to lowest terms.

1. $\dfrac{-54x^3yz^2}{3xz}$ **2.** $\dfrac{7x + 14}{7}$ **3.** $\dfrac{x^2 - 1}{x^2 + 5x + 4}$

Problems 4–6: Build each rational expression to an equivalent expression with the given denominator.

4. $\dfrac{-3ab}{2xy^2} = \dfrac{?}{-18x^3y^2}$ **5.** $\dfrac{5x}{x(x + 3)} = \dfrac{?}{x(x + 3)(x - 2)}$

6. $\dfrac{-4x}{(x - 1)} = \dfrac{?}{3x(x - 1)^2(x + 1)}$

Problems 7–8: Determine the restricted value(s) of the variable in each rational expression.

7. $\dfrac{5x}{x + 9}$ **8.** $\dfrac{25}{x^2 - 15x + 50}$

SECTION 7.2

Problems 9–14: Perform the operations. Reduce each answer to lowest terms.

9. $\dfrac{-8}{75} \cdot \dfrac{-5}{24} \cdot \dfrac{1}{3}$ **10.** $\dfrac{(-2)^2}{2^2} \cdot \dfrac{6}{(-3)^3}$

11. $\dfrac{-2x}{15y^2} \cdot \dfrac{3x^2y^2}{8}$ **12.** $\dfrac{3xy}{20z^2} \div \dfrac{45xy^2}{-4yz^2}$

13. $\dfrac{2}{3} \cdot \dfrac{x + 1}{x - 3} \div \dfrac{4x + 4}{x - 3}$ **14.** $\dfrac{x^2 - 4}{2x + 2y} \div \dfrac{x + 2}{x^2 - y^2} \cdot \dfrac{-14}{x - 2}$

SECTION 7.3

Problems 15–20: Perform the operations.

15. $\dfrac{5}{x} - \dfrac{15}{x}$

16. $\dfrac{10}{x+1} - \dfrac{2}{x+1} + \dfrac{7}{x+1}$

17. $\dfrac{10}{x} - \dfrac{4}{5x}$

18. $\dfrac{3}{x^2y^2z} + \dfrac{23z}{4x^2y^2}$

19. $\dfrac{1}{y^2 - y} + \dfrac{3y}{y^2 - 1}$

20. $\dfrac{6a}{a^2 - 25} - \dfrac{4}{a-5} - \dfrac{7}{a+5}$

SECTION 7.4

Problems 21–22: Graph each function. Complete the table of values rounding decimals to two decimal places. State any restricted value(s) of the independent variable.

21. $f(x) = \dfrac{1}{x-2}$

x	$f(x)$

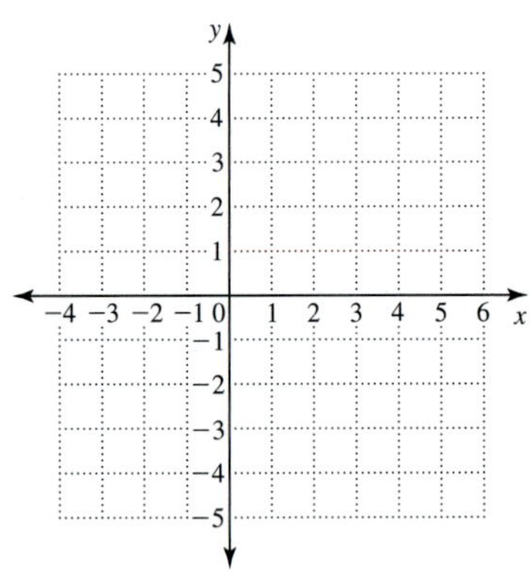

22. $f(x) = \dfrac{1}{x^2 + x - 6}$

x	$f(x)$

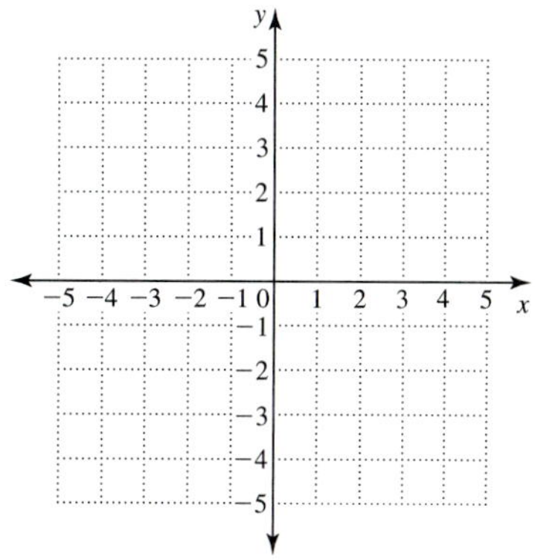

SECTION 7.5

23. Vince and Paul are landscapers. Working alone, Vince can rake a lawn in 3 hours. Paul can rake a lawn in 5 hours. Let t represent the total number of hours required to complete the job when Vince and Paul work together. How many hours will it take to rake the lawn if Vince and Paul work together?

Problems 24–25: Solve each proportion using the Extremes-Means Property of Proportions and Equivalent Fractions.

24. $\dfrac{3}{5} = \dfrac{x}{25}$

25. $\dfrac{x+3}{x-2} = \dfrac{x+2}{x+3}$

Problems 26–29: Solve each rational equation. Confirm the answer(s) analytically. Confirm the solution graphically, if desired.

26. $\dfrac{5}{8} + x = \dfrac{3}{4}$

27. $\dfrac{3}{2x} - \dfrac{1}{x^2} = \dfrac{7}{2x}$

28. $\dfrac{3}{y-2} - \dfrac{2}{y+1} = \dfrac{5}{2y-4}$

29. $\dfrac{2}{x-3} - \dfrac{3}{2x-1} = \dfrac{10}{2x^2 - 7x + 3}$

SECTION 7.6

Problems 30–33: Divide each complex fraction.

30. $\dfrac{\dfrac{72xy^2}{7}}{\dfrac{-24xy^2}{35}}$

31. $\dfrac{\dfrac{x^2 + x - 2}{15xy}}{\dfrac{x^2 - 1}{10x^2}}$

32. $\dfrac{5 + \dfrac{5}{x+1}}{\dfrac{5}{x+1}}$

33. $\dfrac{3 + \dfrac{5x+1}{x^2 - 1}}{1 + \dfrac{1}{x+1}}$

Answers to Getting Started—Chapter 7

SECTION 7.1

1. $\dfrac{12x^2y}{4xy^2} = \dfrac{3x}{y}$

2. Terms are numbers or algebraic expressions that are added (or subtracted). Factors are numbers or expressions that are multiplied.

3. (a) No (b) No. However, if the fraction is rewritten as $\dfrac{x}{5} + \dfrac{5}{5}$, the second fraction could be reduced and the expressions written:
$\dfrac{x+5}{5} = \dfrac{x}{5} + \dfrac{5}{5} = \dfrac{x}{5} + 1.$

4. (a) When $x = 3$: The expression is undefined:
$\dfrac{2(3)+5}{(3)-3} = \dfrac{6+5}{3-3} = \dfrac{11}{0}$. (b) x cannot equal 3 in $\dfrac{2x+5}{x-3}$ because it causes the denominator to equal zero. Division by zero is undefined.

5. (a) = (b) = (c) = (d) =

6. (a) $\dfrac{x}{y} = \dfrac{2x^2}{2xy}$ (b) $\dfrac{4ab}{3cd} = \dfrac{20abcd}{15c^2d^2}$

SECTION 7.2

1. (a) 12 cakes:

6 cups flour	12 eggs	$1\frac{1}{2}$ cups butter
3 cups sugar	4 teaspoons vanilla	1 teaspoon salt
$1\frac{1}{2}$ teaspoon(s) baking powder		$2\frac{1}{2}$ cups milk

(b) $\dfrac{2}{6} = \dfrac{1}{3}$

(c) 2 cakes:

1 cup(s) flour	2 egg(s)	$\frac{1}{4}$ cup(s) butter
$\frac{1}{2}$ cup(s) sugar	$\frac{2}{3}$ teaspoon(s) vanilla	$\frac{1}{6}$ teaspoon(s) salt
$\frac{1}{4}$ teaspoon(s) baking powder		$\frac{5}{12}$ cup(s) milk

(d) To make 12 cakes, multiply by 2. To make 2 cakes, divide by 3 or multiply by $\frac{1}{3}$.

2. (a) $3\dfrac{1}{2}\cdot 4 = \dfrac{7}{2}\cdot\dfrac{4}{1} = \dfrac{28}{2} = 14$ feet

(b) Paul should purchase 1 length of 15 feet.

(c) Multiply $3\frac{1}{2}$ by $4 = 14$ feet. 14 feet is less than 15 feet. Paul needs one length.

3. (a) 4 groups (b) $\dfrac{8}{2} = 4$

(c) 16 (d) $8 \div \dfrac{1}{2} = 8\cdot\dfrac{2}{1} = 16$

(e) Dividing by 2 results in a smaller number.

(f) Dividing by $\dfrac{1}{2}$ results in a larger number.

4. (a) $80{,}000 \div \dfrac{1}{2} = \dfrac{80{,}000}{1}\cdot\dfrac{2}{1} = \$160{,}000$

(b) Don's boss shouldn't have responded so quickly. Dividing Don's salary by $\dfrac{1}{2}$ was equivalent to multiplying the salary by 2, which meant that Don's salary was doubled! While Don was an excellent employee, the boss was surprised to find Don's salary to be \$160,000.

SECTION 7.3

1. (a) $\dfrac{5}{7} + \dfrac{2}{7} = \dfrac{5+2}{7} = \dfrac{7}{7} = 1$

(b) Add or subtract the numerators, the denominator remains the same. Reduce the fraction, if possible.

2. (a) $\dfrac{10}{y} + \dfrac{5}{y} = \dfrac{10+5}{y} = \dfrac{15}{y}$

(b) $\dfrac{12}{x+3} - \dfrac{4}{x+3} = \dfrac{12-4}{x+3} = \dfrac{8}{x+3}$

(c) $\dfrac{7}{x^2} + \dfrac{3}{x^2} = \dfrac{7+3}{x^2} = \dfrac{10}{x^2}$

3. (a) Factors of 12: $2\cdot 2\cdot 3$
Factors of 8: $2\cdot 2\cdot 2$
LCD: $2\cdot 2\cdot 2\cdot 3$

(b) LCD = 24

(c) $\dfrac{5\cdot 2}{12\cdot 2} = \dfrac{10}{24}$ and $\dfrac{3\cdot 3}{8\cdot 3} = \dfrac{9}{24}$

(d) $\dfrac{5}{12} + \dfrac{3}{8} = \dfrac{10}{24} + \dfrac{9}{24} = \dfrac{19}{24}$

4. (a) $\frac{5}{xy} + \frac{7}{2x} = \frac{5 \cdot 2}{xy \cdot 2} + \frac{7 \cdot y}{2x \cdot y} = \frac{10}{2xy} + \frac{7y}{2xy}$

$= \frac{10 + 7y}{2xy}$

(b) $\frac{2x}{4y} - \frac{3xy}{5y^2} = \frac{2x \cdot 5y}{4y \cdot 5y} - \frac{3xy \cdot 4}{5y^2 \cdot 4}$

$= \frac{10xy}{20y^2} - \frac{12xy}{20y^2} = \frac{-2xy}{20y^2} = \frac{-x}{10y}$

SECTION 7.4

1. (a) $x = 0$ is the restricted value and causes the denominator to be equal to zero. $x \neq 0$ in the function $f(x) = \frac{1}{x}$. Division by zero is undefined.

(b)

x	$f(x) = \frac{1}{x}$
−100	−0.01
−10	−0.1
−2	−0.5
−1	−1
−0.4	−2.5
−0.1	−10
−0.01	−100
0	**Undefined**
0.01	100
0.1	10
0.4	2.5
1	1
2	0.5
10	0.1
100	0.01

(c) As x approaches zero from the negative (or left) side, $f(x)$ is decreasing and approaching negative infinity, $f(x) \to -\infty$.

(d) As x approaches zero from the positive (or right) side, $f(x)$ is increasing and approaching positive infinity, $f(x) \to \infty$.

(e) $f(x)$ is approaching zero.

(f) The function $f(x) = \frac{1}{x}$ will never be equal to zero. Since x cannot equal zero and the numerator is 1, the value of $\frac{1}{x}$ will always have a value other than zero.

2.

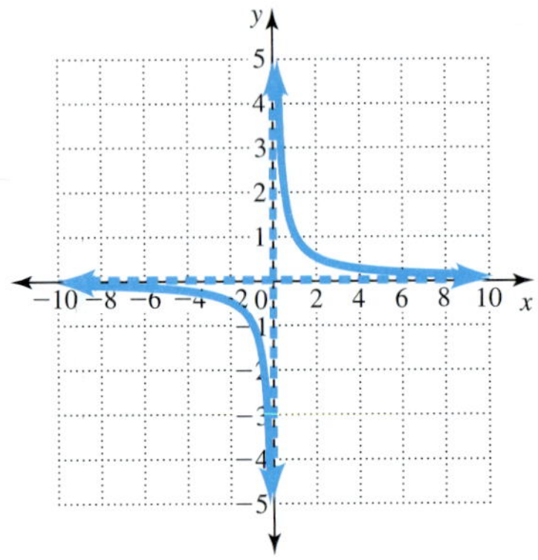

3. $x = 0$ causes division by zero in the function. Division by zero is undefined and is displayed on the calculator as "ERROR" for the dependent variable value.

SECTION 7.5

1. (a)

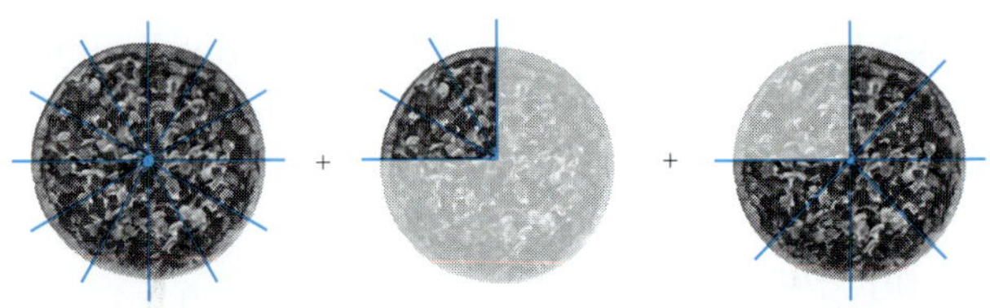

15 pieces of 12-cut pizzas + 6 pieces of 8-cut pizzas form two pizzas:

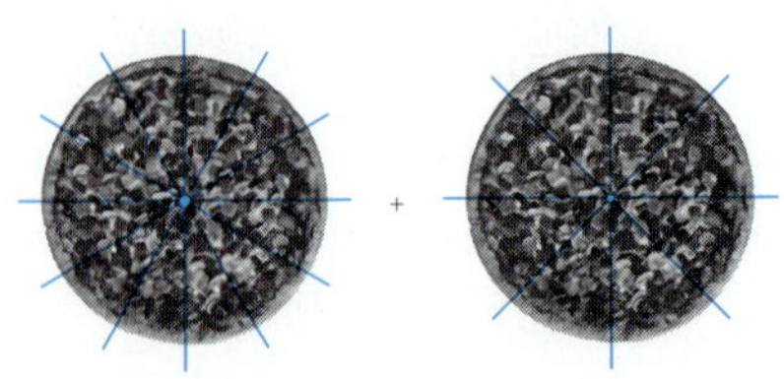

(b) 6 pieces of 8-cut pizza

(c) $\frac{x}{8} + \frac{15}{12} = 2$ or $\frac{x}{8} + \frac{15}{12} = \frac{2}{1}$

$$\frac{x \cdot 3}{8 \cdot 3} + \frac{5 \cdot 6}{4 \cdot 6} = \frac{2}{1} \cdot \frac{24}{24}$$

$$\frac{3x}{24} + \frac{30}{24} = \frac{48}{24}$$

$$\frac{3x}{24} = \frac{48}{24} - \frac{30}{24}$$

$$\frac{3x}{24} = \frac{18}{24}$$

$$3x \cdot 24 = 24 \cdot 18$$

$$72x = 432$$

$$x = \frac{432}{72} = 6$$

(d)

x	$y = \frac{x}{8} + \frac{15}{12} - 2$
−2	−1
0	−0.75
2	−0.5
4	−0.25
6	0
8	0.25

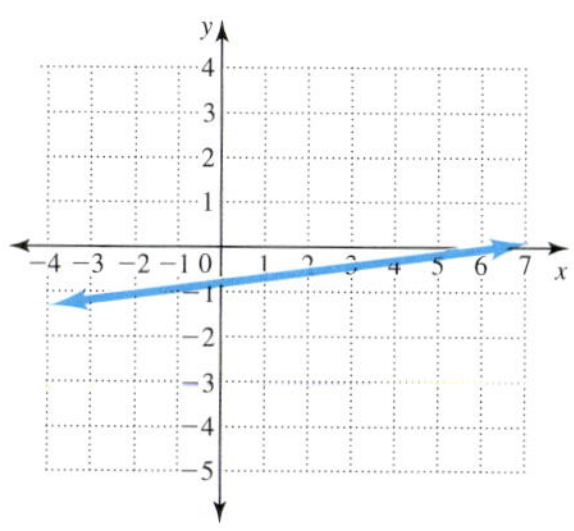

(e) The x-coordinate of the x-intercept is $x = 6$, confirming the analytical solution.

SECTION 7.6

1. (a) 2 people

(b) 3 people

(c) 12 people

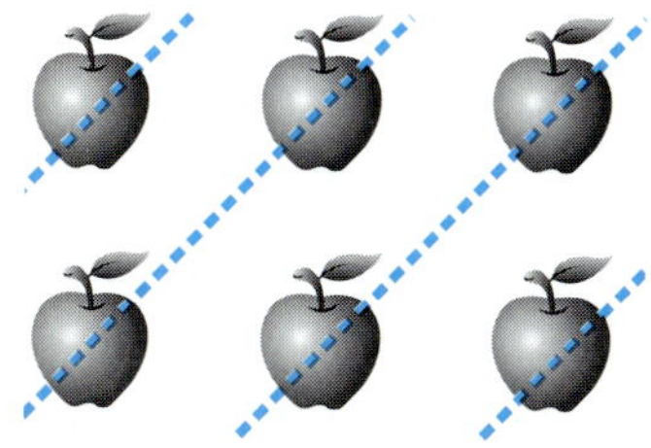

(d) 18 people

(e)

6	$\frac{1}{3}$	$6 \div \frac{1}{3} = 6 \cdot 3$	18
6	$\frac{1}{4}$	$6 \div \frac{1}{4} = 6 \cdot 4$	24
6	$\frac{1}{6}$	$6 \div \frac{1}{6} = 6 \cdot 6$	36

(f) The quotient gets larger.

2. (a) $\dfrac{\frac{3}{4}}{\frac{9}{16}} = \frac{3}{4} \div \frac{9}{16} = \frac{3}{4} \cdot \frac{16}{9} = \frac{4}{3}$

(b) $\dfrac{\frac{y^2}{6}}{\frac{y}{36x}} = \frac{y^2}{6} \div \frac{y}{36x} = \frac{y^2}{6} \cdot \frac{36x}{y} = y \cdot 6x = 6xy$

3. (a) $\dfrac{\frac{y^2}{6}}{\frac{y}{36x}} = \dfrac{\frac{36x}{1} \cdot \left(\frac{y^2}{6}\right)}{\frac{36x}{1} \cdot \left(\frac{y}{36x}\right)} = \frac{6x \cdot y^2}{1 \cdot y} = 6xy$

(b) The numerator and denominator of the complex fraction are each multiplied by the LCD of the individual fractions. The resulting fraction is not a complex fraction.

CHAPTER

8

Roots and Radical Expressions

Nicholas Chuquet (1445–1488) wrote the earliest French algebra text called Triparty en la science des nombres *in 1484. In his work on calculating square and cube roots, he is known for leaving an answer in radical form, when an integer did not result. He had his own notation for roots,* $R^2 6 = \sqrt{6}$ *or* $R^3 12 = \sqrt[3]{12}$*, and used an underline to indicate grouping of the radicand,* $R^2\underline{14\bar{p}R^2 180} = \sqrt{14 + \sqrt{180}}$.

8.1 Roots and Radicals

OBJECTIVES

- Identify perfect powers of integers.
- Identify the radical, radicand, and index of a radical expression.
- Find roots when radicands contain variables.

1 2 3

Getting Started

In this chapter, powers and roots of numbers and algebraic expressions are discussed. The process of taking a root is the inverse process of raising a base to an exponent. "Finding the square root" of a number is the process of "undoing" the squaring of a base. The **radical symbol** $\sqrt{}$ denotes the non-negative square root (or second root) of an expression. The expression under the radical symbol is called the **radicand**.

EXAMPLE 1: The square root (or second root) of 4 is written:

$\sqrt{4}$ ← 4 is the radicand.

$\sqrt{4} = 2$ because $2 \cdot 2 = 2^2 = 4$.

The **square root** of an expression is the quantity that, when used as a factor two times, produces the radicand: $\sqrt{4} = \sqrt{2^2} = 2$.

1. Find the square roots.

(a) $\sqrt{49}$ (b) $\sqrt{81}$ (c) $\sqrt{100}$

2. Before we consider other roots, let's familiarize ourselves with the powers of 2, 3, 4, and 5. Complete the table, writing powers of 3, 4, and 5.

Base	2	3	4	5
Second Power	$2^2 = 2 \cdot 2 = 4$	$3^2 =$	$4^2 =$	$5^2 =$
Third Power	$2^3 = 2 \cdot 2 \cdot 2 = 8$	$3^3 =$	$4^3 =$	$5^3 =$
Fourth Power	$2^4 = 2 \cdot 2 \cdot 2 \cdot 2 = 16$	$3^4 =$	$4^4 =$	$5^4 =$
Fifth Power	$2^5 = 2 \cdot 2 \cdot 2 \cdot 2 \cdot 2 = 32$	$3^5 =$	$4^5 =$	$5^5 =$

➢ *The result of squaring is a* ***perfect square:*** 2^2 *is the perfect square 4. The result of cubing is a* ***perfect cube:*** 2^3 *is the perfect cube 8. The result of raising an expression to the fourth power is called a* ***perfect fourth power:*** 3^4 *is the perfect fourth power 81.*

Finding the root of an expression is the process of "undoing" the raising of a base to an exponent. To denote a root other than a square root, a small number called the **index of the radical** is written as a superscript on the left of the radical, $\sqrt[\text{index}]{\text{radicand}}$. The third root (or cube root) has an index of 3, $\sqrt[3]{\text{radicand}}$. The fourth root is written with an index of 4, $\sqrt[4]{\text{radicand}}$, and so on. When no index is written, it is understood to mean a square root or the second root: $\sqrt{\text{radicand}}$.

EXAMPLE 2: (a) The index is 3. → $\sqrt[3]{8} = 2$ because $2 \cdot 2 \cdot 2 = 2^3 = 8$.
The radical is 8. ↗

The **cube root** of an expression is the quantity, which when used as a factor three times, produces the radicand: $\sqrt[3]{8} = \sqrt[3]{2^3} = 2$.

(b) The index is 4. → $\sqrt[4]{81} = 3$ because $3 \cdot 3 \cdot 3 \cdot 3 = 3^4 = 81$.
The radicand is 81. ↗

The **fourth root** of an expression is the quantity, which when used as a factor four times, produces the radicand: $\sqrt[4]{81} = \sqrt[4]{3^4} = 3$.

3. Factor each radicand and find the root.

		Factors of the radicand	Root
(a)	$\sqrt[3]{27}$	$\sqrt[3]{27} = \sqrt[3]{3 \cdot 3 \cdot 3} = \sqrt[3]{3^3}$	
(b)	$\sqrt[4]{16}$	$\sqrt[4]{16} = \sqrt[4]{2 \cdot 2 \cdot 2 \cdot 2} = \sqrt[4]{2^4}$	
(c)	$\sqrt[5]{3125}$	$\sqrt[5]{3125} = \sqrt[5]{5 \cdot 5 \cdot 5 \cdot 5 \cdot 5} = \sqrt[5]{5^5}$	

(d) What pattern do you notice in the number of factors of a radicand that is a perfect power, the index of the radical, and the answer to a root problem?

4. Answer the following.

(a) If a radicand is a perfect square, how many identical factors exist when the radicand is factored?

(b) If a radicand is a perfect cube, how many identical factors exist when the radicand is factored?

(c) If a radicand is a perfect fourth power, how many identical factors exist when the radicand is factored?

(d) To generalize: If a radicand is a perfect n^{th} power, how many identical factors exist when the radicand is factored?

5. Apply the generalization above to radicands that are algebraic expressions. Complete the table. Assume that all variables are non-negative.

		Factors of the radicand	**Root**
Ex.	$\sqrt{x^2}$	$= \sqrt{x \cdot x}$	$= x$
(a)	$\sqrt{y^2}$	$= \sqrt{y \cdot y}$	$=$
(b)	$\sqrt[3]{x^3}$	$= \sqrt[3]{x \cdot x \cdot x}$	$=$
(c)	$\sqrt[3]{(x+5)^3}$	$= \sqrt[3]{(x+5)(x+5)(x+5)}$	$=$

6. The root of a radicand that is not a perfect power is an irrational number. Calculators are used to find a decimal approximation of these irrational roots. Use a calculator to find each root. Round answers to four decimal places.

(a) $\sqrt{11}$ (b) $\sqrt[3]{10}$ (c) $\sqrt[4]{60}$

7. A calculator returned an error message of "NONREAL ANS" for $\sqrt{-25}$ as shown. Why?

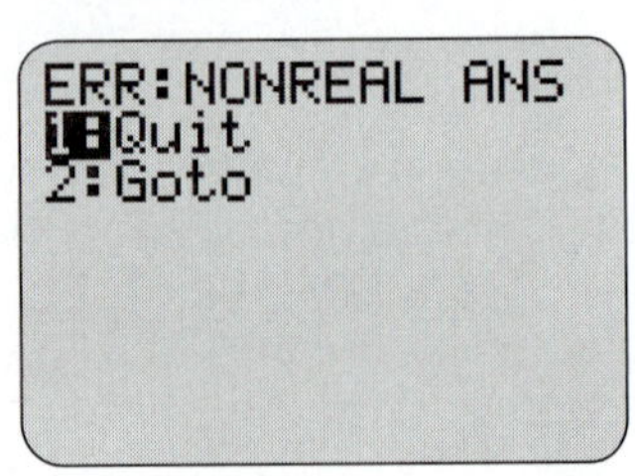

1 2 3

Discussing Mathematics

"Finding a root" is "undoing" the operation of raising a base to an exponent. The process involves identifying the factors of the radicand and looking for perfect powers.

EXAMPLE 3: Factor the radicand and look for perfect powers.

The second root or square root of 4 is 2: $\sqrt{4} = \sqrt{2^2} = 2$.

The third root or cube root of 125 is 5: $\sqrt[3]{125} = \sqrt[3]{5^3} = 5$.

The fourth root of 81 is 3: $\sqrt[4]{81} = \sqrt[4]{3^4} = 3$.

➢ *Understanding the meaning of a root and writing the answer in the appropriate form is essential. Calculators are useful tools when used appropriately, but we should not reach for our calculators to find the root of a radicand that is a small or familiar number. When possible, compute roots mentally. The calculator is used to confirm answers and to find roots that are irrational numbers or not easily identified.*

ROOTS WITH EVEN INDICES

The algebraic sign in front of the radical of a root with an EVEN index must be considered carefully.

EXAMPLE 4: There are two real numbers whose square is 25. There is no real number whose square is -25.

$\sqrt{25} = 5$	$-\sqrt{25} = -5$	$\sqrt{-25}$
The non-negative square root is called the **principal square root**.	The **negative square root** of 25 is -5.	There is **no real number** whose square is -25.
$(+5)^2 = (+5)(+5) = 25$	$(-5)^2 = (-5)(-5) = 25$	
The principal square root of 25 is $+5$.	The negative square root of 25 is -5.	We will write "nonreal" as the answer.

EXAMPLE 5: There are two real numbers whose fourth power is 16. There is no real number whose fourth root is -16.

$\sqrt[4]{16} = +2$	$-\sqrt[4]{16} = -2$	$\sqrt[4]{-16}$
The non-negative fourth root is called the **principal fourth root**.	The **negative fourth root** of 16 is -2.	There is **no real number** whose fourth power is -16.
$(+2)^4 = (+2)(+2)(+2)(+2) = 16$	$(-2)^4 = (-2)(-2)(-2)(-2) = 16$	
The principal fourth root of 16 is $+2$.	The negative fourth root of 16 is -2.	We will write "nonreal" as the answer.

➢ *A calculator in "Real" mode will display an error message, such as "NONREAL ANS," when computing* $\sqrt{-25}$ *or the* $\sqrt[4]{-16}$.

DEFINITION OF AN nth ROOT WHEN n IS ANY POSITIVE EVEN INTEGER

Given n is a positive EVEN integer, the real number a is the nth root of x if $a^n = x$.

- If a is a non-negative real number and $a^n = x$, then $\sqrt[n]{x} = a$ is called the principal n^{th} root of x.
- If a is a negative real number and $a^n = x$, then $-\sqrt[n]{x} = a$ is called the negative n^{th} root of x.
- If x is a negative number, $\sqrt[n]{x}$ is not a real number and the answer will be written "nonreal."

ROOTS WITH ODD INDICES

When the index of a root is a positive ODD integer, a real answer exists, even when the radicand is a negative number.

EXAMPLE 6: Find the roots.

(a) $\sqrt[3]{27} = \sqrt[3]{3\cdot 3\cdot 3} = \sqrt[3]{3^3} = 3$

The non-negative cube root is 3.

(b) $-\sqrt[3]{27} = \sqrt[3]{3\cdot 3\cdot 3} = -\sqrt[3]{3^3} = -3$

The negative cube root is -3.

(c) $\sqrt[3]{-27} = \sqrt[3]{(-3)\cdot(-3)\cdot(-3)} = \sqrt[3]{(-3)^3} = -3$

The cube root of -27 is -3.

(d) $-\sqrt[5]{32} = -\sqrt[5]{2\cdot 2\cdot 2\cdot 2\cdot 2} = -\sqrt[5]{2^5} = -2$

(e) $\sqrt[5]{-243} = \sqrt[5]{(-3)\cdot(-3)\cdot(-3)\cdot(-3)\cdot(-3)} = \sqrt[5]{(-3)^5} = -3$

DEFINITION OF AN nth ROOT WHEN n IS ANY POSITIVE ODD INTEGER

Given n is a positive ODD integer, the real number a is the nth root of x if $a^n = x$.

- If x is a positive real number, then $\sqrt[n]{x} = a$, and a is a positive real number.
- If x is a negative real number, then $\sqrt[n]{x} = a$, and a is a negative real number.

RADICANDS INVOLVING VARIABLES

The definitions of roots presented apply to radicands involving variable expressions. A perfect square has two identical factors. A perfect cube has three identical factors. A perfect fourth power has four identical factors and so forth.

EXAMPLE 7: Find the roots. Variables in radicands with EVEN indices represent non-negative numbers.

(a) $\sqrt{x^2} = x$ (b) $\sqrt[3]{y^3} = y$ (c) $\sqrt[4]{x^4} = x$

(d) $\sqrt{z\cdot z\cdot z\cdot z\cdot z\cdot z} = \sqrt{z^3\cdot z^3} = \sqrt{(z^3)^2} = z^3$

(e) $\sqrt{(x-3)(x-3)} = \sqrt{(x-3)^2} = (x-3)$

(f) $\sqrt[3]{y\cdot y\cdot y\cdot y\cdot y\cdot y} = \sqrt[3]{y^2\cdot y^2\cdot y^2} = \sqrt[3]{(y^2)^3} = y^2$

(g) $\sqrt[3]{(y+5)(y+5)(y+5)} = \sqrt[3]{(y+5)^3} = (y+5)$

(h) $\sqrt[4]{x\cdot x\cdot x\cdot x\cdot x\cdot x\cdot x\cdot x} = \sqrt[4]{x^2\cdot x^2\cdot x^2\cdot x^2} = \sqrt[4]{(x^2)^4} = x^2$

ROOTS THAT ARE IRRATIONAL NUMBERS

Roots are not always rational numbers. If the radicand of a square root is not a perfect square, the result is an irrational number, a decimal that neither terminates nor repeats. If the radicand of a cube root is not a perfect cube, the result is an irrational number. Irrational roots can be left in radical form or as a decimal approximation.

> **IRRATIONAL ROOTS**
>
> - If x is a positive real number that is not a perfect square, then $\sqrt{x}$ is an irrational number.
> - If x is a real number that is not a perfect cube, then $\sqrt[3]{x}$ is an irrational number.

EXAMPLE 8: Irrational numbers computed with a calculator may be rounded to a specified place-value.

(a) $\sqrt{5} \approx 2.24$ Rounded to the nearest hundredth.

(b) $\sqrt[3]{-50} \approx -3.684$ Rounded to the nearest thousandth.

(c) $-\sqrt[4]{13} \approx -1.8988$ Rounded to the nearest ten-thousandth.

> **CONCLUSIONS ABOUT ROOTS AND RADICALS**
>
> 1. Roots may be real numbers (irrational numbers or rational numbers).
> 2. A perfect square has two identical factors. A perfect cube has three identical factors. A perfect fourth power has four identical factors. A perfect n^{th} power has n identical factors. Simplifying a radical with index n means finding perfect n^{th} powers in the radicand.
> 3. The following are true when the index, n, is an EVEN number.
>   (a) The non-negative n^{th} root is called the principal n^{th} root $(\sqrt[n]{\ })$.
>   (b) The negative n^{th} root is indicated when a negative sign will be placed in front of the radical $(-\sqrt[n]{\ })$.
>   (c) When the radicand is negative, then $\sqrt[n]{x}$ $(x < 0)$ cannot be represented as a real number. Write "nonreal" answer.
> 4. The following are true when the index, n, is an ODD number.
>   (a) One real root exists.
>   (b) If the radicand is positive, the root is positive.
>   (c) If the radicand is negative, the root is negative.

1 2 3 Doing Mathematics

- Write the definitions of the words and symbols listed for this section in the Dictionary in Appendix A.
- On a 3 × 5-inch card, write the definition of an nth when n is a positive EVEN and ODD root for future reference.

Problems 1–30: Find each root. If a root is not a real number, write "nonreal" as the answer.

1. $-\sqrt{25}$ **2.** $-\sqrt{16}$ **3.** $\sqrt{-25}$

4. $\sqrt{-4}$ **5.** $\sqrt{-5^2}$ **6.** $-\sqrt{\frac{16}{49}}$

7. $-\sqrt{\frac{100}{81}}$ **8.** $\sqrt{0}$ **9.** $-\sqrt[3]{8}$

10. $-\sqrt[3]{27}$ **11.** $\sqrt[3]{-8}$ **12.** $\sqrt[3]{-125}$

13. $\sqrt[4]{16}$ **14.** $\sqrt[4]{81}$ **15.** $\sqrt[4]{-16}$

16. $\sqrt[4]{-81}$ **17.** $-\sqrt[4]{81}$ **18.** $-\sqrt[4]{16}$

19. $\sqrt[5]{32}$ **20.** $\sqrt[5]{243}$ **21.** $\sqrt[5]{-32}$

22. $\sqrt[5]{-243}$ **23.** $-\sqrt[5]{-32}$ **24.** $-\sqrt[5]{-243}$

25. $-\sqrt[5]{32}$ **26.** $-\sqrt[5]{243}$ **27.** $\sqrt[3]{(25)^3}$

28. $\sqrt[4]{(23)^4}$ **29.** $\sqrt[4]{(11)^4}$ **30.** $\sqrt[5]{(1.7)^5}$

Problems 31–40: Find each root. Assume all variables represent non-negative numbers.

31. $\sqrt{x^2}$ **32.** $\sqrt{y^2}$ **33.** $\sqrt{(x+5)^2}$

34. $\sqrt{(x-1)^2}$ **35.** $\sqrt{x^2-4x+4}$ **36.** $\sqrt{x^2+2x+1}$

37. $\sqrt[5]{x^5}$ **38.** $\sqrt[3]{a^3}$ **39.** $\sqrt[4]{(x-5)^4}$

40. $\sqrt[4]{(y+2)^4}$

Problems 41–56: Use a calculator to find each root. Round answers to four decimal places. If the root is not a real number, write "nonreal" as the answer.

41. $\sqrt{14}$ **42.** $\sqrt{135}$ **43.** $\sqrt{234.67}$

44. $\sqrt{67.11}$ **45.** $-\sqrt{\frac{5}{9}}$ **46.** $-\sqrt{\frac{1}{6}}$

47. $\sqrt{-7^2}$ **48.** $\sqrt{-3^2}$ **49.** $\sqrt[3]{10}$

50. $\sqrt[3]{70}$ **51.** $\sqrt[4]{100}$ **52.** $\sqrt[4]{99}$

53. $\sqrt[4]{-17}$ **54.** $\sqrt[4]{-1}$ **55.** $-\sqrt[4]{18}$

56. $-\sqrt[4]{34}$

Problems 57–60: Rewrite each radicand as a perfect square and find the square root.

	Problem	Rewrite the radicand as a perfect square	Square root
Ex.	$\sqrt{x^4}$	$=\sqrt{(x^2)^2}$	$= x^2$
57.	$\sqrt{x^6}$		
58.	$\sqrt{x^8}$		
59.	$\sqrt{x^{10}}$		
60.	$\sqrt{x^{14}}$		

Problems 61–64: Rewrite each radicand as a perfect cube and find the cube root.

	Problem	Rewrite the radicand as a perfect cube	Cube root
Ex.	$\sqrt[3]{x^6}$	$=\sqrt[3]{(x^2)^3}$	$= x^2$
61.	$\sqrt[3]{x^9}$		
62.	$\sqrt[3]{x^{12}}$		
63.	$\sqrt[3]{x^{15}}$		
64.	$\sqrt[3]{x^{18}}$		

Problems 65–68: Insert the appropriate symbol, $=$ or $\neq$, on the line between the expressions. Assume all variables are non-negative real numbers.

65. $\sqrt{5^2}$ ______ 25

66. $\sqrt{x^3}$ ______ x

67. $-\sqrt{4}$ ______ -2

68. $\sqrt[5]{x^5}$ ______ x

69. When a driver applies the brakes and the car skids to a stop, the minimum speed of the car when the brakes were applied can be estimated by the function $S(d) = \sqrt{30fd}$, where S is the speed of the car in miles per hour, f is the drag factor, and d is the length of the skid marks measured in feet. For this problem, assume that the drag factor is $f = 0.83$. A driver claims that he was going the speed limit of 25 mph when he applied his brakes. The police measured the skid marks at the accident and found them to be 60 feet long.

(a) What was the minimum speed of the car when the brakes were applied?

(b) Should the driver be ticketed for speeding?

70. Explain why $\sqrt{-25}$ is not a real number and why $\sqrt[3]{-8}$ is a real number.

8.2 Multiplication and Division of Radical Expressions

OBJECTIVES

- Identify equivalent forms of radical expressions.
- Multiply and divide radical expressions.

1 2 3 Getting Started

In this section, we discuss the operations of multiplication and division of radical expressions. Expressions involving radicals can often be represented in equivalent forms.

Simplifying a radical is identifying all perfect square factors in the radicand of a square root, all perfect cube factors in a cube root, all perfect fourth power factors in a fourth root, and so forth. These perfect power factors are written without a radical. Other factors remain in the radicand. Simplifying a radical expression employs the

Product Rule for Radicals: $\sqrt{a \cdot b} = \sqrt{a} \cdot \sqrt{b}$ (also written $\sqrt{a} \cdot \sqrt{b} = \sqrt{a \cdot b}$), where $a \geq 0$ and $b \geq 0$. The Product Rule is used to write an equivalent form of a radical expression using the factors in the radicand. In a square root, look for two identical factors.

EXAMPLE 1: Simplifying $\sqrt{18}$ is writing the equivalent form $3 \cdot \sqrt{2}$. Factor the radicand and identify the perfect square 9 in the radicand. Write the square root of 9 without a radical. Factors that are not perfect squares remain in the radicand.

	Simplified Form	Decimal Approximation
$\sqrt{18} = \sqrt{9 \cdot 2} = \sqrt{3^2 \cdot 2} = \sqrt{3^2} \cdot \sqrt{2}$	$= 3 \cdot \sqrt{2}$	$\approx$ 4.2426

9 is a perfect square

➢ *$\sqrt{18}$ is equivalent to the product of 3 and $\sqrt{2}$:* $\sqrt{18} = 3 \cdot \sqrt{2}$.

➢ *The multiplication symbol will not be written explicitly:* $3 \cdot \sqrt{2} = 3\sqrt{2}$.

➢ $3\sqrt{2}$ *is the simplified exact form of* $\sqrt{18}$.

➢ 4.2426 *is a decimal approximation of* $\sqrt{18}$.

1. Use a calculator to check for equivalency of the following radical expressions listed as Root 1 and Root 2. Round each decimal approximation to four decimal places.

	Root 1	Decimal approx. of Root 1	Root 2 simplified form	Decimal approx. of Root 2
Ex.	$\sqrt{18}$	4.2426	$3 \cdot \sqrt{2}$	4.2426
(a)	$\sqrt{45}$		$3 \cdot \sqrt{5}$	
(b)	$\sqrt[3]{24}$		$2 \cdot \sqrt[3]{3}$	
(c)	$\sqrt[4]{48}$		$2 \cdot \sqrt[4]{3}$	

(d) Compare Root 1 and Root 2. Are the two expressions equivalent?

EXAMPLE 2: Apply the Product Rule and simplify $\sqrt{63}$.

$\sqrt{63} = \sqrt{3 \cdot 3 \cdot 7} = \sqrt{3^2 \cdot 7}$ — Factor the radicand into prime factors and look for two identical factors.

$= \sqrt{3^2} \cdot \sqrt{7}$ — Write the perfect square in one radical and the factor of 7 in another radical.

$= 3 \cdot \sqrt{7}$ — Write the square root of 3^2. The number 7 is not a perfect square and remains in a radical.

$= 3\sqrt{7}$ — The multiplication symbol is not written in the final answer.

2. Apply the Product Rule and simplify each square root. Confirm, using a calculator, that Root 1 and Root 2 are equivalent expressions by finding a decimal approximation, rounded to 4 decimal places.

	Root 1	Decimal approx. of Root 1	Factored form	Product of radicals	Root 2 simplified form	Decimal approx. of Root 2
Ex.	$\sqrt{200}$	14.1421	$\sqrt{2^2 \cdot 2 \cdot 5^2}$	$= \sqrt{2^2} \cdot \sqrt{5^2} \cdot \sqrt{2}$	$= 2 \cdot 5 \cdot \sqrt{2} = 10\sqrt{2}$	≈ 14.1421
(a)	$\sqrt{75}$					
(b)	$\sqrt{150}$					

➢ *When the index of a radical is three, look for three identical factors or a perfect cube in the radicand. When the index of a radical is four, look for four identical factors or a perfect fourth power under the radical, and so on for higher indices.*

EXAMPLE 3: The Product Rule applies to cube roots. Simplify $\sqrt[3]{24}$.

$\sqrt[3]{24} = \sqrt[3]{2 \cdot 2 \cdot 2 \cdot 3} = \sqrt[3]{2^3 \cdot 3}$ — Factor the radicand and look for three identical factors.

$= \sqrt[3]{2^3} \cdot \sqrt[3]{3}$ — Write the perfect cube in one radical and the factor 3 in another radical.

$= 2 \cdot \sqrt[3]{3}$ — Write the cube root. The number 3 is not a perfect cube and remains in the radicand.

$= 2\sqrt[3]{3}$

3. Apply the Product Rule and simplify each cube root in the table. Confirm, using a calculator, that Root 1 and Root 2 are equivalent expressions by finding a decimal approximation, rounded to 4 decimal places.

	Root 1	Decimal approx. of Root 1	Factored form	Product of radicals	Root 2 simplified form	Decimal approx. of Root 2
Ex.	$\sqrt[3]{54}$	3.7798	$\sqrt[3]{3 \cdot 3 \cdot 3 \cdot 2} = \sqrt[3]{3^3 \cdot 2}$	$= \sqrt[3]{3^3} \cdot \sqrt[3]{2}$	$= 3\sqrt[3]{2}$	≈ 3.7798
(a)	$\sqrt[3]{375}$					
(b)	$\sqrt[3]{120}$					

4. The Product Rule for Radicals is also used to simplify radicals involving variables. Assume that the variables in the table represent non-negative numbers. Simplify each radical expression.

	Original problem	Factored form	Simplified form
Ex.	$\sqrt{25a^2b}$	$=\sqrt{5\cdot 5\cdot a\cdot a\cdot b}=\sqrt{5^2\cdot a^2\cdot b}=\sqrt{5^2}\cdot\sqrt{a^2}\cdot\sqrt{b}$	$=5a\sqrt{b}$
(a)	$\sqrt{50x^2y}$		
(b)	$\sqrt[3]{27y^3z}$		

5. The Product Rule for Radicals is used to multiply radicals too: $\sqrt{a}\cdot\sqrt{b}=\sqrt{a\cdot b}$. Use a calculator to approximate the products of radicals in the first and last columns, rounded to four decimal places.

	Root 1	Decimal approx. of Root 1	Root 2 Product of radicals	Decimal approx. of Root 2
Ex.	$\sqrt{7}\cdot\sqrt{5}$	5.9161	$\sqrt{35}$	5.9161
(a)	$-\sqrt{11}\cdot\sqrt{2}$			
(b)	$\sqrt{10}\cdot\sqrt{3}$			

1 2 3

Discussing Mathematics

Radical expressions can be rewritten in equivalent forms by simplifying and/or performing operations. **Simplifying a radical** is the process in which all perfect square factors, perfect cube factors, perfect fourth power factors, and so on, are identified and written without a radical. Expressions that are not perfect powers remain in the radicand. The final simplified answer is the product of the rational number(s) and the remaining radical.

PRODUCT RULE FOR RADICALS

The Product Rule for Radicals is used to simplify radical expressions and to multiply radicals with LIKE indices.

PRODUCT RULE FOR RADICALS

(stated in two equivalent forms)

$$\sqrt[n]{a\cdot b}=\sqrt[n]{a}\cdot\sqrt[n]{b} \quad\text{or}\quad \sqrt[n]{a}\cdot\sqrt[n]{b}=\sqrt[n]{a\cdot b},$$

where $\sqrt[n]{a}$ and $\sqrt[n]{b}$ are defined.*

To simplify a radical expression, write factors under separate radicals.

To multiply radicals with LIKE indices, write factors under one radical.

* *If n is an even integer, a and b must be non-negative real numbers* $(a \geq 0, b \geq 0)$. *If n is an odd integer, a and b can be any real number.*

USING THE PRODUCT RULE TO SIMPLIFY RADICALS

EXAMPLE 4: Apply the Product Rule and simplify.

(a) $-\sqrt{175} = -\sqrt{5\cdot 5\cdot 7} = -\sqrt{5^2\cdot 7}$

Factor the radicand into prime factors and look for two identical factors: 5^2 is a perfect square.

$= -\sqrt{5^2}\cdot\sqrt{7}$

Write the perfect square, 5^2, under one radical and the factor, 7, under a second radical.

$= -5\cdot\sqrt{7}$

$= -5\sqrt{7}$

Write the square root without a radical. The number 7 is not a perfect square and remains in the radicand.

(b) $\sqrt[3]{-54} = \sqrt[3]{(-3)\cdot(-3)\cdot(-3)\cdot 2} = \sqrt[3]{(-3)^3\cdot 2}$

Factor the radicand into prime factors and look for three identical factors: $(-3)^3$ is a perfect cube.

$= \sqrt[3]{(-3)^3}\cdot\sqrt[3]{2}$

Write the perfect cubes under individual radicals and the factor, 2, under another radical.

$= (-3)\cdot\sqrt[3]{2}$

$= -3\sqrt[3]{2}$

Write the cube roots without a radical. The number 2 is not a perfect cube and remains in the radicand.

(c) $\sqrt{24x^4y^2z} = \sqrt{2\cdot 2\cdot 2\cdot 3\cdot x\cdot x\cdot x\cdot x\cdot y\cdot y\cdot z}$

$= \sqrt{2^2\cdot 2\cdot 3\cdot x^2\cdot x^2\cdot y^2\cdot z}$

Factor the radicand into prime factors and look for two identical factors: 2^2, x^2, and y^2 are perfect squares.

$= \sqrt{2^2}\cdot\sqrt{x^2}\cdot\sqrt{x^2}\cdot\sqrt{y^2}\cdot\sqrt{2\cdot 3\cdot z}$

Write the perfect squares under individual radicals and remaining factors under another radical.

$= 2\cdot x\cdot x\cdot y\cdot\sqrt{6z}$

$= 2x^2y\sqrt{6z}$

Write the square roots without radicals. The expression, $6z$, is not a perfect square and remains in the radicand.

(d) $\sqrt{2x^2 + 20x + 50} = \sqrt{2(x^2 + 10x + 25)}$

$= \sqrt{2(x+5)(x+5)} = \sqrt{2(x+5)^2}$

Factor the radicand into prime factors and look for two identical factors: $(x+5)^2$ is a perfect square.

$= \sqrt{(x+5)^2}\cdot\sqrt{2}$

$= (x+5)\sqrt{2}$

Write the perfect square under one radical and the factor, 2, under another radical. Simplify.

➢ ***Beware.*** *The Product Rule applies only to products in the radicand.*

Product Rule applies:	*Product Rule does not apply:*
$\sqrt{16\cdot 9} = \sqrt{16}\cdot\sqrt{9} = 4\cdot 3 = 12$	$\sqrt{16+9} \neq 4+3$
$\sqrt{9x^2} = \sqrt{9}\cdot\sqrt{x^2} = 3\cdot x = 3x$	$\sqrt{9-x^2} \neq 3-x$

MULTIPLYING RADICAL EXPRESSIONS

The Product Rule for Radicals is used to multiply radical expressions with LIKE indices. When factors appear outside of radical expressions, they are also multiplied. After multiplying, be sure that the resulting radical expression has no perfect powers in the radicand.

EXAMPLE 5: Apply the Product Rule and multiply.

(a) $2\sqrt{21}\cdot 5\sqrt{3} = \underbrace{2\cdot 5}\cdot\underbrace{\sqrt{21}\cdot\sqrt{3}}$ — Use the Commutative Property of Multiplication to rearrange the order of the factors. Write factors without radicals together and radical expressions together.

$10\cdot\sqrt{21\cdot 3}$ — Use the Associative Property of Multiplication to group and multiply factors without radicals together and radical expressions together.

$10\cdot\sqrt{3\cdot 3\cdot 7}$ — Write the radicand in factored form.

$10\cdot\sqrt{3^2}\cdot\sqrt{7}$ — Write the perfect square under an individual radical and the factor 7 under one radical.

$10\cdot 3\cdot\sqrt{7}$ — Write the square root without a radical.

$30\sqrt{7}$ — Simplify.

(b) $7\cdot\sqrt[3]{25x^2}\cdot\sqrt[3]{10x^2} = \sqrt[3]{25x^2\cdot 10x^2} = 7\sqrt[3]{250x^4}$ — Apply the Product Rule and multiply the radicands.

$= 7\sqrt[3]{2\cdot 5^3\cdot x^3\cdot x}$ — Factor the radicand into prime factors and look for perfect cubes.

$= 7\sqrt[3]{2}\cdot\sqrt[3]{5^3}\cdot\sqrt[3]{x^3}\cdot\sqrt[3]{x}$ — Write the perfect cubes under individual radicals and the factors, 2 and x under one radical.

$= 7\cdot\sqrt[3]{5^3}\cdot\sqrt[3]{x^3}\cdot\sqrt[3]{2x}$ — The factors can be written in this order.

$= 7\cdot 5\cdot x\cdot\sqrt[3]{2x}$ — Write cube roots without a radical.

$= 35x\sqrt[3]{2x}$ — Simplify.

(c) $\sqrt{(x^2-1)}\cdot\sqrt{(x+1)} = \sqrt{(x^2-1)(x+1)}$ — Apply the Product Rule and multiply the radicands.

$= \sqrt{(x-1)(x+1)(x+1)} = \sqrt{(x-1)(x+1)^2}$ — Factor the radicand and look for perfect squares. $(x+1)^2$ is a perfect square.

$= \sqrt{(x-1)}\cdot\sqrt{(x+1)^2}$ — Write factors under individual radicals.

$= \sqrt{x-1}\cdot(x+1)$ — Simplify.

$= (x+1)\sqrt{x-1}$ — Expressions not in a radicand are generally written in front of the radical.

QUOTIENT RULE OF RADICALS

Division is the inverse operation of multiplication and the Quotient Rule for Radicals is an extension of the Product Rule.

QUOTIENT RULE FOR RADICALS

(stated in two equivalent forms)

$$\sqrt[n]{\frac{a}{b}} = \frac{\sqrt[n]{a}}{\sqrt[n]{b}} \quad \text{or} \quad \frac{\sqrt[n]{a}}{\sqrt[n]{b}} = \sqrt[n]{\frac{a}{b}}$$

where $\sqrt[n]{a}$ and $\sqrt[n]{b}$ are defined and $b \neq 0$.*

To simplify a radical expression, write under separate radicals.

To divide radicals with LIKE indices, write the numerator and denominator under one radical.

* *If n is an even integer, a and b must be non-negative real numbers* $(a \geq 0, b > 0)$.
If n is an odd integer, a and b can be any real number $(b \neq 0)$.

SIMPLIFYING AND DIVIDING RADICAL EXPRESSIONS

1. (a) If one radical appears in the problem, reduce the radicand to lowest terms.
   - Do not reduce the radicand if the common factor is part of a perfect power in either the numerator or denominator. It may be more efficient to keep the perfect power and proceed to Step 3.

   (b) If two radicals appear, write the radicands as a quotient under one radical and reduce the radicand to lowest terms, if possible.
2. Rewrite the expression as the quotient of two radicals with the same index as the original radical.
3. Simplify each radical.

SIMPLIFYING QUOTIENTS

EXAMPLE 6: Apply the Quotient Rule and simplify (x and $y > 0$).

(a) $\sqrt{\dfrac{14}{8}} = \sqrt{\dfrac{7}{4}}$ Divide out the common factor, 2. The denominator contains a perfect square.

$= \dfrac{\sqrt{7}}{\sqrt{4}}$ Apply the Quotient Rule and write the radical expression as the quotient of two square roots.

$= \dfrac{\sqrt{7}}{2}$ Write the square root without a radical.

(b) $\sqrt[3]{\frac{-24xy}{27xy^4}} = \sqrt[3]{\frac{(-2)\cdot(-2)\cdot(-2)\cdot 3\cdot x\cdot y}{3\cdot 3\cdot 3\cdot y\cdot y\cdot y\cdot y}}$ — Divide out the common factor, y. In this problem, do not divide out the common factor 3 because the denominator is a perfect cube.

$= \sqrt[3]{\frac{(-2)^3\cdot 3\cdot x}{3^3\cdot y^3}}$ — Identify perfect cubes: $(-2)^3$, 3^3, and y^3.

$= \frac{\sqrt[3]{(-2)^3\cdot 3\cdot x}}{\sqrt[3]{3^3\cdot y^3}}$ — Apply the Quotient Rule and write the radical expression as the quotient of two cube roots.

$= \frac{\sqrt[3]{(-2)^3}\cdot\sqrt[3]{3\cdot x}}{\sqrt[3]{3^3}\cdot\sqrt[3]{y^3}} = \frac{-2\sqrt[3]{3}}{3y}$ — Apply the Product Rule with cube roots under individual radicals and the factors $3x$ under another radical. Simplify.

DIVIDING RADICAL EXPRESSIONS

EXAMPLE 7: Apply the Quotient Rule and divide $(x > 0)$.

(a) $\frac{\sqrt[3]{4x^3}}{\sqrt[3]{32}} = \sqrt[3]{\frac{\overset{1}{\cancel{4}}x^3}{\underset{8}{\cancel{32}}}} = \sqrt[3]{\frac{x^3}{8}}$ — Write the radicands as a quotient under one cube root. Divide out the common factor, 4.

$= \frac{\sqrt[3]{x^3}}{\sqrt[3]{8}}$ — Apply the Quotient Rule and write the radical expression as the quotient of two cube roots.

$= \frac{x}{2}$ — Write cube roots without a radical.

(b) $\frac{\sqrt{x}}{\sqrt{x^3}} = \sqrt{\frac{x}{x^3}} = \sqrt{\frac{1}{x^2}}$ — Apply the Quotient Rule and write the radicands as a quotient under one radical. Divide out the common factor x.

$= \frac{\sqrt{1}}{\sqrt{x^2}} = \frac{1}{x}$ — Apply the Quotient Rule. Simplify.

RATIONALIZING THE DENOMINATOR

A rational expression is not in simplified form if a radical appears in the denominator. To simplify such an expression, we must **rationalize the denominator** by creating a perfect power in the denominator. Multiply the numerator and denominator of the fraction by appropriate factors to form perfect powers in the radicand of the denominator.

EXAMPLE 8: Rationalize each denominator.

(a) $\dfrac{\sqrt{5}}{\sqrt{2}} = \dfrac{\sqrt{5}}{\sqrt{2}} \cdot \dfrac{\sqrt{2}}{\sqrt{2}}$ Multiply the fraction by $\dfrac{\sqrt{2}}{\sqrt{2}} = 1$ to form a perfect square in the radicand of the denominator (Multiplicative Identity).

$= \dfrac{\sqrt{5 \cdot 2}}{\sqrt{2 \cdot 2}} = \dfrac{\sqrt{10}}{\sqrt{2^2}} = \dfrac{\sqrt{10}}{2}$ Simplify.

(b) $\dfrac{\sqrt[3]{2y}}{\sqrt[3]{3x^2}} = \dfrac{\sqrt[3]{2y}}{\sqrt[3]{3x^2}} \cdot \dfrac{\sqrt[3]{3 \cdot 3 \cdot x}}{\sqrt[3]{3 \cdot 3 \cdot x}}$ To create perfect cubes in the denominator of the fraction, two factors of 3 and one factor of x are needed in the radicand. Multiply the fraction by $\dfrac{\sqrt[3]{3 \cdot 3 \cdot x}}{\sqrt[3]{3 \cdot 3 \cdot x}} = 1$.

$= \dfrac{\sqrt[3]{2y \cdot 3 \cdot 3 \cdot x}}{\sqrt[3]{3x^2 \cdot 3 \cdot 3 \cdot x}} = \dfrac{\sqrt[3]{18x}}{\sqrt[3]{3^3 \cdot x^3}} = \dfrac{\sqrt[3]{18x}}{3x}$ Simplify.

1 2 3 Doing Mathematics

- Write the definitions of the words and symbols listed for this section in the Dictionary in Appendix A.
- On a 3 × 5-inch card, write the Product and Quotient Rule for Radicals for future reference.
- Assume that all variables and radicands are non-negative and that denominators are not equal to zero.

Problems 1–24: Apply the Product Rule and simplify each root. Confirm, using a calculator, that the two expressions are equivalent expressions by finding a decimal approximation, rounded to four decimal places.

1. $\sqrt{125}$ **2.** $\sqrt{20}$

3. $\sqrt{200}$ **4.** $\sqrt{72}$

5. $\sqrt{180}$ **6.** $\sqrt{450}$

7. $-\sqrt{98}$ **8.** $-\sqrt{80}$

9. $\sqrt[3]{128}$ **10.** $\sqrt[3]{72}$

11. $-\sqrt[3]{125}$ **12.** $-\sqrt[3]{64}$

13. $\sqrt[3]{-40}$ **14.** $\sqrt[3]{-56}$

15. $-\sqrt[3]{-189}$ **16.** $-\sqrt[3]{-135}$

17. $\sqrt[4]{48}$ **18.** $\sqrt[4]{176}$

19. $-\sqrt[4]{1250}$ **20.** $-\sqrt[4]{405}$

21. $\sqrt[5]{-486}$ **22.** $\sqrt[5]{-160}$

23. $-\sqrt[6]{1458}$ **24.** $-\sqrt[6]{192}$

Problems 25–34: Apply the Product Rule and simplify each root.

25. $-\sqrt{49x^2y^2}$ **26.** $-\sqrt{16x^3y^4}$

27. $\sqrt[3]{-64a^4}$ **28.** $\sqrt[3]{-27z^5}$

29. $\sqrt[4]{162x^6y^4}$ **30.** $\sqrt[4]{48x^6y^4}$

31. $\sqrt{x^2 - 2x + 1}$ **32.** $\sqrt{x^2 + 8x + 16}$

33. $\sqrt[3]{(x - 1)^5}$ **34.** $\sqrt[3]{(y - 5)^4}$

Problems 35–52: Multiply and simplify.

35. $-\sqrt{2}\cdot\sqrt{6}$ **36.** $-\sqrt{5}\cdot\sqrt{15}$

37. $2\cdot\sqrt[3]{5}\cdot\sqrt[3]{25}$ **38.** $3\cdot\sqrt[3]{6}\cdot\sqrt[3]{36}$

39. $\sqrt[4]{27}\cdot\sqrt[4]{15}$ **40.** $\sqrt[4]{8}\cdot\sqrt[4]{6}$

41. $\sqrt{20}\cdot\sqrt{5}\cdot\sqrt{4}$ **42.** $\sqrt{50}\cdot\sqrt{6}\cdot\sqrt{12}$

43. $7\sqrt{5x}\cdot 6\sqrt{15x}$ **44.** $3\sqrt{8x}\cdot 5\sqrt{2x}$

45. $4\sqrt[3]{-12x^2}\cdot\sqrt[3]{18x^2y^2}$ **46.** $\sqrt[3]{2x^2}\cdot\sqrt[3]{-20xy}$

47. $\sqrt{3ab}\cdot\sqrt{6a}\cdot\sqrt{14b}$ **48.** $\sqrt{5xy}\cdot\sqrt{10x^2}\cdot\sqrt{2y}$

49. $-\sqrt[4]{28xy^3}\cdot\sqrt[4]{20x^4y}$ **50.** $-\sqrt[4]{2x^2y^2}\cdot\sqrt[4]{24xy^4}$

51. $\sqrt[3]{9a^2}\cdot\sqrt[3]{3ab^2}\cdot\sqrt[3]{5b^2}$ **52.** $\sqrt[3]{4x}\cdot\sqrt[3]{2x^2y^2}\cdot\sqrt[3]{6xy}$

Problems 53–78: Apply the Quotient Rule and simplify. Rationalize denominators, if necessary.

53. $-\sqrt{\dfrac{8}{9}}$

54. $-\sqrt{\dfrac{27}{4}}$

55. $\dfrac{\sqrt{25}}{\sqrt{50}}$

56. $\dfrac{\sqrt{4}}{\sqrt{27}}$

57. $\dfrac{\sqrt{98}}{\sqrt{2}}$

58. $\dfrac{\sqrt{50}}{\sqrt{2}}$

59. $\dfrac{\sqrt[3]{-64}}{\sqrt[3]{5}}$

60. $\dfrac{\sqrt[3]{-24}}{\sqrt[3]{27}}$

61. $\sqrt[3]{\dfrac{24}{40}}$

62. $\sqrt[3]{\dfrac{81}{8}}$

63. $\sqrt[3]{-\dfrac{x^3}{27y^6}}$

64. $\sqrt[3]{-\dfrac{8x^6}{y^3}}$

65. $\sqrt[4]{\dfrac{x^2}{16y^4}}$

66. $\sqrt[4]{\dfrac{81x^2}{y^8}}$

67. $\sqrt[3]{\dfrac{-2y^3}{8}}$

68. $\sqrt[3]{\dfrac{-3x^3}{64}}$

69. $\sqrt[4]{\dfrac{625x^7}{y^4}}$

70. $\sqrt[3]{\dfrac{27x^3}{y^6}}$

71. $-\dfrac{\sqrt{16y}}{\sqrt{18y}}$

72. $-\dfrac{\sqrt{50x}}{\sqrt{8x}}$

73. $\sqrt{\dfrac{4}{x^2 + 12x + 36}}$

74. $\sqrt{\dfrac{x^2 + 6x + 9}{25}}$

75. $\sqrt[3]{\dfrac{3x(x + 2)^3}{81x}}$

76. $\sqrt[3]{\dfrac{2x(x - 4)^3}{16x}}$

77. $\dfrac{\sqrt{3x^2 + 18x + 27}}{\sqrt{3x + 9}}$

78. $\dfrac{\sqrt{2x^2 + 20x + 50}}{\sqrt{2x + 10}}$

Problems 79–88: Insert the appropriate symbol, either = or ≠, between the expressions.

79. $\sqrt{-9}$ _______ 3

80. $\sqrt{-4}$ _______ -2

81. $\sqrt{2}\cdot\sqrt{11}$ _______ $\sqrt{13}$

82. $\sqrt{3}\cdot\sqrt{5}$ _______ $\sqrt{8}$

83. $\sqrt{3}\cdot\sqrt[3]{7}$ ______ $\sqrt[3]{21}$

84. $\sqrt[3]{5}\cdot\sqrt{2}$ ______ $\sqrt[3]{10}$

85. $\dfrac{\sqrt[3]{8}}{\sqrt[3]{3}}$ ______ $\sqrt[3]{\dfrac{8}{3}}$

86. $\dfrac{\sqrt[4]{48}}{\sqrt[4]{3}}$ ______ $\sqrt[4]{\dfrac{48}{3}}$

87. 5.000 ______ $\sqrt{25}$

88. $\sqrt{10}$ ______ 3.1622776...

Problems 89–98: Simplify each expression. Write answers in simplified radical form. *Hint:* Some expressions are already in simplified form.

89. $\sqrt{5^2 + 3^3}$

90. $\sqrt{10^2 + 7^3}$

91. $\sqrt{6^2 - 4^2}$

92. $\sqrt{5^2 - 2^3}$

93. $\sqrt{x^2 - 4}$

94. $\sqrt{x^2 - 25}$

95. $\sqrt{x^2 + 9}$

96. $\sqrt{36 + x^2}$

97. $\sqrt{(x + 1)^2}$

98. $\sqrt{(x - 2)^2}$

Problems 99–100: Find the area of a rectangle with the given dimensions. Write answers in simplified radical form.

99. Length $6\sqrt{5}$ inches;
Width $2\sqrt{5}$ inches

100. Length $2\sqrt{3}$ inches;
Width $\sqrt{3}$ inches

Problems 101–102: Find the length of a rectangle with the given width and area. Write answers in simplified radical form.

101. Width $5\sqrt{2}$ inches;
Area $20\sqrt{3}$ inches

102. Width $20\sqrt{2}$ inches;
Area $50\sqrt{7}$ inches

8.3 Addition and Subtraction of Radical Expressions

OBJECTIVES

- Identify LIKE radicals.
- Add and subtract radical expressions.
- Multiply monomial and binomial radical expressions.
- Multiply a binomial radical expression by its conjugate.

1 2 3 Getting Started

In this section, the operations of addition and subtraction involving radical expressions are discussed, along with multiplication of binomial expressions containing radicals. Adding and subtracting radical expressions is a similar process to collecting LIKE terms with polynomials. Throughout this section, assume that all variables are non-negative.

1. Given the first term in each row, circle the terms that are LIKE terms.

(a) Given $2x$: $3xy$ $\quad -4x$ $\quad 5x^2$ $\quad 5xy^2$ $\quad \frac{1}{3}x$

(b) Given $5yz$: $7y$ $\quad -9yz$ $\quad 0.5yz$ $\quad 2y^2z$ $\quad y^2z^2$

EXAMPLE 1: **LIKE radicals** have the same radicand and index.

UNLIKE radicals have radicands and/or indices that are different.

$\sqrt{5}$ and $7\sqrt{5}$ are LIKE radicals	$\sqrt{5}$ and $\sqrt{6}$ are UNLIKE radicals because the radicands are different.
$2\sqrt[3]{4}$ and $-\sqrt[3]{4}$ are LIKE radicals	$\sqrt{5}$ and $\sqrt[3]{5}$ are UNLIKE radicals because the indices are different.
$\sqrt{2xy}$, $-13\sqrt{2xy}$ are LIKE radicals	$\sqrt{2xy}$, $\sqrt{2xy^2}$, $\sqrt{2x^2y}$ are UNLIKE radicals because the radicands are different.

2. Given the first term in each row, circle the LIKE radicals.

(a) Given $\sqrt{3}$: $\sqrt{4}$ $\quad \frac{3}{4}\sqrt{3}$ $\quad \sqrt[3]{3}$ $\quad -4\sqrt[4]{3}$ $\quad 5\sqrt{3}$

(b) Given $\sqrt[4]{53}$: $1.6\sqrt[4]{53}$ $\quad \sqrt{53}$ $\quad 8\sqrt[4]{53}$ $\quad \sqrt[3]{53}$ $\quad -2\sqrt{53}$

(c) Given $\sqrt{x}$: $6\sqrt{x}$ $\quad \sqrt{x^2}$ $\quad -0.2\sqrt[3]{x}$ $\quad \sqrt[4]{x}$ $\quad 5\sqrt{x}$

(d) Given $\sqrt{2x}$: $\sqrt{4x}$ $\quad 9\sqrt{2x^2}$ $\quad \sqrt[3]{2x}$ $\quad \sqrt{2x}$ $\quad -\sqrt{2x}$

EXAMPLE 2: The Distributive Property is used to add and subtract LIKE radicals, in a process similar to collecting LIKE terms.

(a) Collect LIKE terms: $9x + 1.5x = (9 + 1.5)x = 10.5x$

(b) Collect LIKE radicals: $8\sqrt{5} + 3\sqrt{5} = (8 + 3)\sqrt{5} = 11\sqrt{5}$

(c) Collect LIKE radicals: $2\sqrt[3]{x} - 7\sqrt[3]{x} = (2 - 7)\sqrt[3]{x} = -5\sqrt[3]{x}$

3. Collect LIKE radicals. If the radicals are not LIKE radicals, write "UNLIKE radicals."

(a) $17\sqrt{3} + 5\sqrt{3}$ (b) $2\sqrt{3} + 5\sqrt{2}$ (c) $5\sqrt{3} + 8\sqrt[3]{3}$

(d) $11\sqrt{y} - 7\sqrt{y}$ (e) $-20.7\sqrt{3x} + 9.7\sqrt{3x}$ (f) $5\sqrt{2x} + 8\sqrt{2y}$

(g) $12\sqrt[5]{x^2} + 16\sqrt[5]{x^2}$

EXAMPLE 3: Add $5\sqrt{2} + 7\sqrt{8}$. Simplify individual radical expressions *before* adding.

$$5\sqrt{2} + 7\sqrt{8} = 5\sqrt{2} + 7\sqrt{2\cdot 2\cdot 2}$$ Write prime factors of 8.

$$= 5\sqrt{2} + 7\sqrt{2^2\cdot 2}$$ Identify perfect square in radicand.

$$= 5\sqrt{2} + 7\cdot 2\sqrt{2}$$ Write square root without a radical.

$$= 5\sqrt{2} + 14\sqrt{2}$$

$$= (5 + 14)\sqrt{2} = 19\sqrt{2}$$ Add LIKE radicals.

4. Write radicals in simplified form and then collect LIKE radicals.

(a) $4\sqrt{3} + 6\sqrt{12}$

(b) $24\sqrt[3]{2} - 3\sqrt[3]{16}$

(c) $5\sqrt{12} - 2\sqrt{75}$

5. Summarize the process of adding and/or subtracting radical expressions.

1 2 3 Discussing Mathematics

ADDING AND SUBTRACTING RADICAL EXPRESSIONS

Radical expressions are called **LIKE radicals** when the radicand and the index of the radical are identical. To add and/or subtract radical expressions, use the Distributive Property to collect LIKE radicals.

ADDING AND SUBTRACTING TERMS CONTAINING RADICALS
1. Simplify individual radical expressions, if possible.
2. Identify terms with LIKE radicals. Use the Distributive Property to collect LIKE radicals.
3. Write the sum or difference of the coefficients in front of the LIKE radical(s) as the final answer.

EXAMPLE 4: Add and/or subtract.

(a) $7\sqrt{2} + 5\sqrt{2} - \sqrt{2} =$
$7\sqrt{2} + 5\sqrt{2} - 1\sqrt{2} =$ — Write the coefficient of the third term explicitly as $1\sqrt{2}$.
$(7 + 5 - 1)\sqrt{2} = 11\sqrt{2}$ — Add and subtract numerical coefficients on LIKE radicals.

(b) $2x\sqrt[3]{y} + 3x\sqrt[3]{y} =$
$(2x + 3x)\sqrt[3]{y} = 5x\sqrt[3]{y}$ — Add $2x$ and $3x$ on LIKE radicals.

(c) $5\sqrt{x} - 3\sqrt{x} + 7\sqrt{y} =$
$(5 - 3)\sqrt{x} + 7\sqrt{y} = 2\sqrt{x} + 7\sqrt{y}$ — Subtract numerical coefficients on LIKE radicals.

(d) $3.5\sqrt{2b} - 0.5\sqrt{2b} =$
$(3.5 - 0.5)\sqrt{2b} = 3.0\sqrt{2b}$ — Subtract numerical coefficients on LIKE radicals.

EXAMPLE 5: Add. Simplify each individual radical before adding.

$9\sqrt{5} + \frac{1}{2}\sqrt{20} = 9\sqrt{5} + \frac{1}{2}\sqrt{2\cdot 2\cdot 5}$ — Write the radicand, 20, in factored form.

$= 9\sqrt{5} + \frac{1}{2}\sqrt{2^2}\cdot\sqrt{5}$ — Simplify the radicand in the second term.

$= 9\sqrt{5} + \frac{1}{2}\cdot 2\cdot\sqrt{5}$ — Multiply.

$= 9\sqrt{5} + 1\sqrt{5} = 10\sqrt{5}$ — Collect LIKE radicals.

EXAMPLE 6: Subtract.

$15x\sqrt[3]{3y} - 7x\sqrt[3]{81y^3} = 15x\sqrt[3]{3y} - 7x\sqrt[3]{3^3}\cdot\sqrt[3]{3y}$ — Write $\sqrt[3]{81y}$ as $\sqrt[3]{3^3}\cdot\sqrt[3]{3y}$.

$= 15x\sqrt[3]{3y} - 7\cdot 3x\sqrt[3]{3y}$ — Simplify the radicand in the second term.

$= 15x\sqrt[3]{3y} - 21x\sqrt[3]{3y}$ — Multiply.

$= (15x - 21x)\sqrt[3]{3y} = -6x\sqrt[3]{3y}$ — Collect LIKE radicals.

SIMPLIFYING RADICALS INVOLVING SEVERAL OPERATIONS

The Order of Operations applies to all expressions. The Distributive Property and/or the FOIL method might need to be employed to perform operations.

EXAMPLE 7: Perform the operations.

(a) $\sqrt{3}\cdot\sqrt{12} - \sqrt{5} = \sqrt{3\cdot 12} - \sqrt{5}$ — Multiply radicals in the first term.

$= \sqrt{3\cdot 2\cdot 2\cdot 3} - \sqrt{5}$

$= \sqrt{2^2\cdot 3^2} - \sqrt{5}$ — Write perfect squares without a radical.

$= 2\cdot 3 - \sqrt{5}$ — Multiply.

$= 6 - \sqrt{5}$

(b) $3\sqrt{7}(\sqrt{14} - 2) = 3\sqrt{7}\cdot\sqrt{14} - 3\sqrt{7}\cdot 2$ — Distribute $3\sqrt{7}$ to remove parentheses.

$= 3\sqrt{7\cdot 2\cdot 7} - 3\cdot 2\sqrt{7}$ — Multiply.

$= 3\sqrt{7^2\cdot 2} - 6\sqrt{7}$ — Identify perfect square in radicand.

$= 3\cdot 7\sqrt{2} - 6\sqrt{7}$ — Write the perfect square without a radical.

$= 21\sqrt{2} - 6\sqrt{7}$ — Simplify.

$= 21\sqrt{2} - 6\sqrt{7}$ — UNLIKE radicals cannot be collected.

(c) $(2\sqrt{5} + 1)(\sqrt{5} - 7)$

$= 2\sqrt{5}\cdot\sqrt{5} - 2\sqrt{5}\cdot 7 + 1\cdot\sqrt{5} - 1\cdot 7$ — Use the FOIL method to multiply the binomials.

F O I L

$= 2\sqrt{5^2} - 14\sqrt{5} + 1\sqrt{5} - 7$ — Write the perfect square without a radical.

$= 2\cdot 5 - 14\sqrt{5} + 1\sqrt{5} - 7$ — Multiply and collect LIKE radicals.

$= 10 - 13\sqrt{5} - 7$ — Simplify.

$= 3 - 13\sqrt{5}$ — Subtract.

(d) $(7\sqrt{x} + 3)^2$

$= (7\sqrt{x} + 3)(7\sqrt{x} + 3)$ — Rewrite the problem as the product of two binomials.

$= 7\sqrt{x}\cdot 7\sqrt{x} + 7\sqrt{x}\cdot 3 + 3\cdot 7\sqrt{x} + 3\cdot 3$ — Use the FOIL method to multiply the binomials.

F O I L

$= 49\sqrt{x^2} + 21\sqrt{x} + 21\sqrt{x} + 9$ — Simplify each term.

$= 49x + (21 + 21)\sqrt{x} + 9$ — Write the perfect square without a radical.

$= 49x + 42\sqrt{x} + 9$ — Combine LIKE radicals.

MULTIPLYING CONJUGATES

The expressions $\sqrt{a} + \sqrt{b}$ and $\sqrt{a} - \sqrt{b}$ are called **conjugates**. The first and second terms are the same, but the arithmetic operations are opposites.

$\sqrt{7} + \sqrt{5}$ and $\sqrt{7} - \sqrt{5}$ are conjugates.

$\sqrt{x} - \sqrt{3}$ and $\sqrt{x} + \sqrt{3}$ are conjugates.

When conjugates are multiplied, the result is the difference of two expressions without radicals.

EXAMPLE 8: Multiply $(\sqrt{x} + \sqrt{3})(\sqrt{x} - \sqrt{3})$.

$= \sqrt{x}\cdot\sqrt{x} - \sqrt{x}\cdot\sqrt{3} + \sqrt{x}\cdot\sqrt{3} - \sqrt{3}\cdot\sqrt{3}$ — Multiply.*

F O I L

$= \sqrt{x^2} - \underbrace{\sqrt{3x} + \sqrt{3x}} - \sqrt{3^2}$ — Write perfect squares without a radical.

$= x + (-1 + 1)(\sqrt{3x}) - 3 = x - 3$ — Collect middle terms.

* *Conjugates also can be multiplied using the formula* $(X + Y)(X - Y) = X^2 - Y^2$:

$$\begin{array}{ccc} (X + Y)(X - Y) & = & X^2 - Y^2 \\ \downarrow & & \downarrow \\ (\sqrt{x} + \sqrt{3})(\sqrt{x} - \sqrt{3}) & = & (\sqrt{x})^2 - (\sqrt{3})^2 = x - 3 \end{array}$$

1 2 3 Doing Mathematics

- Write the definitions of the words and symbols listed for this section in the Dictionary in Appendix A.
- On a 3 × 5-inch card, write the steps of the process of adding and subtracting LIKE radicals for future reference.
- Assume that all variables and radicands are non-negative and that denominators are not equal to zero.

Problems 1–4: Given the first term in each row, circle the LIKE radicals.

1. Given $2\sqrt{5}$:	$\sqrt{7}$	$\sqrt{3}$	$\sqrt[3]{3}$	$-\sqrt{5}$	$7.6\sqrt{5}$
2. Given $\sqrt[3]{5}$:	$-2\sqrt{5}$	$\sqrt[3]{5}$	$\sqrt{5}$	$\sqrt{15}$	$\sqrt[4]{5}$
3. Given $-\sqrt{2y}$:	$\sqrt{y}$	$\sqrt{3y}$	$\sqrt[3]{3y}$	$\sqrt{2y}$	$5\sqrt{2y}$
4. Given $-7\sqrt[3]{x}$:	$\sqrt{x}$	$\sqrt{x^2}$	$4\sqrt[3]{x}$	$10\sqrt[3]{x}$	$\sqrt{x^3}$

Problems 5–18: Collect LIKE radicals. If the radicals are UNLIKE, write "UNLIKE."

5. $5\sqrt{2} + 8\sqrt{2}$

6. $6\sqrt{5} + 11\sqrt{5}$

7. $-\sqrt{22} - 8\sqrt{22}$

8. $-9\sqrt{17} - \sqrt{17}$

9. $5\sqrt{2} - 8\sqrt{7}$

10. $1.3\sqrt{7} - 0.2\sqrt{5}$

11. $1.1\sqrt{x} + 6.8\sqrt{x}$

12. $5.4\sqrt{10x} + 8.3\sqrt{10x}$

13. $5\sqrt{a} + 8\sqrt{b}$

14. $5\sqrt[3]{a^2} + 8\sqrt[3]{a}$

15. $5\sqrt[3]{2x} + 8\sqrt[3]{2x}$

16. $3.2\sqrt[3]{5x} + 1.8\sqrt[3]{5x}$

17. $\frac{2}{3}\sqrt{2a} + \frac{1}{3}\sqrt{2a} - 5\sqrt{2a}$

18. $15\sqrt{2a} + 8\sqrt{2a} - \sqrt{2a}$

Problems 19–30: Write the radicals in simplified form and collect LIKE radicals.

19. $-6\sqrt{18} - 5\sqrt{72}$

20. $-7\sqrt{12} + 5\sqrt{75}$

21. $4\sqrt{12} - 5\sqrt{75} + 12\sqrt{48}$

22. $-0.5\sqrt{45y} + 1.5\sqrt{5y} + 6\sqrt{20y}$

23. $0.56\sqrt[3]{2x} - 1.35\sqrt[3]{16x} + 5.24\sqrt[3]{2x}$

24. $\sqrt{4xy^2} - \sqrt{12x^3y^2} + \sqrt{49xy^2}$

25. $\frac{1}{2}\sqrt{20} + \frac{5}{2}\sqrt{20} - \frac{1}{2}\sqrt{20}$

26. $-9\sqrt[3]{24} + 5\sqrt[3]{3} - 15\sqrt[3]{40}$

27. $6\sqrt[3]{x^4} + 5\sqrt[3]{8x^4} - \sqrt[3]{27x^4}$

28. $\frac{1}{3}\sqrt{x^2} - \frac{5}{3}\sqrt{x^2} + \frac{5}{6}\sqrt{x^2}$

29. $2x\sqrt{4z} + 5x\sqrt{9z} - 10x\sqrt{25z}$

30. $\sqrt{\frac{75x^2}{3}} + \sqrt{36x^2} - 2\sqrt{49x^2}$

Problems 31–54: Perform the operations.

31. $\sqrt{5}\cdot\sqrt{30} + 2\sqrt{6}$

32. $\sqrt{5}\cdot\sqrt{15} + 7\sqrt{3}$

33. $\sqrt{2}\cdot\sqrt{6} + 5\sqrt{3} - 7\sqrt{12}$

34. $\sqrt{3}\cdot\sqrt{6} + 5\sqrt{2} - 7\sqrt{18}$

35. $10\sqrt{x^2} + 3x - 5\sqrt{x^3} + 9x\sqrt{x}$

36. $16.25\sqrt[3]{x^3} + 0.75x - 5\sqrt[3]{x^3} + \sqrt[3]{x^3}$

37. $\sqrt{2}(\sqrt{10} + 3)$

38. $\sqrt{5}(\sqrt{35} + 1)$

39. $\sqrt{x}(\sqrt{x^3} - \sqrt{2x})$

40. $\sqrt{2y}(\sqrt{2y^3} - \sqrt{8y})$

41. $(\sqrt{3} + 2)(\sqrt{3} - 5)$

42. $(\sqrt{7} + 3)(\sqrt{7} - 1)$

43. $5\sqrt[3]{x}(\sqrt[3]{x^2} - 21)$

44. $7\sqrt[3]{2x^2}(\sqrt[3]{4x} - 1)$

45. $(\sqrt{x} + 4)(\sqrt{x} - 5)$

46. $(\sqrt{y} + 3)(\sqrt{y} - 7)$

47. $(\sqrt{x} + 6)^2$

48. $(\sqrt{b} + 1)^2$

49. $(\sqrt{a} - \sqrt{b})^2$

50. $(\sqrt{x} - \sqrt{y})^2$

51. $(\sqrt{5} + \sqrt{2})(\sqrt{5} - \sqrt{2})$

52. $(\sqrt{8.5} + \sqrt{1.5})(\sqrt{8.5} - \sqrt{1.5})$

53. $(\sqrt{x} + \sqrt{11})(\sqrt{x} - \sqrt{11})$

54. $(\sqrt{x} + \sqrt{7})(\sqrt{x} - \sqrt{7})$

Problems 55–60: Insert the appropriate symbol, either = or ≠ , on the line between the expressions.

55. $(\sqrt{5})^2$ ______ 5

56. $(\sqrt{5} - 1)^2$ ______ $(\sqrt{5})^2 - 1^2$

57. $\sqrt[3]{4}(\sqrt{5} + 1)$ ______ $\sqrt[3]{20} + \sqrt[3]{4}$

58. $(\sqrt{7})^3$ ______ 21

59. $(3 + \sqrt{2})^2$ ______ $3^2 + (\sqrt{2})^2$

60. $\sqrt[3]{2}(\sqrt[3]{3} + 1)$ ______ $\sqrt[3]{6} + \sqrt[3]{2}$

Problems 61–62: Write the perimeter and area of the rectangle with given dimensions. Perimeter formula: $P = 2l + 2w$; Area formula: $A = l \cdot w$.

61. Length $= 3\sqrt{2}$ inches;
Width $= \sqrt{2}$ inches

62. Length $= 5\sqrt[3]{2}$ inches;
Width $= \sqrt[3]{2}$ inches

Problems 63–64: Write the area of the isosceles triangle with given dimensions.
Area formula: $A = \frac{1}{2}b \cdot h$, where b = base and h = height

63. Height $= 2\sqrt[3]{5}$ cm;
Base $= 7\sqrt[3]{5}$ cm

64. Height $= 1.2\sqrt{6}$ cm;
Base $= 3.8\sqrt{6}$ cm

Chapter 8 Summary

SECTION 8.1:

Definition of an nth Root

Given n is a positive integer, the real number a is the nth root of x if $a^n = x$.

Conclusions about Roots and Radicals

1. Roots may be real numbers (irrational numbers or rational numbers).
2. Simplifying a radical with index n means making sure there are no perfect n^{th} powers under the radical.
3. When the index, n, is an EVEN number:
   (a) The non-negative n^{th} root is called the principal n^{th} root $\left(\sqrt[n]{\ }\right)$
   (b) The negative n^{th} root is indicated when a negative sign will be placed in front of the radical $\left(-\sqrt[n]{\ }\right)$
   (c) If the radicand is negative, then $\sqrt[n]{x}$ $(x < 0)$ cannot be represented as a real number; write "nonreal" answer
4. When the index, n, is an ODD number, one real root exists:
   (a) If the radicand is positive, the root is positive
   (b) If the radicand is negative, the root is negative

SECTION 8.2:

Product Rule for Radicals

For real numbers a and b and n a positive integer:

$$\sqrt[n]{a \cdot b} = \sqrt[n]{a} \cdot \sqrt[n]{b}, \text{ where } \sqrt[n]{a} \text{ and } \sqrt[n]{b} \text{ are defined.*}$$

**If n is an even integer, a and b must be non-negative real numbers.*
If n is an odd integer, a and b can be any real number.

Quotient Rule for Radicals

For real numbers a and b and n a positive integer:

$$\sqrt[n]{\frac{a}{b}} = \frac{\sqrt[n]{a}}{\sqrt[n]{b}}, \text{ where } \sqrt[n]{a} \text{ and } \sqrt[n]{b} \text{ are defined.*}$$

**If n is an even integer, a and b must be non-negative real numbers.*
If n is an odd integer, a and b can be any real number.

Rationalizing Denominators

The denominator of a rational expression should not be left in a form that contains a radical. Rationalize denominators containing radicals by multiplying the numerator and denominator by the appropriate factors to form a perfect power.

SECTION 8.3:

Adding and Subtracting Terms Containing Radicals

1. Simplify individual radical expressions, if possible.
2. Identify all terms with LIKE radicals. Radical expressions are called LIKE radicals when the radicand and the index of the radical are identical. Use the Distributive Property to collect LIKE radicals.
3. Write the sum or difference of the coefficients in front of the LIKE radical(s) as the final answer.

Chapter 8 Review

- Review the definitions you have written for this chapter in the Dictionary.
- Review any properties and/or processes you have written on 3 × 5-inch cards.
- Assume all variables and radicands are non-negative and that denominators are not equal to zero.

SECTION 8.1

Problems 1–14: Find each root. If a root is not a real number, write "nonreal" as the answer.

1. $-\sqrt{36}$

2. $-\sqrt{-36}$

3. $\sqrt[3]{-27}$

4. $-\sqrt[3]{-27}$

5. $\sqrt[4]{-256}$

6. $-\sqrt[4]{256}$

7. $-\sqrt{\dfrac{9}{25}}$

8. $-\sqrt{\dfrac{64}{49}}$

9. $\sqrt{-6^2}$

10. $-\sqrt[5]{243}$

11. $\sqrt[5]{-243}$

12. $\sqrt[3]{x^3}$

13. $\sqrt{-x^2}$

14. $\sqrt[4]{(x-1)^4}$

Problems 15–22: Use a calculator to find each root. Round answers to four decimal places. If the root is not a real number, write "nonreal" as the answer.

15. $\sqrt{34.67}$

16. $-\sqrt{40}$

17. $-\sqrt{\dfrac{2}{5}}$

18. $\sqrt[3]{12}$

19. $\sqrt[3]{-100}$

20. $\sqrt[4]{200}$

21. $\sqrt[4]{-81}$

22. $\sqrt[5]{-19}$

Problems 23–25: Rewrite each radicand as a perfect square and find the square root.

23. $\sqrt{x^{12}}$

24. $\sqrt{x^{16}}$

25. $\sqrt{x^{20}}$

Problems 26–28: Rewrite each radicand as a perfect cube and find the cube root.

26. $\sqrt[3]{x^9}$ **27.** $\sqrt[3]{x^{21}}$ **28.** $\sqrt[3]{x^{30}}$

SECTION 8.2

Problems 29–40: Apply the Product Rule and simplify each root. Express the simplified answer in radical form. Using a calculator, confirm that the original expression and simplified form represent an equivalent decimal.

29. $-\sqrt{60}$

30. $\sqrt{216}$

31. $\sqrt{162}$

32. $-\sqrt{72}$

33. $\sqrt[3]{-48}$

34. $-\sqrt[3]{64}$

35. $\sqrt[3]{54}$

36. $\sqrt[3]{108}$

37. $\sqrt[5]{-224}$

38. $\sqrt[4]{80}$

39. $\sqrt[5]{-96}$

40. $-\sqrt[6]{3645}$

Problems 41–46: Apply the Product Rule and simplify each root.

41. $-\sqrt{25x^4y^5}$

42. $\sqrt[3]{-64z^7}$

43. $\sqrt[4]{16x^8y^{10}}$

44. $\sqrt{x^2 + 6x + 9}$

45. $\sqrt{12(x-1)^3}$

46. $\sqrt[3]{(2x+3)^4}$

Problems 47–52: Multiply and simplify.

47. $-4\sqrt{10}\cdot\sqrt{15}$

48. $\sqrt[3]{9}\cdot\sqrt[3]{9}$

49. $\sqrt[4]{24}\cdot\sqrt[4]{10}$

50. $\sqrt{6}\cdot\sqrt{15}\cdot\sqrt{10}$

51. $\sqrt{6x}\cdot\sqrt{30x}$

52. $\sqrt[3]{60x^2}\cdot\sqrt[3]{-18x^2}\cdot\sqrt[3]{25x^2}$

Problems 53–60: Apply the Quotient Rule and simplify. Rationalize denominators, if necessary.

53. $\sqrt{\dfrac{20}{9}}$

54. $-\sqrt{\dfrac{50}{8}}$

55. $\sqrt[3]{-\dfrac{27x^{10}}{y^6}}$

56. $\sqrt{\dfrac{x^2 - 10x + 25}{36}}$

57. $\sqrt[4]{\dfrac{256x^8}{y^4}}$

58. $\sqrt[3]{\dfrac{27x^6}{y^3}}$

59. $\dfrac{\sqrt{15}}{\sqrt{30}}$

60. $\sqrt{\dfrac{12x^2y}{15xy^2}}$

Problems 61–64: Insert the appropriate symbol, either = or ≠, between the expressions.

61. $\sqrt[4]{100} \cdot \sqrt{25}$ ______ $\sqrt[4]{2500}$

62. $\sqrt{\dfrac{15}{3}}$ ______ $\sqrt{5}$

63. $\sqrt[3]{15^3}$ ______ 15

64. $\dfrac{\sqrt{4}}{\sqrt[3]{2}}$ ______ $\sqrt{2}$

SECTION 8.3

Problems 65–72: Add and/or subtract. If the radicals are UNLIKE, write "UNLIKE."

65. $9\sqrt{2} - 5\sqrt{2}$

66. $10\sqrt{22} - 8\sqrt{22}$

67. $8\sqrt{2} - 5\sqrt{3}$

68. $3\sqrt[3]{5x} + 8\sqrt[3]{5x}$

69. $2\sqrt{7xy} + 6\sqrt{2xy}$

70. $2.07\sqrt{3x} + 1.56\sqrt{3x} - 0.14\sqrt{3x}$

71. $5\sqrt{11x} + 8\sqrt{11x}$

72. $-9\sqrt[4]{x} + 2\sqrt[3]{x}$

Problems 73–93: Perform the operations and simplify.

73. $10\sqrt{12} - 5\sqrt{75} + 12\sqrt{27}$

74. $-10\sqrt{18} + 5\sqrt{72}$

75. $-3.5\sqrt{45y} + 1.5\sqrt{5y} + 6\sqrt{80y}$

76. $9.56\sqrt[3]{2x} + 1.35\sqrt[3]{54x} - 5.24\sqrt[3]{2x}$

77. $\sqrt{4xy^2} - \sqrt{12x^3y^2} + 5x\sqrt{49xy^2}$

78. $\frac{1}{2}\sqrt{4x} + \frac{5}{2}\sqrt{4x} - \frac{3}{2}\sqrt{4x}$

79. $9\sqrt[3]{24} + 5\sqrt[3]{3} - 15\sqrt[3]{32}$

80. $7\sqrt[3]{x^5} - 3\sqrt[3]{125x^5} + 5\sqrt[3]{8x^5}$

81. $\frac{4}{3}\sqrt{x^2} - \frac{5}{3}\sqrt{x^2} + \frac{5}{6}\sqrt{x^2}$

82. $20x\sqrt{4z} + 5x^2\sqrt{9z} - 10x\sqrt{25z}$

83. $2\sqrt{5}\cdot\sqrt{15} + 7\sqrt{3}$

84. $5\sqrt{2}\cdot\sqrt{6} + 5\sqrt{3} - 7\sqrt{12}$

85. $10\sqrt{x^2} - 3x^2 + 5\sqrt{x^3} + 9\sqrt{x^4}$

86. $\sqrt{\frac{125}{5}} + \sqrt{100} - 2\sqrt{64}$

87. $7\sqrt{x}\cdot\sqrt{x^3} + 5x^2 - \sqrt{2x}\cdot\sqrt{2x} + \sqrt{2x}\cdot\sqrt{8x}$

88. $\sqrt{2}(\sqrt{2} + 3)$

89. $4\sqrt{x}(\sqrt{x} - 10)$

90. $(\sqrt{3} + 4)(\sqrt{3} - 6)$

91. $(\sqrt{x} - 4)(\sqrt{x} - 5)$

92. $(\sqrt{x} - 7)(\sqrt{x} + 7)$

93. $(\sqrt{x} - 4)^2$

Practice Test—Chapter 8

- Assume that the variables and radicands are non-negative and that denominators are not equal to zero.

SECTION 8.1

Problems 1–8: Find each root. If a root is not a real number, write "nonreal" answer.

1. $\sqrt[3]{-64}$

2. $\sqrt[4]{-81}$

3. $\sqrt{100}$

4. $\sqrt{\dfrac{4}{49}}$

5. $-\sqrt{500}$

6. $\sqrt[3]{-24}$

7. $2\sqrt{625}$

8. $5\sqrt[4]{48}$

Problems 9–12: Use a calculator to find each root. Round answers to four decimal places.

9. $-\sqrt{\dfrac{2}{3}}$

10. $\sqrt[3]{25}$

11. $\sqrt[4]{50}$

12. $\sqrt[5]{-200}$

SECTION 8.2

Problems 13–23: Apply the Product and Quotient Rules and simplify each root. Rationalize denominators, if necessary.

13. $-\sqrt{75x^5y^2}$

14. $\sqrt[3]{8y^4}$

15. $\sqrt[3]{-40z^3}$

16. $\sqrt{x^2 - 6x + 9}$

17. $\sqrt[5]{(x-1)^6}$

18. $\sqrt{3}\cdot\sqrt{15}$

19. $-\sqrt{2x}\cdot\sqrt{18x^2}$

20. $\sqrt[3]{4y^2}\cdot\sqrt[3]{6y}\cdot\sqrt[3]{45y^2}$

21. $\sqrt[3]{-\dfrac{125x^4}{y^6}}$

22. $\sqrt{\dfrac{x^2 + 10x + 25}{100}}$

23. $\sqrt[3]{\dfrac{5x^9}{40}}$

SECTION 8.3

Problems 24–33: Perform the operations and write answers in simplified form. If the radicals are UNLIKE, write "UNLIKE."

24. $-7\sqrt{72} + 5\sqrt{72}$

25. $5\sqrt{3b} + 8\sqrt{b} - \sqrt{2b}$

26. $19.56\sqrt[3]{16x^4} + 1.35x\sqrt[3]{54x} - 5.24\sqrt[3]{2x^4}$

27. $2x\sqrt{9z} + 5x^2\sqrt{9z} - 10x\sqrt{36z}$

28. $\sqrt{2}(\sqrt{2} + \sqrt{5})$

29. $-4\sqrt{5}\cdot\sqrt{15} + 5\sqrt{3} + 15\sqrt{12}$

30. $18x\sqrt{x^2} - 7x^2 + 5\sqrt{x^3} - 9\sqrt{x^4}$

31. $8\sqrt{x^3}\cdot\sqrt{x} + 4x^2 + \sqrt{3x}\cdot\sqrt{3x} + \sqrt{5x}\cdot\sqrt{5x}$

32. $4\sqrt{x}(\sqrt{x} - \sqrt{y})$

33. $(3\sqrt{x} + 2)(5\sqrt{x} - 1)$

Answers to Getting Started—Chapter 8

SECTION 8.1

1. (a) $\sqrt{49} = 7$

(b) $\sqrt{81} = 9$

(c) $\sqrt{100} = 10$

2.

Base	3	4	5
Second Power	$3^2 = 9$	$4^2 = 16$	$5^2 = 25$
Third Power	$3^3 = 27$	$4^3 = 64$	$5^3 = 125$
Fourth Power	$3^4 = 81$	$4^4 = 256$	$5^4 = 625$
Fifth Power	$3^5 = 243$	$4^5 = 1024$	$5^5 = 3125$

3.

(a)	$\sqrt[3]{27} = 3$
(b)	$\sqrt[4]{16} = 2$
(c)	$\sqrt[5]{3125} = 5$

(d) In a perfect power, the factors of the radicand are identical. When identical factors occur the same number of times as the index of the radical, the factor is the root of the radical.

4. (a) If a radicand is a perfect square, there are two identical factors.

(b) If a radicand is a perfect cube, there are three identical factors.

(c) If a radicand is a perfect fourth power, there are four identical factors.

(d) If a radicand is a perfect n^{th} power, there are n identical factors.

5.

(a)	$\sqrt{y^2}$	y
(b)	$\sqrt[3]{x^3}$	x
(c)	$\sqrt[3]{(x+5)^3}$	$x + 5$

6. (a) $\sqrt{11} = 3.3166$

(b) $\sqrt[3]{10} = 2.1544$

(c) $\sqrt[4]{60} = 2.7832$

7. The calculator returned an error message of "NONREAL ANS" for $\sqrt{-25}$ because there is no real number whose square is -25.

SECTION 8.2

1.

	Decimal approx. of Root 1	Decimal approx. of Root 2
(a)	6.7082	6.7082
(b)	2.8845	2.8845
(c)	2.6321	2.6321
(d)	The expressions are equivalent with the roots of perfect powers written outside the radicand.	

2.

(a)	8.6603	$\sqrt{3\cdot 5^2}$	$\sqrt{5^2}\cdot\sqrt{3}$	$5\sqrt{3}$	8.6603
(b)	12.2474	$\sqrt{2\cdot 3\cdot 5^2}$	$\sqrt{5^2}\cdot\sqrt{2\cdot 3}$	$5\sqrt{6}$	12.2474

3.

(a)	7.2112	$\sqrt[3]{5^3\cdot 3}$	$\sqrt[3]{5^3}\cdot\sqrt[3]{3}$	$5\sqrt[3]{3}$	7.2112
(b)	4.9324	$\sqrt[3]{2^3\cdot 3\cdot 5}$	$\sqrt[3]{2^3}\cdot\sqrt[3]{3\cdot 5}$	$2\sqrt[3]{15}$	4.9324

4.

(a)	$\sqrt{2\cdot 5\cdot 5\cdot x\cdot x\cdot y}$ $= \sqrt{2\cdot 5^2\cdot x^2\cdot y}$ $= \sqrt{5^2}\cdot\sqrt{x^2}\cdot\sqrt{2}\cdot\sqrt{y}$ $= 5\cdot x\cdot\sqrt{2y}$	$= 5x\sqrt{2y}$
(b)	$\sqrt[3]{3\cdot 3\cdot 3\cdot y\cdot y\cdot y\cdot z}$ $= \sqrt[3]{3^3\cdot y^3\cdot z}$ $= \sqrt[3]{3^3}\cdot\sqrt[3]{y^3}\cdot\sqrt[3]{z}$ $= 3\cdot y\cdot\sqrt[3]{z}$	$= 3y\sqrt[3]{z}$

5.

(a)	$-\sqrt{11}\cdot\sqrt{2}$	-4.6904	$-\sqrt{22}$	-4.6904
(b)	$\sqrt{10}\cdot\sqrt{3}$	5.4772	$\sqrt{30}$	5.4772

SECTION 8.3

1. (a) $-4x$ $\quad \frac{1}{3}x$

   (b) $-9yz$ $\quad 0.5yz$

2. (a) $\frac{3}{4}\sqrt{3}$ $\quad 5\sqrt{3}$

   (b) $1.6\sqrt[4]{53}$ $\quad 8\sqrt[4]{53}$

   (c) $6\sqrt{x}$ $\quad 5\sqrt{x}$

   (d) $\sqrt{2x}$ $\quad -\sqrt{2x}$

3. (a) $(17 + 5)\sqrt{3} = 22\sqrt{3}$

   (b) UNLIKE radicals

   (c) UNLIKE radicals

   (d) $(11 - 7)\sqrt{y} = 4\sqrt{y}$

   (e) $(-20.7 + 9.7)\sqrt{3x} = -11\sqrt{3x}$

   (f) UNLIKE radicals

   (g) $(12 + 16)\sqrt[5]{x^2} = 28\sqrt[5]{x^2}$

4. (a) $4\sqrt{3} + 6\sqrt{12} = 4\sqrt{3} + 6\cdot 2\sqrt{3}$
   $= (4 + 12)\sqrt{3} = 16\sqrt{3}$

   (b) $24\sqrt[3]{2} - 3\sqrt[3]{16} = 24\sqrt[3]{2} - 3\cdot 2\sqrt[3]{2}$
   $= (24 - 6)\sqrt[3]{2} = 18\sqrt[3]{2}$

   (c) $5\sqrt{12} - 2\sqrt{75} = 5\cdot 2\sqrt{3} - 2\cdot 5\sqrt{3}$
   $= 10\sqrt{3} - 10\sqrt{3} = 0$

5. Simplify each radical. When radicals are LIKE radicals, add or subtract the numbers or expressions in front of the radicals, keeping the radical the same.

Cumulative Test Chapters 7–8

- Assume that the variables and radicands are non-negative and that denominators are not equal to zero.

1. Determine the restricted value(s) of the variable in the rational expression:

$$\frac{2x}{x^2 - x - 6}.$$

Problems 2–6: Perform the operations. Reduce each answer to lowest terms.

2. $\frac{-5x}{2y} \cdot \frac{x-2}{3x+1} \div \frac{x^2 - 2x}{6x + 2}$

3. $\frac{3x - 3y}{x^2 - 1} \div \frac{x - y}{x + 1} \cdot \frac{x - 1}{12}$

4. $\frac{5x}{2x + 1} - \frac{2x}{2x + 1} + \frac{3x}{2x + 1}$

5. $\frac{5}{2ab^2c} - \frac{7b}{6ac}$

6. $\frac{9x}{x^2 - 9} - \frac{3}{x + 3} + \frac{6}{x - 3}$

7. Graph $f(x) = \frac{1}{x^2 + x - 6}$.

State the restricted values of the independent variable in the function.

x	$f(x)$

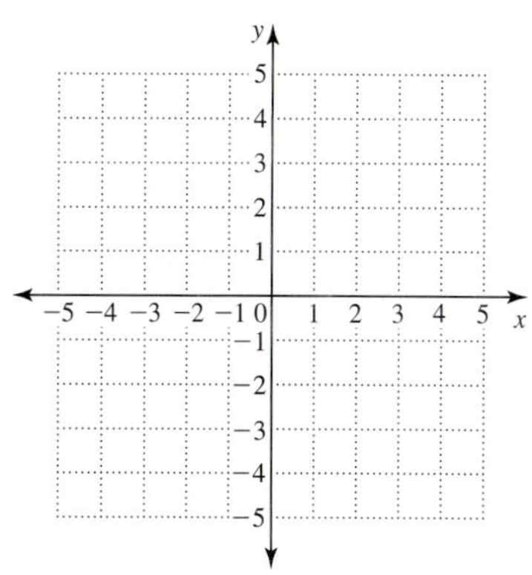

Problems 8–9: Solve each rational equation. Confirm the solution analytically. Confirm the solution graphically, if desired.

8. $\dfrac{8}{y^2 + y - 6} = \dfrac{6}{y^2 - 9}$

9. $\dfrac{3}{x - 2} - \dfrac{5}{x + 3} = \dfrac{1}{x^2 + x - 6}$

10. Divide the complex fraction $\dfrac{\dfrac{x^2 - x - 6}{2x^2}}{\dfrac{x^2 - 9}{8x}}$.

Problems 11–14: Find each root. If a root is not a real number, write "nonreal" answer.

11. $-\sqrt{200}$

12. $\sqrt{\dfrac{9y^{10}}{z^{12}}}$

13. $\sqrt[3]{-125x^6}$

14. $\sqrt[4]{-x^6}$

Problems 15–18: Perform the operations and write answers in simplified form. If the radicals are UNLIKE, write "UNLIKE."

15. $\sqrt{18x^2} + x\sqrt{50}$

16. $(2\sqrt{5} - \sqrt{6})^2$

17. $\dfrac{6 - \sqrt{12}}{-2}$

18. $-3\sqrt{x}(\sqrt{x} + 1)$

CHAPTER

R

Arithmetic Review

The simplest mathematical idea of counting organized around base 10, in words and in symbols, can be traced to the Egyptians and the Babylonians (3000–2000 B.C.) and the Chinese (1000–500 B.C.). The symbols used for the first nine numbers of our present base 10 number system have their origins in the Brahmi system of writing in India. Around the year 600, the Indians dropped earlier symbols and began to use the modern numerals.

R.1 Place-Value, Rounding and Estimating, Order of Operations

OBJECTIVES

- Review place-value.
- Round a number to a specified number of decimal places.
- Estimate the result of arithmetic operations.
- Apply Order of Operations to perform operations with non-negative numbers.

1 2 3 Reviewing

This chapter reviews basic arithmetic concepts necessary for success in an introductory algebra course. Many of these topics are discussed in other chapters of this textbook with arithmetic concepts extended to algebraic expressions containing variables.

PLACE-VALUE

In our numbering system, each place-value is assigned a value representing a power of 10. This numbering system is called the **base-10 numbering system** or **decimal system**. The decimal point is used as a symbol to separate whole units from fractional parts of units. As you move to the left in a number, each place-value increases by a power of ten. As you move to the right in a number, each place-value decreases by a power of ten.

EXAMPLE 1: The number 729.4763 is in **decimal form**, is equivalent to the sum of each digit multiplied by its place-value, and can be rewritten in **expanded fractional form**:

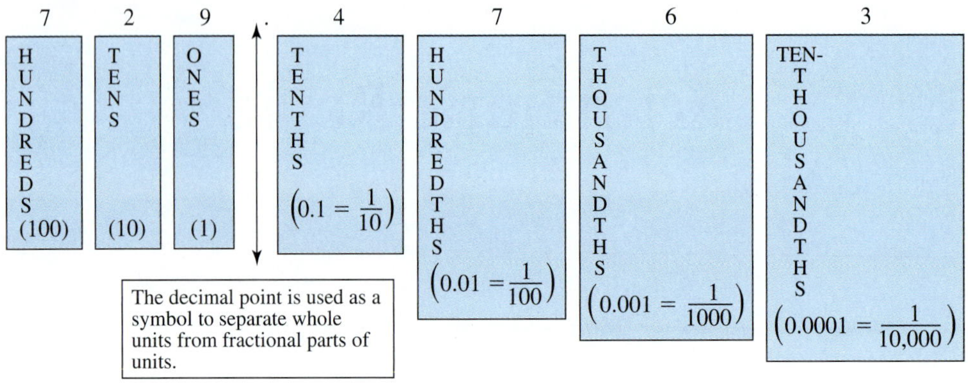

7 hundreds + 2 tens + 9 ones + 4 tenths + 7 hundredths + 6 thousandths + 3 ten-thousandths

$$700 + 20 + 9 + \frac{4}{10} + \frac{7}{100} + \frac{6}{1000} + \frac{3}{10,000} = 729.4763$$

1. Write each decimal in expanded fractional form.

(a) 45.652 (b) 1007.84

ROUNDING AND ESTIMATING

An exact answer is not always necessary. Numbers may be rounded and results of operations estimated.

2. Sandi used a computer software program to calculate the amount of federal income tax she owed. As she made the monetary entries into different categories, she noticed that the computer recorded her entries differently than she typed them in.

Sandi's entry	Computer display for the same number
$1342.56	$1343
$ 426.15	$ 426
$ 221.01	$ 221
$5743.98	$5744
$ 0.89	$ 1

Describe the mathematical process used to determine the computer display.

3. A number can be rounded to a specified place-value. To round a number to the specified place-value, look at the digit to the right of the specified place-value. If this digit is less than 5, round down by dropping all digits to the right of the specified place-value. If the digit is 5 or greater, round up by increasing the digit in the specified place-value position by 1 and dropping digits to the right. Complete the table.

	Original number	Specified place-value	Rounded number
Ex. 2	1568.3579	Nearest hundreds place	1600
Ex. 3	1568.3579	Nearest thousandths place	1568.358
(a)	1568.3579	Nearest tens place	
(b)	1568.3579	Nearest ones place	
(c)	1568.3579	Nearest tenths place	
(d)	1568.3579	Nearest hundredths place	

EXAMPLE 4: It is often useful to estimate the result of arithmetic operations mentally. The goal of estimating is NOT to find an exact answer but, rather, to find an answer that is close to the exact answer. Is $231.45 - 102.55$ closer to 100 or closer to 10?

$200 - 100 = 100$ Round each number to the nearest hundred and subtract.

The difference is closer to 100.

4. Estimate the results of each operation. Do not perform the actual operations. Circle the most reasonable answer.

(a) Is 3.14×5 closer to 15 or closer to 150?

(b) Is $\frac{156}{5}$ closer to 30 or closer to 3?

(c) Is $53.576 + 47.999$ closer to 100 or closer to 1000?

(d) Is $\sqrt{50}$ closer to 7 or closer to 10?

ORDER OF OPERATIONS

In a problem with more than one operation, the order in which addition, subtraction, multiplication, and division are performed can produce different answers. To standardize the results, the order in which operations are performed is governed by a set of rules called the **Order of Operations**. Arithmetic operations should be performed in this order: **P**arentheses, **E**xponents, then **M**ultiplication OR **D**ivision (left to right), and **A**ddition OR **S**ubtraction (left to right). The Order of Operations is programmed into many calculators and should be memorized. The acronym "PEMDAS" or the sentence "**P**lease **E**xcuse **M**y **D**ear **A**unt **S**ally," can be used as memory tools to help remember the Order of Operations.

➢ *An* **exponent** *is a symbol that denotes repeated multiplication. In the expression 3^4, the exponent, 4, means to multiply the* **base**, *3, four times:* $3^4 = \underbrace{3 \cdot 3 \cdot 3 \cdot 3}_{\text{4 times}} = 81$.

EXAMPLE 5: Perform the operations $1 + 3(5 - 2)^2$.

$1 + 3(5 - 2)^2$

$= 1 + 6 \cdot (3)^2$ First: Perform subtraction in parentheses.

$= 1 + 6 \cdot 9$ Second: Simplify exponent.

$= 1 + 54$ Third: Multiply.

$= 55$ Fourth: Add.

5. Use the Order of Operations to perform the operations.

(a) $5 + 2 \cdot 3$ (b) $40 \div 5 \cdot (6 - 4)^2$

(c) $6 + 2 \cdot 7 - 10 \div 2$

1 2 3 Discussing Mathematics

PLACE-VALUE

Our base-10 numbering system (page 550) uses powers of 10 to represent each place-value. Numbers may be written in decimal form or expanded fractional form.

EXAMPLE 6: The decimal form of 15.4078 can be written in expanded fractional form:

$$\begin{aligned} 15.4078 &= 1 \cdot 10 + 5 \cdot 1 + 4 \cdot \frac{1}{10} + 0 \cdot \frac{1}{100} + 7 \cdot \frac{1}{1000} + 8 \cdot \frac{1}{10{,}000} \\ &= 10 + 5 + \frac{4}{10} + \frac{0}{100} + \frac{7}{1000} + \frac{8}{10{,}000} \\ &= 10 + 5 + \frac{4}{10} + \frac{7}{1000} + \frac{8}{10{,}000} \end{aligned}$$

ROUNDING AND ESTIMATING

An exact value for a number is often desirable, but there are many problem situations where numbers are rounded and/or estimated. **Rounding** a number to a specified decimal place is a process determining a number close to, but not equal to, the original number.

SUMMARY OF RULES FOR ROUNDING

1. Identify the digit in the specified place-value to which the number is to be rounded.
2. Look at the digit to the right of the specified place-value.
3. If this digit is less than 5, round down by dropping all digits to the right of the specified place-value.
4. If the digit is greater than or equal to 5, round up by increasing the digit in the specified place-value position by 1 and dropping all digits to the right.

EXAMPLE 7: The number 6534.53789 rounded to the place-values indicated.

(a) Nearest Thousands: 7000

(b) Nearest Hundreds: 6500

(c) Nearest Tens: 6530

(d) Nearest Ones: 6535

(e) Nearest Tenths: 6534.5

(f) Nearest Hundredths: 6534.54

(g) Nearest Thousandths: 6534.538

EXAMPLE 8: When estimating, an exact answer is not necessary. Round to the nearest place-value that makes sense for the given problem and perform the operations mentally.

(a) Estimate the sum to the nearest ones place.

$125.11 + 100.2322 \approx 125 + 100$ Round each number to the nearest ones place.

$= 225$ Add.

(b) Estimate the difference to the nearest hundred.

$5489 - 2197 \approx 5500 - 2200$ Round each number to the nearest hundred.

$= 3300$ Subtract.

(c) Estimate the quotient.

$\frac{210.38}{51} \approx \frac{200}{50}$ Round 210.38 to the nearest hundred.

Round 51 to the nearest tens place.

$= 4$ Divide.

ORDER OF OPERATIONS

The order in which operations are performed can change the answer to a problem. The Order of Operations standardize the results.

ORDER OF OPERATIONS

	Operation to be performed	Process—Rewrite the problem after each step	Acronym memory tool	Sentence memory tool
1.	**P**arentheses	Perform any operations enclosed in parentheses. A fraction bar may act as grouping symbol. Simplify the numerator or denominator of a fraction, if possible.	**P**	**P**lease
2.	**E**xponents	Expand and simplify exponents.	**E**	**E**xcuse
3.	**M**ultiplication *or* **D**ivision (left to right)	Start at the left side of the expression and identify the first multiplication *or* division operation in the problem. If division appears first, perform the division. If multiplication appears first, perform the multiplication.	**M** **D**	**M**y **D**ear
4.	**A**ddition *or* **S**ubtraction (left to right)	Start at the left side of the expression and identify the first addition *or* subtraction operation in the problem. If subtraction appears first, perform subtraction. If addition appears first, perform the addition.	**A** **S**	**A**unt **S**ally

EXAMPLE 9: Perform the operations.

(a) $6 \cdot 2 + 7 =$ First: Multiply.

$12 + 7 = 19$ Second: Add.

(b) $6(2 + 7) =$ First: Perform addition in parentheses.

$6 \cdot 9 = 54$ Second: Multiply.

➢ *Multiplication or division is always performed before addition. When both multiplication and division appear in the problem, perform multiplication* or *division, whichever operation appears first when viewing the problem from left to right.*

(c) $4 + 12 \div 3 \cdot 2 =$ First: Divide.

$4 + 4 \cdot 2 =$ Second: Multiply.

$4 + 8 =$ Third: Add.

12

➢ *When both addition and subtraction appear in the problem, perform addition* or *subtraction left to right.*

(d) $10 - (9 - 7)^2 + 3 - 5 + 7 =$ First: Perform subtraction in parentheses.

$10 - (2)^2 + 3 - 5 + 7 =$ Second: Simplify exponent.

$10 - 4 + 3 - 5 + 7 =$ Third: Subtract 4 from 10.

$6 + 3 - 5 + 7 =$ Fourth: Add 6 and 3.

$9 - 5 + 7 =$ Fifth: Subtract 5 from 9.

$4 + 7 =$ Sixth: Add 4 and 7.

11

1 2 3 Doing Mathematics

Problems 1–5: Write each number in expanded fractional form.

1. 503.79

2. 25.1

3. 1.23

4. 1524.5

5. 1,003,523

6. David and Roger own Trim & Fit, a sporting goods store. They need to determine how many cases of tennis balls to order for the next year. Last year, their store sold 2440 cans of tennis balls. Each case of balls contains 24 cans of balls. David divided $\frac{\text{2440 cans}}{\text{24 cans per case}}$ on his calculator which displayed this answer: 101.6666667. How many cases of tennis balls should be ordered?

7. Michael keeps track of his business travel expenses and requests reimbursement from his employer. His employer asks Michael to round all expenses to the nearest dollar. Complete the table by rounding Michael's expenses to the nearest dollar.

	Expense	Actual amount	Amount rounded to the nearest dollar
(a)	Airfare	$285.75	
(b)	Hotel (5 days)	$789.10	
(c)	Taxi and Tips	$ 43.50	
(d)	Meals	$198.99	
(e)	Parking	$ 71.20	
(f)	Phone Calls	$ 14.49	

8. Round each number to the specified place-value.

	Number	Specified place-value	Rounded value
(a)	2134.1235	Nearest thousands place	
(b)	149.29	Nearest tens place	
(c)	0.50	Nearest ones place	
(d)	149.2115	Nearest tenths place	
(e)	637.288	Nearest hundredths place	
(f)	54.123456	Nearest ten-thousandths place	

9. Jan needs to review 300 pages in her psychology book for the next test. Estimate the number of pages she must read each day if the test is in 9 days.

10. Anne needs to type 33 pages of a report within the next 5 days. Estimate the number of pages she needs to type each day to complete the report.

Problems 11–14: Estimate the result of each operation. Circle the number that is closest to the given number. Do not perform the operations on paper or use your calculator.

11. Is 5.25×6 closer to 30 or closer to 300?

12. Is $\frac{21.7}{3}$ closer to 7 or closer to 0.7?

13. Is $83.576 - 43.999$ closer to 40 or closer to 100?

14. Is $\sqrt{91}$ closer to 7 or closer to 10?

15. Write each expression in expanded form without an exponent. Multiply the factors.

(a) 10^2 (b) 4^3 (c) 3^5

Problems 16–35: Use the Order of Operations to perform the operations without the use of a calculator.

16. $14 + 6 - 6 \cdot 2$

17. $35 - 5 - 4 \cdot 3$

18. $20 - (5 - 2)^2$

19. $35 - (20 - 15)^2$

20. $[3(10 - 8)]^2$

21. $[2(7 - 3)]^2$

22. $6 - 6 \div 6 \cdot 6 + 6^2$

23. $40 - 10 \div 5 \cdot 2 + 2^2$

24. $\dfrac{10^2 + 5}{5}$

25. $\dfrac{7^2 - 7}{7}$

26. $\dfrac{4^3 - 8 \cdot 3}{4 + 6}$

27. $\dfrac{5^2 + 3 \cdot 3}{5 + 12}$

28. $15 + [9(8 \div 4) - 10]$

29. $3 \cdot 20 + [15 - 2(6 \div 3)]$

30. $60 \div 30 \cdot 2 + 7 - 3$

31. $75 \div 3 \cdot 5 + 10 - 2$

32. $5[2(3 \cdot 10 - 25)]^2$

33. $2[3(20 \div 5 - 2)]^2$

34. $12 \cdot 6 \div 2 + 5^2$

35. $12 \div 6 \cdot 2 + 5^2$

36. Describe a real-world problem situation where estimating is useful.

R.2 Operations with Fractions

OBJECTIVES

- Build and reduce fractions.
- Change improper fractions to mixed numbers and vice versa.
- Multiply, divide, add, and subtract fractions.
- Apply the Order of Operations to problems involving fractions.

1 2 3 Reviewing

In this section, we focus on equivalent forms created by building or reducing fractions, and operations with fractions.

PRINCIPLES OF FRACTIONS

A fraction is a mathematical notation that denotes parts of a whole. For example, if a pizza was cut into 8 pieces and Bob eats three pieces of pizza, the portion eaten can be represented by the fraction $\frac{3}{8}$. The top number, 3, is called the **numerator** of the fraction.

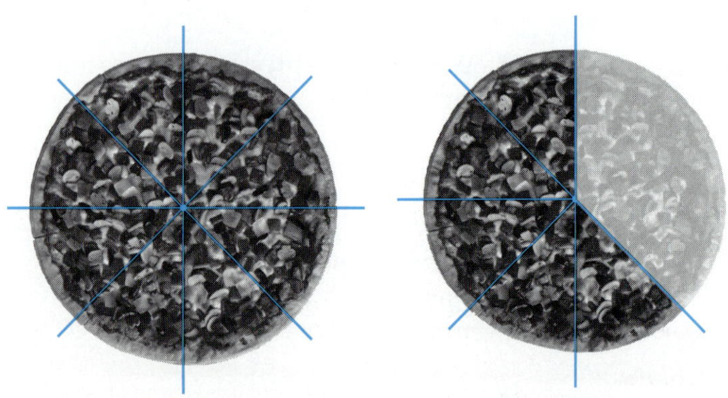

Eight total pieces

Bob eats $\frac{3}{8}$ of the pizza.

The bottom number, 8, is called the **denominator** of the fraction: $\frac{3 \text{ (numerator)}}{8 \text{ (denominator)}}$.

Fractions can be written in equivalent forms. Half a pizza can be written in many equivalent fractions, depending on how many pieces are cut:

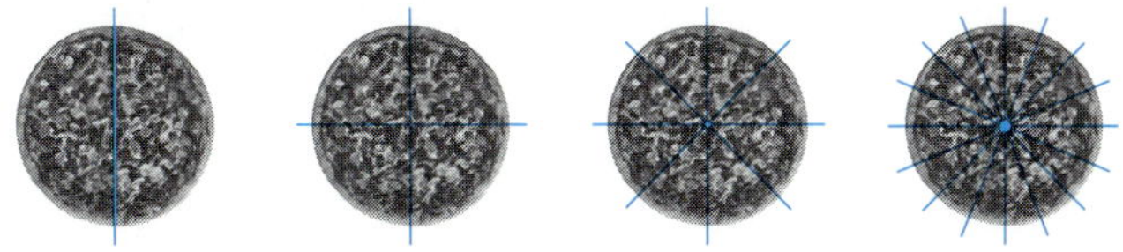

$\frac{1}{2} = \frac{2}{4} = \frac{4}{8} = \frac{8}{16}$

All four fractions are equivalent, $\frac{1}{2} = \frac{2}{4} = \frac{4}{8} = \frac{8}{16}$.

➢ **Reducing a fraction** *to an equivalent form is the process of dividing out common factors from the numerator and denominator.*

EXAMPLE 1: Reduce $\frac{8}{16}$ to lowest terms.

$$\frac{8}{16} = \frac{1 \cdot \cancel{8}^{1}}{2 \cdot \cancel{8}_{1}} = \frac{1}{2}.$$ Divide out common factor 8.

$\frac{8}{16}$ is equivalent to $\frac{1}{2}$.

1. Reduce each fraction to lowest terms by dividing out common factors.

(a) $\frac{10}{16}$ (b) $\frac{25}{45}$

➢ **Building a fraction** *to an equivalent form is the process of multiplying the numerator and denominator of a fraction by the same nonzero real number.*

EXAMPLE 2: Build the fraction to an equivalent fraction with the given denominator.

$$\frac{3}{5} = \frac{?}{20}.$$

$$\frac{3}{5} = \frac{3}{5} \cdot \frac{4}{4} = \frac{12}{20}$$ Multiply the numerator and denominator by 4.

$\frac{3}{5}$ is equivalent to $\frac{12}{20}$.

2. Build each fraction to an equivalent fraction with the given denominator.

(a) $\frac{5}{7} = \frac{?}{21}$ (b) $\frac{6}{11} = \frac{?}{44}$

IMPROPER FRACTIONS AND MIXED NUMBERS

An **improper fraction** is a fraction where the numerator is larger than or equal to the denominator. A **mixed number** is a number comprised of an integer part and a fractional part. An improper fraction can be written as an equivalent mixed number and vice versa.

EXAMPLE 3: The improper fraction $\frac{16}{3}$ is equivalent to the mixed number $5\frac{1}{3}$.

$$\frac{16}{3} = 3\overline{)16}\;(5) = 5 \text{ remainder of } 1 = 5\frac{1}{3}$$
$$\begin{array}{r} 15 \\ \hline 1 \end{array}$$

An improper fraction is changed to a mixed number by dividing the denominator into the numerator and representing the remainder as a fraction.

3. Change each improper fraction to a mixed number.

(a) $\frac{27}{5}$ (b) $\frac{7}{4}$

EXAMPLE 4: The mixed number $4\frac{3}{5}$ is equivalent to $\frac{23}{5}$.

$$4\frac{3}{5} = \frac{5\cdot 4 + 3}{5} = \frac{23}{5}$$

Multiply the denominator of the fraction by the integer part of the number and add the numerator of the fraction. The denominator of the resulting fraction is the same as the original denominator.

4. Change each mixed number to an improper fraction.

(a) $3\frac{4}{5}$ (b) $10\frac{1}{4}$

MULTIPLYING FRACTIONS

To multiply fractions:
1. Multiply the numerators of the fractions.
2. Multiply the denominators of the fractions.
3. Divide out common factors, if possible.
4. Multiply remaining factors. Write improper fractions as mixed numbers.

EXAMPLE 5: Multiply.

$$\frac{1}{2}\cdot\frac{6}{5} = \frac{1\cdot 6}{2\cdot 5}$$ Multiply numerators and denominators.

$$= \frac{6}{10} = \frac{\overset{1}{\cancel{2}}\cdot 3}{\underset{1}{\cancel{2}}\cdot 5}$$ Divide out the common factor of 2. Multiply remaining factors.

$$= \frac{3}{5}$$ Dividing out may also be written: $\frac{1\cdot\overset{3}{\cancel{6}}}{\underset{1}{\cancel{2}}\cdot 5}$ or $\frac{\overset{3}{\cancel{6}}}{\underset{5}{\cancel{10}}} = \frac{3}{5}$.

5. Multiply the fractions.

(a) $\frac{4}{5}\cdot\frac{35}{2}$ (b) $\frac{22}{3}\cdot\frac{6}{55}$

DIVIDING FRACTIONS

Multiplication and division are inverse operations. Every division problem can be rewritten as a multiplication problem. To divide fractions:

1. Rewrite division as multiplication. (Multiply the dividend by the reciprocal of the divisor.) Perform the multiplication.
2. Divide out common factors, if possible.
3. Multiply remaining factors. Write improper fractions as mixed numbers.

EXAMPLE 6: Divide.

$$\frac{3}{4} \div \frac{15}{11} = \frac{3}{4} \cdot \frac{11}{15}$$ Rewrite division as multiplication.

$$= \frac{\overset{1}{\cancel{3}}}{4} \cdot \frac{11}{\underset{5}{\cancel{15}}} = \frac{11}{20}$$ Divide out common factor 3. Multiply remaining factors.

6. Divide.

(a) $\frac{4}{5} \div \frac{2}{45}$ (b) $\frac{12}{7} \div \frac{15}{14}$

ADDING (OR SUBTRACTING) FRACTIONS WITH DENOMINATORS THAT ARE THE SAME

To add (or subtract) fractions with LIKE denominators:

1. Add (or subtract) the numerators of the fractions.
2. Write the sum (or difference) over the original denominator.
3. Reduce the resulting fraction, if necessary. Write improper fractions as mixed numbers.

EXAMPLE 7: Subtract.

$$\frac{27}{10} - \frac{3}{10} = \frac{27 - 3}{10} = \frac{24}{10}$$ Subtract the numerators. Write the difference over the original denominator.

$$= \frac{12 \cdot \overset{1}{\cancel{2}}}{\underset{1}{\cancel{2}} \cdot 5} = \frac{12}{5} = 2\frac{2}{5}$$ Divide out common factor 2. Write improper fraction as a mixed number.

7. Add or subtract.

(a) $\frac{21}{8} + \frac{3}{8} = \frac{\quad}{8}$ (b) $\frac{10}{7} - \frac{5}{7} = \frac{\quad}{7}$

ADDING (OR SUBTRACTING) FRACTIONS WITH DENOMINATORS THAT ARE DIFFERENT

To add or subtract fractions, fractions must be written with LIKE denominators. The **least common multiple** (LCM) is a number that is the smallest multiple of all denominators under consideration. The **least common denominator** (LCD) is the least common multiple of all denominators in the fractions being considered.

To add (or subtract) fractions with denominators that are different:

1. Identify the least common multiple of the denominators (LCM).
2. Use the LCM as the LCD and build the original fractions into equivalent fractions with the LCD as the denominator.
3. Add (or subtract) the numerators. Write the sum (or difference) using the LCD as the denominator.
4. If possible, reduce the resulting fraction. Write improper fractions as mixed numbers.

8. Use the following process to identify the LCM/LCD and add $\frac{4}{9} + \frac{7}{15}$.

(a) The least common multiple can be found by writing the prime factors of the numbers under consideration. Factor 9 and 15 into the product of prime factors, writing like factors in vertical columns. A positive number greater than 1 whose only factors are 1 and itself, is called a prime number. Draw an arrow through each factor in each vertical column and write the number as a factor of the LCD.

$9 =$

$15 =$

$\text{LCD} =$

(b) Multiply the factors listed as the LCD.

(c) Build each fraction to an equivalent fraction with a denominator of LCD = 45 to add $\frac{4}{9} + \frac{7}{15}$.

9. Find the LCD and subtract $\frac{3}{10} - \frac{1}{16}$.

1 2 3

Discussing Mathematics

EQUIVALENT FRACTIONS

The Fundamental Principle of Equivalent Fractions is used to reduce a fraction to lowest terms and build a fraction to an equivalent fraction with a different denominator.

FUNDAMENTAL PRINCIPLE OF EQUIVALENT FRACTIONS

When the numerator (p) and denominator (q) of a fraction are multiplied or divided by the same nonzero number (n), the result is an equivalent fraction.

Reducing a fraction: $\frac{p}{q} = \frac{p \div n}{q \div n}$ **Building a fraction:** $\frac{p}{q} = \frac{p \cdot n}{q \cdot n}$

EXAMPLE 8:

Reducing a fraction:

$$\frac{25}{35} = \frac{25 \div 5}{35 \div 5} = \frac{5}{7}^*$$

$\frac{25}{35}$ is equivalent to $\frac{5}{7}$

*Dividing the numerator and denominator by the same number is the same as dividing out a common factor.

Building a fraction:

$$\frac{5}{6} = \frac{5 \cdot 3}{6 \cdot 3} = \frac{15}{18}^*$$

$\frac{5}{6}$ is equivalent to $\frac{15}{18}$

*Multiplying the numerator and denominator by the same number is the same as multiplying by 1.

MULTIPLYING FRACTIONS

To multiply fractions, multiply the numerators and denominators of the fraction. Divide out any common factors.

MULTIPLYING FRACTIONS

When multiplying fractions, multiply the numerators of the fractions and multiply the denominators of the fractions:

$$\frac{A}{B}\cdot\frac{C}{D}=\frac{A\cdot C}{B\cdot D}, \text{ where } B\neq 0 \text{ and } D\neq 0.$$

Reduce the resulting product by dividing out common factors, if possible.

➢ *It may be necessary to change the form of a mixed number to an improper fraction before performing the multiplying. During multiplication, fractions can be reduced to lowest terms. Divide common factors at any time. Repeat the reducing process until there are no common factors remaining.*

EXAMPLE 9: Multiply.

$$5\cdot\left(4\frac{3}{7}\right)=\frac{5}{1}\cdot\frac{31}{7}$$ Rewrite the whole number 5 as a fraction. Rewrite the mixed number as an improper fraction.

$$=\frac{155}{7}$$ Multiply the numerators and denominators.

$$=22\frac{1}{7}$$ Reduce and/or write as a mixed number.

DIVIDING OF FRACTIONS

Every division problem can be rewritten as a multiplication problem. Division is equivalent to multiplying by the reciprocal of the divisor.

DIVIDING FRACTIONS

The operation of division is equivalent to multiplying the dividend by the reciprocal of the divisor:

$$\frac{A}{B}\div\frac{C}{D}=\frac{A}{B}\cdot\frac{D}{C}, \text{ where } B\neq 0,\ D\neq 0, \text{ and } C\neq 0.$$

If possible, reduce the resulting product by dividing out common factors.

EXAMPLE 10: Divide.

$$\frac{5}{18}\div\frac{15}{4}=\frac{5}{18}\cdot\frac{4}{15}$$ Rewrite division as multiplication.

$$=\frac{\overset{1}{\cancel{5}}}{\underset{9}{\cancel{18}}}\cdot\frac{\overset{2}{\cancel{4}}}{\underset{3}{\cancel{15}}}$$ Multiply and divide out common factors 3 and 5.

$$=\frac{1\cdot 2}{9\cdot 3}=\frac{2}{27}$$ Multiply remaining factors.

ADDING OR SUBTRACTING FRACTIONS

When adding or subtracting fractions, the denominators of the fractions must be the same.

ADDING (OR SUBTRACTING) FRACTIONS WITH DENOMINATORS THAT ARE THE *SAME*

Given the fractions, $\frac{A}{B}$ and $\frac{C}{B}$, then

Sum: $\frac{A}{B} + \frac{C}{B} = \frac{A + C}{B}, (B \neq 0)$; **Difference:** $\frac{A}{B} - \frac{C}{B} = \frac{A - C}{B}, (B \neq 0)$.

EXAMPLE 11: Perform the operations.

$$\frac{5}{8} + \frac{15}{8} - \frac{3}{8} = \frac{5 + 15 - 3}{8}$$ Add and subtract the numerators.

$$= \frac{17}{8} = 2\frac{1}{8}$$ Write the result over the original denominator. Write improper fraction as a mixed number.

When adding or subtracting fractions, the denominators must have LIKE denominators. If the denominators are not the same, a common multiple, a number that is a multiple of all denominators under consideration, is identified. The least common multiple (LCM) is used as the **least common denominator (LCD)** to add or subtract fractions.

ADDING (OR SUBTRACTING) FRACTIONS WITH DENOMINATORS THAT ARE *DIFFERENT*

1. Find the Least Common Multiple (LCM):
   a) Write the prime factorizations of each denominator. Write LIKE factors in vertical columns.
   b) Draw an arrow through each factors and vertical column. Write each factor in the LCM. (Each factor should be included the greatest number of times it appears in any one factorization.)
   c) Multiply the factors together to form the LCM.
2. Use the LCM as the Least Common Denominator (LCD) and build the original fractions into equivalent fractions with the LCD as the denominator.
3. Add (or subtract) the fractions.
4. If possible, reduce the resulting fraction. Write improper fractions as mixed numbers.

EXAMPLE 12: Add $\frac{5}{8} + \frac{7}{12}$.

$$8 = 2 \cdot 2 \cdot 2$$
$$12 = 2 \cdot 2 \quad \cdot 3$$

Write the prime factorizations of each denominator with LIKE factors in vertical columns.

Factors of LCM: $2 \cdot 2 \cdot 2 \cdot 3 = 24$

The LCM is 24.

Draw an arrow through each factor and vertical column. Write each factor in the LCM and multiply.

Multiply the first fraction by 3. Multiply the second fraction by 2.

$$\frac{5\cdot3}{8\cdot3}+\frac{7\cdot2}{12\cdot2}=\frac{15}{24}+\frac{14}{24}$$

Use the LCM as the LCD. Build each fraction to an equivalent fraction with LCD of 24.

$$=\frac{29}{24}=1\frac{5}{24}$$

Add the numerators and keep the donominator. Write as a mixed number.

EXAMPLE 13: Prior to 2000, stock prices were written in fractional form. The opening price of a given stock can be calculated with this equation:

Opening Price = Last Price (closing) − Change (change from opening).

Using this equation, the Opening Price of Microsoft stock on December 15, 1999 was: Opening Price of Intel = Last − Change

$$=72\frac{13}{16}-\left(-1\frac{1}{4}\right)$$

Method 1: Each mixed number is converted into an improper fraction, fractions with LIKE denominators are combined, and the final answer converted to a mixed number.

$$72\frac{13}{16}-\left(-1\frac{1}{4}\right)$$

$$=\frac{1165}{16}-\left(\frac{-5}{4}\right)=\frac{1165}{16}+\frac{5}{4}$$

$$=\frac{1165}{16}+\frac{20}{16}$$

$$=\frac{1185}{16}=74\frac{1}{16}$$

Method 2: Mixed numbers and fractional parts are written as individual sums. Terms are rearranged according to the Commutative and Associative Properties. The whole numbers are combined, the fractional parts with LIKE denominators are combined, and the result is simplified.

$$72\frac{13}{16}-\left(-1\frac{1}{4}\right)$$

$$=72+\frac{13}{16}-\left(-1-\frac{1}{4}\right)=72+\frac{13}{16}+1+\frac{1}{4}$$

$$=72+1+\frac{13}{16}+\frac{4}{16}=73+\frac{13+4}{16}$$

$$=73\frac{17}{16}=74\frac{1}{16}$$

ORDER OF OPERATIONS

The Order of Operations discussed in Section R.1 also should be applied when performing operations with fractions.

EXAMPLE 14: Perform the operations.

(a) $\frac{1}{2}+\underbrace{\frac{3}{2}\cdot\frac{2}{5}}$ Multiply first.

$=\frac{1}{2}+\frac{3}{5}$ Identify the LCD = 10.

Multiply first fraction by 5. Multiply second fraction by 2. Build each fraction to equivalent fractions with the LCD of 10.

$$\frac{1}{2} + \frac{3}{5} = \frac{1}{2}\cdot\frac{5}{5} + \frac{3}{5}\cdot\frac{2}{2}$$ Multiply each fraction.

$$= \frac{5}{10} + \frac{6}{10}$$

$$= \frac{11}{10} = 1\frac{1}{10}$$ Add the numerators and keep the denominator. Write the improper fraction as a mixed number.

(b) $$\left(\frac{1}{2} + \frac{3}{2}\right)\cdot\frac{2}{5} = \left(\frac{1+3}{2}\right)\cdot\frac{2}{5}$$ Add inside the parentheses first.

$$= \left(\frac{\overset{2}{\cancel{4}}}{\underset{1}{\cancel{2}}}\right)\cdot\frac{2}{5}$$ Divide out common factors.

$$= \frac{2}{1}\cdot\frac{2}{5} = \frac{4}{5}$$ Multiply.

1 2 3 Doing Mathematics

Problems 1–6: Reduce each fraction to lowest terms.

1. $\frac{5}{20}$

2. $\frac{100}{500}$

3. $\frac{48}{64}$

4. $\frac{32}{48}$

5. $\frac{35}{150}$

6. $\frac{52}{200}$

Problems 7–12: Build each fraction to an equivalent fraction with the given denominator.

7. $\frac{2}{5} = \frac{\quad}{10}$

8. $\frac{3}{4} = \frac{\quad}{20}$

9. $\frac{1}{2} = \frac{\quad}{50}$

10. $\frac{2}{25} = \frac{\quad}{100}$

11. $\frac{16}{3} = \frac{\quad}{18}$

12. $\frac{2}{7} = \frac{\quad}{28}$

Problems 13–18: Rewrite each mixed number as an improper fraction

13. $2\frac{1}{3}$

14. $10\frac{7}{9}$

15. $16\frac{4}{5}$

16. $9\frac{7}{8}$

17. $245\frac{3}{8}$

18. $11\frac{3}{5}$

Problems 19–24: Rewrite each improper fraction as a mixed number.

19. $\frac{10}{3}$

20. $\frac{21}{4}$

21. $\frac{152}{21}$

22. $\frac{54}{5}$

23. $\frac{101}{10}$

24. $\frac{37}{8}$

Problems 25–30: Multiply. Reduce each fraction to lowest terms. Write improper fractions as mixed numbers.

25. $\frac{10}{3} \cdot \frac{2}{5}$

26. $\frac{5}{6} \cdot \frac{15}{7}$

27. $2 \cdot 3\frac{2}{5}$

28. $2\frac{1}{7} \cdot 5\frac{1}{4}$

29. $\frac{10}{3} \cdot \frac{2}{5} \cdot \frac{12}{7} \cdot \frac{14}{5}$

30. $\frac{15}{4} \cdot \frac{50}{3}$

Problems 31–36: Divide. Reduce each fraction to lowest terms. Write improper fractions as mixed numbers.

31. $\frac{10}{3} \div \frac{2}{5}$

32. $\frac{2}{3} \div \frac{4}{3}$

33. $\frac{5}{6} \div 6\frac{2}{3}$

34. $2 \div \frac{2}{5}$

35. $2\frac{1}{2} \div 5\frac{1}{4}$

36. $\frac{25}{4} \div \frac{5}{2}$

Problems 37–48: Add or subtract. Reduce each fraction to lowest terms. Write improper fractions as mixed numbers.

37. $\frac{151}{4} + \frac{61}{4}$

38. $\frac{10}{3} + \frac{2}{3}$

39. $\frac{5}{6} + \frac{7}{24} - \frac{1}{4}$

40. $\frac{1}{3} + \frac{2}{5} + \frac{5}{6}$

41. $4\frac{5}{12} + 9\frac{7}{18}$

42. $2\frac{1}{2} + 3\frac{1}{4} + 5\frac{3}{8}$

43. $\frac{10}{3} - \frac{2}{3}$

44. $\frac{153}{6} - \frac{59}{6}$

45. $\frac{10}{3} - \frac{7}{5}$

46. $\frac{5}{12} - \frac{7}{18}$

47. $12\frac{5}{6} - 8\frac{7}{24}$

48. $20\frac{3}{5} - 1\frac{4}{5}$

49. Jan and Bob are putting a fence completely around their property, which is a pentagon with a $4\frac{1}{4}$-foot gate on one side. Each side of the property is $50\frac{3}{4}$ feet long.

How many feet of fencing are needed to enclose the property, not counting the gate?

50. Dave is replacing the molding around all four sides of three windows. The windows are each $2\frac{1}{2}$ feet wide and $5\frac{1}{4}$ feet high.

(a) What is the total length of molding that he needs to purchase?

(b) If the molding comes in lengths of 8 feet, how many lengths should he purchase?

Problems 51–52: When stock prices are given in fractional form, the Opening Price of a given stock can be calculated with this equation:

Opening Price = Last Price (closing) − Change (change from opening).

51. Compute the Opening Price of the following stock.

Nasdaq: December 15, 1999[1]				
	Trade name	Last price	Change	Opening Price = Last Price − Change
(a)	Learn2.cm	$4\frac{31}{32}$	$+1\frac{15}{16}$	
(b)	3Com	$50\frac{5}{16}$	$+5\frac{13}{16}$	
(c)	SunMic s	$75\frac{9}{16}$	$-4\frac{7}{8}$	

52. Compute the Opening Price of the following stock.

Amex: December 21, 1999[1]				
	Trade name	Last price	Change	Opening Price = Last Price − Change
(a)	Nasd100T n	$158\frac{47}{64}$	$-3\frac{1}{4}$	
(b)	SPDR	$140\frac{3}{4}$	$-1\frac{3}{8}$	
(c)	InterDig	$19\frac{3}{8}$	$-\frac{5}{8}$	

Problems 53–60: Perform the operations. Be sure to follow Order of Operations. Reduce each fraction to lowest terms. Write improper fractions as mixed numbers.

53. $\frac{10}{3} + \frac{2}{3} - \frac{5}{3}$

54. $\frac{10}{3} - \frac{2}{3} + \frac{5}{3}$

55. $\frac{205}{4} - \frac{61}{4} \cdot \frac{8}{3}$

56. $\left(\frac{73}{4} - \frac{61}{4}\right) \cdot \frac{8}{3}$

57. $\frac{10}{7} + \frac{4}{27} \div \frac{2}{9}$

58. $\frac{10}{3} + \frac{7}{24} \div \frac{21}{4}$

59. $\frac{5}{12} - \frac{7}{18} + \frac{2}{3}$

60. $\frac{5}{14} - \frac{3}{10} + \frac{4}{7}$

[1]*Palm Beach Post,* December 15, 1999, p. 9B.

R.3 Decimals and Percents

OBJECTIVES

- Write equivalent forms of decimals, fractions, and percents.
- Perform operations with percents and decimals.

1 2 3 Reviewing

In this section, we review equivalency between decimals, fractions, and percents, and use them to solve real-life problems.

DECIMALS, FRACTIONS, AND PERCENTS

A fraction can be written as an equivalent decimal. The fraction $\frac{3}{8}$ is the result of dividing 3 by 8 and is equivalent to the decimal 0.375. In performing the division, the numerator is written as the dividend with the decimal point written to the right of the number 3. Zeros are written to create place-values needed during the division process.

$$\begin{array}{r} .375 \\ 8\overline{)3.000} \\ \underline{2\,4} \\ 60 \\ \underline{56} \\ 40 \\ \underline{40} \\ 0 \end{array}$$

If the numerator and denominator of a fraction are integers, the fraction is called a **rational number**. A rational number is any number that can be written either as a terminating or repeating decimal. A **terminating decimal** is a decimal with a finite number of decimal places. A **repeating decimal** is a decimal where a sequence of numbers repeat.

EXAMPLE 1: (a) $\frac{1}{2} = 0.5$ Terminating Decimal.

(b) $\frac{1}{11} = 0.0909\ldots = 0.\overline{09}$ Repeating Decimal.

1. Use long division to write each fraction as a decimal. State whether it is a terminating decimal or a repeating decimal.

(a) $\frac{3}{4}$ (b) $\frac{5}{9}$

Percent denotes the number of parts under consideration of a total of one hundred parts. The percent symbol (%) is a notation for $\frac{1}{100}$.

EXAMPLE 2: Each individual block in the grid of 100 blocks represents $\frac{1}{100}$ of the whole. 30% is equivalent to 30 parts of a total of 100 parts.

30 of the 100 blocks are shaded.

$30 \cdot \frac{1}{100} = 30\%$ of the total 100 blocks are shaded.

2. Write the percent of blocks that are shaded.

CONVERTING A PERCENT TO A DECIMAL

EXAMPLE 3: Convert 30% to a decimal by either of two methods.

Method 1: Replace the % symbol with $\frac{1}{100}$ and multiply.

30%
↓

$$30 \cdot \frac{1}{100} = \frac{30}{100} = 0.30$$

Method 2: Move the decimal point two places to the left and omit the percent symbol.

$$30\% = 30.0\% = 0.30$$

A zero is often written to the left of the decimal point to emphasize the location of the decimal point.

3. Convert each percent to a decimal.

(a) 45%

(b) 2%

CONVERTING A DECIMAL TO A PERCENT

EXAMPLE 4: Convert 0.67 to a percent by either of two methods.

Method 1: Multiply the number by 100%.

$$0.67 = 0.67 \cdot 100\% = 67\%$$

Method 2: Move the decimal point two places to the right and insert percent symbol.

$$0.67 = 0.67 = 67\%$$

4. Convert each decimal to a percent.

(a) 0.89 (b) 0.04

CONVERTING A FRACTION TO A PERCENT

EXAMPLE 5: Convert $\frac{1}{2}$ to a percent.

First, express the fraction $\frac{1}{2}$ as a decimal: $\frac{1}{2} = 0.5$. Then use either Method 1 or 2.

Method 1: Multiply the decimal by 100%.

$$0.5 = 0.5 \cdot 100\% = 50\%$$

Method 2: Move the decimal point two places to the right and insert the percent symbol.

$$0.5 = 0.50 = 50\%$$

5. Convert each fraction to a percent.

(a) $\frac{3}{4}$ (b) $\frac{4}{5}$

ADDING AND SUBTRACTING DECIMALS

To add or subtract decimals, align the decimal points and corresponding place-values vertically. Add or subtract each place-value vertically and place the decimal point in the answer directly below the decimal point in the problem.

6. A checking account is a good example of adding or subtracting decimals. Add deposits or subtract the checks in the checking account to complete the table without a calculator.

Account activity: (deposits and checks)	Arithmetic operations:	Account balance ($):
Balance brought forward		$925.17
Check for apartment rent $575.00	$925.17 − 575.00 = $350.17	$350.17
Deposit of paycheck $478.87		
Check for gas $12.35		
Check for car payment $106.37		
Check for groceries of $43.78		
Deposit of paycheck $478.87		Ending balance

MULTIPLYING DECIMALS

To multiply decimals, multiply the numbers and place the decimal in the appropriate place in the product. The number of decimal places in the final product is equal to the sum of the number of decimal places in each factor.

EXAMPLE 6: Multiply 14.9×2.53.

$$\begin{array}{r} 14.9 \\ \times\ \underline{2.53} \\ 447 \\ 745 \\ \underline{298} \\ 37.697 \end{array}$$

Align the right-most digits.

14.9 has ONE digit to the right of the decimal point.
2.53 has TWO digits to the right of the decimal point.

The product of 14.9 and 2.53 has THREE digits to the right of the decimal point.

EXAMPLE 7: Multiply 245.66×12.831.

$$\begin{array}{r} 245.66 \\ \times\ \underline{12.831} \\ 24566 \\ 73698 \\ 196528 \\ 49132 \\ \underline{24566} \\ 3152.06346 \end{array}$$

Align the right-most digits.

245.66 has TWO digits to the right of the decimal point.
12.831 has THREE digits to the right of the decimal point.

The product of 245.66 and 12.831 has FIVE digits to the right of the decimal point.

7. Multiply

(a) 135.46×25.7 (b) 435.19×6.723

DIVIDING DECIMALS

EXAMPLE 8: Divide $25.6 \div 0.5$.

Method 1: Multiply the numerator and denominator of the fraction by an appropriate power of 10 to form an integer in the denominator.

$$25.6 \div 0.5 = \frac{25.6}{0.5} \cdot \frac{10}{10} = \frac{256}{5} = 51.2$$

Method 2: Long division.

$$\frac{25.6}{0.5} = 0\,5.\overline{)25\,6.} = 5\overline{)256.}$$

$$\begin{array}{r} 51.2 \\ 5\overline{)256.0} \\ \underline{25} \\ 6 \\ \underline{5} \\ 10 \\ \underline{10} \\ 0 \end{array}$$

8. Divide.

(a) $33.28 \div 6.5$ (b) $10.3 \div 0.002$

1 2 3 Discussing Mathematics

PERCENTS AND DECIMALS

PERCENT

Percent means the number of parts of the whole.

The % symbol is a notation for $\frac{1}{100}$.

CONVERTING A PERCENT TO A DECIMAL	
Method 1: Replace the % symbol with $\frac{1}{100}$ and multiply.	**Method 2:** Move the decimal point two places to the left and omit the percent symbol.

EXAMPLE 9: Convert 62.5% to a decimal.

Method 1: Replace the % symbol with $\frac{1}{100}$ and multiply.

62.5%

$$62.5 \cdot \frac{1}{100} = \frac{62.5}{100} = 0.625$$

Method 2: Move the decimal point two places to the left and omit the percent symbol.

$62.5\% = 62.5\% = 0.625$

➢ *A zero is often written to the left of the decimal point to emphasize the location of the decimal point.*

CONVERTING A DECIMAL TO A PERCENT	
Method 1: Multiply the number by 100%.	**Method 2:** Move the decimal point two places to the right and insert the percent symbol.

EXAMPLE 10: Convert 1.25 to a percent.

Method 1:

$1.25 = 1.25 \cdot 100\% = 125\%$

Method 2:

$1.25 = 1.25 = 125\%$

CONVERTING A FRACTION TO A PERCENT	
Method 1: Express the fraction as a decimal. Multiply the decimal by 100%.	**Method 2:** Express the fraction as a decimal. Move the decimal point two places to the right and insert the percent symbol.

EXAMPLE 11: Convert $16\frac{3}{4}$ to a percent.

Method 1:

$16\frac{3}{4} = 16.75$

$16.75 = 16.75 \cdot 100\% = 1675\%$

Method 2:

$16\frac{3}{4} = 16.75$

$16.75 = 16.75 = 1675\%$ (move the decimal point two places to the right)

USING PERCENTS AND DECIMALS IN REAL-WORLD PROBLEMS

To find a percent of a number, change the percent to an equivalent decimal and multiply.

EXAMPLE 12: (a) Find 33% of 150.

0.33×150 — Change 33% to 0.33.

$= 49.5$ — Multiply.

(b) Find 2.5% of 550.

0.025×550 — Change 2.5% to 0.025.

$= 13.75$ — Multiply.

EXAMPLE 13: The regular price of the dress was \$120 and the sale price was 40% off. Find the sale price of the dress.

$$\begin{aligned}
\text{Amount of discount} &= 40\% \text{ of } \$120 \\
&= 0.40 \cdot \$120 \\
&= \$48 \\
\text{Sale price of dress} &= \text{Original Price} - \text{Discount} \\
&= \$120 - \$48 = \$72
\end{aligned}$$

ADDITION AND SUBTRACTION OF DECIMALS

TO ADD OR SUBTRACT DECIMALS

1. Align the decimal points and corresponding place-values vertically.
2. Add or subtract each place-value vertically.
3. Place the decimal point in the answer directly below the decimal point in the problem.

EXAMPLE 14: Add 7665.349 + 10.5.

$$\begin{array}{r}
7665.349 \\
+\ \ 10.500 \\
\hline
7675.849
\end{array}$$

Align the decimal points and corresponding place-values vertically. Add two zeros to the right of 10.5 to align columns.
Add each place-value vertically.
Place the decimal point in the answer directly below the decimal point in the problem.

EXAMPLE 15: Subtract 125.3 − 37.56.

$$\begin{array}{r}
125.30 \\
+-\ 37.56 \\
\hline
87.74
\end{array}$$

Align the decimal points and corresponding place-values vertically. Add one zero to the right of 125.3 to align columns.
Subtract each place-value vertically.
Place the decimal point in the answer directly below the decimal point in the problem.

MULTIPLYING DECIMALS

TO MULTIPLY DECIMALS

1. Multiply the numbers.
2. Place the decimal in the appropriate place in the product. The number of decimal places in the product is equal to the sum of the number of decimal places in each factor.

EXAMPLE 16: Multiply.

$$
\begin{array}{r}
325.31 \\
\times\ \underline{47.562} \\
65062 \\
195186 \\
162655 \\
227717 \\
\underline{130124} \\
15472.39422
\end{array}
$$

325.31 has TWO digits to the right of the decimal point.
47.562 has THREE digits to the right of the decimal point.

The product has FIVE digits to the right of the decimal point.

DIVIDING DECIMALS

TO DIVIDE DECIMALS

1. Write the quotient using a division symbol: divisor$\overline{)\text{dividend}}$
2. Move the decimal in the divisor to the right to form a whole number.
3. Move the decimal in the dividend the same number of places.
4. Perform the division.
5. Place the decimal point in the quotient directly above the decimal in the dividend.

EXAMPLE 17: Divide.

$$13.5\overline{)1697.625}$$

$$135.\overline{)16976.25} = 135.\overline{)16976.25}$$

Write the quotient using a division symbol.
Move the decimal in the divisor to the right to form a whole number. Move the decimal in the dividend the same number of places.

$$
\begin{array}{r}
125.75 \\
135\overline{)16976.25} \\
\underline{135} \\
347 \\
\underline{270} \\
776 \\
\underline{675} \\
1012 \\
\underline{945} \\
675 \\
\underline{675} \\
0
\end{array}
$$

Perform the division.

Place the decimal point in the quotient directly above the decimal in the dividend.

1 2 3 Doing Mathematics

Problems 1–4: Write each fraction as a decimal.

1. $\frac{7}{8}$

2. $\frac{2}{3}$

3. $\frac{16}{9}$

4. $\frac{5}{2}$

Problems 5–8: Write each decimal as a fraction. If an answer is an improper fraction, do not change it.

5. 23.5

6. 6.21

7. 1.235

8. 0.942

9. In August 28, 1999, the Plain Dealer newspaper in Cleveland, Ohio announced that it would be simplifying its stock tables by listing stock prices in dollars and cents (rounded decimal form), instead of fractions. A stock that ended the day at a price of $\$35\frac{27}{32}$ would then be listed as \$35.84 (35.84375 rounded to the nearest cent). Write each stock price as a decimal, rounded to the nearest cent (nearest hundredths place).

(a) $\$68\frac{1}{32}$

(b) $\$5\frac{1}{8}$

(c) $\$12\frac{5}{32}$

(d) $\$\frac{3}{8}$

(e) $\$9\frac{11}{32}$

(f) $\$76\frac{5}{8}$

(g) $\$3\frac{21}{32}$

(h) $\$10\frac{7}{8}$

10. Identify each decimal as terminating, repeating, or neither.

(a) 1.55

(b) $1.\overline{5}$

(c) 2.23606797...

(d) 0.777777...

(e) 25.01

(f) 3.141592654...

Problems 11–14: Convert each percent to a decimal.

11. 20%

12. 68.25%

13. 100%

14. 70%

Problems 15–18: Convert each decimal to a percent.

15. 0.50

16. 0.75

17. 1.253

18. 0.125

Problems 19–22: Convert each fraction to a percent.

19. $\frac{3}{5}$

20. $\frac{7}{2}$

21. $\frac{5}{8}$

22. $\frac{1}{3}$

Problems 23–30: Add or subtract.

23. $25.36 + 15.1$

24. $205.123 + 130.5$

25. $0.41662 + 23.11$

26. $0.9999 + 0.035$

27. $125.836 - 55.4$

28. $1.0036 - 0.9151$

29. $1275.123 - 10.56$

30. $0.123 - 0.1105$

Problems 31–34: Multiply.

31. 5.036×7.12

32. 6.18×15.45

33. 101.1×1.782

34. 0.542×0.31

Problems 35–38: Divide.

35. $205.875 \div 15.25$

36. $2406.72 \div 9.6$

37. $125.8 \div 0.005$

38. $0.625 \div 0.0025$

39. A carpenter has been hired to replace the baseboards in a living room having the dimensions shown in the figure.

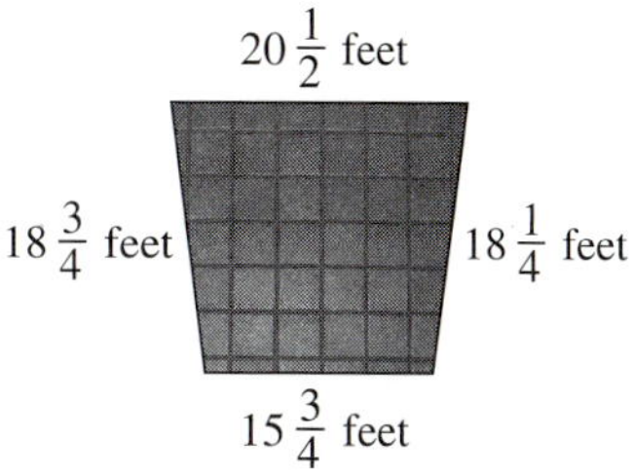

(a) Write each measurement as a decimal.

$20\frac{1}{2} =$ ______ $18\frac{3}{4} =$ ______ $18\frac{1}{4} =$ ______ $15\frac{3}{4} =$ ______

(b) Find the perimeter of the living room.

Hint: The distance around the outside of the room is called the perimeter.

(c) Find the total cost of the baseboards if the price is \$3.52 per foot.

40. An X-ray technologist uses the product (P) of the milliamperage (mA) and the time (t) to produce appropriate radiographic exposures.

Product is milliamperage multiplied by time

$$P = \text{mA} \times t$$

Find the constant (P) for the following milliamperage (mA) and time (t) combinations. Write all answers in decimal form. Round to the nearest hundredth, if necessary.

EXAMPLE: Find the product P, given mA = 300 and $t = 0.035$ seconds.

$$P = 300 \times 0.035 = 10.5$$

	Milliamperage = mA	Time = t	$P = \text{mA} \times t$
(a)	400 mA	$\frac{1}{120}$ sec	
(b)	600 mA	$\frac{1}{30}$ sec	
(c)	200 mA	$\frac{3}{10}$ sec	
(d)	100 mA	$\frac{3}{20}$ sec	
(e)	100 mA	0.4 sec	
(f)	400 mA	0.025 sec	
(g)	500 mA	0.4 sec	
(h)	50 mA	0.035 sec	

41. An X-ray technologist can compute the milliamperage (mA) if the product (P) and the number of seconds of exposure (t) are known by dividing the product by time.

Milliamperage	is	Product divided by time
mA	=	$\frac{P}{t}$

Find the milliamperage (mA) for the given values of the product (P) and time (t). Write all answers in decimal form.

EXAMPLE: Find mA, given $P = 10.5$ and $t = 0.035$ seconds.

$$\text{mA} = \frac{10.5}{0.035} = 300$$

	Product = P	Time = t	Milliamperage $mA = \frac{P}{t}$
(a)	20 mA	$\frac{1}{30}$ sec	
(b)	60 mA	$\frac{3}{10}$ sec	
(c)	15 mA	$\frac{3}{20}$ sec	
(d)	10 mA	0.025 sec	
(e)	40 mA	0.4 sec	
(f)	14 mA	0.035 sec	

42. It is occasionally necessary to change the degree of blackness of an X-ray image by altering the product (P). The amount of increase or decrease desired is added or subtracted from the product (P) as shown using the formulas:

Increase (or decrease) = Previous product (P) × Percent increase (or decrease)

New P value = Original P value + Increase

OR

New P value = Original P value − Decrease

EXAMPLE: If the original product is 100 and the increase in degree of darkness is 40%, compute the increase and the New P value.

Increase	= Original P value	× Percent Increase
	= 100	× 40%
	= 100	× .40 = 40
New P value	= Original P value	+ Increase
	= 100	+ 40
New P value	= 140	

	Original *P*	Percent	Increase or decrease	New *P* value
(a)	160 mA	40% Decrease		
(b)	18 mA	50% Increase		
(c)	5.6 mA	200% Increase		
(d)	25 mA	30% Decrease		

Problems 43–46: Perform the operations.

43. Find 20% of 90.

44. Find 75% of 250.

45. Find 125% of 500.

46. Find 0.05% of 500.

47. Lydell wants to buy a motorcycle that costs $7500. The dealer announced a sale of 15% off all cycles. What is the sale price of the motorcycle?

48. Computers4U advertised a computer for sale at 35% off the original price of $999. What is the sale price of the computer?

Problems 49–50: Writing Exercises

49. Create a real-world problem involving multiplication of decimals. Find the answer to your problem.

50. Create a real-world problem involving percent. Find the answer to your problem.

R.4 Ratio, Proportion, and Conversions

OBJECTIVES

- Write ratios and rates.
- Use ratio and proportion to convert one unit of measure to another.
- Solve proportions.

1 2 3 Reviewing

RATIOS AND RATES

A **ratio** is a quotient of two quantities with the same unit of measure. A **rate** is a quotient of two quantities with different units of measure.

EXAMPLE 1: A ratio may be written as a fraction, a word statement, or with a colon.

	Ratio as a Fraction	Word Statement	Colon
(a)	$\dfrac{\text{5 pieces of pizza eaten by Tom}}{\text{3 pieces of pizza eaten by Bill}}$	5 pieces to 3 pieces	5:3
(b)	$\dfrac{\text{9 pages read on Monday}}{\text{3 pages read on Tuesday}} = \dfrac{\text{3 pages}}{\text{1 pages}}$	3 pages to 1 page	3:1

EXAMPLE 2: A rate is written as a fraction or word statement.

	Rate as a Fraction	Word Statement
(a)	$\dfrac{\text{500 miles}}{\text{10 hours}} = \dfrac{\text{50 miles}}{\text{1 hour}}$	50 miles per hour
(b)	$\dfrac{\text{20 feet}}{\text{3 seconds}} = \dfrac{6.\overline{6}\text{ feet}}{\text{1 second}}$	$6.\overline{6}$ feet per second

1. Before Super Bowl XXXIX in 2005, statistics were published about the New England Patriots and the Philadelphia Eagles offensive teams' record in the 2004 regular season.

(a) What is the ratio of total net yards gained rushing by the Patriots to the total net yards gained rushing by the Eagles?

$$\frac{\text{Yards gained rushing by Patriots}}{\text{Yards gained rushing by Eagles}} = \underline{\qquad}$$

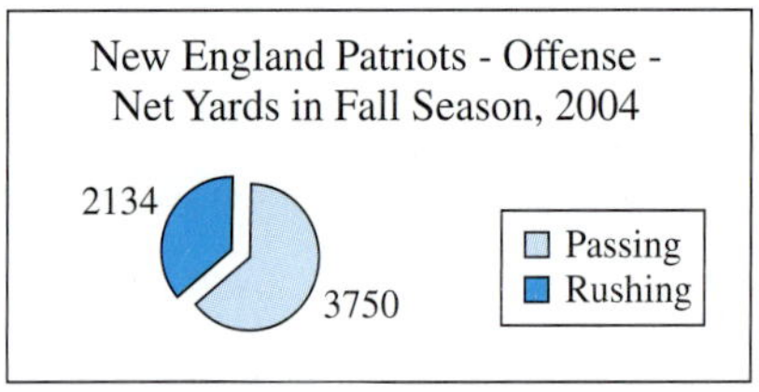

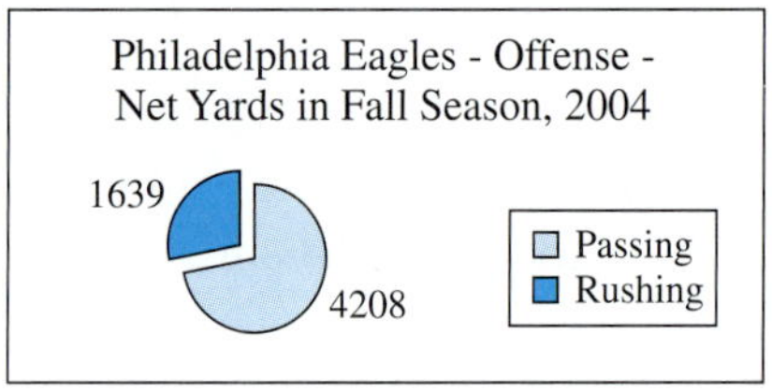

(b) Divide the denominator of the ratio into the numerator. Round to the nearest hundredth. What does the ratio tell you about the two offensive teams?

(c) What is the ratio of total first downs earned rushing by the Patriots to the total first downs earned rushing by the Eagles?

$$\frac{\text{First downs earned rushing by Patriots}}{\text{First downs earned rushing by Eagles}} = \underline{\qquad}$$

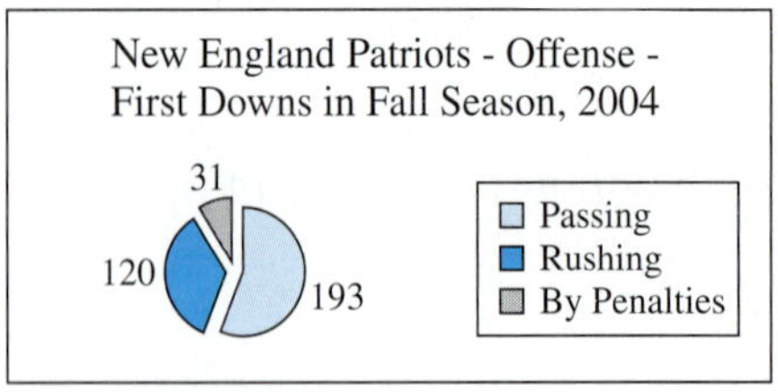

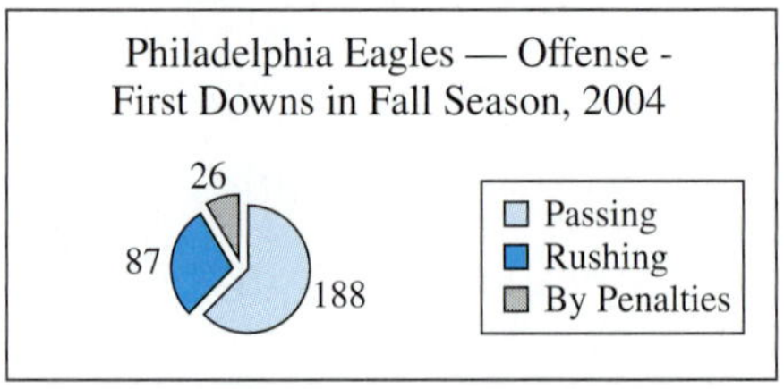

(d) Divide the denominator of the ratio into the numerator. Round to the nearest hundredth. What does the ratio tell you about the two offensive teams?

(e) Reflecting on the ratios and statistics above, which team had the better offense in the 2004 regular season?

PROPORTIONS

Two equal ratios form an equation called a **proportion**. The quantities in a proportion have the specific names of **extremes** and **means**:

In the proportion: $\frac{1}{2} = \frac{4}{8}$ 1 is to 2 as 4 is to 8

Means (2 and 4)

Extremes (1 and 8)

In any proportion, the product of the extremes is equal to the product of the means.

Extremes

$$\frac{1}{2} \times \frac{4}{8}$$

Means

The product of the extremes and the means can be visualized with two crossed arrows called the **cross product**.

Product of Extremes = Product of Means

$$1 \cdot 8 = 2 \cdot 4$$
$$8 = 8$$

An equation can be formed with the products indicated by each arrow, illustrating the equivalency.

Proportions may involve variables and the product of the extremes and the means can be used to solve equations.

EXAMPLE 3: Find the product of the extremes and means to solve the proportion $\frac{3}{5} = \frac{x}{20}$ for x.

$$\frac{3}{5} \times \frac{x}{20}$$ Write an equation with the product of the extremes equal to the product of the means.

$$3 \cdot 20 = 5 \cdot x$$
$$60 = 5x$$
$$12 = x$$ Solve the equation for x.

2. Solve each proportion.

(a) $\frac{6}{11} = \frac{x}{55}$ (b) $\frac{2}{x} = \frac{1}{14}$

Discussing Mathematics

RATIOS AND RATES

Ratios and rates may be written in different forms. It is common to divide the denominator of a ratio or rate into the numerator, rewriting the fraction with 1 in the denominator.

EXAMPLE 4: The ratio of $\frac{\text{2059 yards rushing}}{\text{1811 yards rushing}}$ can be written in these forms:

Ratio as a fraction	Word statement	Colon
$\frac{\text{2059 yards rushing}}{\text{1811 yards rushing}} = \frac{\text{1.14 yards rushing}}{\text{1 yard rushing}}$	1.14 yards rushing to 1 yard rushing	1.14:1

EXAMPLE 5: The rate $\frac{200 \text{ feet}}{50 \text{ seconds}}$ can be written in these forms:

Rate as a fraction	Word statement
$\frac{200 \text{ feet}}{50 \text{ seconds}} = \frac{4 \text{ feet}}{1 \text{ second}}$	4 feet per second.

PROPORTIONS

An equation where two ratios are equal is called a **proportion**. The Extremes-Means Property of Proportions is a useful tool for showing equivalency of fractions or solving for an unknown.

EXTREMES-MEANS PROPERTY OF PROPORTIONS

In any proportion, the product of the extremes equals the product of the means:

If $\frac{a}{b} = \frac{c}{d}$ then $a \cdot d = b \cdot c$ $(b \neq 0, d \neq 0)$. Product of Extremes = Product of Means

EXAMPLE 6: Use the Extremes-Means Property of Proportions to determine if the two fractions $\frac{7}{8}$ and $\frac{42}{48}$ are equivalent.

$$\frac{7}{8} \stackrel{?}{=} \frac{42}{48}$$

$7 \cdot 48 \stackrel{?}{=} 8 \cdot 42$ The product of the extremes equals the product of the means.

$336 = 336$ The original fractions are equivalent.

EXAMPLE 7: Solve the proportion $\frac{x}{105} = \frac{1}{7}$.

$\frac{x}{105} = \frac{1}{7}$ Write an equation with the product of the extremes equal to the product of the means.

$x \cdot 7 = 105 \cdot 1$ Multiply.

$7x = 105$ Solve for x.

$x = 15$

CONVERTING UNITS OF MEASURE

Equivalent measures can be written as rates and used to convert from one unit of measurement to another.

- The equivalent measurement of 1 foot (ft) = 12 inches (in) can be written as either of the rates: $\frac{1 \text{ ft}}{12 \text{ in}}$ or $\frac{12 \text{ in}}{1 \text{ ft}}$.
- The approximate measurement of 1 mile (mi) $\approx$ 1.61 kilometer (km) can be written as either of the rates: $\frac{1 \text{ mi}}{1.61 \text{ km}}$ or $\frac{1.61 \text{ km}}{1 \text{ mi}}$.

Multiplying by an equivalent measure is equivalent to multiplying by 1. The value is not changed, only its unit of measure. When converting, the rates must be written so that similar units divide out, leaving the desired unit of measurement. It may be necessary to use approximate measures for some conversions.

EXAMPLE 8: Convert 246 inches to feet. Use the equivalent measure $\frac{1\text{ ft}}{12\text{ in}}$.

$$246\text{ inches} = \frac{\overset{41}{\cancel{246}}\ \cancel{\text{inches}}}{1}\cdot\frac{1\text{ ft}}{\underset{2}{\cancel{12}}\ \cancel{\text{inches}}}$$

Inch units (inches) and 6 divide out as common factors.

$$= \frac{41\text{ ft}}{2} = 20.5\text{ ft}$$

Multiply. The desired unit of measurement, feet (ft), remain.

EXAMPLE 9: Convert 480 miles to kilometers. Use the approximate measure $\frac{1.61\text{ km}}{1\text{ mi}}$.

$$480\text{ miles} \approx \frac{480\ \cancel{\text{mi}}}{1}\cdot\frac{1.61\text{ km}}{1\ \cancel{\text{mi}}}$$

Mile units (mi) divide out as common factors.

$$\approx 772.8\text{ km}$$

Multiply. The desired unit of measurement, kilometer (km), remains.

1 2 3 Doing Mathematics

1. The bar graph shows the projected number of tourists in the year 2020 to visit five countries.[1]

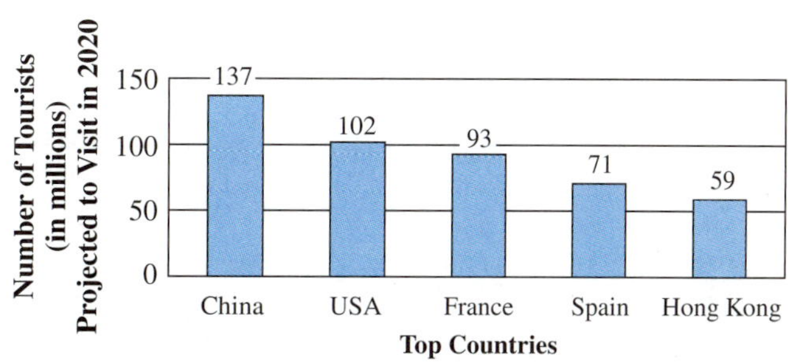

(a) What is the ratio of the number of tourists (in millions) projected to visit China to the number of tourists (in millions) projected to visit the United States?

(b) Divide the denominator of the ratio in part (a) into the numerator. Round the decimal to the nearest hundredth. Express the ratio using a colon.

(c) What is the ratio of the number of tourists (in millions) projected to visit Spain to the number of tourists (in millions) projected to visit the United States?

(d) Divide the denominator of the ratio in part (c) into the numerator. Round the decimal to the nearest hundredth. Express the ratio using a colon.

[1] *USA Today*, January 20, 2000, Sec. D, p. 1.

2. Before Super Bowl XXXIX in 2005, the following statistics were published about the New England Patriots and the Philadelphia Eagles defensive teams' record in the 2004 regular season.

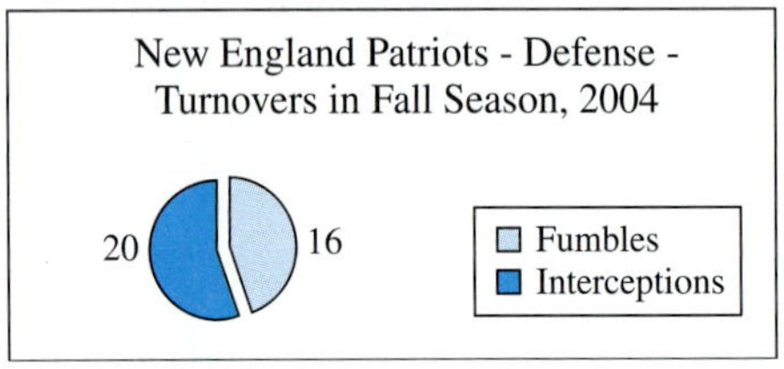

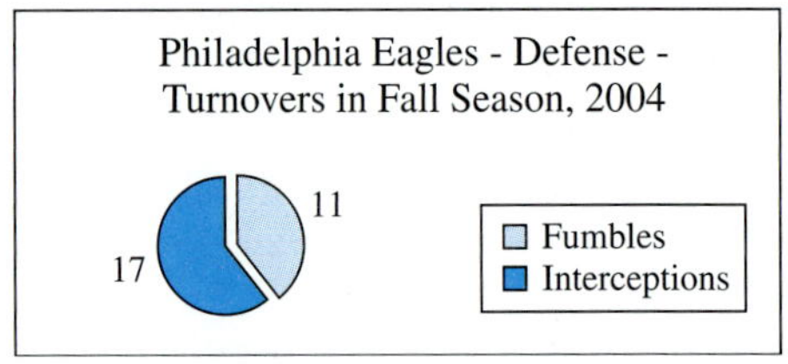

(a) What is the ratio of interceptions made by the Eagles' defense to the interceptions made by the Patriots' defense?

$$\frac{\text{Interceptions made by Eagles' defense}}{\text{Interceptions made by Patriots' defense}} = \underline{\qquad\quad}$$

(b) What is the ratio of fumbles recovered by the Eagles' defense to fumbles recovered by the Patriots' defense?

$$\frac{\text{Fumbles recovered by Eagles' defense}}{\text{Fumbles recovered by Patriots' defense}} = \underline{\qquad\quad}$$

(c) Reflecting on the ratios and statistics above, which team had the better defense in the 2004 regular season?

Problems 3–6: Write each rate.

3. Leslie traveled 420 miles in 8 hours.

4. The rocket traveled 125 meters in 3 seconds.

5. The airplane traveled 600 miles in 2 hours.

6. Pierre traveled 220 kilometers in 2.5 hours.

Problems 7–16: Use the Extremes-Means Property of Proportions to determine whether the given fractions are equivalent. Insert the appropriate symbol, either = or ≠, on the line between the fractions.

7. $\frac{3}{4}$ ______ $\frac{6}{8}$

8. $\frac{10}{20}$ ______ $\frac{1}{2}$

9. $\frac{2}{3}$ ______ $\frac{7}{9}$

10. $\frac{4}{5}$ ______ $\frac{12}{15}$

11. $\frac{5}{9}$ ______ $\frac{15}{27}$

12. $\frac{5}{6}$ ______ $\frac{15}{20}$

13. $\frac{12}{18}$ ______ $\frac{2}{3}$

14. $\frac{2}{5}$ ______ $\frac{18}{45}$

15. $\frac{18}{12}$ ______ 6

16. 3 ______ $\frac{65}{20}$

Problems 17–26: Solve each proportion. Express answers as rational numbers. Do not change fractions to decimals.

17. $\frac{3}{4} = \frac{x}{32}$

18. $\frac{1}{5} = \frac{x}{130}$

19. $\frac{x}{3} = \frac{5}{9}$

20. $\frac{x}{4} = \frac{2}{17}$

21. $\frac{x}{9} = 6$

22. $\frac{x}{2} = 8$

23. $\frac{5}{x} = \frac{2}{3}$

24. $\frac{2}{x} = \frac{3}{4}$

25. $\frac{3}{2x} = \frac{2}{3}$

26. $\frac{5}{3x} = \frac{1}{6}$

Problems 27–34: Use the following equivalent and approximate measures to convert the given measurements to the specified units. Choose the appropriate equivalent measure and write it as a rate. Multiply to convert to the specified units. Round each answer to the nearest hundredth.

1 inch (in) = 2.54 centimeters (cm)

12 inches (in) = 1 foot (ft)

1 yard (yd) = 3 feet (ft)

1.09 yard (yd) ≈ 1 meter (m)

1 mile (mi) ≈ 1.61 kilometers (km)

1 gallon (gal) ≈ 3.79 liters (L)

27. Convert 150 miles to kilometers.

28. Convert 500 kilometers to miles.

29. Convert 5 liters to gallons.

30. Convert 15 gallons to liters.

31. Convert 12 yards to meters.

32. Convert 20 feet to yards.

33. (a) Convert 150 centimeters to inches.

(b) Convert the answer in (a) to feet.

34. (a) Convert 3.5 meters to yards.

(b) Convert the answer in (a) to feet.

35. Years ago, gasoline cost \$1.50 per gallon in the United States and \$6.00 per liter in Europe. Was gasoline less expensive in the US or in Europe? How did you arrive at your answer?

36. Judy and Jim rented a car on their vacation in France. The distance from Paris to Marseilles was 790 kilometers (km). What is the distance rounded to the nearest mile?

Problems 37–40: Which product is a more economical purchase?

37. Topsoil that costs \$60 for 3 cubic yards or topsoil that costs \$105 for 5 cubic yards

38. A cup of coffee that costs \$0.79 for 20 ounces or a cup of coffee that costs \$0.99 for 32 ounces

39. A glass of soda that costs \$0.89 for 20 ounces or an 8-ounce soda for \$0.48

40. An 8-foot-long wood plank for \$10.50 or an 11-foot-long plank for \$13.99

Problems 41–44: In radiology, "Conventional Units" (measured in rads and rems) are used to measure the amount of radiation in an x-ray image in the United States. Other countries use "International Units" (measured in Grays and Sieverts). Use the following equivalent measures to convert Conventional Units to International Units.

Conventional units:		International Units:
0.01 Radiation Absorbed Dose (rad)	=	1 Gray (Gy)
0.01 Radiation Equivalent Man (rem)	=	1 Sievert (Sv)

EXAMPLE: Convert 50 rad to the equivalent Gray (Gy) units.

$$\frac{50\ \cancel{\text{rad}}}{1}\cdot\frac{1\text{ Gy}}{0.01\ \cancel{\text{rad}}}=\frac{50\text{ Gy}}{0.01}=5000\text{ Gy}$$

41. Convert 40 rad to the equivalent Gray (Gy) units:

42. Convert 200 rad to the equivalent Gray (Gy) units:

43. Convert 500 rem to the equivalent Sievert (Sv) units:

44. Convert 600 rem to the equivalent Sievert (Sv) units:

45. Create and solve a real-world problem involving ratios, proportions, or conversions.

R.5 Exponents and Scientific Notation

OBJECTIVES

- Find products of expressions with positive and negative exponents.
- Perform multiplication and division of expressions with exponents.
- Write numbers in scientific notation.
- Multiply and divide numbers written in scientific notation.
- Understand and use scientific notation displayed on a calculator.

1 2 3 Reviewing

EXPONENTS

A full discussion of exponents is presented in Chapter 5 of this textbook. The purpose of this discussion is to review the definition of an exponent, multiply and divide expressions with exponents, and write numbers in scientific notation.

An **exponent** is a shorthand notation used to indicate that a number (called the base) is multiplied by itself a given number of times (called the exponent). A number written with an exponent is in **exponential notation**. The number written without an exponent as a product of factors is in **expanded factor notation**. The product of the factors results in the **standard form** of the number.

Exponential Notation	Expanded Factor Notation	Standard Form
$10^4 =$	$\underbrace{10 \cdot 10 \cdot 10 \cdot 10}$	$= 10{,}000$
10 is the **base**. 4 is the **exponent**.	The base 10 multiplied 4 times.	

A **negative exponent** is a shorthand notation for reciprocating.

$$10^{-3} = \frac{1}{10^3} = \frac{1}{10 \cdot 10 \cdot 10} = 0.001 \quad \text{and} \quad \frac{1}{10^{-3}} = \frac{10^3}{1} = \frac{10 \cdot 10 \cdot 10}{1} = 1000$$

➢ *A negative exponent does not mean that the final product is negative.*

1. Write each expression in expanded factor notation and standard form.

	Exponential notation	Expanded factor notation	Standard form
(a)	$10^4 =$		
(b)	$10^6 =$		
(c)	$10^{-2} =$		
(d)	$10^{-4} =$		

SCIENTIFIC NOTATION

Scientific notation uses exponents to write very large or very small numbers. A number written in **scientific notation** consists of the product of two components: (1) a number greater than or equal to 1 and less than 10 and (2) an integer power of 10. The power of 10 is positive on large numbers and negative on small numbers.

To change a number given in **scientific notation to standard form**, look at the exponent:

➢ *When the exponent of 10 is positive: move the decimal point to the right the number of decimal places equal to the absolute value of the exponent;*

➢ *When the exponent of 10 is negative: move the decimal point to the left the number of decimal places equal to the absolute value of the exponent.*

EXAMPLE 1: (a) The radius of the planet Mercury written in scientific notation is

$$2.43 \times 10^6 \text{ meters}$$

Number between 1 and 10 — Power of 10 where exponent is positive

To write 2.43×10^6 in standard form, move the decimal point six places to the right:
2.4‸3‸0‸0‸0‸0. = 2,430,000 m.

(b) The length of a housefly written in scientific notation is

$$5.0 \times 10^{-3} \text{ meters}$$

Number between 1 and 10 — Power of 10 where exponent is negative

To write 5.0×10^{-3} in standard form, move the decimal point three places to the left:
0.0‸0‸5.0 = 0.005 m.

2. Write each measurement in standard form.

	Item	Measurement	Move the decimal	Standard form
(a)	Radius of Venus	6.06×10^6 m.		
(b)	Radius of Saturn	5.85×10^7 m.		
(c)	Size of a dust particle	1.0×10^{-4} m.		
(d)	Mass of a mosquito	1.0×10^{-5} kg.		

To change a number given in **standard form to scientific notation**:

- When the given number is greater than 10:
    1. Move the decimal point to the left until the resulting number is between 1 and 10.i
    2. Write the number with the decimal point in the new location.
    3. Multiply by 10 raised to a positive exponent equal to the number of places the decimal was moved.
- When the given number is less than 1:
    1. Move the decimal point to the right so that the resulting number is between 1 and 10.
    2. Write the number with the decimal point in the new location.
    3. Multiply by 10 raised to a negative exponent equal to the number of places the decimal was moved.

EXAMPLE 2: Write each number in scientific notation.

(a) 947,000

9.4 7 0 0 0 — Move the decimal point five places to the left.

$947{,}000 = 9.47 \times 10^5$ — Multiply by 10 to the positive exponent 5.

(b) 0.000425

0 0 0 0 4 .25 — Move the decimal point four places to the right.

$0.000425 = 4.25 \times 10^{-4}$ — Multiply by 10 to the negative exponent 4.

3. Write each number in scientific notation.

(a) 5,600,000 (b) 0.00325

(c) 1,230 (d) 0.0000198

MULTIPLICATION AND DIVISION INVOLVING EXPONENTS

Multiplying expressions with LIKE bases and exponents involves multiplying factors or adding exponents.

EXAMPLE 3: Multiply $10^4 \cdot 10^5$.

Method 1: Write the expression in expanded factor notation and multiply.

$$10^4 \cdot 10^5 = \underbrace{10 \cdot 10 \cdot 10 \cdot 10}_{4 \text{ factors}} \cdot \underbrace{10 \cdot 10 \cdot 10 \cdot 10 \cdot 10}_{5 \text{ factors}}$$

$$= \underbrace{10 \cdot 10 \cdot 10 \cdot 10 \cdot 10 \cdot 10 \cdot 10 \cdot 10 \cdot 10}_{9 \text{ factors}}$$

$$= 10^9 = 1{,}000{,}000{,}000$$

Method 2: Write the base and add the exponents.

$$10^4 \cdot 10^5 = 10^{4+5}$$

$$= 10^9 = 1{,}000{,}000{,}000$$

4. Multiply using Method 2.

(a) $10^2 \cdot 10^4$ (b) $10^3 \cdot 10^2$

Dividing expressions with LIKE bases and exponents involves dividing out common factors or subtracting exponents.

EXAMPLE 4: Divide $\dfrac{10^5}{10^3}$.

Method 1: Write the expression in expanded factor notation and divide.

$$\frac{10^5}{10^3} = \frac{\overbrace{\cancel{10} \cdot \cancel{10} \cdot \cancel{10} \cdot 10 \cdot 10}^{5 \text{ factors}}}{\underbrace{\cancel{10} \cdot \cancel{10} \cdot \cancel{10}}_{3 \text{ factors}}} = \frac{10 \cdot 10}{1}$$

$$= \frac{10^2}{1} = 10^2 = 100$$

Method 2: Write the base and subtract the exponents.

$$\frac{10^5}{10^3} = 10^{5-3}$$

$$= 10^2 = 100$$

5. Divide using Method 2.

(a) $\dfrac{10^4}{10^3}$ (b) $\dfrac{10^6}{10^4}$

1 2 3

Discussing Mathematics

EXPONENTS

An exponent is a shorthand notation to indicate that a number (called the base) is being multiplied by itself a given number of times (the exponent).

EXPONENT

For any real number a and any counting number n:

$$a^n = \underbrace{a \cdot a \cdot a \ldots \cdot a}_{n \text{ factors}}$$

For any real number a, except $a \neq 0$, and any counting number n:

$$a^{-n} = \frac{1}{\underbrace{a \cdot a \cdot a \ldots \cdot a}_{n \text{ factors}}}$$

also $$\frac{1}{a^{-n}} = \frac{\overbrace{a \cdot a \cdot a \ldots \cdot a}}{1}$$

EXAMPLE 5: The definition of an exponent governs how numbers may be written in equivalent forms.

Exponential notation	Expanded factor notation	Standard form
4^3	$= 4 \cdot 4 \cdot 4$	$= 64$
4^{-3}	$= \frac{1}{4^3} = \frac{1}{4 \cdot 4 \cdot 4}$	$= \frac{1}{64}$
$\frac{1}{4^{-3}}$	$= \frac{4^3}{1} = \frac{4 \cdot 4 \cdot 4}{1}$	$= 64$

SCIENTIFIC NOTATION

Scientific notation is shorthand notation using exponents and powers of 10 to express very large and very small numbers.

> **SCIENTIFIC NOTATION:** A number between 1 and 10 × the base 10 raised to an integer power.
>
> To change a number given in scientific notation to standard form:
>
> - When the exponent of 10 is positive: move the decimal point to the right the number of decimal places equal to the absolute value of the exponent.
> - When the exponent of 10 is negative: move the decimal point to the left the number of decimal places equal to the absolute value of the exponent.
>
> To change a number given in standard form to scientific notation:
>
> - When the given number is greater than 10:
>   1. Move the decimal point to the left until the resulting number is between 1 and 10.
>   2. Write the number with the decimal point in the new location.
>   3. Multiply by 10 raised to a positive exponent equal to the number of places the decimal was moved.
> - When the given number is less than 1:
>   1. Move the decimal point to the right so that the resulting number is between 1 and 10.
>   2. Write the number with the decimal point in the new location.
>   3. Multiply by 10 raised to a negative exponent equal to the number of places the decimal was moved.

EXAMPLE 6: (a) The speed of light (in meters per second) written in scientific notation is

$$2.99 \times 10^8$$

Number between 1 and 10 — Power of 10

To write 2.99×10^8 in standard form, move the decimal eight places to the right:

$$2\,9\,9\,0\,0\,0\,0\,0\,0. = 299{,}000{,}000 \text{ m/s}$$

(b) The number of cubic meters in one US gallon written in scientific notation is

$$3.785 \times 10^{-3}$$

Number between 1 and 10 — Power of 10

To write 3.785×10^{-3} in standard form, move the decimal three places to the left:

$$0.0\,0\,3\,785 = 0.003785 \text{ m}^3/\text{gal}$$

EXAMPLE 7: Write each number in scientific notation.

(a) 25,300,000

$= 2.5_{\wedge}3_{\wedge}0_{\wedge}0_{\wedge}0_{\wedge}0_{\wedge}0$ — Move the decimal point seven places to the left so that the resulting number is 2.53 (a number between 1 and 10).

$= 2.53 \times 10^7$ — Write the number with the decimal point in the new location. Multiply by 10 to the positive exponent 7.

$2.53 \times 10^7 = 2.53 \times 10{,}000{,}000$ — Check the answer by writing the number in standard form. ✓ It checks.

$= 25{,}300{,}000$

(b) 0.00000179

$= 0\ 0_{\wedge}0_{\wedge}0_{\wedge}0_{\wedge}0_{\wedge}1.79$ — Move the decimal point six places to the right so that the resulting number is 1.79 (a number between 1 and 10).

$= 1.79 \times 10^{-6}$ — Write the number with the decimal point in the new location. Multiply by 10 to the negative exponent 6.

$1.79 \times 10^{-6} = 1.79 \times 0.000001$ — Check the answer by writing the number in standard form. ✓ It checks.

$= 0.00000179$

EXAMPLE 8: Calculators display scientific notation in different ways. The TI-83 calculator displays E and a number to denote the power of 10.

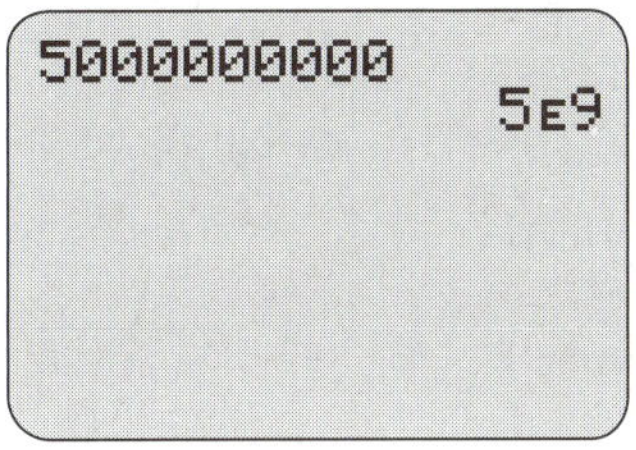

$5{,}000{,}000{,}000 = 5 \times 10^9$

$0.000027 = 2.7 \times 10^{-5}$

MULTIPLICATION AND DIVISION OF EXPRESSIONS INVOLVING EXPONENTS

Multiplication of LIKE bases with integer exponent is generalized in the following Property:

> **MULTIPLICATION PROPERTY OF EXPONENTS**
> For any real number a, and integers m and n: $a^m \cdot a^n = a^{m+n}$

EXAMPLE 9: Multiply.

(a) $6^3 \cdot 6^4 = 6^{3+4} = 6^7 = 279{,}936$ — The bases are ALIKE. Add exponents.

(b) $10^{-3} \cdot 10^8 = 10^{-3+8} = 10^5 = 100{,}000$ — The bases are ALIKE. Add exponents.

➢ ***Watch out!*** *Exponents can be added ONLY if the bases are identical and the operation is multiplication.*

EXAMPLE 10: Perform the operations.

(a) $10^4 + 10^5 =$ — Do *not* add exponents.

$\underbrace{10 \cdot 10 \cdot 10 \cdot 10}_{10,000} + \underbrace{10 \cdot 10 \cdot 10 \cdot 10 \cdot 10}_{100,000} =$ — Multiply each term.

$10,000 + 100,000 = 110,000$ — Add numbers.

(b) $2^3 \cdot 7^4 =$ — The bases are *not* identical.

$2 \cdot 2 \cdot 2 \cdot 7 \cdot 7 \cdot 7 \cdot 7 = 8 \cdot 2401 = 19,208$ — Do *not* add exponents. Multiply factors.

Division of LIKE bases with integer exponent is generalized in the following Property:

DIVISION PROPERTY OF EXPONENTS

For any real number a $(a \neq 0)$ and integers m and n: $\dfrac{a^m}{a^n} = a^{m-n}$

EXAMPLE 11: Divide.

$\dfrac{4^6}{4^5} = 4^{6-5} = 4^1$ — Subtract exponents.

➢ ***Watch out!*** *The exponents can be subtracted ONLY if the bases are identical and the operation is division.*

EXAMPLE 12: Perform the operations.

(a) $10^4 - 10^3 = 10 \cdot 10 \cdot 10 \cdot 10 - 10 \cdot 10 \cdot 10$ — Multiply each term.

$= 10,000 - 1000$ — Subtract.

$= 9000$

(b) $\dfrac{10^5}{2^3} = \dfrac{10 \cdot 10 \cdot 10 \cdot 10 \cdot 10}{2 \cdot 2 \cdot 2}$ — Do *not* subtract exponents since the bases are different.

$= \dfrac{100,000}{8} = 12,500$ — Multiply in the numerator and denominator.

PERFORMING OPERATIONS WITH NUMBERS IN SCIENTIFIC NOTATION

It may be necessary to perform operations with numbers written in scientific notation. Use the Commutative and Associative Properties to rearrange and group decimals together and powers of 10 together to make calculations easier.

EXAMPLE 13: Multiply.

$$(5\cdot 3 \times 10^2)\cdot(1.2 \times 10^5) = (5.3\cdot 1.2) \times (10^2\cdot 10^5)$$

Write decimals together and powers of 10 together.

$$= (6.36) \times (10^{2+5})$$

Multiply decimals. Multiply powers of 10 by adding exponents.

$$= 6.36 \times 10^7 = 63{,}600{,}000$$

Write in standard form, if requested.

EXAMPLE 14: Divide.

$$\frac{(2.5 \times 10^{-5})}{(0.5 \times 10^2)} = \frac{2.5}{0.5} \times \frac{10^{-5}}{10^2}$$

Write the quotient of the decimals in one fraction and the quotient of the powers of 10 in another fraction.

$$= \frac{\overset{5}{\cancel{2.5}}}{\underset{1}{\cancel{0.5}}} \times 10^{-5-2}$$

Divide decimals. Divide powers of 10 by subtracting exponents.

$$= 5 \times 10^{-7} = 0.0000005$$

Write in standard form, if requested.

EXAMPLE 15: Perform the operations.

$$\frac{(25{,}000{,}000)(420{,}000{,}000)}{(5{,}000)(7{,}000{,}000)}$$

$$= \frac{(2.5 \times 10^7)}{(5 \times 10^3)}\cdot\frac{(4.2 \times 10^8)}{(7 \times 10^6)}$$

Write each number in scientific notation.

$$= \frac{2.5 \times 4.2 \times 10^7 \times 10^8}{5 \times 7 \times 10^6 \times 10^3}$$

Write decimal factors and powers of 10 together in the numerator and denominator.

$$= \frac{\overset{0.5}{\cancel{2.5}} \times \overset{0.6}{\cancel{4.2}} \times 10^{7+8}}{\underset{1}{\cancel{5}} \times \underset{1}{\cancel{7}} \times 10^{6+3}}$$

Divide out common factors in decimal factors.

$$= \frac{0.5 \times 0.6 \times 10^{15}}{1 \times 1 \times 10^9} = \frac{0.300 \times 10^{15}}{1 \times 10^9}$$

Multiply powers of 10 by adding exponents in the numerator and denominator. Multiply remaining decimals.

$$= \frac{0.30 \times 10^{15-9}}{1}$$

Divide powers of 10 by subtracting exponents.

$$= 0.30 \times 10^6 = 300{,}000$$

Write in standard form.

1 2 3

Doing Mathematics

Problems 1–4: Write each expression in expanded factor notation and standard form.

Exponential notation	Expanded factor notation	Standard form
1. 10^3		
2. 4^5		
3. 10^{-5}		
4. 6^{-1}		

Problems 4–8: Write each measurement in standard form.[2]

5. Typical altitude of a satellite orbiting the earth (meters): 2×10^5

6. Radius of the earth (meters): 6×10^6

7. Approximate mass of a frog (kilograms): 1×10^{-1}

8. Size of cells of most living organisms (meters): 1×10^{-5}

Problems 9–12: A "triple point" is a unique temperature and pressure point where the solid, liquid, and vapor state of a substance exist in equilibrium measured in Newton/meters^2.

Convert each number from scientific notation to standard form.

	Substance	Triple point[3]	Standard form
9.	Carbon Dioxide	5.16×10^5	
10.	Ammonia	6.06×10^3	
11.	Nitrogen	1.25×10^4	
12.	Oxygen	1.52×10^2	

Problems 13–16: The heat conduction of a substance is affected by its "thermal conductivity constant," measured in $kcal/s \cdot m \cdot C°$. Write each constant in standard form.

	Substance	Thermal conductivity[4]	Standard form
13.	Copper	9.2×10^{-2}	
14.	Glass	1.9×10^{-4}	
15.	Water	1.4×10^{-4}	
16.	Air	5.5×10^{-6}	

[2]Serway and Faughn, p. 4.

[3]Giancoli, *Physics*, p. 388.

[4]Giancoli, *Physics*, p. 412; Beiser, *Physics*, p. 362.

Problems 17–32: Write each number in scientific notation.

17. 3,500,000

18. 1400

19. 125

20. 451,000

21. The approximate age of a college student in seconds:[5] 600,000,000

22. The number of minutes in one year: 525,600

23. The length of a football field (in yards): 100

24. The number of pennies in $1000: 100,000

25. 0.00021

26. 0.000000825

27. 0.00000764

28. 0.00392

29. The approximate mass of a bacterium (in kg):[6] 0.000000000000001

30. Number of miles in one kilometer: 0.621

31. The number of seconds between normal heartbeats:[7] 0.08

32. The number of seconds in a period of an audible sound wave:[8] 0.01

33. Pierce multiplied 3,500,000 and 12,500 on his graphing calculator with this display.

Write the calculator answer in standard form.

34. Susan multiplied 0.00000015 and 0.000028 on her graphing calculator with this display.

Write the calculator answer in standard form.

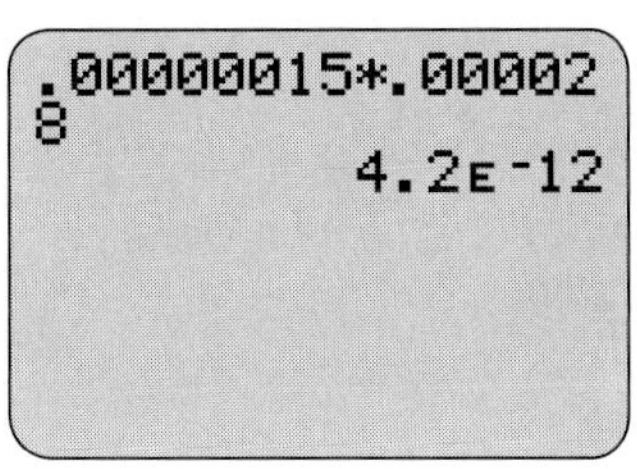

[5]Serway and Faughn, p. 5.
[6]Serway and Faughn, p. 5.
[7]Serway and Faughn, p. 5.
[8]Serway and Faughn, p. 5.

Problems 35–46: Perform the operations. Write the answers in exponential form with positive exponents.

35. $10^2 \cdot 10^7$

36. $10^3 \cdot 10^4$

37. $10^{-5} \cdot 10^{-3}$

38. $10^{-12} \cdot 10^{-5}$

39. $5^8 \cdot 5^6$

40. $2^8 \cdot 2^6$

41. $\dfrac{10^8}{10^3}$

42. $\dfrac{10^{15}}{10^5}$

43. $\dfrac{10^4}{10^7}$

44. $\dfrac{10^7}{10^1}$

45. $\dfrac{4^{10}}{4^7}$

46. $\dfrac{3^8}{3^6}$

Problems 47–54: Perform the operations. Write answers in scientific notation. Confirm answers using a calculator.

47. $(4.5 \times 10^2) \cdot (0.3 \times 10^4)$

48. $(9.1 \times 10^3) \cdot (0.3 \times 10^2)$

49. $(2.21 \times 10^2) \cdot (11 \times 10^4)$

50. $(9 \times 10^{11}) \cdot (6 \times 10^6)$

51. $\dfrac{(22 \times 10^8)}{(11 \times 10^4)}$

52. $\dfrac{(3.6 \times 10^{14})}{(1.8 \times 10^6)}$

53. $(0.55 \times 10^4) \div (1.1 \times 10^2)$

54. $(9.6 \times 10^7) \div (0.8 \times 10^3)$

Problems 55–57: Convert each number to scientific notation and perform the operations. Write answers in scientific notation. Confirm answers using a calculator.

55. (1,250,000,000)(5,000,000)

56. $\dfrac{(14{,}000)(28{,}000{,}000)}{(7000)(40{,}000)}$

57. $\dfrac{(1{,}250{,}000{,}000)(5{,}000{,}000)}{(250{,}000)(50{,}000)}$

58. Find a problem situation in a course other than mathematics in which scientific notation is used.

Answers to Reviewing—Chapter R

SECTION R.1

1. (a) $45.652 = 4\cdot 10 + 5\cdot 1 + 6\cdot\frac{1}{10} + 5\cdot\frac{1}{100} + 2\cdot\frac{1}{1000} = 40 + 5 + \frac{6}{10} + \frac{5}{100} + \frac{2}{1000}$

(b) $1007.84 = 1\cdot 1000 + 7\cdot 1 + 8\cdot\frac{1}{10} + 4\cdot\frac{1}{100} = 1000 + 7 + \frac{8}{10} + \frac{4}{100}$

2. Sandi's computer software program rounded to the *nearest dollar*. If the number of cents was less than 50, the cents were dropped. The entry was rounded DOWN. If the number of cents was greater than or equal to 50, the dollar amount was rounded UP (increased by 1) and the cents were dropped.

3. (a) 1570

(b) 1568

(c) 1568.4

(d) 1568.36

4. (a) 3.14×5 is closer to 15 than to 150. To perform this operation in your head, round 3.14 to the whole number 3 and multiply by 5 $(3 \times 5 = 15)$. The estimated answer is 15 (exact answer is 15.7).

(b) $\frac{156}{5}$ is closer to 30 than to 3. To perform this operation in your head, round 156 to 150 and divide by 5: $\left(\frac{150}{5} = 30\right)$. The estimated answer is 30 (exact answer is 31.2).

(c) $53.576 + 47.999$ is closer to 100 than to 1000. Round 53.576 to 50 and round 47.999 to 50 and add $50 + 50$ in your head to get 100. The exact answer is 101.575.

(d) $\sqrt{50}$ is closer to 7 than to 10. The radicand, 50, is very close to the perfect square 49. The square root of 49 is 7. The answer is 7.07106....

5. (a) $5 + 2\cdot 3 = 5 + 6 = 11$

(b) $40 \div 5\cdot(6 - 4)^2 = 40 \div 5\cdot(2)^2 = 40 \div 5\cdot 4 = 8\cdot 4 = 32$

(c) $6 + 2\cdot 7 - 10 \div 2 = 6 + 14 - 5 = 20 - 5 = 15$

SECTION R.2

1. (a) $\frac{5}{8}$

(b) $\frac{5}{9}$

2. (a) $\frac{5}{7} = \frac{15}{21}$

(b) $\frac{6}{11} = \frac{24}{44}$

3. (a) $5\frac{2}{5}$

(b) $1\frac{3}{4}$

4. (a) $\frac{19}{5}$

(b) $\frac{41}{4}$

5. (a) 14

(b) $\frac{4}{5}$

6. (a) 18

(b) $\frac{8}{5} = 1\frac{3}{5}$

7. (a) 3

(b) $\frac{5}{7}$

8. (a)
$$9 = 3\cdot 3$$
$$15 = \quad 3\cdot 5$$

(b) $\text{LCD} = 3\cdot 3\cdot 5 = 45$

(c) $\frac{4\cdot 5}{9\cdot 5} + \frac{7\cdot 3}{15\cdot 3} = \frac{20}{45} + \frac{21}{45} = \frac{41}{45}$

9.
$$10 = \qquad 2\cdot 5$$
$$16 = 2\cdot 2\cdot 2\cdot 2$$
$$\text{LCD} = 2\cdot 2\cdot 2\cdot 2\cdot 5 = 80$$

$$\frac{3\cdot 8}{10\cdot 8} - \frac{1\cdot 5}{16\cdot 5} = \frac{24}{80} - \frac{5}{80} = \frac{19}{80}$$

SECTION R.3

1. (a) 0.75; terminating decimal

   (b) $0.\overline{5}$; repeating decimal

2. 10%

3. (a) 0.45

   (b) 0.02

4. (a) 89%

   (b) 4%

5. (a) 75%

   (b) 80%

6.

Activity on account	Account balance
Balance brought forward	\$925.17
Check written for apartment rent of \$575.00	\$350.17
Deposit of paycheck in the amount of \$478.87	\$829.04
Check written for gas in the amount of \$12.35	\$816.69
Check written for car payment in the amount of \$106.37	\$710.32
Check written for groceries in the amount of \$43.78	\$666.54
Deposit of paycheck in the amount of \$478.87	\$1145.41 Ending balance

7. (a) 3481.322

   (b) 2925.78237

8. (a) 5.12

   (b) 5150

SECTION R.4

1. (a) $\frac{\text{Yards gained rushing by Patriots}}{\text{Yards gained rushing by Eagles}} = \frac{2134}{1639}$

   (b) $\frac{2134}{1639} \approx \frac{1.30}{1}$ The ratio tells us that the Patriots' total yards gained rushing was greater than the Eagles'. The Patriots gained 1.30 yards for each yard gained by the Eagles.

   (c) $\frac{\text{First downs earned rushing by Patriots}}{\text{First downs earned rushing by Eagles}} = \frac{120}{87}$

   (d) $\frac{120}{87} \approx \frac{1.38}{1}$ The ratio tells us that the Patriots' had more first downs earned rushing than the Eagles'. The Patriots had 1.38 first downs for each first down of the Eagles.

   (e) The given statistics and ratios do not provide enough information to determine which team had the better offense. However, based on total net yards and first downs earned rushing, the Patriots had a stronger record.

2. (a) $x = 30$

   (b) $x = 28$

SECTION R.5

1.

	Expanded factor notation		Standard form
(a)	$10 \cdot 10 \cdot 10 \cdot 10$	=	10,000
(b)	$10 \cdot 10 \cdot 10 \cdot 10 \cdot 10 \cdot 10$	=	1,000,000
(c)	$\frac{1}{10 \cdot 10}$	=	0.01
(d)	$\frac{1}{10 \cdot 10 \cdot 10 \cdot 10}$	=	0.0001

2.

	Standard form
(a)	6,060,000
(b)	58,500,000
(c)	0.0001
(d)	0.00001

3. (a) 5.6×10^6

   (b) 3.25×10^{-3}

   (c) 1.23×10^3

   (d) 1.98×10^{-5}

4. (a) $10^{2+4} = 10^6 = 1{,}000{,}000$

   (b) $10^{3+2} = 10^5 = 100{,}000$

5. (a) $10^{4-3} = 10^1 = 10$

   (b) $10^{6-4} = 10^2 = 100$

APPENDIX

A

Dictionary

Section 1.1

Base:

Equation:

Exponent:

Radicand:

Real number line:

Two uses for "=" symbol:

1.

2.

Two uses for "+" symbol:

1.

2.

Three uses for “$-$” symbol:

1.

2.

3.

“$<$” symbol:

“$>$” symbol:

“$\leq$” symbol:

“$\geq$” symbol:

“$=$” symbol:

“$\neq$” symbol:

“$\approx$” symbol:

“$-\infty$ and $+\infty$” symbol:

Section 1.2

Additive Identity:

Associative Property of Addition:

Associative Property of Multiplication:

Commutative Property of Addition:

Commutative Property of Multiplication:

Counting numbers (natural numbers) (C):

Distributive Property of Multiplication over Addition (or Subtraction):

Elements (or members) of a set:

Ellipsis:

Finite set:

Identity Property of Addition:

Identity Property of Multiplication:

Infinite set:

Integers (I):

Irrational numbers (P):

Multiplicative Identity:

Order of Operations:

Rational numbers (Q):

Real numbers:

Repeating decimal:

Repetend:

Set of numbers:

Subset:

Terminating decimal:

Variables:

Vinculum:

Whole numbers (W):

Section 1.3

Absolute value:

Addends:

Difference:

Minuend:

Signed numbers:

Subtrahend:

Section 1.4

Dividend:

Divisor:

Quotient:

Section 1.5

Polya's Problem-Solving Method:

I.

II.

III.

IV.

List three problem-solving strategies:

1.

2.

3.

Section 2.1

Ascending order:

Binomial:

Constant:

Degree of a polynomial:

Descending order:

Factors:

Monomial:

Numerical coefficient:

Polynomial:

Term:

Trinomial:

Section 2.2

Collecting LIKE Terms:

LIKE Terms:

Section 2.3

Write an example of the "scale method" for solving equations:

Write an example of the "symbolic method" for solving equations:

Equality Principles of Equations:

Addition Principle:

Subtraction Principle:

Multiplication Principle:

Division Principle:

Solving an equation:

Section 2.4

Addition (and subtraction) Principle of Inequalities:

Closed dots or brackets:

Division Principle of Inequalities:

Interval notation:

Multiplication Principle of Inequalities:

Open dots or parentheses:

Reminder square:

Set-builder notation:

Section 2.5

Hypotenuse:

Literal equation:

Pythagorean Theorem:

Right triangle:

Section 3.1

Cartesian plane:

First coordinate:

Graphing an equation:

Ordered pair:

Origin:

Plotting an ordered pair:

Quadrants:

Second coordinate:

x-axis:

y-axis:

XMIN:

XMAX:

YMIN:

YMAX:

Section 3.2

Dependent variable:

Evaluating:

Function:

Function notation:

Independent variable:

Input value:

Output value:

Relation:

Solving:

Vertical line test:

Section 3.3

Function form of a horizontal line:

General form of a vertical line:

Linear function:

Linear relation:

x-intercept:

y-intercept:

Zero of a function (or relation):

List the steps involved in graphing a function using the Intercept Method:

1.

2.

3.

4.

5.

Section 3.4

Horizontal displacement:

Ratio:

Slope Formula:

Slope of a decreasing line:

Slope of a horizontal line:

Slope of an increasing line:

Slope of a vertical line:

Slope-Intercept form of a line:

Vertical displacement:

Section 3.5

General Form of a linear equation:

Point-Slope Formula of a line:

Section 4.1

Intersection-of-Graphs Graphical Method of solving an equation:

X-Intercept Graphical Method of solving an equation:

Zero of a relation or function:

Section 4.2

Consistent linear system:

Dependent linear system:

Inconsistent linear system:

Solution of a system of equations:

System of equations:

Section 4.3

List the steps in the Method of Substitution for solving a linear system:

1.

2.

3.

4.

5.

List the steps in the Method of Elimination for solving a linear system:

1.

2.

3.

4.

Section 4.4

Linear inequality in two variables:

List the steps in graphing the solution of a linear inequality in two variables:

1.

2.

3.

4.

5.

Solution of an inequality in two variables:

Section 5.1

Expanded form:

Exponential form:

Properties of Exponents:

I.

II.

III.

IV(A).

IV(B).

V.

VI.

Section 5.2

Properties of Exponents:

VII(A).

VII(B).

Reciprocal:

Section 5.3

What do the letters in FOIL represent when multiplying two binomials?

Section 5.4

Formula for the product of the sum and difference of two terms:

Formulas for the square of a binomial (sum or difference):

Middle term:

Section 5.5

None

Section 6.1

Factored form:

Factoring:

Greatest common factor (GCF):

Section 6.2

An example of a perfect square trinomial:

An example of the difference of two squares:

Formula for factoring a perfect square trinomial:

Formula for factoring the difference of two squares:

Section 6.3

An example of a trinomial of the form $ax^2 + bx + c$ $(a \neq 1)$:

Section 6.4

Degree of a polynomial equation:

Zero-Product Property:

Zeros of a polynomial:

Section 6.5

Formula for factoring the sum of two cubes:

Formula for factoring the difference of two cubes:

List the steps in the Multi-Step Factoring Strategy:

1.

2.

3.

4.

Section 7.1

Building a Fraction:

Extremes-Means Property of Proportions:

Greatest Common Factor:

Proportion:

Rate:

Ratio:

Rational Expression:

Reducing the Fraction:

Restricted values:

Section 7.2

List the steps involved in multiplying rational expressions:

1.

2.

List of steps involved in dividing rational expressions:

1.

2.

3.

4.

5.

Section 7.3

Least Common Denominator (LCD):

List the steps in the process for finding the LCD of the denominators of fractions:

1.

2.

3.

Write the generalization for adding or subtracting rational expressions with denominators that are the *same*:

Sum:

Difference:

List the steps in the process for adding or subtracting rational expressions with denominators that are *different*:

1.

2.

3.

4.

Section 7.4

Horizontal asymptote:

List the steps for graphing a rational function:

1.

2.

3.

4.

5.

6.

7.

Rational function:

Vertical asymptote:

Section 7.5

Clearing Fractions:

List the steps for solving a rational equation by building fractions to the LCD:

List the steps for solving a rational equation by clearing fractions:

Rational equation:

Section 7.6

Complex Fraction:

List the steps for simplifying a complex fraction by performing the operations and multiplying by the reciprocal of the divisor (Method 1):

1.

2.

3.

4.

List the steps for simplifying a complex fraction by multiplying by the LCD (Method 2):

1.

2.

3.

4.

Section 8.1

Cube root:

Fourth root:

Index of a radical:

Irrational root:

Negative n^{th} root when n is any positive even integer:

Negative square root: n^{th} root when n is any positive odd integer:

Perfect cube:

Perfect fourth power:

Perfect square:

Principal n^{th} root when n is any positive even integer:

Principal square root:

Radical symbol:

Radicand:

Section 8.2

Product Rule for Radicals:

Quotient Rule for Radicals:

Rationalizing the denominator:

Simplifying a radical:

Section 8.3

Conjugate:

LIKE Radicals:

List the steps for adding and subtracting terms containing radicals:

1.

2.

3.

Write an example of an expression with UNLIKE radicals:

APPENDIX

B

Measures of Central Tendency

OBJECTIVES

- Find the mean, median, and mode of a set of data and draw conclusions.

1 2 3

Getting Started

Mathematics is more than performing operations or solving equations. It involves collecting data, organizing it, thinking about it, finding patterns, and drawing conclusions. In this section, we discuss **measures of central tendency**, which include:

- The **mean** is the average of a set of numbers (also called the arithmetic mean).
- The **median** is the middle number of a set of ordered numbers.
- The **mode** is the number(s) or data point(s) that occurs most frequently.

1. Consider the following data set listing the test scores from an Introductory Algebra class.

Test scores				
58	71	95	71	99
70	93	85	81	89
63	87	79	71	61

(a) To calculate the mean, add all the test scores together and divide the sum by the total number of scores. Find the mean of the test scores.

$$\frac{\text{Sum of all scores}}{\text{Total number of scores}} = \frac{\quad}{15}$$

(b) To find the median arrange the test scores from lowest to highest.

The median is the middle score (such that there is an equal number of data items below and above that score). Find the median.

(c) Find the most frequent score, the mode.

(d) What do the values of the mean, median, and mode tell you about these test scores?

Test scores	

2. Rhonda was excited to accept a job as a luggage inspector. She had been told that the average weight of the luggage she would have to lift was 35.5 pounds and she could lift around 36 pounds. Here are the weights (in pounds) of the first 10 pieces of luggage that Rhonda had to lift: 48, 15, 48, 20, 18, 48, 48, 16, 45, 49.

(a) Find the mean, median, and mode.

(b) Which measure of central tendency would have been the best descriptor of the weights of the luggage for Rhonda?

(c) If she had been given that measure, do you think Rhonda should have accepted the job? Why or why not?

Discussing Mathematics

MEAN, MEDIAN, AND MODE

Measures of central tendency, the mean, median, and mode, are useful descriptors of a data set.

FINDING THE MEAN, MEDIAN, AND MODE

- To calculate the **mean**: find the sum of all elements in the data set and divide by the total number of elements in the data set.
- To calculate the **median**: arrange the data in either ascending or descending order. Identify the "middle" number of the data set, such that there is an equal number of data points below and above that number. If *one* number does not appear in the middle, find the average of the two middle data points.
- The **mode** is the most frequent data point(s). There may be one mode, more than one mode, or no mode.

EXAMPLE 1: The weights of the players on the men's basketball team at Smart College are listed in this table.

Player	1	2	3	4	5	6	7	8	9	10
Weight (pounds)	200	189	202	176	198	193	210	202	189	208

(a) Calculate the mean.

$$\frac{200 + 189 + 202 + 176 + 198 + 193 + 210 + 202 + 189 + 208}{10} = \frac{1967}{10} = 196.7$$

(b) Calculate the median.

← 5 numbers → ← 5 numbers →

176 189 189 193 198 200 202 202 208 210

The median is the middle data point.

A data point exactly in the middle of this data set does not exist. The median is the average of 198 and 200: median $= \frac{198 + 200}{2} = 199$.

(c) 189 and 202 appear twice in the data. In this case, the data set is called **bimodal**. Both 189 and 202 are modes of the data set.

(d) The mean, median, and mode in this problem are close to one another. One measure does not provide much more information than the other.

➢ *When the mean and median are not close to the same value, use caution when drawing conclusions about a data set.*

EXAMPLE 2: Brandon can afford a house that costs no more than \$86,000. He read in the paper that the average price of a house (mean) in a certain community was \$99,000. He concluded that he should not consider homes in that community. Did Brandon make a wise decision?

Brandon should ask more questions before making his decision. Consider this list of home prices:

Home prices:		Measures of Central Tendency:
\$75,000	\$86,000	Mean: \$99,000
\$80,000	\$88,000	Median: \$86,000
\$83,000	\$200,000	Mode: None
\$85,000		

The mean was much higher than what Brandon could afford, but the median was below his price level. Therefore, at least half of the homes in that community were affordable. Brandon should have considered the homes in that community in his price range.

1 2 3 Doing Mathematics

- Write the definitions of the words and symbols listed for this section in the Dictionary in Appendix A.
- On a 3 × 5-inch card, write the processes of finding the mean, median, and mode for future reference.

1. The table lists the temperatures recorded every hour for a 24-hour period at Cleveland (Ohio) Hopkins International Airport beginning July 14, 2001.[1]

July 14, 2001: 6 P.M.	75	6 A.M.	60
7 P.M.	74	7 A.M.	63
8 P.M.	71	8 A.M.	73
9 P.M.	68	9 A.M.	77
10 P.M.	63	10 A.M.	80
11 P.M.	62	11 A.M.	82
Midnight: July 15, 2001	61	Noon	83
1 A.M.	61	1 P.M.	83
2 A.M.	59	2 P.M.	84
3 A.M.	57	3 P.M.	83
4 A.M.	56	4 P.M.	83
5 A.M.	60	5 P.M.	82

(a) Find the mean, median, and mode. Round the mean to the nearest tenth.

(b) What can you conclude from the measures of central tendency?

2. The weights (in pounds) of fifteen persons are recorded in this table.

Person	1	2	3	4	5	6	7	8	9	10	11	12	13	14	15
Weight	180	52	131	153	75	210	128	218	146	132	60	131	175	131	192

(a) Find the mean, median, and mode. Round the mean to the nearest tenth.

(b) What can you conclude from the measures of central tendency?

3. Find the mean, median, and mode of the list of test scores from an Introductory Algebra class.

(a) Find the mean, median, and mode. Round the mean to the nearest tenth.

(b) What can you conclude from the measures of central tendency?

Test scores		
52	76	86
61	78	87
65	79	89
68	81	92
73	82	94
75	85	98

[1]"The Plain Dealer" (newspaper), July 16, 2001, page B8.

4. The total number of credits taken by each student in an Algebra class during fall semester table is listed in the table.

Total number of credits taken by each student during fall semester									
12	10	9	4	9	10	12	11	6	12
9	7	5	12	3	11	12	8	12	8

(a) Find the mean median, and mode.

(b) What can you conclude from the measures of central tendency?

5. Record the age (in years) of eleven adults (21 years and older).

Person	1	2	3	4	5	6	7	8	9	10	11
Age											

(a) Find the mean, median, and mode.

(b) What can you conclude from the measures of central tendency?

6. Record the height (in inches) of thirteen people.

Person	1	2	3	4	5	6	7	8	9	10	11	12	13
Height													

(a) Find the mean, median, and mode.

(b) What can you conclude from the measures of central tendency?

7. Leslie and Dave want to buy a home that costs no more than $150,000. Their realtor informed them that the average price of homes in a certain community was $175,000 with a median of $140,000. Should they consider looking at homes in that community? Why or why not?

8. Ignacio is looking for a new job. The mean and median salaries at three companies are shown in the table. Which job should he consider first if he hopes to earn at least $55,000 per year? Why?

Company	CDs4U	VideoesRUs	Movies2See
Mean	$50,000	$43,000	$40,000
Median	$42,000	$43,000	$60,000

9. Use the Internet or newspaper to find a data set of an event, survey, or other information.

Record your data here:

(a) Find the mean, median, and mode.

(b) What can you conclude from the measures of central tendency?

Problems 10–18: Writing Exercises

10. If the mean is the same for two data sets, are the two data sets identical? Why or why not?

11. If two students earn a perfect score of 100 on a test in a class of 30 students, how do the two perfect scores influence the value of the mean, median, and mode?

12. Which measure of central tendency is most affected by one very large or very small value in the data set?

13. What can you conclude about a data set where the mean is larger than the median?

14. What can you conclude about a data set where the median is larger than the mean?

15. Why are the mean, median, and mode called measures of central tendency?

16. Describe a real-world situation where the mean is the best descriptor of the data set.

17. Describe a real-world situation where the median is the best descriptor of the data set.

18. Describe a real-world situation where the mode is the best descriptor of the data set.

Answers to Getting Started—Appendix B

1. (a) Mean: 78.2

(b) Median: 79

Test scores	
58	81
61	85
63	87
70	89
71	93
71	95
71	99
79	

(c) Mode: 71

(d) The average test score was 78.2. There were an equal number of test scores above and below 79. The most frequent score was 71 (three people received 71).

2. (a) Mean: 35.5; Median: 46.5; Mode: 48.

(b) The median would have been a better descriptor for Rhonda. She would then know that half of the pieces of luggage would be more than 46.5 pounds.

(c) No, if Rhonda knew the median was 46.5, she would have realized that half of the luggage she would have to lift would be more than 36 pounds.

APPENDIX

C

Books to Improve Study Skills

Burns, M. (1998). *Math: Facing an American Phobia*. Sausalito, CA: Math Solutions Publications.

Davidson, R., and Levitov, E. (1993). *Overcoming Math Anxiety*. New York: Harper Collins College Publishers.

Mitchell, C. (1987). *Math Anxiety: What it is and What to do about it*. Tempe, AZ: Action Press.

Nolting, P. D. (1997). *Winning at Math: Your Guide to Learning Mathematics Through Successful Study Skills*. Bradenton, FL: Academic Success Press.

Polya, G. (1973). *How to Solve It: A New Aspect of Mathematical Method*. Princeton, NJ: Princeton University Press.

Ruedy, E. and Nirenberg, S. (1990). *Where Do I Put the Decimal Point?* New York: Barnes and Noble Books.

Smith, R. M. (1998). *Mastering Mathematics: How to Be a GREAT Math Student*. Pacific Grove, CA: Brooks/Cole Publishing Company.

Tobias, S. (1995). *Overcoming Math Anxiety*. New York: W. W. Norton.

Answers to Exercises and Tests

ARITHMETIC REVIEW SKILLS TEST ANSWERS

1. b **2.** b **3.** a **4.** b **5.** a **6.** c **7.** a **8.** a **9.** c **10.** d **11.** d **12.** c **13.** b **14.** c **15.** c **16.** b **17.** c **18.** d **19.** a **20.** b **21.** c **22.** c **23.** b **24.** d **25.** c **26.** b **27.** a **28.** b **29.** c **30.** d

CHAPTER 1

DOING MATHEMATICS SECTION 1.1

1. Answers will vary: $10 \div 5 \times 3 + 7 - 3 = 10$
2. Answers will vary: $5 \times 8 \div \boxed{10} + 3 + 3 = 10$
3. = **4.** > **5.** < **6.** = **7.** < **8.** < **9.** > **10.** < **11.** = **12.** = **13.** = **14.** > **15.** > **16.** < **17.** < **18.** < **19.** > **20.** < **21.** > **22.** > **23.** < **24.** > **25.** = **26.** = **27.** = **28.** = **29.**

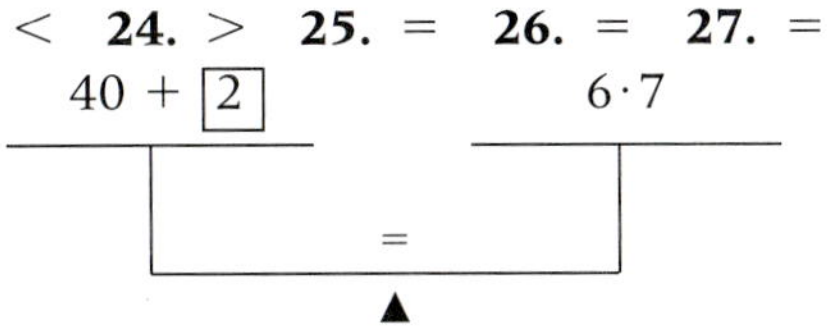

30.

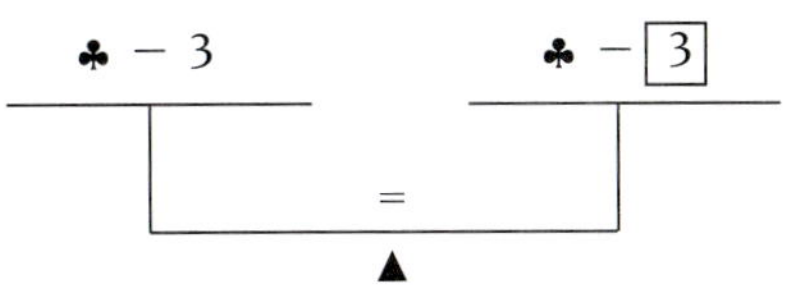

31.

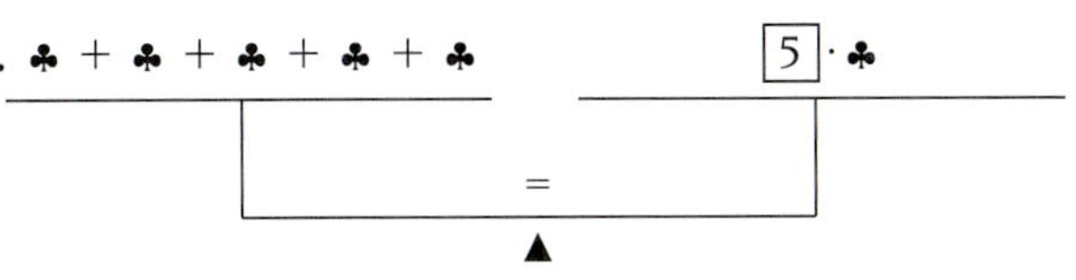

32.

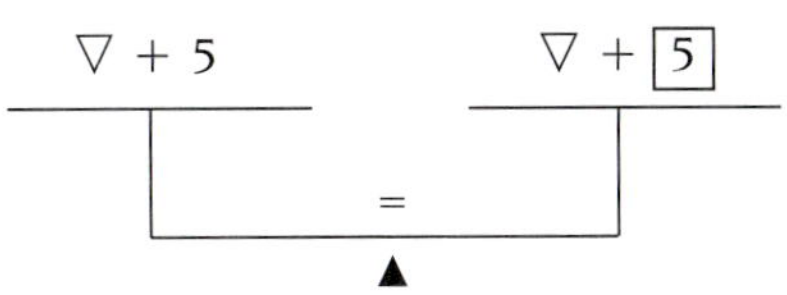

33.

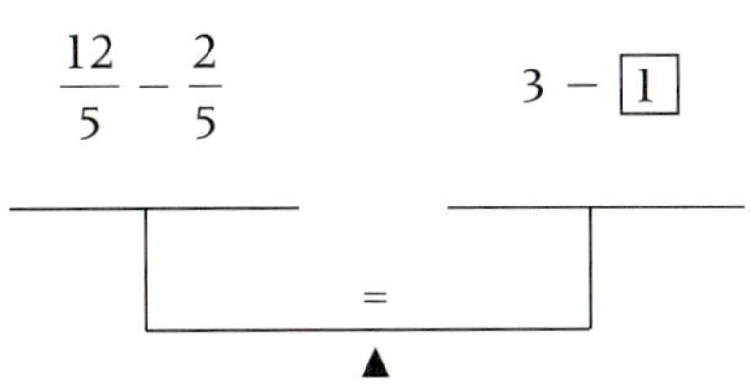

34. 4 **35.** 2 **36.** 3 **37.** 5 **38.** 3 **39.** $\frac{1}{2}$
40. Answers will vary: (a) $2 < 3$ (b) $5 > 3$ (c) $6 \le 10$ (d) $7 \ge 2$ **41.** (a) P (b) A (c) P (d) A (e) Answers will vary: A + symbol included next to a number within parentheses denotes the number's positive value. A + symbol between two values is an instruction to add. **42.** (a) S (b) N (c) N (d) S (e) Answers will vary: A − symbol included next to a number within parentheses or next to a number that is being divided denotes the number's negative value. A − symbol between two numbers is an instruction to subtract. **43.** The − key is the subtraction key. The (−) key is the negative key. **44.** $5 \cdot 5 = 25$
45. $6 \cdot 6 \cdot 6 = 216$ **46.** $3.5 \cdot 3.5 \cdot 3.5 = 42.875$
47. $2 \cdot 2 \cdot 2 \cdot 2 \cdot 2 = 32$ **48.** $3 \cdot 3 \cdot 3 = 27$
49. $2.3 \cdot 2.3 = 5.29$

DOING MATHEMATICS SECTION 1.2

1. > **2.** > **3.** < **4.** < **5.** > **6.** < **7.** < **8.** > **9.** W, I, Q **10.** P **11.** C, W, I, Q **12.** Q **13.** Q **14.** P **15.** C, W, I, Q **16.** I, Q **17.** P **18.** Q **19.** Q **20.** Q **21.** Q **22.** Q
23. (a) 4.582575695... (b) −10.34408043... (c) 1.054092553... (d) 0.7071067812...
24. (a) 7.416198487... (b) −4.123105626... (c) 0.7824607964... (d) 0.4082482905...
25. True **26.** False: All rational numbers are real numbers. **27.** False: Repeating and terminating decimals are rational numbers. **28.** True
29. False: Every rational number can be written as a repeating or terminating decimal. **30.** True
31. False: The number 1 is the multiplicative identity. **32.** False: The set of whole numbers is a subset of the rational numbers. **33.** $6 + 51$
34. $b \cdot a$ **35.** $\frac{1}{3} \cdot \frac{17}{4}$ **36.** $\frac{5}{2} + \frac{4}{3}$ **37.** $y + x$
38. $42 \cdot 125$ **39.** $(2 + 3) + 5$ **40.** $(9 \cdot 5) \cdot 7$
41. $(a \cdot b) \cdot c$ **42.** $x + (y + z)$ **43.** $7 \cdot (6 \cdot 3)$
44. $(5.32 + 7.67) + 9.11$ **45.** $10 \cdot 2 + 10 \cdot 6$
46. $7 \cdot 5 - 7 \cdot 2$ **47.** $15 \cdot 2 + 78 \cdot 2$ **48.** $\frac{2}{3} \cdot 9 - \frac{2}{3} \cdot \frac{6}{5}$
49. $a \cdot b + a \cdot c$ **50.** $y \cdot x - z \cdot x$ **51.** $2(5 + 3)$
52. $(b - c)a$ **53.** $537(6 - 2)$
54. $1.5(2.35 + 6.43)$ **55.** $\frac{1}{2}(x - y)$ **56.** $x(y + z)$
57. Commutative—Addition **58.** Commutative—Addition **59.** Distributive—Addition
60. Associative—Addition **61.** Associative—Multiplication **62.** Identity—Multiplication
63. Commutative—Multiplication
64. Commutative—Multiplication
65. Distributive—Addition **66.** Distributive—Subtraction **67.** Distributive—Subtraction

68. Identity—Multiplication **69.** Commutative—Addition **70.** Associative—Addition
71. Identity—Addition **72.** Distributive—Addition
73. $6.59 + 20.8 = 27.39$

74. $\frac{4}{49} + \frac{45}{49} = \frac{49}{49} = 1$

75. $8 \cdot 2 + 7$
$16 + 7 = 23$

76. $6 + 20 \div 5$
$6 + 4 = 10$

77. $6 + 2 \cdot 2$
$6 + 4 = 10$

78. $26.01 + 10 - 3 + 2 \cdot 6$
$26.01 + 10 - 3 + 12$
$36.01 - 3 + 12$
$33.01 + 12 = 45.01$

79. $3(7) - 2^3 + 5(4)$
$3(7) - 8 + 5(4)$
$21 - 8 + 20$
$13 + 20 = 33$

80. $\frac{55}{5} = 11$

81. $16 \div 4 \cdot 2$
$4 \cdot 2 = 8$

82. $6 + 12 \div 3 \cdot 4 - 2 + 16$
$6 + 4 \cdot 4 - 2 + 16$
$6 + 16 - 2 + 16$
$22 - 2 + 16$
$20 + 16 = 36$

83. $(64 - 4) \div 10$
$(60) \div 10 = 6$

84. $(2)^2 \div 2 \cdot 5$
$4 \div 2 \cdot 5 = 2 \cdot 5 = 10$

85. $\frac{27 - 2}{3 + 2} = \frac{25}{5} = 5$

86. $\frac{6 + 3 \cdot 2}{36 - 30} = \frac{6 + 6}{36 - 30} = \frac{12}{6} = 2$

87. $100 - 2 + 3 - 50 \div 25$
$100 - 2 + 3 - 2$
$98 + 3 - 2 = 101 - 2 = 99$

88. $15 - 2 + 12 \div 6 - 3$
$15 - 2 + 2 - 3$
$13 + 2 - 3 = 15 - 3 = 12$

89. $2(3)^2 + 3^2 - 2^2$
$2(9) + 9 - 4$
$18 + 9 - 4$
$27 - 4 = 23$

90. $4(2)^2 + 5^2 - 3^2$
$4(4) + 25 - 9$
$16 + 25 - 9$
$41 - 9 = 32$

91. $10^2 - (3)^2 + 5$
$100 - 9 + 5 = 91 + 5 = 96$

92. $6^2 - (2)^2 + 7$
$36 - 4 + 7 = 32 + 7 = 39$

93. 2117.120199 **94.** 737.2104 **95.** (b) 20 and 8 must be added before dividing by 4. Parentheses are necessary to denote addition before division.

DOING MATHEMATICS SECTION 1.3

1.

10/1	−35.00	−35.00
10/3	−55.30	−90.30
10/4	−90.65	−180.95
10/15	−(−55.30)	−125.65
10/31	+100.00	−25.65
11/9	−120.85	−146.50
11/20	−10.50	−156.55
11/30	+100.00	−56.55

2. (a) $25 - 5 + 3(2) = 26$ (b) 26° F
3. (a) $-6 - (-16) = 10$ (b) 10° C
4. 6 **5.** −6 **6.** −2 **7.** 2 **8.** 6 **9.** −2 **10.** −6
11. 5 **12.** 7 **13.** −10 **14.** −2 **15.** 1 **16.** 2
17. −15 **18.** −22 **19.** 12.1 **20.** 8.2 **21.** 0
22. 0 **23.** 1 **24.** −4 **25.** −3.2 **26.** 6.2
27. −13 **28.** −50 **29.** 15 **30.** 17 **31.** −12
32. −13 **33.** −2 **34.** −9 **35.** 0 **36.** 65
37. −12.3 **38.** 4 **39.** −14 **40.** > **41.** >
42. > **43.** > **44.** < **45.** > **46.** > **47.** >
48. < **49.** < **50.** 6 **51.** −10 **52.** 10 **53.** −7
54. 1 **55.** Answers may vary (a) Balancing your checkbook (b) Losing weight (c) Finding temperature changes

DOING MATHEMATICS SECTION 1.4

1. 8 **2.** −8 **3.** −8 **4.** 8 **5.** 2 **6.** −2 **7.** −2
8. 2 **9.** 12 **10.** −11.172 **11.** 36 **12.** −36
13. −36 **14.** −8 **15.** $\frac{-12}{15} = \frac{-4}{5}$ **16.** −2
17. $\frac{3}{4}$ **18.** 0 **19.** $\frac{-1}{8}$ **20.** 67 **21.** −33 **22.** $-\frac{4}{3}$
23. −630 **24.** −400 **25.** 9 **26.** 0 **27.** 3
28. −16 **29.** 79 **30.** −33 **31.** 60 **32.** 2
33. −2 **34.** 7 **35.** 9/625 **36.** 0.77 **37.** −26
38. 6 **39.** −512 **40.** 5 **41.** > **42.** < **43.** >
44. < **45.** > **46.** > **47.** > **48.** < **49.** 6
50. 6 **51.** 1 **52.** 11 **53.** 13 **54.** −36 **55.** 36

DOING MATHEMATICS SECTION 1.5

1. (a) $120 (b) 1100 miles (c) Answers will vary. Mathematical expression to describe the problem situation: Cost = $45 + 0.15x$, where x = miles driven. **2.** (a) $1700 (b) 200 people (c) Answers will vary. Mathematical expression to describe the problem situation: Cost = $500 + 12x$, where x = number of persons. **3.** (a) 20 feet (b) Answers will vary: find the square root of the length squared. **4.** (a) 8 feet (b) Answers will vary: "15 times something" had to equal 120. So divide 120 by 15. **5.** (a) 66 feet (b) $610.50 (c) Answers will vary: add the lengths around the outer boundary of the store, then multiply the total by the cost per foot. **6.** (a) 4 small boxes and 6 large boxes. (b) Answers will vary. Create system of equations and solve the system: $10 = x + y$ and $64 = 8x + 4y$, where y = small boxes and x = large boxes. **7.** (a) 1000 minutes local, 750 minutes long distance (b) Answers will vary. Create system of equations and solve the system: $1750 = x + y$ and $25 = 0.01x + 0.02y$ where x = number of local calls and y = number of long distance calls. **8.** (a) $330 (b) $1000 (c) Answers will vary. Mathematical expression to describe the problem situation: Salary = $300 + 0.06x$, where x = total monthly sales. **9.** (a) $450 (b) $500 (c) Answers will vary. Mathematical expression to describe the problem situation: Salary = $250 + 0.25x$, where x = number of calls completed. **10.** (a) 64 (b) Answers will vary. Mathematical expression to describe the problem situation: $93 + 87 + 76 + x = 320$ **11.** (a) 92 (b) Answers will vary. Mathematical expression to describe the problem situation: $98 + 89 + 81 + x = 360$ **12.** (a) 2 (b) 7 (c) Answers will vary. Mathematical expression to describe the problem situation: Profit = $2000x - 4000$ where x = number of computers sold

13.

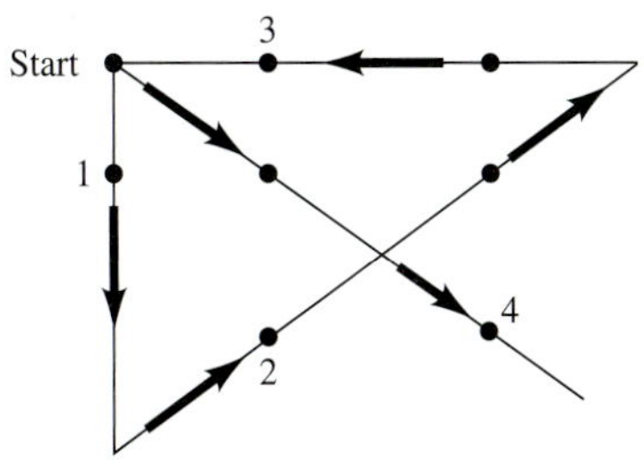

Strategy: Use the hint.

14. Answers may vary: (a) Mathematical expression (b) It was easy to see how all of the numbers in each problem were used. (c) Fewer steps are more efficient.

CHAPTER 1 REVIEW

1. Answers will vary: $6 \div 2 \times 9 + 8 + 6/2 = 2 + 2 \times 2 \times 9$
2. Answers will vary: $6 \times 6 - 2 \cdot 6 = 11 - 11 + 8 \div 2 \times 6$
3. 0.65 **4.** 3 **5.** 5 **6.** 2/7 **7.** 4 **8.** 3 **9.** 64 **10.** 625 **11.** 2.197 **12.** 243 **13.** 16 **14.** 1.4641 **15.** $<$ **16.** $>$ **17.** $<$ **18.** $<$ **19.** $>$ **20.** $<$ **21.** $>$ **22.** $>$ **23.** C, W, I, Q **24.** Q **25.** P **26.** C, W, I, Q **27.** Q **28.** Q **29.** C, W, I, Q **30.** P **31.** Q **32.** I, Q **33.** True **34.** False: All integers are real numbers. **35.** True **36.** False: Every whole number is an integer. **37.** False: Every repeating decimal is a rational number. **38.** True **39.** True **40.** True **41.** Associative Property of Addition **42.** Commutative Property of Multiplication **43.** Distributive Property over Subtraction **44.** Identity Property of Multiplication **45.** Commutative Property of Addition **46.** Associative Property of Multiplication **47.** Identity Property of Addition **48.** Distributive Property over Addition **49.** Commutative Property of Multiplication **50.** Associative Property of Addition **51.** 14 **52.** 14 **53.** 30 **54.** 34 **55.** 21 **56.** 4 **57.** 10 **58.** 37 **59.** 54 **60.** 26 **61.** 3 **62.** 4 **63.** $-\frac{7}{2}$ **64.** $\frac{31}{4}$

65. $-6 + 2 = -4$

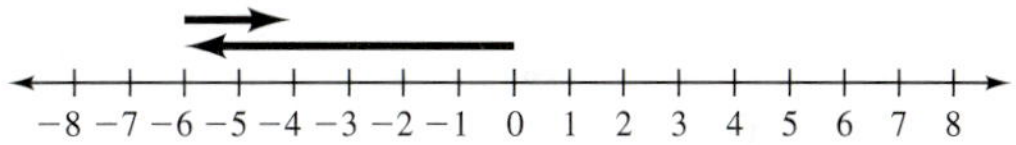

66. $-6 + (-2) = -8$

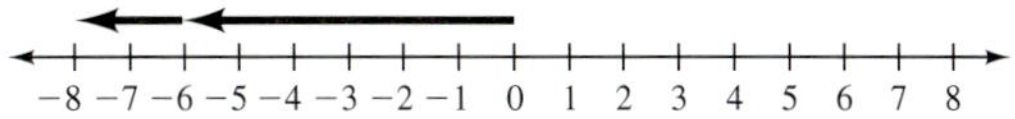

67. $-6 - 2 = -8$

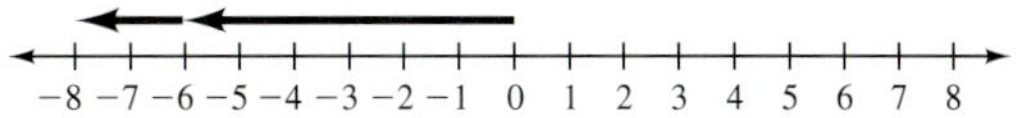

68. $-6 - (-2) = -4$

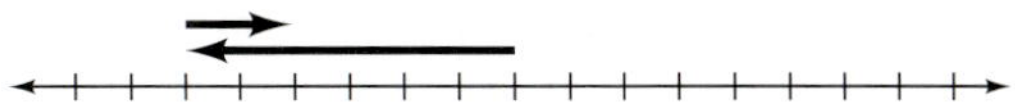

69. $-(-6) - 2 = 4$

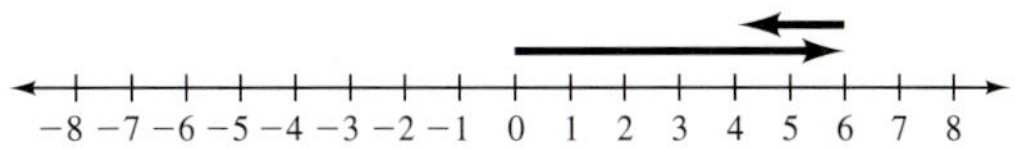

70. 25 **71.** 10 **72.** −22 **73.** 18 **74.** −7.9 **75.** 0 **76.** −4 **77.** −4.0 **78.** −9 **79.** 27 **80.** 62 **81.** −85 **82.** 0 **83.** −3 **84.** 30 **85.** −20 **86.** $>$ **87.** $<$ **88.** $>$ **89.** $<$ **90.** 2 **91.** −4 **92.** −6 **93.** 4 **94.** −4 **95.** −8 **96.** 4 **97.** −4 **98.** −8 **99.** −8 **100.** −8 **101.** 48

102. 16 **103.** −16 **104.** −21 **105.** −394 **106.** −24 **107.** 638 **108.** 3 **109.** −4 **110.** 32 **111.** 1/6 **112.** −19 **113.** −31 **114.** −1944 **115.** −117 **116.** > **117.** < **118.** > **119.** < **120.** < **121.** < **122.** > **123.** > **124.** −10 **125.** 30 **126.** 5 **127.** 4 **128.** −6 **129.** 14 **130.** −3 **131.** −4 **132.** (a) \$5.50 (b) 65 miles (c) Answers will vary. Mathematical expression to describe the problem situation: Cost = 3.50 + 0.10x, where x = miles driven. **133.** (a) 20 feet (b) Answers will vary. Draw picture and write an expression: $25x = 500$, where x = the width of the pen. **134.** (a) 576 feet (with gate) (b) \$5184 (c) Answers will vary. Draw picture, multiply perimeter by the cost of fencing and add the cost of the gate: $571(9) + 45 = 5184$. **135.** (a) 102 (b) Answers will vary. Mathematical expression to describe the problem situation: $80 + 86 + 92 + x = 360$. **136.** (a) \$4000 (b) \$116,666.67 (c) Answers will vary. Mathematical expression to describe the problem situation: Salary = 1000 + 0.03x, where x = total monthly sales.

CHAPTER 1 PRACTICE TEST ANSWERS

1. (a) = (b) < (c) < (d) > **2.** (a) Q (b) P (c) W, I, Q (d) Q (e) I, Q (f) Q (g) Q (h) C, W, I, Q **3.** (a) 30 + 25 (b) $b + a$ (c) 10×9 (d) $y \cdot x$ **4.** (a) $(1 + 2) + 7$ (b) $a + (b + c)$ (c) $3 \times (5 \times 10)$ (d) $x \cdot (y \cdot z)$ **5.** (a) $20 \cdot 8 + 20 \cdot 6$ (b) $a \cdot b - a \cdot c$ (c) $3(2 + 5)$ (d) $x(y + z)$

6. (a) $4 - (-2) = 6$

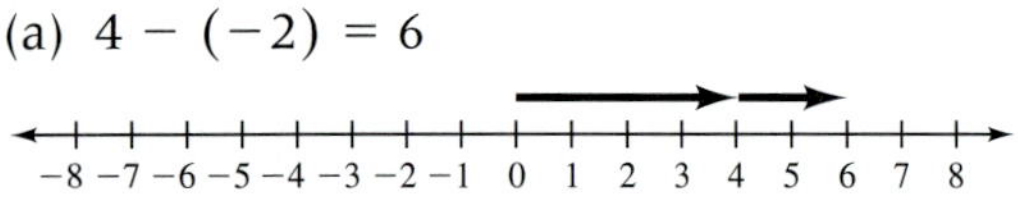

(b) $-4 + (-2) = -6$

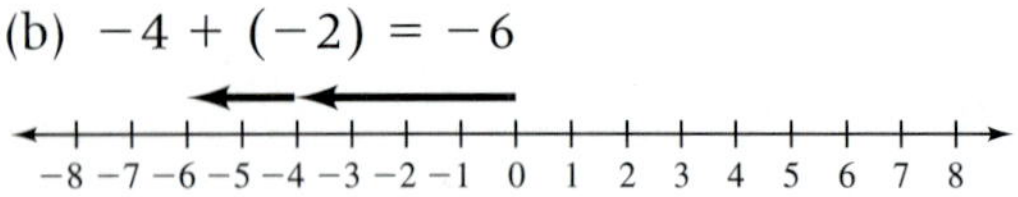

7. (a) 12 (b) 375 **8.** (a) Negative (b) Negative (c) Negative (d) Positive **9.** (a) −6 (b) 13 (c) −36 (d) 5 **10.** −16 **11.** −42 **12.** 15 **13.** −17 **14.** 512 **15.** Loss of \$16 or −16 **16.** 144 beats per minute

CHAPTER 2

DOING MATHEMATICS SECTION 2.1

1. (a) V (b) II (c) IV (d) VI (e) III (f) I **2.** (a) Answers will vary: $x = 1$ and $y = 3$ (b) $3 + 2 + y = 3 + x + 4$ **3.** = **4.** = **5.** < **6.** > **7.** = **8.** < **9.** > **10.** = **11.** = **12.** < **13.** $x = 8$ **14.** $y = 17$ **15.** $x = 2$ **16.** $y = -40$ **17.** $y = 15$ **18.** $x = -25$ **19.** $y = 4$ **20.** $x = 23$ **21.** −13 **22.** −16 **23.** 68 **24.** −7 **25.** 51 **26.** −144 **27.** 56 **28.** 0 **29.** 40 **30.** 24

31.	Binomial	$-5x$ -7	-5 Constant	1
32.	Monomial	$1.5x^2$	1.5	2
33.	Trinomial	$2x^3$ $-4x^2$ 5	2 -4 Constant	3
34.	Polynomial	$7y^3$ $-5y^2$ $9y$ -10	7 -5 9 Constant	3
35.	Trinomial	$\frac{1}{2}x$ -5 $\frac{2}{3}x^2$	$\frac{1}{2}$ Constant $\frac{2}{3}$	2
36.	Monomial	100	Constant	0
37.	Monomial	$-3x$	-3	1

38. $3x^5 - 9x^4 + x^3 - 4x^2 + 10x - 16$
39. $18y^6 + 2y^3 + 3y^2 - 5y - 15$
40. $4.25x^3 - 5x^2 + 1.5x - 2.75$
41. $2x^3 + \frac{3}{2}x^2 - \frac{1}{2}x - 5$
42. $16 + 10x - 4x^2 + x^3 - 9x^4 + 3x^5$
43. $-15 - 5y + 3y^2 + 2y^3 + 18y^6$
44. $2.75 + 1.5x - 5x^2 + 4.25x^3$
45. $\frac{16}{3} + x - \frac{2}{3}x^2 - x^4$ **46.** 24 feet **47.** 60 feet **48.** 18.8 feet **49.** 40 feet **50.** Answers will vary: (a) $P = 2l + 2w$, where P = perimeter of a rectangle, l = length, and w = width (b) $A = l \times w$, where A = area of a rectangle, l = length, and w = width (c) $A = s \times s$, where A = area of a square and s = length of one side Answers will vary: How much more money needs to be earned to purchase textbooks, if the books cost \$300 and you have \$150. Let x represent the amount to be earned: $300 = 150 + x$. Solve for x.

DOING MATHEMATICS SECTION 2.2

1. $5x$ **2.** $-14xy + 43x = 43x - 14xy$ **3.** $-3.5x^2 - 4.98x$ **4.** $-21x$ **5.** $\frac{-1}{2}y$ **6.** $-2x^3 + 7x^2$ **7.** $10.3ab + 5.2a - 6.7b$ **8.** x **9.** $75xyz - 5xy + 7xz + 21x$ **10.** $7.5cd + c + 4d$

11. $13x$ **12.** $5x - 3y$ **13.** $=$ **14.** $\neq$ **15.** $\neq$
16. $=$ **17.** $5x - 30$ **18.** $7x + 35$
19. $-x^2 - 2x - 1$ **20.** $6y^2 - 15y + 12$
21. $-10x^2 + 12$ **22.** $-4y + 6$ **23.** $25x + 4$
24. $11x - 8$ **25.** $-3x^2 + 7$ **26.** $3xy - x - 3$
27. $2x - 1$ **28.** $-10x - 23$ **29.** $6x^2 - 8x - 16$
30. $x^2 + 3x + 22$ **31.** $x^2 + 5x - 21$
32. $5x^4 - 6x^2 - 3x + 4$ **33.** $5x + 2$ **34.** 0
35. $3ab - b + 1$ **36.** $-3x^2 - 5x + 4$
37. $-4x^2 - 18x + 4$ **38.** $10x^5 - 2x^3 - x + 1$
39. $10x^3 - 3x^2 - 1$ **40.** $6x^3 + 2x^2 + x - 1$
41. $-2x^4 - 7x^2 - 2x - 17$
42. $2x^5 - 8x^3 - 6x^2 + 8$
43. $-6x^3 - 7x^2 + 2x - 17$
44. $-8x^3 - 2x^2 + 3x + 21$
45. $-2x^4 - 3x^2 + 4x - 1$
46. $2x^5 + 8x^3 - 4x^2 + 2x - 8$
47. Answers will vary. LIKE terms have identical variables and exponents.

DOING MATHEMATICS SECTION 2.3

1. (a) No: $5 + 4 = 9 \neq 19$ (b) Yes: $3(2) + 1 = 7$
2. (a) No: $-4(-3) = 12 \neq 0$
(b) No: $6400\left(\frac{1}{4}\right) = 1600 \neq 4000$ **3.** 5 **4.** 4
5. 19 **6.** 22 **7.** -20 **8.** -7 **9.** -5 **10.** -6
11. 40 **12.** 10 **13.** -12 **14.** -60 **15.** -6
16. -8 **17.** 10 **18.** 5 **19.** 6 **20.** 5 **21.** $-\frac{1}{4}$
22. $\frac{4}{3}$ **23.** 78 **24.** 16 **25.** 32 **26.** 20
27. 27 **28.** 10 **29.** $\frac{1}{5}$ **30.** 10 **31.** 8 **32.** 2
33. 25 **34.** 2 **35.** 2 **36.** 9 **37.** 29 **38.** $-\frac{3}{2}$
39. 4 **40.** 16 **41.** 2 **42.** 2 **43.** -4 **44.** 5
45. $-\frac{6}{7}$ **46.** 3 **47.** 1 **48.** -10 **49.** 9 **50.** -5
51. 7 **52.** -4.2 **53.** $-\frac{4}{3}$
54. (a) $T = 300 + 0.06x$ (b) \$330 (c) \$6000
55. (a) $P = 1500c - 600c - 2700$ (b) 3 computers (c) at least 15 computers **56.** (a) $T = 200 + 0.20c$ (b) \$360 (c) 9000 calls **57.** (a) $P = 3.5g - 875$ (b) 250 games (c) 400 games
58. (a) $0.03c + 0.03(2c) = 27$ (b) 300 minutes of long distance calls; 600 minutes of local calls (c) 900 minutes total

DOING MATHEMATICS SECTION 2.4

1. $x > -1$
(a) 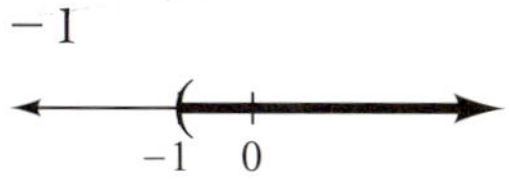
(b) $(-1, \infty)$

2. $x \geq -1$
(a) 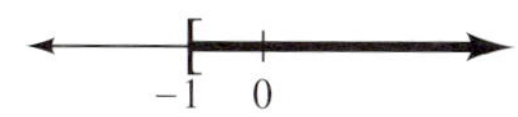
(b) $[-1, \infty)$

3. $x < -1$
(a) 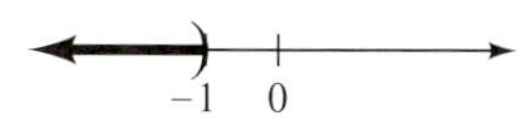
(b) $(-\infty, -1)$

4. $x \leq -1$
(a) 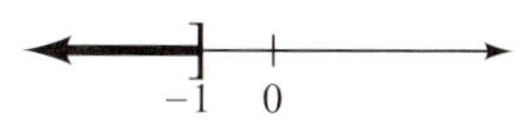
(b) $(-\infty, -1]$

5. $[-2, \infty)$
(a) 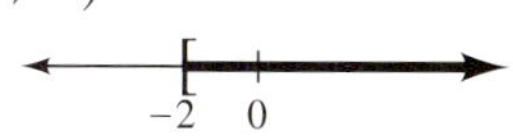
(b) $\{x | x \geq -2\}$

6. $(5, \infty)$
(a) 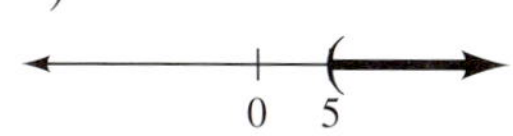
(b) $\{x | x > 5\}$

7. $(-\infty, -3]$
(a) 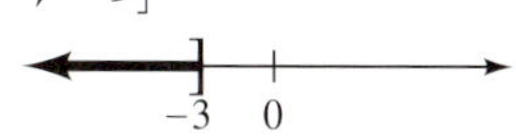
(b) $\{x | x \leq -3\}$

8. $(-\infty, 0)$
(a) 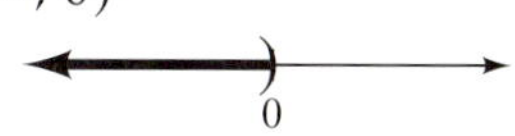
(b) $\{x | x < 0\}$

9. $x > 7$
(a) 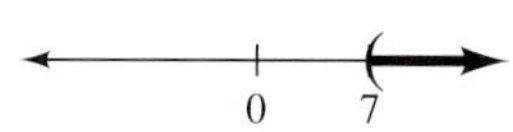
(b) $\{x | x > 7\}$ (c) $(7, \infty)$

10. $y < 24$
(a) 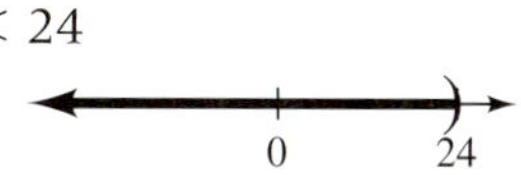
(b) $\{y | y < 24\}$ (c) $(-\infty, 24)$

11. $x \geq 3$
(a) 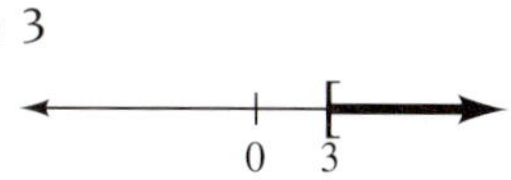
(b) $\{x | x \geq 3\}$ (c) $[3, \infty)$

12. $z \geq -3$
(a) 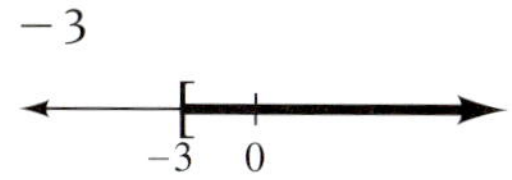
(b) $\{z | z \geq -3\}$ (c) $[-3, \infty)$

13. $x \geq -4$
(a) 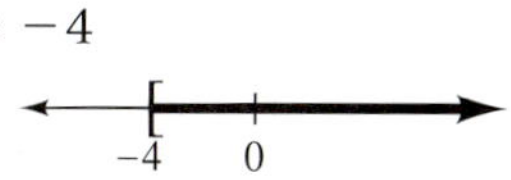
(b) $\{x | x \geq -4\}$ (c) $[-4, \infty)$

14. $y > 2$

(a) 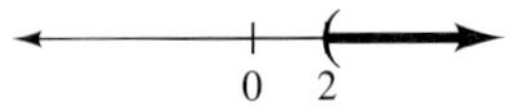

(b) $\{y|y > 2\}$ (c) $(2, \infty)$

15. $x \geq -\frac{1}{2}$

(a) 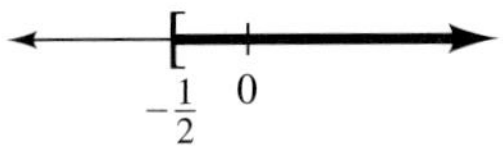

(b) $\left\{x|x \geq -\frac{1}{2}\right\}$ (c) $\left[-\frac{1}{2}, \infty\right)$

16. $x \geq -\frac{1}{4}$

(a) 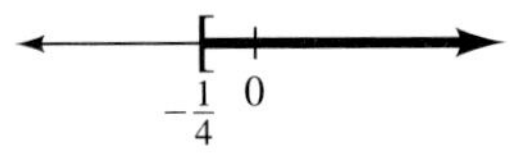

(b) $\left\{x|x \geq -\frac{1}{4}\right\}$ (c) $\left[-\frac{1}{4}, \infty\right)$

17. $x > \frac{1}{4}$

(a) 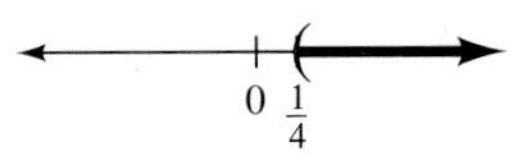

(b) $\left\{x|x > \frac{1}{4}\right\}$ (c) $\left(\frac{1}{4}, \infty\right)$

18. $x > 2$

(a) 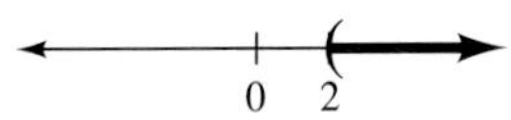

(b) $\{x|x > 2\}$ (c) $(2, \infty)$

19. $a > 3$

(a) 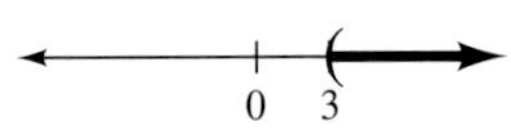

(b) $\{a|a > 3\}$ (c) $(3, \infty)$

20. $x \geq 2$

(a) 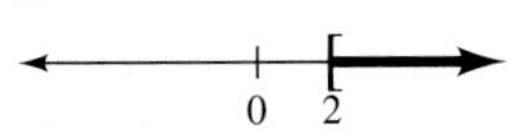

(b) $\{x|x \geq 2\}$ (c) $[2, \infty)$

21. $y \leq 4$

(a) 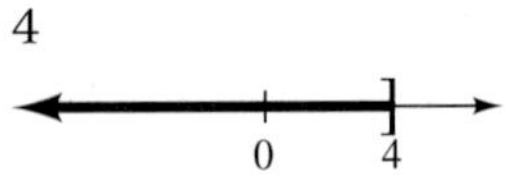

(b) $\{y|y \leq 4\}$ (c) $(-\infty, 4]$

22. $y \geq 0$

(a) 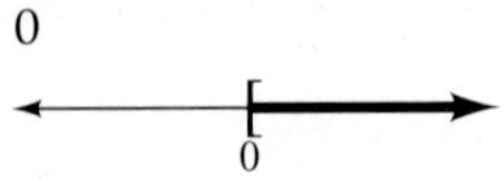

(b) $\{y|y \geq 0\}$ (c) $[0, \infty)$

23. $x \geq 0$

(a) 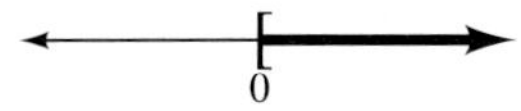

(b) $\{x|x \geq 0\}$ (c) $[0, \infty)$

24. $x < 7$

(a) 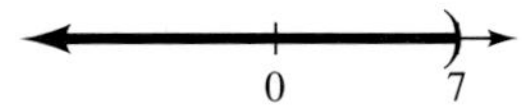

(b) $\{x|x < 7\}$ (c) $(-\infty, 7)$

25. $x \leq -\frac{3}{2}$

(a) 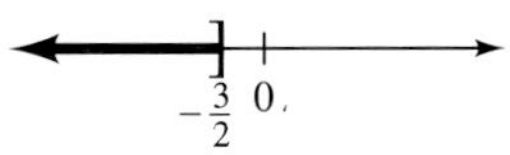

(b) $\left\{x|x \leq -\frac{3}{2}\right\}$ (c) $\left(-\infty, -\frac{3}{2}\right]$

26. $y < -3$

(a) 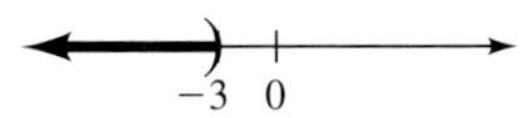

(b) $\{y|y < -3\}$ (c) $(-\infty, -3)$

27. $x \leq 2$

(a) 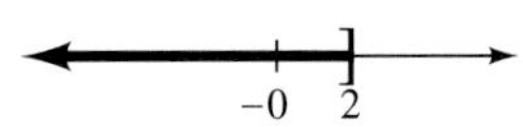

(b) $\{x|x \leq 2\}$ (c) $(-\infty, 2]$

28. $x \geq 1$

(a) 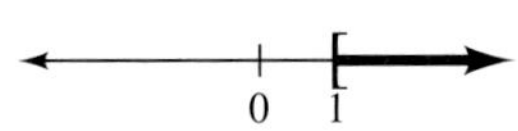

(b) $\{x|x \geq 1\}$ (c) $[1, \infty)$

29. $x > 2$

(a) 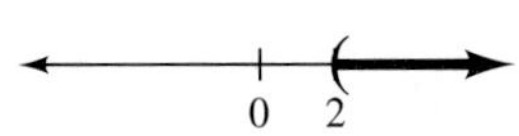

(b) $\{x|x > 2\}$ (c) $(2, \infty)$

30. $x > 6$

(a) 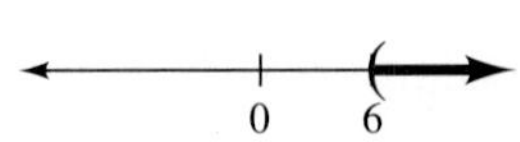

(b) $\{x|x > 6\}$ (c) $(6, \infty)$

31. (a) 87 (b) $273 + x \geq 360$ (c) $x \geq 87$

(d) 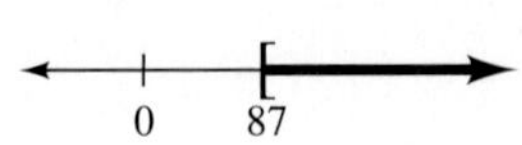 (e) $\{x|x \geq 87\}$

(f) $[87, \infty)$ (g) Yes. Each test is a maximum of 100 points, so the solution set must have a maximum of 100.

32. (a) as of 5/19/2004

Symbolic	Mailing cost
$3 < x \le 4$	\$5.30
$4 < x \le 5$	\$5.85
$x > 5$	Need ZIP code

(b) 4 lb (c) \$5.85 **33.** (a) $421 + x > 600$ (b) $x > 179$, $\{x \mid x > 179\}$ (c) 180 phones (d) No. 180 is more than she has ever sold in a one-month period and she only has one month remaining.
34. (a) $R - 300 \ge 500$ (b) $R \ge 800$, $[800, \infty)$
35. (a) as of 5/19/2004 (b) \$24.20

Symbolic	Mailing cost
$\frac{1}{2} < x \le 1$	\$13.65
$1 < x \le 2$	\$17.85
$2 < x \le 3$	\$21.05
$3 < x \le 4$	\$24.20
$4 < x \le 5$	\$27.30

36. Answers will vary: If you need at least 50 game tickets to purchase a stuffed animal at Chuck E. Cheese and you only have 32, we can write $32 + x \ge 50$. In solving, $x \ge 18$ means 18 or more tickets are needed to purchase a stuffed animal.

DOING MATHEMATICS SECTION 2.5

1. (a) 5 ft (b) 10 ft (c) 6.7 ft (d) 50 ft (e) 20 ft (f) 8.3 ft (g) There are an infinite number of values for the length and width such that the area = length·width = 500 square feet.
2. $l = \frac{P - 2w}{2}$ **3.** $a = P - b - c$ **4.** (a) 75 ft (b) 50 ft (c) 20 ft (d) 90 ft (e) 70 ft (f) 25 ft (g) There are an infinite number of values for the length and width such that the perimeter is 200 feet.
5. (a) 15 feet (b) 36 feet **6.** 20 **7.** 4 **8.** 3
9. 6 **10.** 31.4 feet of fence, Area = 78.5 sq ft
11. 43.96 feet of fence, Area = 153.86 sq ft
12. (a) $C = R - P$ (b) \$400
13. (a) $C = \frac{5(F - 32)}{9}$

(b)

Fahrenheit	Celsius
−20	−28.9
40	4.4
45	7.2
60	15.6

14. $r = \frac{d}{t}$ **15.** $r = \frac{I}{Pt}$ **16.** $d = rt$ **17.** $w = \frac{V}{lh}$
18. $r = \frac{C}{2\pi}$ **19.** $m = \frac{y - b}{x}$ **20.** $x = \frac{y - b}{m}$
21. $r = \frac{d}{2}$ **22.** $h = \frac{V}{\pi r^2}$ **23.** $A = 180° - B - C$
24. $t = \frac{A - P}{Pr}$ **25.** $x = 5 - y$ **26.** $y = x - 14$
27. $x = \frac{10 - 5y}{3}$ **28.** $y = 12 - 4x$
29. $y = 2x + 14$ **30.** $x = \frac{11 - y}{3}$ **31.** $x = \frac{12 + 4y}{5}$
32. $y = \frac{20 - 3x}{4}$ **33.** $x = \frac{4y + 20}{3}$

CHAPTER 2 REVIEW

1. (a) Answers will vary. $x = 8$, $y = 5$ (b) Infinite (c) $10 + 5 + y = 10 + 2 + x$ **2.** −336 **3.** 79
4. 96 **5.** 28 **6.** 50 **7.** 76

8.	$5x^2 - 7z$	Binomial	$5x^2$ $-7z$	5 −7	2
9.	$11.5a$	Monomial	$11.5a$	11.5	1
10.	$2x^2 - 4x + 5$	Trinomial	$2x^2$ $-4x$ 5	2 −4 Constant	2
11.	$7x^3 + x - 5$	Trinomial	$7x^3$ x −5	7 1 Constant	3
12.	$8y + 10$	Binomial	$8y$ 10	8 Constant	1
13.	$x + y + z + w$	Polynomial	x y z w	1 1 1 1	1
14.	$-26x$	Monomial	$-26x$	−26	1
15.	$x^3 - 6x^2 + 9x - 1$	Polynomial	x^3 $-6x^2$ $9x$ −1	1 −6 9 Constant	3

16. $7x^6 - 9x^4 + x^3 - 16x^2 - 4x$
17. $4.25y^5 - 5y^3 + 3.75y^2 + 1.5y$
18. $-11x - 10x^2 - 8x^3 + 7x^4 + 2x^5$
19. $-15 - 5y + 3y^2 + 2y^3 + 18y^6$ **20.** 5 ft
21. 48 sq ft **22.** 20 **23.** 17 **24.** -4 **25.** 30
26. $-8x$ **27.** $20x - 18xy$ **28.** $-3.8x^2 - 2x$
29. $-z$ **30.** $10.3ab + 5.2a - 6.7b$ **31.** $-x$
32. $xyz - 5xy + 9xz$ **33.** $=$ **34.** $\neq$ **35.** $\neq$
36. $=$ **37.** $x^2 + 17x + 4$ **38.** $4y$ **39.** $4x - 2$
40. $-21x + 4$ **41.** $6x^2 - 8x - 16$
42. $-4x^4 + 7x^2 - x + 4$ **43.** $5x + 2$
44. $-6x^2 - 2x + 3$ **45.** $3x^2 + 12x + 37$
46. $28x^3 - 5x^2 - 3x - 1$ **47.** $3x^3 - 3x^2 + x - 23$
48. $x^3 - 7x^2 + 5x - 1$ **49.** $-8x^3 - 2x^2 + 3x + 21$
50. $0.25q + 0.10d + 0.05n + 0.01p = 1.48$
51. $T = 5c + 26d + 9f + 32p$
52. Yes: $-3 - 9 = -12$ **53.** No: $10 + 6 = 16 \neq 11$
54. No: $2(0) = 0 \neq 2$ **55.** Yes: $\frac{-12}{-2} = 6$ **56.** -2
57. 2 **58.** 27 **59.** 3 **60.** $\frac{34}{22} = \frac{17}{11}$ **61.** 5 **62.** 20
63. -36 **64.** (a) $S = 500 + 0.02x$ (b) \$1700
(c) \$230,000
65. $x > -3$
(a) 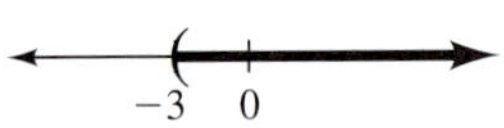
(b) $(-3, \infty)$
66. $x \geq -3$
(a) 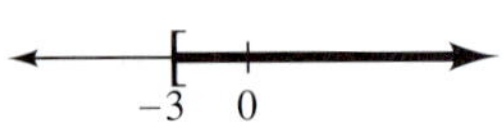
(b) $[-3, \infty)$
67. $x < -3$
(a) 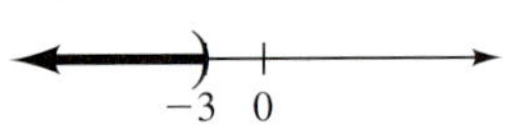
(b) $(-\infty, -3)$
68. $x \leq -3$
(a) 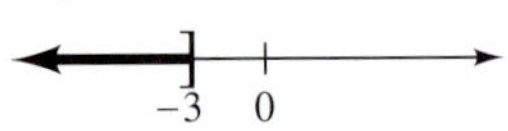
(b) $(-\infty, -3]$
69. $(-1, \infty)$
(a) 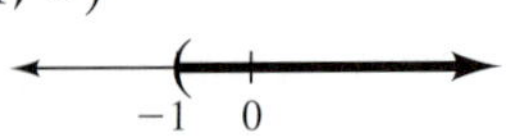
(b) $\{x | x > -1\}$
70. $[0, \infty)$
(a) 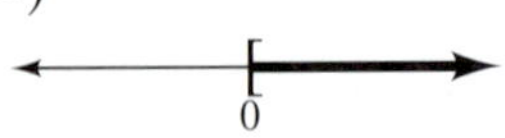
(b) $\{x | x \geq 0\}$
71. $(-\infty, 1)$
(a) 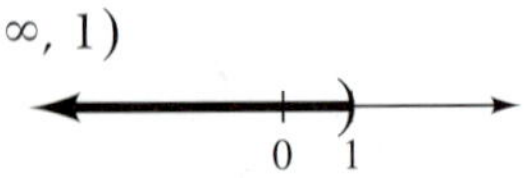
(b) $\{x | x < 1\}$
72. $(-\infty, 3)$
(a) 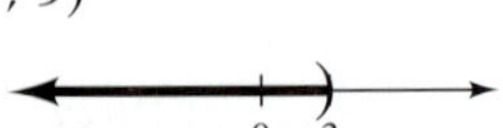
(b) $\{x | x < 3\}$

Symbolic representation of the weight	Cost to mail a package from the UPS office in your city to Anchorage, Alaska
$0 < x \leq 1$	Answers will vary.
$1 < x \leq 2$	
$2 < x \leq 3$	
$3 < x \leq 4$	
$4 < x \leq 5$	

74. (a) $24 + x > 25$ (b) 2 or more (c) Yes: He has sold more than 2 cars per month for 5 months in a row.
75. $x < 6$
(a) 
(b) $\{x | x < 6\}$ (c) $(-\infty, 6)$
76. $z \geq -6$
(a) 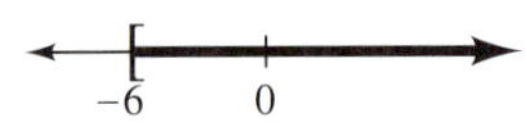
(b) $\{z | z \geq -6\}$ (c) $[-6, \infty)$
77. $x < 13$
(a) 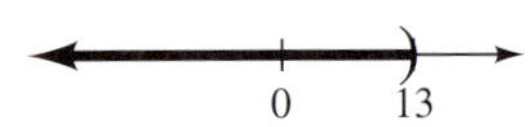
(b) $\{x | x < 13\}$ (c) $(-\infty, 13)$
78. $y < -6$
(a) 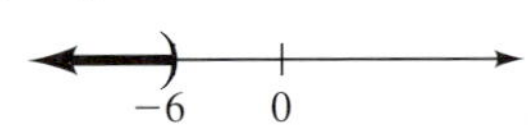
(b) $\{y | y < -6\}$ (c) $(-\infty, -6)$
79. $x < 3$
(a) 
(b) $\{x | x < 3\}$ (c) $(-\infty, 3)$
80. $x < 2$
(a) 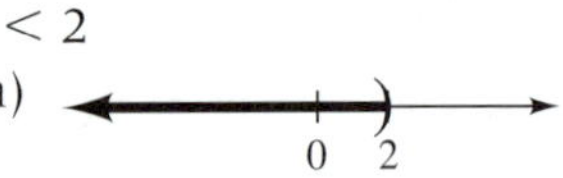
(b) $\{x | x < 2\}$ (c) $(-\infty, 2)$
81. $x \leq -\frac{4}{3}$
(a) 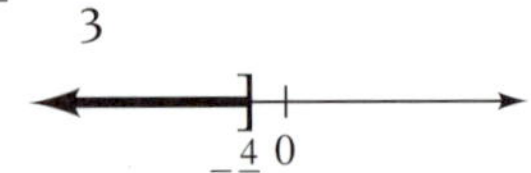
(b) $\{x | x \leq -4/3\}$ (c) $(-\infty, -4/3]$

82. $x \geq -1$

(a) 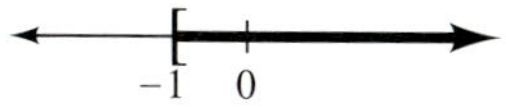

(b) $\{x|x \geq -1\}$ (c) $[-1, \infty)$
83. (a) 3 ft (b) 6 ft (c) 12 ft (d) 30 ft (e) 6 ft (f) 10 ft (g) There are an infinite number of values for the length and width such that the area is 300 square feet. **84.** (a) 50 ft (b) 100 ft (c) 125 ft (d) 140 ft (e) 100 ft (f) 120 ft (g) There are an infinite number of values for the length and width such that the perimeter is 300 feet.

85. $t = \dfrac{I}{Pr}$ **86.** $h = \dfrac{V}{lw}$ **87.** $y = \dfrac{C - Ax}{B}$
88. $C = 180° - A - B$ **89.** $b = P - a - c$
90. $x = 7 - 2y$ **91.** $x = \dfrac{15 - 5y}{4}$ **92.** $y = \dfrac{12 - 8x}{5}$
93. $x = \dfrac{5 - y}{2}$ **94.** $y = \dfrac{7x - 12}{5}$
95. $x = 4y + 20$

CHAPTER 2 PRACTICE TEST ANSWERS

1. (a) 148 (b) 14 (c) -50 (d) 14

2.

(a) $125x + y^2 + z$	Trinomial	$125x$ y^2 z	125 1 1	2
(b) $25a^3$	Monomial	$25a^3$	25	3
(c) $4x^5 - y$	Binomial	$4x^5$ $-y$	4 -1	5
(d) $x^4 + 2x^3 - 3x + 9$	Polynomial	x^4 $2x^3$ $-3x$ 9	1 2 -3 Constant	4

3. (a) $2x^5 + x^4 - 9x^3 + x^2 - 3x + 1$
(b) $5y^4 - 2y^3 + 7y^2 - y$
4. (a) $7 + 3x + 8x^2 - 4x^3 - 2x^4$
(b) $10 - 5y + 2y^3 + 3y^5$ **5.** (a) $x = 14$ (b) $y = 8$
(c) $x = 6$ (d) $y = 42$ **6.** (a) $8x + 5y$
(b) $11x^2y^2 + 4xy$ (c) $11xy + x - 6y$
(d) $-9xy + xz - 4zy$ **7.** (a) $3x^2 + 6x + 3$
(b) $-14x - 2$ (c) $3x + 12$ (d) $4x^2 + 4x + 34$
8. (a) $20x^3 + 3x^2 - 9x + 11$ (b) $x^3 - 6x^2 + 8x + 4$
9. $y = 3$ **10.** $x = -4$ **11.** $y = -13$ **12.** $x = -5$
13. $x = -0.4$ **14.** (a) $S = 400 + 0.04x$
(b) $S = \$3600$ (c) $x = \$170{,}000$

15. (a) 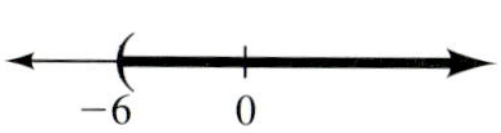 $(-6, \infty)$

(b) 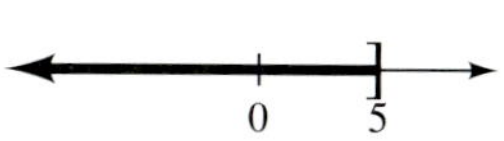 $(-\infty, 5]$

16. (a) 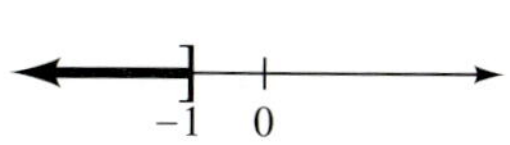 $\{x|x \leq -1\}$

(b) 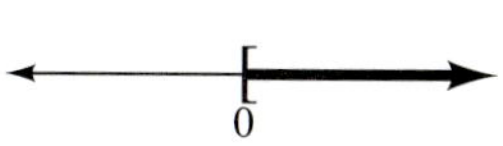 $\{x|x \geq 0\}$

17. (a) 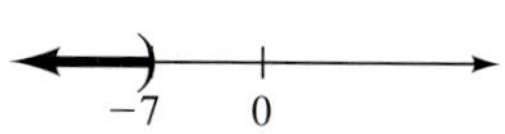

(b) $\{z|z < -7\}$ (c) $(-\infty, -7)$

18. (a) 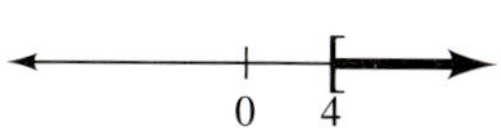

(b) $\{x|x \geq 4\}$ (c) $[4, \infty)$

19. (a) 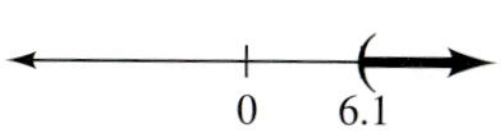

(b) $\{x|x > 6.1\}$ (c) $(6.1, \infty)$

20. (a) 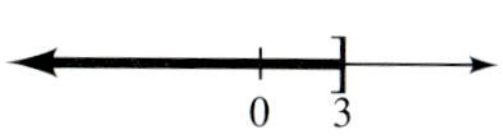

(b) $\{x|x \leq 3\}$ (c) $(-\infty, 3]$

21. $P = \dfrac{I}{rt}$ **22.** $d = \dfrac{C}{\pi}$ **23.** $y = \dfrac{C - Ax}{B}$

24. $x = \dfrac{y + 17}{3}$ **25.** $y = \dfrac{5x - 10}{2}$

CHAPTER 3

DOING MATHEMATICS SECTION 3.1

1. (a) I (b) III (c) IV (d) III (e) II (f) III (g) IV (h) I (i) II (j) I (k) IV (l) None **2.** (a) IV (b) II (c) I (d) None (e) III (f) IV (g) III (h) II (i) None (j) IV (k) I (l) III

1.

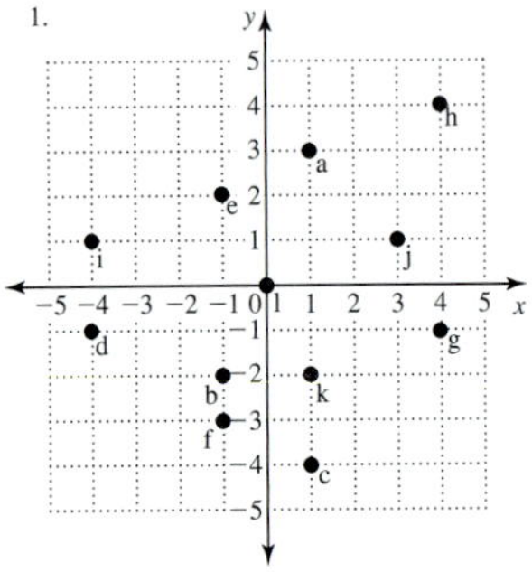

2.

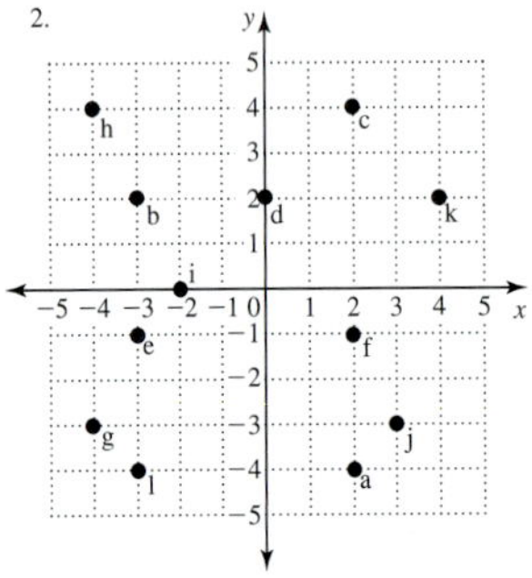

3. −4 **4.** −2 **5.** −3 **6.** 4 **7.** (a) 10,400 (b) 11,000 (c) 10,800 (d) 10,500 (e) 9500 **8.** (a) 9 million (b) 7 million (c) Answers may vary: More and more women are going to be attending college. The number of women in college seems to be increasing more rapidly than the number of men in college. **9.** (a) (2, 15); (4, 20); (6, 25) (b) 0, 6, 2 (c) 10, 25, 5

(d)

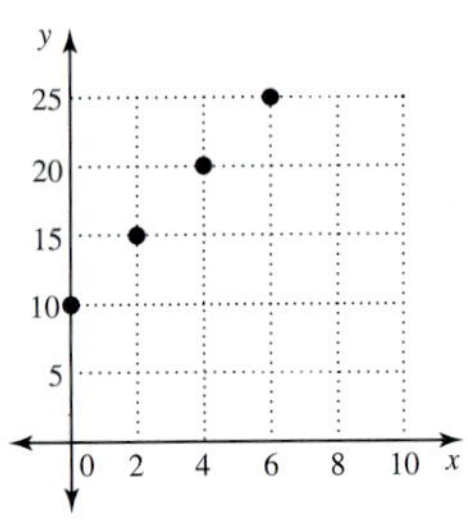

10. (a) (−50, −15); (0, 0); (50,15); (100, 30) (b) −100, 100, 50 (c) −30, 30, 15

(d)

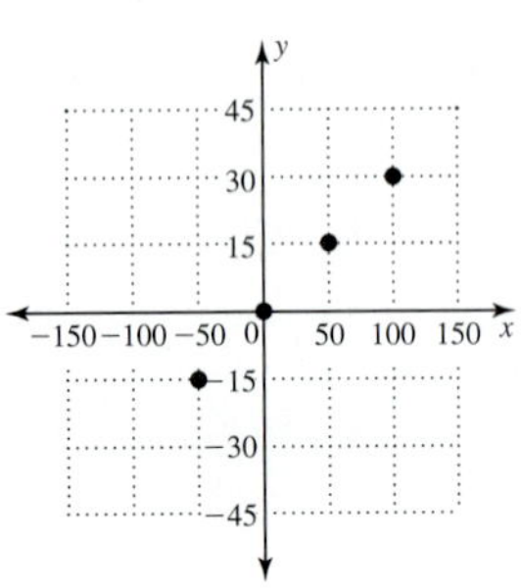

11. (a)

n = Number of credit hours taken	C = Tuition cost
9	405
15	675

(b) 3, 15, 3 (c) 135, 675, 135

(d)

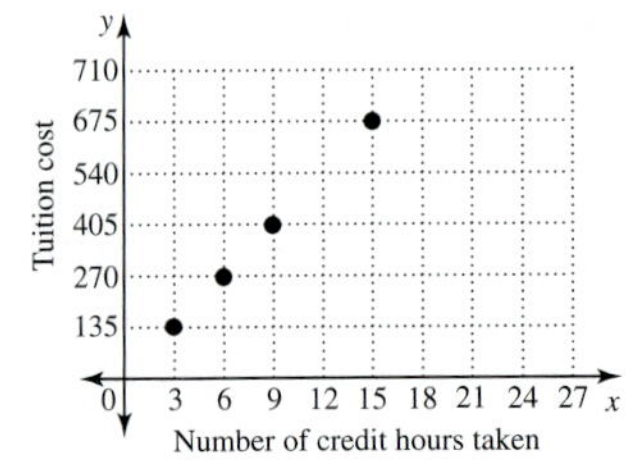

12. (a) 0, 100, 10 (b) 10, 242, 30

(c)

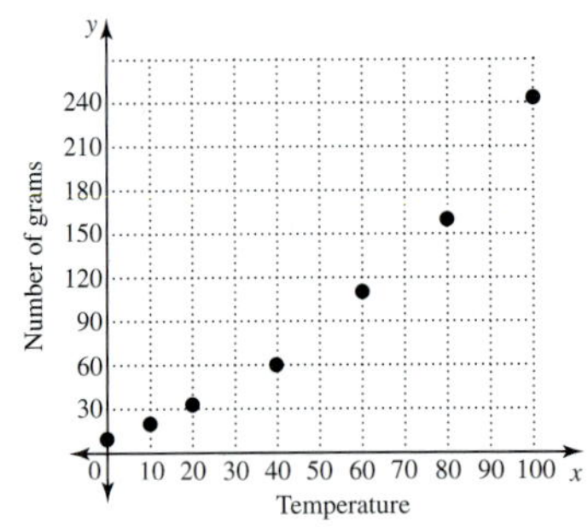

13. (a) 0.5, 5.7, 0.7 (b) 4, 60, 8

(c)

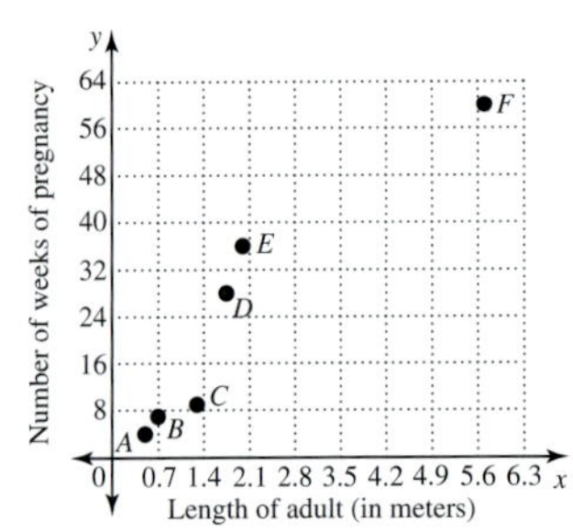

(d) The longer the pregnancy, the longer the adult.

14. (a)

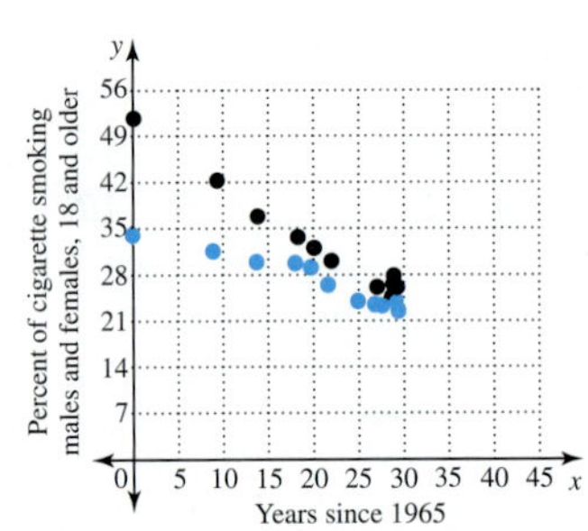

(b) The percentages of male and female smokers are declining. Still, the number of male smokers remains higher. **15.** Answers will vary. **16.** No **17.** No **18.** Yes **19.** Yes

20.

x	$y = x - 4$	Ordered pairs
−2	−6	(−2, −6)
−1	−5	(−1, −5)
0	−4	(0, −4)
1	−3	(1, −3)
2	−2	(2, −2)

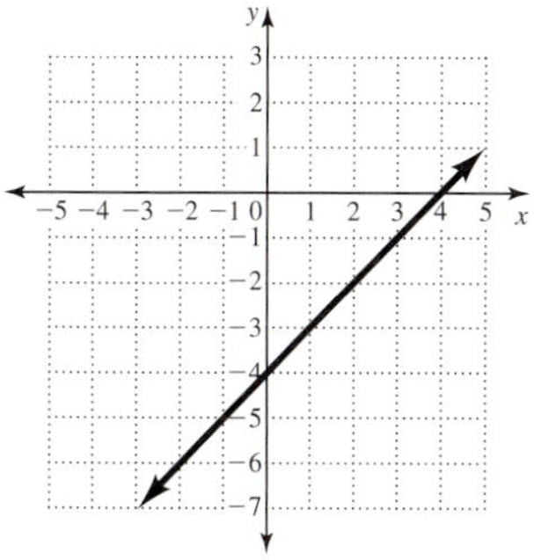

21.

x	$y = x - 1$	Ordered pairs
−2	−3	(−2, −3)
−1	−2	(−1, −2)
0	−1	(0, −1)
1	0	(1, 0)
2	1	(2, 1)

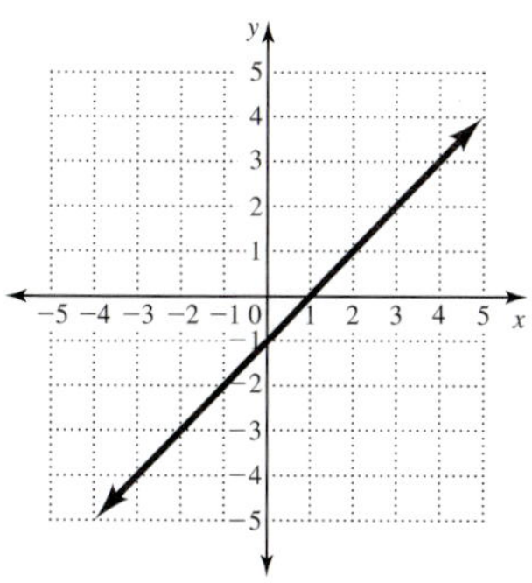

22.

x	$y = -2x + 3$	Ordered pairs
−2	7	(−2, 7)
−1	5	(−1, 5)
0	3	(0, 3)
1	1	(1, 1)
2	−1	(2, −1)

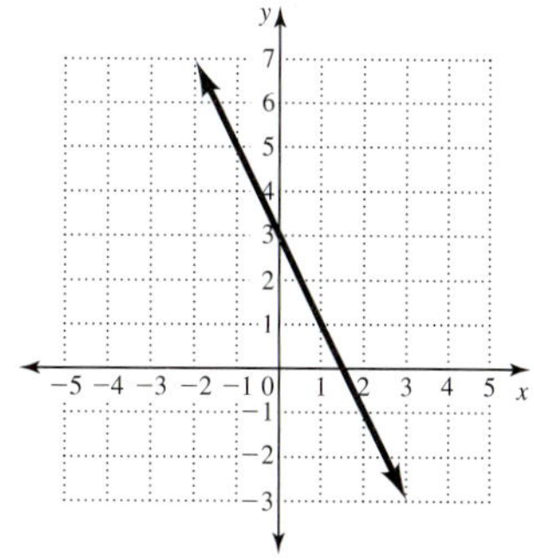

23.

x	$y = -3x + 1$	Ordered pairs
−2	7	(−2, 7)
−1	4	(−1, 4)
0	1	(0, 1)
1	−2	(1, −2)
2	−5	(2, −5)

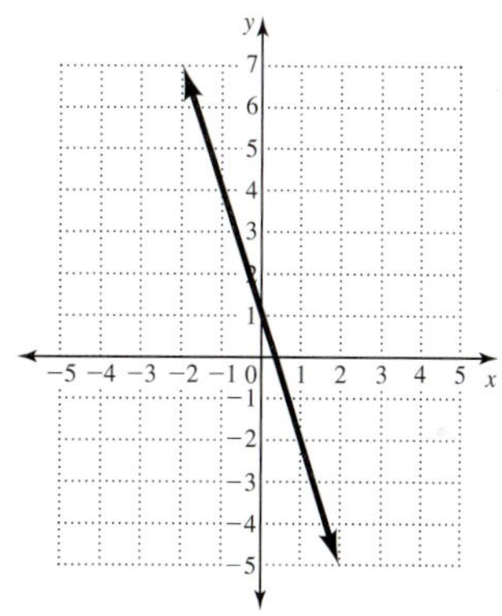

24.

x	$y = -x + 4$	Ordered pairs
−2	6	(−2, 6)
−1	5	(−1, 5)
0	4	(0, 4)
1	3	(1, 3)
2	2	(2, 2)

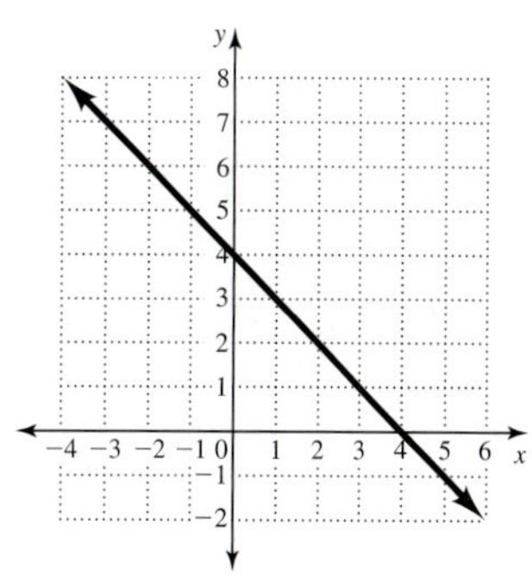

25.

x	$y = -x + 6$	Ordered pairs
-2	8	$(-2, 8)$
-1	7	$(-1, 7)$
0	6	$(0, 6)$
1	5	$(1, 5)$
2	4	$(2, 4)$

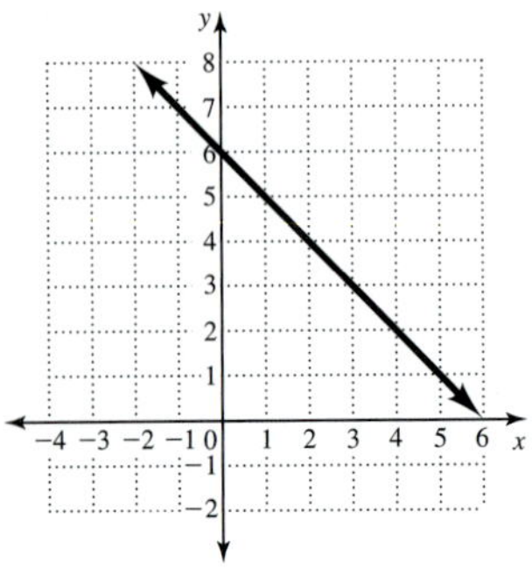

26.

x	$y = 4 + 2x$	Ordered pairs
-2	0	$(-2, 0)$
-1	2	$(-1, 2)$
0	4	$(0, 4)$
1	6	$(1, 6)$
2	8	$(2, 8)$

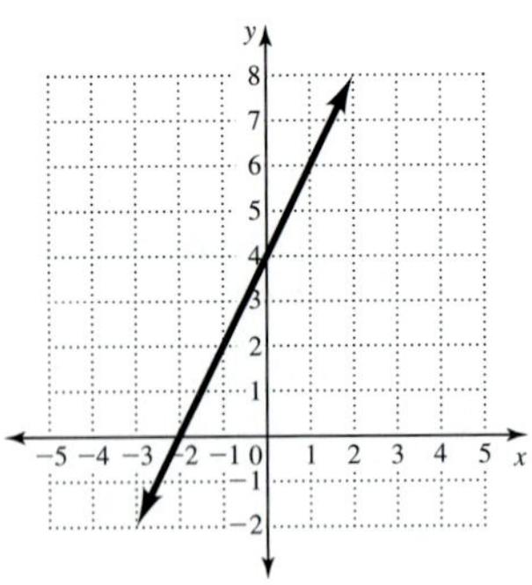

27.

x	$y = 6 + 3x$	Ordered pairs
-2	0	$(-2, 0)$
-1	3	$(-1, 3)$
0	6	$(0, 6)$
1	9	$(1, 9)$
2	12	$(2, 12)$

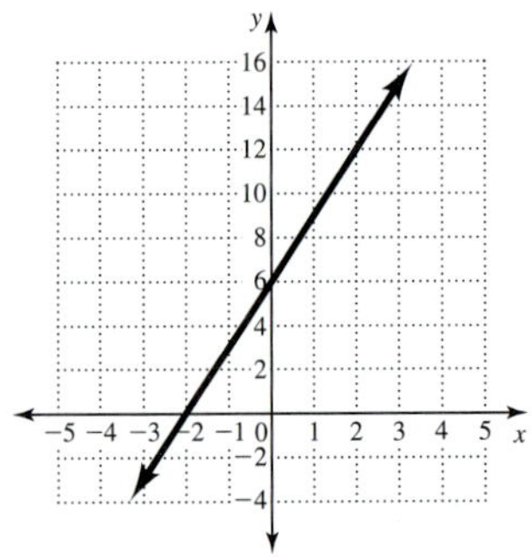

DOING MATHEMATICS SECTION 3.2

1. $\{(0, 0), (0, 1), (1, -2), (1, 3), (3, -2), (3, 4)\}$
Not a function

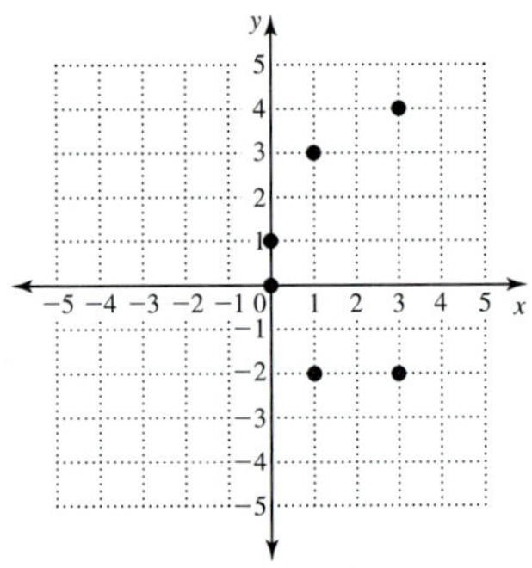

2. $\{(-1, -4), (-2, 1), (0, 3), (1, 2)\}$ Function

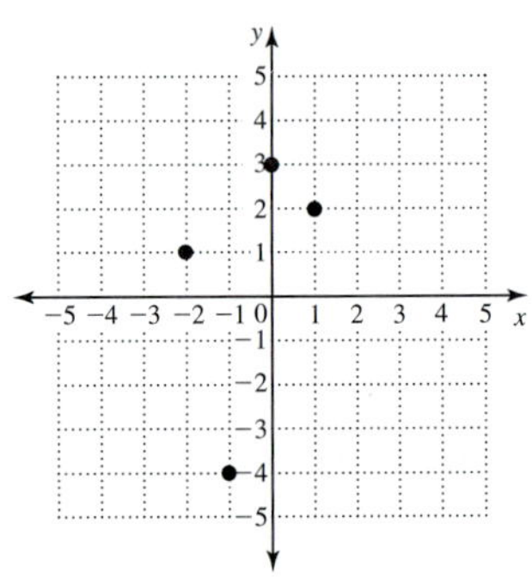

3. Not a Function

Input	Output
3	4
3	5
3	6
3	7
3	8

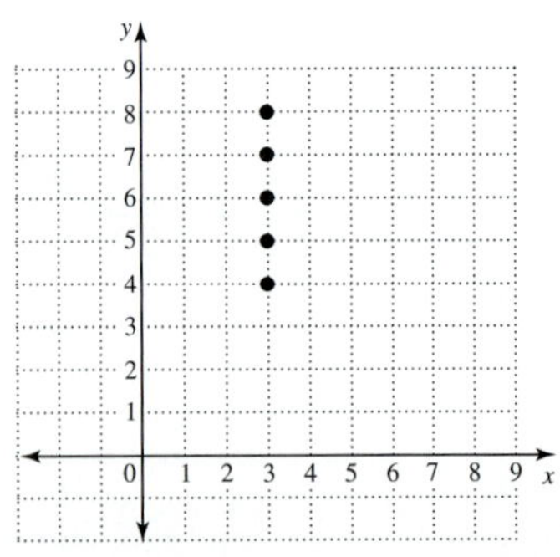

4. Function

Input	Output
4	3
5	3
6	3
7	3

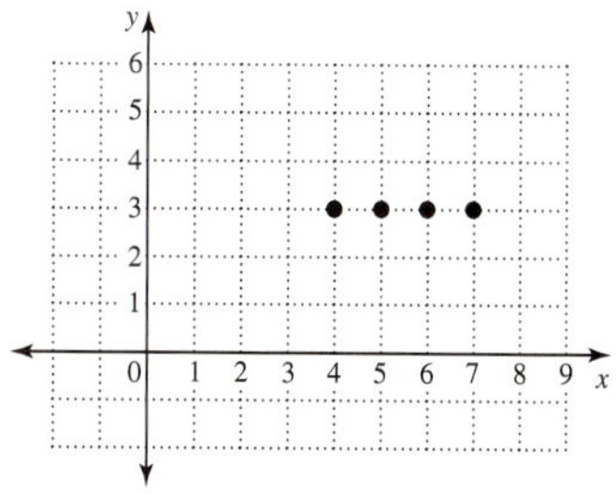

5. Function **6.** Function **7.** Not a function

8. Function

9. $f(x) = 2x - 1$

x	$f(x)$	Ordered pairs
-2	-5	$(-2, -5)$
-1	-3	$(-1, -3)$
0	-1	$(0, -1)$
1	1	$(1, 1)$
2	3	$(2, 3)$

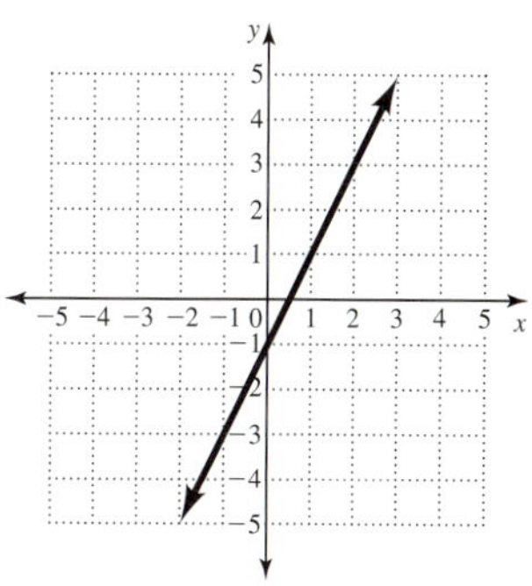

10. $f(x) = x + 2$

x	$f(x)$	Ordered pairs
-2	0	$(-2, 0)$
-1	1	$(-1, 1)$
0	2	$(0, 2)$
1	3	$(1, 3)$
2	4	$(2, 4)$

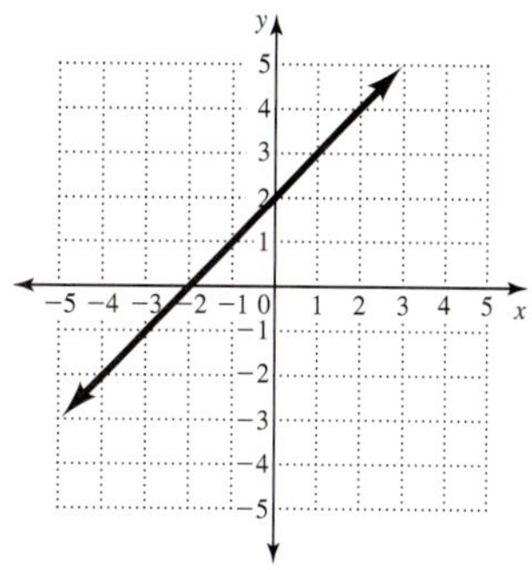

11. $f(x) = 2x$

x	$f(x)$	Ordered pairs
-2	-4	$(-2, -4)$
-1	-2	$(-1, -2)$
0	0	$(0, 0)$
1	2	$(1, 2)$
2	4	$(2, 4)$

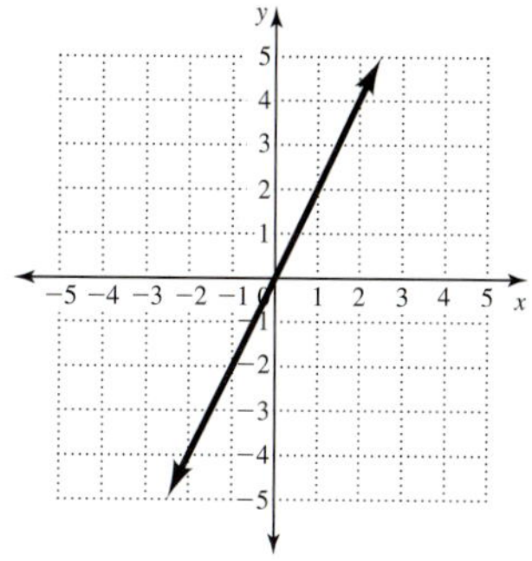

12. $f(x) = 3x$

x	$f(x)$	Ordered pairs
−2	−6	(−2, −6)
−1	−3	(−1, −3)
0	0	(0, 0)
1	3	(1, 3)
2	6	(2, 6)

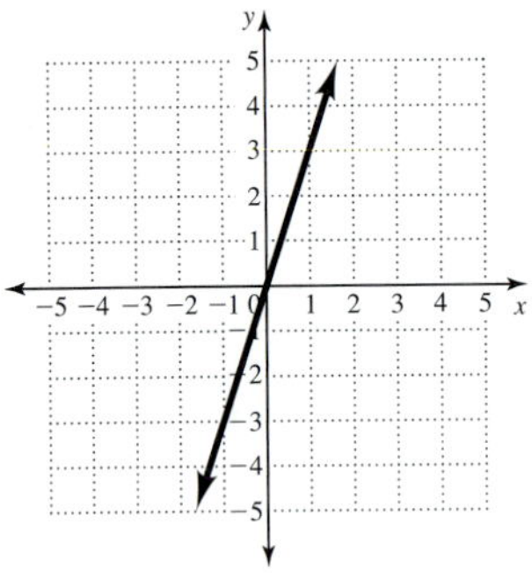

13. $f(x) = -x$

x	$f(x)$	Ordered pairs
−2	2	(−2, 2)
−1	1	(−1, 1)
0	0	(0, 0)
1	−1	(1, −1)
2	−2	(2, −2)

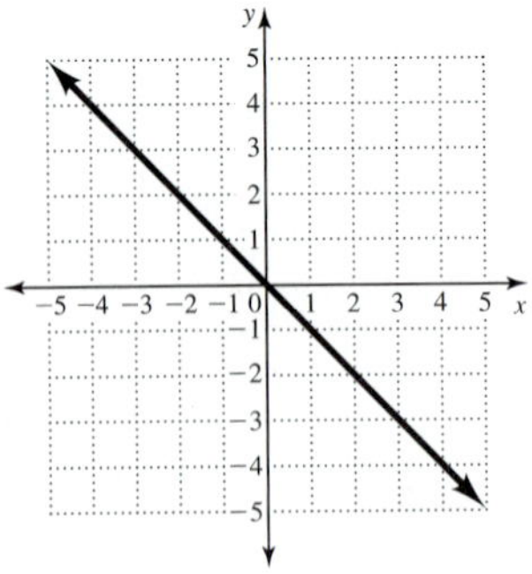

14. $f(x) = 1 - x$

x	$f(x)$	Ordered pairs
−2	3	(−2, 3)
−1	2	(−1, 2)
0	1	(0, 1)
1	0	(1, 0)
2	−1	(2, −1)

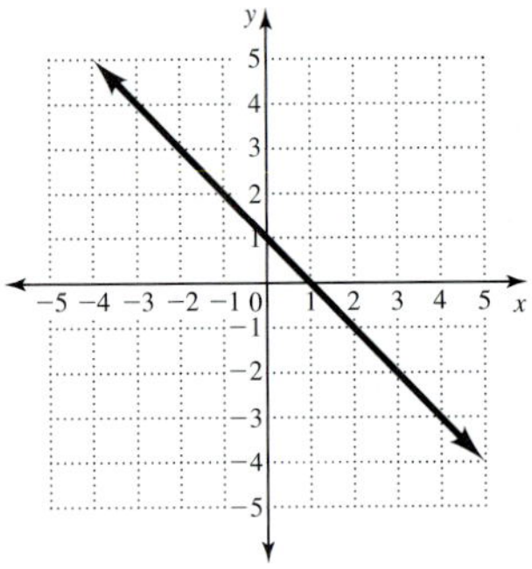

15. $f(x) = 6$

x	$f(x)$	Ordered pairs
−2	6	(−2, 6)
−1	6	(−1, 6)
0	6	(0, 6)
1	6	(1, 6)
2	6	(2, 6)

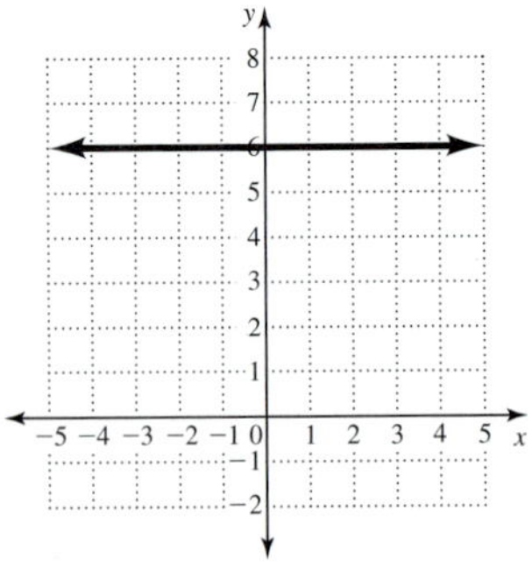

16. $f(x) = -2$

x	$f(x)$	Ordered pairs
-2	-2	$(-2, -2)$
-1	-2	$(-1, -2)$
0	-2	$(0, -2)$
1	-2	$(1, -2)$
2	-2	$(2, -2)$

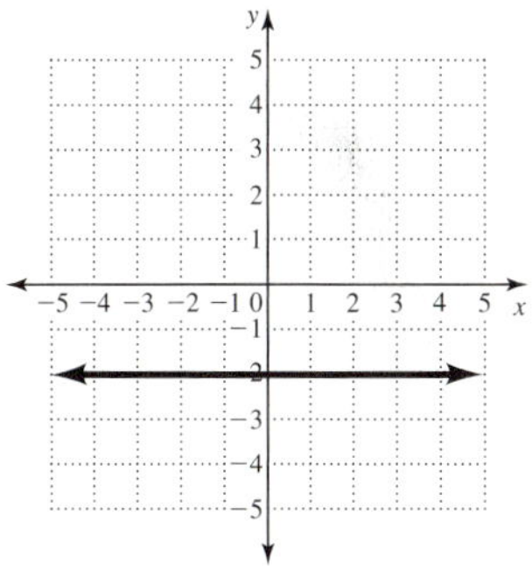

17.

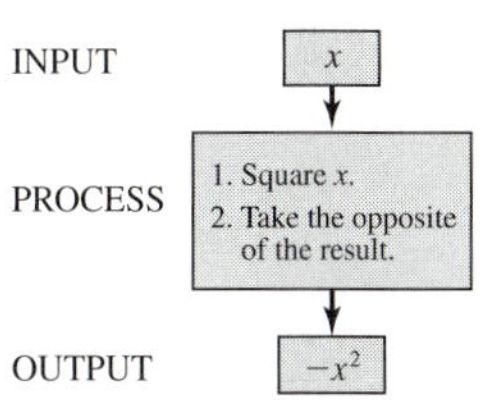

(a) the output value is -9 (b) the output value is -9 (c) the output value is -400 (d) the output value is -400

18.

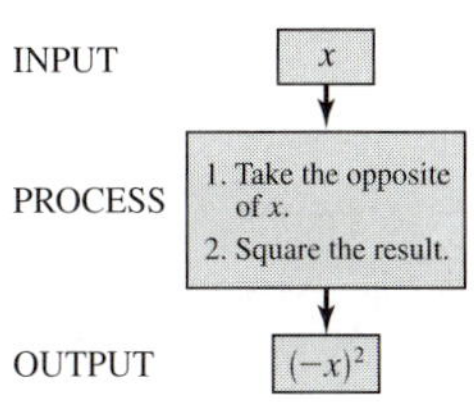

(a) the output value is 9 (b) the output value is 9 (c) the output value is 400 (d) the output value is 400

19.

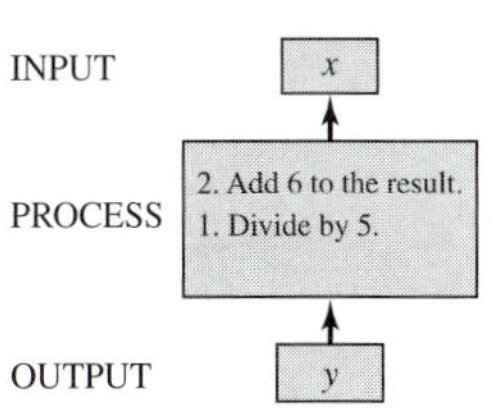

(a) the input value is 8 (b) the input value is 10 (c) the input value is 3 (d) the input value is 0

20.

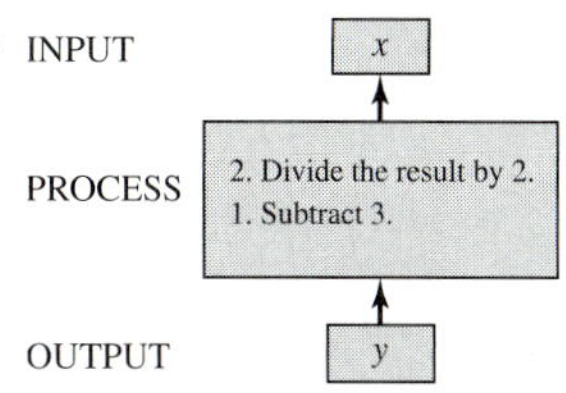

(a) the input value is 5 (b) the input value is 12 (c) the input value is -4 (d) the input value is -10

21. (a) 11 (b) -3 **22.** (a) 26 (b) 1

23. (a) 103 (b) 7 **24.** (a) -21 (b) -17

25. $x = 2$ **26.** $x = 0$ **27.** $x = -1$ **28.** $x = -3$

29. (a) Dollar amount of sales (b) Total salary

(c) See table

s = Dollar amount of sales	T = Total salary (in dollars)	Ordered pairs
5000	16,750	(5000, 16750)
6000	17,100	(6000, 17100)
10,000	18,500	(10000, 18500)
s		$T(s) = 15{,}000 + 0.35s$

(d)

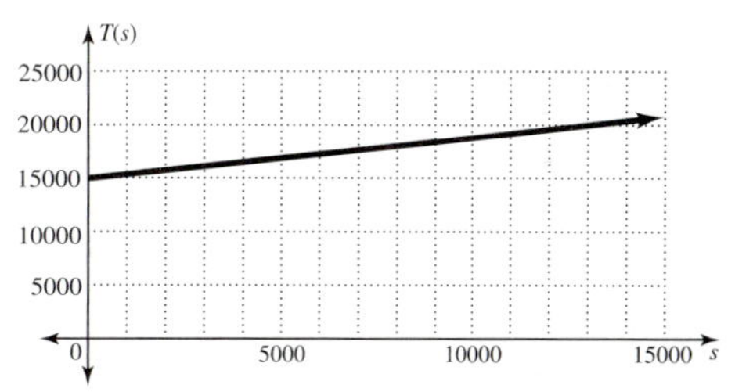

(e) (7500, 18000) (f) $18,000

(g) $T(7500) = \$17{,}625$ (h) $T(9500) = \$18{,}325$

(i) $s = \$8200$ (j) $s = \$15{,}000$

30. (a) Miles driven (b) Cost to rent car

(c) See table

m = Number of miles driven	C = Cost to rent the car
100	$35
200	$45
500	$75
700	$95
m	$C = 25 + (0.10)m$

(d)

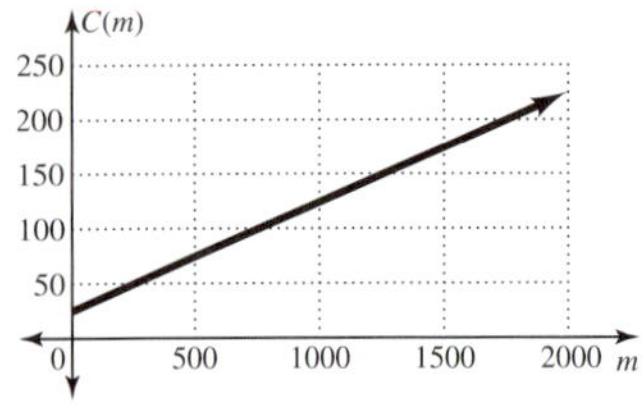

(e) (300, 55) (f) \$55 (g) $C(300) = \$55$
(h) $C(1000) = \$125$ (i) $m = 1100$ miles
(j) $m = 1750$ miles

31. (a)

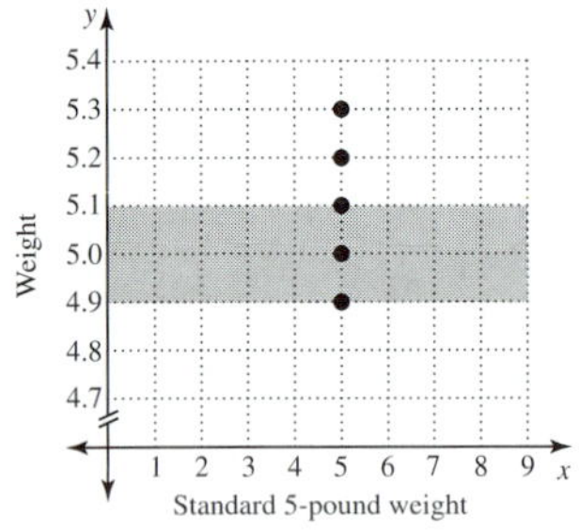

(b) No. For the input value of 5 pounds, there are six corresponding output values. (c) 4.9, 4.9, 5.0, 5.1.

32. (a)

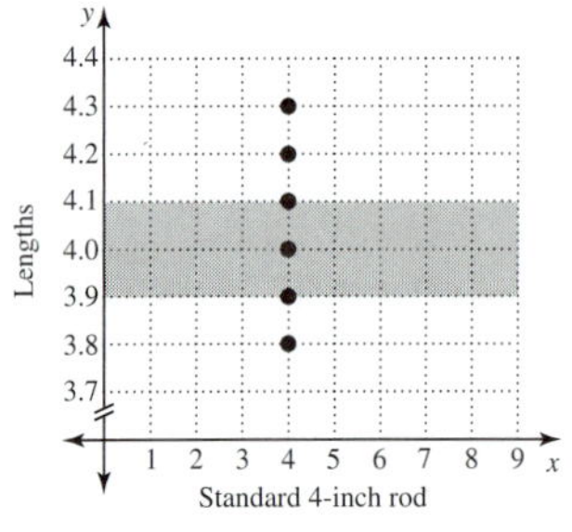

(b) No. For the input value of 4 inches, there are six corresponding output values. (c) 3.9, 4.0, 4.1 inches
33. Answers will vary. **34.** Answers will vary.

DOING MATHEMATICS SECTION 3.3

1. (a), (c) **2.** (a), (d) **3.** (a) (−2, 0) (b) (0, 4) **4.** (a) (−3, 0) (b) (0, −9) **5.** (a) None (b) (0, 5)

6.

x	$f(x) = 3x + 2$
0	2
$-\frac{2}{3}$	0
Two additional points to verify the graph:	
−1	−1
−2	−4

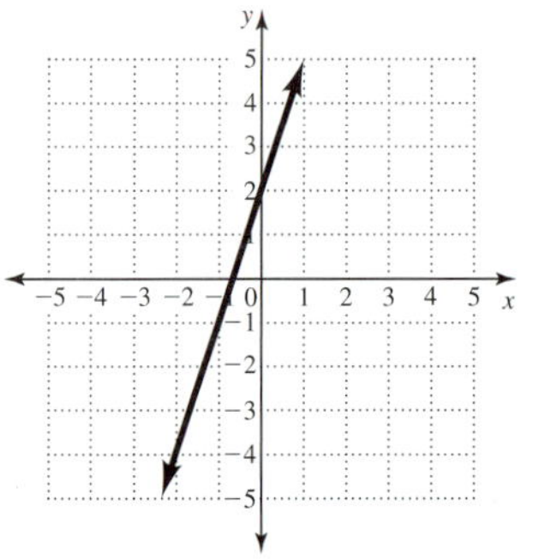

7.

x	$f(x) = 2x - 1$
0	−1
$\frac{1}{2}$	0
Two additional points to verify the graph:	
−2	−5
1	1

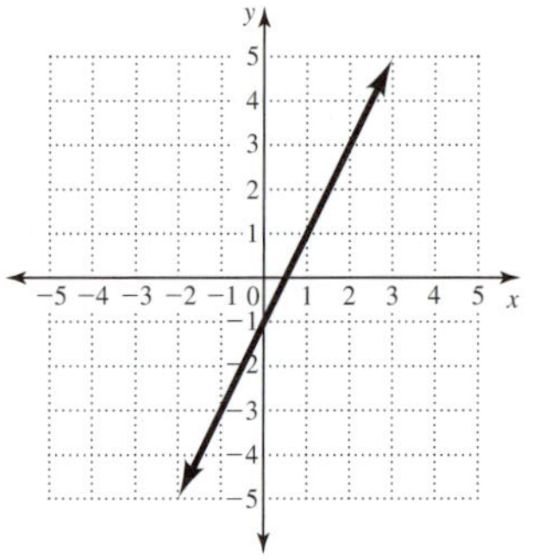

8.

x	$f(x) = -x - 3$
0	−3
−3	0
Two additional points to verify the graph:	
−1	−2
2	−5

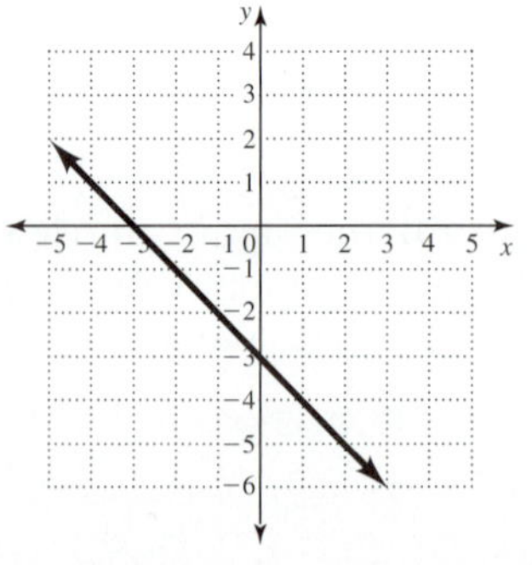

9.

x	$f(x) = -x + 4$
0	4
4	0
Two additional points to verify the graph:	
-1	5
2	2

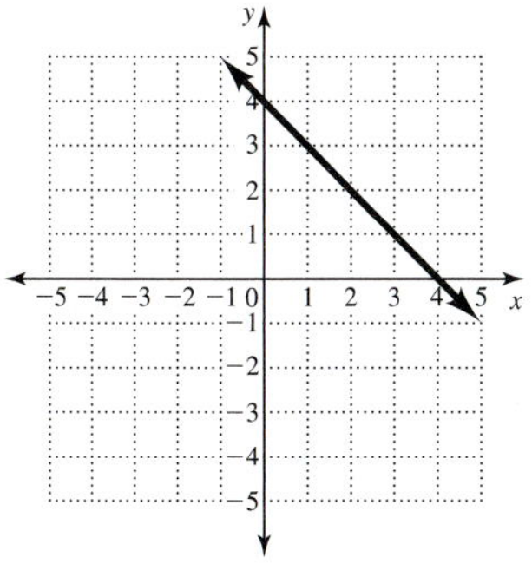

10.

x	$y = \frac{12 - 4x}{-3} = \frac{4x - 12}{3}$
0	-4
3	0
Two additional points to verify the graph:	
-2	$-\frac{20}{3}$
1	$-\frac{8}{3}$

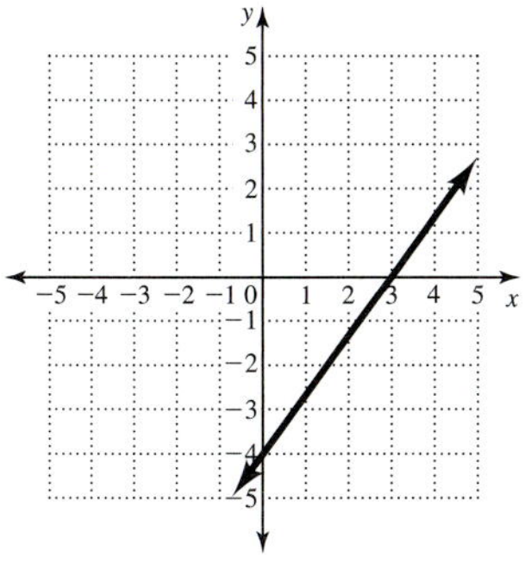

11.

x	$y = \frac{2x - 6}{3}$
0	-2
3	0
Two additional points to verify the graph:	
-2	$-\frac{10}{3}$
1	$-\frac{4}{3}$

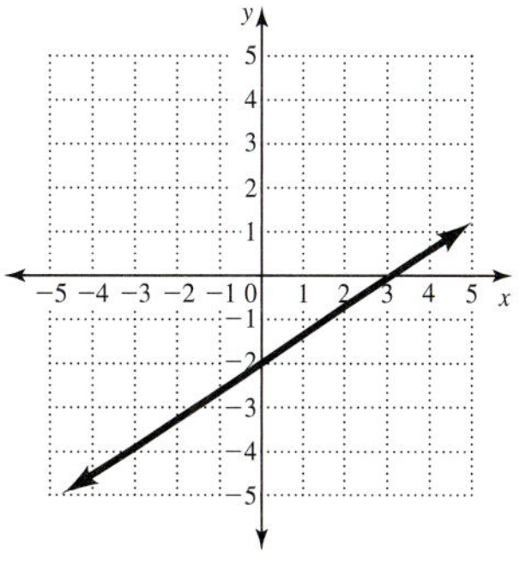

12.

x	$y = 2x - 8$
0	-8
4	0
Two additional points to verify the graph:	
1	-6
3	-2

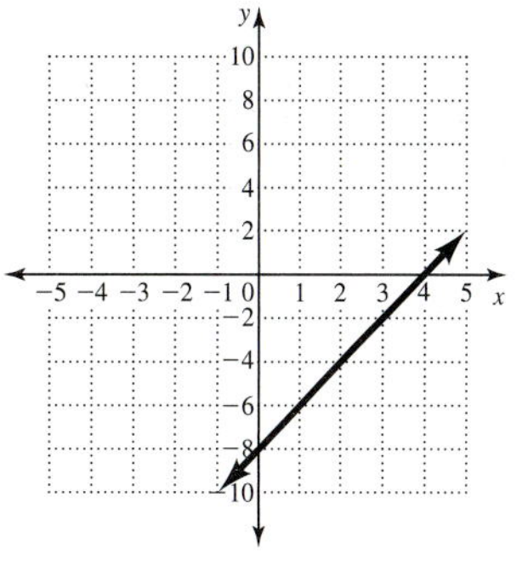

13.

x	$y = -2x + 2$
0	2
1	0
Two additional points to verify the graph:	
-1	4
2	-2

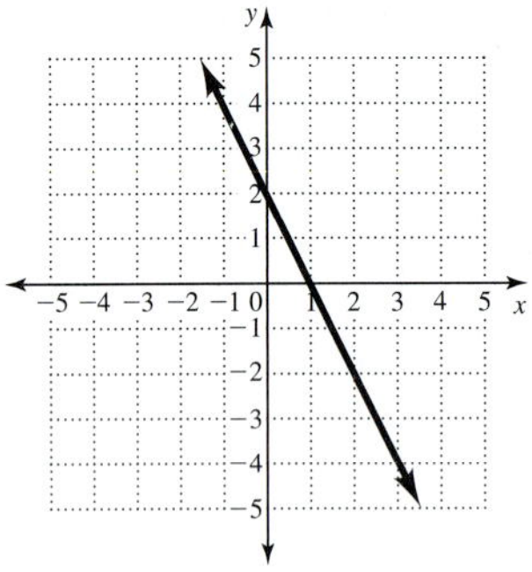

14.

x	$y = \frac{1}{2}x - 1$
0	-1
2	0
Two additional points to verify the graph:	
-2	-2
1	$-\frac{1}{2}$

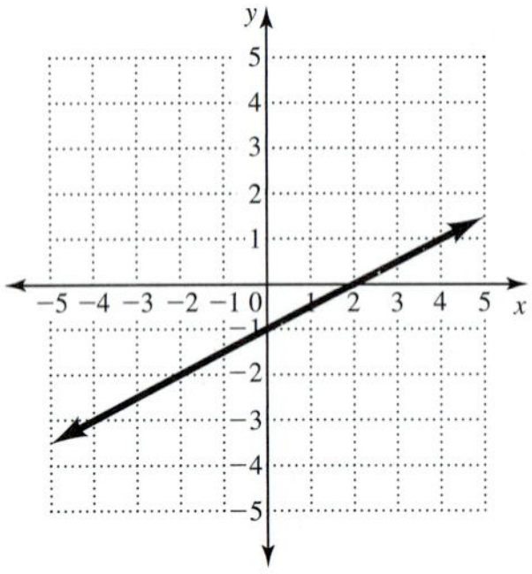

15.

x	$y = -\frac{3}{4}x$
0	0
Two additional points to verify the graph:	
-4	3
4	-3

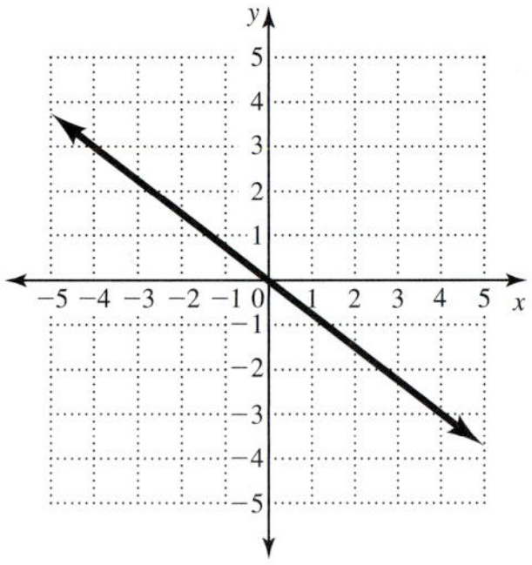

16.

x	$y = 2$
-3	2
-2	2
-1	2
0	2
1	2
2	2
3	2

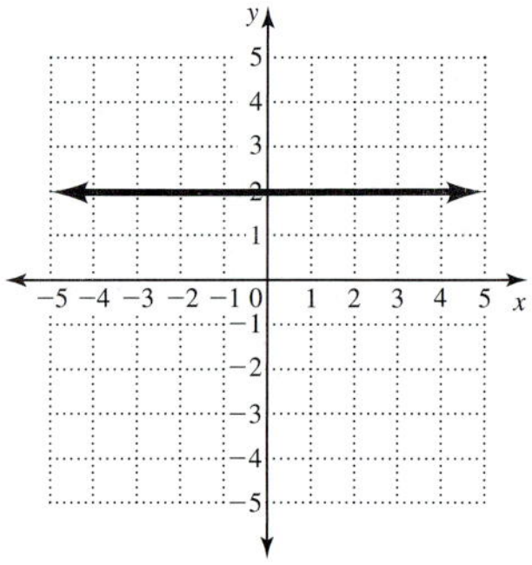

(a) Yes. Passes the vertical line test. (b) $f(x) = 2$

17.

x	$y = -1$
-3	-1
-2	-1
-1	-1
0	-1
1	-1
2	-1
3	-1

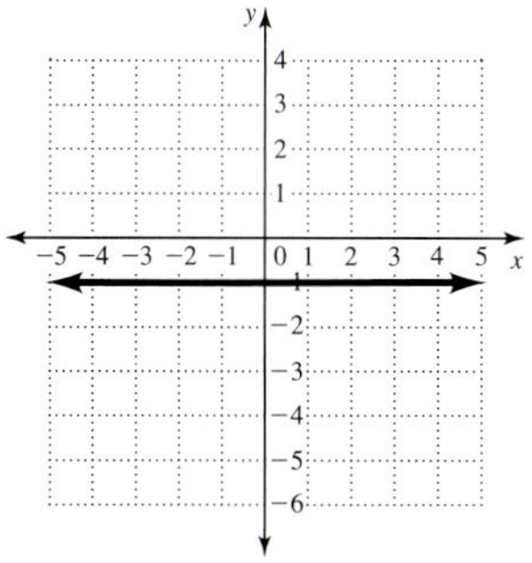

(a) Yes. Passes the vertical line test. (b) $f(x) = -1$

18.

$x = -3$	y
-3	-3
-3	-2
-3	-1
-3	0
-3	1
-3	2
-3	3

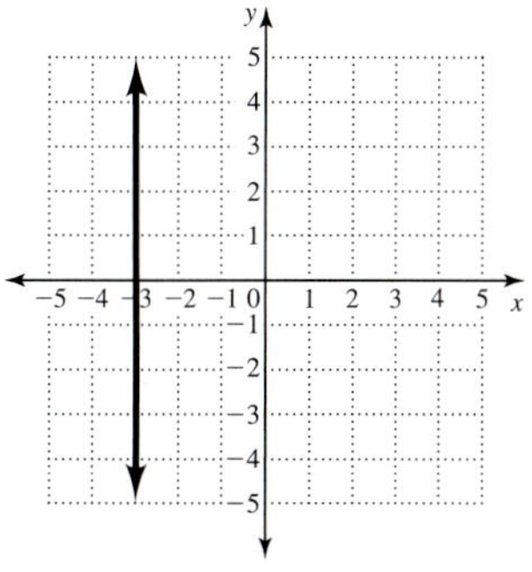

(a) No. Does not pass the vertical line test.
(b) Does not apply.

19.

$x = 0$	y
0	-3
0	-2
0	-1
0	0
0	1
0	2
0	3

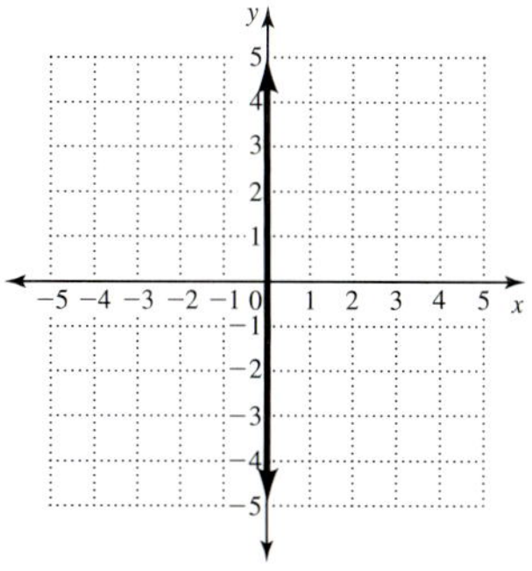

(a) No. Does not pass the vertical line test.
(b) Does not apply.

20. (a)

n	Computations	P
0	$200(0) - 1500$	-1500
10	$200(10) - 1500$	500
15	$200(15) - 1500$	1500
20	$200(20) - 1500$	2500
25	$200(25) - 1500$	3500
n	$200(n) - 1500$	$P(n) = 200n - 1500$

(b)

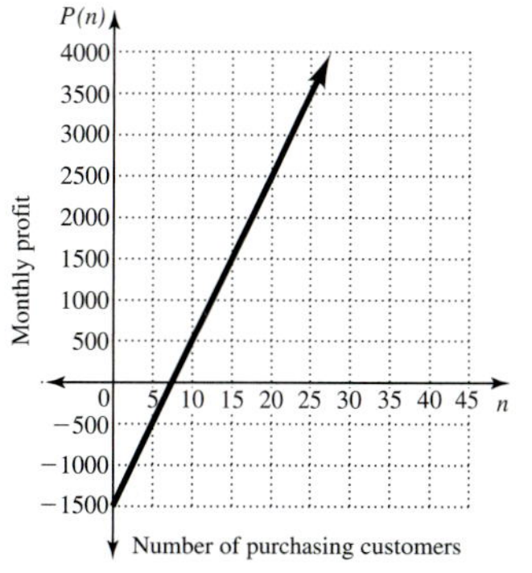

(c) 8 customers, impossible to have 7.5 customers
(d) \$850 (e) $P(12) = \$900$

21. (a)

n	Computations:	S
0	$25(0)$	\$0
5	$25(5)$	\$125
8	$25(8)$	\$200
10	$25(10)$	\$250
12	$25(12)$	\$300
n	$25(n)$	$S(n) = 25n$

(b)

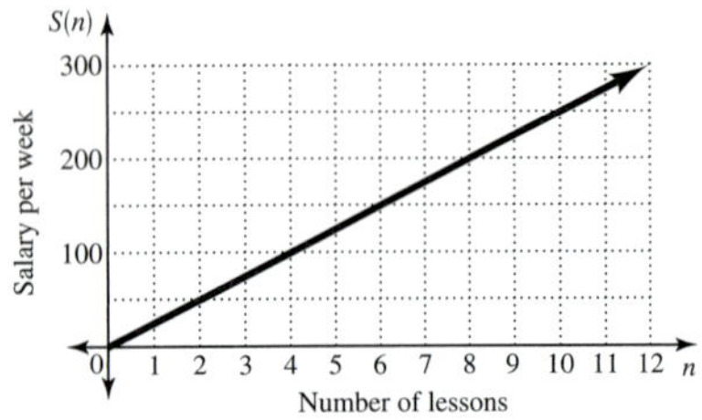

(c) 0 lesson (d) \$500 (e) 11 lessons
(f) 30 lessons

22. (a)

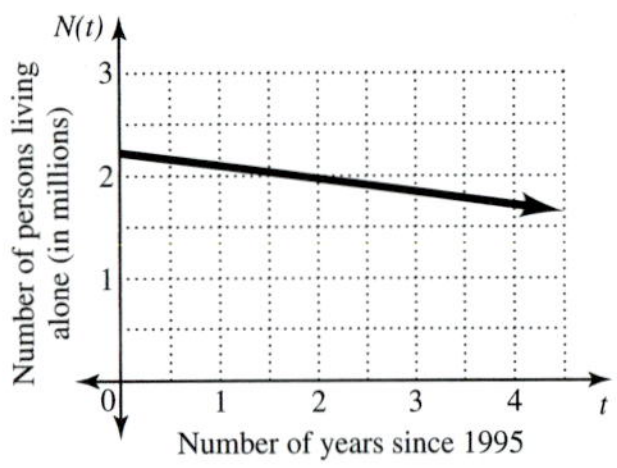

(b) 1.938 million (c) $N(10) = 1.74$ million
23. The graph of a horizontal line is a function because for each input value there is one and only one output value. It passes the vertical line test.
24. The graph of a vertical line is not a function, since the single input value has an infinite number of output values. It does not pass the vertical line test.

DOING MATHEMATICS SECTION 3.4

1. Killington, Stowe, Sugarbush **2.** (a) 1678.24 feet (b) $\frac{3160}{1678.24} \approx 1.88$ **3.** (a) 52,747.68 feet (b) $\frac{2350}{52{,}747.68} \approx 0.04$ **4.** 0.27 **5.** 0.40 **6.** 0.65 **7.** 0.75 **8.** The slope is not very steep. **9.** The slope is very steep. **10.** (a) Horizontal (b) Increasing (c) Vertical (d) Decreasing **11.** (a) Increasing (b) Horizontal (c) Decreasing (d) Vertical **12.** (a) Undefined (b) Negative (c) Positive (d) Zero **13.** (a) Positive (b) Undefined (c) Zero (d) Negative
14. (a) y-intercept of the line: (0, 3)

(b) Slope of the line: 1

x	$f(x)$
−2	1
2	5

14.

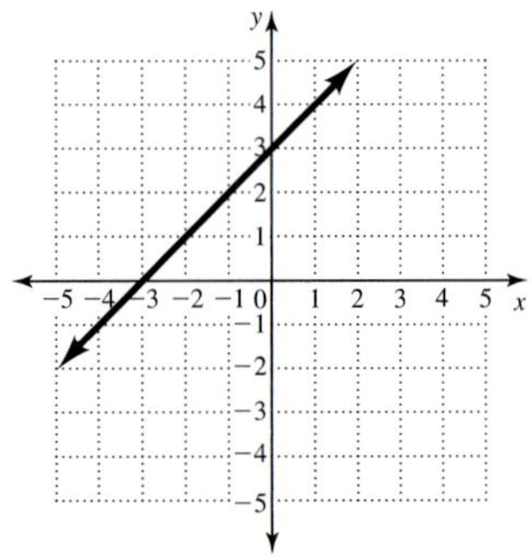

15. (a) y-intercept of the line: (0, −1)

(b) Slope of the line: 1

x	$f(x)$
−1	−2
2	1

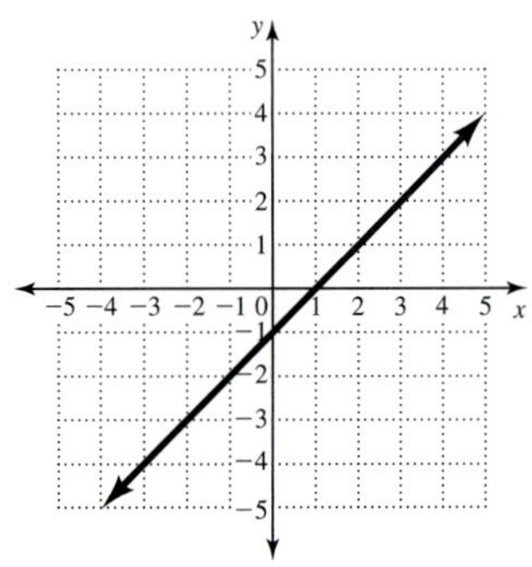

16. (a) y-intercept of the line: (0, 1)

(b) Slope of the line: 2

x	y
−2	−3
2	5

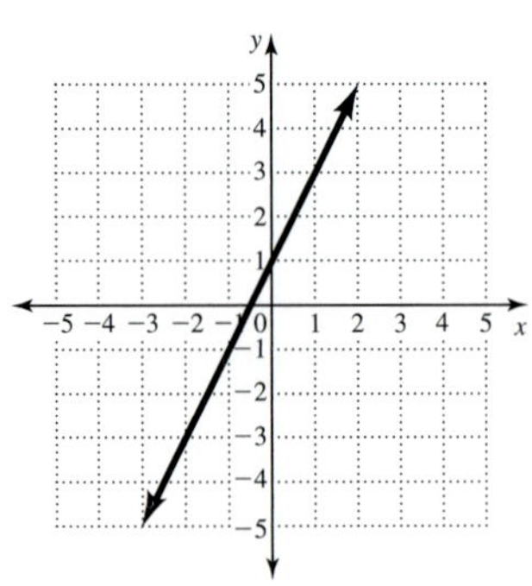

17. (a) y-intercept of the line: (0, 2)
(b) Slope of the line: 3

x	y
-2	-4
-1	-1

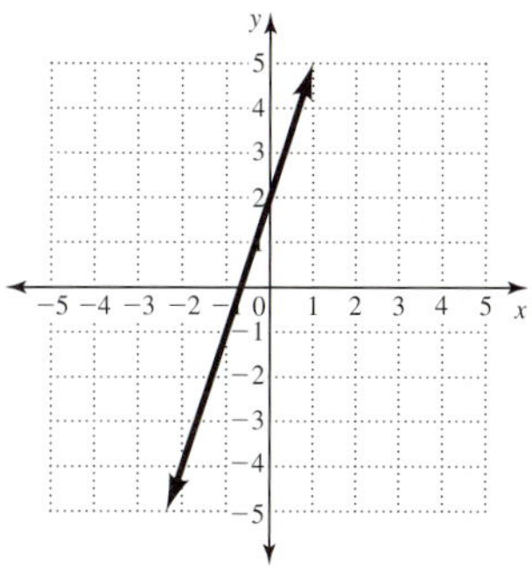

18. (a) y-intercept of the line: (0, 1)
(b) Slope of the line: -3

x	y
-1	4
2	-5

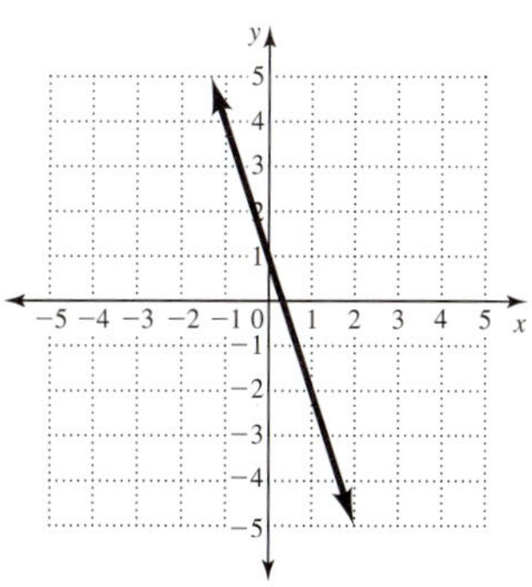

19. (a) y-intercept of the line: (0, -3)
(b) Slope of the line: -2

x	y
-2	1
1	-5

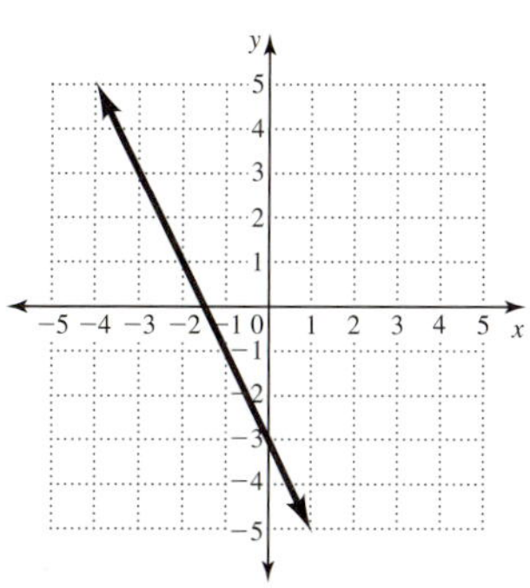

20. (a) y-intercept of the line: (0, 4)
(b) Slope of the line: 0

x	y
-1	4
2	4

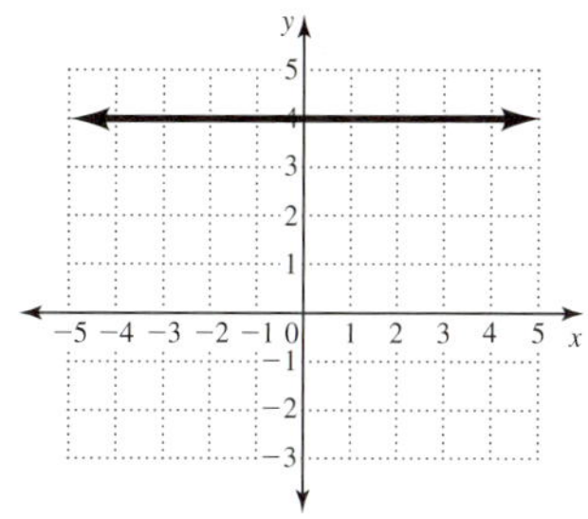

21. (a) y-intercept of the line: (0, -2)
(b) Slope of the line: 0

x	y
-1	-2
2	-2

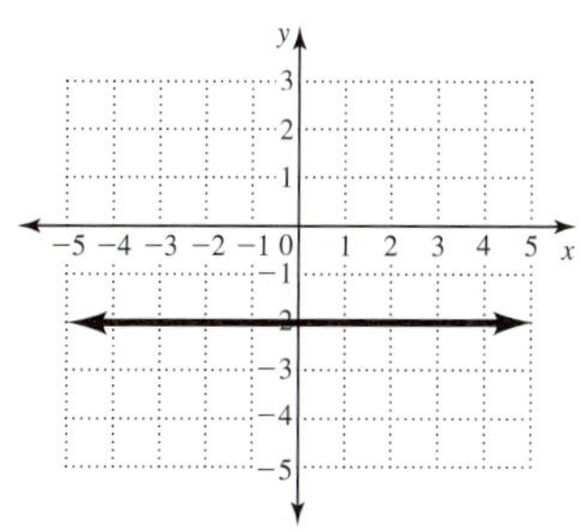

22. (a) y-intercept of the line: None
(b) Slope of the line: Undefined

x	y
-2	1
-2	2

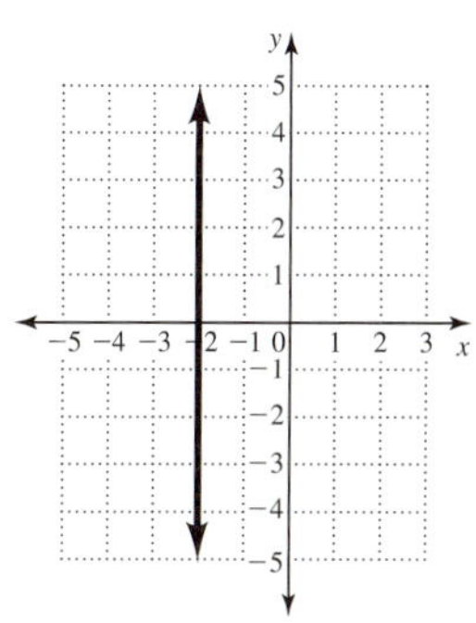

23. (a) y-intercept of the line: None
(b) Slope of the line: Undefined

x	y
4	−1
4	2

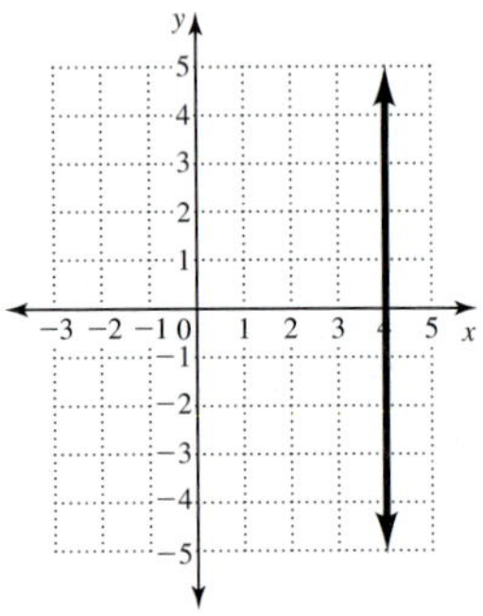

24. (a) y-intercept of the line: $(0, -1)$
(b) Slope of the line: $\frac{2}{3}$

x	y
−3	−3
3	1

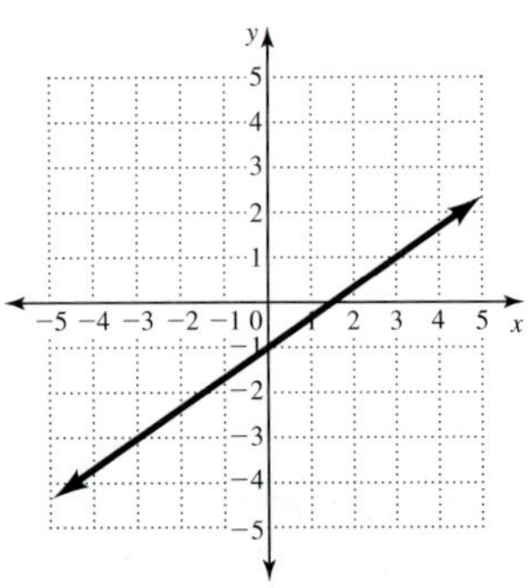

25. (a) y-intercept of the line: $(0, -2)$
(b) Slope of the line: $\frac{3}{4}$

x	y
−4	−5
4	1

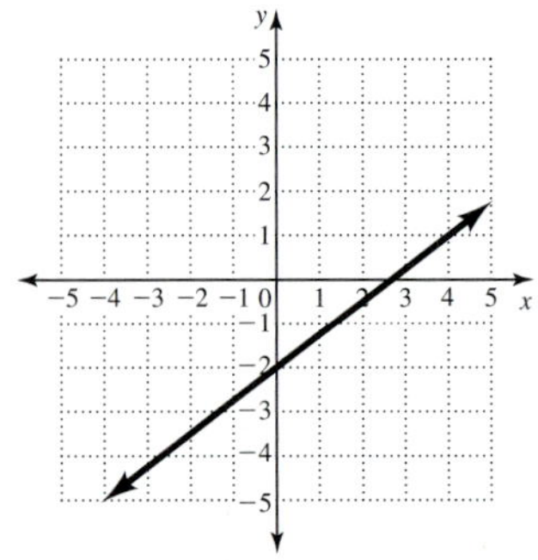

26. (a) y-intercept of the line: $(0, 3)$
(b) Slope of the line: $-\frac{2}{5}$

x	y
−1	$\frac{17}{5}$
5	1

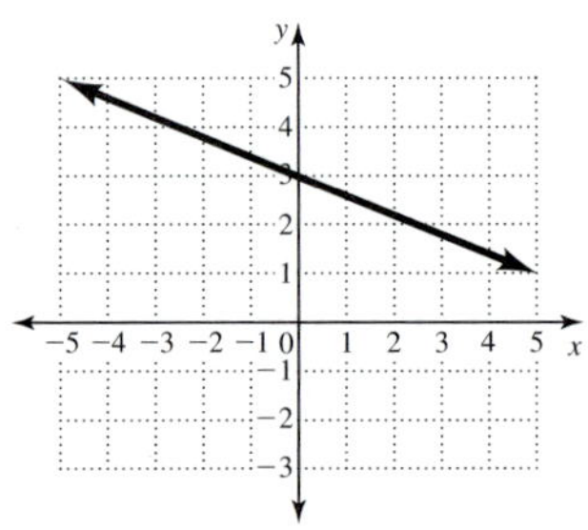

27. (a) y-intercept of the line: $(0, 1)$
(b) Slope of the line: $-\frac{3}{2}$

x	y
−1	$\frac{5}{2}$
2	−2

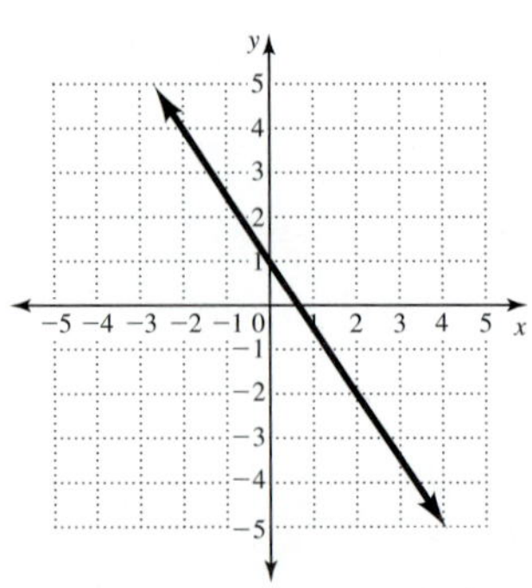

28. (a) y-intercept of the line: (0, 2)
(b) Slope of the line: -3

x	$f(x)$
-1	5
2	-4

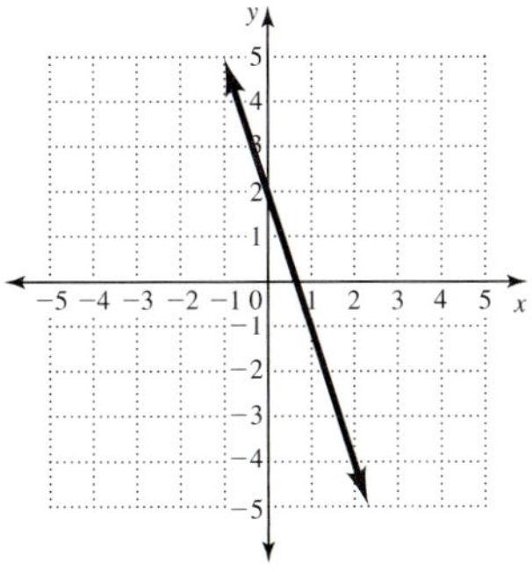

29. (a) y-intercept of the line: (0, 1) (b) Slope of the line: -2

x	$f(x)$
-1	3
2	-3

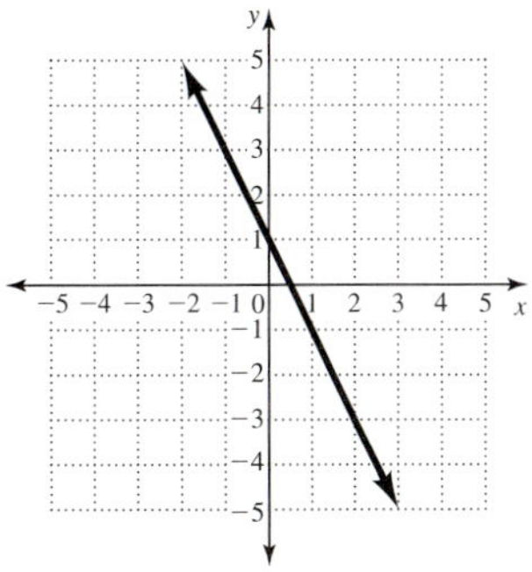

30. $y = 5x - 2$ **31.** $y = 4x - 1$ **32.** $y = \frac{2}{3}x + 1$
33. $y = \frac{2}{7}x + 3$ **34.** $y = 2.3x - 4.2$
35. $y = 1.75x - 3.5$ **36.** $y = -6x + 5$
37. $y = -4x + 3$ **38.** $y = -x + \frac{1}{2}$ **39.** $y = -x - \frac{5}{7}$
40. $y = -x$ **41.** $y = 3x$ **42.** (a) 1.8 (b) (0, 32) (c) $y = 1.8x + 32$ or $F = 1.8C + 32$ (d) 55.4°F **43.** (a) $m = 65$ (b) (0, 0) (c) $y = 65x$ (d) 390 miles **44.** (a) $m = 0.1$ (b) (0, 4.2) (c) $y = 0.1x + 4.2$ (d) 5.2 thousand **45.** (a) $m = 0.1$ (b) (0, 2.3) (c) $y = 0.1x + 2.3$ (d) 3.3 thousand **46.** (a) $m = -0.7$ (b) (0, 14.8) (c) $y = -0.7x + 14.8$ (d) 7.8 thousand

47. (a) $m = -0.1$ (b) (0, 7.8) (c) $y = -0.1x + 7.8$ (d) 6.8 thousand **48.** Slope is the ratio of vertical displacement (RISE) to horizontal displacement (RUN).

DOING MATHEMATICS SECTION 3.5

1–4. Answers will vary.

1. $\frac{\text{60 miles}}{\text{2 hours}}$ **2.** $\frac{\text{20 meters}}{\text{3 seconds}}$

3. $\frac{\text{3 Free throws made}}{\text{4 Free throws attempted}}$ **4.** $\frac{\text{20 feet}}{\text{1 second}}$

5. (a) Decreasing (b) -3 (c) -4 **6.** (a) Increasing (b) 2 (c) $+5$ **7.** (a) Increasing (b) 3 (c) -4 **8.** (a) Decreasing (b) -2 (c) $+5$ **9.** $y = 5x + 1$ **10.** $y = 4x + 3$
11. $y = -5x - 1$ **12.** $y = -3x - 2$
13. $y = 5x - 5$ **14.** $y = 4x - 10$ **15.** $y = -2x + 7$
16. $y = -3x + 10$ **17.** $y = 6x + 5$
18. $y = 2x - 1$ **19.** $y = \frac{1}{2}x + \frac{3}{2}$ **20.** $y = \frac{1}{3}x + \frac{7}{3}$
21. $y = -\frac{5}{3}x - \frac{4}{3}$ **22.** $y = -\frac{2}{5}x - 3$ **23.** $y = -6x + 4$
24. $y = -x + 9$ **25.** $y = 1$ **26.** $y = 2$
27. $y = x - 2$ **28.** $y = -x + 5$ **29.** $y = -1$
30. $y = 5$ **31.** $y = \frac{3}{2}x + 8$ **32.** $y = -\frac{5}{2}x + 9$
33. $x = -5$ **34.** $x = -2$ **35.** $y = 7x + 21$
36. $y = \frac{3}{5}x$ **37.** $-5x + y = 1$ or $5x - y = -1$
38. $-4x + y = 3$ or $4x - y = -3$ **39.** $5x + y = -1$
40. $3x + y = -2$ **41.** $-5x + y = -5$ or $5x - y = 5$
42. $-4x + y = -10$ or $4x - y = 10$ **43.** (a) 3 (b) 4 (c) 3 (d) -6 (e) They are equal.
44. (a) -2 (b) -3 (c) -2 (d) 5 (e) They are equal. **45.** (a) (2008, 244) (b) $\frac{1}{1}$ (c) $y = x - 1764$ (d) 248 thousand
46. (a) (2008, 433) (b) $\frac{3}{1}$ (c) $y = 3x - 5591$ (d) 445 thousand **47.** (a) $m \approx 1.4$ (b) $y = 1.4x + 325.3$ (c) 374.3 ppm
48. (a)

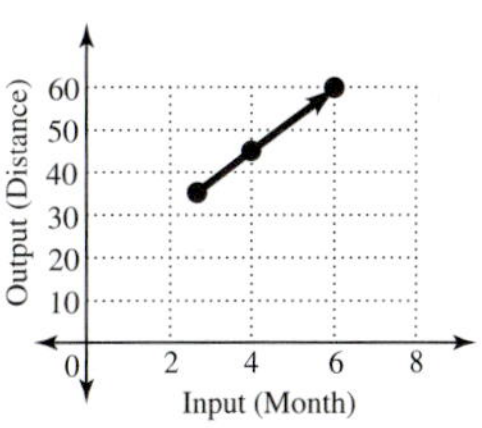

(b) $m = 7$ (c) $f(x) = y = 7x + 17$ (d) $f(7) = 66$ feet (e) No
49. (a) $m = 0.25$ (b) $y = 0.25x$ (c) 5%

50. (a) $m = -0.0325$ (b) $y = -0.0325x + 2.065$ (c) 1.74 million **51.** (a) $m = 0.0975$ (b) $y = 0.0975x + 3.875$ (c) 4.85 million

CHAPTER 3 REVIEW

1. (a) I (b) I (c) II (d) III (e) III (f) II (g) IV (h) IV (i) III

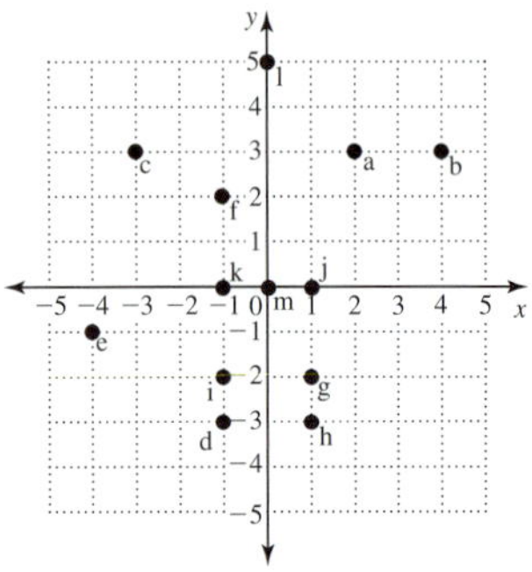

2. −2 and 4 **3.** −2 **4.** 3 **5.** 1 and 4

6. (a)

x = Number of credit hours	y = Tuition cost	Ordered pairs
1	185	(1, 185)
5	925	(5, 925)
10	1850	(10, 1850)
15	2775	(15, 2775)

(b) 1, 15, 5 (c) 185, 2775, 350

(d)

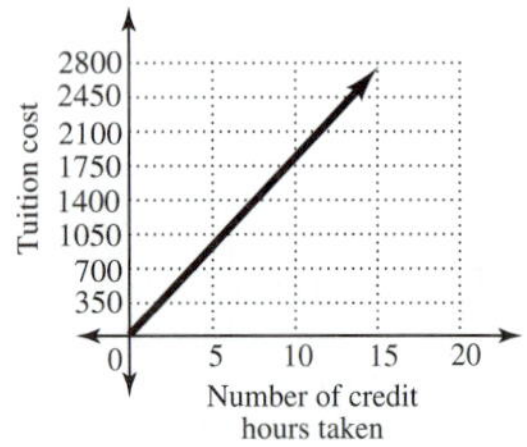

7. (a)

Ordered pairs
(10, 120.5)
(15, 92.4)
(20, 40.2)
(25, 28.5)
(30, 13.6)

(b) 10, 30, 5 (c) 13.6, 120.5, 15

(d)

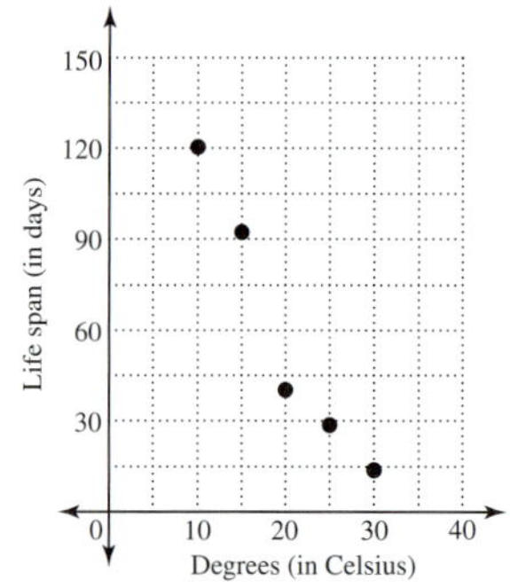

8. (a) Dollar sales of equipment (b) Total salary

(c)

s = Sales	Computations	T = Total salary
1000	20,000 + 0.30(1000)	\$20,300
5000	20000 + 0.30(5000)	\$21,500
10,000	20000 + 0.30(10000)	\$23,000
15,000	20000 + 0.30(15000)	\$24,500
20,000	20000 + 0.30(20000)	\$26,000
s	$20000 + 0.30(s)$	$T(s) = 20,000 + 0.30s$

(d)

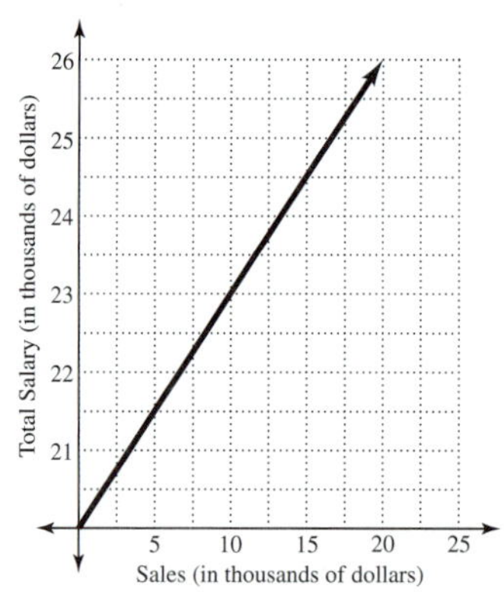

(e) \$25,000 (f) $T(16,000) = \$24,800$ (g) \$7000

9. (a) Miles driven (b) Cost of renting the car

(c)

m = Number of miles driven	C = Cost of renting the car
50	\$55
100	\$60
150	\$65
200	\$70
250	\$75
m	$C(m) = 50 + 0.10m$

(d)

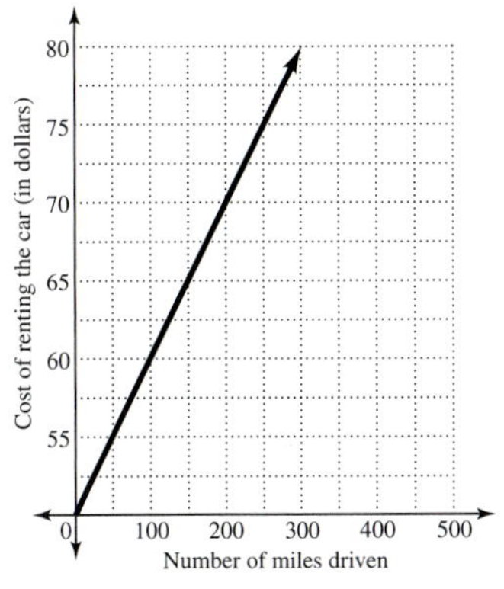

(e) \$80 (f) $C(300) = \$80$ (g) 180 miles

10. Not a function.

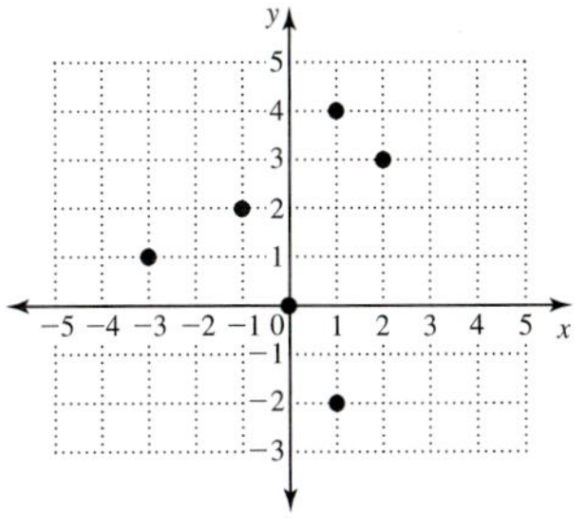

11. Yes, a function.

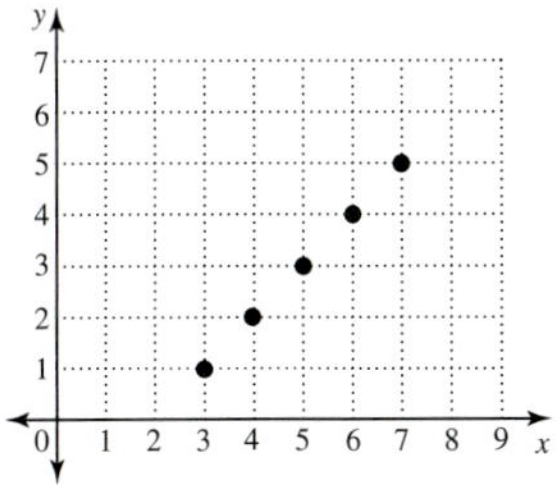

12. $f(x) = 2x - 1$.

x	$f(x)$	Ordered pairs
−2	−5	(−2, −5)
−1	−3	(−1, −3)
0	−1	(0, −1)
1	1	(1, 1)
2	3	(2, 3)

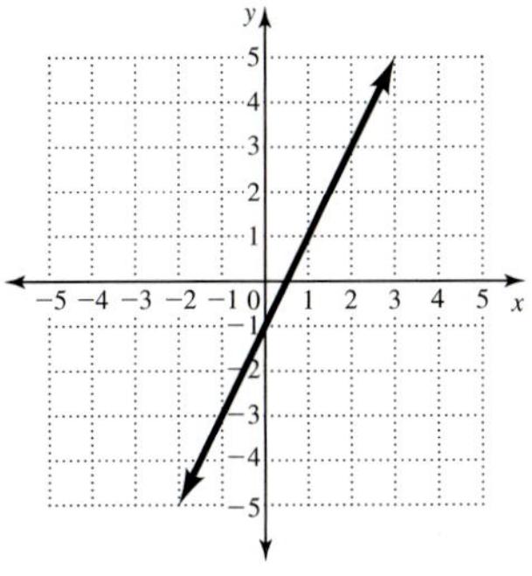

13. (a) 25 (b) 4 **14.** (a) −2 (b) 7 **15.** (a) 4 (b) 39 **16.** (a) 50 (b) 23 **17.** (b) **18.** (a) and (c) **19.** (a) x-intercept at (1.5, 0) (b) y-intercept at (0, 3) **20.** (a) x-intercept at (4, 0) (b) y-intercept at (0, −4)

21. (a)

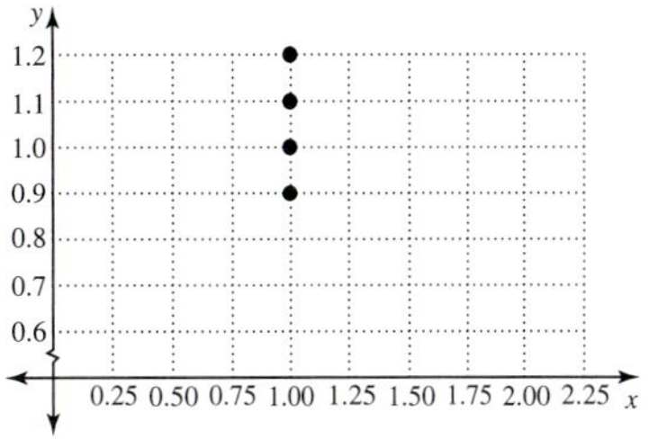

(b) No. For the input value of 1.0, there are four output values.

22. $3x - y = 3$

x	y
0	−3
1	0
Two additional points to verify the graph:	
−1	−6
2	3

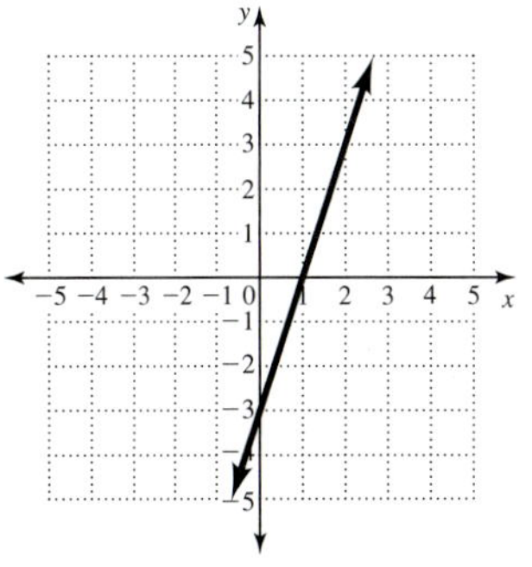

23. $3x + y = -6$

x	y
0	−6
−2	0
Two additional points to verify the graph:	
−1	−3
−2	0

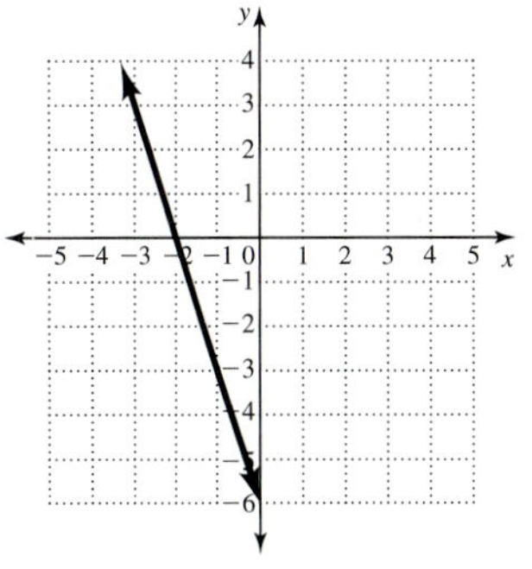

24. $y = -2$

x	y
−2	−2
−1	−2
0	−2
1	−2
2	−2

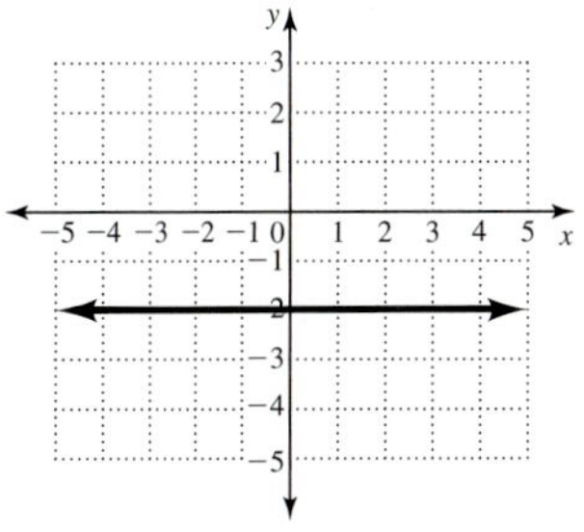

(a) The equation and graph represent a function. For each input value (x) there is one and only one output value (y). (b) $f(x) = -2$

25. $x = 4$

x	y
4	−2
4	−1
4	0
4	1
4	2

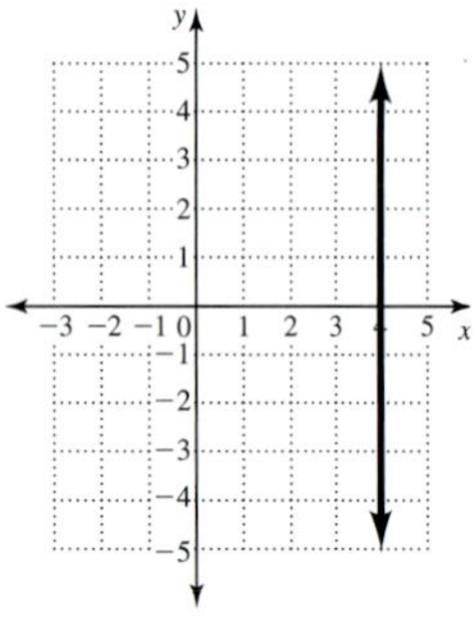

(a) No. For the input value of 4, there are an infinite number of output values. (b) N/A

26. $\frac{1080}{4100} \approx 0.26$ **27.** $\frac{1090}{2500} \approx 0.44$ **28.** $\frac{1310}{2115} \approx 0.62$ **29.** $\frac{820}{999} \approx 0.82$ **30.** It has the greatest slope. **31.** (a) Decreasing (b) Increasing (c) Zero (d) Undefined **32.** (a) $\frac{5}{9}$ (b) −17.78 (c) $y = \frac{5}{9}x - 17.78$ (d) 17.22° C

33. (a)

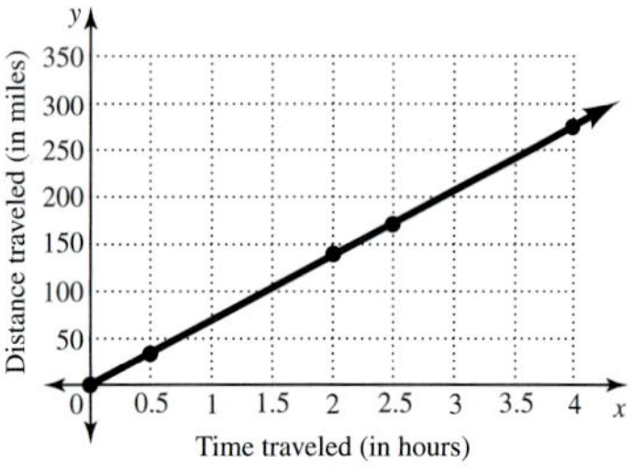

(b) 68 (c) 0 (d) $y = 68x$ (e) 408 miles
34. (a) $m = 0.8$ (b) (0, 17.0) (c) $y = 0.8x + 17.0$ (d) 25 thousand **35.** (a) $m = -0.1$ (b) (0, 6.6) (c) $y = -0.1x + 6.6$ (d) 5.6 thousand

36. (a) y-intercept of the line: (0, −3)

x	$y = 2x - 3$
−1	−5
2	1

(b) Slope of the line: 2

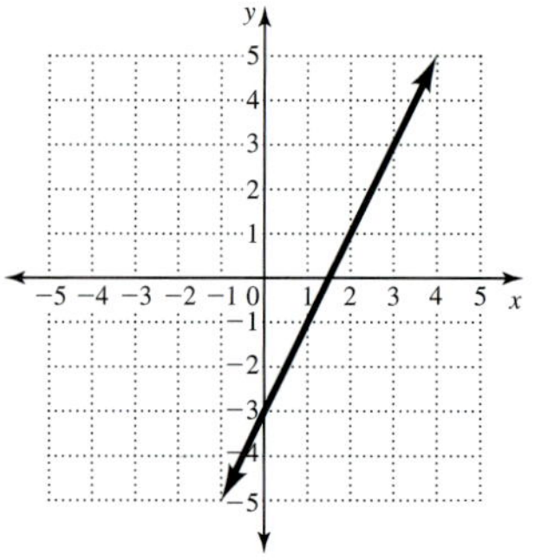

37. (a) y-intercept of the line: (0, 1)

x	$y = -x + 1$
−1	2
2	−1

(b) Slope of the line: −1

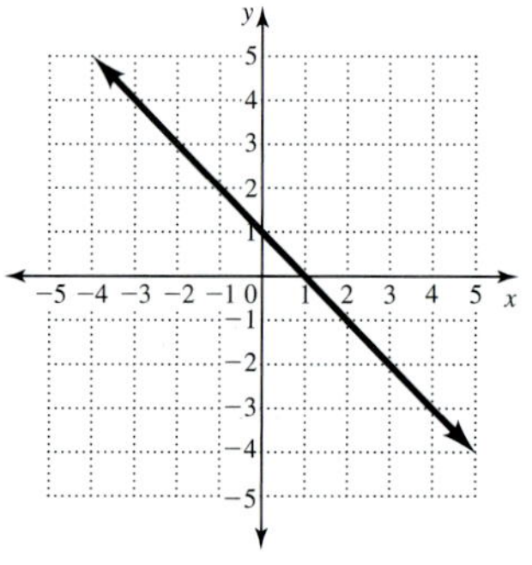

38. (a) y-intercept of the line: (0, 5)

x	$y = 5$
−1	5
2	5

(b) Slope of the line: 0

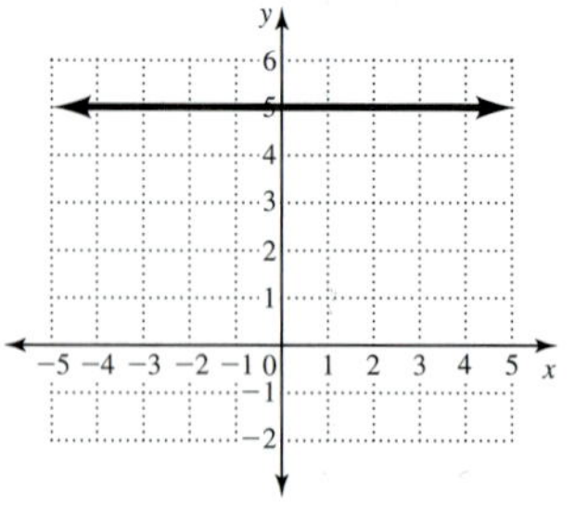

39. (a) y-intercept of the line: None

x	y
2	-1
2	3

(b) Slope of the line: Undefined

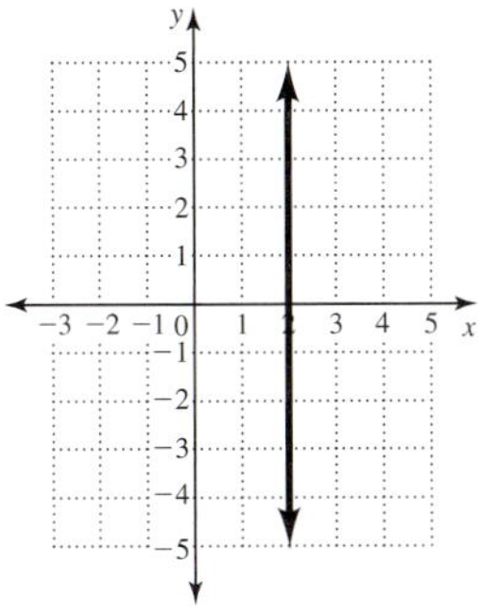

40. (a) 20.2 (b) $y = 20.2x + 197$ (c) 661.6 billion miles **41.** (a) (2010, 1456) (b) $\dfrac{15}{1}$ (c) $y = 15x - 28{,}694$ (d) 1501 thousand

42. $y = x$ **43.** $y = \dfrac{-10}{3}x + \dfrac{1}{3}$ **44.** $y = x - 5$

45. $y = 7x - 1$ **46.** $y = \dfrac{1}{2}x + 4$

47. $y = \dfrac{-2}{3}x - 3$ **48.** $y = \dfrac{-5}{7}x$

49. $y = -6x + 54$ **50.** $x - y = 0$

51. $10x + 3y = 1$ **52.** $x - y = 5$ **53.** $7x - y = 1$

CHAPTER 3 PRACTICE TEST ANSWERS

1. (a) 3 (b) -3 (c) -1 (d) -2 and 4

2. (a) Smallest first coordinate value: 1
Largest first coordinate value: 8
Scale for horizontal axis: 1

(b) Smallest second coordinate value: 0
Largest second coordinate value: 1000
Scale for vertical axis: 100

(c)

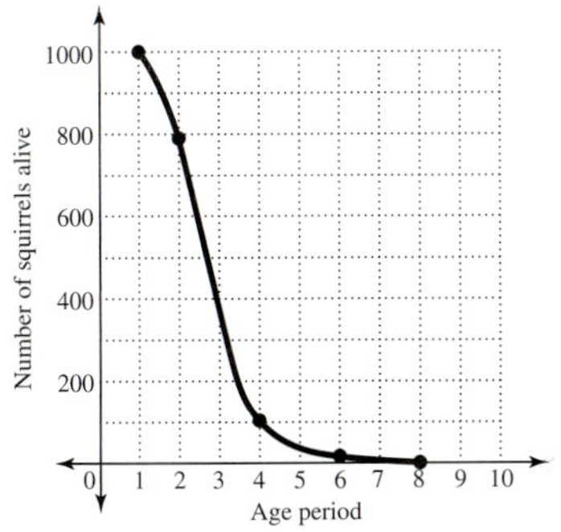

(d) 400 squirrels **3.** (a) Input: the number of tickets sold. (b) Output: the profit from sale of tickets.

(c)

$n =$ Number of raffle tickets sold	$P =$ Total profit
20	75
40	175
60	275
80	375
100	475
n	$P(n) = 5n - 25$

(c–d)

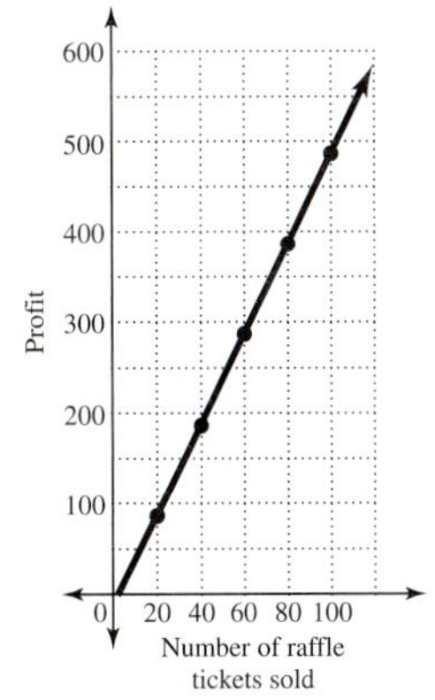

(e) \$430 (f) $P(90) = \$425$ (g) 85 tickets

4. Not a function

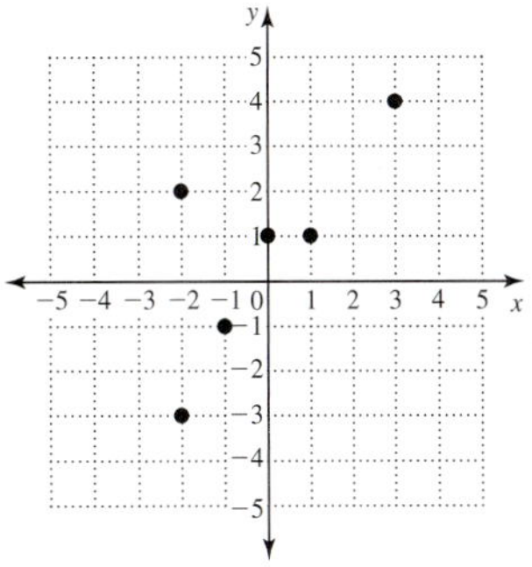

5. Function

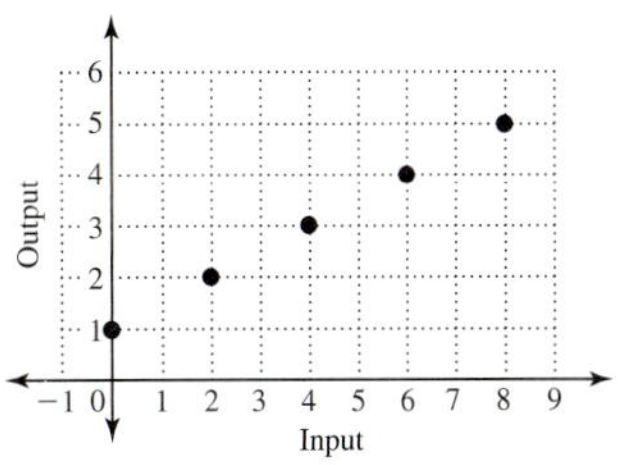

6. (a) $f(3) = 17$ (b) $f(-1) = -3$ (c) $g(-3) = 16$ (d) $g(0) = -5$ **7.** (a), (c), and (f)

8.

x	$f(x)$
0	3
−3	0
Two additional points:	
−1	2
1	4

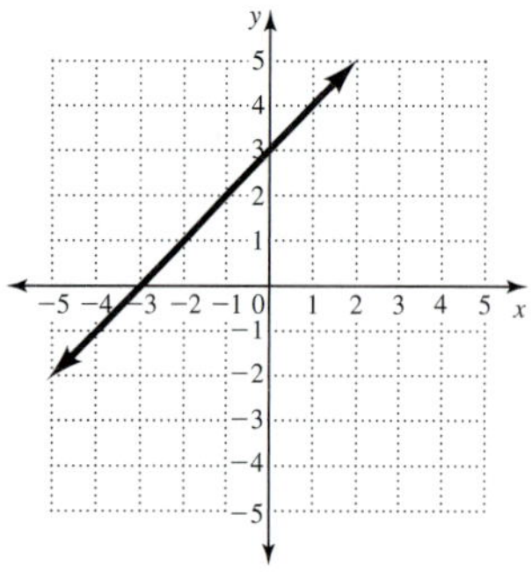

9.

x	$y = 3$
−2	3
−1	3
0	3
1	3
2	3

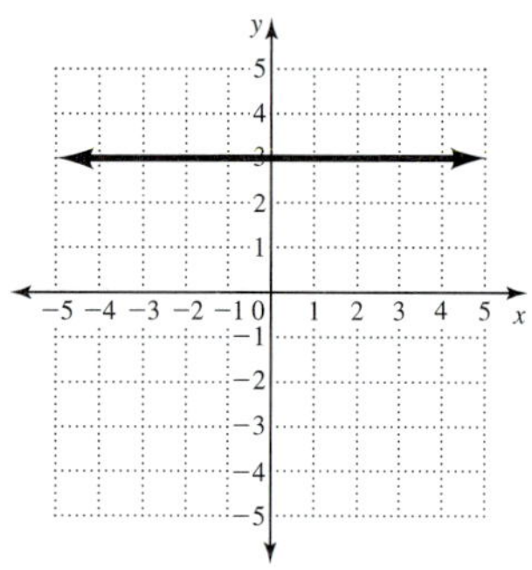

10. (a) y-intercept of the line: $(0, -1)$

(b) Slope of the line: 3

x	$f(x) = 3x - 1$
−1	−4
2	5

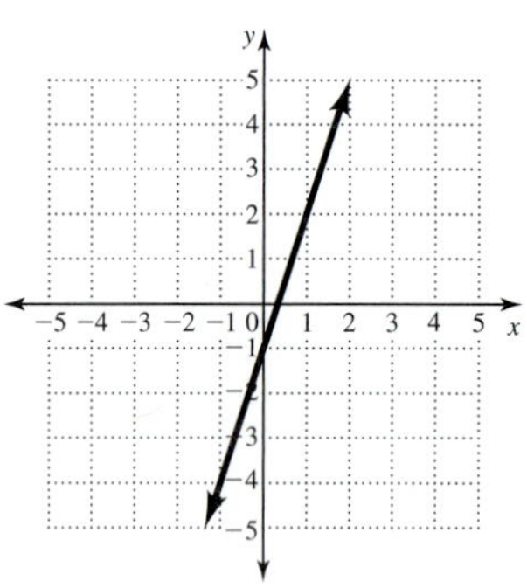

11. (a) y-intercept of the line: $(0, -1)$

(b) Slope of the line: $-\frac{3}{2}$

x	$f(x) = -\frac{3}{2}x - 1$
−2	2
2	−4

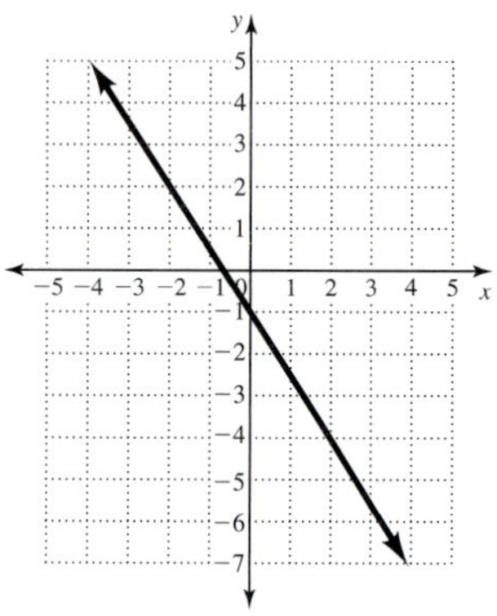

12. (a) y-intercept of the line: 0

(b) Slope of the line: −2

x	$y = -2x$
−1	2
2	−4

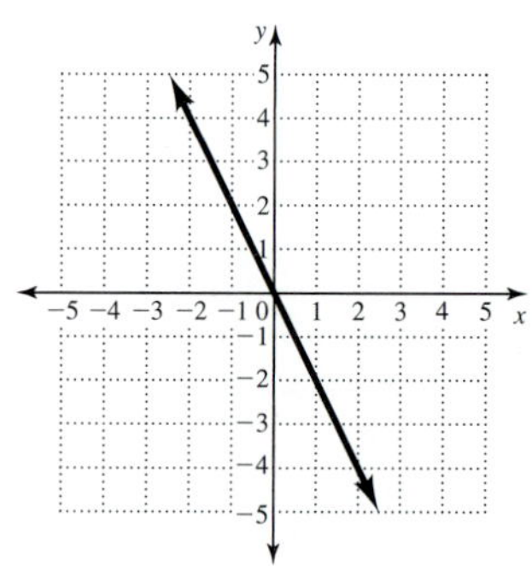

13. $y = 5x + 2$ **14.** $y = \frac{3}{2}x - 1$ **15.** $x = -1$

16. $y = -6$ 17. (a) $m = \frac{1}{4}$ (b) $y = \frac{1}{4}x$ (c) 3%

18. $y = -\frac{1}{2}x + \frac{5}{2}$, $x + 2y = 5$

19. $y = \frac{4}{5}x - \frac{11}{5}$, $4x - 5y = 11$

20. $y = \frac{4}{3}x + \frac{8}{3}$, $4x - 3y = -8$

21. $y = \frac{1}{2}x + 1$, $x - 2y = -2$

CUMULATIVE TEST CHAPTER 1–3 ANSWERS

1. (a) $-5 - (-3) = -2$

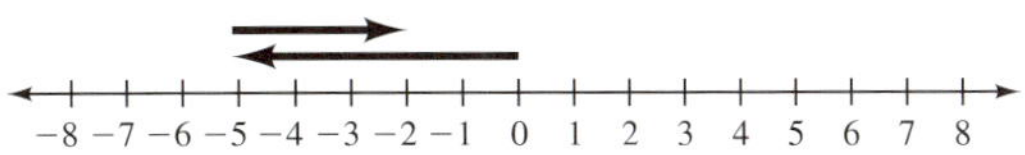

(b) $-6 + 4 = -2$

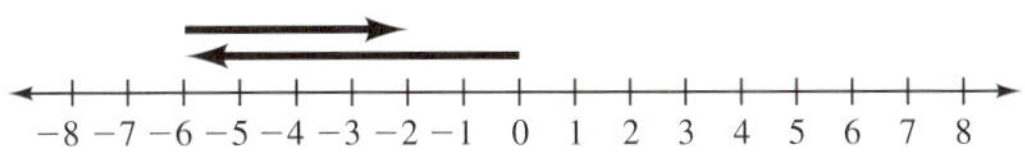

2. (a) Positive (b) Negative (c) Positive (d) Negative **3.** (a) -12 (b) -64 **4.** -5 **5.** -22 **6.** -16 **7.** 2187 **8.** $111 profit **9.** (a) -44 (b) 62 **10.** $x^2 - 14x - 13$ **11.** $-x^2 - 6x + 7$ **12.** $20x^3 + 3x^2 - 9x + 11$ **13.** $y = -13$ **14.** $x = 9$

15. (a) Graph: 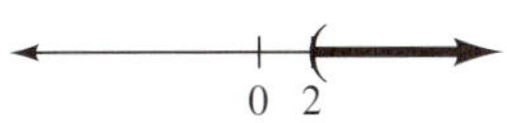

(b) $\{z | z > 2\}$ (c) $(2, \infty)$

16. (a) Graph: 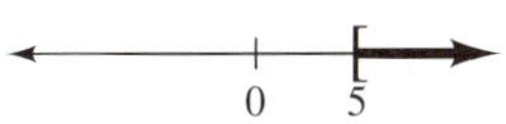

(b) $\{x | x \geq 5\}$ (c) $[5, \infty)$ **17.** $y = \dfrac{C - Ax}{B}$

18. $y = 3x - 17$

19. (a)

n = Number of t-shirts sold	P = Total profit
20	0
40	20
60	40
80	60
100	80
n	$P(n) = n - 20$

(b)

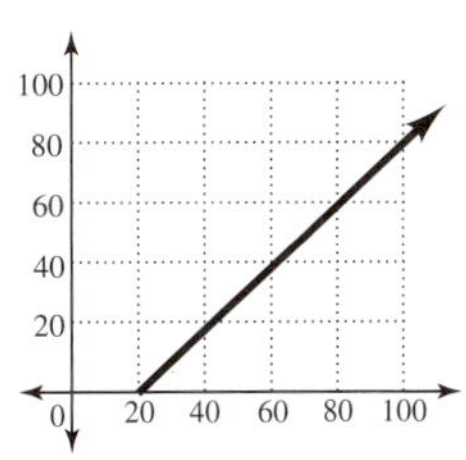

(c) $35 (d) $P(55) = 35$ (e) Must sell 20 shirts to break even **20.** (a) $f(-3) = 1$ (b) $f(1) = 1$ (c) $g(-2) = 5$ (d) $g(0) = -5$

21.

x	$f(x) = -x + 3$
0	3
3	0
Two additional points to verify the graph:	
-1	4
1	2

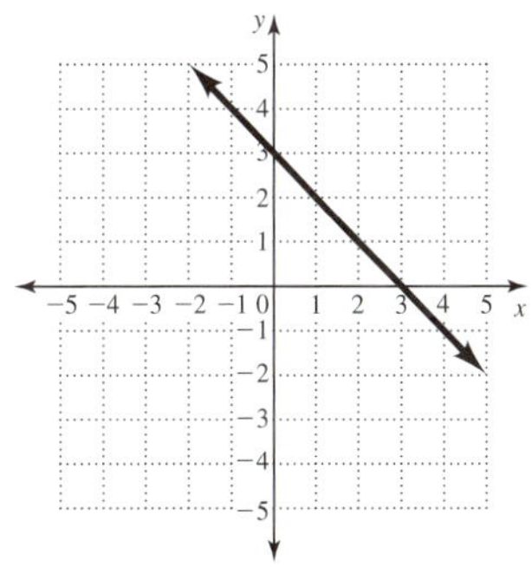

22. $y = -\dfrac{5}{2}x + \dfrac{1}{2}$; $5x + 2y = 1$
23. $y = -3x - 2$; $3x + y = -2$
24. $y = 6x - 2$; $6x - y = 2$
25. $y = -\dfrac{1}{2}x - \dfrac{1}{2}$; $x + 2y = -1$
26. (a) $m = 0.5$ (b) $(0, 4.2)$ (c) $y = 0.5x + 4.2$ (d) 9.2 thousand

CHAPTER 4

DOING MATHEMATICS SECTION 4.1

1. (a) $x = -1$

(b)

x	$y = 2(x + 1)$
-3	-4
-2	-2
-1	0
0	2
1	4
2	6
3	8

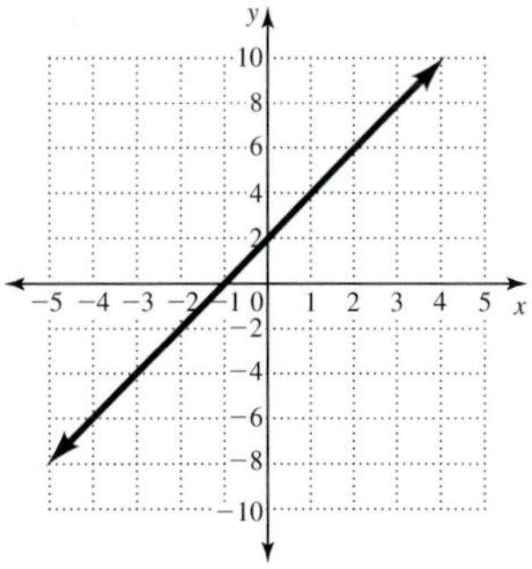

(c) $x = -1$ (d) Yes; $x = -1$ checks

2. (a) $x = 2$

(b)

x	$y = 3(x - 2)$
−3	−15
−2	−12
−1	−9
0	−6
1	−3
2	0
3	3

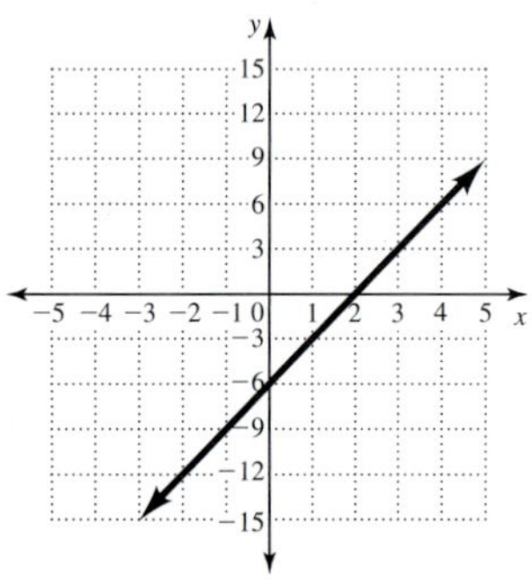

(c) $x = 2$ (d) Yes; $x = 2$ checks

3. (a) $x = 1$

(b)

x	$y = 2x - 5 + 3$
−3	−8
−2	−6
−1	−4
0	−2
1	0
2	2
3	4

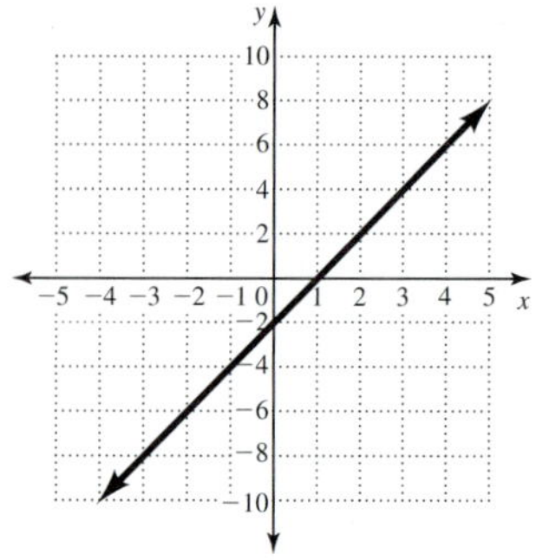

(c) $x = 1$ (d) Yes; $x = 1$ checks

4. (a) $x = -2$

(b)

x	$y = 3x + 4 + 2$
−3	−3
−2	0
−1	3
0	6
1	9
2	12
3	15

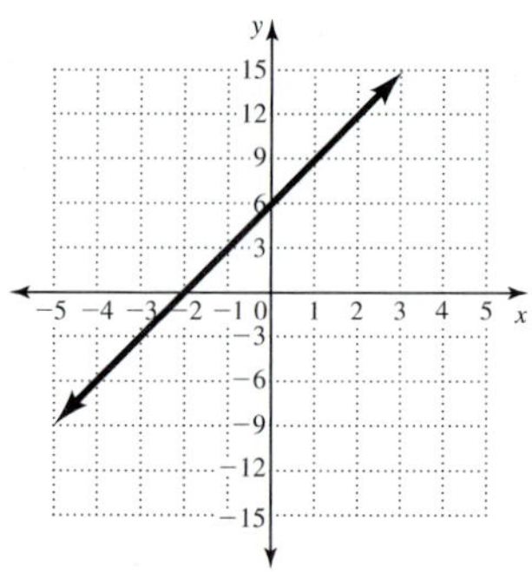

(c) $x = -2$ (d) Yes; $x = -2$ checks

5. (a) $x = 0$

(b)

x	$y = -(x + 1) + 1$
−3	3
−2	2
−1	1
0	0
1	−1
2	−2
3	−3

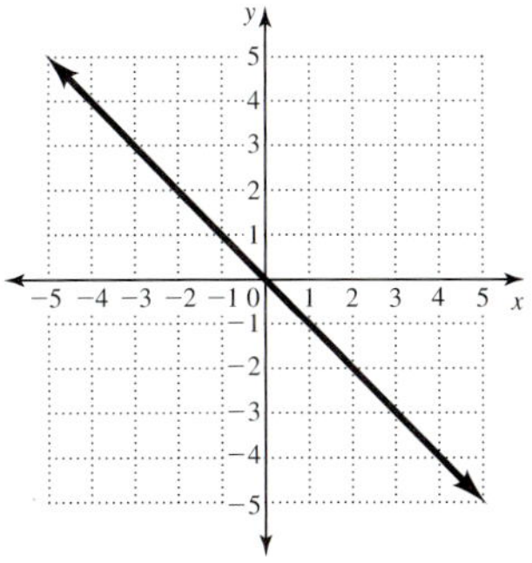

(c) $x = 0$ (d) Yes; $x = 0$ checks

6. (a) $x = 5$

(b)

x	$y = -2(x - 3) + 4$
−1	12
1	8
3	4
5	0
7	−4
9	−8

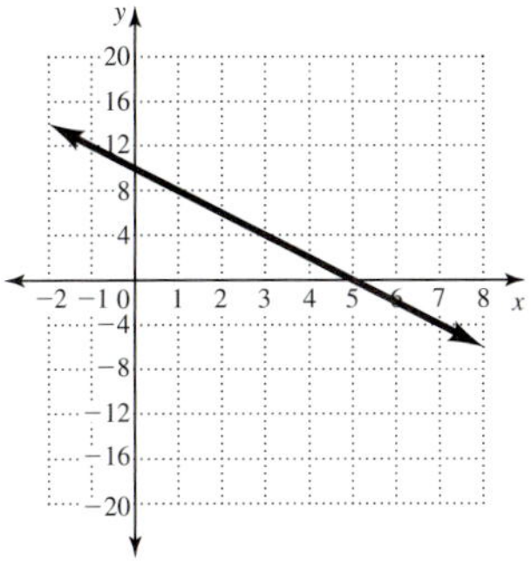

(c) $x = 5$ (d) Yes; $x = 5$ checks

7. (a) $x = -3$

(b)

x	$y = -17 - 6x - 1$
−5	12
−3	0
−1	−12
1	−24
3	−36

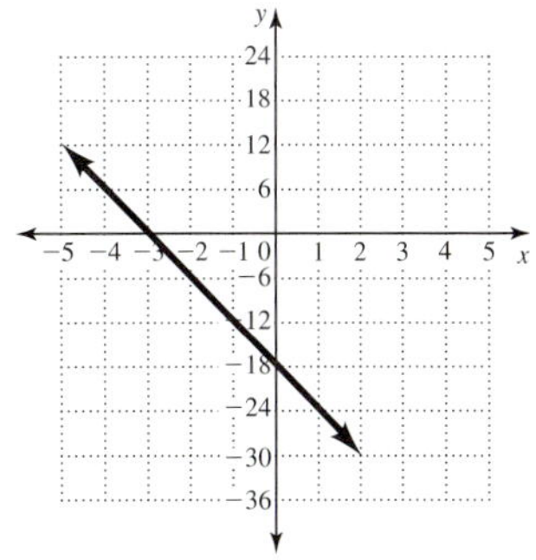

(c) $x = -3$ (d) Yes; $x = -3$ checks

8. (a) $x = 5$

(b)

x	$y = -7 + 4x - 13$
2	−12
3	−8
4	−4
5	0
6	4
7	8

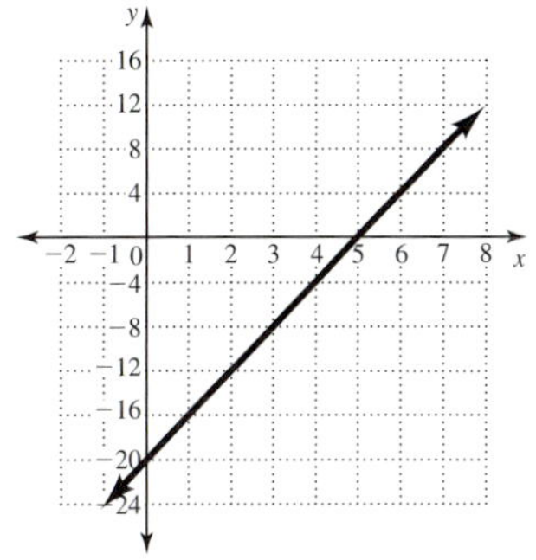

(c) $x = 5$ (d) Yes; $x = 5$ checks

9. (a) $x = 4$

(b)

x	$y_1 = 8x - 3$	$y_2 = 3(2x + 3) - x$
−1	−11	4
0	−3	9
1	5	14
2	13	19
3	21	24
4	29	29
5	37	34

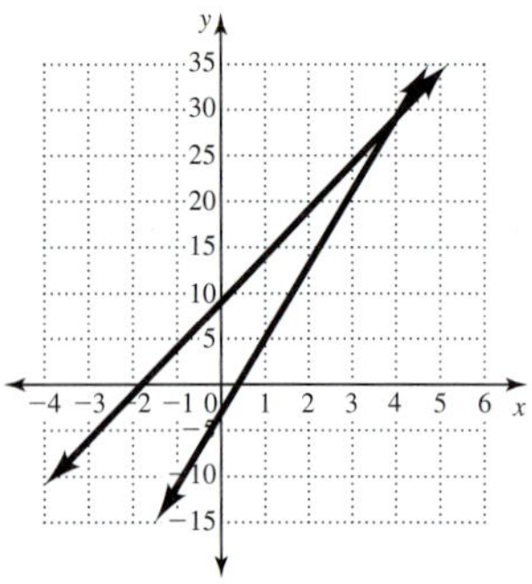

(c) $x = 4$ (d) Yes; $x = 4$ checks

10. (a) $x = 4$

(b)

x	$y_1 = 2(6x - 1)$	$y_2 = 2(3 + 4x) + 2x$
2	22	26
3	34	36
4	46	46
5	58	56
6	70	66

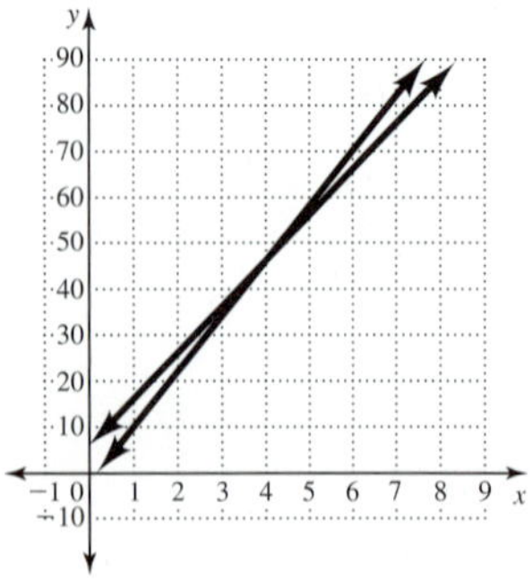

(c) $x = 4$ (d) Yes; $x = 4$ checks

11. (a) $x = -3$

(b)

x	$y_1 = 2(3x + 5)$	$y_2 = 2(x - 1)$
−4	−14	−10
−3	−8	−8
−2	−2	−6
−1	4	−4
0	10	−2

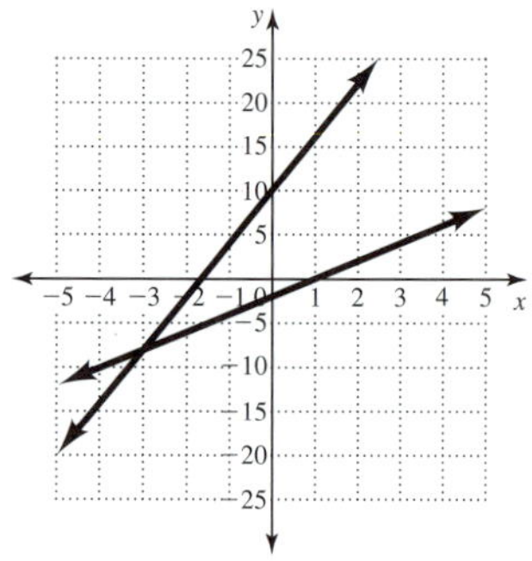

(c) $x = -3$ (d) Yes; $x = -3$ checks

12. (a) $x = 3$

(b)

x	$y_1 = 6.3x - (1.3x + 2)$	$y_2 = 4x + 1$
−1	−7	−3
0	−2	1
1	3	5
2	8	9
3	13	13
4	18	17
5	23	21

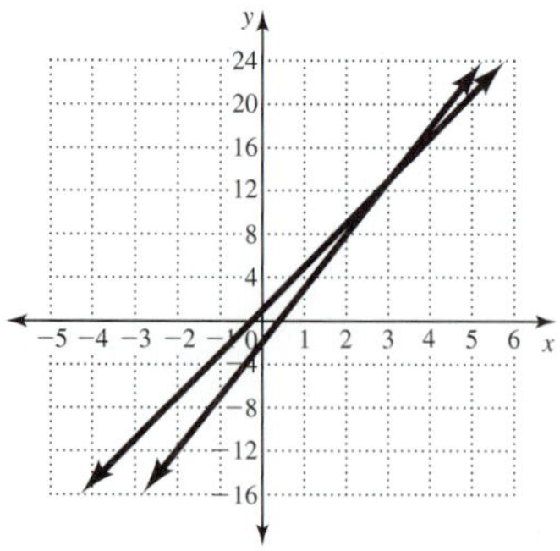

(c) $x = 3$ (d) Yes; $x = 3$ checks

13. (a) $x = -5$

(b)

x	$y_1 = -2.1x - (4.2x + 3.5)$	$y_2 = -3x + 13$
−7	40.6	34
−6	34.3	31
−5	28	28
−4	21.7	25
−3	15.4	22

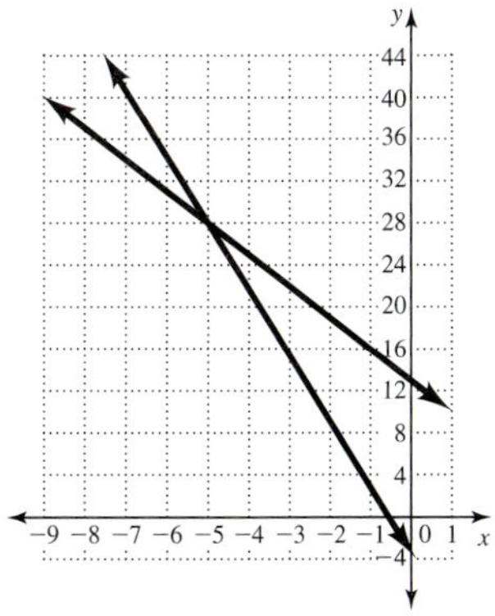

(c) $x = -5$ (d) Yes; $x = -5$ checks

14. (a) $x = -3$

(b)

x	$y_1 = 5x + 2$	$y_2 = 2(x - 3.5)$
−5	−23	−17
−4	−18	−15
−3	−13	−13
−2	−8	−11
−1	−3	−9

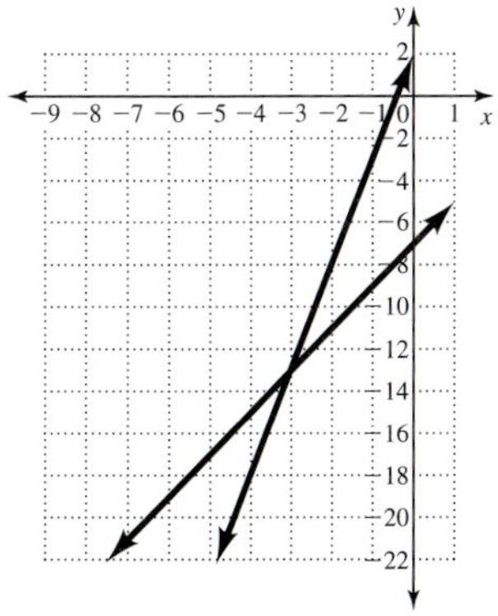

(c) $x = -3$ (d) Yes; $x = -3$ checks

15. (a) $x = 5$

(b)

x	$y_1 = (x + 3) - (x - 4)$	$y_2 = 2x - 3$
1	7	−1
2	7	1
3	7	3
4	7	5
5	7	7
6	7	9
7	7	11

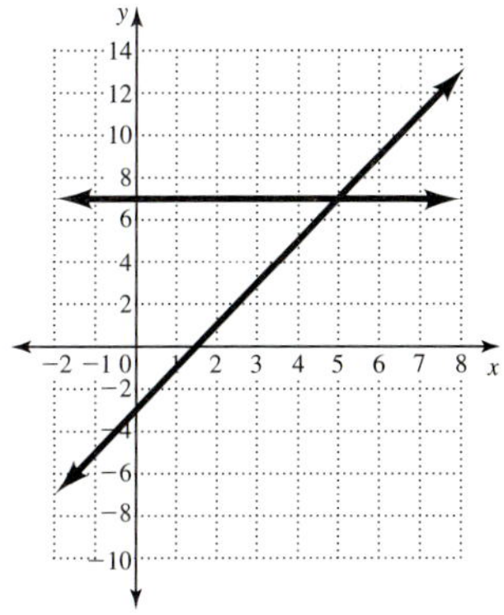

(c) $x = 5$ (d) Yes; $x = 5$ checks

16. (a) $x = 3$

(b)

x	$y_1 = (x - 2) - (x + 3)$	$y_2 = 2x - 11$
1	−5	−9
2	−5	−7
3	−5	−5
4	−5	−3
5	−5	−1

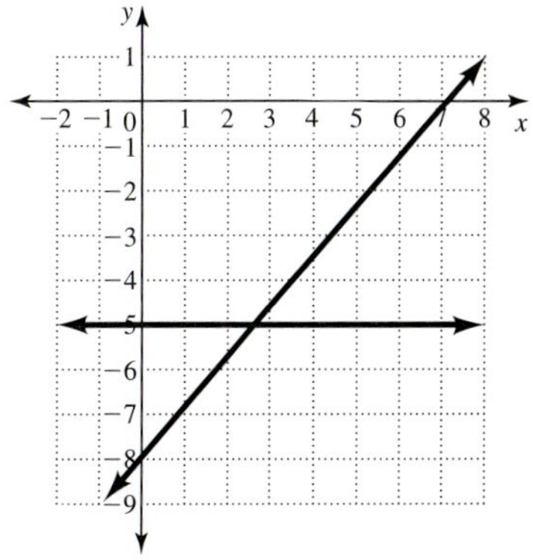

(c) $x = 3$ (d) Yes; $x = 3$ checks

17. (a) $x = -1$ (b) Solving the equation graphically using the Intersection-of-Graphs Method:

x	$y_1 = 3(0.5x + 0.2)$	$y_2 = 0.5x - 0.4$
-2	-2.4	-1.4
-1	-0.9	-0.9
0	0.6	-0.4
1	2.1	0.1
2	5.1	0.6

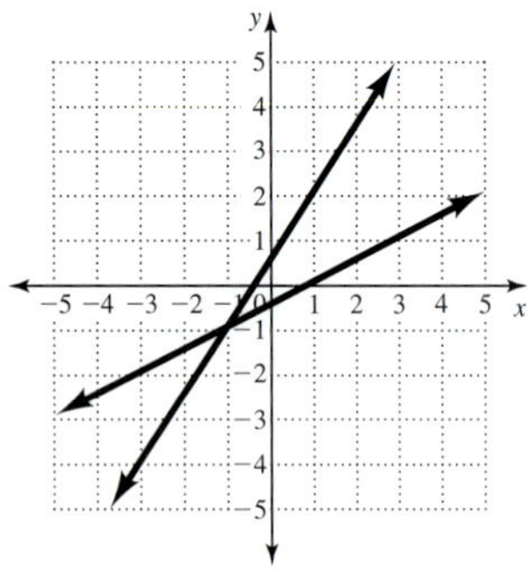

(c) $x = -1$ (d) Yes; $x = -1$ checks

18. (a) $x = 2$ (b) Solving the equation graphically using the Intersection-of-Graphs Method:

x	$y_1 = 3(1.5x - 2)$	$y_2 = 2.5x - 2$
-2	-15	-7
-1	-10.5	-4.5
0	-6	-2
1	-1.5	0.5
2	3	3
3	7.5	5.5

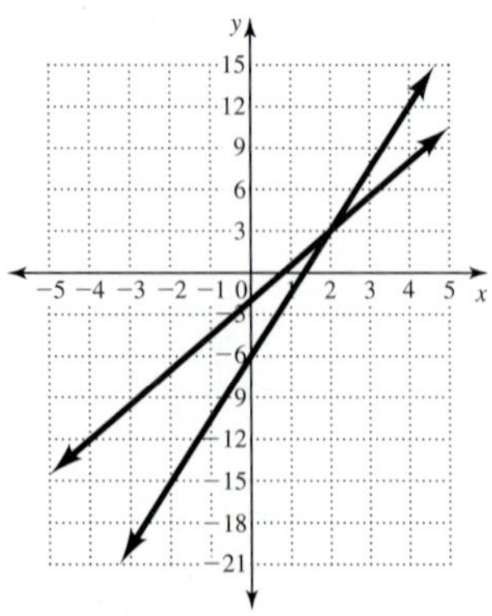

(c) $x = 2$ (d) Yes; $x = 2$ checks

19. (a) $x = -3$ (b) Solving the equation graphically using the Intersection-of-Graphs Method:

x	$y_1 = 21 - 2(x - 4)$	$y_2 = 5(1 - 2x)$
-4	37	45
-3	35	35
-2	33	25
-1	31	15
0	29	5
1	27	-5

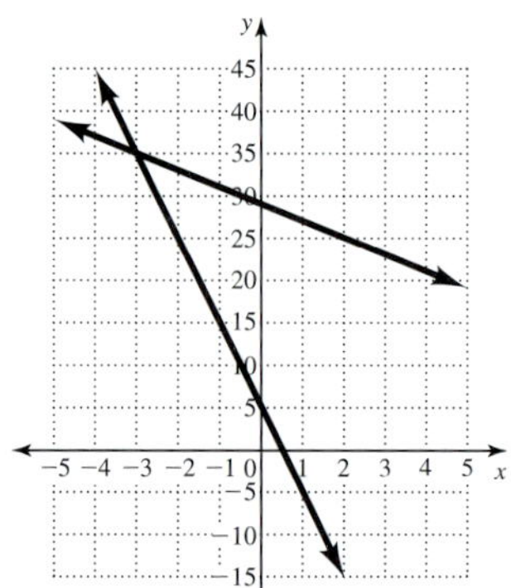

(c) $x = -3$ (d) Yes; $x = -3$ checks

20. (a) $x = -1$ (b) Solving the equation graphically using the X-Intercept Graphical Method:

x	$y = 10 - 3(x + 2) - 2(6 - x) + 7$
-3	2
-2	1
-1	0
0	-1
1	-2
2	-3

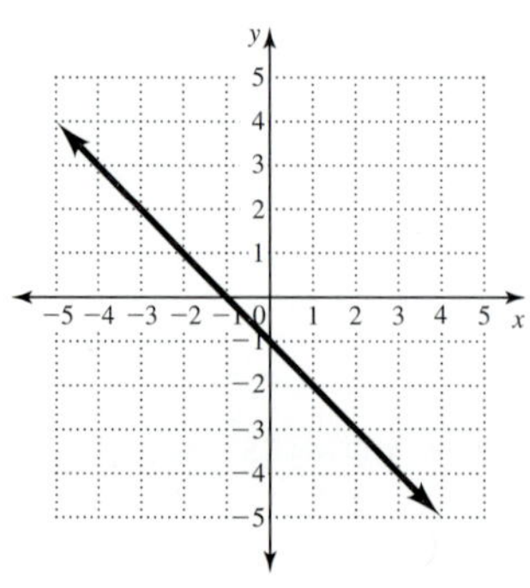

(c) $x = -1$ (d) Yes; $x = -1$ checks

21. (a) $x = 2$ (b) Solving the equation using the X-Intercept Graphical Method:

x	$y = -0.4x + 2.6 - 1.6x + 1.4$
-1	6
0	4
1	2
2	0
3	-2

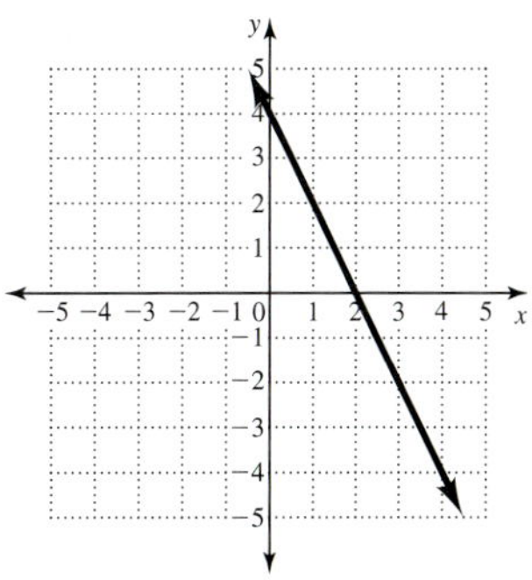

(c) $x = 2$ (d) Yes; $x = 2$ checks

22. (a) $x = 3$ (b) Solving the equation using the X-Intercept Graphical Method:

x	$y = \frac{2}{3}x + \frac{1}{5} - x + \frac{4}{5}$
-1	$\frac{4}{3}$
0	1
1	$\frac{2}{3}$
2	$\frac{1}{3}$
3	0
4	$-\frac{1}{3}$

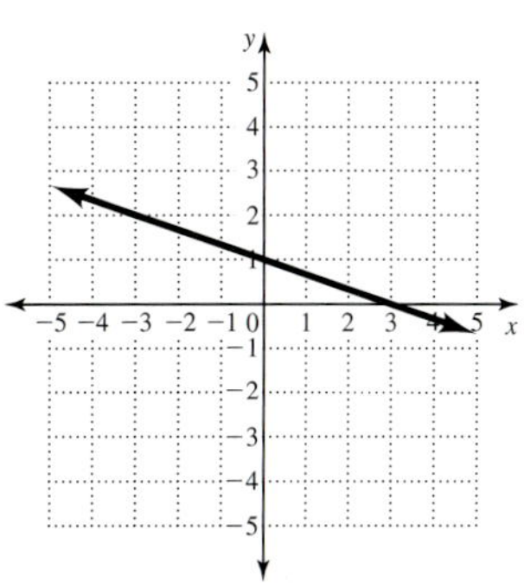

(c) $x = 3$ (d) Yes; $x = 3$ checks

23. (a)

Number of magazines sold	Profit function
0	-5000
2000	-1000
2500	0
3000	1000
4000	3000
5000	5000
x	$P(x) = 2x - 5000$

(b) $0 = 2x - 5000$, $x = 2500$ (c) 2500 magazines
(d) Xscl 1000 Yscl 1000
(e)

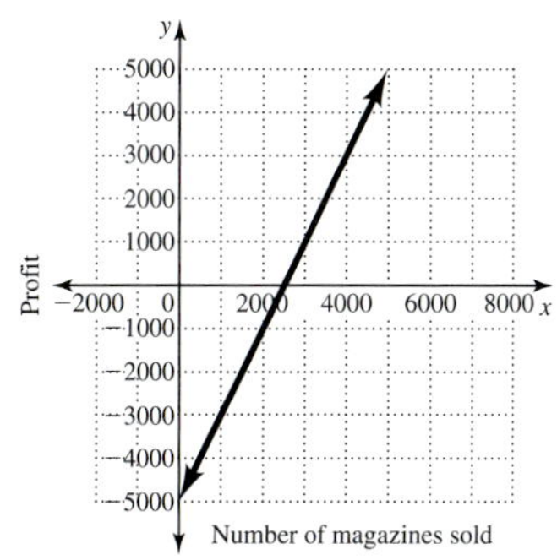

(f) x-intercept: $x = 2500$ (g) The x-intercept of the graph and the breakeven point of the profit function are the same value.

24. (a)

Number of CDs sold	Profit function
1000	$-30,000$
2000	$-15,000$
3000	0
4000	15,000
5000	30,000
x	$P(x) = 15x - 45,000$

(b) $0 = 15x - 45,000$, $x = 3000$ (c) 3000 CDs
(d) Xscl 1000 Yscl 10,000

(e)

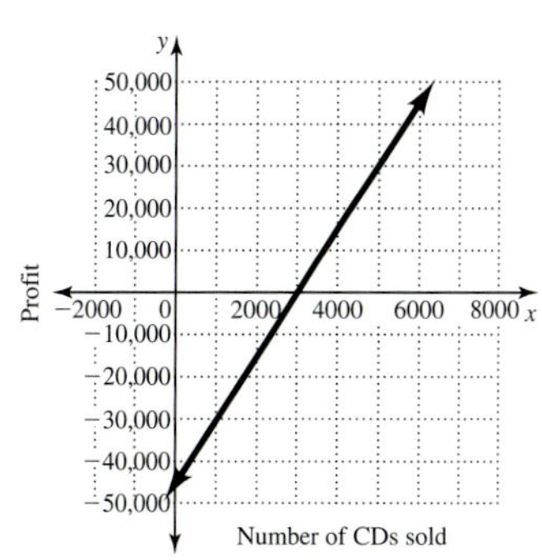

(f) x-intercept: $x = 3000$ (g) The x-intercept of the graph and the breakeven point of the profit function are the same value.

25. (a)

Number of candy bars sold	Profit function
0	−381
200	−81
300	69
400	219
500	369
x	$P(x) = x(1.50) - 500(.75) - 6$

(b) $0 = x(1.50) - 500(.75) - 6, x = 254$
(c) 254 candy bars (d) Xscl 100 Yscl 100
(e)

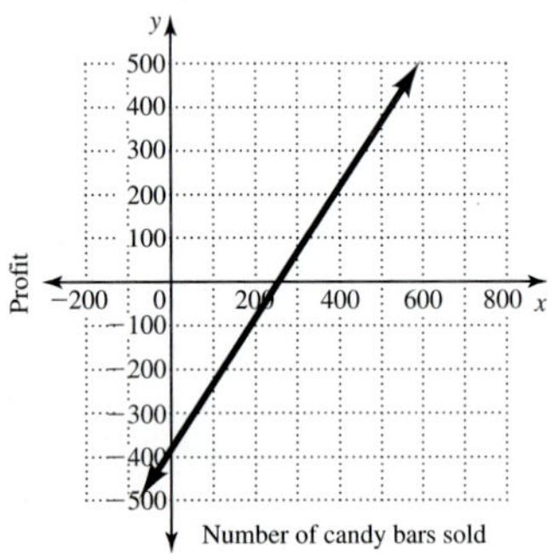

(f) x-intercept: $x = 254$ (g) The x-intercept of the graph and the breakeven point of the profit function are the same value.

26. (a)

Number of t-shirts sold	Profit function
0	−12360
315	−4800
515	0
815	7200
1000	11640
x	$P(x) = 24x - 12000 - 360$

(b) $0 = 24x - 12{,}000 - 360, x = 515$
(c) 515 t-shirts (d) Xscl 100 Yscl 4000

(e)

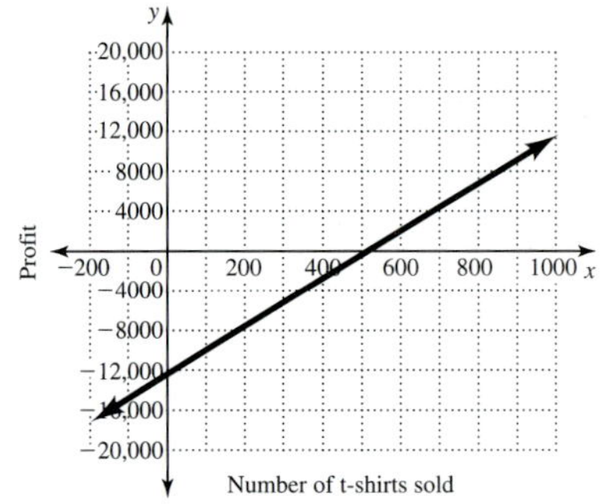

(f) The x-intercept is $x = 515$. (g) The x-intercept of the graph and the breakeven point of the profit function are the same value.

27. (a) $x = 15$ hours (b) and (c)

x	$y_1 = 6x + 40$	$y_2 = 130$
0	40	130
5	70	130
10	100	130
15	130	130
20	160	130
30	190	130

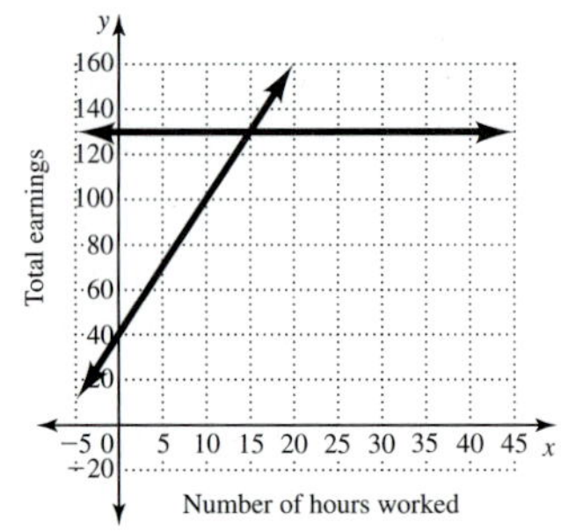

(d) Point of intersection: (15, 130) (e) Beth worked 15 hours. (f) The x-coordinate of the point of intersection is the solution of the equation and represents the number of hours Beth worked.

28. (a) $x = 8$ hours
(b) and (c)

x	y_1	y_2
0	21	71
2	33.5	71
4	46	71
6	58.5	71
8	71	71
10	83.5	71

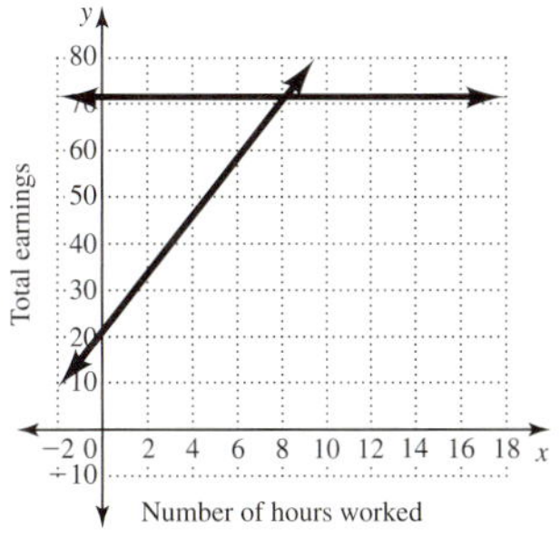

(d) Point of intersection: (8, 71) (e) Carl worked 8 hours. (f) The x-coordinate of the point of intersection is the solution of the equation and represents the number of hours Carl worked. **29.** (a) $x + 10$ (b) $x + (x + 10) = 210$ (c) $x = 100$ (d) Don spent \$100. (e) Bill spent \$100 + 10 = \$110. (f) Using the X-Intercept Method, graph $y = 2x + 10 - 210 = 2x - 200$. The x-intercept is $x = 100$. The x-intercept of the graph and the solution of the equation are the same value and represents the amount Don spent \$100. **30.** (a) $x + 150$ (b) $x + (x + 150) = 950$ (c) $x = 400$ (d) Leah spent \$400. (e) Crystal spent \$400 + 150 = \$550. (f) Using the Intersection-of-Graphs Method, graph $y_1 = x + (x + 150)$ and $y_2 = 950$. The analytical solution is the x-coordinate of the point of intersection. The point of intersection is (400, 950). The amount spent by Leah was \$400. **31.** Answers will vary. The zero of an equation is the x-intercept of the graph of the equation when $y = 0$.
32. The analytical solution of an equation is the same value as the x-coordinate of the x-intercept of the associated function. **33.** The analytical solution of an equation is the same value as the x-coordinate of the point of intersection of the graphs of the two associated functions.

DOING MATHEMATICS SECTION 4.2

1. Yes **2.** No **3.** No **4.** Yes **5.** Yes **6.** No **7.** No **8.** Yes **9.** (2, 4) **10.** (2, 2) **11.** Inconsistent **12.** Dependent **13.** Inconsistent **14.** (3, −3) **15.** Dependent **16.** (2, −1) **17.** (−2, 10) **18.** Dependent **19.** Dependent **20.** Inconsistent **21.** Inconsistent **22.** (2, 0) **23.** (1, −6) **24.** Dependent **25.** Dependent **26.** (3, 7) **27.** (−2, −2) **28.** (1, 4) **29.** (−1, 5) **30.** (3, 4) **31.** (5, 7) **32.** Inconsistent **33.** Inconsistent **34.** (9, 4) **35.** (2, 1) **36.** (4, −2)

37. (a) Equation 1: $\boxed{x} + \boxed{y} = \boxed{15}$

Equation 2: $\boxed{8x} + \boxed{5y} = \boxed{99}$

(b) $y_1 = 15 - x$ (c) $y_2 = \dfrac{99 - 8x}{5}$

(d)

x	y_1	y_2
0	15	19.8
3	12	15
6	9	10.2
9	6	5.4
12	3	.6
15	0	−4.2
18	−3	−9

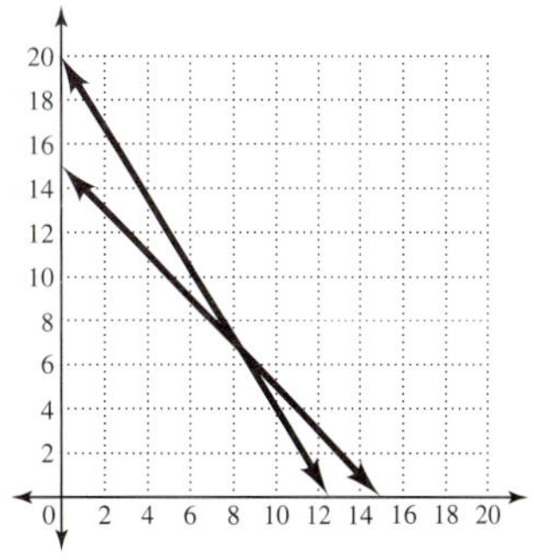

(e) Coordinates of the point of intersection: (8, 7) (f) T-shirts to be purchased: 8 (g) Baseball caps to be purchased: 7 (h) (8, 7) checks

38. (a) Equation 1: $\boxed{x} + \boxed{y} = \boxed{28}$

Equation 2: $\boxed{7x} + \boxed{10y} = \boxed{244}$

(b) Equation 1: $x + y = 28$ (c) $y_1 = 28 - x$
(d) Equation 2: $7x + 10y = 244$
(e) $y_2 = \dfrac{244 - 7x}{10}$

(f)

x	y_1	y_2
0	28	24.4
3	25	22.3
6	22	20.2
9	19	18.1
12	16	16
15	13	13.9
18	10	11.8

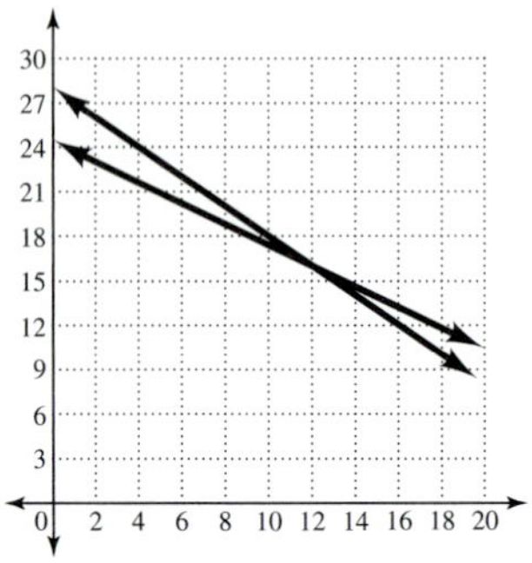

(g) Coordinates of the point of intersection: (12, 16)
(h) Sandi worked 12 hours (i) Sharon worked 16 hours (j) (12, 16) checks

39. (a) Equation 1: x + y = 500

Equation 2: $10x$ + $5y$ = 4000

(b) Equation 1: $x + y = 500$ (c) $y_1 = 500 - x$
(d) Equation 2: $10x + 5y = 4000$

(e) $y_2 = \dfrac{4000 - 10x}{5}$

(f)

x	y_1	y_2
0	500	800
50	450	700
100	400	600
150	350	500
200	300	400
250	250	300
300	200	200

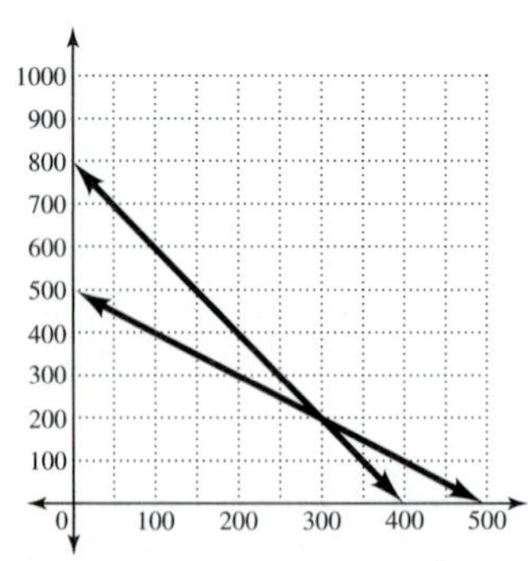

(g) Coordinates of the point of intersection: (300, 200) (h) 300 adult tickets sold (i) 200 children's tickets sold (j) (300, 200) checks

40. (a) Equation 1: x + y = 50

Equation 2: $4x$ + $6y$ = 240

(b) Equation 1: $x + y = 50$ $y_1 = 50 - x$

(c) Equation 2: $4x + 6y = 240$ $y_2 = \dfrac{240 - 4x}{6}$

(d)

x	y_1	y_2
0	50	40
10	40	23.333
20	30	26.667
30	20	20
40	10	13.333
50	0	6.6667
60	−10	0

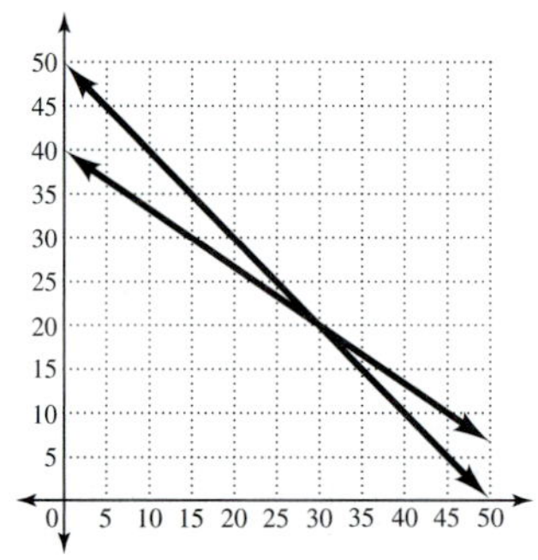

(e) Coordinates of the point of intersection: (30, 20)
(f) 30 oil filters purchased (g) 20 air filters purchased (h) (30, 20) checks

41. (a) Equation 1: x + y = 40

Equation 2: $30x$ + $20y$ = 960

(b) Equation 1: $x + y = 40$ (c) $y_1 = 40 - x$
(d) Equation 2: $30x + 20y = 960$

(e) $y_2 = \dfrac{960 - 30x}{20}$

(f)

x	y_1	y_2
0	40	48
8	32	36
16	24	24
24	16	12
32	8	0
40	0	−12
48	−8	−24

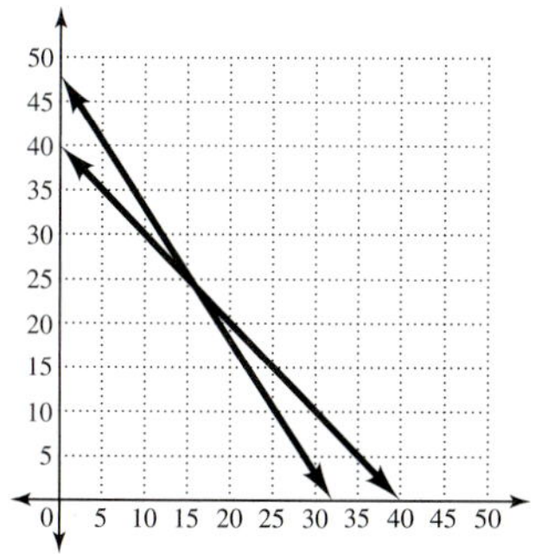

(g) Coordinates of the point of intersection: (16, 24) (h) Damien worked 16 hours (i) Jondarr worked 24 hours (j) (16, 24) checks

42. (a) Equation 1: $\boxed{x} + \boxed{y} = \boxed{300}$

Equation 2: $\boxed{2x} + \boxed{y} = \boxed{500}$

(b) Equation 1: $x + y = 300$ (c) $y_1 = 300 - x$

(d) Equation 2: $2x + y = 500$ (e) $y_2 = 500 - 2x$

(f)

x	y_1	y_2
0	300	500
50	250	400
100	200	300
150	150	200
200	100	100
250	50	0
300	0	−100

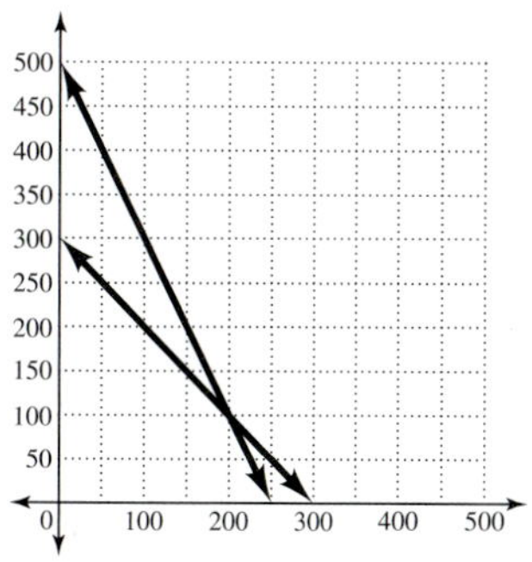

(g) Coordinates of the point of intersection: (200, 100) (h) 200 hot dogs sold (i) 100 orders of french fries sold (j) (200, 100) checks

43. When a system is consistent, there is one solution and one point of intersection. When a system is inconsistent, there is no solution and the lines do not intersect. When a system is dependent, there is an infinite number of solutions and the lines coincide.

DOING MATHEMATICS SECTION 4.3

1. (9, 8) **2.** (4, 9) **3.** (−2, −4) **4.** (8, 2) **5.** Dependent **6.** Inconsistent **7.** Inconsistent **8.** Dependent **9.** Inconsistent **10.** Dependent **11.** Inconsistent **12.** Inconsistent **13.** $\left(\frac{3}{4}, \frac{1}{4}\right)$ **14.** (0, 2) **15.** (3, −3) **16.** $\left(\frac{1}{2}, 1\right)$ **17.** (2, 4) **18.** (1, 2) **19.** $\left(\frac{5}{3}, -\frac{7}{3}\right)$ **20.** $\left(\frac{1}{4}, -\frac{1}{2}\right)$ **21.** Dependent **22.** Dependent **23.** Dependent **24.** Inconsistent **25.** Inconsistent **26.** Dependent **27.** Inconsistent **28.** Inconsistent **29.** (2, 2) **30.** (1, 5) **31.** (5, 4) **32.** (3, 4)

33. (a) and (b) Equation 1: $x + y = 4500$; Equation 2: $0.05x + 0.11y = 375$ (c) (2000, 2500) (d) \$2000 invested at 5% (e) \$2500 invested at 11%

(f) Equation 1: $2000 + 2500 \stackrel{?}{=} 4500$
$4500 = 4500$

Equation 2: $0.05(2000) + 0.11(2500) \stackrel{?}{=} 375$
$375 = 375$

34. (a) and (b) Equation 1: $x + y = 100{,}000$; Equation 2: $0.03x + 0.07y = 4000$
(c) (75,000, 25,000) (d) \$75,000 invested at 3%
(e) \$25,000 invested at 7%

(f) Equation 1: $75{,}000 + 25{,}000 \stackrel{?}{=} 100{,}000$
$100{,}00 = 100{,}000$

Equation 2:
$0.03(75{,}000) + 0.07(25{,}000) \stackrel{?}{=} 4000$
$4000 = 4000$

35. (a) and (b) Equation 1: $x + y = 7$; Equation 2: $5x + 10y = 55$ (c) (3, 4) (d) 3 five-dollar bills (e) 4 ten-dollar bills

(f) Equation 1: $3 + 4 \stackrel{?}{=} 7$
$7 = 7$

Equation 2: $5(3) + 10(4) \stackrel{?}{=} 55$
$55 = 55$

36. (a) and (b) Equation 1: $x + y = 1100$; Equation 2: $0.25x + 0.01y = 131$ (c) (500, 600) (d) 500 quarters (e) 600 pennies

(f) Equation 1: $500 + 600 \stackrel{?}{=} 1100$
$1100 = 1100$

Equation 2: $0.25(500) + 0.01(600) \stackrel{?}{=} 131$
$131 = 131$

37. (a) Equation 1: $x + y = 9$; Equation 2: $0.50x + 0.75y = 5.25$ (b) (6, 3) (c) 6 blue pens (d) 3 green pens

(e) Equation 1: $6 + 3 \stackrel{?}{=} 9$
$9 = 9$

Equation 2: $0.50(6) + 0.75(3) \stackrel{?}{=} 5.25$
$5.25 = 5.25$

38. (a) Equation 1: $x + y = 8$; Equation 2: $1.25x + 0.75y = 9.00$ (b) (6, 2) (c) 6 Superduper batteries (d) 2 Reallygood batteries
(e) Equation 1: $6 + 2 \stackrel{?}{=} 8$
$8 = 8$
Equation 2: $1.25(6) + 0.75(2) \stackrel{?}{=} 9.00$
$9.00 = 9.00$

39. (a) Equation 1: $x + y = 75$; Equation 2: $0.25x + 0.10y = 0.20(75)$ (b) (50, 25)
(c) 50 cubic cm (d) 25 cubic cm
(e) Equation 1: $50 + 25 \stackrel{?}{=} 75$
$75 = 75$
Equation 2: $0.25(50) + 0.10(25) \stackrel{?}{=} 0.20(75)$
$15 = 15$

40. (a) Equation 1: $x + y = 250$; Equation 2: $0.20x + 0.05y = 0.15(250)$ (b) $\left(166\frac{2}{3}, 83\frac{1}{3}\right)$
(c) $166\frac{2}{3}$ cubic cm of 20% (d) $83\frac{1}{3}$ cubic cm of 5%
(e) Equation 1: $166\frac{2}{3} + 83\frac{1}{3} \stackrel{?}{=} 250$
$250 = 250$

Equation 2:
$0.2\left(166\frac{2}{3}\right) + 0.05\left(83\frac{1}{3}\right) \stackrel{?}{=} 0.15(250)$
$37.5 = 37.5$

41. (a) Equation 1: $x + y = 500$; Equation 2: $0.10x + 0.05y = 30$ (b) (100, 400) (c) 100 peak minutes (d) 400 off-peak minutes
(e) Equation 1: $100 + 400 \stackrel{?}{=} 500$
$500 = 500$
Equation 2: $0.10(100) + 0.05(400) \stackrel{?}{=} 30$
$30 = 30$

42. (a) Equation 1: $x + y = 500$; Equation 2: $0.09x + 0.05y = 30$ (b) (125, 375) (c) 125 peak minutes (d) 375 off-peak minutes
(e) Equation 1: $125 + 375 \stackrel{?}{=} 500$
$500 = 500$
Equation 2: $0.09(125) + 0.05(375) \stackrel{?}{=} 30$
$30 = 30$
(f) The best choice of telephone plan depends on whether Jessica needs or wants more peak minutes or more off-peak minutes. **43.** $(2, -4)$
44. $(5, -5)$ **45.** Inconsistent **46.** Inconsistent
47. $\left(-\frac{1}{2}, 4\right)$ **48.** (1, 2) **49.** $\left(4, \frac{1}{2}\right)$ **50.** (6, 6)
51. $\left(-\frac{3}{2}, 11\right)$ **52.** $\left(2, \frac{3}{4}\right)$ **53.** Inconsistent
54. Dependent **55.** $(2.8, -2.4)$ **56.** (4, 0)
57. Inconsistent **58.** $(-2, -7)$ **59.** Dependent
60. Inconsistent

DOING MATHEMATICS SECTION 4.4

1. (a) Yes (b) Yes (c) Yes **2.** (a) Yes (b) Yes (c) No **3.** (a) Yes (b) No (c) Yes **4.** (a) No (b) Yes (c) Yes **5.** (a) No (b) No (c) No **6.** (a) No (b) No (c) Yes

7.

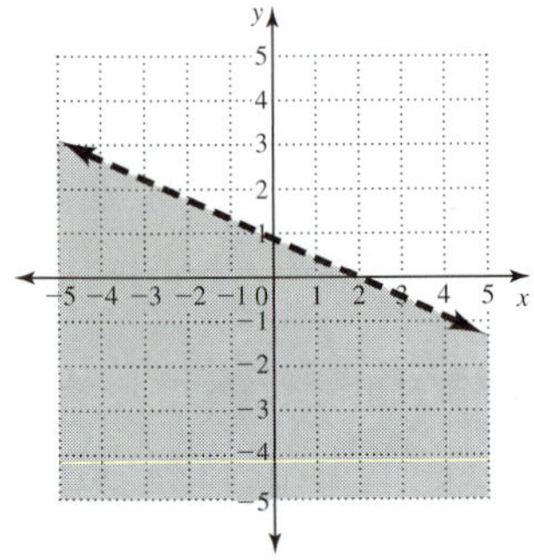

8.

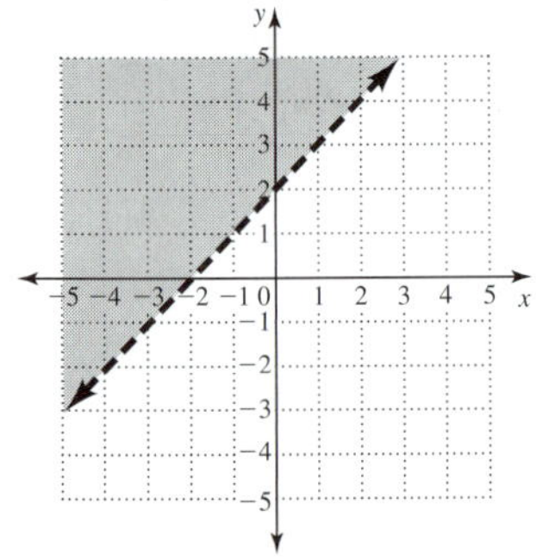

9.

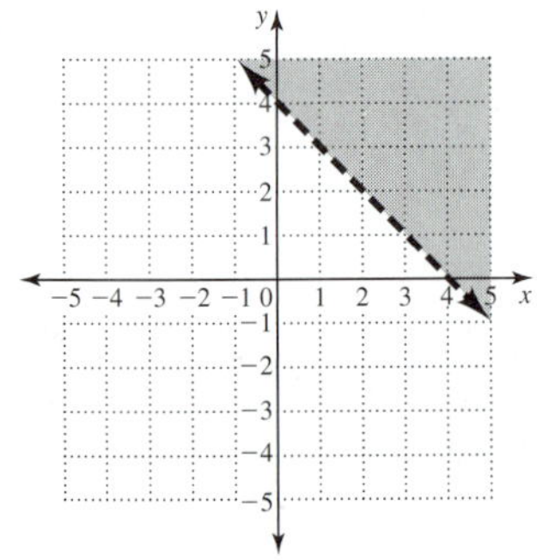

10.

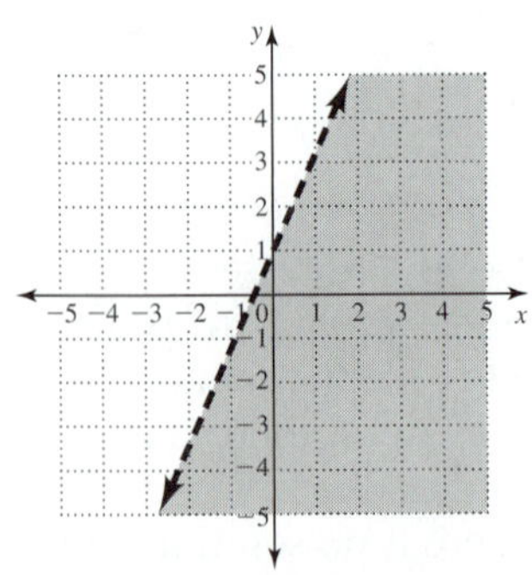

11.

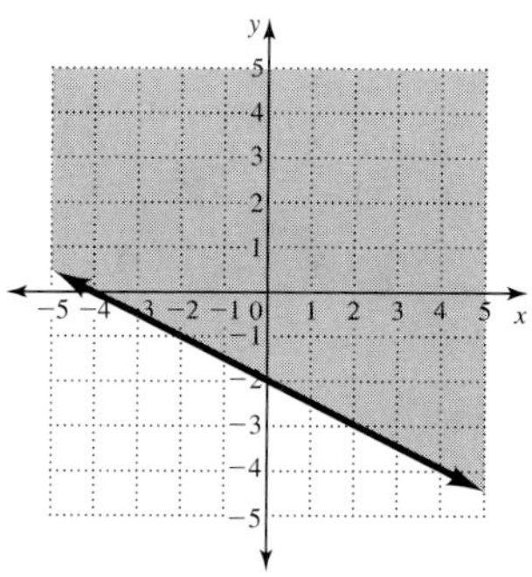

12.

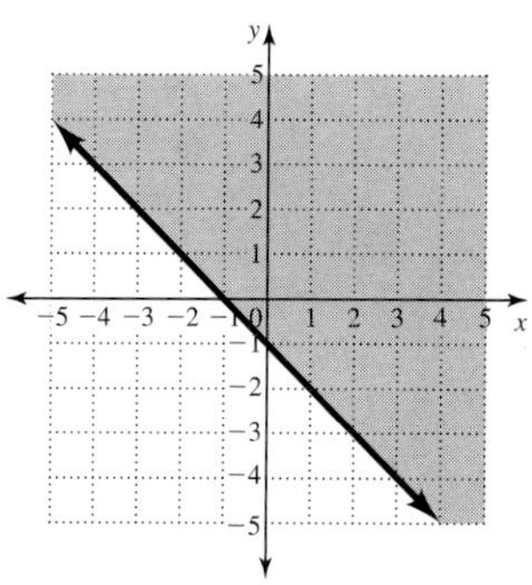

13.

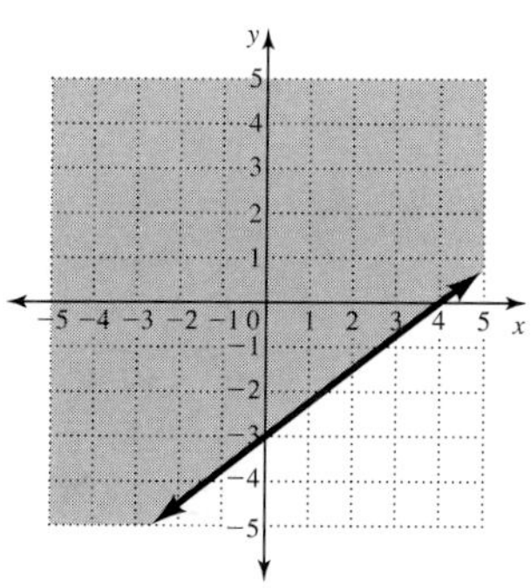

14.

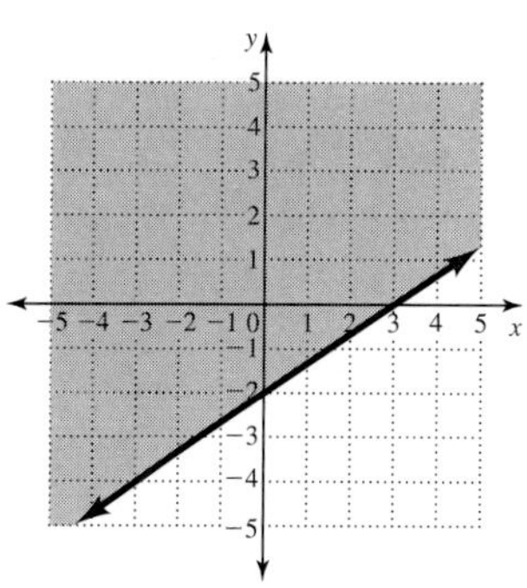

15.

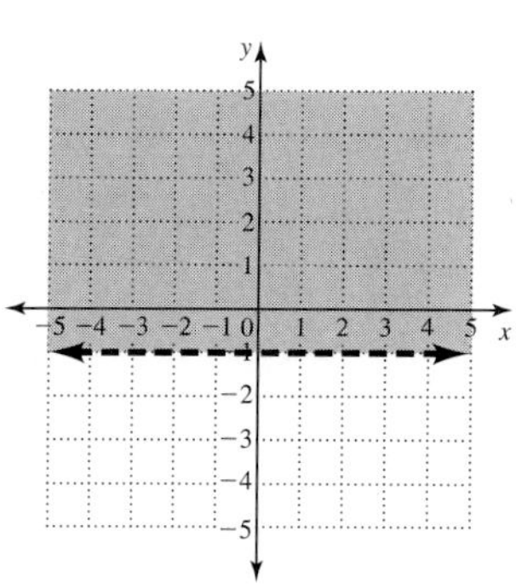

16.

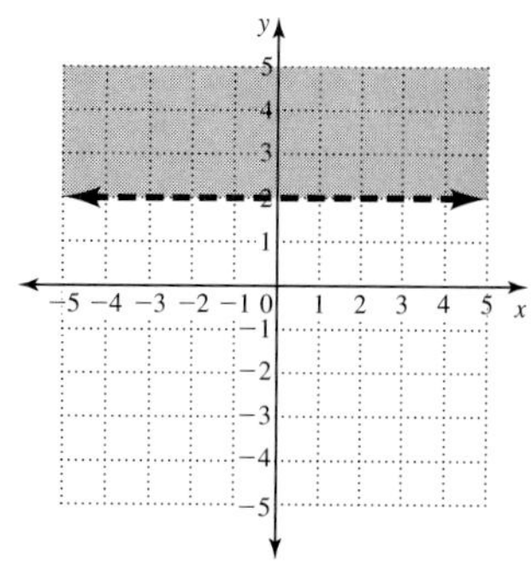

17.

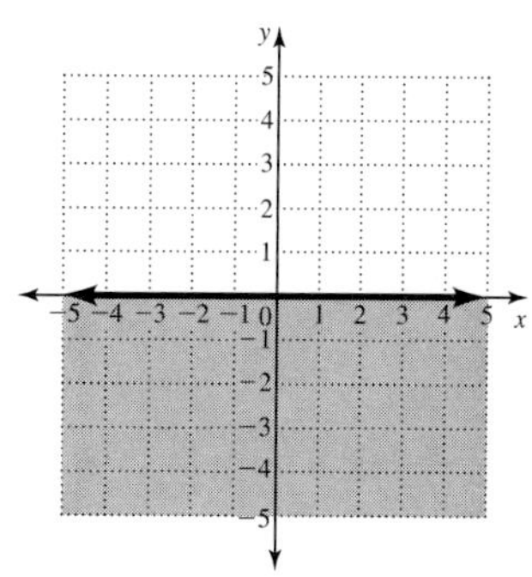

18.

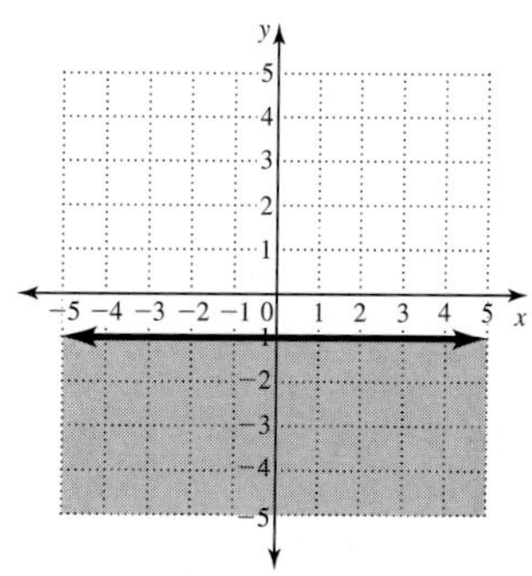

19.

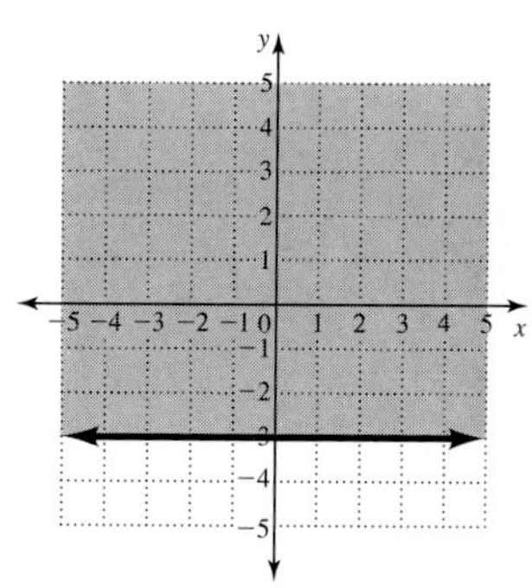

20.

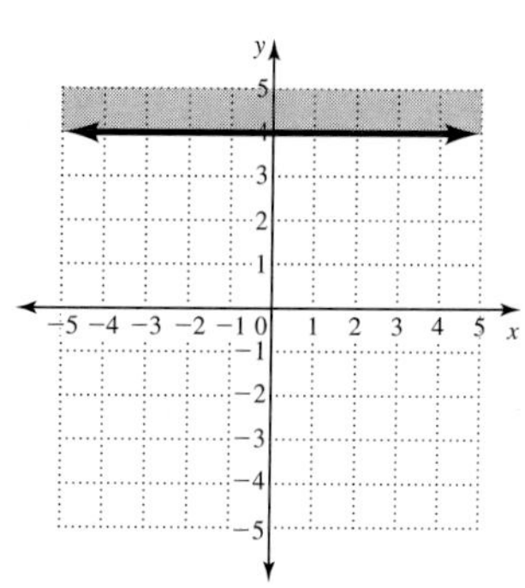

21.

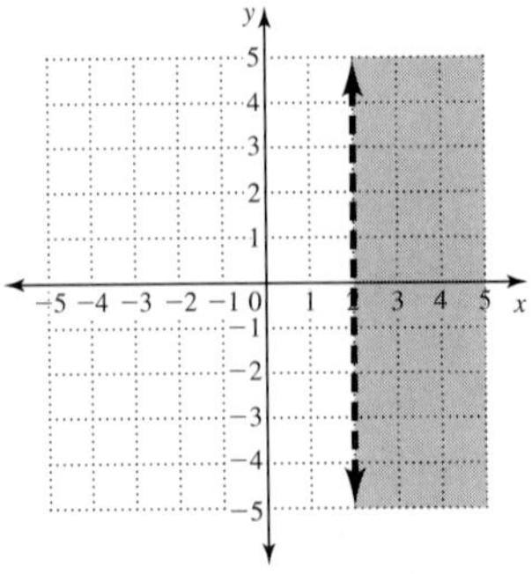

22.

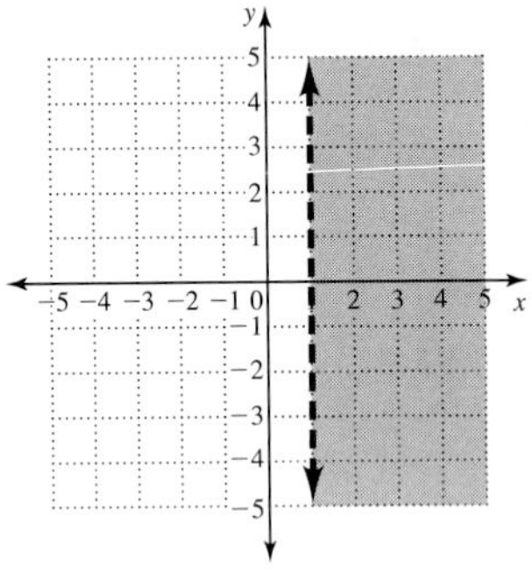

23.

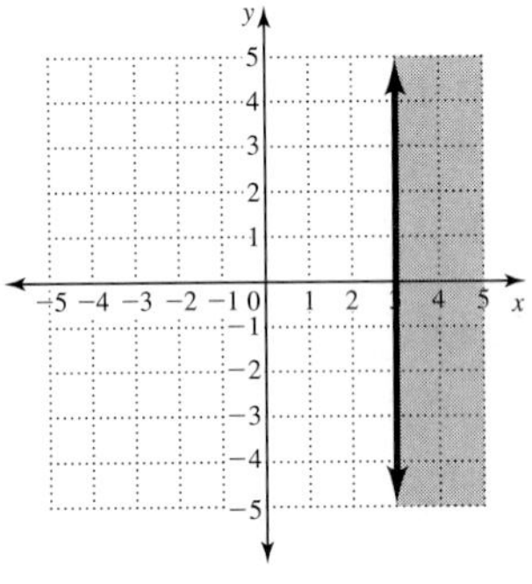

24.

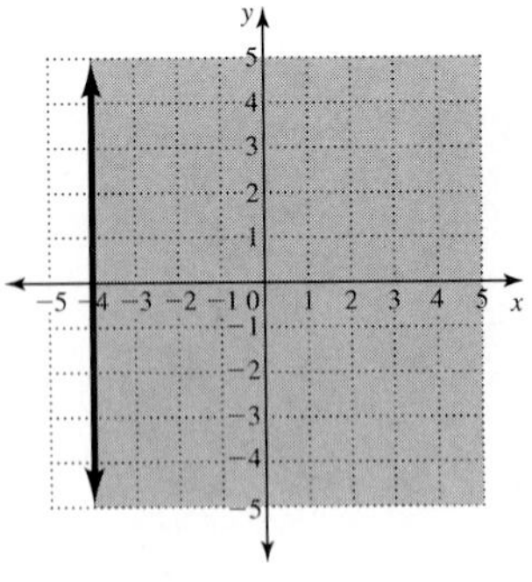

25.

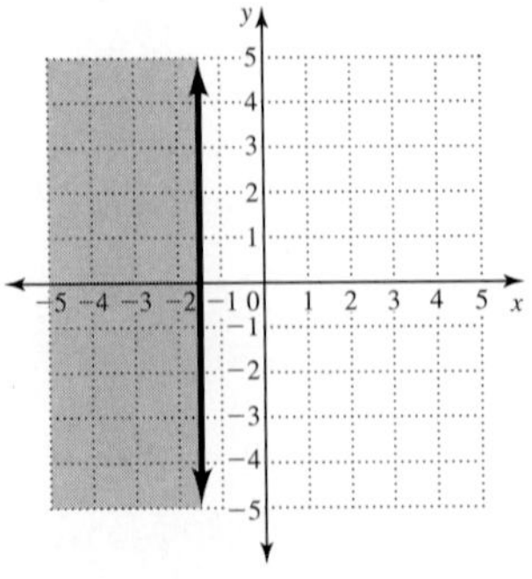

26.

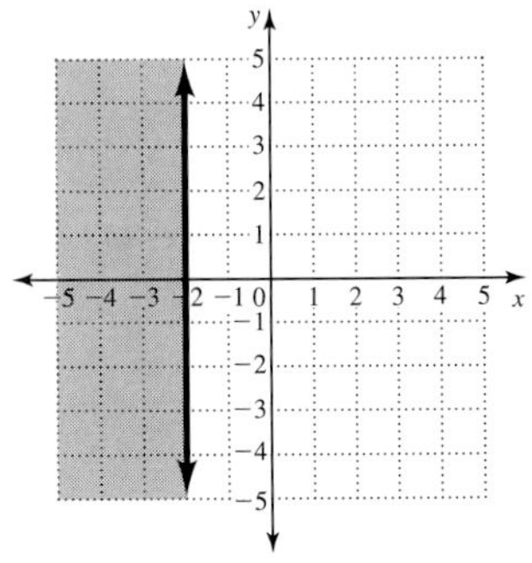

27.

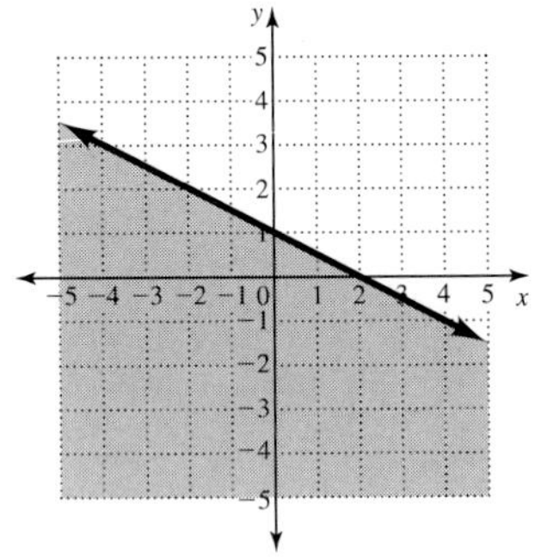

28.

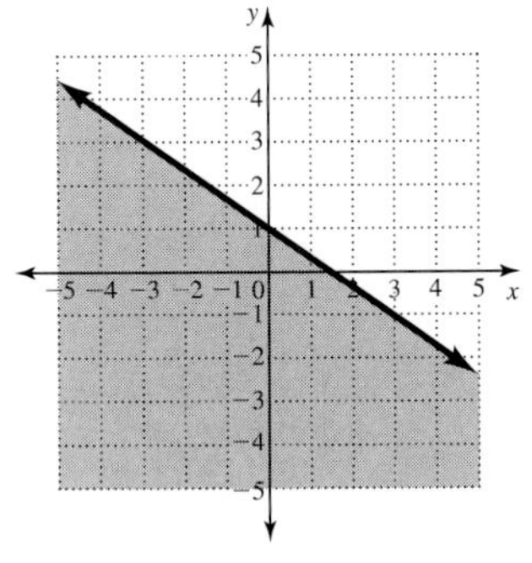

29.

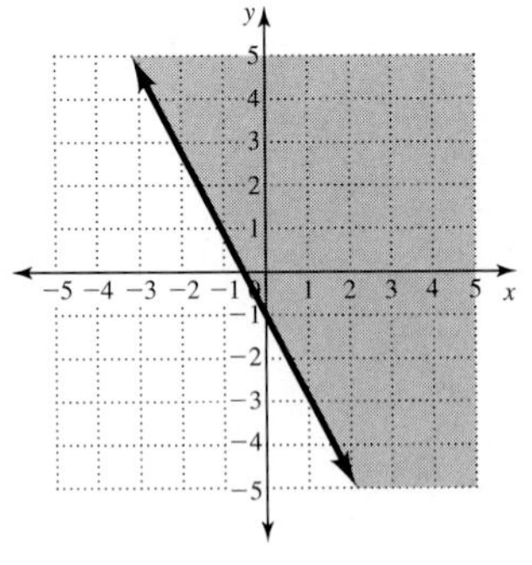

30.

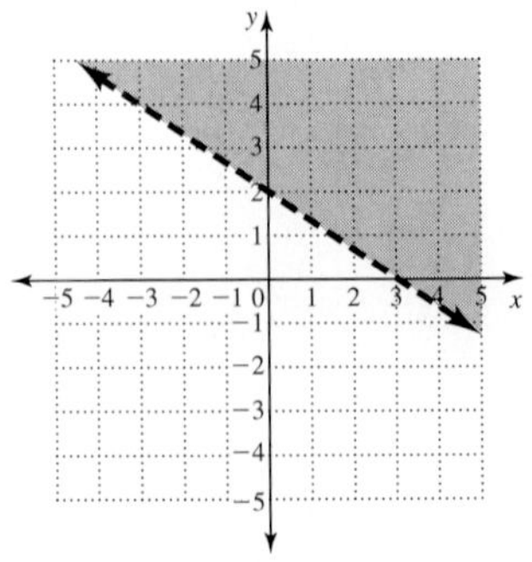

31.

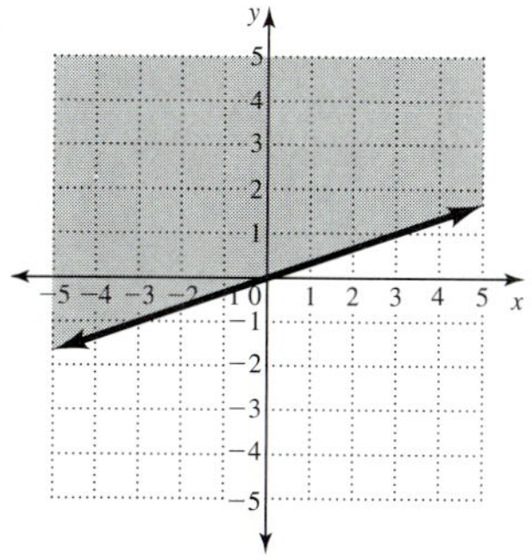

CHAPTER 4 REVIEW

1. (a) $x = 6$

(b)

x	$y = -\frac{1}{2}x + 3$
0	3
2	2
4	1
6	0
8	−1

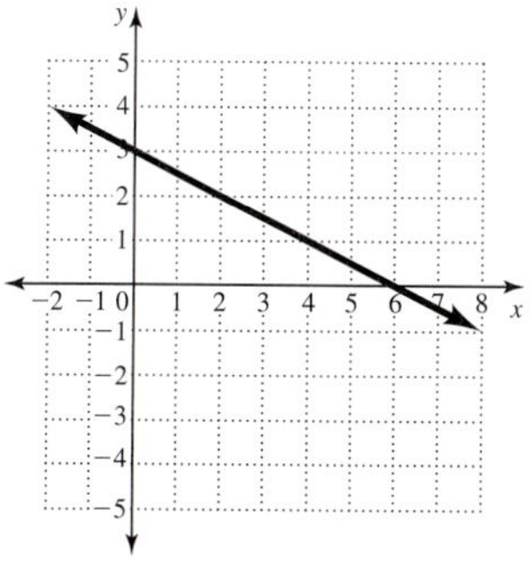

(c) $x = 6$ (d) Yes; $x = 6$ checks

2. (a) $x = 1$ (b) $y = 2x + 2(2x - 1) - 4$

x	y
−1	−12
0	−6
1	0
2	6
3	12

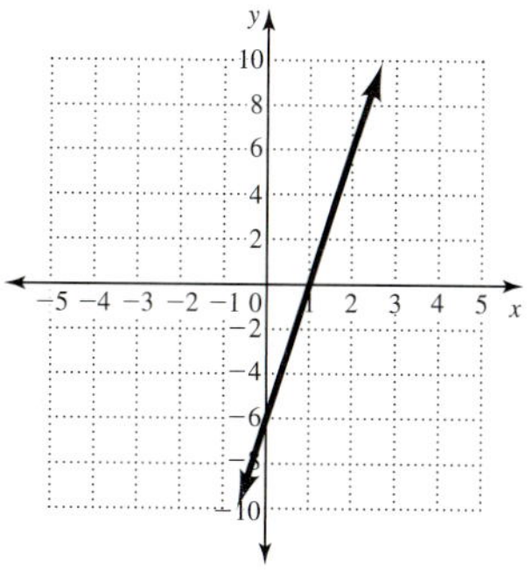

(c) $x = 1$ (d) Yes; $x = 1$ checks

3. (a) $x = 2$

(b) $y_1 = 9.3x - (1.3x + 2)$ and $y_2 = 4x + 6$

x	y_1	y_2
−1	−10	2
0	−2	6
1	6	10
2	14	14
3	22	18

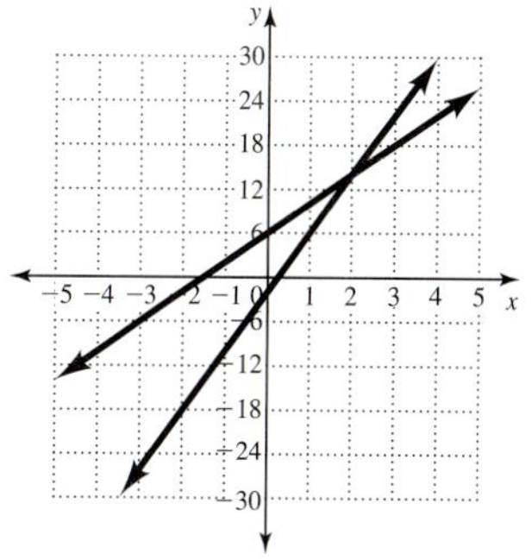

(c) (2, 14) (d) Yes; $x = 2$ checks

4. (a) $x = 2$

(b) $y_1 = 5.7x - 3.6(x - 1)$ and $y_2 = 7.8$

x	y_1	y_2
−1	1.5	7.8
0	3.6	7.8
1	5.7	7.8
2	7.8	7.8
3	9.9	7.8

4.

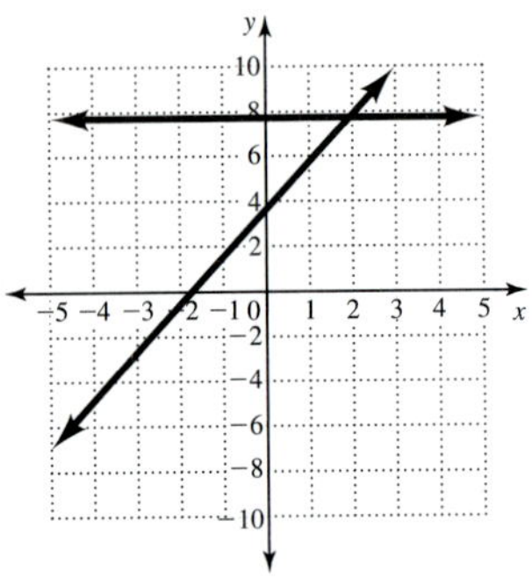

(c) (2, 7.8) (d) Yes; $x = 2$ checks

5. (a) $x = 7$

(b) X-Intercept Method:

$y_1 = (2x + 3) - (x - 6) - 3x + 5$

x	y_1
3	8
5	4
7	0
9	−4

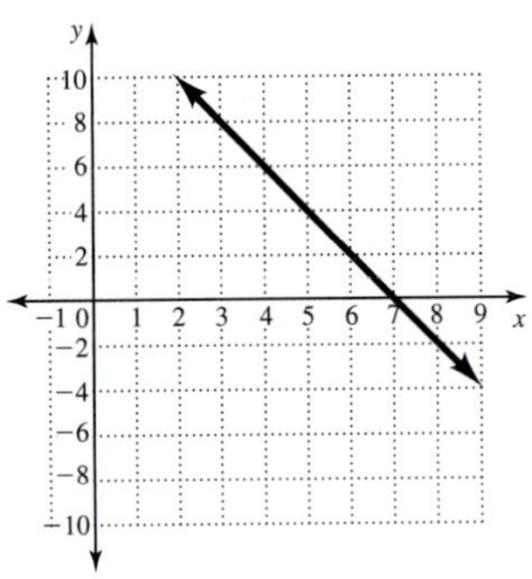

(c) $x = 7$ (d) Yes; $x = 7$ checks

6. (a)

x	$P(x)$ = Profit function
30	−375
40	875
50	2125
x	$P(x) = 125x - 55(75) = 125x - 4125$

(b) $0 = 125x - 4125$, $x = 33$ (c) 33 racquets

(d) Xscl 10 Yscl 500

(e)

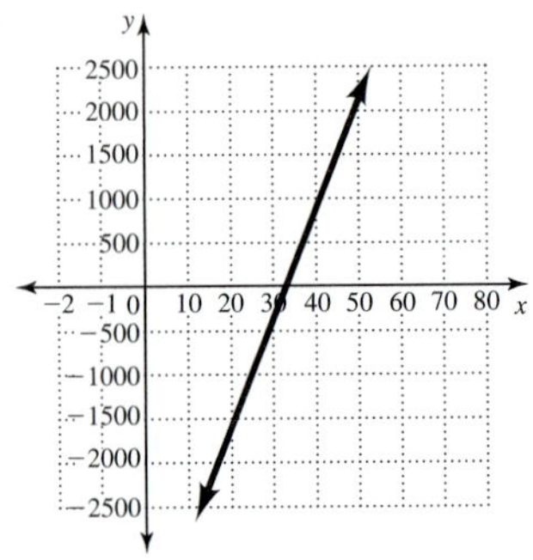

(f) x-intercept of the graph: (33, 0) (g) 53 racquets

(h) They are the same.

7. (a)

x	$P(x)$ = Profit function
0	−500
50	−250
100	0
150	250
200	500
300	1000
400	1500
x	$P(x) = 5x - 500$

(b) $0 = 5x - 500$, $x = 100$ (c) 100 tickets

(d) Xscl 50 Yscl 500

(e)

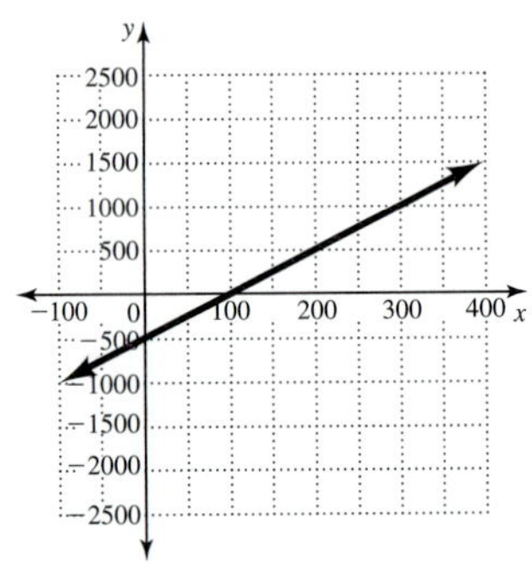

(f) x-intercept of the graph: (100, 0) (g) 500 tickets to make a profit of \$2000

8. (a) $x + 50$ (b) $x + (x + 50) = 530$ (c) $x = 240$ (d) Stephanie spent \$240 (e) Karen spent \$290 (f) Using the X-Intercept Method and the equation $y = x + (x + 50) - 530$:

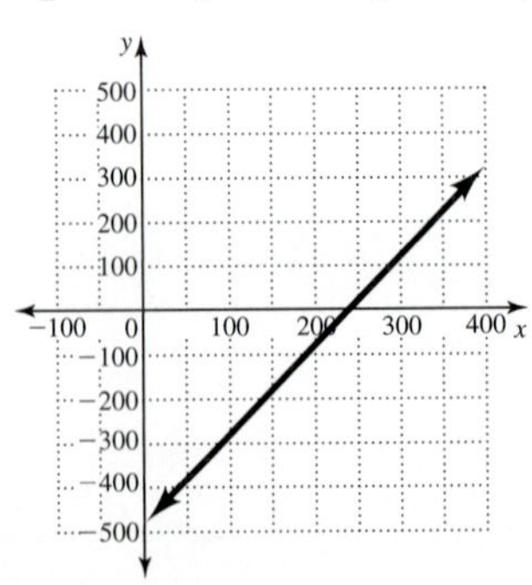

The x-intercept at (240, 0) confirms the analytical solution of $x = 240$.

9. (a) Equation 1: $x + y = 28$
Equation 2: $10x + 8y = 256$
(b) $y_1 = 28 - x$ (c) $y_2 = \frac{256 - 10x}{8} = \frac{128 - 5x}{4}$
(d)

x	y_1	y_2
0	28	32
8	20	22
12	16	17
16	12	12
18	10	9.5

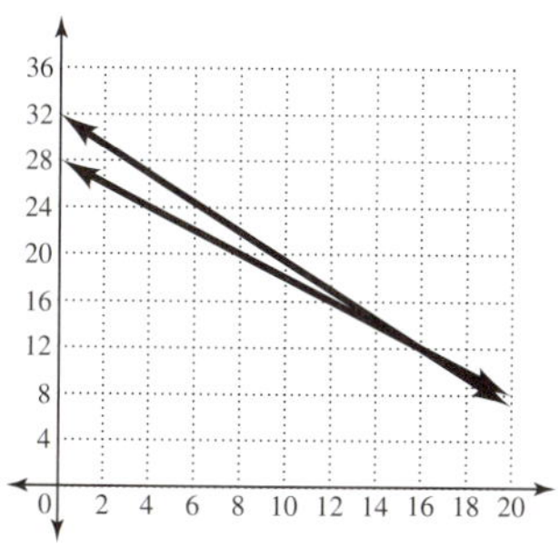

(e) (16, 12) (f) 16 golf gloves (g) 12 golf towels (h) $x = 16$ checks
10. (a) Picture (b) $x + y = 600$
(c) $y_1 = 600 - x$ (d) $4x + 6y = 2700$
(e) $y_2 = \frac{2700 - 4x}{6}$

x	y_1	y_2
100	500	383.33
200	400	316.67
300	300	250
400	200	183.33
450	150	150
500	100	116.67

(f)

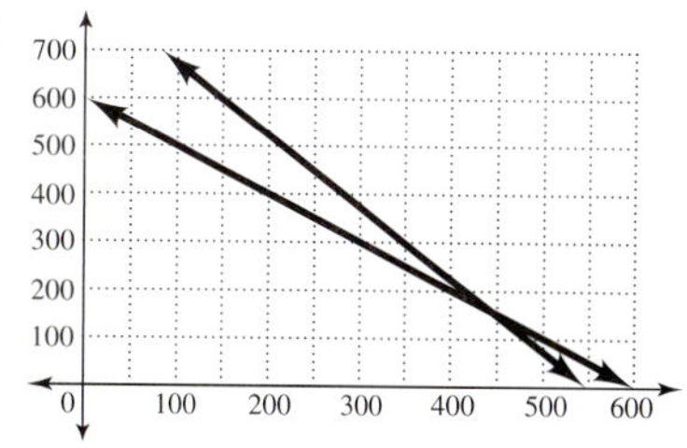

(g) (450, 150) (h) 450 student tickets (i) 150 non-student tickets (j) (450, 150) checks
11. (1, 5) **12.** Inconsistent **13.** (2, 2)
14. (0, −5) **15.** Inconsistent **16.** Dependent
17. Dependent **18.** (a) Equation 1: $x + y = 6000$; Equation 2: $0.06x + 0.15y = 585$ (b) (3500, 2500) (c) \$3500 invested at 6% (d) \$2500 invested at 15%
(e) Equation 1: $3500 + 2500 \stackrel{?}{=} 6000$
$6000 = 6000$
Equation 2: $0.06(3500) + 0.15(2500) \stackrel{?}{=} 585$
$585 = 585$
19. (a) Equation 1: $x + y = 10$; Equation 2: $5x + 20y = 155$ (b) (3, 7) (c) 3 five-dollar bills (d) 7 twenty-dollar bills
(e) Equation 1: $3 + 7 \stackrel{?}{=} 10$
$10 = 10$
Equation 2: $5(3) + 20(7) \stackrel{?}{=} 155$
$155 = 155$
20. (a) Equation 1: $x + y = 100$; Equation 2: $0.50x + 0.0y = 0.30(100)$ (b) (60, 40) (c) 60 cc of 50% dextrose solution (d) 40 cc of 0% dextrose solution
(e) Equation 1: $60 + 40 \stackrel{?}{=} 100$
$100 = 100$
Equation 2: $0.50(60) + 0.0(40) \stackrel{?}{=} 0.30(100)$
$30 = 30$

21. (2, 1) **22.** $\left(\frac{7}{3}, -\frac{1}{3}\right)$ **23.** Dependent
24. Dependent **25.** Inconsistent **26.** Inconsistent
27. (4, 2) **28.** (3, −1) **29.** Inconsistent
30. Inconsistent **31.** Dependent **32.** Dependent
33. (a) No (b) Yes (c) Yes **34.** (a) Yes (b) Yes (c) No **35.** (a) No (b) No (c) No
36. (a) No (b) Yes (c) No
37.

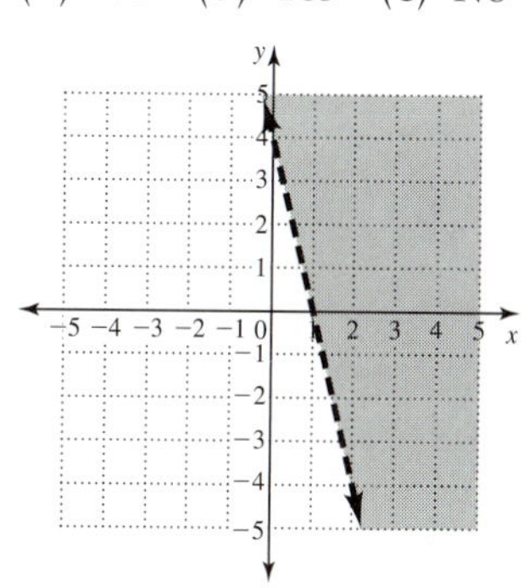

38.

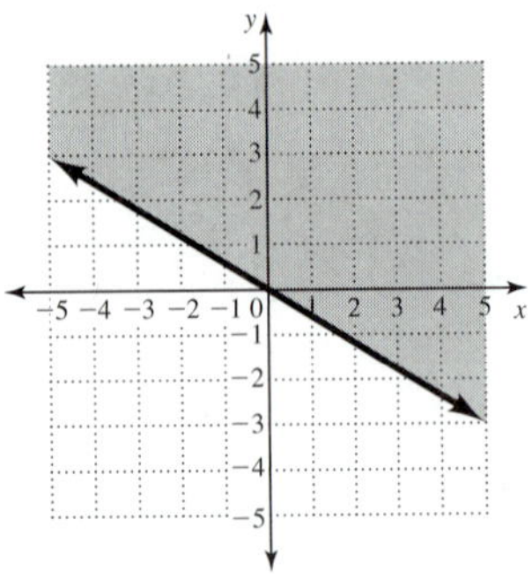

39.

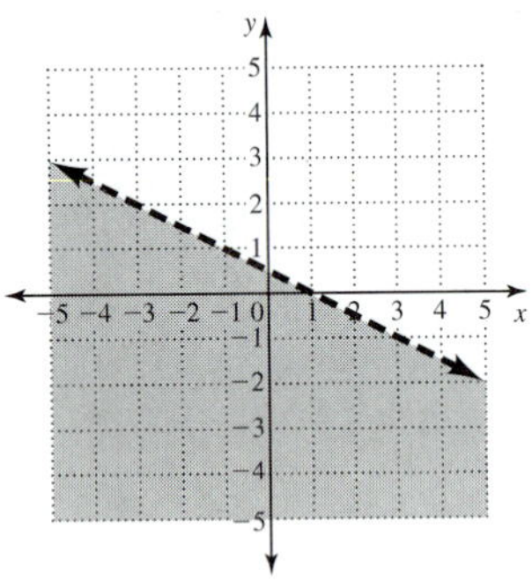

40.

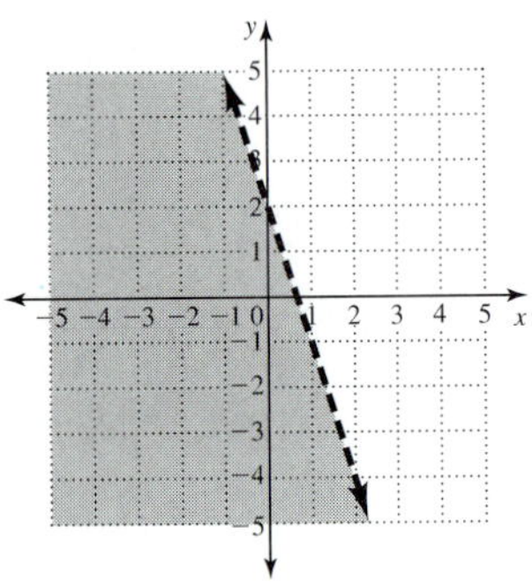

41.

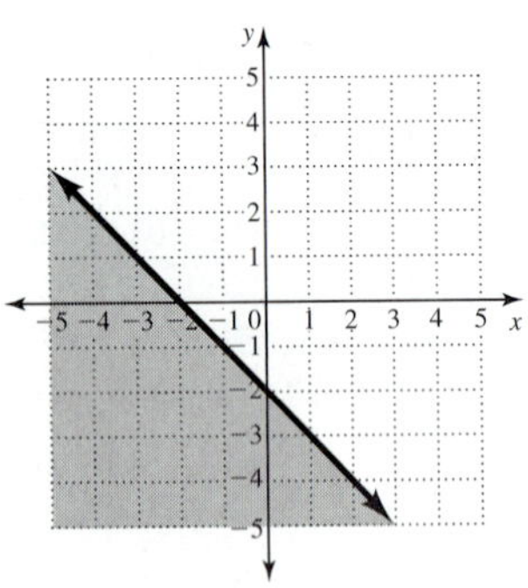

42.

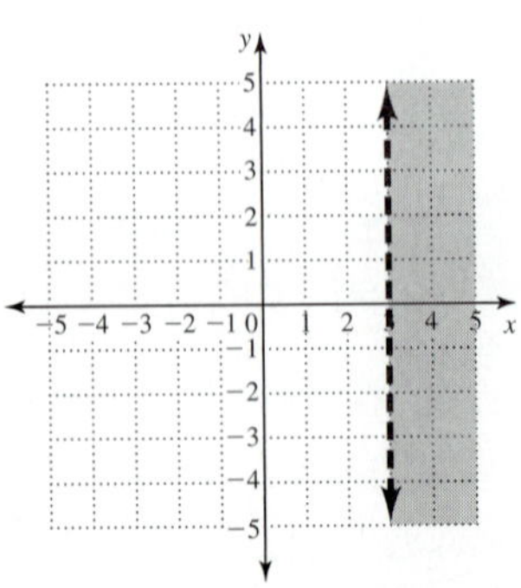

43.

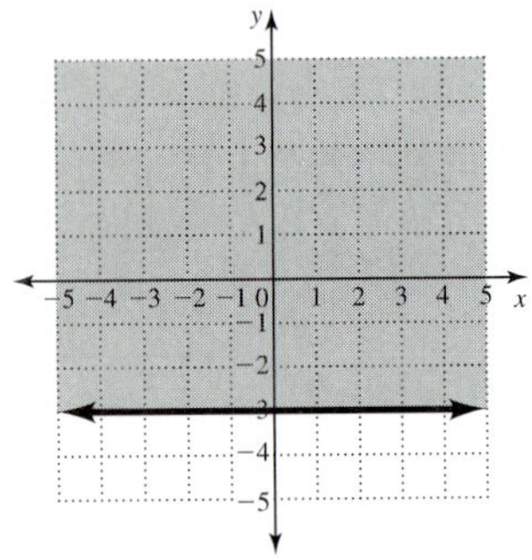

44.

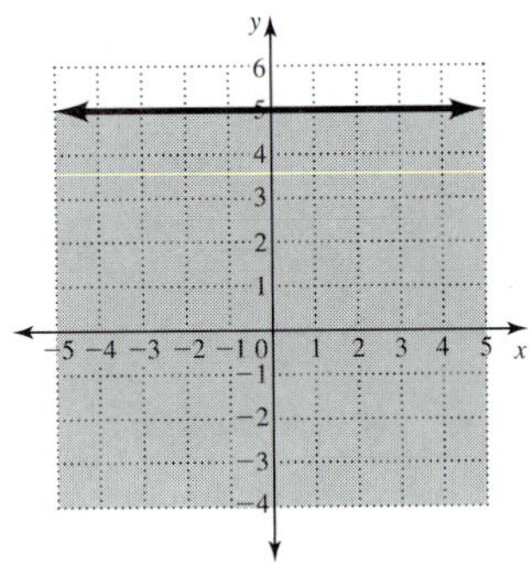

CHAPTER 4 PRACTICE TEST ANSWERS

1. (a)

x = Tickets sold	$P(x)$ = Profit function
100	−300
150	−100
200	100
300	500
Generalize x	$P(x) = 4x - 700$

(b) 175 tickets (c) Xscl 50 Yscl 100

(d)

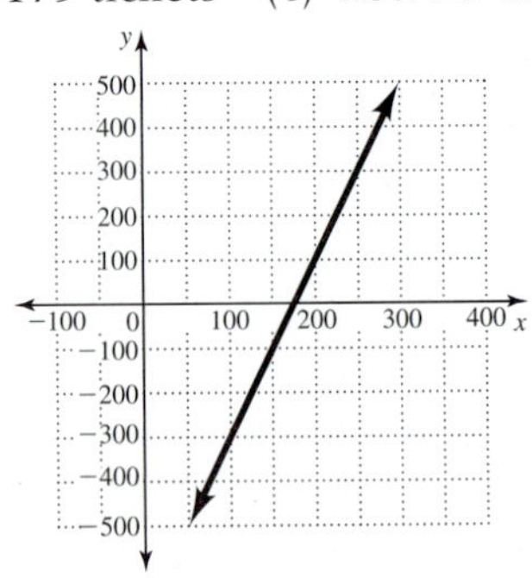

(e) x-intercept approx. 175 (f) Yes

(g) $x = 320$ tickets

2. (a) $x = 4$
(b)

x	$y = 2x - 8$
2	−4
3	−2
4	0
5	2

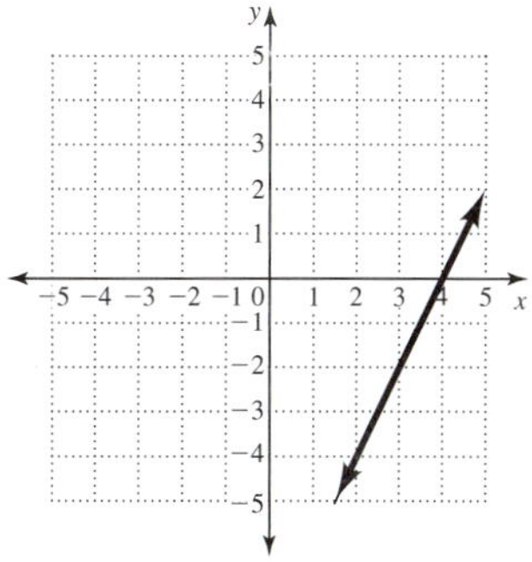

(c) $x = 4$ (d) Yes (e) $x = 4$ checks

3. (a) $x = 4$
(b)

x	$y_1 = 7x - 2$	$y_2 = 4x + 10$
2	12	18
3	19	22
4	26	26
5	33	30

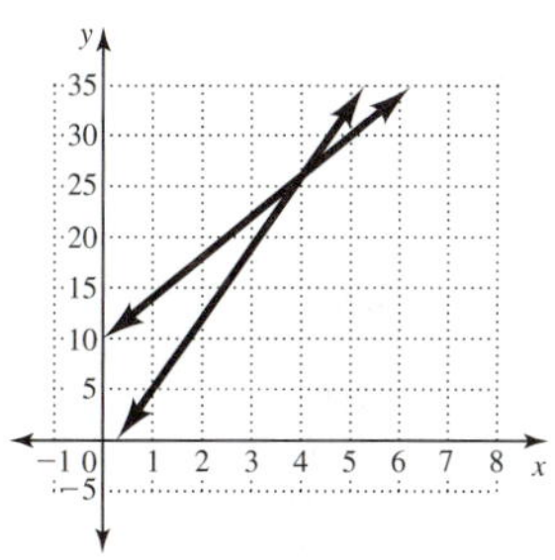

(c) x-coordinate is 4 (d) Yes (e) $x = 4$ checks

4. (a) $x \approx 1.05$
(b)

x	$y_1 = 4(1.56x - 1)$	$y_2 = 2.43x$
0	−4	0
1	2.24	2.43
2	8.48	4.86
3	14.72	7.29

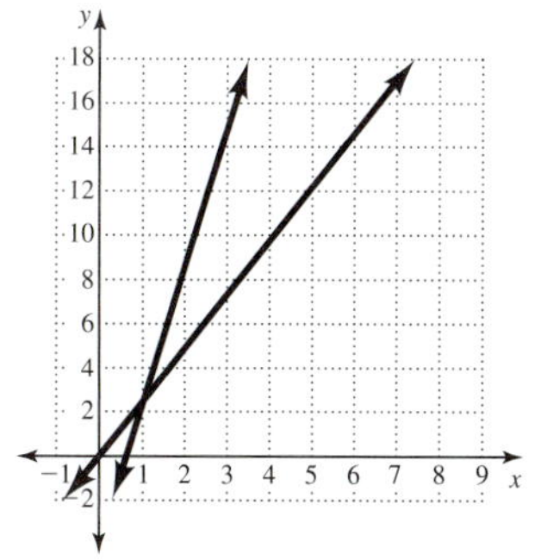

(c) x-coordinate is approx. 1 (d) Yes
(e) $x = 1.05$ checks

5. (a) Picture (b) $x + y = 65$ (c) $y_1 = 65 - x$
(d) $4x + 1.5y = 185$ (e) $y_2 = \dfrac{185 - 4x}{1.5}$
(f)

x	$y_1 = 65 - x$	$y_2 = \dfrac{185 - 4x}{1.5}$
15	50	83.3
30	35	43.3
35	30	30
45	20	3.3
60	5	−36.7

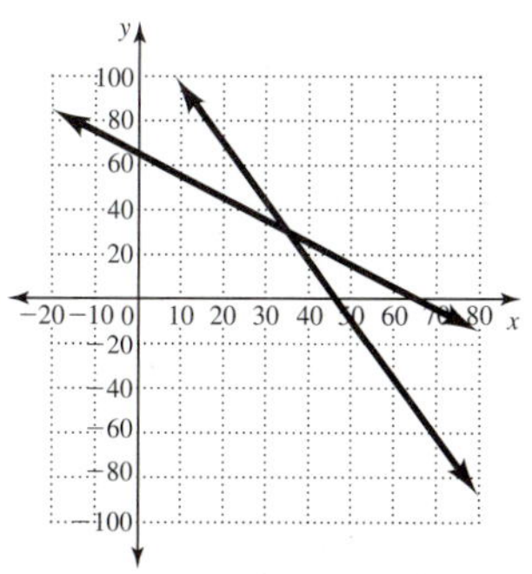

(g) (35, 30) (h) 35 notebooks (i) 30 dividers
(j) (35, 30) check **6.** (2, 4) **7.** Inconsistent
8. Dependent **9.** (−1, −2)
10. (a)

$$x + y = 10{,}000$$
$$0.05x + 0.12y = 815$$

(b) (5500, 4500) (c) \$5500 at 5% (d) \$4500 at 12% (e) (5500, 4500) checks
11. (a)

$$x + y = 2$$
$$0.5x + 0.1y = 0.35(2)$$

(b) (1.25, 0.75) (c) 1.25 liters of 50% solution
(d) 0.75 liters of 10% solution (e) (1.25, 0.75) checks **12.** (1, 5) **13.** Inconsistent **14.** (−2, 4)
15. Dependent **16.** $\left(0, -\dfrac{3}{2}\right)$

17. $x + y > 1$

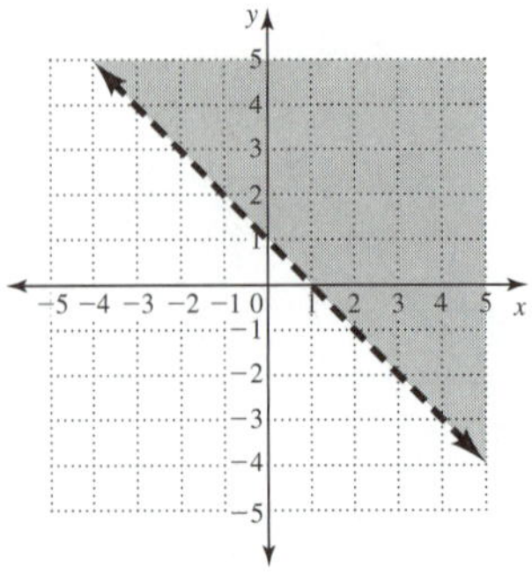

18. $y \le 2$

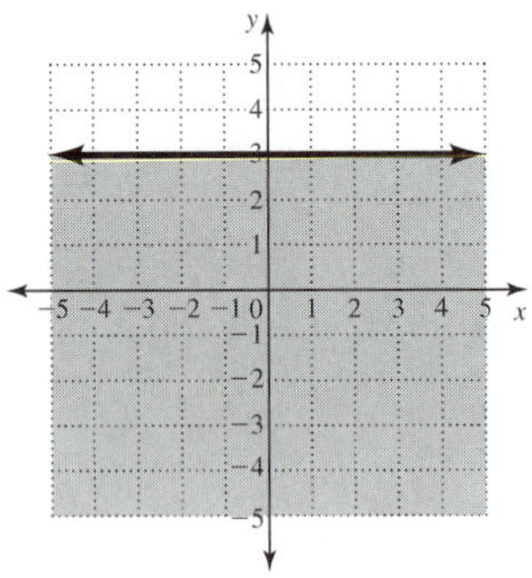

19. $3x - y \ge 4$

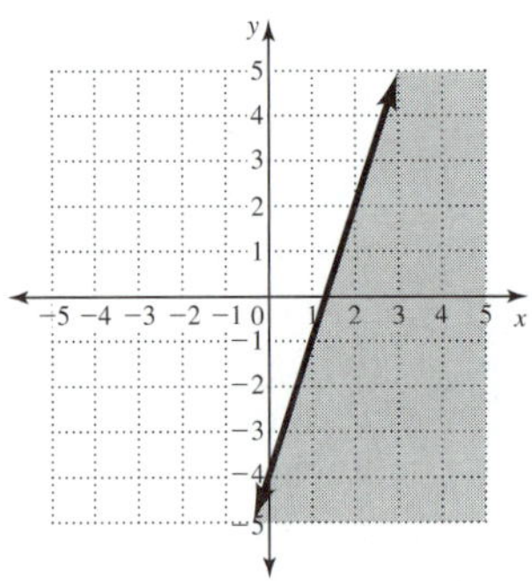

20. $x > 2.5$

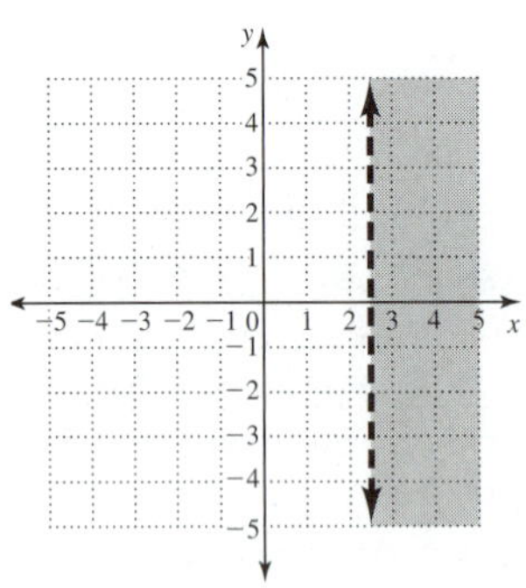

CHAPTER 5

DOING MATHEMATICS SECTION 5.1

1. x^{10} **2.** x^{12} **3.** $4^5 = 1024$ **4.** $5^5 = 3125$ **5.** z^7
6. y^{13} **7.** a^7 **8.** x^3 **9.** n^{15} **10.** y^{20} **11.** y^{14}
12. x^6 **13.** $2^{10} = 1024$ **14.** $10^6 = 1{,}000{,}000$
15. b^{36} **16.** z^{30} **17.** a^4b^4 **18.** x^3y^3 **19.** $9c^2d^2$
20. $4x^2y^2$ **21.** $16c^4d^4$ **22.** $100x^2y^2$ **23.** $-27a^3$
24. $-8x^3$ **25.** $6^3 = 216$ **26.** $5^3 = 125$ **27.** $\frac{1}{p}$
28. $\frac{1}{y^{15}}$ **29.** $\frac{-6y}{x}$ **30.** $\frac{-3x^4}{y}$ **31.** $\frac{4q}{p^2r^4}$ **32.** $\frac{4bc^3}{a^2}$
33. $\frac{4}{25}$ **34.** $\frac{9}{16}$ **35.** $\frac{x^3}{y^3}$ **36.** $\frac{x^2}{z^2}$ **37.** $\frac{-27a^3}{64b^3}$
38. $\frac{-64x^3}{y^3}$ **39.** $\frac{16x^4y^4}{z^4}$ **40.** $\frac{81x^4}{y^4z^4}$ **41.** 1 **42.** 1
43. 1 **44.** 1 **45.** $-7x^2$ **46.** 2 **47.** 1 **48.** 1
49. b^4 **50.** x^6 **51.** $\frac{c^8}{2}$ **52.** $2x^2$ **53.** $25x^{20}y^{10}$
54. $36x^{16}y^6$ **55.** $-7x$ **56.** $-2x$
57. $2a^{36}$ **58.** $5y^{29}$ **59.** $16x^{12}y^8$ **60.** $81x^8y^{12}$
61. $\frac{1}{x^{12}}$ **62.** y^{10} **63.** $\frac{81a^{12}}{b^8}$ **64.** $\frac{125x^6}{y^{21}}$ **65.** $8x^{15}z^{18}$
66. $81b^4c^6$
67. (a) $f(x) = (-3x)^2$

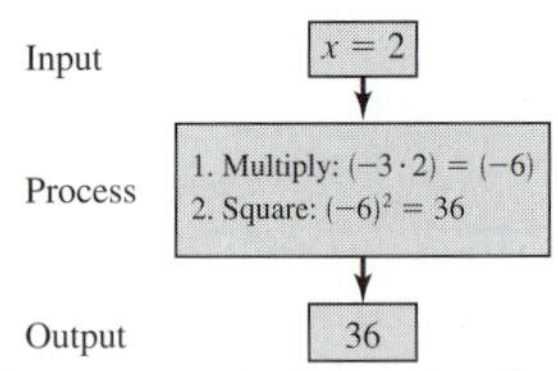

$f(2) = (-3\cdot 2)^2 = (-6)^2 = 36$

(b) $g(x) = 9x^2$

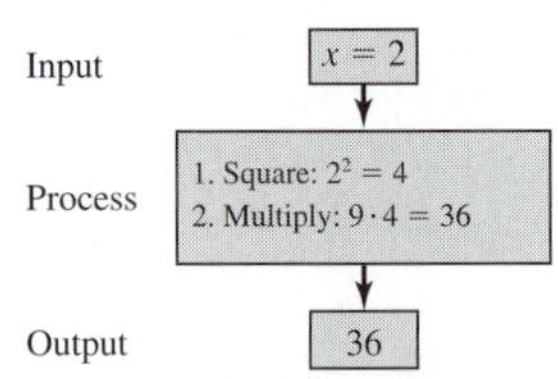

$g(2) = 9(2)^2 = 9\cdot 4 = 36$

(c) Yes

68. (a) $f(x) = (5x)^2$

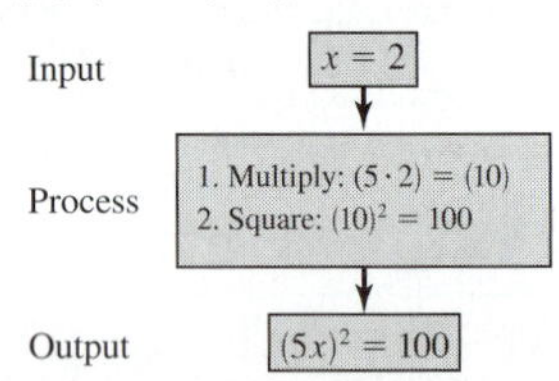

$f(2) = (5\cdot 2)^2 = (10)^2 = 100$

(b) $g(x) = 25x^2$

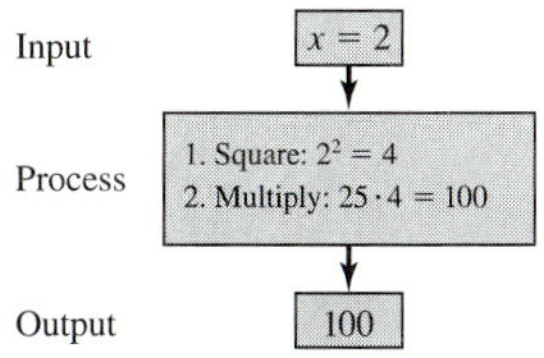

$g(2) = 25(2)^2 = 25 \cdot 4 = 100$

(c) Yes

69. $\neq$ **70.** $\neq$ **71.** $\neq$ **72.** $=$ **73.** $=$ **74.** $\neq$
75. $=$ **76.** $\neq$ **77.** $\neq$ **78.** $\neq$ **79.** $=$ **80.** $=$

DOING MATHEMATICS SECTION 5.2

1. $\frac{1}{a^5}$ **2.** $\frac{1}{x^2}$ **3.** $\frac{1}{25}$ **4.** $\frac{1}{81}$ **5.** $\frac{3x^4}{y^2}$ **6.** $\frac{4a}{b^3}$
7. $\frac{-4x^3}{y^2}$ **8.** $\frac{-25x}{y^3}$ **9.** $\frac{15}{c^3d^6}$ **10.** $\frac{20}{a^4b^2}$ **11.** 27
12. 32 **13.** $-m^3$ **14.** $-x^2$ **15.** pq^3 **16.** a^2b^3
17. $\frac{-2y^3}{7x^5}$ **18.** $\frac{x^3z}{15y^2}$ **19.** $\frac{wyz^4}{-5x^2} = \frac{-wyz^4}{5x^2}$
20. $\frac{2yz^4}{-7x^3} = \frac{-2yz^4}{7x^3}$ **21.** 9 **22.** 16 **23.** n^6 **24.** a^8
25. $\frac{1}{27x^{12}}$ **26.** $\frac{1}{4x^8}$ **27.** $\frac{c^8d^2}{64}$ **28.** $\frac{b^6}{81a^4}$ **29.** $\frac{27q^{15}}{8p^6}$
30. $\frac{16y^2}{9x^4}$ **31.** -32 **32.** -125 **33.** $4x^7$ **34.** $2x^3$
35. $\frac{6y^5}{z}$ **36.** $\frac{5}{x^2y}$ **37.** 1 **38.** 1 **39.** $\frac{1}{z^6}$ **40.** $\frac{1}{x^3}$

41. (a) $f(x) = -4x^{-2}$

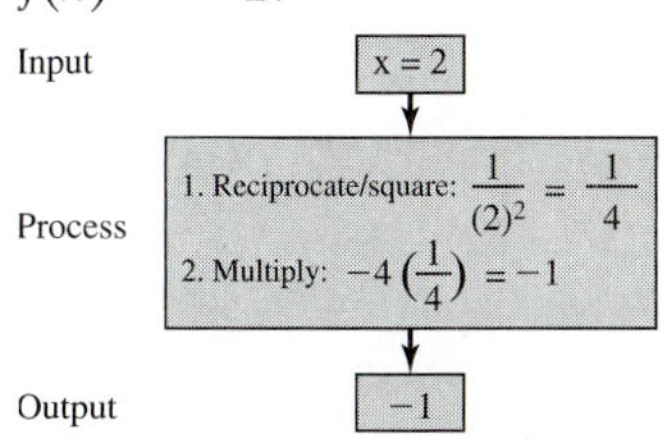

$$f(2) = -4(2)^{-2} = \frac{-4}{(2)^2} = \frac{-4}{4} = -1$$

(b) $g(x) = \frac{-4}{x^2}$

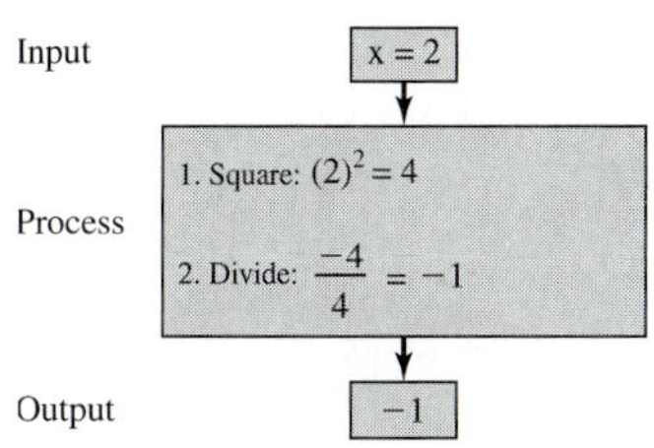

$$g(2) = \frac{-4}{(2)^2} = \frac{-4}{4} = -1$$

(c) Yes

42. (a) $f(x) = 16x^{-3}$

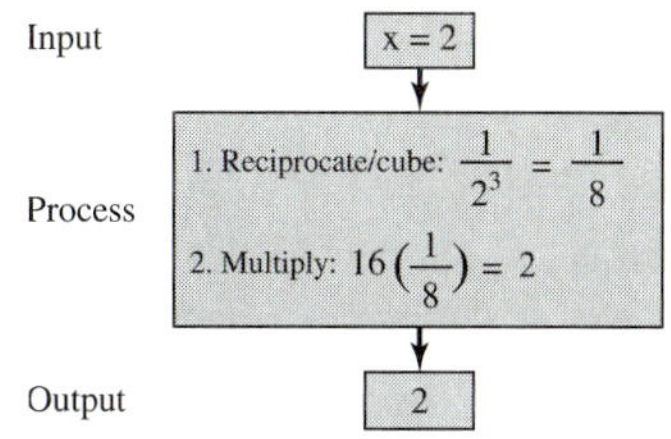

$$f(2) = 16(2)^{-3} = \frac{16}{(2)^3} = \frac{16}{8} = 2$$

(b) $g(x) = \frac{16}{x^3}$

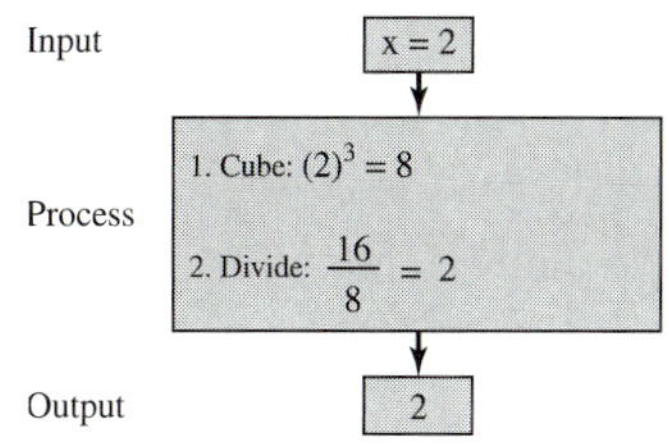

$$g(2) = \frac{16}{(2)^3} = \frac{16}{8} = 2$$

(c) Yes

43. $f(2) = \frac{3}{8}$ **44.** $f(4) = \frac{3}{64}$ **45.** $g(-2) = 16$
46. $g(-3) = 36$ **47.** $h(-2) = \frac{1}{36}$ **48.** $h(-4) = \frac{1}{64}$
49. $\neq$ **50.** $\neq$ **51.** $\neq$ **52.** $\neq$ **53.** $=$ **54.** $=$
55. $\neq$ **56.** $\neq$ **57.** $=$ **58.** $=$ **59.** $\neq$ **60.** $\neq$

DOING MATHEMATICS SECTION 5.3

1. (a) 2 (b) 3 (c) 1 (d) 2 (e) 3 (f) 4 (g) 1 (h) 4 **2.** (a) 3 (b) 1 (c) 2 (d) 1 (e) 4 (f) 3 (g) 4 (h) 2 **3.** $50z^6$ **4.** $-6x^6$
5. $-5.473a^4b^4c^3$ **6.** $11.5x^3y^5$ **7.** $-48y^6$ **8.** $-24x^4$
9. $-\frac{1}{48}p^3q^3r^2$ **10.** $-\frac{1}{6}a^3b^3c^3$ **11.** $-3.75z^4$
12. $-35y^7$ **13.** $15x^4 + 5x^3 - 10x^2$
14. $-5x^3 + 5x^2 + 5x$ **15.** $-64t^5 + 48t^3 - 16t^2$
16. $-6x^3 + 10x^2 + 20x$ **17.** $2l^3w + l^2w^2 + lw^3$
18. $6x^3y + 2x^2y^2 - 2xy^3$ **19.** $18b^2c^3d^2 - 48bc^3d^3$
20. $-18a^2b^2 - 45a^2$
21. $-10k^4 + 15k^3 + 35k^2 + 25k$
22. $-8x^4 + 10x^3 - 20x^2 + 16x$ **23.** $v^2 + 10v + 16$
24. $x^2 + 14x + 45$ **25.** $t^2 - 8t + 7$ **26.** $x^2 - 14x + 45$
27. $y^2 - y - 12$ **28.** $x^2 - 4x - 5$
29. $21s^2 + 41s + 10$ **30.** $10x^2 + 23x + 12$
31. $3a^2 - 10ab + 3b^2$ **32.** $5x^2 - 16xy + 3y^2$
33. $u^2 - 100$ **34.** $x^2 - 25$ **35.** $2a^2 + 5ab - 3b^2$
36. $x^2 - 4xy - 5y^2$ **37.** $w^2 - 5w - 36$
38. $y^2 + y - 42$ **39.** $21x^2 + 11x - 2$
40. $10x^2 - 7x - 12$
41. $3y^4 + 7y^3 - 17y^2 - 18y + 28$

42. $2x^4 - 7x^2 - 22x - 15$
43. $2x^4 - x^3 - 4x^2 + 8x - 5$
44. $20x^4 - 17x^3 + 40x^2 - 20x - 16$
45. $8x^5 + 22x^4 - 9x^3 - 4x^2 + 4x - 1$
46. $5x^5 - 38x^4 + 73x^3 - 39x^2 + 17x + 2$
47. $-2x^6 + x^5 - 2x^4 - 4x^3 + x^2 - 3x - 3$
48. $4x^6 + 2x^5 - 4x^4 - 3x^3 + 4x^2 + 9x - 18$
49. $x^4 + x^3 - x^2 + 1$ **50.** $-3x^3 + 46x^2 - 21x + 2$
51. $2x^4 - 7x^3 - 18x^2 - 11x - 2$
52. $2x^4 - 11x^2 - 14x - 5$
53. $6x^5 + x^4 - 21x^3 - 45x^2 - 27x - 10$
54. $5x^5 - 38x^4 + 73x^3 - 39x^2 + 17x + 2$
55. $-8x^6 - 12x^5 - 4x^4 + 12x^3 - 3x^2 + 10x + 5$
56. $4x^6 + 2x^5 - 4x^4 - 3x^3 + 4x^2 + 9x - 18$
57. (a) Area $= (2x + 9)(3x - 2) = 6x^2 + 23x - 18$
(b) 247 square inches
58. (a) Area $= (4x + 5)(2x - 1) = 8x^2 + 6x - 5$
(b) 225 square inches
59. (a) Area $= (6x)(6x) = 36x^2$
(b) 900 square inches
60. (a) Area $= (4x)(4x) = 16x^2$
(b) 400 square inches
61. (a) Area $= (3x + 7)(x + 4) - (x)(x + 9) = 2x^2 + 10x + 28$
(b) 128 square inches
62. (a) Area $= (2x + 3)(x + 2) - (x)(x + 5) = x^2 + 2x + 6$
(b) 41 square inches
63. (a)

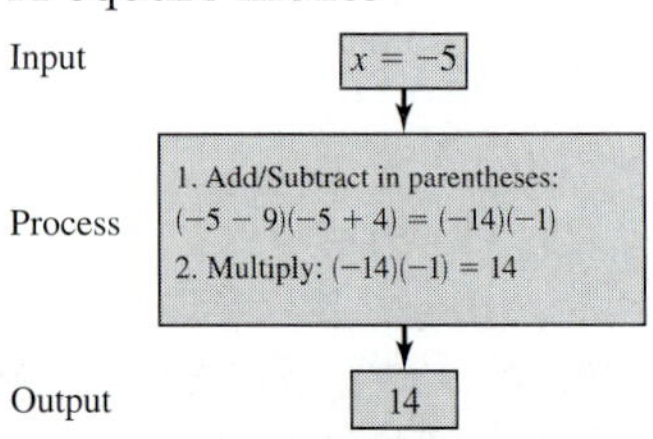

$f(-5) = (-5 - 9)(-5 + 4) = 14$

(b)

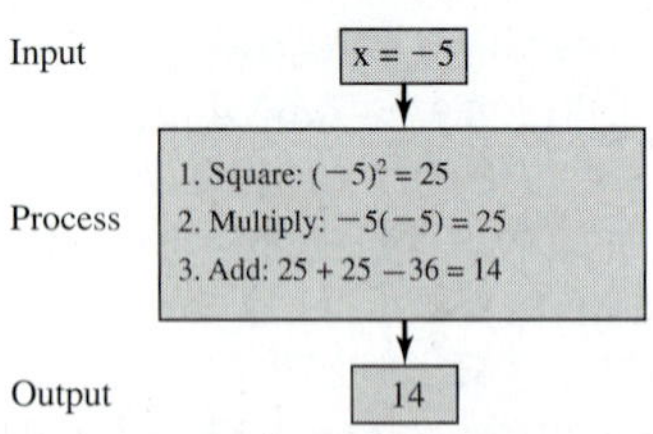

$g(-5) = (-5)^2 - 5(-5) - 36 = 14$
(c) Yes
64. (a)

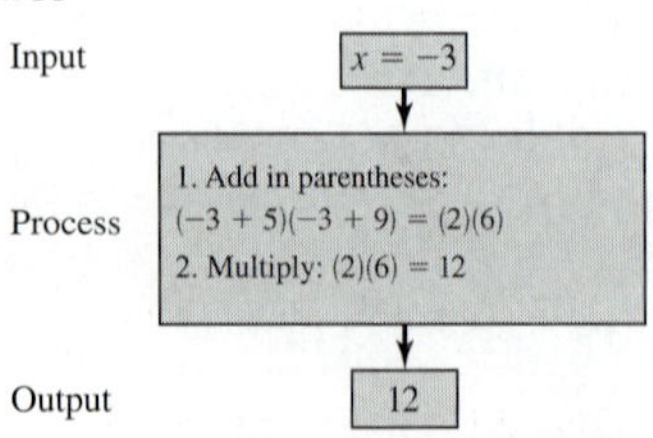

$f(-3) = (-3 + 5)(-3 + 9) = 12$

(b)

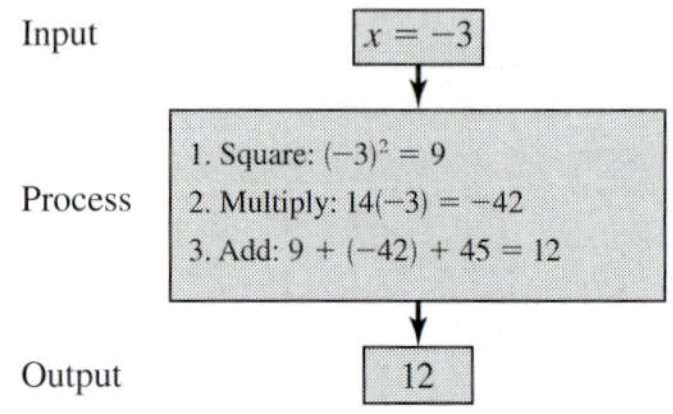

$g(-3) = (-3)^2 + 14(-3) + 45 = 12$
(c) Yes
65. $\neq$ **66.** $\neq$ **67.** $=$ **68.** $=$ **69.** $\neq$ **70.** $\neq$
71. $\neq$ **72.** $\neq$ **73.** $=$ **74.** $=$ **75.** $\neq$ **76.** $\neq$
77. $=$ **78.** $=$

DOING MATHEMATICS SECTION 5.4

1. (a) 2 (b) 1 (c) 1 (d) 2 (e) 1 **2.** (a) 2 (b) 1 (c) 1 (d) 1 (e) 2 **3.** $t^2 - 25$ **4.** $x^2 - 4$
5. $36x^2 - 1$ **6.** $4a^2 - 9$ **7.** $x^2 - 64y^2$ **8.** $a^2 - 9b^2$
9. $4 - x^2$ **10.** $36 - y^2$ **11.** $9x^2 - 16y^2$
12. $4x^2 - 49y^2$ **13.** $w^2 + 8w + 16$ **14.** $x^2 + 2x + 1$
15. $y^2 + 10y + 25$ **16.** $x^2 + 6x + 9$
17. $9x^2 + 24x + 16$ **18.** $4x^2 + 20x + 25$
19. $9p^2 + 6pq + q^2$ **20.** $4x^2 + 4xy + y^2$
21. $r^2 - 2r + 1$ **22.** $x^2 - 4x + 4$
23. $b^2 - 18b + 81$ **24.** $x^2 - 10x + 25$
25. $16 - 8x + x^2$ **26.** $64 - 16y + y^2$
27. $25x^2 - 40xy + 16y^2$ **28.** $4a^2 - 12ab + 9b^2$
29. $-5x^3 - 50x^2 - 125x$ **30.** $-4x^3 + 56x^2 - 196x$
31. $2x^5 - 20x^4 + 50x^3$ **32.** $7x^4 + 28x^3 + 28x^2$
33. $\dfrac{1}{4x^2 - 12x + 9}$ **34.** $\dfrac{1}{x^2 + 4x + 4}$
35. $x^3 - 3x^2 - 9x + 27$ **36.** $x^3 + 7x^2 + 8x - 16$
37. $-75x^5y^2 + 48x^3y^2$ **38.** $-20x^4y + 5x^2y$
39. $-9x^2 + 42xy - 49y^2$ **40.** $-12x^2 - 60xy - 75y^2$
41. $x^4 - 18x^2 + 81$ **42.** $x^4 + 6x^3 + 13x^2 + 12x + 4$
43. (a) Volume $= x(x + 4)(x - 4) = x^3 - 16x$
(b) 120 cubic inches
44. (a) Volume $= x(x + 5)(x - 5) = x^3 - 25x$
(b) 66 cubic inches
45. (a) Volume $= 2x(x + 6)^2 = 2x^3 + 24x^2 + 72x$
(b) 1728 cubic inches
46. (a) Volume $= 2x(x + 1)^2 = 2x^3 + 4x^2 + 2x$
(b) 588 cubic inches
47. (a) Area $= (3x + 1)^2 - (x + 7)^2 = 8x^2 - 8x - 48$
(b) 112 square feet
48. (a) Area $= (2x + 3)^2 - (x + 5)^2 = 3x^2 + 2x - 16$
(b) 69 square feet

49. (a)

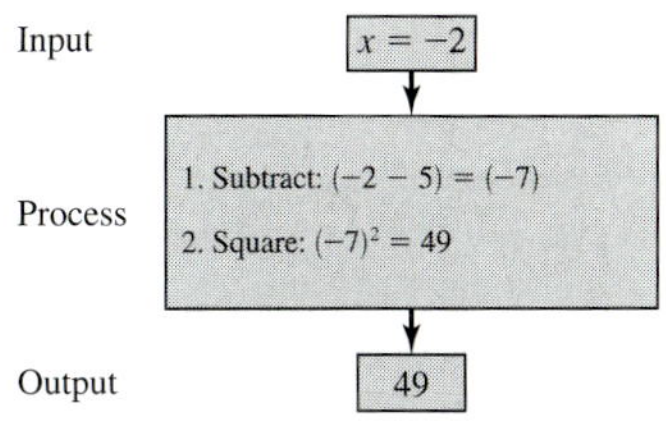

$f(-2) = (-2 - 5)^2 = (-7)^2 = 49$

(b)

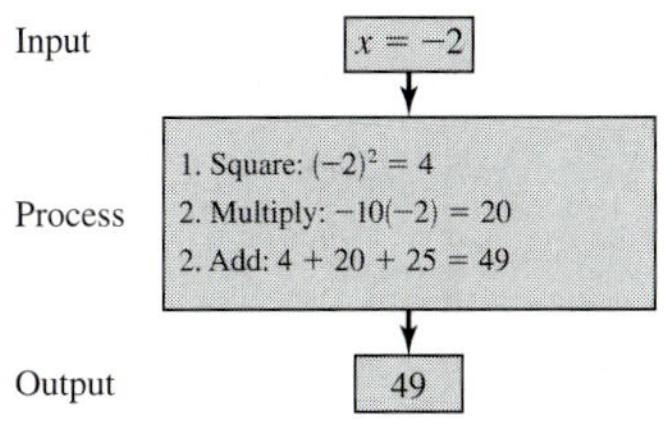

$g(-2) = (-2)^2 - 10(-2) + 25$
$= 4 + 20 + 25 = 49$

(c) Yes

50. (a)

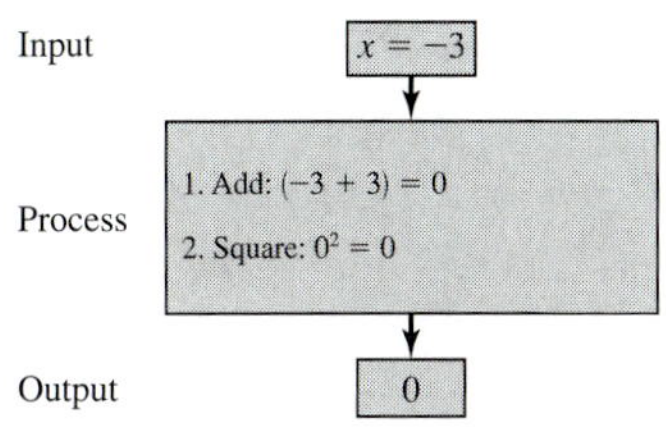

$f(-3) = (-3 + 3)^2 = 0^2 = 0$

(b)

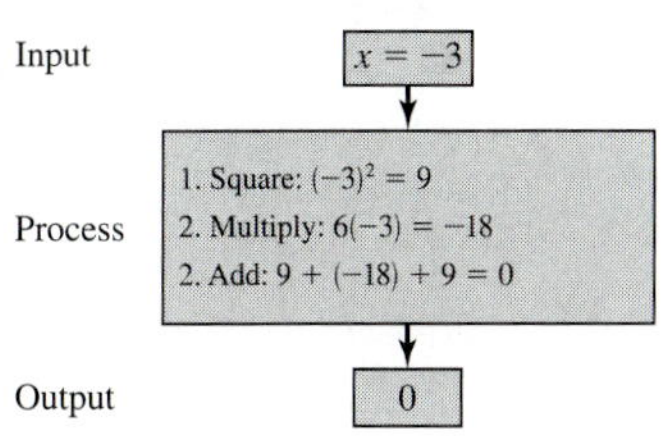

$g(-3) = (-3)^2 + 6(-3) + 9 = 9 - 18 + 9 = 0$

(c) Yes

51. $\neq$ **52.** $\neq$ **53.** $\neq$ **54.** $=$ **55.** $=$ **56.** $=$
57. $=$ **58.** $\neq$ **59.** $\neq$ **60.** $\neq$ **61.** $=$ **62.** $=$
63. $\neq$ **64.** $\neq$

DOING MATHEMATICS SECTION 5.5

1. (a) 2 (b) 1 (c) 3 (d) 2 (e) 3 (f) 1
2. (a) 2 (b) 1 (c) 3 (d) 1 (e) 3 (f) 2
3. $8x^2y$ **4.** $5x^2$ **5.** $\frac{-11}{mn}$ **6.** $\frac{-6}{a^2b^3}$ **7.** $\frac{-13pr^2}{q}$
8. $\frac{-4x^2y^3}{z^3}$ **9.** $\frac{9y^2z}{x}$ **10.** $\frac{11bc^2}{a}$ **11.** $\frac{3w^3}{z^2} - 2$
12. $\frac{5x^2}{y} + 3$ **13.** $1 - \frac{3}{w} + \frac{12}{w^2}$ **14.** $x + 3 - \frac{7}{x}$
15. $\frac{y}{2} - 2 - \frac{14}{y}$ **16.** $\frac{-x}{3} - 2 - \frac{5}{x}$
17. $\frac{2m}{n^2} + \frac{3}{2mn} - \frac{1}{2m^2}$ **18.** $\frac{-5x}{y} + 1 - \frac{2y}{x}$
19. $x + 1 + \frac{-2}{x - 6}$ **20.** $x + 3 + \frac{-7}{x - 1}$
21. $2x + 1 + \frac{2}{3x + 5}$ **22.** $x - 4 + \frac{-18}{2x + 3}$
23. $2x^2 - 4x + 8$ **24.** $2x^2 + 2x + 2 + \frac{-6}{x - 1}$
25. $y^2 - 3$ **26.** $6x^3 + 8x^2 + 8x + 8 + \frac{13}{x - 1}$
27. $\frac{-3j^2}{k^2}$ **28.** $\frac{-3x^2}{y^2}$ **29.** $-2t - 4 + \frac{10}{t}$
30. $3y + \frac{5}{2} - \frac{9}{2y}$ **31.** $2x + 1 + \frac{12}{x - 2}$
32. $x + 4 + \frac{-42}{x + 3}$ **33.** $x^2 - 1 + \frac{2}{x^2 + 1}$
34. $x^2 + 3 + \frac{18}{x^2 - 3}$ **35.** $-2y - 6x + 3z$
36. $-6ab^2 - 7c^2 + 4b$ **37.** $\frac{3s^4}{r^2}$ **38.** $\frac{34ac^2}{b}$
39. (a) Length $= \frac{x^2 + 2x - 8}{x - 2}$
$= x + 4$
(b) 9 feet
40. (a) Length $= \frac{x^2 - x - 6}{x - 3}$
$= x + 2$
(b) 7 feet
41. (a) Length $= \frac{9x^2 - 4}{3x - 2}$
$= 3x + 2$
(b) 17 feet
42. (a) Length $= \frac{4x^2 - 1}{2x + 1}$
$= 2x - 1$
(b) 9 feet
43. $\neq$ **44.** $\neq$ **45.** $=$ **46.** $=$ **47.** $=$ **48.** $=$
49. $\neq$ **50.** $\neq$ **51.** $\neq$ **52.** $\neq$ **53.** $\neq$ **54.** $\neq$
55. $=$ **56.** $=$ **57.** $=$ **58.** $=$
59. An example: $\frac{2x + 2}{2}$
60. An example: $\frac{x^3 - 4x^2 + 7x + 2}{x^2 + 8x + 9}$

CHAPTER 5 REVIEW

1. x^5 **2.** $\frac{1}{5^3} = \frac{1}{125}$ **3.** $-x^{10}$ **4.** 4096
5. $-3.375x^3y^3$ **6.** x^8y^6 **7.** $3^8 = 6561$ **8.** $y^{-5} = \frac{1}{y^5}$
9. $\frac{x^2}{y^2}$ **10.** $\frac{9}{25}$ **11.** 1 **12.** 1 **13.** $\frac{1}{y^3}$
14. $(2x)^3 = 8x^3$ **15.** $\frac{5}{x^4}$ **16.** $3x^7$ **17.** $\frac{x^6}{2}$ **18.** $\frac{9x^4}{y^6}$
19. $1024x^5$ **20.** $5y$ **21.** 343 **22.** $\frac{1}{9x^4y^6}$ **23.** $6y^{15}$

24. $\frac{y^{21}}{8x^6}$ **25.** $\frac{b^4}{25c^6}$ **26.** $\frac{8x^3}{125}$ **27.** $\neq$ **28.** $=$
29. $\neq$ **30.** $\neq$ **31.** $2x^7$ **32.** $48x^4$ **33.** $\frac{a^3b^3c^3}{6}$
34. $-5x^3 + 10x^2 - 15x$ **35.** $-8x^3 - 10x^2 + 20x$
36. $-3a^3b + 10a^3b^2$ **37.** $x^2 + x - 30$ **38.** $x^2 - y^2$
39. $x^2 + 14x + 49$ **40.** $18x^2 - 37x - 20$
41. $4x^2 - 1$ **42.** $5x^2 + 2xy - 3y^2$
43. $y^2 - 8y + 16$ **44.** $9x^2 + 12x + 4$
45. $a^2 - 2ab + b^2$ **46.** $x^2 - 10xy + 25y^2$
47. $10x^2 + 23x + 12$
48. $3x^4 - 14x^3 - 6x^2 + 6x - 5$
49. $5x^4 - 31x^3 - 39x^2 - 31x + 8$
50. $8x^4 - 28x^3 + 15x^2 + 11x - 5$
51. $7x^5 - 17x^4 + 74x^3 - 19x^2 - 23x - 2$
52. $3x^4 + 5x^3 + 32x^2 - 4x - 16$
53. $25x^5 - 70x^4 + 109x^3 - 89x^2 + 27x - 2$
54. (a) $A = (3x - 5)(6x + 2) = 18x^2 - 24x - 10$
(b) 320 square inches
55. (a) $A = (2x - 5)(2x - 5) = 4x^2 - 20x + 25$
(b) 25 square inches
56. (a) $A = \pi(3x + 2)^2 - (x + 1)^2$
$A = \pi(9x^2 + 12x + 4) - (x^2 + 2x + 1)$
$A = 9\pi x^2 + 12\pi x + 4\pi - x^2 - 2x - 1$
(b) 871.46 square inches
57. $\neq$ **58.** $\neq$ **59.** $=$ **60.** $=$ **61.** $=$ **62.** $\neq$
63. $-6a^3b^2$ **64.** $\frac{y^2}{z^4}$ **65.** $x + 3 - \frac{7}{x^2}$
66. $\frac{5x}{y} - 2 - \frac{5y}{3x}$ **67.** $x + 4 + \frac{-14}{x + 1}$
68. $x^2 + x + 1 + \frac{-7}{x - 1}$ **69.** $\frac{-8x^2}{y^7}$
70. $3y + \frac{5}{2} - \frac{4}{y}$ **71.** $x + 1 + \frac{-15}{x + 3}$ **72.** $x + 7$
73. $\frac{9}{b^3} + \frac{11c}{a^2b^2} - \frac{6}{ab}$
74. Volume $= x(x + 2)^2 = x^3 + 4x^2 + 4x$; 1089 cubic feet
75. Height $= \frac{3x^3 - 75x}{(x + 5)(x - 5)} = \frac{3x(x + 5)(x - 5)}{(x + 5)(x - 5)} = 3x$
76. $\neq$ **77.** $=$ **78.** $\neq$ **79.** $=$

CHAPTER 5 PRACTICE TEST ANSWERS

1. $7x^8$ **2.** $\frac{5}{x^6}$ **3.** $-27x^3y^6$ **4.** $\frac{-125x^6}{y^6}$ **5.** $-\frac{bc^2}{9a^2}$
6. $5xy^6$ **7.** 64 **8.** a^{12} **9.** $\frac{y^2}{49x^2z^2}$ **10.** $\frac{x^{12}}{16y^4z^4}$
11. $54x^4$ **12.** $-2x^4 - 2x^3 + 60x^2$
13. $14x^2 - 3x - 5$ **14.** $-3x^3y + 6x^2y^2$
15. $4x^2 - 12x + 9$ **16.** $-15a^3b^4c^2 + 12a^2b^2c^3$
17. $x^3 - 3x^2 - 6x + 8$ **18.** $-5x^3y + 5xy^3$
19. $2x^4 + 5x^3 - 12x^2 + 22x - 35$ **20.** $-3b$
21. $6x - 3$ **22.** $1 - 3ab - 4b$ **23.** $x - 2$
24. $x^2 + 3x + 9$
25. (a) $A = (4x - 1)(x + 3) - (x + 5)(x)$
$= 3x^2 + 6x - 3$
(b) 141 square inches

CHAPTER 6

DOING MATHEMATICS SECTION 6.1

1. (a) -7 (b) $(2x - 3)$ (c) 3 (d) $(5c + 1)$ (e) None (f) $3y$ **2.** (a) 3 (b) $(x + 1)$ (c) 2 (d) $(2x - 3)$ (e) None (f) $2y$ **3.** $3(x - 2)$
4. $5(x + 6)$ **5.** $-3(x - 2)$ **6.** $-3(4y + 3)$
7. $-5x(x + 1)$ **8.** $-4x(1 + x)$ **9.** $-3x^3(x^2 - 2)$
10. $-2m^3(m - 4)$ **11.** $2x(x^2 + 2x + 5)$
12. $3x(x^3 + 3x^2 + 8x + 12)$ **13.** $-x(2x^2 + x + 1)$
14. $-y(2y^2 - 3y + 5)$ **15.** $3abc(5ab^2 - 4c)$
16. $7pqr^3(p + 4q)$ **17.** $-2x^2(2x^2 - x + 4)$
18. $-3x(2x^2 + x + 3)$ **19.** $(y + 7)(x + 2)$
20. $(4a + 3)(b + 3)$ **21.** $(x - 2)(3 - y)$
22. $(n - 3)(5m - 2)$ **23.** $(x - 1)(5 - y)$
24. $(u + 2)(3 - v)$ **25.** $(c + 4)(ab + 1)$
26. $(z - 3)(xy + 1)$ **27.** $(2x - 3)(5x - 4)$
28. $(2x + 3)(3x - 2)$ **29.** $(3a + 1)(2a + 5)$
30. $(3y + 4)(5y + 2)$ **31.** $(x + 6)(x - 5)$
32. $(x + 2)(x - 3)$ **33.** $(x - 5)(2x + 1)$
34. $(3x - 5)(7x + 2)$ **35.** 2 **36.** 3 **37.** $\frac{5}{2}$
38. $\frac{1}{2}$ **39.** $\frac{x + 2}{4}$ **40.** $\frac{x - 7}{2}$ **41.** $\frac{2a + 1}{3}$
42. $\frac{3u + 4}{2}$ **43.** $\frac{x - 3}{2x + 1}$ **44.** $\frac{4x - 5}{x - 4}$ **45.** $\frac{-3x}{5}$
46. $-3y$ **47.** $=$ **48.** $=$ **49.** $\neq$ **50.** $\neq$

DOING MATHEMATICS SECTION 6.2

1. (a) 1 and 3 (b) 1 (c) 4 (d) 2 (e) 1 (f) 1 and 3 (g) 4 (h) 2 (i) 4 (j) 1 **2.** (a) 1 and 3 (b) 2 (c) 1 and 3 (d) 1 and 3 (e) 1 (f) 4 (g) 4 (h) 2 (i) 1 (j) 1 and 3 **3.** $(y + 5)(y + 2)$
4. $(x + 7)(x + 3)$ **5.** $(m + 10)(m + 3)$
6. $(x + 5)(x + 4)$ **7.** $(n + 4)(n - 8)$
8. $(x + 1)(x - 5)$ **9.** $(y + 7)(y - 6)$
10. $(x - 4)(x + 3)$ **11.** $(x + 2)(x + 3)$
12. $(x + 5)(x + 1)$ **13.** $(x - 6)(x - 4)$
14. $(x - 8)(x - 3)$ **15.** $(c + 5)(c - 8)$
16. $(x - 3)(x + 7)$ **17.** $(x + 8)(x - 7)$
18. $(x - 4)(x + 5)$ **19.** $(x + 9)(x - 9)$
20. $(x + 2)(x - 2)$ **21.** $(m + 1)(m - 1)$
22. $(y + 3)(y - 3)$ **23.** $(z + 12)(z - 12)$

24. $(x + 10)(x - 10)$ **25.** $(b + 4c)(b - 4c)$
26. $(a + 6b)(a - 6b)$ **27.** $(x + 7y)(x - 7y)$
28. $(x + 4y)(x - 4y)$ **29.** $(1 + 10a)(1 - 10a)$
30. $(1 + 6y)(1 - 6y)$ **31.** $(a + 13b)(a - 13b)$
32. $(y + 7z)(y - 7z)$ **33.** $(x + 2y)(x - 2y)$
34. $(x + 9y)(x - 9y)$ **35.** $(y - 9)^2$ **36.** $(x - 2)^2$
37. $(x + 5)^2$ **38.** $(x - 5)^2$ **39.** $(a + 10)^2$
40. $(x + 4)^2$ **41.** $(n + 6)^2$ **42.** $(y + 7)^2$
43. $(p - 2)^2$ **44.** $(x + 1)^2$ **45.** $(x + 11y)^2$
46. $(x + 5y)^2$ **47.** $(x - 12)^2$ **48.** $(x - 9)^2$
49. $(c - 4d)^2$ **50.** $(x - 6y)^2$ **51.** $(x - 8)(x + 3)$
52. $(x - 4)(x + 1)$ **53.** Prime **54.** Prime
55. $(x + y)^2$ **56.** $(x + 3y)^2$ **57.** $(s + 7)(s + 11)$
58. $(x - 7)^2$ **59.** Prime **60.** Prime
61. $(a + 1)(a - 1)$ **62.** $(y + 25)(y + 2)$
63. $(x - 4)^2$ **64.** $(a - 3)^2$ **65.** Prime
66. $(x + 9y)(x - 9y)$ **67.** $\frac{x - 5}{x - 10}$ **68.** $\frac{x - 3}{x + 5}$
69. $\frac{r + 7}{r + 3}$ **70.** $\frac{y + 1}{y - 3}$ **71.** $\frac{z + 2}{z - 4}$ **72.** $\frac{a - 5}{a - 7}$
73. $\frac{x - 6}{x - 9}$ **74.** $\frac{y + 5}{y - 6}$ **75.** $\frac{x + 7}{x + 1}$ **76.** $\frac{x + 3}{x - 1}$
77. $\frac{m + 8}{m - 8}$ **78.** $\frac{a + 3}{a - 3}$ **79.** $\frac{x - 6}{x + 6}$ **80.** $\frac{y + 1}{y - 1}$
81. $\frac{x + 4}{x - 3}$ **82.** $\frac{x - 8}{x + 2}$ **83.** $=$ **84.** $=$ **85.** $\neq$
86. $\neq$

DOING MATHEMATICS SECTION 6.3

1. (a) 2 and 4 (b) 2 (c) 3 (d) 1 (e) 5 (f) 3 (g) 2 and 4 (h) 2 **2.** (a) 1 (b) 2 and 4 (c) 5 (d) 2 and 4 (e) 2 (f) 3 (g) 5 (h) 2 and 4
3. $(3y + 1)(y - 3)$ **4.** $(2x + 1)(x - 5)$
5. $(5z - 1)(z - 4)$ **6.** $(3x - 2)(x - 4)$
7. $(4a + 3)(3a + 4)$ **8.** $(3y + 2)(2y - 3)$
9. $(2a + 1)(2a - 5)$ **10.** $(6x - 1)(x + 3)$
11. $(3p - 4)(2p + 3)$ **12.** $(4x + 7)(x - 2)$
13. $(6m - 1)(2m + 1)$ **14.** $(4m + 1)(3m - 1)$
15. $(2x - 1)(2x - 9)$ **16.** $(2x + 1)(3x + 4)$
17. $(7x + 2)(2x - 5)$ **18.** $(4x - 3)(x + 3)$
19. $(3x - 2)(3x + 2)$ **20.** $(5x - 7)(5x + 7)$
21. $(4x - 5y)(4x + 5y)$ **22.** $(3a - 5b)(3a + 5b)$
23. $(2x + 3)^2$ **24.** $(2x - 1)^2$ **25.** $(4x - y)^2$
26. $(2x + 5y)^2$ **27.** $(3x + y)(2x + y)$
28. $(3x + 5y)(2x - y)$ **29.** $(8x + 9)(3x + 4)$
30. $(5a - 2)(4a - 1)$ **31.** $(2x - 5)(6x + 7)$
32. $(7y + 1)(3y + 2)$ **33.** $(4x + 11)^2$
34. $(2x - 5)(x - 5)$ **35.** $(3x + 5)(9x - 2)$
36. $(7x - 3)(x + 4)$ **37.** $\frac{3v + 1}{v + 1}$ **38.** $\frac{3x + 5}{x + 1}$
39. $\frac{3x - 10}{x - 8}$ **40.** $\frac{x - 4}{x + 1}$ **41.** $\frac{2x + 1}{x - 2}$ **42.** $\frac{3x + 2}{x + 3}$
43. $\frac{z - 3}{3z + 1}$ **44.** $\frac{3y - 2}{7y - 3}$ **45.** $\frac{3x - 1}{2x - 1}$ **46.** $\frac{2x + 1}{x - 1}$
47. $\frac{7x - 1}{3x - 2}$ **48.** $\frac{2x + 1}{2x - 5}$ **49.** $\frac{3x + 5}{3x + 1}$ **50.** $\frac{2x - 9}{4x - 3}$
51. $\frac{x + 5y}{4x - 3y}$ **52.** $\frac{5x + y}{x + 2y}$ **53.** $=$ **54.** $=$ **55.** $\neq$
56. $=$

DOING MATHEMATICS SECTION 6.4

1. $(x + 5)(x - 6) = 0, x = -5, x = 6$
2. $(x + 1)(x - 3) = 0, x = -1, x = 3$
3. $(y + 3)(y + 3) = 0, y = -3$
4. $(y + 1)(y - 4) = 0, y = -1, y = 4$
5. $(x + 7)(x + 1) = 0, x = -7, x = -1$
6. $(x + 5)(x + 1) = 0, x = -5, x = -1$
7. $(y - 5)(y - 4) = 0, y = 5, y = 4$
8. $(y - 4)(y - 3) = 0, y = 4, y = 3$
9. $(x - 9)(x - 5) = 0, x = 9, x = 5$
10. $(x - 2)(x - 7) = 0, x = 2, x = 7$
11. $(a - 1)(a - 1) = 0, a = 1$
12. $(a + 3)(a - 1) = 0, a = -3, a = 1$
13. $(x + 2)(x - 6) = 0, x = -2, x = 6$
14. $(x - 3)(x - 3) = 0, x = 3$
15. $4x(x + 1) = 0, x = 0, x = -1$
16. $3x(x - 1) = 0, x = 0, x = 1$
17. $2x(x + 5) = 0, x = 0, x = -5$
18. $5x(x - 3) = 0, x = 0, x = 3$
19. $-9x(x - 4) = 0, x = 0, x = 4$
20. $-7x(x + 4) = 0, x = 0, x = -4$
21. $y(y - 1) = 0, y = 0, y = 1$
22. $6x(x + 2) = 0, x = 0, x = -2$
23. $(3x + 1)(x - 5) = 0, x = -\frac{1}{3}, x = 5$
24. $(x + 5)(2x - 1) = 0, x = -5, x = \frac{1}{2}$
25. $(x - 1)(x - 1) = 0, x = 1$
26. $(x - 5)(x - 5) = 0, x = 5$
27. $(7x + 2)(3x - 4) = 0, x = -\frac{2}{7}, x = \frac{4}{3}$
28. $(2x - 3)(5x - 4) = 0, x = \frac{3}{2}, x = \frac{4}{5}$
29. $(x + 4)(3x - 2) = 0, x = -4, x = \frac{2}{3}$
30. $(2x + 1)(x - 5) = 0, x = -\frac{1}{2}, x = 5$
31. $(2y - 9)(y - 4) = 0, y = \frac{9}{2}, y = 4$
32. $(3y - 2)(y - 4) = 0, y = \frac{2}{3}, y = 4$

33. $(2x - 3)(2x - 3) = 0, x = \frac{3}{2}$

34. $(3x - 1)(3x - 1) = 0, x = \frac{1}{3}$

35. $(3y + 4)(2y - 9) = 0, y = -\frac{4}{3}, y = \frac{9}{2}$

36. $(3y + 5)(2y - 1) = 0, y = -\frac{5}{3}, y = \frac{1}{2}$

37. $(2x + 1)(3x + 2) = 0, x = -\frac{1}{2}, x = -\frac{2}{3}$

38. $(3x + 2)(4x - 9) = 0, x = -\frac{2}{3}, x = \frac{9}{4}$

39. $(5a + 9)(5a - 9) = 0, a = -\frac{9}{5}, a = \frac{9}{5}$

40. $(3a - 7)(3a + 7) = 0, a = \frac{7}{3}, a = -\frac{7}{3}$

41. Length = 30 feet; Width = 21 feet
42. Length = 200 feet; Width = 100 feet
43. Length = 25 feet; Width = 20 feet
44. 5 seconds **45.** Height = 3 feet; Base = 4 feet
46. Height = 5 feet; Base = 12 feet
47. (a) x-coordinate of vertex is -2.5
(b) $f(-2.5) = -0.25$; $(-2.5, -0.25)$

DOING MATHEMATICS SECTION 6.5

1. (a) 2 (b) 1 (c) 2 (d) 3 (e) 5 (f) 1 (g) 4 (h) 5 **2.** (a) 1 (b) 2 (c) 1 (d) 3 (e) 4 (f) 3 (g) 4 (h) 5
3. $(x)^3 + (6)^3 = (x + 6)(x^2 - 6x + 36)$
4. $(y)^3 + (3)^3 = (y + 3)(y^2 - 3y + 9)$
5. $(x)^3 + (1)^3 = (x + 1)(x^2 - x + 1)$
6. $(x)^3 + (4)^3 = (x + 4)(x^2 - 4x + 16)$
7. $(x)^3 + (y)^3 = (x + y)(x^2 - xy + y^2)$
8. $(3x)^3 + (2y)^3 = (3x + 2y)(9x^2 - 6xy + 4y^2)$
9. $(2y)^3 + (5)^3 = (2y + 5)(4y^2 - 10y + 25)$
10. $(4x)^3 + (3)^3 = (4x + 3)(16x^2 - 12x + 9)$
11. $(2a)^3 + (3b)^3 = (2a + 3b)(4a^2 - 6ab + 9b^2)$
12. $(a)^3 - (3b)^3 = (a - 3b)(a^2 + 3ab + 9b^2)$
13. $(y)^3 - (3)^3 = (y - 3)(y^2 + 3y + 9)$
14. $(x)^3 - (6)^3 = (x - 6)(x^2 + 6x + 36)$
15. $(x)^3 - (4)^3 = (x - 4)(x^2 + 4x + 16)$
16. $x^3 - 1^3 = (x - 1)(x^2 + x + 1)$
17. $(3x)^3 - (2y)^3 = (3x - 2y)(9x^2 + 6xy + 4y^2)$
18. $x^3 - y^3 = (x - y)(x^2 + xy + y^2)$
19. $(4y)^3 - (z)^3 = (4y - z)(16y^2 + 4yz + z^2)$
20. $(5x)^3 - (3)^3 = (5x - 3)(25x^2 + 15x + 9)$

21. $2^3(x + 1)^3 + (y - 2)^3 = (2(x + 1))^3 + (y - 2)^3 = (2x + 2)^3 + (y - 2)^3 = [(2x + 2) + (y - 2)][(2x + 2)^2 - (2x + 2)(y - 2) + (y - 2)^2] = [2x + y][(4x^2 + 8x + 4) - (2xy - 4x + 2y - 4) + (y^2 - 4y + 4)] = [2x + y][4x^2 + 8x + 4 - 2xy + 4x - 2y - 4 + y^2 - 4y + 4] = (2x + y)(4x^2 + y^2 - 2xy + 12x - 6y + 12)$
22. $(2(x - 5))^3 - (y + 1)^3 = (2x - 10)^3 - (y + 1)^3 = ((2x - 10) - (y + 1))((2x - 10)^2 + (2x - 10)(y + 1) + (y + 1)^2) = [2x - 10 - y - 1][(4x^2 - 40x + 100) + (2xy + 2x - 10y - 10) + (y^2 + 2y + 1)] = (2x - y - 11)(4x^2 + y^2 + 2xy - 38x - 8y + 91)$
23. $3xy(x - 1)$ **24.** $5x(x^2 + 2x + 4)$
25. $5(x + 10y)(x - 10y)$ **26.** $3(2x + 5z)(2x - 5z)$
27. $-2x(x - 5y)^2$ **28.** $x(7x + 5y)(x - 2y)$
29. $y(x - 15)(x + 4)$ **30.** $2x(x + 5)(x - 5)$
31. $-3x(x + 1)^2$ **32.** $-5x(x + 2)^2$
33. $-x(3x - 1)(2x + 5)$ **34.** $4x(2x + 3y)^2$
35. $4(x + 3)(x^2 - 3x + 9)$
36. $-2y(3y - 2)(9y^2 + 6y + 4)$
37. $2(x^3 - xy^2 + x^2 - y^2) = 2[x(x^2 - y^2) + 1(x^2 - y^2)] = 2(x^2 - y^2)(x + 1) = 2(x + y)(x - y)(x + 1)$
38. $-3(x^4 + xy^3 - x^3y - y^4) = -3[x(x^3 + y^3) - y(x^3 + y^3)] = -3(x^3 + y^3)(x - y) = -3(x + y)(x^2 - xy + y^2)(x - y)$

39. $\frac{(x - 3)(x^2 + 3x + 9)}{2(x^2 + 3x + 9)} = \frac{x - 3}{2}$

40. $\frac{(x + 4)(x^2 - 4x + 16)}{3(x^2 - 4x + 16)} = \frac{x + 4}{3}$

41. $\frac{6(2x + y)(2x - y)}{3(2x + y)(3x - y)} = \frac{2(2x - y)}{(3x - y)} = \frac{4x - 2y}{3x - y}$

42. $\frac{2(3x + y)(3x - y)}{4(3x - y)(x + y)} = \frac{(3x + y)}{2(x + y)} = \frac{3x + y}{2x + 2y}$

43. $\frac{5x(x + 3)(x + 3)}{5xy(x + 3)(x^2 - 3x + 9)} = \frac{1}{y(x + 3)} = \frac{1}{xy + 3y}$

44. $\frac{x(x + 2)(x + 2)}{x(x + 2)(x^2 - 2x + 4)} = \frac{x + 2}{x^2 - 2x + 4}$

45. $\frac{-1(4x^2 - 2x + 1)}{3(2x - 1)} = \frac{-4x^2 + 2x - 1}{6x - 3}$

46. $\frac{-6(x+3)(x-2)}{3(x+3)(x-3)} = \frac{-2(x-2)}{x-3} = \frac{-2x+4}{x-3}$

47. $\frac{(x-y)(x+y)}{-1(x-y)} = -(x+y) = -x-y$

48. $\frac{-1(a-b)}{(a-b)(a+b)} = \frac{-1}{a+b}$

49. $\frac{-2(x-2)(x+2)}{(x-2)(x+3)} = \frac{-2(x+2)}{x+3} = \frac{-2x-4}{x+3}$

50. $\frac{-3(x+3)}{2x+1} = \frac{-3x-9}{2x+1}$

51. $3x^2(x-8) = 0, x = 0, x = 8$

52. $4x^2(x-125) = 0, x = 0, x = 125$

53. $(x+2)(x-2)(2x+1) = 0$
$x = -2, x = 2, x = -\frac{1}{2}$

54. $(x+1)(x+3)(x-3) = 0$
$x = -1, x = -3, x = 3$

55. $3x(5x-2)(5x-2) = 0, x = 0, x = \frac{2}{5}$

56. $2x(x-5)(x-5) = 0, x = 0, x = 5$

57. $5x(3x-4)(2x-3) = 0$
$x = 0, x = \frac{4}{3}, x = \frac{3}{2}$

58. $x(5x+1)(x-3) = 0, x = 0, x = -\frac{1}{5}, x = 3$

59. $x(3x+5)(x+5) = 0, x = 0, x = -5, x = -\frac{5}{3}$

60. $x(2x-1)(4x+3) = 0, x = 0, x = \frac{1}{2}, x = -\frac{3}{4}$

61. $x(2x-3)(x-1) = 0, x = 0, x = 1, x = \frac{3}{2}$

62. $3x(2x+1)(2x+1) = 0, x = 0, x = -\frac{1}{2}$

63. Calculator

CHAPTER 6 REVIEW

1. Answers will vary: (a) $2a + 6b = 2(a+3b)$
(b) $2a + 2b + a^2 + ab = 2(a+b) + a(a+b)$
$= (a+b)(2+a)$ (c) $x^2 + 2x + 5$
(d) $4x^2 - 2x + 5$ (e) $x^2 + 12x + 36 = (x+6)^2$
(f) $x^2 - 100 = (x)^2 - (10)^2$
(g) $x^3 + 8 = (x)^3 + (2)^3$
(h) $x^3 - 27 = (x)^3 - (3)^3$

2. $2xy(x-2)$ **3.** $-5y(y+6x)$

4. $-3y(y^2+3y-1) = -3y(y+5)(y-2)$

5. $-24x^6(x+2)$ **6.** $6y(x^3-2y^2)$

7. $-2xy(x^3 - 2x^2y + 4xy^2 - 10y^3)$

8. $2(a+b) + a(a+b) = (a+b)(2+a)$

9. $x(y-2) + 3(y-2) = (y-2)(x+3)$

10. $5(x-1) - y(x-1) = (x-1)(5-y)$

11. $a(a-7) + b(a-7) = (a-7)(a+b)$

12. $3x(4x+5) - 2(4x+5) = (4x+5)(3x-2)$

13. $3x(x-5) - 4(x-5) = (x-5)(3x-4)$

14. $(x+3)(x-4)$ **15.** $(x-2)(x-1)$

16. $(x+10)(x+2)$ **17.** $(x-6)(x-5)$

18. $(x-4)(x+2)$ **19.** $(x+3)(x+6)$

20. $(x+7)(x-7)$ **21.** $(x+10)(x-10)$

22. $(3x+7)(3x-7)$ **23.** $(4x+9y)(4x-9y)$

24. $(x+y)(x-y)$ **25.** $(2x+5y)(2x-5y)$

26. $(y-7)^2$ **27.** $(x+5)^2$ **28.** $(x-4y)^2$

29. $(5x-3)^2$ **30.** $(2x+5)^2$ **31.** $(3x+y)^2$

32. $(2x+5)(x-1)$ **33.** $(4x+1)(x-5)$

34. $(3x+4)(2x-1)$ **35.** $(2y-5)(y-5)$

36. $(3y+5)(2y-1)$ **37.** $(5x+2)(3x+1)$

38. $(x+1)^2 = 0, x = -1$

39. $(x+3)(x+4) = 0, x = -3, x = -4$

40. $(x+1)(x-1) = 0, x = -1, x = 1$

41. $(x+2)(x-2) = 0, x = -2, x = 2$

42. $(2x+1)(2x-3) = 0, x = -\frac{1}{2}, x = \frac{3}{2}$

43. $(3x-1)(x+5) = 0, x = \frac{1}{3}, x = -5$

44. $(x)^3 + (2)^3 = (x+2)(x^2-2x+4)$

45. $(2x)^3 + (3y)^3 = (2x+3y)(4x^2-6xy+9y^2)$

46. $(1)^3 + (x)^3 = (1+x)(1-x+x^2)$

47. $(a)^3 + (b)^3 = (a+b)(a^2-ab+b^2)$

48. $(6x)^3 + (5)^3 = (6x+5)(36x^2-30x+25)$

49. $(5x)^3 + (4y)^3 = (5x+4y)(25x^2-20xy+16y^2)$

50. $(a)^3 - (b)^3 = (a-b)(a^2+ab+b^2)$

51. $(x)^3 - (1)^3 = (x-1)(x^2+x+1)$

52. $(2)^3 - (5x)^3 = (2-5x)(4+10x+25x^2)$

53. $(3a)^3 - (b)^3 = (3a-b)(9a^2+3ab+b^2)$

54. $(6x)^3 - (5y)^3 = (6x-5y)(36x^2+30xy+25y^2)$

55. $(3x)^3 - (y)^3 = (3x-y)(9x^2+3xy+y^2)$

56. $2(x^3-64y^3) = 2((x)^3-(4y)^3) =$
$2(x-4y)(x^2+4xy+16y^2)$

57. $2(x^2-4y^2) = 2(x+2y)(x-2y)$

58. $3(2a+3b)^2$ **59.** $-4x(2x^2+3xy-5y^2) =$
$-4x(2x+5y)(x-y)$ **60.** $-10x^2y^2(3x+2)^2$

61. $-3xy(y^3+1) = -3xy(y+1)(y^2-y+1)$

62. $3y(2x+3)(2x-3)$ **63.** $-5x^3(3x+4)^2$

64. $(4x^2-9)(y^3-8) =$
$(2x+3)(2x-3)(y-2)(y^2+2y+4)$

65. $-2x(x^3-8) + 2y(x^3-8) =$
$(-2x+2y)(x^3-8) =$
$-2(x-y)(x-2)(x^2+2x+4)$

66. $-3x(x^3+y^3) + 3y(x^3+y^3) =$
$(-3x+3y)(x^3+y^3) =$
$-3(x-y)(x+y)(x^2-xy+y^2)$ **67.** $\frac{x-3}{8}$

68. $\dfrac{x+3}{10x-5}$ **69.** $\dfrac{y(3x-y)}{3x(2x-y)} = \dfrac{3xy-y^2}{6x^2-3xy}$

70. $\dfrac{x-4y}{4x^2+2xy+y^2}$ **71.** $\dfrac{4(a-b)}{4a^2-6ab+9b^2}$

72. $\dfrac{x+3}{2x-14}$ **73.** $\dfrac{-3(x+5)}{(x+3)} = \dfrac{-3x-15}{x+3}$

74. $\dfrac{-(a+2b)}{4b^2+2ab+a^2} = \dfrac{-a-2b}{4b^2+2ab+a^2}$

75. $5x(x-27)=0, x=0, x=27$

76. $3x(4x-1)(4x-1)=0, x=0, x=\dfrac{1}{4}$

77. $x(3x-5)(x+5)=0, x=0, x=\dfrac{5}{3}, x=-5$

78. $y(y-7)(y-7)=0, y=7, y=0$

79. $(x+7)(x+3)=0, x=-7, x=-3$

80. $(2x+3)(x-5)=0, x=-\dfrac{3}{2}, x=5$

81. $2x(x+4)(x-4)=0, x=0, x=-4, x=4$

82. $3y(y-6)(y-4)=0, y=0, y=6, y=4$

83. Calculator **84.** $x(x-6)=280$; Length: 20 feet; Width: 14 feet **85.** $(x+5)x=300$; Length: 20 feet Width: 15 feet **86.** (a) $(x+4)(x+4)=256$ (b) x = length and width of swimming pool = 12 feet (c) Area of pool = $12 \times 12 = 144$ sq. ft. (d) Area of sidewalk = $256 - 144 = 112$ sq. ft.

CHAPTER 6 PRACTICE TEST ANSWERS

1. Answers will vary: (a) $2x^2+10x-20 = 2(x^2+5x-10)$ (b) $x(y+2)-5(y+2)$ (c) x^2-x-6 (d) $3x^2-20x+12$ (e) $x^2+10x+25$ (f) x^2-16 (g) x^3+8 (h) x^3-27 (i) x^2+16 (j) x^2+x-5

2. $-2xy(x+5)(x-5)$ **3.** $6x(x+10)(x-1)$

4. $-18a^2b^2c^2(2+3abc)$ **5.** $(x-y)(z+3)$

6. $(a+5)(a-b)$ **7.** $(x-9)(x+3)$

8. $(y-5)^2$ **9.** $(3x-2)(x-6)$

10. $(2x+3y)(2x-3y)$ **11.** $(2x-y)(3x+4y)$

12. $5x(x+4)(x-4)$ **13.** $(y+7)(x-2)(x+2)$

14. $-3(3a-5b)(a+2b)$

15. $(x-3)(x^2+3x+9)$

16. $\dfrac{x+3}{x+5}$ **17.** $\dfrac{x-3}{2(x+2)}$ **18.** $\dfrac{a+4}{2(a-3)}$

19. $\dfrac{2x-1}{3x+5}$ **20.** $x=9, x=-9, x=0$

21. $x=\dfrac{3}{4}, x=-1$ **22.** $y=\dfrac{7}{4}, y=-1$

23. $y=-\dfrac{5}{2}, y=1, y=0$

24. Length: 19 feet; Width: 14 feet

CHAPTERS 4–6 CUMULATIVE TEST ANSWERS

1. $(-1, 3)$ **2.** Inconsistent

3. (a)
$$x + y = 3$$
$$0.40x + 0.08y = 0.32(3)$$
(b) $(2.25, 0.75)$ (c) 2.25 liters (d) 0.75 liters

4. $(2, -3)$ **5.** Inconsistent **6.** Dependent

7.

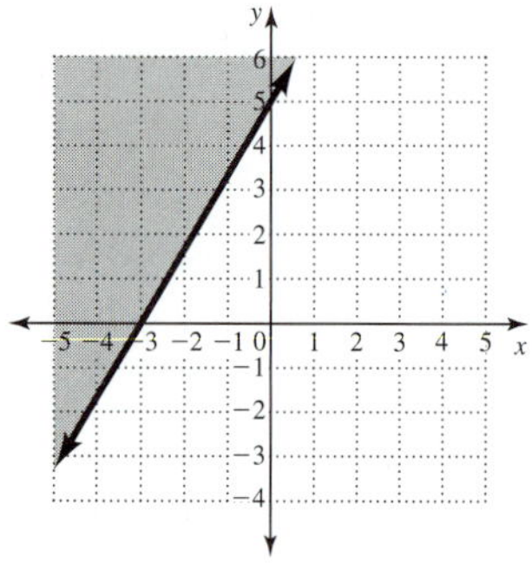

8. $-125x^6y^6$ **9.** $\dfrac{-a^2b^2}{6c}$ **10.** $10y^6$ **11.** 1 **12.** $\dfrac{4y^2}{x^4}$

13. $\dfrac{y^8}{1296x^8z^4}$ **14.** $36x^7$

15. $-20x^4y-20x^3y-5x^2y$ **16.** $10x^2-27x+18$

17. $-4x^5+2x^4-4x^3+24x^2+70x$ **18.** $\dfrac{1}{8y}$

19. $\dfrac{1}{2}-\dfrac{3ab}{2}+2b$ **20.** $x^2+x+1+\dfrac{-4}{x-1}$

21. (a) $A(x)=(3x-1)(2x)-x(x+4)$ or $A(x)=5x^2-6x$ (b) Area = 144 square inches

22. $-10a^2b^3c^2(5+abc^3)$ **23.** $(a-7)(a-b)$

24. $3x(x-2)(x^2+2x+4)$ **25.** $2(x^2+25)$

26. $(2a-b)(2a+3b)$ **27.** $\dfrac{5a-2}{4a+3}$

28. $x=12, x=-12$ **29.** $x=0, x=6, x=-5$

30. $y=\dfrac{1}{2}, y=-\dfrac{3}{2}$ **31.** $y=0, y=4, y=-4$

32. (a) Length = 18 feet (b) Width = 13 feet (c) Area = 108 square feet

CHAPTER 7

DOING MATHEMATICS SECTION 7.1

1. $\dfrac{1}{2}$ **2.** -13 **3.** $\dfrac{1}{4}$ **4.** $\dfrac{4b}{3}$ **5.** $\dfrac{9c^2}{8b}$ **6.** $\dfrac{-4x}{y}$

7. $x+2$ **8.** $a-4$ **9.** $\dfrac{5}{6}$ **10.** $2x$ **11.** $\dfrac{x-2}{x+7}$

12. $\dfrac{x+3}{x-4}$ **13.** $\dfrac{y-3}{y}$ **14.** $\dfrac{x-y}{x+3y}$ **15.** $\dfrac{a^2+1}{a+1}$

16. $? = 7$ **17.** $? = 15$ **18.** $? = 2x^2$ **19.** $? = -8a^2$

20. $? = 10x^2y$ **21.** $? = -30abx$ **22.** $? = 6ab + 6a$

23. $? = xy - x$ **24.** $? = -4x^2 - 12x$

25. $? = 2x^2 - 3x - 20$ **26.** $? = 9x$ **27.** $? = 20a$
28. $? = 20x - 45$ **29.** $? = 7x^3 + 14x^2 + 28x$
30. $? = -a^2b + ab$ **31.** $x \neq 1$ **32.** $x \neq -\dfrac{1}{2}$
33. $x \neq 1, x \neq -2$ **34.** $x \neq 2, x \neq 5$ **35.** $x \neq -\dfrac{1}{7}$
36. $x \neq \dfrac{5}{3}$ **37.** $x \neq -3, x \neq -4$
38. $x \neq 3, x \neq -2$ **39.** Answers will vary.
40. Answers will vary.
41. (a) $\dfrac{60 \text{ meters}}{2.5 \text{ seconds}} = \dfrac{24 \text{ meters}}{1 \text{ second}}$ $= 24$ meters per second
(b) $\dfrac{3 \text{ cups}}{1 \text{ batch}} = 3$ cups per batch
(c) $\dfrac{\$1.98}{2 \text{ cantaloupes}} = \dfrac{\$0.99}{1 \text{ cantaloupe}}$ $= \$0.99$ per cantaloupe
42. (a) $\dfrac{100 \text{ miles}}{2 \text{ hours}} = \dfrac{50 \text{ miles}}{1 \text{ hour}} = 50$ miles per hour
(b) $\dfrac{240 \text{ words}}{3 \text{ minutes}} = \dfrac{80 \text{ words}}{1 \text{ minute}} = 80$ words per minute
(c) $\dfrac{250 \text{ CDs}}{10 \text{ workers}} = \dfrac{25 \text{ CDs}}{1 \text{ worker}} = 25$ CDs per worker
43. $=$ **44.** $=$ **45.** $\neq$ **46.** $\neq$ **47.** $=$ **48.** $=$
49. $\neq$ **50.** $=$ **51.** $=$ **52.** $=$ **53.** $x = 4$
54. $x = 30$ **55.** $x = 88$ **56.** $x = 140$
57. $x = 242$ **58.** $x = 84$ **59.** (a) Biggie Market is cheaper. 12 coconuts cost \$19.80 at Biggie and \$19.96 at Friendly. (b) Friendly Market is a better buy if the coconuts were much larger there.
60. (a) Brand Y is cheaper per ounce.
(b) Brand X costs \$0.09 per load. Brand Y costs 2(\$0.08) per load. Brand X is most economical.
61. \$37.50 **62.** 42 minutes **63.** (a) $\dfrac{6}{1.5}$ (b) $\dfrac{13}{x}$
(c) $\dfrac{6}{1.5} = \dfrac{13}{x}$, $x = 3.25$ (d) 3.25 liters of water
64. The calculator displays "ERROR" when division by zero occurs: $x = 1$ creates division by zero. $x = 1$ is a restricted value of the function.

DOING MATHEMATICS SECTION 7.2

1. 12 eggs, $1\frac{1}{2}$ cups water, $2\frac{1}{4}$ teaspoons salt, $1\frac{1}{2}$ cups milk, 3 cups flour, 3 teaspoons vanilla extract, 6 tablespoons clarified butter, 6 tablespoons brandy (optional), 3–6 teaspoons sugar (for dessert crepes)

2. 3 eggs, $\frac{3}{8}$ cup water, $\frac{9}{16}$ teaspoon salt, $\frac{3}{8}$ cup milk, $\frac{3}{4}$ cup flour, $\frac{3}{4}$ teaspoon vanilla extract, $1\frac{1}{2}$ tablespoons clarified butter, $1\frac{1}{2}$ tablespoons brandy (optional), $\frac{3}{4}$–$1\frac{1}{2}$ teaspoons sugar (for dessert crepes)

3. (a) 60 miles per hour (b) 40 minutes (c) 4:40 P.M.
4. $\dfrac{15}{2} = 7\dfrac{1}{2}$ **5.** $\dfrac{23}{2} = 11\dfrac{1}{2}$ **6.** 5 **7.** $-\dfrac{10}{17}$
8. $\dfrac{3}{2} = 1\dfrac{1}{2}$ **9.** 12 **10.** $\dfrac{-5}{2a}$ **11.** $\dfrac{-3}{10xy}$
12. $\dfrac{3az}{2y}$ **13.** $\dfrac{-9b^2}{a}$ **14.** $\dfrac{1}{2r}$ **15.** $\dfrac{1}{6x}$
16. $\dfrac{-5x}{x+2}$ **17.** $\dfrac{(2x-1)(x+3)}{5}$ **18.** 14
19. $-\dfrac{5}{2}$ **20.** $\dfrac{2bc}{a}$ **21.** $\dfrac{5}{6b^4}$ **22.** $\dfrac{3}{2xy}$
23. $\dfrac{1}{abxyz^2}$ **24.** $-10a + 10b$
25. $-2x - 2y$ **26.** $-\dfrac{2(x-y)(x-z)}{5}$ **27.** $-\dfrac{1}{6}$
28. 1 **29.** $-\dfrac{9(x+1)}{4(x+7)}$ **30.** 5
31. $\dfrac{(x-3)(x+1)}{(x-4)(x+3)}$ **32.** $-\dfrac{x-1}{2x+1}$ **33.** -3
34. $\dfrac{6(x+5)(x-y)}{(x-1)^2}$ **35.** $\dfrac{3(x+y)(2x-1)}{(x+1)}$
36. $\dfrac{3x+2}{x+1}$ **37.** $\dfrac{2x-1}{8x}$ **38.** 1
39. $A(x) = \dfrac{4x^2}{x+5}$, $A(5) = 10$ sq. in.
40. $A(x) = \dfrac{x^2+3x}{2}$, $A(5) = 20$ sq. in.
41. $A(x) = x + 2$, $A(5) = 7$ sq. in.
42. $A(x) = \dfrac{x+2}{2}$, $A(5) = \dfrac{7}{2}$ sq. in.

DOING MATHEMATICS SECTION 7.3

1. Factors of 9: $3 \cdot 3$
Factors of 15: $3 \cdot 5$
LCD: $3 \cdot 3 \cdot 5 = 45$

2. Factors of 12: $2 \cdot 2 \cdot 3$
Factors of 15: $3 \cdot 5$
LCD: $2 \cdot 2 \cdot 3 \cdot 5 = 60$

3. Factors of xy: $x \cdot y$
Factors of $2x^2$: $2 \cdot x \cdot x$
LCD: $2 \cdot x \cdot x \cdot y = 2x^2y$

4. Factors of $3b^2$: $3 \cdot b \cdot b$
Factors of ab: $a \cdot b$
LCD: $3 \cdot a \cdot b \cdot b = 3ab^2$

5. Factors of $3xy$: $3 \cdot x \cdot y$
Factors of $14x^2y$: $2 \cdot 7 \cdot x \cdot x \cdot y$
LCD: $2 \cdot 3 \cdot 7 \cdot x \cdot x \cdot y = 42x^2y$

6. Factors of $30xy$: $2 \cdot 3 \cdot 5 \cdot x \cdot y$
Factors of $7xy^3$: $7 \cdot x \cdot y \cdot y \cdot y$
LCD: $2 \cdot 3 \cdot 5 \cdot 7 \cdot x \cdot y \cdot y \cdot y = 210xy^3$

7. Factors of x: x
Factors of $x(x + 1)$: $x(x + 1)$
LCD: $x(x + 1)$

8. Factors of $2y$: $2 \cdot y$
Factors of $y(y - 1)$: $y(y - 1)$
LCD: $2y(y - 1)$

9. Factors of $(y - 2)^2$: $(y - 2)(y - 2)$
Factors of $3(y - 2)$: $3(y - 2)$
LCD: $3(y - 2)(y - 2)$

10. Factors of $(y + 3)^2$: $(y + 3)(y + 3)$
Factors of $5(y + 3)$: $5(y + 3)$
LCD: $5(y + 3)(y + 3)$

11. Factors of $y^2 - 1$: $(y + 1)(y - 1)$
Factors of $y^2 - y - 2$: $(y + 1)$ $(y - 2)$
LCD: $(y + 1)(y - 1)(y - 2)$

12. Factors of $y^2 + 5y + 6$: $(y + 2)(y + 3)$
Factors of $y^2 - 4$: $(y + 2)$ $(y - 2)$
LCD: $(y + 2)(y + 3)(y - 2)$

13. 1 **14.** 8 **15.** $\frac{100}{21}$ **16.** $-\frac{1}{24}$ **17.** $\frac{11x}{10}$

18. $\frac{26x}{75}$ **19.** $\frac{15}{y}$ **20.** $\frac{14}{x}$ **21.** $\frac{6}{x}$ **22.** $\frac{30}{x}$

23. $\frac{6}{y - 5}$ **24.** $\frac{9}{x + 3}$ **25.** $\frac{1}{6x}$ **26.** $\frac{13}{10x}$

27. $\frac{2c + 8ab}{abc}$ **28.** $\frac{3y + 4x}{xy}$ **29.** $\frac{8x^2y + 15}{12x^2y^2}$

30. $\frac{15x^3y + 46}{10x^2y^2}$ **31.** $\frac{-10x - 28}{(x - 2)(x + 3)}$

32. $\frac{5x + 7}{(x + 1)(x + 2)}$ **33.** $\frac{3x + 3y + 2xy}{(x - y)(x + y)}$

34. $\frac{a + ab - b^2}{(a - b)(a + b)}$

35. $\frac{5x - 2}{(x + 1)(x - 1)}$

36. $\frac{8y + 5}{(y + 1)(y - 1)}$

37. $\frac{-5x^2 + 10x - 2}{(x - 2)(x + 2)}$ **38.** $\frac{9 - a}{(a - 3)(a + 3)}$

39. $\frac{2x}{(3x - 1)(x + 1)(x - 1)}$

40. $\frac{8x - 4}{(2x + 1)(x + 1)(x - 3)}$

41. $57\frac{1}{2}$ feet **42.** $112\frac{1}{2}$ inches **43.** 60 hours

44. $P(x) = \frac{5x}{3}$;
$P(4) = \frac{20}{3} = 6\frac{2}{3}$ feet

45. $P(x) = \frac{9x}{10}$;
$P(4) = \frac{18}{5} = 3\frac{3}{5}$ feet

46. $P(x) = \frac{20x - 8}{(x + 1)(x - 1)}$, $(x \neq -1, x \neq 1)$;
$P(4) = \frac{24}{5} = 4\frac{4}{5}$ feet

47. $P(x) = \frac{28x - 16}{(x + 2)(x - 2)}$, $(x \neq -2, x \neq 2)$;
$P(4) = 8$ feet

DOING MATHEMATICS SECTION 7.4

1. $x \neq 3$

x	$f(x) = \frac{1}{x - 3}$
−100	−0.01
−10	−0.08
−1	−0.25
0	−0.33
1	−0.5
2	−1
2.5	−2
3	Undefined
3.5	2
4	1
5	0.5
6	0.33
10	0.14
100	0.01

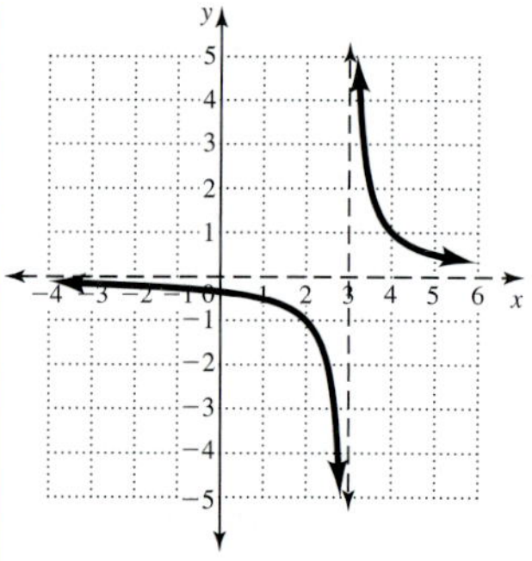

2. $x \neq 2$

x	$f(x) = \frac{1}{x-2}$
−100	−0.01
−10	−0.08
0	−0.5
1	−1
1.5	−2
2	Undefined
2.5	2
3	1
4	0.5
6	0.25
10	0.13
100	0.01

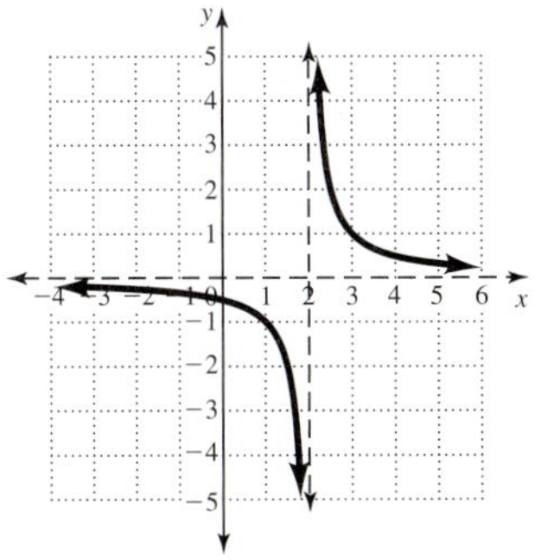

3. $x \neq 1$

x	$f(x) = \frac{3}{x-1}$
−100	−0.03
−10	−0.27
−2	−1
−1	−1.5
0	−3
0.5	−6
1	Undefined
1.5	6
2	3
3	1.5
4	1
10	0.33
100	0.03

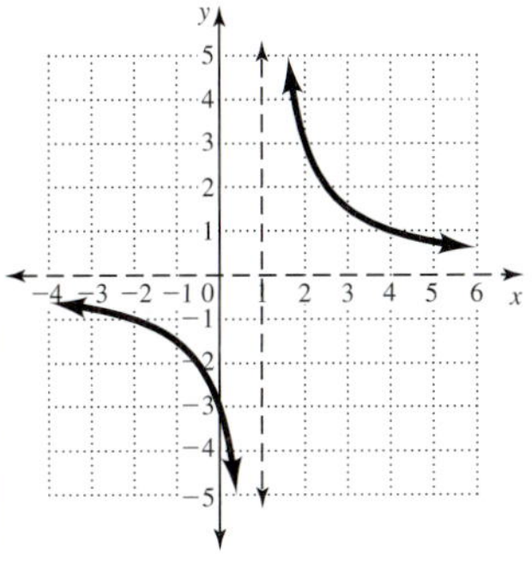

4. $x \neq -3$

x	$f(x) = \frac{2}{x+3}$
−100	−0.02
−10	−0.29
−6	−0.67
−5	−1
−4	−2
−3.5	−4
−3	Undefined
−2.5	4
−2	2
−1	1
0	0.67
10	0.15
100	0.02

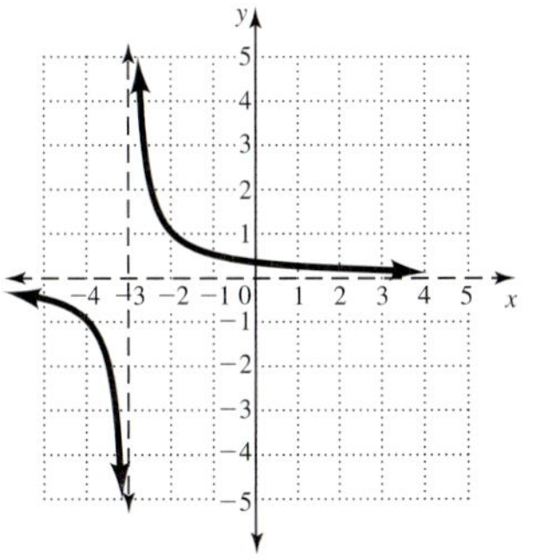

5. $x \neq -1, x \neq 2$

x	$f(x) = \frac{1}{x^2 - x - 2}$
−100	0.0001
−10	0.01
−4	0.06
−3	0.1
−2	0.25
−1.5	0.57
−1	Undefined
−0.5	−0.8
0	−0.5
1	−0.5
1.5	−0.8
2	Undefined
2.5	0.57
3	0.25
4	0.1
5	0.56
6	0.04
10	0.01
100	0.0001

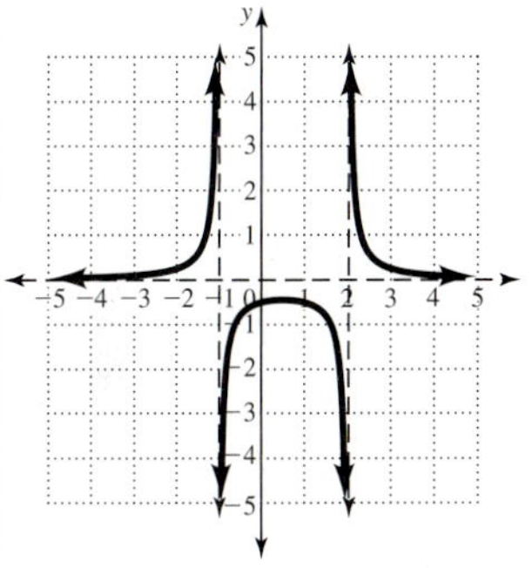

6. $x \neq 1, x \neq -3$

x	$f(x) = \frac{2}{x^2 + 2x - 3}$
−100	0.0002
−10	0.03
−5	0.17
−4	0.4
−3.5	0.89
−3	Undefined
−2.5	−1.14
−2	−0.67
−1	−0.5
0	−0.67
0.5	−1.14
1	Undefined
1.5	0.89
2	0.4
3	0.17
4	0.01
5	0.06
10	0.02
100	0.0002

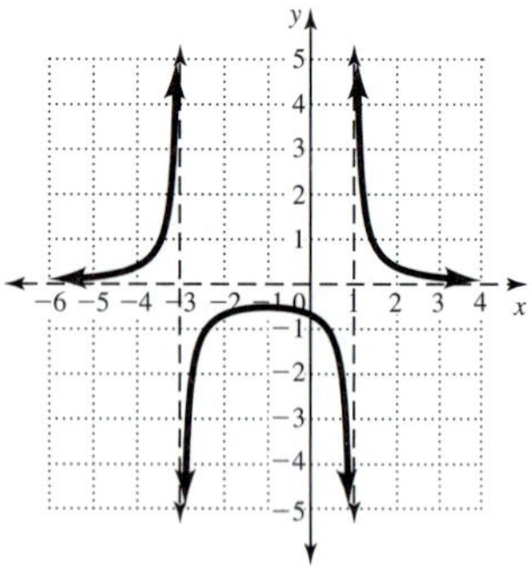

7. $x \neq 1, x \neq 0$

x	$f(x) = \frac{1}{x^2 - x}$
−100	0.0001
−10	0.01
−4	0.05
−3	0.08
−2	0.17
−1	0.5
0	Undefined
0.2	−6.25
0.4	−4.17
0.6	−4.17
0.8	−6.25
1	Undefined
2	0.5
3	0.17
4	0.08
5	0.05
6	0.03
10	0.01
100	0.0001

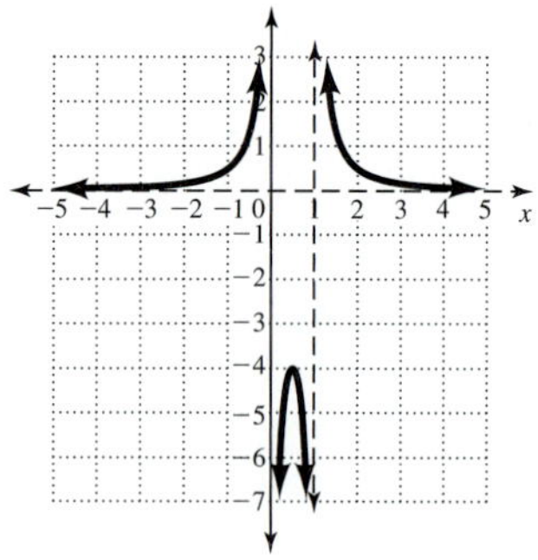

8. $x \neq -1, x \neq 0$

x	$f(x) = \frac{1}{x^2 + x}$
−100	0.0001
−10	0.11
−4	0.08
−3	0.17
−2	0.5
−1.5	1.33
−1	Undefined
−0.8	−6.25
−0.6	−4.17
−0.4	−4.17
−0.2	−6.25
0	Undefined
1	0.5
2	0.17
3	0.08
4	0.05
10	0.01
100	0.0001

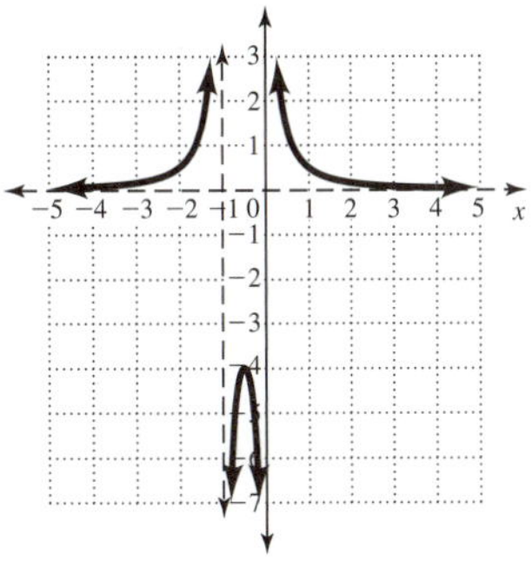

9. $x \neq -2, x \neq 4$

x	$f(x) = \frac{1}{x^2 - 2x - 8}$
−100	0.0001
−10	0.01
−4	0.06
−3	0.14
−2	Undefined
−1	−0.2
0	−0.13
1	−0.11
2	−0.13
3	−0.2
4	Undefined
5	0.14
6	0.06
10	0.01
100	0.0001

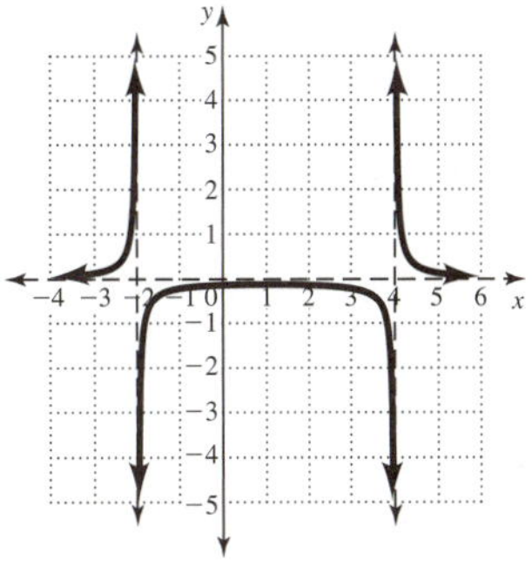

10. $x \neq -2, x \neq 2$

x	$f(x) = \frac{1}{x^2 - 4}$
−100	0.0001
−10	0.01
−5	0.48
−4	0.08
−3	0.2
−2.5	0.44
−2	Undefined
−1.5	−0.57
−1	−0.33
0	−0.25
1	−0.33
1.5	−0.57
2	Undefined
2.5	0.44
3	0.2
4	0.08
10	0.01
100	0.0001

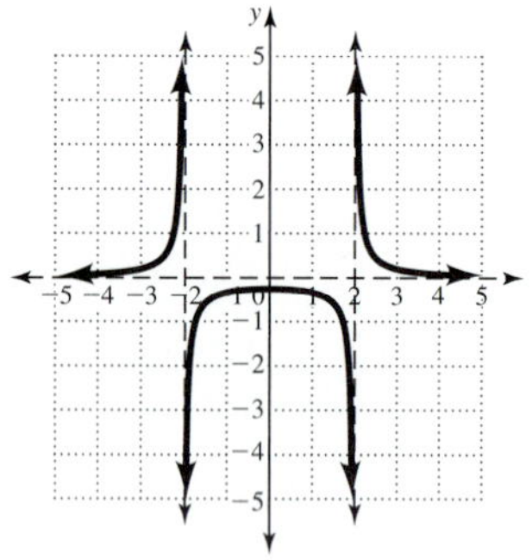

11. Restrictions in the domain are created by values of the independent variable that cause the denominator to equal zero (undefined). Horizontal asymptotes are restrictions in the dependent variable, where the values of the dependent variable approach, but never equal, a restricted value.
12. Zero divided by any nonzero number is zero because it is not possible to divide zero into any number of parts. **13.** It is not possible to divide a number into zero parts.

DOING MATHEMATICS SECTION 7.5

1. $x = 72$ **2.** $x = 6$ **3.** $x = 3, x = -8$ **4.** $x = 8, x = -5$ **5.** $x = -\frac{17}{9}$ **6.** $x = -5$ **7.** $y = 2$ **8.** $x = 18$ **9.** $x = 6$ **10.** $x = 10$ **11.** $x = -\frac{13}{24}$ **12.** $x = \frac{1}{3}$ **13.** $y = 20$ **14.** $y = 30$ **15.** $x = -\frac{1}{2}$ **16.** $x = -1$ **17.** $x = 3, x = -2$ **18.** $x = -3, x = -4$ **19.** $x = 5$ **20.** $x = 2$ **21.** $x = -\frac{1}{2}$ **22.** $x = 4$ **23.** $x = -8$ **24.** $y = 25$ **25.** $x = 6$ **26.** $x = 2$ **27.** No solution, $x \neq 2$ **28.** No solution, $x \neq 1$ **29.** $x = 2$ **30.** $x = -1$ **31.** $y = \frac{1}{3}$ **32.** $x = 3$ **33.** $x = 0$ **34.** $x = -\frac{2}{5}$ **35.** No solution, $x \neq 3$ **36.** No solution, $x \neq -1$ **37.** $x = -\frac{11}{13}$ **38.** $x = 10$ **39.** $a = 4$, $(a \neq 1)$ **40.** $x = -1$ **41.** $x = 4, x = \frac{3}{2}$ **42.** $x = 4, x = 5$ **43.** $x = -\frac{8}{5}$ **44.** $x = 9$ **45.** No solution, $x \neq \frac{2}{3}$

46. $\frac{x}{8} + \frac{9}{12} = 2$
10 pieces of 8-cut pizza are needed.

47. $\frac{x}{12} + \frac{2}{8} = 1$
9 pieces of 12-cut chocolate cheesecake are needed.

48. $\frac{1}{7} + \frac{1}{t} = \frac{1}{4}$
$t = \frac{28}{3}$ hours

49. $\frac{1}{40} + \frac{1}{60} = \frac{1}{t}$
$t = 24$ minutes

50. Width $= w$;
Length $= w + 8$;
$\frac{w + 8}{w} = \frac{4}{3}$
Width $= 24$ feet
Length $= 32$ feet

51. Length $= l$;
Width $= l - 3$;
$\frac{l - 3}{l} = \frac{2}{5}$
Length $= 5$ feet
Width $= 2$ feet

52. Answers will vary.

DOING MATHEMATICS SECTION 7.6

1. b, c, e **2.** b, d, e **3.** Dividing the number 10 by 2 means that 10 is divided into 2 groups, each group with 5 elements. Dividing 10 by $\frac{1}{2}$ means that each of the 10 units is divided into 2 parts and the result is 20 elements. **4.** (a) True (b) False **5.** $\frac{5}{4}$ **6.** $\frac{3}{2}$ **7.** $\frac{x}{2}$ **8.** y^2 **9.** $-\frac{ab}{6y}$ **10.** $-\frac{5wyz}{7x}$

11. $\dfrac{2(x-2)}{y(x-1)} = \dfrac{2x-4}{xy-y}$ **12.** $\dfrac{2(x+1)}{x} = \dfrac{2x+2}{x}$

13. $\dfrac{x+y}{x}$ **14.** $\dfrac{b(a-b)}{a^2} = \dfrac{ab-b^2}{a^2}$ **15.** $x-1$

16. $x-3$ **17.** $\dfrac{-2y(x+1)}{3x(x-3)} = \dfrac{-2xy-2y}{3x^2-9x}$

18. $\dfrac{2(x+5)}{xyz(x+4)} = \dfrac{2x+10}{x^2yz+4xyz}$

19. $\dfrac{2(x-2)}{y(x-1)} = \dfrac{2x-4}{xy-y}$ **20.** $-\dfrac{x+3}{2x}$ **21.** $\dfrac{1}{2}$

22. -17 **23.** $\dfrac{2y-4x}{xy}$ **24.** $\dfrac{3b-5a}{3ab}$

25. $\dfrac{3x+2y}{y+2x}$ **26.** $\dfrac{5y-3x}{4y-2x}$ **27.** $\dfrac{2+3x}{9}$

28. $\dfrac{3x-4}{4}$ **29.** $\dfrac{x^2+1}{x^2+5}$ **30.** $2a-b$

31. x **32.** $\dfrac{5+2x}{10x}$ **33.** $\dfrac{2a+b}{3ab-5b}$ **34.** x

35. $a-3$ **36.** $\dfrac{2x+1}{x+1}$ **37.** $\dfrac{y+1}{y-2}$ **38.** $-\dfrac{6y}{5}$

39. Method 1:

Step 1: Step 2:

$$\frac{\dfrac{1}{xy^2}-\dfrac{3y}{xy^2}}{\dfrac{3y}{x^2y^2}+\dfrac{x}{x^2y^2}} = \frac{\dfrac{1-3y}{xy^2}}{\dfrac{3y+x}{x^2y^2}}$$

Step 3:

$$\frac{1-3y}{xy^2} \div \frac{3y+x}{x^2y^2} =$$

Step 4:

$$\frac{1-3y}{xy^2} \cdot \frac{x^2y^2}{3y+x} = \frac{(1-3y)x}{3y+x} =$$

Step 5: $\dfrac{x-3xy}{3y+x}$

Method 2:

LCD: x^2y^2

Step 1:

$$\frac{\dfrac{x^2y^2}{1}}{\dfrac{x^2y^2}{1}}\left(\frac{\dfrac{1}{xy^2}-\dfrac{3}{xy}}{\dfrac{3}{x^2y}+\dfrac{1}{xy^2}}\right) =$$

Step 2:

$$\frac{x-3xy}{3y+x}$$

Method 2 has fewer steps.

CHAPTER 7 REVIEW

1. $\dfrac{3}{8}$ **2.** $x+2$ **3.** $7yz$ **4.** $\dfrac{x+1}{x+4}$ **5.** $\dfrac{x-1}{3}$

6. $x-1$ **7.** $? = 4xy + 8y$ **8.** $? = -27x^2 + 27x$

9. $? = -x^3 - x^2$ **10.** $? = 3x^2 + 15x$ **11.** $x \neq -1$

12. $x \neq \frac{1}{2}$ **13.** $x \neq -4, x \neq 3$ **14.** $x \neq -\frac{1}{3}, x \neq 2$

15. 54 drinks **16.** 67.2 minutes

17. The pine mulch costs \$2.50 per cubic yard. The hardwood mulch costs \$2.37 per cubic yard. Hardwood mulch is less expensive.

18. The value $x = -\dfrac{1}{2} = -0.5$ causes the denominator to be zero. The calculator records division by zero as "ERROR" because division by zero is undefined.

19. $-\dfrac{1}{68}$ **20.** $-\dfrac{125}{4}$ **21.** $-\dfrac{1}{10x}$ **22.** $\dfrac{z}{6y}$

23. $-x^2-3x$ **24.** $\dfrac{14}{5}$ **25.** $-\dfrac{z}{15y}$ **26.** $\dfrac{5a^2b^2}{4}$

27. $\dfrac{2}{3}$ **28.** $-5x+5y$

29. $\dfrac{(x-4)(x+6)}{(x-5)(x-3)} = \dfrac{x^2+2x-24}{x^2-8x+15}$ **30.** $\dfrac{4x+9}{x-3}$

31. $38\frac{3}{8}$ feet **32.** $\dfrac{10}{x}$ **33.** $\dfrac{15}{x-1}$ **34.** $\dfrac{4x}{2x+3}$

35. $\dfrac{7x}{60}$ **36.** $\dfrac{71}{5x}$ **37.** $\dfrac{13y+14x}{xy}$ **38.** $\dfrac{10xy-x}{25y^2}$

39. $\dfrac{2x-8}{(x+3)(x-2)}$ **40.** $\dfrac{3x-3y+10xy}{(x+y)(x-y)}$

41. $\dfrac{6y-2}{(y-1)(y+1)}$

42. $x \neq -3$

x	$f(x) = \dfrac{1}{x+3}$
−100	−0.01
−10	−0.14
−5	−0.5
−4	−1
−3.5	−2
−3	Undefined
−2.5	2
−2	1
−1	0.5
0	0.33
1	0.25
2	0.2
10	0.08
100	0.01

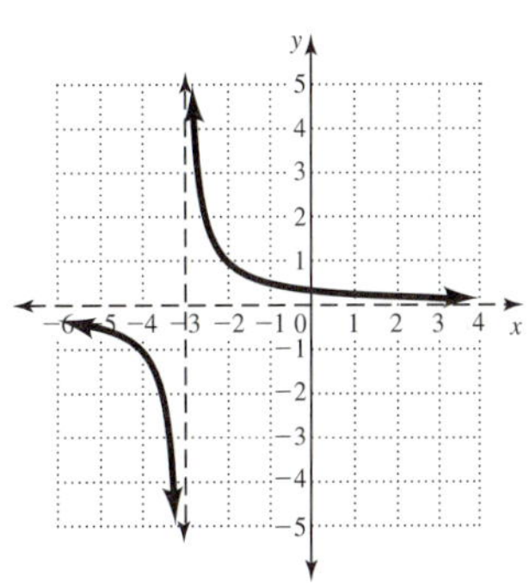

43. $x \neq -2$ and $x \neq 3$

x	$f(x) = \dfrac{1}{x^2 - x - 6}$
−100	0.0001
−10	0.01
−4	0.07
−3	0.17
−2.5	0.36
−2	Undefined
−1.5	−0.44
−1	−0.25
0	−0.17
1	−0.17
2	−0.25
2.5	−0.44
3	Undefined
3.5	0.36
4	0.17
5	0.07
6	0.04
10	0.01
100	0.0001

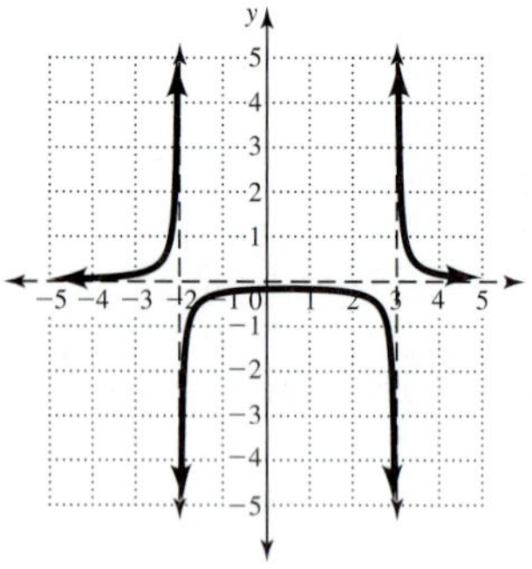

44. $x \neq 0$ and $x \neq 1$

x	$f(x) = \dfrac{3}{2x^2 - 2x}$
−100	0.0002
−10	0.01
−4	0.08
−3	0.13
−2	0.25
−1	0.75
−0.5	2
0	Undefined
0.2	−9.38
0.4	−6.25
0.6	−6.25
0.8	−9.38
1	Undefined
1.5	2
2	0.75
3	0.25
4	0.13
5	0.08
6	0.05
10	0.02
100	0.0002

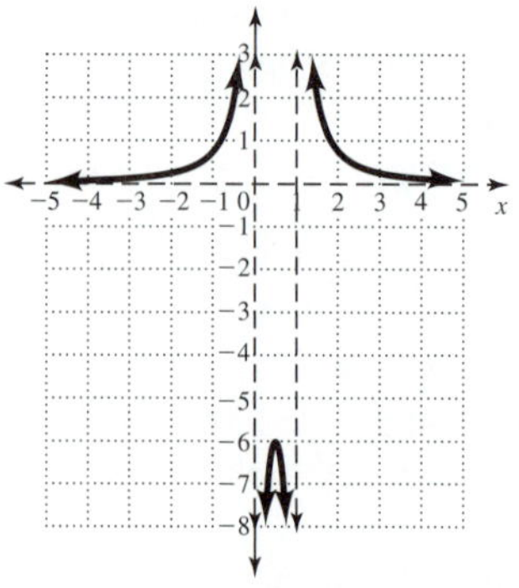

45. 2.4 hours **46.** Width = 8 feet; Length = 10 feet **47.** $x = 12$

48. $x = -\dfrac{17}{9}$ **49.** $x = 77$ **50.** $y = 4$ **51.** $x = -\dfrac{2}{3}$

52. $x = -3$ **53.** $y = -30$ **54.** $x = 2$ **55.** $x = -4$

56. $x = 3$ **57.** $x = -\frac{10}{7}$ **58.** $x = \frac{5}{4}, \left(x \neq -\frac{1}{2}\right)$
59. $x = 2, \left(x \neq -\frac{1}{2}\right)$ **60.** No solution $\left(x \neq \frac{1}{2}\right)$
61. b, d, f **62.** $\frac{27}{16}$ **63.** $-\frac{9}{7}$ **64.** $-108x$
65. $\frac{2x-4}{xy-y}$ **66.** $\frac{5x^2y+5xy}{x-2}$ **67.** $2a-b$ **68.** x
69. $\frac{2a+4}{3a-1}$ **70.** $\frac{2y-4}{3y+5}$

CHAPTER 7 PRACTICE TEST ANSWERS

1. $-18x^2yz$ **2.** $x+2$ **3.** $\frac{x-1}{x+4}$ **4.** $27abx^2$
5. $5x^2-10x$ **6.** $-12x^4+12x^2$ **7.** $x \neq -9$
8. $x \neq 5, x \neq 10$ **9.** $\frac{1}{135}$ **10.** $-\frac{2}{9}$ **11.** $-\frac{x^3}{20}$
12. $-\frac{1}{75}$ **13.** $\frac{1}{6}$ **14.** $-7x+7y$ **15.** $\frac{-10}{x}$
16. $\frac{15}{x+1}$ **17.** $\frac{46}{5x}$ **18.** $\frac{12+23z^2}{4x^2y^2z}$
19. $\frac{3y^2+y+1}{y(y+1)(y-1)}$ **20.** $\frac{-5a+15}{(a-5)(a+5)}$
21. $x \neq 2$

x	$f(x) = \frac{1}{x-2}$
−100	−0.01
−10	−0.08
−1	−0.33
0	−0.5
1	−1
1.5	−2
2	Undefined
2.5	2
3	1
4	0.5
5	0.33
10	0.13
100	0.01

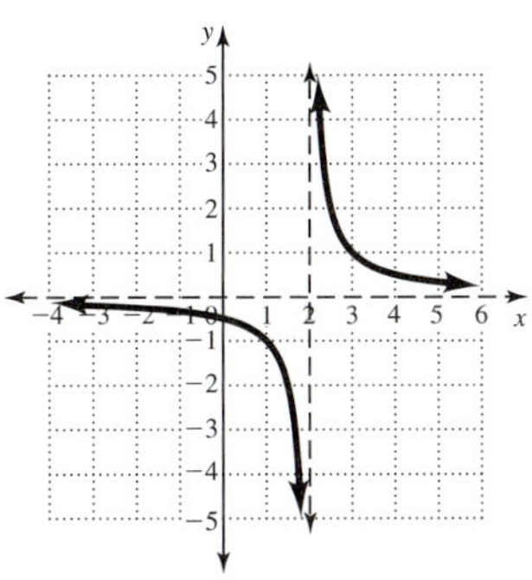

22. $x \neq -3, x \neq 2$

x	$f(x) = \frac{1}{x^2+x-6}$
−100	0.0001
−10	0.01
−4	0.17
−3.5	0.36
−3	Undefined
−2.5	−0.44
−2	−0.25
−1	−0.17
0	−0.17
1	−0.25
1.5	−0.44
2	Undefined
2.5	0.36
3	0.17
10	0.01
100	0.000099

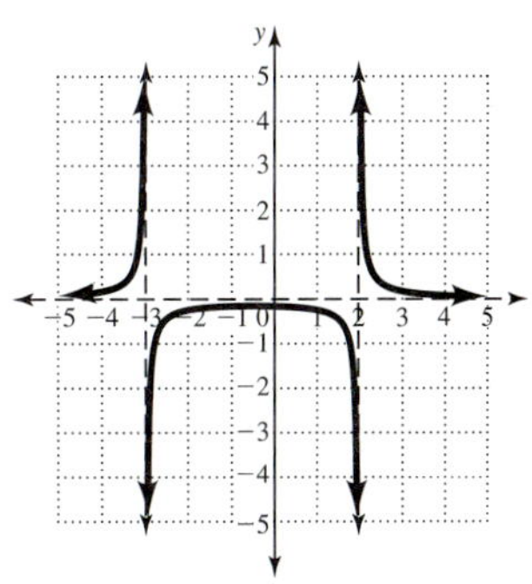

23. 1.875 hours **24.** $x = 15$ **25.** $x = -\frac{13}{6}$
26. $x = \frac{1}{8}$ **27.** $x = -\frac{1}{2}$ **28.** $y = 3$ **29.** No solution, $x \neq 3$ **30.** -15 **31.** $\frac{2x^2+4x}{3xy+3y}$ **32.** $x+2$
33. $\frac{3x-1}{x-1}$

CHAPTER 8

DOING MATHEMATICS SECTION 8.1

1. -5 **2.** -4 **3.** Nonreal **4.** Nonreal
5. Nonreal **6.** $-\frac{4}{7}$ **7.** $-\frac{10}{9}$ **8.** 0 **9.** -2
10. -3 **11.** -2 **12.** -5 **13.** 2 **14.** 3
15. Nonreal **16.** Nonreal **17.** -3 **18.** -2
19. 2 **20.** 3 **21.** -2 **22.** -3 **23.** 2 **24.** 3
25. -2 **26.** -3 **27.** 25 **28.** 23 **29.** 11 **30.** 1.7
31. x **32.** y **33.** $x + 5$ **34.** $x - 1$
35. $\sqrt{(x-2)^2} = x - 2$ **36.** $\sqrt{(x+1)^2} = x + 1$
37. x **38.** a **39.** $x - 5$ **40.** $y + 2$ **41.** 3.7417
42. 11.6190 **43.** 15.3189 **44.** 8.1921
45. -0.7454 **46.** -0.4082 **47.** Nonreal
48. Nonreal **49.** 2.1544 **50.** 4.1213 **51.** 3.1623
52. 3.1543 **53.** Nonreal **54.** Nonreal
55. -2.0598 **56.** -2.4147 **57.** $\sqrt{x^6} = \sqrt{(x^3)^2} = x^3$
58. $\sqrt{x^8} = \sqrt{(x^4)^2} = x^4$ **59.** $\sqrt{x^{10}} = \sqrt{(x^5)^2} = x^5$
60. $\sqrt{x^{14}} = \sqrt{(x^7)^2} = x^7$ **61.** $\sqrt[3]{x^9} = \sqrt[3]{(x^3)^3} = x^3$
62. $\sqrt[3]{x^{12}} = \sqrt[3]{(x^4)^3} = x^4$
63. $\sqrt[3]{x^{15}} = \sqrt[3]{(x^5)^3} = x^5$
64. $\sqrt[3]{x^{18}} = \sqrt[3]{(x^6)^3} = x^6$ **65.** $\sqrt{5^2} \neq 25$
66. $\sqrt{x^3} \neq x$ **67.** $-\sqrt{4} = -2$ **68.** $\sqrt[5]{x^5} = x$
69. (a) 38.65 mph (b) Yes, the driver was driving over the speed limit. **70.** There is no real number whose square is -25. Therefore, $\sqrt{-25}$ does not exist as a real number. When -2 is raised to the third power, the result is -8. Therefore, $\sqrt[3]{-8} = \sqrt[3]{(-2)^3} = -2$. It is possible to find the root of a negative number when the index is an odd number. It is not possible to find a real root of a negative number when the index is an even number.

DOING MATHEMATICS SECTION 8.2

1. $\sqrt{125} = \sqrt{5^2}\cdot\sqrt{5} = 5\sqrt{5} \approx 11.1803$
2. $\sqrt{20} = \sqrt{2^2}\cdot\sqrt{5} = 2\sqrt{5} \approx 4.4721$
3. $\sqrt{200} = \sqrt{2^2}\cdot\sqrt{5^2}\cdot\sqrt{2} = 2\cdot 5\sqrt{2} = 10\sqrt{2} \approx 14.1421$ **4.** $\sqrt{72} = \sqrt{2^2}\cdot\sqrt{3^2}\cdot\sqrt{2} = 2\cdot 3\sqrt{2} = 6\sqrt{2} \approx 8.4853$ **5.** $\sqrt{180} = \sqrt{2^2}\cdot\sqrt{3^2}\cdot\sqrt{5} = 2\cdot 3\sqrt{5} = 6\sqrt{5} \approx 13.4164$ **6.** $\sqrt{450} = \sqrt{3^2}\cdot\sqrt{5^2}\cdot\sqrt{2} = 3\cdot 5\sqrt{2} = 15\sqrt{2} \approx 21.2132$ **7.** $-\sqrt{98} = -\sqrt{7^2}\cdot\sqrt{2} = -7\sqrt{2} \approx -9.8995$ **8.** $-\sqrt{80} = -\sqrt{2^2}\cdot\sqrt{2^2}\cdot\sqrt{5} = -2\cdot 2\sqrt{5} = -4\sqrt{5} \approx -8.9443$ **9.** $\sqrt[3]{128} = \sqrt[3]{2^3}\cdot\sqrt[3]{2^3}\cdot\sqrt[3]{2} = 2\cdot 2\cdot\sqrt[3]{2} = 4\sqrt[3]{2} \approx 5.0397$
10. $\sqrt[3]{72} = \sqrt[3]{2^3}\cdot\sqrt[3]{9} = 2\sqrt[3]{9} \approx 4.1602$
11. $-\sqrt[3]{125} = -\sqrt[3]{5^3} = -5$
12. $-\sqrt[3]{64} = -\sqrt[3]{4^3} = -4$
13. $\sqrt[3]{-40} = \sqrt[3]{(-2)^3}\cdot\sqrt[3]{5} = -2\sqrt[3]{5} \approx -3.4200$
14. $\sqrt[3]{-56} = \sqrt[3]{(-2)^3}\cdot\sqrt[3]{7} - 2\sqrt[3]{7} \approx -3.8259$
15. $-\sqrt[3]{-189} = -\sqrt[3]{(-3)^3}\cdot\sqrt[3]{7} = -(-3)\sqrt[3]{7} = 3\sqrt[3]{7} \approx 5.7388$ **16.** $-\sqrt[3]{-135} = -\sqrt[3]{(-3)^3}\cdot\sqrt[3]{5} = -(-3)\sqrt[3]{5} = 3\sqrt[3]{5} \approx 5.1299$
17. $\sqrt[4]{48} = \sqrt[4]{2^4}\cdot\sqrt[4]{3} = 2\sqrt[4]{3} \approx 2.6321$
18. $\sqrt[4]{176} = \sqrt[4]{2^4}\cdot\sqrt[4]{11} = 2\sqrt[4]{11} \approx 3.6423$
19. $-\sqrt[4]{1250} = -\sqrt[4]{5^4}\cdot\sqrt[4]{2} = -5\sqrt[4]{2} \approx -5.9460$
20. $-\sqrt[4]{405} = -\sqrt[4]{3^4}\cdot\sqrt[4]{5} = -3\sqrt[4]{5} \approx -4.4860$
21. $\sqrt[5]{-486} = \sqrt[5]{(-3)^5}\cdot\sqrt[5]{2} = -3\sqrt[5]{2} \approx -3.4461$
22. $\sqrt[5]{-160} = \sqrt[5]{(-2)^5}\cdot\sqrt[5]{5} = -2\sqrt[5]{5} \approx -2.7595$
23. $-\sqrt[6]{1458} = -\sqrt[6]{3^6}\cdot\sqrt[6]{2} = -3\sqrt[6]{2} \approx -3.3674$
24. $-\sqrt[6]{192} = -\sqrt[6]{2^6}\cdot\sqrt[6]{3} = -2\sqrt[6]{3} \approx -2.4019$
25. $-7xy$ **26.** $-4xy^2\sqrt{x}$ **27.** $-4a\sqrt[3]{a}$
28. $-3z\sqrt[3]{z^2}$ **29.** $3xy\sqrt[4]{2x^2}$ **30.** $2xy\sqrt[4]{3x^2}$
31. $x - 1$ **32.** $x + 4$ **33.** $(x-1)\sqrt[3]{(x-1)^2}$
34. $(y-5)\sqrt[3]{(y-5)}$ **35.** $-2\sqrt{3}$ **36.** $-5\sqrt{3}$
37. 10 **38.** 18 **39.** $3\sqrt[4]{5}$ **40.** $2\sqrt[4]{3}$ **41.** 20
42. 60 **43.** $210x\sqrt{3}$ **44.** $60x$ **45.** $-24x\sqrt[3]{xy^2}$
46. $-2x\sqrt[3]{5y}$ **47.** $6ab\sqrt{7}$ **48.** $10xy\sqrt{x}$
49. $-2xy\sqrt[4]{35x}$ **50.** $-2y\sqrt[4]{3x^3y^2}$
51. $3ab\sqrt[3]{5b}$ **52.** $2xy\sqrt[3]{6x}$ **53.** $-\frac{2\sqrt{2}}{3}$ **54.** $-\frac{3\sqrt{3}}{2}$
55. $\frac{\sqrt{2}}{2}$ **56.** $\frac{2\sqrt{3}}{9}$ **57.** 7 **58.** 5 **59.** $\frac{-4\sqrt[3]{25}}{5}$
60. $\frac{-2\sqrt[3]{3}}{3}$ **61.** $\frac{\sqrt[3]{75}}{5}$ **62.** $\frac{3\sqrt[3]{3}}{2}$ **63.** $\frac{-x}{3y^2}$
64. $\frac{-2x^2}{y}$ **65.** $\frac{\sqrt[4]{x^2}}{2y}$ **66.** $\frac{3\sqrt[4]{x^2}}{y^2}$ **67.** $\frac{-y\sqrt[3]{2}}{2}$
68. $\frac{-x\sqrt[3]{3}}{4}$ **69.** $\frac{5x\sqrt[4]{x^3}}{y}$ **70.** $\frac{3x}{y^2}$ **71.** $-\frac{2\sqrt{2}}{3}$
72. $-\frac{5}{2}$ **73.** $\frac{2}{x+6}$ **74.** $\frac{x+3}{5}$ **75.** $\frac{x+2}{3}$
76. $\frac{x-4}{2}$ **77.** $\sqrt{x+3}$ **78.** $\sqrt{x+5}$ **79.** $\neq$
80. $\neq$ **81.** $\neq$ **82.** $\neq$ **83.** $\neq$ **84.** $\neq$
85. $=$ **86.** $=$ **87.** $=$ **88.** $=$ **89.** $2\sqrt{13}$
90. $\sqrt{443}$ **91.** $2\sqrt{5}$ **92.** $\sqrt{17}$ **93.** $\sqrt{x^2 - 4}$
94. $\sqrt{x^2 - 25}$ **95.** $\sqrt{x^2 + 9}$ **96.** $\sqrt{36 + x^2}$
97. $x + 1$ **98.** $x - 2$ **99.** Area $=$ 60 square inches

100. Area = 6 square inches
101. Length = $2\sqrt{6}$ inches
102. Length = $\frac{5\sqrt{14}}{4}$ inches

DOING MATHEMATICS SECTION 8.3

1. $-\sqrt{5}$ $7.6\sqrt{5}$ **2.** $\sqrt[3]{5}$ **3.** $\sqrt{2y}$ $5\sqrt{2y}$
4. $4\sqrt[3]{x}$ $10\sqrt[3]{x}$ **5.** $13\sqrt{2}$ **6.** $17\sqrt{5}$ **7.** $-9\sqrt{22}$
8. $-10\sqrt{17}$ **9.** UNLIKE **10.** UNLIKE **11.** $7.9\sqrt{x}$
12. $13.7\sqrt{10x}$ **13.** UNLIKE **14.** UNLIKE
15. $13\sqrt[3]{2x}$ **16.** $5\sqrt[3]{5x}$ **17.** $-4\sqrt{2a}$ **18.** $22\sqrt{2a}$
19. $-48\sqrt{2}$ **20.** $11\sqrt{3}$ **21.** $31\sqrt{3}$ **22.** $12\sqrt{5y}$
23. $3.1\sqrt[3]{2x}$ **24.** $9y\sqrt{x} - 2xy\sqrt{3x}$ **25.** $5\sqrt{5}$
26. $-13\sqrt[3]{3} - 30\sqrt[3]{5}$ **27.** $13x\sqrt[3]{x}$ **28.** $-\frac{1}{2}x$
29. $-31x\sqrt{z}$ **30.** $-3x$ **31.** $7\sqrt{6}$ **32.** $12\sqrt{3}$
33. $-7\sqrt{3}$ **34.** $-13\sqrt{2}$ **35.** $13x + 4x\sqrt{x}$
36. $13x$ **37.** $2\sqrt{5} + 3\sqrt{2}$ **38.** $5\sqrt{7} + \sqrt{5}$
39. $x^2 - x\sqrt{2}$ **40.** $2y^2 - 4y$ **41.** $-7 - 3\sqrt{3}$
42. $4 + 2\sqrt{7}$ **43.** $5x - 105\sqrt[3]{x}$ **44.** $14x - 7\sqrt[3]{2x^2}$
45. $x - \sqrt{x} - 20$ **46.** $y - 4\sqrt{y} - 21$
47. $x + 12\sqrt{x} + 36$ **48.** $b + 2\sqrt{b} + 1$
49. $a - 2\sqrt{ab} + b$ **50.** $x - 2\sqrt{xy} + y$ **51.** 3
52. 7.0 **53.** $x - 11$ **54.** $x - 7$ **55.** = **56.** ≠
57. ≠ **58.** ≠ **59.** ≠ **60.** =
61. Perimeter = $8\sqrt{2}$ inches
Area = 6 square inches
62. Perimeter = $12\sqrt[3]{2}$ inches
Area = $5\sqrt[3]{4}$ square inches
63. Area = $7\sqrt[3]{25}$ square cm.
64. Area = 13.68 square cm.

CHAPTER 8 REVIEW

1. -6 **2.** Nonreal **3.** -3 **4.** 3 **5.** Nonreal
6. -4 **7.** $-\frac{3}{5}$ **8.** $-\frac{8}{7}$ **9.** Nonreal **10.** -3
11. -3 **12.** x **13.** Nonreal **14.** $x - 1$
15. 5.8881 **16.** -6.3246 **17.** -0.6325
18. 2.2894 **19.** -4.6416 **20.** 3.7606 **21.** Nonreal
22. -1.8020 **23.** x^6 **24.** x^8 **25.** x^{10} **26.** x^3
27. x^7 **28.** x^{10} **29.** $-2\sqrt{15}$ **30.** $6\sqrt{6}$ **31.** $9\sqrt{2}$
32. $-6\sqrt{2}$ **33.** $-2\sqrt[3]{6}$ **34.** -4 **35.** $3\sqrt[3]{2}$
36. $3\sqrt[3]{4}$ **37.** $-2\sqrt[5]{7}$ **38.** $2\sqrt[4]{5}$ **39.** $-2\sqrt[5]{3}$
40. $-3\sqrt[6]{5}$ **41.** $-5x^2y^2\sqrt{y}$ **42.** $-4z^2\sqrt[3]{z}$
43. $2x^2y^2\sqrt[4]{y^2}$ **44.** $x + 3$ **45.** $2(x - 1)\sqrt{3(x - 1)}$
46. $(2x + 3)\sqrt[3]{2x + 3}$ **47.** $-20\sqrt{6}$ **48.** $3\sqrt[3]{3}$
49. $2\sqrt[4]{15}$ **50.** 30 **51.** $6x\sqrt{5}$ **52.** $-30x^2$
53. $\frac{2\sqrt{5}}{3}$ **54.** $-\frac{5}{2}$ **55.** $-\frac{3x^3\sqrt[3]{x}}{y^2}$ **56.** $\frac{x - 5}{6}$
57. $\frac{4x^2}{y}$ **58.** $\frac{3x^2}{y}$ **59.** $\frac{\sqrt{2}}{2}$ **60.** $\frac{2\sqrt{5xy}}{5y}$ **61.** ≠
62. = **63.** = **64.** ≠ **65.** $4\sqrt{2}$ **66.** $2\sqrt{22}$
67. UNLIKE **68.** $11\sqrt[3]{5x}$ **69.** UNLIKE
70. $3.49\sqrt{3x}$ **71.** $13\sqrt{11x}$ **72.** UNLIKE
73. $31\sqrt{3}$ **74.** 0 **75.** $15\sqrt{5y}$ **76.** $8.37\sqrt[3]{2x}$
77. $2y\sqrt{x} - 2xy\sqrt{3x} + 35xy\sqrt{x} =$
$(2y + 35xy)\sqrt{x} - 2xy\sqrt{3x}$
78. $3\sqrt{x}$ **79.** $23\sqrt[3]{3} - 30\sqrt[3]{4}$ **80.** $2x\sqrt[3]{x^2}$ **81.** $\frac{1}{2}x$
82. $15x^2\sqrt{z} - 10x\sqrt{z} = (15x^2 - 10x)\sqrt{z}$ **83.** $17\sqrt{3}$
84. $\sqrt{3}$ **85.** $10x + 6x^2 + 5x\sqrt{x}$ **86.** -1
87. $12x^2 + 2x$ **88.** $2 + 3\sqrt{2}$ **89.** $4x - 40\sqrt{x}$
90. $-21 - 2\sqrt{3}$ **91.** $x - 9\sqrt{x} + 20$ **92.** $x - 49$
93. $x - 8\sqrt{x} + 16$

CHAPTER 8 PRACTICE TEST ANSWERS

1. -4 **2.** Nonreal **3.** 10 **4.** $\frac{2}{7}$ **5.** $-10\sqrt{5}$
6. $-2\sqrt[3]{3}$ **7.** 50 **8.** $10\sqrt[4]{3}$ **9.** -0.8165
10. 2.9240 **11.** 2.6591 **12.** -2.8854
13. $-5x^2y\sqrt{3x}$ **14.** $2y\sqrt[3]{y}$ **15.** $-2z\sqrt[3]{5}$
16. $x - 3$ **17.** $(x - 1)\sqrt[5]{(x - 1)}$ **18.** $3\sqrt{5}$
19. $-6x\sqrt{x}$ **20.** $6y\sqrt[3]{5y^2}$ **21.** $-\frac{5x}{y^2}\sqrt[3]{x}$ **22.** $\frac{x + 5}{10}$
23. $\frac{x^3}{2}$ **24.** $-12\sqrt{2}$ **25.** UNLIKE
26. $37.93x\sqrt[3]{2x}$ **27.** $15x^2\sqrt{z} - 54x\sqrt{z} =$
$(15x^2 - 54x)\sqrt{z}$
28. $2 + \sqrt{10}$ **29.** $15\sqrt{3}$ **30.** $2x^2 + 5x\sqrt{x}$
31. $12x^2 + 8x$ **32.** $4x - 4\sqrt{xy}$ **33.** $15x + 7\sqrt{x} - 2$

CHAPTERS 7–8 CUMULATIVE TEST ANSWERS

1. $x \neq 3, x \neq -2$ **2.** $-\frac{5}{y}$ **3.** $\frac{1}{4}$ **4.** $\frac{6x}{2x + 1}$
5. $\frac{15 - 7b^3}{6ab^2c}$ **6.** $\frac{12x + 27}{x^2 - 9}$

7. $x \neq -3, x \neq 2$

x	$f(x) = \dfrac{1}{x^2 + x - 6}$
-100	0.0001
-10	0.01
-4	0.17
-3.5	0.36
-3	Undefined
-2.5	-0.44
-2	-0.25
-1	-0.17
0	-0.17
1	-0.25
2	Undefined
2.5	0.36
3	0.16
10	0.01
100	0.00009

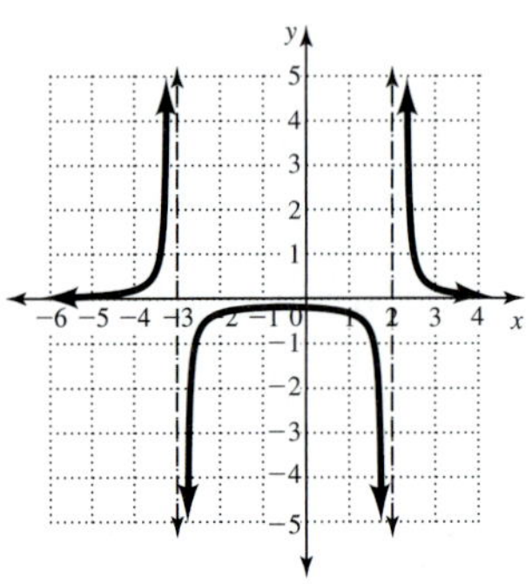

8. $y = 6$ **9.** $x = 9$ **10.** $\dfrac{4x + 8}{x^2 + 3x}$

11. $-10\sqrt{2}$ **12.** $\dfrac{3y^5}{z^6}$ **13.** $-5x^2$ **14.** Nonreal

15. $8x\sqrt{2}$ **16.** $26 - 4\sqrt{30}$ **17.** $-3 + \sqrt{3}$
18. $-3x - 3\sqrt{x}$

CHAPTER R

DOING MATHEMATICS SECTION R.1

1. $500 + 3 + \dfrac{7}{10} + \dfrac{9}{100}$ **2.** $20 + 5 + \dfrac{1}{10}$

3. $1 + \dfrac{2}{10} + \dfrac{3}{100}$ **4.** $1000 + 500 + 20 + 4 + \dfrac{5}{10}$

5. 1,000,000 + 3000 + 500 + 20 + 3 **6.** At least 102 cases **7.** (a) Airfare \$286 (b) Hotel \$789 (c) Taxi and Tips \$44 (d) Meals \$199 (e) Parking \$71 (f) Phone Calls \$14 **8.** (a) 2000 (b) 150 (c) 1 (d) 149.2 (e) 637.29 (f) 54.1235
9. Estimating, Jan needs to review more than 30 pages each day. The exact answer is $33.\overline{3}$. If she reviews 34 pages the first 8 days, she will have 28 pages left to review on the 9th day. **10.** Estimating, Anne needs to type more than 6 pages per day. If she types 7 pages the first 4 days, she will have 5 pages left on the 5th day. **11.** 30 **12.** 7 **13.** 40
14. 10 **15.** (a) $10^2 = 10 \cdot 10 = 100$
(b) $4^3 = 4 \cdot 4 \cdot 4 = 64$ (c) $3^5 = 3 \cdot 3 \cdot 3 \cdot 3 \cdot 3 = 243$
16. 8 **17.** 18 **18.** 11 **19.** 10 **20.** 36 **21.** 64
22. 36 **23.** 40 **24.** 21 **25.** 6 **26.** 4 **27.** 2
28. 23 **29.** 71 **30.** 8 **31.** 133 **32.** 500 **33.** 72
34. 61 **35.** 29 **36.** Answers will vary. Estimating the amount of money needed for vacation. Estimating the cost of purchases at a store to be sure you have enough money.

DOING MATHEMATICS SECTION R.2

1. $\dfrac{1}{4}$ **2.** $\dfrac{1}{5}$ **3.** $\dfrac{3}{4}$ **4.** $\dfrac{2}{3}$ **5.** $\dfrac{7}{30}$ **6.** $\dfrac{13}{50}$ **7.** $\dfrac{4}{10}$

8. $\dfrac{15}{20}$ **9.** $\dfrac{25}{50}$ **10.** $\dfrac{8}{100}$ **11.** $\dfrac{96}{18}$ **12.** $\dfrac{8}{28}$ **13.** $\dfrac{7}{3}$

14. $\dfrac{97}{9}$ **15.** $\dfrac{84}{5}$ **16.** $\dfrac{79}{8}$ **17.** $\dfrac{1963}{8}$ **18.** $\dfrac{58}{5}$

19. $3\dfrac{1}{3}$ **20.** $5\dfrac{1}{4}$ **21.** $7\dfrac{5}{21}$ **22.** $10\dfrac{4}{5}$ **23.** $10\dfrac{1}{10}$

24. $4\dfrac{5}{8}$ **25.** $1\dfrac{1}{3}$ **26.** $1\dfrac{11}{14}$ **27.** $6\dfrac{4}{5}$ **28.** $11\dfrac{1}{4}$

29. $6\dfrac{2}{5}$ **30.** $62\dfrac{1}{2}$ **31.** $8\dfrac{1}{3}$ **32.** $\dfrac{1}{2}$ **33.** $\dfrac{1}{8}$ **34.** 5

35. $\dfrac{10}{21}$ **36.** $2\dfrac{1}{2}$ **37.** 53 **38.** 4 **39.** $\dfrac{7}{8}$

40. $1\dfrac{17}{30}$ **41.** $13\dfrac{29}{36}$ **42.** $11\dfrac{1}{8}$ **43.** $2\dfrac{2}{3}$ **44.** $15\dfrac{2}{3}$

45. $1\dfrac{14}{15}$ **46.** $\dfrac{1}{36}$ **47.** $4\dfrac{13}{24}$ **48.** $18\dfrac{4}{5}$

49. $249\dfrac{1}{2}$ feet **50.** (a) $46\dfrac{1}{2}$ feet (b) Purchase 6 lengths that are 8 feet long. **51.** (a) $3\dfrac{1}{32}$

(b) $44\dfrac{1}{2}$ (c) $80\dfrac{7}{16}$ **52.** (a) $161\dfrac{63}{64}$ (b) $142\dfrac{1}{8}$

(c) 20 **53.** $2\dfrac{1}{3}$ **54.** $4\dfrac{1}{3}$ **55.** $10\dfrac{7}{12}$ **56.** 8

57. $2\dfrac{2}{21}$ **58.** $3\dfrac{7}{18}$ **59.** $\dfrac{25}{36}$ **60.** $\dfrac{22}{35}$

DOING MATHEMATICS SECTION R.3

1. 0.875 **2.** $0.\overline{6}$ **3.** $1.\overline{7}$ **4.** 2.5 **5.** $\frac{47}{2}$ **6.** $\frac{621}{100}$ **7.** $\frac{247}{200}$ **8.** $\frac{471}{500}$ **9.** (a) \$68.03 (b) \$5.13 (c) \$12.16 (d) \$0.38 (e) \$9.34 (f) \$76.63 (g) \$3.66 (h) \$10.88 **10.** (a) Terminating (b) Repeating (c) Neither (d) Repeating (e) Terminating (f) Neither **11.** 0.20 **12.** 0.6825 **13.** 1.00 **14.** 0.70 **15.** 50% **16.** 75% **17.** 125.3% **18.** 12.5% **19.** 60% **20.** 350% **21.** 62.5% **22.** $33\frac{1}{3}$% or $33.\overline{3}$% **23.** 40.46 **24.** 335.623 **25.** 23.52662 **26.** 1.0349 **27.** 70.436 **28.** 0.0885 **29.** 1264.563 **30.** 0.0125 **31.** 35.85632 **32.** 95.481 **33.** 180.1602 **34.** 0.16802 **35.** 13.5 **36.** 250.7 **37.** 25160 **38.** 250 **39.** (a) 20.5 feet, 18.75 feet, 18.25 feet, 15.75 feet (b) 73.25 feet (c) \$257.84 **40.** (a) $3.\overline{3}$ (b) 20 (c) 60 (d) 15 (e) 40 (f) 10 (g) 200 (h) 1.75 **41.** (a) 600 (b) 200 (c) 100 (d) 400 (e) 100 (f) 400 **42.** (a) 64; 96 (b) 9; 27 (c) 11.2; 16.8 (d) 7.5; 17.5 **43.** 18 **44.** 187.5 **45.** 625 **46.** 0.25 **47.** \$6375 **48.** \$649.35 **49.** Answers will vary. **50.** Answers will vary.

DOING MATHEMATICS SECTION R.4

1. (a) $\frac{137}{102}$ (b) 1.34 : 1 (c) $\frac{71}{102}$ (d) 0.70 : 1 **2.** (a) $\frac{17}{20} = 0.85$ (b) $\frac{11}{16} = 0.6875$ (c) It appears from these statistics that the Patriots may have had a stronger defense. In 2005, the Patriots beat the Eagles in the Super Bowl.

3. $\frac{420 \text{ miles}}{8 \text{ hours}} = 52.5$ miles per hour
4. $\frac{125 \text{ meters}}{3 \text{ seconds}} = 41.\overline{6}$ meters per second
5. $\frac{600 \text{ miles}}{2 \text{ hours}} = 300$ miles per hour
6. $\frac{220 \text{ kilometers}}{2.5 \text{ hours}} = 88$ kilometers per hour
7. = **8.** = **9.** ≠ **10.** = **11.** = **12.** ≠ **13.** = **14.** = **15.** ≠ **16.** ≠ **17.** $x = 24$ **18.** $x = 26$ **19.** $x = \frac{5}{3}$ **20.** $x = \frac{8}{17}$ **21.** $x = 54$ **22.** $x = 16$ **23.** $x = \frac{15}{2}$ **24.** $x = \frac{8}{3}$ **25.** $x = \frac{9}{4}$ **26.** $x = 10$ **27.** 241.5 km **28.** 310.56 mi **29.** 1.32 gal **30.** 56.85 L **31.** 11.01 m **32.** 6.67 yd **33.** (a) 59.06 in (b) 4.92 ft **34.** (a) 3.82 yd (b) 11.46 ft **35.** Gasoline was less expensive in the US. \$1.50 per gallon is equivalent to \$0.40 per liter. **36.** 491 miles **37.** Topsoil that costs \$60 for 3 cubic yards is \$20 per cubic yard. Topsoil that costs \$105 for 5 cubic yards is \$21 per cubic yard. Topsoil that costs \$60 for 3 cubic yards is less expensive. **38.** A cup of coffee that costs \$0.79 for 20 ounces is \$0.0395 per ounce. A cup of coffee that costs \$0.99 for 32 ounces is \$0.0309 per ounce. A cup of coffee that costs \$0.99 for 32 ounces is less expensive. **39.** A glass of soda that costs \$0.89 for 20 ounces is \$0.0445 per ounce. An 8-ounce soda for \$0.48 is \$0.06 per ounce. A glass of soda that costs \$0.89 for 20 ounces is less expensive. **40.** An 8-foot long wood plank for \$10.50 is \$1.3125 per foot. An 11-foot long plank for \$13.99 is \$1.2718 per foot. An 11-foot long plank for \$13.99 is less expensive. **41.** 4,000 Gy **42.** 20,000 Gy **43.** 50,000 Sv **44.** 60,000 Sv **45.** Answers will vary.

DOING MATHEMATICS SECTION R.5

1. $10^3 = 10 \cdot 10 \cdot 10 = 1000$
2. $4^5 = 4 \cdot 4 \cdot 4 \cdot 4 \cdot 4 = 1024$
3. $10^{-5} = \frac{1}{10 \cdot 10 \cdot 10 \cdot 10 \cdot 10} = \frac{1}{100{,}000} = 0.00001$
4. $6^1 = \frac{1}{6} = \frac{1}{6} = 0.1\overline{6}$ **5.** 200,000 **6.** 6,000,000 **7.** 0.1 **8.** 0.00001 **9.** 516,000 **10.** 6060 **11.** 12,500 **12.** 152 **13.** 0.092 **14.** 0.00019 **15.** 0.00014 **16.** 0.0000055 **17.** 3.5×10^6 **18.** 1.4×10^3 **19.** 1.25×10^2 **20.** 4.51×10^5 **21.** 6×10^8 **22.** 5.256×10^5 **23.** 1×10^2 **24.** 1×10^5 **25.** 2.1×10^{-4} **26.** 8.25×10^{-7} **27.** 7.64×10^{-6} **28.** 3.92×10^{-3} **29.** 1×10^{-15} **30.** 6.21×10^{-1} **31.** 8×10^{-2} **32.** 1×10^{-2} **33.** 43,750,000,000 **34.** 0.0000000000042 **35.** 10^9 **36.** 10^7 **37.** $10^{-8} = \frac{1}{10^8}$ **38.** $10^{-17} = \frac{1}{10^7}$ **39.** 5^{14} **40.** 2^{14} **41.** 10^5 **42.** 10^{10} **43.** $10^{-3} = \frac{1}{10^3}$ **44.** 10^6 **45.** 4^3 **46.** 3^2 **47.** 1.35×10^6 **48.** 2.73×10^5 **49.** 2.431×10^7 **50.** 5.4×10^{18} **51.** 2×10^4 **52.** 2×10^8 **53.** 5×10^1 **54.** 1.2×10^5 **55.** 6.25×10^{15} **56.** 1.4×10^3 **57.** 5×10^5 **58.** Answers will vary.

APPENDIX B

DOING MATHEMATICS

1. (a) Mean: 70.8; Median: 72; Mode: 83
(b) The mean and median are close to the same value. The average temperature in July in Cleveland, Ohio is 70–72 degrees. Nights are cool and the most frequent temperature of 83 (mode) occurs between noon and 4 P.M.

2. (a) Mean: 140.9; Median: 132; Mode: 131
(b) The mean is greater than the median. Some large weights are influencing the value of the mean.

3. (a) Mean: 78.9; Median: 80; Mode: None
(b) The mean and median are close to the same value and scores are distributed between 52 and 98, somewhat uniformly, with each score occurring once.

4. (a) Mean: 9.1; Median: 9.5; Mode: 12
(b) The mean and median are close to the same value. The mode of 12 occurs for six persons and shows the most frequent and popular number of credit hours taken by students.

5. Answers will vary.

6. Answers will vary.

7. Since the median is less than the mean, at least half of the homes in that community were in Leslie and Dave's price range and they should have looked at them.

8. Movies2See would be a good first choice. The mean is less than the median. Half of the salaries are greater than $60,000 and more than Ignacio's desired salary of at least $55,000.

9. Answers will vary.

10. No, the two sets would not be identical. Data sets that are composed of very different elements may have the same mean.

11. Two large scores will raise the mean, and have less effect on the median and mode.

12. Mean.

13. If the mean is larger than the median then some of the data are large compared to the other data.

14. If the median is larger than the mean then some of the data are small compared to the other data.

15. The mean, median, and mode provide information about the middle or the center of a data set.

16. Problem 1 in Getting Started.

17. Problem 2 in Getting Started.

18. Problem 4 in Getting Started.

Index